DYNAMIC SEDIMENTARY ENVIRONMENTS OF MANGROVE COASTS

DYNAMIC SEDIMENTARY ENVIRONMENTS OF MANGROVE COASTS

Edited by

FRIDA SIDIK
Institute for Marine Research and Observation
Ministry of Marine Affairs and Fisheries
Indonesia

DANIEL A. FRIESS
Department of Geography
National University of Singapore, Singapore

ELSEVIER

Elsevier
Radarweg 29, PO Box 211, 1000 AE Amsterdam, Netherlands
The Boulevard, Langford Lane, Kidlington, Oxford OX5 1GB, United Kingdom
50 Hampshire Street, 5th Floor, Cambridge, MA 02139, United States

Notices
Knowledge and best practice in this field are constantly changing. As new research and experience broaden our understanding, changes in research methods, professional practices, or medical treatment may become necessary.

Practitioners and researchers must always rely on their own experience and knowledge in evaluating and using any information, methods, compounds, or experiments described herein. In using such information or methods they should be mindful of their own safety and the safety of others, including parties for whom they have a professional responsibility.

To the fullest extent of the law, neither the Publisher nor the authors, contributors, or editors, assume any liability for any injury and/or damage to persons or property as a matter of products liability, negligence or otherwise, or from any use or operation of any methods, products, instructions, or ideas contained in the material herein.

Library of Congress Cataloging-in-Publication Data
A catalog record for this book is available from the Library of Congress

British Library Cataloguing-in-Publication Data
A catalogue record for this book is available from the British Library

ISBN: 978-0-12-816437-2

Abstract and keywords available online

Publisher: Charlotte Cockle
Acquisitions Editor: Louisa Munro
Editorial Project Manager: Liz Heijkoop
Production Project Manager: Kumar Anbazhagan
Cover Designer: Christian J. Bilbow

Typeset by TNQ Technologies

Contents

Contributors ix
Foreword xiii
List of reviewers xv
Introduction xvii

I

Biogeomorphological processes

1. Biogeomorphic evolution and expansion of mangrove forests in New Zealand's sediment-rich estuarine systems

Andrew Swales, Mark Pritchard and Graham B. McBride

1. Introduction 4
2. Biogeomorphology 4
3. Mangroves in New Zealand 6
4. Estuary evolution 11
5. Case study—sand flat to mangrove forest 14
6. Future shock—what's the fate of New Zealand's mangrove forests? 30
7. Summary and conclusions 35

Acknowledgments 38
References 39

2. Mangroves: a natural early-warning system of erosion on open muddy coasts in French Guiana

Christophe Proisy, Romain Walcker, Elodie Blanchard, Antoine Gardel and Edward J. Anthony

1. Introduction 48
2. Mud dynamics and vulnerability of the wave-exposed mangrove coast of French Guiana 49
3. Status and dynamics of mangroves on a rapidly changing muddy coast 52
4. Mangrove-based early warning of coastal vulnerability to mud erosion 55
5. Concluding remarks 62

Acknowledgments 62
References 63

3. Groundwater research in mangrove coastal ecosystems—new prospects

Paul Macklin, Judith Rosentreter, Virni Budi Arifanti and I Gusti Ngurah Agung Suryaputra

1. Introduction 68
2. Linkage between the hydrogeological and carbon cycle in mangrove ecosystems 68
3. Mangrove forest degradation and loss 71
4. Available techniques in SGD research 72
5. Summary 77

References 78

4. Flow and sediment dynamics around structures in mangrove ecosystems—a modeling perspective

Marine Le Minor, Martin Zimmer, Véronique Helfer, Lucy Gwen Gillis and Katrin Huhn

1. Introduction 84
2. General hydro- and sediment dynamics in mangrove ecosystems 86
3. State-of-the-art numerical modeling on hydro-/sediment dynamics in mangrove ecosystems 89
4. Disparate treatment of spatial scales 107
5. Conclusions and outlook 108

Acknowledgments 110
References 110

5. Morphological plasticity and survival thresholds of mangrove plants growing in active sedimentary environments

Alejandra G. Vovides, Uta Berger and Thorsten Balke

1. Introduction 121
2. Colonization and biogeomorphic succession 124
3. Wind resistance and mechanical stability in unstable sediments 127
4. Root responses for mechanical stability 129

5. Plastic responses to sediment dynamics in mature forests 134
6. Concluding remarks 135
References 135

6. Microbial communities in mangrove sediments

Véronique Helfer and Christiane Hassenrück

1. Microbial communities in mangrove sediments—setting the scene 142
2. Microbial communities, their drivers, and their role in element cycling 143
3. Microbial communities in mangroves 151
4. Techniques to depict microbial communities and the processes they drive 157
5. Conclusion and perspectives 165
Appendix 1: Excerpts from the R scripts (full script available under: https://gitlableibniz-zmt.de/vhe/microbial-communities-in-mangroves) used for the bibliographic review, showing the different syntaxes used to group the studies by microorganism types, habitat types and applied analytical approaches 167
Acknowledgments 169
References 169

II

Long-term sedimentary processes and sea-level rise

7. The history of surface-elevation paradigms in mangrove biogeomorphology

Daniel A. Friess and Karen L. McKee

1. Introduction 180
2. Surface elevation determines mangrove species distribution and growth (Paradigm 1) 181
3. Mangroves can modify their surface elevations (Paradigm 2) 188
4. Conclusions 194
Acknowledgments 195
References 195

8. Radiocarbon dating of mangrove sediments

Juliet Sefton, Sarah Woodroffe and Philippa Ascough

1. Introduction 199
2. Brief introduction to the principles of radiocarbon dating 200
3. Radiocarbon dating of mangrove sediments 204
4. Other quantitative dating approaches 209
5. Conclusions 210
References 210

9. Australian mangroves through the Holocene: interactions between sea level, mangrove extent, and carbon sequestration

Neil Saintilan, Jeff Kelleway and Kerrylee Rogers

1. Introduction 218
2. Early phases of the postglacial marine transgression: 13,000–9000 BP 219
3. Late phases of the postglacial marine transgression: 9000–6000 BP 221
4. Estuary infill in the mid-late Holocene 224
5. Contemporary patterns of mangrove–saltmarsh interaction 226
6. 21st century sea-level rise acceleration and the challenges of coastal planning 227
7. Conclusions 230
References 230

10. Responses of mangrove ecosystems to sea level change

Raymond D. Ward and Luiz Drude de Lacerda

1. Introduction 236
2. Factors contributing to mangrove resilience to sea level rise 237
3. Landward migration and sediment availability in arid and semiarid areas 238
4. New biogeochemical setting following mangrove migration 242
5. Plant productivity 243
6. Measuring mangrove resilience to sea level rise 244
7. Differential responses of mangrove types (estuarine/deltaic, fringing) to sea level rise 245
8. Regions with mangroves under threat from sea level rise 246
9. Conclusions 248
References 249

11. Does geomorphology determine vulnerability of mangrove coasts to sea-level rise?

Karen L. McKee, Ken W. Krauss and Donald R. Cahoon

1. Introduction 256
2. Methods 260
3. Results and discussion 263

4. Conclusions 269
Acknowledgments 270
References 270

III

Blue carbon

12. Environmental drivers of blue carbon burial and soil carbon stocks in mangrove forests

Richard MacKenzie, Sahadev Sharma and Andre S. Rovai

1. Introduction 276
2. Abiotic factors 279
3. Biotic factors 283
4. Conclusion 287
References 288

13. Gaps, challenges, and opportunities in mangrove blue carbon research: a biogeographic perspective

Andre S. Rovai and Robert R. Twilley

1. Introduction 295
2. Mangroves as blue carbon sinks: stocks versus sequestration 298
3. Biogeography of blue carbon research in mangroves 307
4. Global trends in blue C in mangroves 315
5. Challenges and opportunities to improve mangrove blue C inventories and mapping 325
References 329

14. State of biogeochemical blue carbon in South Asian mangroves

Raghab Ray, Sahadev Sharma and Tapan Kumar Jana

1. Introduction 336
2. Overview of past and current research in South Asia 341
3. Mangrove blue C in the Indian Sundarbans 348
4. Research gaps and future directions 357
References 360

15. Quantity and quality of organic matter in mangrove sediments

Martin Zimmer and Véronique Helfer

1. Organic matter in mangrove sediments—setting the scene 369
2. Quantity versus quality of SOM 374
3. Origin of SOM 376
4. Mangroves worldwide—alike but distinct 382
5. Conclusion 388
References 388

16. Relevance of allochthonous input from an agriculture-dominated hinterland for "Blue Carbon" storage in mangrove sediments in Java, Indonesia

Tim C. Jennerjahn

1. Introduction 394
2. Material and methods 395
3. Results 399
4. Discussion 400
5. Conclusions 410
References 410

17. Potential carbon loss in sediment through methane production during early development stage of mangrove regeneration in restored mangroves

Yaya I. Ulumuddin, Sara Beavis, Michael Roderick, Stephen Eggins, Irawan Sugoro and Sukristijono Sukardjo

1. Introduction 416
2. Study area and sampling sites 418
3. Sample collection and analytical methods 420
4. Results 422
5. Discussion 427
6. Conclusion 436
Appendix 1: soil surface CH_4 fluxes 436
Appendix 2: CH_4 fluxes from tree stem and soil surface in the literature 438
Acknowledgments 439
References 439

18. Blue carbon storage comparing mangroves with saltmarsh and seagrass habitats at a warm temperate continental limit

Sinegugu P. Banda, Janine B. Adams, Anusha Rajkaran, Jaime Leigh Johnson and Jacqueline L. Raw

1. Introduction 448
2. Materials and methods 450

3. Results 455
4. Discussion 460
5. Conclusion 466
References 466

19. Mangrove carbon sequestration and sediment deposition changes under cordgrass invasion
Luzhen Chen, Yining Chen, Yihui Zhang and Hongyu Feng

1. Introduction 474
2. Cordgrass invasion status in Chinese mangroves 475
3. Biological factors affecting the sediment carbon flux 480
4. Vegetation–sediment dynamic interactions and the consequent sedimentary processes 488
5. Conceptual model and evaluation of regional carbon storage 496
6. Conclusions and perspectives 500
Acknowledgments 501
References 501

IV

Mangrove management and restoration

20. A framework for the quantitative assessment of mangrove resilience
Wei Jian Ong and Joanna C. Ellison

1. Introduction 513
2. Methods 517
3. Results and discussion 521
4. Conclusions 530
References 532

21. Assessment of typhoon impacts and post-typhoon recovery in Philippine mangroves: lessons and challenges for adaptive management
Severino G. Salmo III

1. Introduction 540
2. Materials and methods 541
3. Results and discussion 544
4. Summary: vulnerability, research/management implications, and recommendations 556
Acknowledgments 558
References 559

22. Managing sediment dynamics through reintroduction of tidal flow for mangrove restoration in abandoned aquaculture ponds
Frida Sidik, Denny Wijaya Kusuma, Bayu Priyono, Christophe Proisy and Catherine E. Lovelock

1. Introduction 564
2. Mangrove–sediment–hydrodynamics interactions 565
3. Estuarine ecohydrology in mangrove restoration and rehabilitation 567
4. Implications of mangrove restoration: carbon abatement opportunity and climate change mitigation 568
5. Case study: hydrological recovery in abandoned ponds in Perancak, Bali 568
6. Lesson learned from the case study 576
Acknowledgments 578
References 578

23. Impacts of forestry on mangrove sediment dynamics
Mériadec Sillanpää, Massimo Lupascu, Ruhuddien Pandu Yudha, Sigit D. Sasmito, Susan Vulpas, Rui Xiang Teo and Sartji Taberima

1. Introduction 584
2. Global overview of soil dynamics and recovery in mangrove harvest regimes 587
3. Postharvest impacts on mangrove soils in different harvest regimes 589
4. Forestry case study: soil dynamics in a large-scale selective logging regime in the Bintuni Bay mangrove, West Papua, Indonesia 594
5. Overview and future considerations 602
Appendix 603
Acknowledgments 604
References 604
Further reading 607

Conclusions 609
Index 615

Contributors

Janine B. Adams Nelson Mandela University, Port Elizabeth, Eastern Cape, South Africa

Edward J. Anthony LEEISA, CNRS, IFREMER, Univ Guyane, Cayenne, French Guiana, France; CEREGE, Univ Aix Marseille, CNRS, Collège de France, INRAE, IRD, Aix-en-Provence, France

Virni Budi Arifanti Center for Research and Development of Social, Economy, Policy and Climate Change (P3SEKPI), Ministry of Environment and Forestry, Bogor, Indonesia

Philippa Ascough Scottish Universities Environmental Research Centre, NERC Radiocarbon Facility, East Kilbride, United Kingdom

Thorsten Balke School of Geographical and Earth Sciences, University of Glasgow, Scotland, United Kingdom

Sinegugu P. Banda Nelson Mandela University, Port Elizabeth, Eastern Cape, South Africa

Sara Beavis Fenner School of Environment and Society, Australian National University, Acton, ACT, Australia

Uta Berger Institute of Forest Growth and Forest Computer Sciences, Technische Universität Dresden, Tharandt, Germany

Elodie Blanchard AMAP, University of Montpellier, IRD, CIRAD, CNRS, INRAE, Montpellier, France

Donald R. Cahoon U.S. Geological Survey, Patuxent Wildlife Research Center, Beltsville, MD, United States

Luzhen Chen Key Laboratory of the Ministry of Education for Coastal and Wetland Ecosystems, College of the Environment and Ecology, Xiamen University, Xiamen, Fujian, China

Yining Chen Second Institute of Oceanography, Ministry of Natural Resources, Hangzhou, Zhejiang, China

Luiz Drude de Lacerda Instituto de Ciências do Mar, Universidade Federal do Ceará, Fortaleza, Ceará, Brazil

Stephen Eggins Research School of Earth Sciences, Australian National University, Acton, ACT, Australia

Joanna C. Ellison School of Geography, Planning and Spatial Sciences, University of Tasmania, Launceston, Tasmania, Australia

Hongyu Feng Key Laboratory of the Ministry of Education for Coastal and Wetland Ecosystems, College of the Environment and Ecology, Xiamen University, Xiamen, Fujian, China

Daniel A. Friess Department of Geography, National University of Singapore, Singapore

Antoine Gardel LEEISA, CNRS, IFREMER, Univ Guyane, Cayenne, French Guiana, France

Lucy Gwen Gillis ZMT - Leibniz Centre for Tropical Marine Research, Bremen, Germany

Christiane Hassenrück ZMT - Leibniz Centre for Tropical Marine Research, Bremen, Germany

Véronique Helfer ZMT - Leibniz Centre for Tropical Marine Research, Bremen, Germany

Katrin Huhn MARUM - Center for Marine Environmental Sciences, University of Bremen, Bremen, Germany

Tapan Kumar Jana Department of Marine Science, The University of Calcutta, Kolkata, West Bengal, India

Tim C. Jennerjahn Leibniz Centre for Tropical Marine Research, Bremen, Germany; Faculty of Geoscience, University of Bremen, Bremen, Germany

Jaime Leigh Johnson University of the Western Cape, Bellville, Cape Town, South Africa

Jeff Kelleway School of Earth, Atmospheric and Life Sciences, University of Wollongong, NSW, Australia

Ken W. Krauss U.S. Geological Survey, Wetland and Aquatic Research Center, Lafayette, LA, United States

Denny Wijaya Kusuma Institute for Marine Research and Observation, Ministry of Marine Affairs and Fisheries, Bali, Indonesia

Marine Le Minor MARUM - Center for Marine Environmental Sciences, University of Bremen, Bremen, Germany; Géosciences Rennes, Université de Rennes 1, CNRS, UMR 6118, Campus de Beaulieu, Rennes, France

Catherine E. Lovelock School of Biological Sciences, University of Queensland, St Lucia, QLD, Australia

Massimo Lupascu Department of Geography, National University of Singapore, Singapore

Richard MacKenzie Institute of Pacific Islands Forestry, Pacific Southwest Research Station, USDA Forest Service, Hilo, HI, United States

Paul Macklin National Marine Science Centre, Southern Cross University, Coffs Harbour, NSW, Australia; Department of Chemical Analysis, Faculty of Mathematics and Natural Sciences, Universitas Pendidikan Ganesha, Singaraja, Bali, Indonesia

Graham B. McBride National Institute of Water and Atmospheric Research (NIWA), Hamilton, Waikato, New Zealand

Karen L. McKee U.S. Geological Survey, Wetland and Aquatic Research Center, Lafayette, LA, United States

Wei Jian Ong School of Geography, Planning and Spatial Sciences, University of Tasmania, Launceston, Tasmania, Australia

Mark Pritchard National Institute of Water and Atmospheric Research (NIWA), Hamilton, Waikato, New Zealand

Bayu Priyono Institute for Marine Research and Observation, Ministry of Marine Affairs and Fisheries, Bali, Indonesia

Christophe Proisy French Institute of Pondicherry, Pondicherry, India; AMAP, University of Montpellier, IRD, CIRAD, CNRS, INRAE, Montpellier, France

Anusha Rajkaran University of the Western Cape, Bellville, Cape Town, South Africa

Jacqueline L. Raw Nelson Mandela University, Port Elizabeth, Eastern Cape, South Africa

Raghab Ray Atmosphere and Ocean Research Institute, The University of Tokyo, Kashiwa, Chiba, Japan

Michael Roderick Research School of Earth Sciences, Australian National University, Acton, ACT, Australia

Kerrylee Rogers School of Earth, Atmospheric and Life Sciences, University of Wollongong, NSW, Australia

Judith Rosentreter Centre for Coastal Biogeochemistry, School of Environment, Science and Engineering Southern Cross University, Lismore, NSW, Australia

Andre S. Rovai Department of Oceanography and Coastal Sciences, College of Coast and Environment, Louisiana State University, Baton Rouge, LA, United States; Departamento de Oceanografia, Universidade Federal de Santa Catarina, Florianópolis, Santa Catarina, Brazil

Neil Saintilan Department of Environmental Sciences, Macquarie University, NSW, Australia

Severino G. Salmo III Institute of Biology, College of Science, University of the Philippines Diliman, Quezon City, Philippines

Sigit D. Sasmito Department of Geography, National University of Singapore, Singapore; Center for International Forestry Research (CIFOR), Jalan CIFOR, Situgede, Bogor, Indonesia; NUS Environmental Research Institute, National University of Singapore, Singapore

Juliet Sefton Department of Geography, Durham University, Durham, United Kingdom

Sahadev Sharma Institute of Ocean and Earth Sciences, University of Malaya, Kuala Lumpur, Selangor, Malaysia

Frida Sidik Institute for Marine Research and Observation, Ministry of Marine Affairs and Fisheries, Bali, Indonesia; School of Biological

Sciences, University of Queensland, St Lucia, QLD, Australia

Mériadec Sillanpää Department of Geography, National University of Singapore, Singapore; Green Forest Product and Tech. Pte. Ltd., Research Department, Singapore

Irawan Sugoro National Nuclear Energy Agency of Indonesia, Jakarta, Indonesia

Sukristijono Sukardjo Research Center for Oceanography, Indonesian Institute of Sciences., Jakarta, Indonesia

I Gusti Ngurah Agung Suryaputra Department of Chemical Analysis, Faculty of Mathematics and Natural Sciences, Universitas Pendidikan Ganesha, Singaraja, Bali, Indonesia

Andrew Swales National Institute of Water and Atmospheric Research (NIWA), Hamilton, Waikato, New Zealand

Sartji Taberima Department of Soil Science and Land Resource, Faculty of Agriculture, Papua University, Manokwari, West Papua, Indonesia

Rui Xiang Teo Green Forest Product and Tech. Pte. Ltd., Research Department, Singapore

Robert R. Twilley Department of Oceanography and Coastal Sciences, College of Coast and Environment, Louisiana State University, Baton Rouge, LA, United States

Yaya I. Ulumuddin Research Center for Oceanography, Indonesian Institute of Sciences., Jakarta, Indonesia; Fenner School of Environment and Society, Australian National University, Acton, ACT, Australia

Alejandra G. Vovides School of Geographical and Earth Sciences, University of Glasgow, Scotland, United Kingdom

Susan Vulpas Department of Environmental Science, American University, Washington D.C., United States

Romain Walcker Laboratoire Ecologie Fonctionnelle et Environnement, Univ Toulouse, CNRS, Toulouse, France

Raymond D. Ward Centre for Aquatic Environments, School of the Environment and Technology, University of Brighton, Brighton, United Kingdom; Institute of Agriculture and Environmental Sciences, Estonian University of Life Sciences, Tartu, Estonia

Sarah Woodroffe Department of Geography, Durham University, Durham, United Kingdom

Ruhuddien Pandu Yudha Forestry Department, PT. Bintuni Utama Murni Wood Industries, Jakarta, Indonesia

Yihui Zhang Key Laboratory of the Ministry of Education for Coastal and Wetland Ecosystems, College of the Environment and Ecology, Xiamen University, Xiamen, Fujian, China

Martin Zimmer ZMT - Leibniz Centre for Tropical Marine Research, Bremen, Germany; University of Bremen, Faculty of Biology/ Chemistry, Bremen, Germany; IUCN-SSC Mangrove Specialist Group

Foreword

People see mangroves as a forest, as coastal vegetation that can withstand saline conditions and are well adapted to the harsh and dynamic intertidal environment. More generally, mangroves are an ecosystem with strong geomorphic, sedimentary, and biogeochemical links with adjacent coastal systems, such as mudflats. The long-term functioning of the mangrove ecosystem involves biogeomorphic interactions between mangroves, hydrodynamics, and sediments. The mangrove sedimentary system is very dynamic, creating an ecosystem that is able to act as a coastal stabilizer, carbon sink, and a nutrient-rich habitat that supports other ecosystems across the wider coastal landscape.

Recently, a growing awareness of mangrove in coastal protection and climate change has risen among policy makers and the general population to protect and restore mangroves. However, these laudable efforts often neglect the interaction between hydromorphodynamics and vegetation and other natural processes controlling the development of mangroves. The long-term survival of mangroves and the sustained provision of their ecosystem services are dependent on their ability to adapt to sea-level rise and other climate change impacts. Scientific knowledge of these key processes will help provide the evidence base from which policy makers and coastal managers can make suitable management and restoration decisions.

This book summarizes the latest conceptual and empirical research on sedimentary dynamics and their importance for managing mangrove forests. Part I covers the background knowledge of mangrove ecosystems and associated hydrological and sedimentary dynamics. These fundamental biogeomorphological processes have consequences on the role of mangroves in adapting to and mitigating the changing climate. Part II includes chapters on long-term sedimentary dynamics operating along mangrove coasts in response to local rises in sea level, focusing on the capacity of mangrove soil surfaces to potentially increase in elevation. Part III explores the "blue carbon" potential of mangrove sedimentary environments as a mitigation strategy for climate change. Both parts address the importance of the complexities of mangrove sedimentary dynamics for coastal management in the tropics and subtropics to ensure the future of mangroves. This understanding is linked to the successes and failures in mangrove management in addressing the issues of mangrove degradation and restoration (Part IV).

As the editors, we are grateful to many colleagues worldwide in mangrove science who contributed their ideas and work presenting the state of mangrove regions in this book. The *Dynamic Sedimentary Environments of Mangrove Coasts* encompasses a broad and multidisciplinary range of topics in mangroves around the globe, which would not be possible without the tremendous work and cooperation of the contributors. We thank them for their efforts that have remarkably shaped the direction and key messages of this book. We also greatly appreciate the reviewers who have provided valuable reviews and feedback that have strengthened the work presented here.

We hope that this book will help readers better understand the state of the art of research in mangrove sedimentary environments, and best utilize this knowledge and lessons learned to improve mangrove management in the face of global change.

Frida Sidik and Daniel A. Friess

List of reviewers

Fernanda Adame
Daniel M. Alongi
Aplena E. Bless
Clint Cameron
Luzhen Chen
Nicole Cormier
Cheryl Doughty
Daniel Friess
Glenn Guntenspergen
Erik Horstman
Mahmood Hossain
Catherine E. Lovelock
Cyril Marchand
David Neil
Jacqueline Raw
Tim Shaw
Frida Sidik
Lili Wei
Pim Willemsen
Erik Yando

Introduction

Frida Sidik[1], *Daniel A. Friess*[2]

[1]Institute for Marine Research and Observation, Ministry of Marine Affairs and Fisheries, Bali, Indonesia; [2]Department of Geography, National University of Singapore, Singapore

Coastal mangrove forests serve as an important interface between the land and sea. Located in the intertidal zone, mangroves are generally distributed from mean sea level to highest spring tide, often found in the upper intertidal zone on many low-energy coasts, and can coexist with other coastal ecosystems such as seagrasses, salt marshes, and tidal flats (Alongi and Brinkman, 2011; Wolanski et al., 2009).

Their location in the intertidal zone means that mangrove forests interact with a range of processes that cross the terrestrial, coastal, and marine domains. Mangroves are a dynamic ecosystem, interacting with biophysical factors that operate over temporal scales from seconds to millennia, and at spatial scales from the individual tree to the global scale. In *Dynamic Sedimentary Environments of Mangrove Coasts*, we highlight key frontiers in mangrove sedimentary environments. Part I summarizes the state of the art of our knowledge of sedimentary processes influencing this important intertidal ecosystem, from microbial communities in mangrove sediments to the large-scale biogeomorphic evolution of mangrove coastlines. Part II takes a long-term and geological view of sedimentary processes in mangroves, and how they influence the vulnerability or resilience of mangroves to sea-level rise. Part III focuses on an increasingly important aspect of mangroves; how sedimentary processes contribute to the sequestration, accumulation, and storage of "blue" carbon. Finally, Part IV presents case studies that highlight the importance of sedimentary understanding for mangrove management, sustainable use, conservation, and restoration.

Mangrove distribution is a function of climate, biogeomorphology, and sedimentary processes

Mangroves are controlled by tidal patterns that shape mangrove forests into five principal geomorphic types: fringe, riverine, overwash, basin, and dwarf systems (Lugo and Snedaker, 1974). Woodroffe and Davies (2009) further defined mangroves based on morphological settings, which are river-dominated, tide-dominated, wave-dominated, composite river and wave-dominated, drowned bedrock valley, and carbonate settings. These classifications represent the linkages between mangroves and environmental settings (Lugo and Snedaker, 1974; Woodroffe and Davies, 2009).

The distribution of mangroves is dependent on a range of geomorphic and sedimentary processes, coastal geography, and climate (Duke et al., 1998; Osland et al., 2017). The global distribution of mangroves is constrained primarily by temperature, that limits the mangrove belt primarily (but not exclusively) to the tropical and subtropical latitudes, where air temperature are at least 20°C in the coldest month (Duke, 1992). Mangrove-temperature linkages are prominent in most subtropical mangrove coasts (e.g. America, eastern Australia, New Zealand, eastern Asia, and southeast Africa) which can result in expansion or contraction the extent of mangrove forests (Osland et al., 2017). Temperature, interplaying with rainfall, salinity, and tidal variation, contributes to mangrove species richness and biomass at the regional scale (Duke et al., 1998; Osland et al., 2017). Mangroves in wet tropics, especially in Indo-West Pacific, are often greatest in diversity and biomass, whereas arid mangrove areas are characterized by low species richness, scattered with specific tree architecture (Duke et al., 1998).

Within the broader boundary conditions provided by temperature, biophysical and sedimentary factors and their gradients become important factors driving mangrove distribution, as mangroves show species-specific tolerances and sensitivity to environmental changes (Duke et al., 1998; Woodroffe, 1992). Each mangroves species has a specific tolerance and response to gradients in salinity, soil type and chemistry, nutrient content, sediment supply, and hydrological regime, that enables them to survive in harsh environment (Alongi, 2009; Woodroffe and Davies, 2009). This suggests that, as coastal vegetation, mangroves are not a single group of species but rather a diverse range of mangrove plants structuring a forest with different levels of adaptation to a dynamic and changing environment (Smith, 1992). Vegetation community composition can thus change through time in association with the morphological evolution of shorelines (Woodroffe, 1992). The zonation also often matches with topographic contours that indicates the importance of tidal inundation as factor controlling the zonation (Duke et al., 1998). To this end, we currently focus on hydrogeomorphological processes of mangroves to place this book in context of sedimentary coasts.

Hydrogeomorphological processes can both facilitate and constrain the development of mangroves. For example, mangrove loss and regrowth have been observed simultaneously along the French Guiana mangrove coastline, which is attributed to natural processes of erosion and retreat (Anthony et al., 2013). This periodic morphological change represents the natural process of a mangrove shoreline reaching equilibrium with hydrodynamics and sediment transport (Anthony et al., 2013; Woodroffe and Davies, 2009). In most cases, mangroves actively contribute to the stabilization of land by capturing sediments, resulting in coastal accretion in the low-energy zone between the mud bank and the shore (Furukawa and Wolanski, 1996). In contrast, when mangroves receive a reduced sediment supply, they are not able to establish adequate coastal accretion due to lack of sediment trapping capacity, leading to coastline erosion (Willemsen et al., 2016).

Mangroves provide important ecosystem services to society

The natural processes controlling the interaction between morphodynamics and vegetation generate their potential as natural barriers that stabilize and protect the coastline from coastal hazards (Spalding et al., 2014). It has been claimed that mangroves can provide varying levels of protection against waves (Sánchez-Núñez et al., 2020), storm surges (Zhang et al., 2012), cyclones (Das and Vincent, 2009), and potentially some tsunamis (Danielsen et al., 2005). Functioning can only be understood by considering the interaction between trees, sediments, and hydrodynamic forcing, and their combined effects. For example, for the ecosystem service of land consolidation, the sediment, which is brought in from outside to the system (allochthonous) or produced within the ecosystem (autochthonous), is carried in suspension, trapped, and accumulated (Furukawa and Wolanski, 1996; Woodroffe, 1992). Over the course of time, the new sediments consolidate and become more compact, thus enabling mangrove propagules to colonize in the new land (Toorman et al., 2018). When mangroves grow, like all trees, they adapt and find their stability by developing an adaptive root system in the soil (McKee, 1996), spreading in horizontal direction to divide their weight over a larger area (Komiyama et al., 2008). The roots do not grow deeply due to anaerobic conditions and the low permeability of the mud, which is usually characterized by a dense sediment–water mixture dominated by clay particles (Furukawa et al., 1997; Gill and Tomlinson, 1977).

Mangrove forests also contribute to global climate regulation, due to the capacity of mangroves to sequester carbon (Howard et al., 2014; Sidik et al., 2019). Mangroves are productive coastal forests and have a large capacity for carbon storage, termed as "blue" carbon, which is captured from the atmosphere, transported materials from upstream and adjacent coastal environments, and organic material produced within the system (Alongi, 2014; Donato et al., 2011). Estimations of global mangrove carbon stocks vary from 4.19 to 10 Pg (Alongi, 2018; Hamilton and Friess, 2018). Unlike terrestrial forests, the largest carbon pool in mangroves is found in the soil, thus enhanced sediment accumulation is the key component to keep largest contribution of soil carbon to the total carbon stock of mangrove ecosystem, particularly minerogenic mangrove site (Alongi, 2018). Accumulation rates, sediment sources (allochthonous vs. autochthonous), and carbon storage capacity vary between and within sedimentary and geomorphic settings, resulting in a range of soil carbon stocks from <50 tons/ha to >500 tons/ha (Alongi, 2018; Lovelock et al., 2017; Rovai et al., 2018; Twilley et al., 2018). Recent estimates reveal that mangrove soils could store 2.6–6.4 Pg C in the top meter of the soil column, with Indonesia as the country with the greatest soil carbon stock (Atwood et al., 2017; Jardine and Siikamäki, 2014; Sanderman et al., 2018). These examples show the importance of understanding key sedimentary processes to better understand the creation of mangrove ecosystem services.

Threats to mangroves and their ecosystem services

While mangroves provide various ecosystem services to coastal communities, they continue to decline due to natural forces and human disturbances (Alongi, 2002). A number of studies have presented the evidence that the loss and degradation of mangroves over the last decades is due in large part to mangrove conversion to aquaculture, particularly in Southeast Asia (Thomas et al., 2017) where the largest proportion of mangrove is found globally (Bunting et al., 2018). Clearing mangroves for aquaculture ponds is also reported in South America, South Asia, and Africa, as this practice has been widespread in these regions (Feka and Ajonina, 2011; Ferreira and Lacerda, 2016; Paul et al., 2017; Rahman et al., 2010). However, mangroves also experience a number of other drivers of land cover conversion that vary regionally in importance, include oil palm in Southeast Asia, rice in Southeast Asia and West Africa, and overextraction in East Africa (Friess et al., 2019) These anthropogenic drivers of mangrove loss often interact with natural processes, such as erosion and climate change impacts, that can exacerbate mangrove loss (Gopal, 2012).

Mangrove loss and degradation have multiple potential effects, ranging from loss of protection to adjacent ecosystems, such coral reefs (Hayden and Granek, 2015), and coastal communities (Badola and Hussain, 2005; Barbier, 2016), depleted seafood resources (Barbier, 2000; Ruitenbeek, 1992) to increases in CO_2 emissions (Lovelock et al., 2017; Pendleton et al., 2012). In recent years, the latter has become a concern in climate change discussions as widespread mangrove deforestation could result in the release of 0.45 Pg CO_2e yr^{-1} to the atmosphere (Pendleton et al., 2012). Having mangrove forests with the greatest soil carbon stock, Indonesia is also the country with the highest potential CO_2 emissions due to mangrove loss, followed by Malaysia, United States, and Brazil (Atwood et al., 2017). Therefore, estimations of soil carbon stocks have become increasingly important to bring attention to the threats of deforestation, which could yield potential CO_2 emissions from soils of ~7.0 Tg CO_2e yr^{-1}(Alongi, 2018; Atwood et al., 2017).

The role of mangroves in shaping geomorphological processes also influences their potential resilience to climate change. The capacity of mangroves to accelerate their rate of soil elevation represents one potential adaptation to the rising sea level (Cahoon and Guntenspergen, 2010). Sediment trapping can result in soil accretion that, in combination with biotic processes, contributes to surface and subsurface soil change (Cahoon et al., 2006; Friess et al., 2019). Contemporary changes of this vertical movement of coastal surfaces can be measured by rod surface elevation table–marker horizon method (RSET-MH), which has been used globally across a range of wetland ecosystems such as mangroves and salt marshes. Over longer time scales, the process of sediment deposition associated with sea-level fluctuation in mangrove coasts contributes to broader geomorphological evolution and is commonly detected by paleoecological reconstruction and radioisotope analysis (Wolanski et al., 2009; Woodroffe, 1992).

Using the sedimentary properties of mangroves to promote their conservation and restoration

Unlike terrestrial forests, although the main carbon in mangroves is found in the soil, it is often excluded from carbon credits (Wylie et al., 2016). The inclusion of the soil component in conservation and restoration programmes can assist in prioritizing management of sites in ways to preserve existing carbon sinks or increase carbon sequestration (Macreadie et al., 2017; Sidik et al., 2018, 2019). Recognizing the value of blue carbon, a growing awareness has risen among policymakers, stakeholders, and communities to protect and restore mangroves (Thomas, 2014). Mangrove restoration for carbon abatement is plausible but tends to fail if the underlying factors that cause the initial degradation or death of the habitat are not first addressed (Lewis et al., 2016). Selecting a successful approach for mangrove restoration requires knowledge of ecohydrology and other processes operating in this dynamic sedimentary environment (Anthony and Goichot, 2019; Lewis, 2005). Hydrodynamic processes can alter the establishment and early development of mangroves, from propagule dispersal to recruitment of sapling stage (Krauss et al., 2008); however, this knowledge is often neglected in restoration and plantation projects, leading to low levels of success (Iftekhar, 2008). Therefore, a key challenge is the translation of science into a format that the public and policymakers can understand and implement in mangrove management.

References

Alongi, D.M., 2002. Present state and future of the world's mangrove forests. Environ. Conserv. 29 (3), 331–349. https://doi.org/10.1017/S0376892902000231.

Alongi, D.M., 2009. The Energetics of Mangrove Forests. Springer Netherlands. https://doi.org/10.1007/978-1-4020-4271-3.

Alongi, D.M., 2014. Carbon cycling and storage in mangrove forests. Annu. Rev. Mar. Sci. 6, 195–219. https://doi.org/10.1146/annurev-marine-010213-135020.

Alongi, D.M., 2018. Blue Carbon: Coastal Sequestration for Climate Change Mitigation. Springer.

Alongi, D.M., Brinkman, R., 2011. Hydrology and biogeochemistry of mangrove forests. In: Levia, D.F., Carlyle-Moses, D., Tanaka, T. (Eds.), Forest Hydrology and Biogeochemistry: Synthesis of Past Research and Future Direction, vol. 216. Springer Netherlands, pp. 203–219. https://doi.org/10.1007/978-94-007-1363-5.

Anthony, E.J., Gardel, A., Proisy, C., Fromard, F., Gensac, E., Peron, C., Walcker, R., Lesourd, S., 2013. The role of fluvial sediment supply and river-mouth hydrology in the dynamics of the muddy, Amazon-dominated Amapá–Guianas coast, South America: A three-point research agenda. J. S. Am. Earth Sci. 44, 18–24. https://doi.org/10.1016/j.jsames.2012.06.005.

Anthony, E.J., Goichot, M., 2019. Sediment Flow in the Context of Mangrove and Restoration: A Rapid Assessment Guidance Manual.

Atwood, T.B., Connolly, R.M., Almahasheer, H., Carnell, P.E., Duarte, C.M., Lewis, C.J.E., Irigoien, X., Kelleway, J.J., Lavery, P.S., Macreadie, P.I., Serrano, O., Sanders, C.J., Santos, I., Steven, A.D.L., 2017. Global patterns in mangrove soil carbon stocks and losses. Nat. Clim. Change 7. https://doi.org/10.1038/NCLIMATE3326.

Badola, R., Hussain, S.A., 2005. Valuing ecosystem functions: an empirical study on the storm protection function of Bhitarkanika mangrove ecosystem, India. Environ. Conserv. 32 (1), 85–92. https://doi.org/10.1017/S0376892905001967.

Barbier, E.B., 2000. Valuing the environment as input: review of applications to mangrove-fishery linkages. Ecol. Econ. 35, 47–61.

Barbier, E.B., 2016. The protective service of mangrove ecosystems: A review of valuation methods. Mar. Pollut. Bull. 109 (2), 676–681. https://doi.org/10.1016/j.marpolbul.2016.01.033.

Bunting, P., Rosenqvist, A., Lucas, R.M., Rebelo, L.-M., Hilarides, L., Thomas, N., Hardy, A., Itoh, T., Shimada, M., Finlayson, C.M., 2018. The global mangrove watch – a new 2010 global baseline of mangrove extent. Remote Sens. 10 (10), 1669.

Cahoon, D.R., Guntenspergen, G.R., 2010. Climate change, sea-level rise, and coastal wetlands. Natl. Wetlands Newsl. 32, 8–12.

Cahoon, D.R., Hensel, P.F., Spencer, T., Reed, D.J., Mckee, K.L., Saintilan, N., 2006. Coastal wetland vulnerability to relative sea-level rise: wetland elevation trends and process controls. In: Verhoeven, J.T.A., Beltman, B., Bobbink, R., Whigham, D.F. (Eds.), Ecological Studies, vol. 190. Springer Berlin Heidelberg, pp. 271–292.

Danielsen, F., Sørensen, M.K., Olwig, M.F., Selvam, V., Parish, F., Burgess, N.D., Hiraishi, T., Karunagaran, V.M., Rasmussen, M.S., Hansen, L.B., Quarto, A., 2005. The Asian tsunami: a protective role for coastal vegetation. Science 310, 643.

Das, S., Vincent, J.R., 2009. Mangroves protected villages and reduced death toll during Indian super cyclone. PNAS 106 (18), 7357–7360.

Donato, D.C., Kauffman, J.B., Murdiyarso, D., Kurnianto, S., Stidham, M., Kanninen, M., 2011. Mangroves among the most carbon-rich forests in the tropics. Nat. Geosci. 4 (5), 293–297. https://doi.org/10.1038/ngeo1123.

Duke, N.C., 1992. Mangrove floristics and biogeography. In: Robertson, A.I., Alongi, D.M. (Eds.), Tropical Mangrove Ecosystems. American Geophysical Union, pp. 63–100.

Duke, N.C., Ball, M.C., Ellison, J.C., 1998. Factors influencing biodiversity and distributional gradients in mangroves. Global Ecol. Biogeography Lett. 7 (1), 27–47.

Feka, N.Z., Ajonina, G.N., 2011. Drivers causing decline of mangrove in West-Central Africa : a review. Int. J. Biovidersity Sci. Manag. 7 (3), 217–230. https://doi.org/10.1080/21513732.2011.634436.

Ferreira, A.C., Lacerda, L.D., 2016. Degradation and conservation of Brazilian mangroves, status and perspectives. Ocean Coastal Manag. 125, 38–46. https://doi.org/10.1016/j.ocecoaman.2016.03.011.

Friess, D.A., Rogers, K., Lovelock, C.E., Krauss, K.W., Hamilton, S.E., Lee, S.Y., Lucas, R., Primavera, J.H., Rjkaran, A., Shi, S., 2019. The state of the world's mangrove forests: past, present and future. Annu. Rev. Environ. Resour. 44, 89–115.

Furukawa, K., Wolanski, E., 1996. Sedimentation in mangrove forests. Mangroves Salt Marshes 1, 3–10.

Furukawa, K., Wolanski, E., Mueller, H., 1997. Currents and sediment transport in mangrove forests. Estuarine, Coastal Shelf Sci. 44 (3), 301–310. https://doi.org/10.1006/ecss.1996.0120.

Gill, A.M., Tomlinson, P.B., 1977. Studies on the growth of red mangrove (*Rhizophora mangle* L .) 4. The adult root system. Biotropica 9 (3), 145–155.

Gopal, B., 2012. Future of wetlands in tropical and subtropical Asia, especially in the face of climate change. Aquat. Sci. 75 (1), 39–61. https://doi.org/10.1007/s00027-011-0247-y.

Hamilton, S.E., Friess, D.A., 2018. Global carbon stocks and potential emissions due to mangrove deforestation from 2000 to 2012. Nat. Clim. Change 8, 240–244. https://doi.org/10.1038/s41558-018-0090-4.

Hayden, H.L., Granek, E.F., 2015. Coastal sediment elevation change following anthropogenic mangrove clearing. Estuarine, Coastal Shelf Sci. 165, 70–74. https://doi.org/10.1016/j.ecss.2015.09.004.

Howard, J., Hoyt, S., Isensee, K., Pidgeon, E., Telszewski, M., 2014. Coastal blue carbon. Methods for assessing carbon stocks and emissions factors in mangroves, tidal salt marshes, and seagrass meadows. In: Conservation International, Intergovernmental Oceanographic Commission of UNESCO, International Union for Conservation of Nature.

Iftekhar, M., 2008. Functions and development of reforested mangrove areas: a review. Int. J. Biovidersity Sci. Manag. 4 (1), 1–14.

Jardine, S., Siikamäki, J., 2014. A global predictive model of carbon in mangrove soils. Environ. Res. Lett. 9 https://doi.org/10.1088/1748-9326/9/10/104013.

Komiyama, A., Ong, J.E., Poungparn, S., 2008. Allometry, biomass, and productivity of mangrove forests: a review. Aquat. Bot. 89 (2), 128–137. https://doi.org/10.1016/j.aquabot.2007.12.006.

Krauss, K.W., Lovelock, C.E., McKee, K.L., López-Hoffman, L., Ewe, S.M., Sousa, W.P., 2008. Environmental drivers in mangrove establishment and early development: a review. Aquat. Bot. 89 (2), 105–127. https://doi.org/10.1016/j.aquabot.2007.12.014.

Lewis, R.R., 2005. Ecological engineering for successful management and restoration of mangrove forests. Ecol. Eng. 24 (4), 403–418. https://doi.org/10.1016/j.ecoleng.2004.10.003.

Lewis, R.R., Milbrandt, E.C., Brown, B., Krauss, K.W., Rovai, A.S., Beever, J.W., Flynn, L.L., 2016. Stress in mangrove forests: early detection and preemptive rehabilitation are essential for future successful worldwide mangrove forest management. Mar. Pollut. Bullet. https://doi.org/10.1016/j.marpolbul.2016.03.006.

Lovelock, C.E., Atwood, T., Baldock, J., Duarte, C.M., Hickey, S., Lavery, P.S., Masque, P., Macreadie, P.I., Ricart, A.M., Serrano, O., Steven, A., 2017. Assessing the risk of carbon dioxide emissions from blue carbon ecosystems. Front. Ecol. Environ. 15 (5), 257–265. https://doi.org/10.1002/fee.1491.

Lugo, A.E., Snedaker, S.C., 1974. The ecology of mangroves. Annu. Rev. Ecol. Syst. 5, 39–64.

Macreadie, P.I., Nielsen, D.A., Kelleway, J.J., Atwood, T.B., Seymour, J.R., Petrou, K., Connolly, R.M., Thomson, A.C.G., Trevathan-tackett, S.M., Ralph, P.J., 2017. Can we manage coastal ecosystems to sequester more blue carbon ? Front. Ecol. Environ. 15 (4), 206–213. https://doi.org/10.1002/fee.1484.

McKee, K.L., 1996. Growth and physiological responses of neotropical mangrove seedlings to root zone hypoxia. Tree Physiol. 16, 883–890.

Osland, M., Feher, L., Griffith, K., Cavanaugh, K., Enwright, N., 2017. Climatic controls on the global distribution, abundance , and species richness of mangrove forests. Ecol. Monogr. 87 (2), 341–359.

Paul, A.K., Ray, R., Kamila, A., Jana, S., 2017. Mangrove degradation in the Sundarbans. In: Finkl, C.W., Makowski, C. (Eds.), Coastal Wetlands: Alteration and Remediation. Springer, pp. 357–392. https://doi.org/10.1007/978-3-319-56179-0.

Pendleton, L., Donato, D.C., Murray, B.C., Crooks, S., Jenkins, W.A., Sifleet, S., Craft, C., Fourqurean, J.W., Kauffman, J.B., Marbà, N., Megonigal, P., Pidgeon, E., Herr, D., Gordon, D., Baldera, A., 2012. Estimating global "blue carbon" emissions from conversion and degradation of vegetated coastal ecosystems. PloS One 7 (9). https://doi.org/10.1371/journal.pone.0043542.

Rahman, M.M., Rahman, M.M., Islam, K.S., 2010. The causes of deterioration of Sundarban mangrove forest ecosystem of Bangladesh. Aquacult., Aquarium, Conserv. Legis., Bioflux 3 (2), 77–90.

Rovai, A.S., Twilley, R.R., Castañeda-moya, E., Riul, P., Cifuentes-jara, M., Manrow-villalobos, M., Horta, P.A., Simonassi, J.C., Fonseca, A.L., Pagliosa, P.R., 2018. Global controls on carbon storage in mangrove soils. Nat. Clim. Change 8 (6), 534–538. https://doi.org/10.1038/s41558-018-0162-5.

Ruitenbeek, J., 1992. Mangrove Management: An Economic Analysis of Management Options with a Focus on Bintuni Bay, Irian Jaya..

Sánchez-Núñez, D., Pineda, J., Osorio, A., 2020. From local- to global-scale control factors of wave attenuation in mangrove environments and the role of indirect mangrove wave attenuation. Estuarine, Coastal Shelf Sci. 245, 106926.

Sanderman, J., Hengl, T., Fiske, G., Solvik, K., Adame, M.F., Benson, L., Bukoski, J.J., Carnell, P., Cifuentes-jara, M., Donato, D., Duncan, C., Eid, E.M., Ermgassen, P., Lewis, C.J.E., Macreadie, P.I., Glass, L., Gress, S., Jardine, S.L., Jones, T.G., Landis, E., 2018. A global map of mangrove forest soil carbon at 30 m spatial resolution. Environ. Res. Lett. 13 https://doi.org/10.1088/1748-9326/aabe1c.

Sidik, F., Adame, M.F., Lovelock, C.E., 2019. Carbon sequestration and fluxes of restored mangroves in abandoned aquaculture ponds. J. Indian Ocean Reg. 1–16. https://doi.org/10.1080/19480881.2019.1605659.

Sidik, F., Supriyanto, B., Krisnawati, H., Muttaqin, M.Z., 2018. Mangrove conservation for climate change mitigation in Indonesia. Wiley Interdiscip. Rev. Clim. Change 9 (5), e529. https://doi.org/10.1002/wcc.529.

Smith, T.J., 1992. Forest structure. In: Robertson, A.I., Alongi, D.M. (Eds.), Tropical Mangrove Ecosystems. American Geophysical Union, p. 329.

Spalding, M.D., Ruffo, S., Lacambra, C., Meliane, I., Hale, L.Z., Shepard, C.C., Beck, M.W., 2014. The role of ecosystems in coastal protection: adapting to climate change and coastal hazards. Ocean Coastal Manag. 90, 50–57. https://doi.org/10.1016/j.ocecoaman.2013.09.007.

Thomas, N., Lucas, R., Bunting, P., Hardy, A., Rosenqvist, A., Simard, M., 2017. Distribution and drivers of global mangrove forest change, 1996 – 2010. PloS One 12 (6), e0179302.

Thomas, S., 2014. Blue carbon : Knowledge gaps, critical issues, and novel approaches. Ecol. Econ. 107, 22–38. https://doi.org/10.1016/j.ecolecon.2014.07.028.

Toorman, E.A., Anthony, E.J., Augustinus, P.G.E.F., Gardel, A., Gratiot, N., Homenauth, O., Huybrechts, N., Monbaliu, J., Moseley, K., Naipal, S., 2018. Interaction of mangroves, coastal hydrodynamics and morphodynamics along the coastal fringes of the Guianas. In: Makowski, C., Finkl, C. (Eds.), Threats to Mangrove Forests. Springer, pp. 429–473. https://doi.org/10.1007/978-3-319-73016-5_20.

Twilley, R.R., Rovai, A.S., Riul, P., 2018. Coastal morphology explains global blue carbon distributions. Front. Ecol. Environ. 16 (9), 1–6. https://doi.org/10.1002/fee.1937.

Willemsen, P.W.J.M., Horstman, E.M., Borsje, B.W., Friess, D.A., Dohmen-janssen, C.M., 2016. Geomorphology Sensitivity of the sediment trapping capacity of an estuarine mangrove forest. Geomorphology 273, 189–201. https://doi.org/10.1016/j.geomorph.2016.07.038.

Wolanski, E., Brinson, M.M., Cahoon, D.R., Perillo, G.M., 2009. Coastal wetlands: a synthesis. In: Perillo, G.M.E., Wolanski, E., Cahoon, D.R., Brinson, M.M. (Eds.), Coastal Wetlands an Integrated Ecosystem Approach. Elsevier, pp. 1–62.

Woodroffe, C.D., 1992. Mangrove sediments and geomorphology. In: Robertson, A.I., Alongi, D.M. (Eds.), Tropical Mangrove Ecosystems. American Geophysical Union, p. 329.

Woodroffe, C.D., Davies, G., 2009. The morphology and development of tropical coastal wetlands. In: Perillo, G.M.E., Wolanski, E., Cahoon, D.R., Brinson, M.M. (Eds.), Coastal Wetlands an Integrated Ecosystem Approach. Elsevier, pp. 65–84.

Wylie, L., Sutton-grier, A.E., Moore, A., 2016. Keys to successful blue carbon projects : lessons learned from global case studies. Mar. Pol. 65, 76–84. https://doi.org/10.1016/j.marpol.2015.12.020.

Zhang, K., Liu, H., Li, Y., Xu, H., Shen, J., Rhome, J., Smith, T.J., 2012. The role of mangroves in attenuating storm surges. Estuarine, Coastal Shelf Sci. 102–103, 11–23. https://doi.org/10.1016/j.ecss.2012.02.021.

PART I

Biogeomorphological processes

CHAPTER

1

Biogeomorphic evolution and expansion of mangrove forests in New Zealand's sediment-rich estuarine systems

Andrew Swales, Mark Pritchard, Graham B. McBride

National Institute of Water and Atmospheric Research (NIWA), Hamilton, Waikato, New Zealand

OUTLINE

1. Introduction 4

2. Biogeomorphology 4

3. Mangroves in New Zealand 6

3.1 Geographic distribution, environmental settings, and forest types 6

3.2 Historical patterns 9

3.3 Biophysical drivers 10

4. Estuary evolution 11

5. Case study—sand flat to mangrove forest 14

5.1 Environmental setting 15

5.2 Mangrove-seedling recruitment 16

5.3 Biogeomorphic evolution 18

5.4 Surface-elevation dynamics 23

6. Future shock—what's the fate of New Zealand's mangrove forests? 30

6.1 Elevation capital and future fate 30

6.2 Estuaries—predicting the shape of things to come 30

6.3 Legacy sediments 35

7. Summary and conclusions 35

Acknowledgments 38

References 39

Dynamic Sedimentary Environments of Mangrove Coasts
https://doi.org/10.1016/B978-0-12-816437-2.00013-6

1. Introduction

Mangrove forests form a major component of intertidal habitats on river deltas, muddy shores, and estuaries of the tropical Indo-Pacific, Atlantic-East Pacific Regions, and in the high-latitude temperate estuaries of Australasia and southern Africa (Giri et al., 2011; Woodroffe et al., 2016). The largest mangrove forests occur in the fluvial sediment–rich and meso- and macrotidal settings where accumulating fine sediment has formed extensive deltas and low-gradient mudflats (Woodroffe et al., 2016). These fluvial sediment mangrove forests contrast markedly with forests that develop in autochthonous systems, such as oceanic low islands and karst coasts where peat development and/or carbonate production are the major sedimentary processes controlling mangrove forest evolution (Snedaker, 1995; McKee et al., 2007).

Mangrove forest loss intensified during the 20th century primarily due to activities associated with aquaculture, land reclamation, and overutilization by growing human populations and today occupy an estimated 85–138 × 10^3 km^2 (Giri et al., 2011; Hamilton and Casey, 2016; Richards and Friess, 2016). There is some evidence that rates of mangrove forest loss have declined particularly in Southeast Asia over the last decade through improved management and restoration efforts (Richards and Friess, 2016). The extent of remaining mangrove forests, however, represents less than half of that which existed ~50 years ago (Spiers, 1999; Spalding et al., 2010; Giri et al., 2011).

Although the world's largest mangrove forests occur in fluvial sediment–rich settings, studies based on observations of the biogeomorphic evolution of these systems over annual to decadal time scales are rare (Walsh and Nittrouer, 2004; Swales et al., 2015, 2019; Woodroffe et al., 2016; Sidik et al., 2016). Many of the sediment-rich mangrove forests of New Zealand's northern estuaries have developed relatively recently (i.e., during the 20th century) so that we have accurate records of the environmental changes. These data include mapping of mangrove-habitat expansion and sedimentary records for a range of estuarine geomorphologies. Research undertaken over the last decade has identified key processes controlling mangrove ecology and provided new insights into the role of mangrove forests in the geomorphic evolution of these estuaries.

In this chapter, we identify the key biophysical processes controlling mangrove ecology and the role of mangrove forests in the geomorphic evolution of New Zealand's sediment-rich northern estuaries. We draw on the historical record of mangrove-habitat expansion over the last several decades and insights from sedimentary records for a range of estuary types based on their geomorphology. A case study of the biogeomorphic evolution of a mangrove forest that has occurred in a large coastal embayment (Firth of Thames) since the early 1960s is presented. Finally, using data and models, we discuss how New Zealand's mangrove forests are likely to respond to future changes in sediment supply and rates of sealevel rise (SLR).

2. Biogeomorphology

Biogeomorphology (or ecogeomorphology) focuses on the interface between the earth and biological sciences, which considers the two-way feedbacks between ecological and geomorphic processes and how these processes influence ecosystems and landform evolution (Viles, 1988; Naylor et al., 2002; Wheaton et al., 2011). Although the term "biogeomorphology" was first coined by Viles (1988), early ecologists *"recognised the need for the documentation of geomorphic form and process"* (e.g., Cowles, 1899, 1901; Clements, 1916) to explain species distributions (Hupp et al., 1995). Integrated studies explicitly exploring the two-way interactions between ecology and geomorphology began in the 1950s (Hall and Smith, 1955; Olsen, 1958; Hack and Goodlett, 1960; Imeson, 1976; Alpert, 1985).

The role of biology in landscape evolution in terrestrial (Kirby, 1995; Phillips, 1995; Goodson et al., 2002; Dietrich and Perron, 2006; Osterkamp and Hupp, 2010) as well as marine environments (D'Alpaos et al., 2007; Murray et al., 2009; Corenblit et al., 2011; Marani et al., 2010, 2013; Coco et al., 2013; Van Maanen et al., 2015) is an emerging theme in contemporary geomorphology. Dietrich and Perron (2006) posed the following question in relation to landscape evolution: "*If life had not arisen, would the tectonic and climatic processes that drive uplift and erosion of landscapes be significantly different*?" Bird (1986) posed this question on the role of mangroves in the geomorphic evolution of estuaries: "*whether mangroves promote sedimentation, and thereby the evolution of depositional landforms, or whether they simply occupy sites that have become ecologically suitable, moving in to colonise (and possibly thereafter to stabilise and protect) an intertidal morphology that would have formed independently in their absence*" (Bird, 1986). Geomorphologists and ecologists have sought to address this question for over a century (Vaughan, 1909; Davis, 1940; Chapman and Ronaldson, 1958; Scoffin, 1970; Thom et al., 1975; Vann, 1980; Bird, 1986; Walsh and Nittrouer, 2004; McKee, 2011). The prevailing scientific consensus regarding the biogeomorphic evolution of mangrove forests in fluvial sediment–rich systems is that mangroves only colonize intertidal flats when local environmental conditions become "ecologically suitable" (Bird, 1986), with substrate elevation and stability being key factors (Balke et al., 2015; Oh et al., 2017). Once established, do mangroves in turn enhance sedimentation relative to unvegetated tidal flats through biophysical feedbacks (Fig. 1.1)?

Over the last decade or so, computational models of coastal wetland biogeomorphology have been developed that explicitly account for the two-way coupling of coastal wetlands with physical processes (i.e., hydrodynamics, sediment transport, and deposition) and resulting landscape evolution occurring over a range of time scales (D'Alpaos et al., 2007; Murray et al., 2009; Marani et al., 2010, 2013; Horstman et al., 2015; Van Maanen et al., 2015; Willemsen et al., 2016). Exploratory/abstracted modeling approaches are increasingly being used to overcome the limitations of deterministic (bottom-up) numerical models. A critical issue that remains to be solved, however, is how to accurately simulate long-term geomorphological evolution over meaningful spatial and temporal scales (Coco et al., 2013). Various biophysical processes may also operate over a range of time scales (Phillips, 1995). Plant communities respond to seasonal environmental drivers, whereas landscape evolution may be strongly influenced by infrequent episodic events. This begs the question to what extent do plant communities and geomorphology interact (Phillips, 1995). In the case of mangroves, Woodroffe et al. (2016) have argued that there is still insufficient understanding of the ecology and morphodynamics of mangrove forest sedimentary systems to confidently predict their long-term evolution using existing models. The experimental and observational data, including measurements of processes and long-term geomorphic evolution that are required to underpin and validate these biogeomorphic models of mangrove systems, are still rare. These fundamental questions and knowledge gaps about the role of mangroves in the geomorphic evolution of estuaries have been explored for a New Zealand estuarine system (Firth of Thames) using observations and modeling across a range of time scales (i.e., from discrete storm events to decades) (Swales et al., 2015, 2016, 2019). The major findings of this research are described in the case study section.

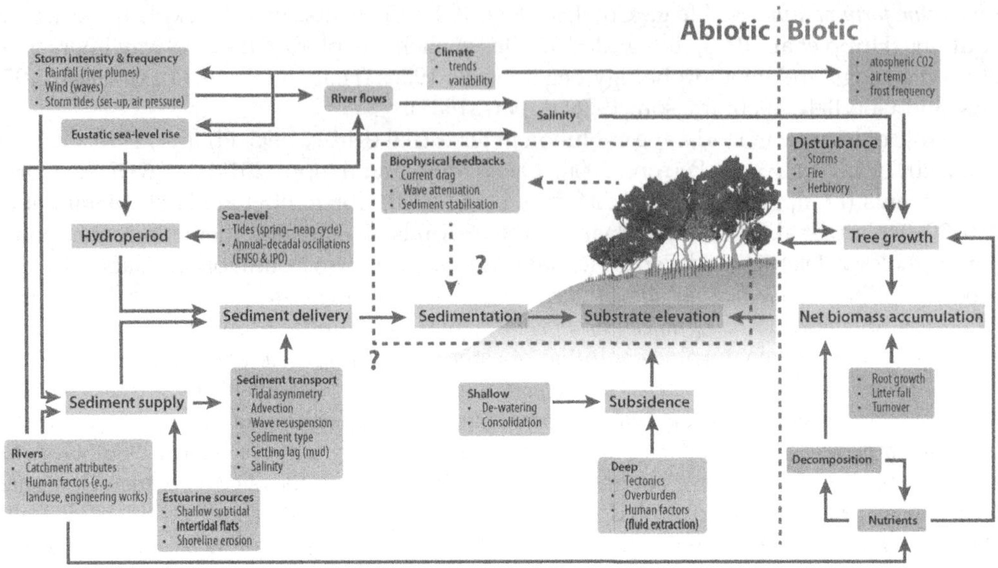

FIGURE 1.1 Conceptual model of processes influencing mangrove forest biogeomorphology in fluvial sediment–rich estuaries. Key processes controlling sedimentation, surface elevation gain, and mangrove forest development (*blue* (*light gray in print version*) *boxes*) and associated environmental drivers (*green* (*dark gray in print version*) *boxes*). *Biotic components of this diagram adapted from Cahoon, D.R., Guntenspergen, G.R., January–February 2010. Climate change, sea-level rise and coastal wetlands. Natl. Wetl. Newsl., 8–12.*

3. Mangroves in New Zealand

3.1 Geographic distribution, environmental settings, and forest types

Mangrove forests composed of monospecific stands of the gray mangrove *Avicennia marina* var *australasica* occur in the numerous estuaries of northern New Zealand. This mangrove is also known as Mānawa, which is the name given to *A. marina* by the indigenous Māori people who colonized New Zealand around 1300 AD (Wilmshurst et al., 2008). This subspecies also occurs in the estuaries and on the coasts of southeastern Australia, New Caledonia, and Lord Howe Island (Duke, 1991). The southern latitudinal limit of mangroves in New Zealand is at 38.08°S (de Lange and de Lange, 1994), which is close to the southernmost limit for mangroves globally (i.e., 38.9°S, Corner Inlet, South Victoria, Australia; Bird, 1986). The present southern latitudinal limit for mangroves in New Zealand occurs at the Kawhia (west coast) and Ohiwa Harbours (east coast) (Fig. 1.2). The southern limit for mangroves in Australasia is controlled by low winter air temperatures and the frequency of lethal frosts (e.g., Chapman and Ronaldson, 1958; Saintilan et al., 2009). Physiological limitations at low air temperatures may also be causal factors (Walbert, 2002; Beard, 2006). The North Island's biogeography and oceanography also appear to limit the dispersal of mangrove propagules along the coast (de Lange and de Lange, 1994). Although studies of *A. marina* propagule dispersal show that they can travel tens of kilometers or more (e.g., Clarke, 1993), they may not remain viable for more than several days (de Lange and de Lange, 1994). Furthermore, the relatively small number of, and distance

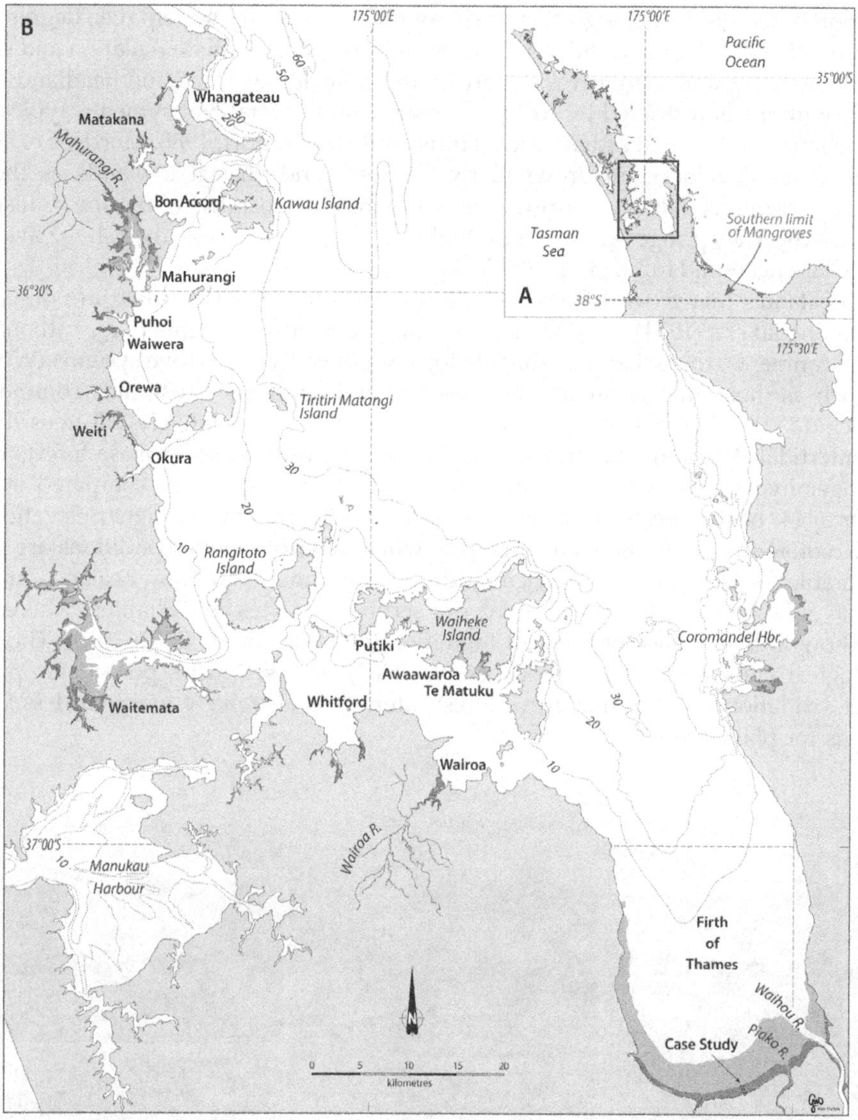

FIGURE 1.2 Distribution of *Avicennia. marina* mangroves in the estuaries of New Zealand's upper North Island: (A) The present southern latitudinal limit of mangroves is 38°S and occurs at Kawhia Harbour on the west coast and Ohiwa Harbour on the east coast; (B) location of estuaries on Auckland's east coast and the Firth of Thames.

between, estuaries located south of mangroves present southern limit suitable for colonization are effective barriers to mangroves increasing their range in New Zealand.

New Zealand's *A. marina* mangrove forests occur in a range of estuarine geomorphic settings (Fig. 1.3). The Roy et al. (2001) classification scheme for south-eastern Australian estuaries is also broadly applicable to New Zealand estuaries and describes mangrove distribution based on the geological maturity of the sedimentary system (i.e., degree of infilling) and relative dominance of tides and waves in controlling water circulation and sediment transport. New Zealand mangroves mainly occur in barrier-enclosed and headland-enclosed estuaries, with inlets restricted by rocky headlands, and coastal embayments, typically with small catchments relative to estuary area (Hume and Herdendorf, 1988; Morrisey et al., 2010). These estuaries developed in drowned river valleys and coastal lowlands as they were flooded by rising sea levels during the most recent postglacial marine transgression (Hume, 2003). They range in size (i.e., high-tide area) from less than 1—~750 km^2 (NZ Estuary Classification, Hume et al., 2007).

New Zealand's mangrove forests include several functional types that are described by Lugo and Snedaker's (1974) classification. Although developed originally for carbonate-sediment environments, this scheme is suitable for low-diversity mangrove systems (Woodroffe, 1992), such as those that occur in New Zealand and southeast Australia. Common forest types include (1) fringe and (2) riverine forests composed of relatively tall trees that flank mid-to-intertidal flats and river/tidal channels, respectively. Trees in these forests typically have longer hydroperiods and receive more nutrients and sediment compared to interior (3) scrub or (4) basin forests. Trees in these interior forests grow at higher elevations and/or more remote from tidal flats and channels where environmental conditions are typically less favorable for tree growth. Nutrient-addition experiments in New Zealand's mangrove forests have shown that tree growth at interior scrub forest sites are N limited, whereas fringe forests show little evidence of nutrient limitation (Lovelock et al., 2007, 2010). The reduced "hydrological connectivity" (Lovelock et al., 2007) and sediment/pore-water properties (e.g., pH, oxidation state) of mangrove forest interiors is also likely to impart less favorable conditions for plant growth.

FIGURE 1.3 Examples of New Zealand's *Avicennia. marina* mangrove estuaries: (A) southern Firth of Thames, Waikato Region (coastal embayment, 37.25°S, 175.40°E); and (B) Waiwera estuary (headland/barrier-enclosed 37.54°S, 174.69°E), (C) Whangapoua (barrier enclosed, 36.14°S, 175.41°E), and (D) Orewa estuary (barrier enclosed, 36.60°S, 174.69°E), Auckland Region.

3.2 Historical patterns

Mangrove forest loss occurred in New Zealand estuaries due to human activities over the last century or so associated with land reclamation for port infrastructure, urban development and agriculture, road and causeway construction, landfills, and stock grazing. Historical changes (both losses and gains) in the extent of New Zealand's mangrove forests have not been accurately quantified because substantial habitat change occurred prior to systematic aerial photographic surveys that began in the 1930s (Morrisey et al., 2010).

Contrary to the global pattern of ongoing mangrove forest loss, rapid expansion of mangrove forests has occurred in many New Zealand estuaries. Rates of mangrove-habitat expansion since the 1930s have averaged 4.1% yr^{-1} (range −0.2%−20.2% yr^{-1}) in terms of the intertidal area occupied. This rate of habitat expansion is some twofold higher than for temperate *Avicennia* mangrove forests of southeastern Australia (average 2.1% yr^{-1}; range 0.7%−9.1% yr^{-1}) (Morrisey et al., 2010). In New Zealand, mangrove-habitat expansion has primarily occurred by colonization of unvegetated intertidal flats seaward of existing forests (Burns and Ogden, 1985; Ellis et al., 2004; Morrisey et al., 2010; Lundquist et al., 2014; Swales et al., 2015; Horstman et al., 2018). Mangrove forests presently occupy about 260 km^2 of intertidal flats in New Zealand estuaries (Spalding et al., 2010). This habitat expansion has coincided with large-scale catchment deforestation from the mid-1800s, resulting in increased soil erosion and formation of intertidal flats as the rate of estuary infilling accelerated (e.g., Swales et al., 1997, 2002a,b).

Estuary infilling, associated with a shift from sand- to mud-dominated systems, resulted in the reduction in the areal extent of subtidal habitats and loss or degradation of ecosystems. This degradation has been characterized by loss of plants and animals sensitive to increased water turbidity, reduced light levels, and fine sediment deposition (e.g., seagrass meadows, filter-feeding bivalves) (Inglis, 2003; Thrush et al., 2004). For example, extensive areas of mature mangrove forest had developed in the Kaipara Harbour (Northland NZ) by the 1920s and infrequently inundated by spring tides (Ferrar, 1934). Historical accounts of settlers dating back to the 1860s describe white-sand beaches fringing the shoreline. These accounts indicate that mangroves were not widespread in the large tidal rivers of the northern Kaipara. Slash and burn agriculture resulting in catchment disturbance was practiced by Māori over the several hundred years prior to European arrival. Sedimentary records suggest that Māori land use activities had a minor impact on these northern estuaries. Background sediment accumulation rates (SAR) over the several thousand years prior to European settlement averaged 0.1−1 mm/yr (e.g., Hume and McGlone, 1986; Sheffield et al., 1995; Swales et al., 1997, 2002a,b; Thrush et al., 2004; Oldman et al., 2009). These background SAR were an order of magnitude lower than the sedimentation that followed large-scale catchment disturbance due to timber extraction, mining, and the development of pastoral agriculture from the mid-1800s. Historical records suggest that mangrove forests formed in estuaries as they infilled with eroded soils in the decades following large-scale catchment deforestation (Swales et al., 2011) (Fig. 1.4).

This rapid expansion of mangrove habitat in New Zealand estuaries has also polarized coastal communities over the last decade or so (Harty, 2009; Horstman et al., 2018). Environmental groups have advocated protection of mangrove habitat, whereas a growing number of community groups have formed with a major purpose being to lobby local government

FIGURE 1.4 Example of tidal-creek infilling and mangrove-habitat expansion (Mangemangeroa, east Auckland), 1863 to 2016. View looking downstream from the Whitford Road Bridge. Major environmental changes include lateral growth of the intertidal flats and associated reduction in channel size, consistent with progressive loss of tidal prism and development of a mangrove forest of the intertidal flats. *Photo credits: Bridge over the Mangemangeroa Creek near Howick, 1860s. John Kinder photographs. MSS & Archives 2009/7, item PH 233. Special Collections, University of Auckland Libraries and Learning Services.*

agencies to manage the expansion and/or reduce the area of mangrove habitat in estuaries (Harty, 2009; Lundquist et al., 2014; Horstman et al., 2018). Mangrove removal campaigns have typically not realized anticipated ecological benefits such as reestablishment of seagrass meadows and shellfish beds (Bulmer et al., 2017; Horstman et al., 2018). Environmental management agencies also increasingly recognize that mangrove-habitat expansion is a symptom of estuary infilling resulting from land use activities that have increased soil erosion in catchments.

3.3 Biophysical drivers

Mangrove forests occupy the intertidal zone above mean sea level (MSL) (Galloway, 1982; Bird, 1986; Ellison, 1993), although the position of the lower elevation of mangrove forests varies about MSL due to local conditions. For example, mangrove forests occupy most of their potential habitat down to MSL in sheltered tidal creeks and smaller fetch-limited estuaries in New Zealand (Auckland Region) estuaries (Swales et al., 2009). The lower elevation limit for *A. marina* forests can be decimeters higher than MSL due to wave exposure and increasing tidal range (e.g., Swales et al., 2009; Balke et al., 2015; Barker et al., 2015).

The approximate MSL minimum for mangrove forest distribution is related to the underlying physiological tolerance of mangroves to emersion. Regular daily exposure at low tide enables seedlings growing in muddy substrates in particular to maintain an adequate oxygen supply to their tissues (Clarke and Hannon, 1970; Curran et al., 1986; Hovenden et al., 1995). A corollary of this physiological constraint on mangrove establishment is that the geomorphic evolution of estuaries and coastal sedimentary landforms exerts a first-order control on mangrove forest ecology through sediment transport, deposition, and resulting vertical accretion of intertidal flats. Mangroves colonize intertidal flats only after they become "ecologically suitable" (Bird, 1986). Estuarine sedimentation builds intertidal flats that provide habitat that is potentially suitable for mangrove colonization. Therefore, although intertidal flat development provides the opportunity for mangrove forest establishment or expansion, seedling recruitment depends on biophysical factors that control the likelihood of propagule survival and substrate disturbance. Key factors include wave exposure and timing of propagule production relative to changes in tidal-flat hydroperiod (i.e., frequency and duration of tidal inundation) that occur during the spring–neap tidal cycle, stability of the substrate and bed-level dynamics (e.g., Lovelock et al., 2010; Balke et al., 2015). Biostabilization of intertidal sediments by microorganisms can reduce the susceptibility of intertidal-flat sediment to erosion (e.g., Widdows et al., 2004), thereby promoting propagule establishment. Predation of propagules (e.g., Clarke and Kerrigan, 2002) will reduce recruitment.

Sea-level variations in estuaries over seasonal to decadal time scales will also influence recruitment success. These variations are also superimposed on the long-term trend of SLR induced by climate warming (Nerem et al., 1998; Church and White, 2011; Watson et al., 2015). In the Pacific Region, major interannual and decadal cycles include El Niño-Southern Oscillation and the Inter-decadal Pacific Oscillation. These large-scale processes result in decimeter-scale variability in MSL and wind climate from year to year (Goring and Bell, 1999; Hannah, 2004). Crucially during the summer months, when mangrove propagules are produced, lower than normal sea levels during El Niño events and calm or offshore winds coinciding with spring-to-neap tidal phases provide disturbance-free "windows of opportunity" over several days or more (Balke et al., 2015) for stranded mangrove propagules to establish on intertidal substrates (Thom, 1967; Chapman, 1976; Clarke and Myerscough, 1993; Balke et al., 2015; Lovelock et al., 2015a). Over longer time scales (i.e., decades), the biogeomorphic evolution of mangrove forests is also influenced by feedbacks that influence sedimentation processes (e.g., negative feedback between surface elevation and hydroperiod, Swales et al., 2015; Horstman et al., 2015), sediment supply, and accommodation space (Walsh and Nittrouer, 2004; McKee et al., 2007; Alongi, 2008; Swales et al., 2016; Willemsen et al., 2016; Woodroffe et al., 2016). Thus, biophysical processes across a range of spatial and temporal scales exert a strong influence on mangrove forest evolution in estuaries and on muddy coasts. Over geological to historical time scales, estuary infilling results in large-scale environmental changes. Subtidal areas and water-depth decrease and as a result hydrodynamic conditions, sediment properties and accumulation rates, geomorphological characteristics, and ecosystems change.

4. Estuary evolution

New Zealand's estuaries have progressively infilled with fluvial and marine sediments over the last 6000–7000 years since their formation following the last ice age. Stages of development range from youthful systems that retain a substantial proportion of their original tidal volume to mature estuaries dominated by river discharge. In semimature estuaries, sediment infilling is usually associated with the expansion of accreting intertidal flats that progressively displace subtidal basins. The rate of estuary infilling primarily reflects the original volume of the tidal basin, rate of sediment supply, and changes in sea level. Evidence from global salt marsh sedimentary records indicate that low rates of sea-level change (i.e., tenths of mm/yr) persisted until as recently as the early 20th century. Both sedimentary and tide-gauge records show that average rates of SLR have increased to the order of mm/yr in the modern era (Church et al., 2013). The biogeomorphic evolution of estuaries can be conceptualized as developmental stages (youth, middle, and old age) that are characterized by a progressive shift from subtidal to intertidal systems (Fig. 1.5).

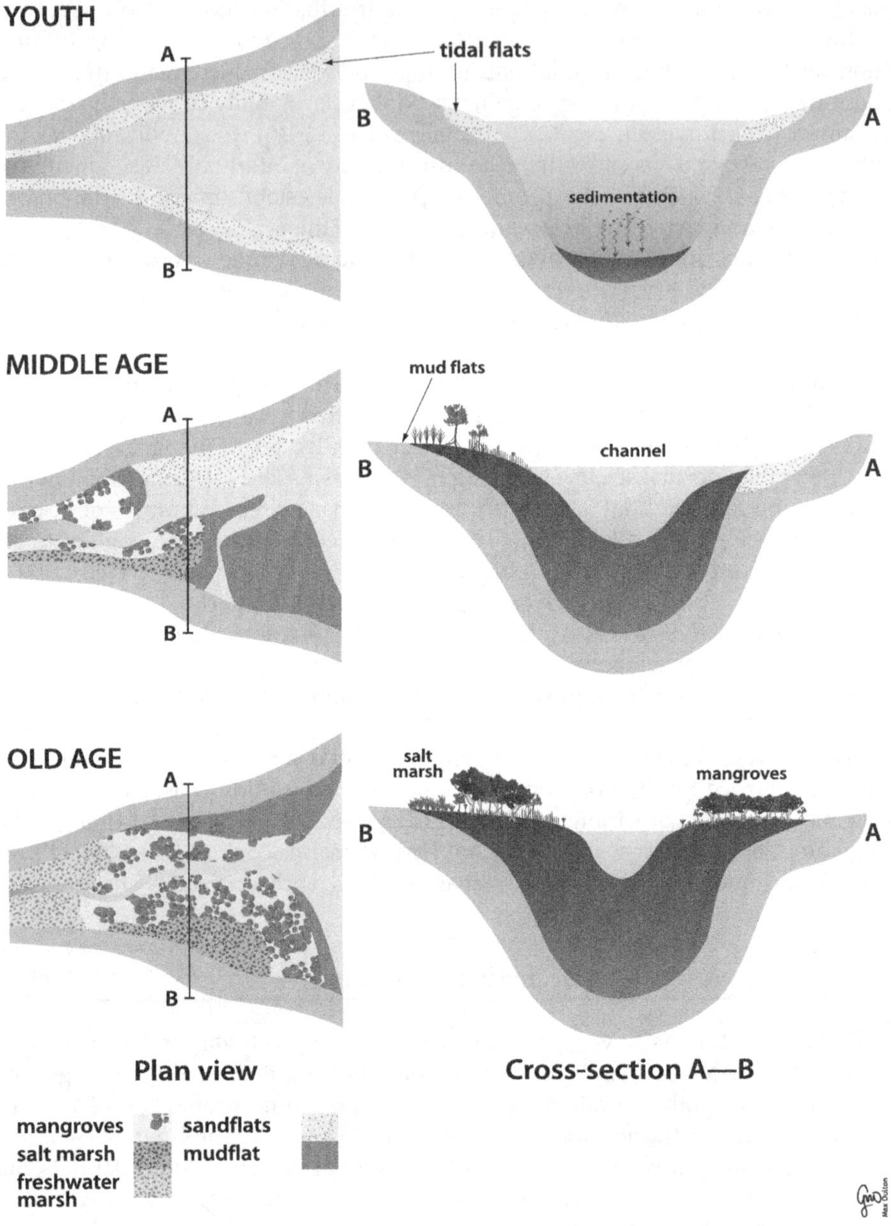

FIGURE 1.5 Biogeomorphic evolution of estuaries. The estuaries that have formed during the Holocene follow a cycle of development determined by the size (volume) of the ancestral basin/flooded river valley, the rate of sediment input, and trapping efficiency of the system. The sediment trapping efficiency typically reduces as estuaries infill due to reduction in accommodation volume and increased efficiency of sediment remobilization by fetch-limited waves.

Tidal ranges in barrier-enclosed estuaries and lagoons are reduced by sand deposits at their inlets, so that tidal currents are weak and wind waves and wind-driven circulation control sediment transport on tidal flats (Roy et al., 2001). In northern New Zealand, many drowned river valley estuaries have reached an advanced stage of infilling since sea levels stabilized about 6500 years ago and sediment transport on intertidal flats is controlled by wind waves rather than tidal currents (e.g., Green et al., 1997; Swales et al., 2004; Green and Coco, 2014).

Given that the geomorphic maturity of estuaries is strongly influenced by sediment supply, then a relationship should exist between the extent of estuary infilling and catchment size. This catchment-estuary infilling relationship has been demonstrated for large drowned-valley systems in New South Wales (Australia) that have high-tide surface areas of tens of km^2 (Fig. 1.7, Roy et al., 2001) but is not consistent for smaller estuaries. Data for a subset of 18 mangrove estuaries located on the east coast of New Zealand's North Island (Auckland and Waikato Regions; Swales et al., 2009, 2015; Hume et al., 2007) have been analyzed to determine relationships between intertidal-flat area above MSL (i.e., potential mangrove habitat) and catchment sediment load. In this analysis, the area of potential mangrove habitat (i.e., intertidal flat above MSL, A_M) as a proportion of the high-tide estuary area (A_E) was compared to the ratio of annual catchment sediment yield to estuary tidal prism volume (Fig. 1.6).

The tidal prism was calculated as the intertidal volume based on an average spring-tide range (Hume et al., 2007). The analysis shows that potential mangrove habitat is well predicted by the ratio of annual catchment sediment yield to the estuary tidal prism volume (SY_V/Ω) ($r^2 = 0.69$, $P = < 0.001$). Coastal embayment–type estuaries are less infilled due to their relatively small catchment sediment supply and consequently have smaller intertidal areas (Fig. 1.6). By contrast, barrier-enclosed estuaries formed in drowned river valleys are substantially more infilled than coastal embayments due to their larger sediment supply.

Barrier-enclosed estuaries therefore have proportionally larger intertidal areas above MSL suitable for mangrove forest development. These estuaries have also experienced the largest percentage increases in mangrove-habitat areal extent since the 1940s, ranging from 0.5% to 8.5% yr^{-1} between the 1940s/1960s and 2007 (Swales et al., 2009). Specific catchment sediment yields for the range of land uses in these largely rural lowland catchment–estuary systems have averaged ~120 t km^2 yr^{-1} and varied from ~80 t km^2 yr^{-1} (funnel-type estuaries) to ~120 t km^2 yr^{-1} (coastal embayments) and up to 160 t km^2 yr^{-1} (barrier-enclosed estuaries) (Swales et al., 2009). Specific sediment yields for these lowland northern catchments that are within the range of New Zealand mangroves are relatively modest. By comparison, major river systems draining the North Island's mountainous central and eastern regions yield hundreds to several thousand tonnes km^2 yr^{-1} of sediment (Hicks et al., 2011). Mangrove-habitat expansion in New Zealand's northern estuaries is consistent with patterns observed in modern temperate and tropical mangrove forests that have increased their distribution on river deltas and muddy coasts where fluvial sediments feed rapid accretion of intertidal flats (Thom, 1967; Panapitukkul et al., 1998; Bedin, 2001; Walsh and Nittrouer, 2004; Carvalho do Amaral et al., 2006; Swales et al., 2015).

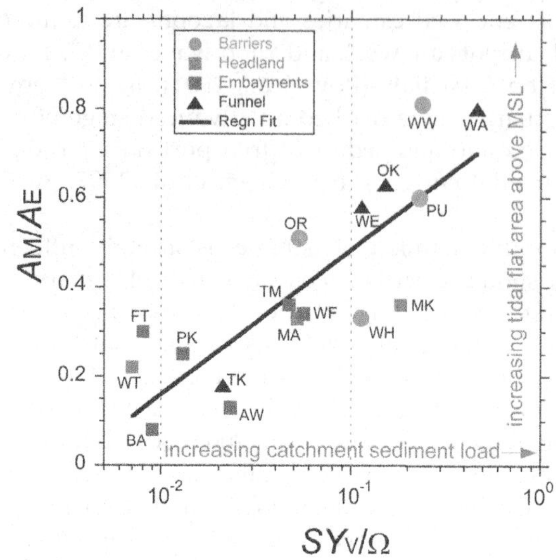

FIGURE 1.6 Auckland east-coast estuaries—relationship between catchment annual sediment yield to estuary tidal prism volume (SY_V/Ω) ratio and potential mangrove habitat (i.e., interidal area above mean sea level) as a proportion of estuary high-tide area (A_M/A_E). **Model fit**: $A_M/A_E = 0.139{*}\text{LN}(SY_V/\Omega) + 0.808$ ($r^2 = 0.69$, $P = < 0.001$). **Geomorphic class** (after Hume and Herdendorf, 1988): (1) Barrier-enclosed estuaries with inlets formed by Holocene spits. Estuaries: Waiwera (WW), Puhoi (PU), Orewa (OR), and Whangateau (WH); (2) Headland-enclosed estuaries with inlets restricted by rocky headlands. Estuaries: Matakana (MK), Mahurangi (MA), and Waitemata (WT); (3) Coastal embayments typically with small catchments. Estuaries: Te Matuku (TM), Whitford (WF), Awaawaroa (AW), Putiki (PK), Firth of Thames (FT), and Bon Accord (BA); and (4) Funnel-shaped estuaries that have no inlet barrier, are simple or branched, and form on low-energy coasts. Estuaries: Wairoa (WA), Weiti (WE), and Okura (OK). **Definitions**: (1) A_M/A_E—ratio of the tidal flat above mean sea level (MSL) to the estuary mean high-tide surface area (km^2), (2) SY_V/Ω—ratio of the estimated annual average catchment sediment yield (m^3) to the spring-tidal prism volume (m^3). The annual load is converted from tonnes to m^3 using a typical wet-bulk density of estuarine sediment deposits of 1.2 t m^3 (Swales et al., 2002a). *Hume, T.M., Snelder, T., Weatherhead, M., Liefting, R., 2007. A controlling factor approach to estuary classification. Ocean Coast Manag. 50, 905–929, Elliot, A.H., Shankar, U., Hicks, D.M., Woods, R.A., Dymond, J.R., 2008. SPARROW regional regression for sediment yields in New Zealand rivers. In: Schmidt, J., Cochrane, T., Phillips, C., Elliot, S., Davies, T., Basher, L. (Eds.), Sediment Dynamics in Changing Environments (Proceedings of a Symposium Held in Christchurch, New Zealand, December 2008. IAHS Publication, vol. 325, Swales, A., Gorman, R., Oldman, J.W., Altenberger, A., Hart, C., Bell, R.G., Claydon, L., Wadha, S., Ovenden, R., 2009. Potential Future Changes in Mangrove Habitat in Auckland's East-coast Estuaries. Auckland Council Technical Report No. 079 (June 2009), ISSN 1179-0512 (online), 219 pp, Hicks, D.M., Shankar, U., McKerchar, A.I., Basher, L., Jessen, M., Lynn, I., Page, M., 2011. Supsended sediment yields from New Zealand rivers. J. Hydrol. (NZ) 50(1), 81–142.*

5. Case study—sand flat to mangrove forest

Understanding of the processes driving the biogeomorphic evolution and rapid expansion of New Zealand's mangrove forest over the last century has been advanced by research conducted over the last decade in the Firth of Thames (Fig. 1.2). This multidisciplinary research has included studies of processes controlling mangrove-seedling recruitment on mudflats (Swales et al., 2007; Balke et al., 2015), nutrient dynamics and forest succession (Lovelock et al., 2010), biogeomorphic evolution of the mangrove forest and role in estuary sedimentation (Swales et al., 2015), the influence of large-scale geological processes (e.g., basin subsidence, Swales et al., 2016), and identifying biophysical processes that drive sedimentation and surface-elevation dynamics in the mangrove forest at seasonal to decadal time scales (Swales et al., 2019).

5.1 Environmental setting

The Firth of Thames is a ~800 km^2 mesotidal estuarine embayment located on the north-east coast of New Zealand's North Island (Fig. 1.2). The Firth is one of numerous estuaries fringing the east coast of New Zealand's upper North Island that have experienced substantial increases in mangrove forest habitat over recent decades. Tides in the Firth are semidiurnal with average spring and neap-tidal ranges of 2.8 and 2 m, respectively (Swales et al., 2007). The rate of relative sea-level rise (RSLR) at the nearby Port of Auckland has averaged 1.5 ± 0.1 mm yr^{-1} over the 20th century (1899–2008 AD, Hannah and Bell, 2012). This rate of RSLR is similar to the increase in global mean sea level of 1.7 mm yr^{-1} (range: 1.5–1.9 mm yr^{-1}) recorded between 1901 and 2010 (IPCC, 2013).

The Firth is located in the tectonically active Hauraki Rift, a ~25 km wide by 250 km long rift-valley system (Hochstein and Nixon, 1979; Hochstein et al., 1986). Steeply dipping (~70 degrees) and roughly parallel NNW striking normal faults form deep sedimentary basins that have infilled with Tertiary–Quaternary sediments to a maximum depth of about 3 km (Hochstein and Nixon, 1979). Unconsolidated Quaternary sediments, which extend to 0.7–1 km depth, suggest subsidence in these basins over geological time scales (Hochstein et al., 1986).

The climate of the Firth is temperate with an average minimum air temperature of 10.3°C in July and an average maximum of 19.8°C in February. Mean monthly rainfall varies from 73 mm in January to 145 mm in July, with a mean annual rainfall of 1211 mm (station B75152, 1936–2011, New Zealand Meteorological Service). Major sources of fluvial sediment accumulating in the southern Firth are the Waihou (1966 km^2) and Piako (1476 km^2) Rivers, which account for ~96% of the 3600 km^2 land catchment. These rivers supply an estimated 160,000 and 30,000 t yr^{-1}, respectively, of suspended sediments to the southern Firth (Hicks et al., 2011).

Peak sediment influxes to the Firth coincided with large-scale catchment deforestation during the mid–late 1800s associated with timber extraction, gold mining, and land conversion to pastoral agriculture by European settlers (Healy, 2002; Swales et al., 2015). These activities substantially increased sediment loads to the Firth. Hydrographic surveys conducted in 1882 and 1918 suggest that ~43 million m^3 was deposited in a 210 km^2 area of the southern Firth over this four-decade period (Swales et al., 2007). Sediment-core data (Swales et al., 2007, 2015) indicate that mass deposition rates in the southern Firth have averaged ~500 kt/yr since the 1950s. An apparent shortfall (~300 kt yr^{-1}) between this mud deposition and river inputs (~190 kt yr^{-1}) has been explained in terms of a shoreward-sediment flux convergence from a large sediment source (Swales et al., 2015). Legacy sediments associated with historical catchment deforestation, mining, and resulting increased soil erosion (e.g., James, 2013), stored in the extensive shallow subtidal zone, are the likely source of these fluvial muds in the Firth of Thames. Physical processes favoring onshore transport and deposition of mud in the upper-intertidal zone include tidal asymmetry, wave-driven resuspension, and onshore tidal advection; shoreward wave-energy dissipation due to shoaling, scour and settling lags, and flocculation (Swales et al., 2015). The combination of these physical processes, relatively large tidal range, contemporary and legacy fluvial sediment supply has subsequently built some 70 km^2 of intertidal mudflats in the southern Firth. The interaction of prevailing southwest and opposing northeast winds with buoyant silt-laden river plumes control storm-event and long-term sedimentation patterns in the Firth of Thames. Mud depocentres on the upper-intertidal flats and in the shallow subtidal zone, several kilometres north of the mangrove forest, are a product of these coupled physical processes (Healy, 2002; Pritchard et al., 2015).

Mangroves have colonized ~11 km^2 of intertidal flats in the southern Firth since the early 1960s that were previously restricted to river deltas until the early 1950s, as noted by Captain James Cook in 1769 (Beaglehole, 1968). These forests occur close to the southernmost limit of mangroves globally (i.e., 38.9°S, Corner Inlet, South Victoria, Australia; Bird, 1986). Three major forest zones are identified (Swales et al., 2019), based on the classification of Lugo and Snedaker (1974). These forest zones represent a temporal sequence of tidal flat colonization: (1) relict fringe forest (trees that were tall seaward fringing trees in the 1970s); (2) scrub forest composed of shorter trees formed within a comparatively hydrologically isolated area (mid-1990s—present); and (3) tall seaward-mudflat fringing trees of the fringe forest (early 1990s) (Lovelock et al., 2010). These forest zones occupy a range of elevations within the tidal frame and distances from the unvegetated mudflat (Lovelock et al., 2010; Swales et al., 2019).

5.2 Mangrove-seedling recruitment

The historical sequence of mangrove forest development in the southern Firth has been reconstructed using aerial photographic records (1944—2006), field observations, and modeling (Swales et al., 2007, 2015; Balke et al., 2015). Historical records indicate that the mangrove forest development since the early 1950s is the result of at least four major seedling recruitment events. These events are characterized by the establishment of mangrove seedlings at spatial densities sufficient to form a near-continuous canopy of mature trees across tens of meters of mudflat immediately seaward of the preexisting fringe forest. Major recruitment events contrast with limited seedling recruitment that occurs in a typical year due to mudflat erosion and resulting high seedling mortality (i.e., >90% in 60 days, Swales et al., 2007) that produces isolated widely spaced individual mature trees. The aerial photographs and radiocarbon dating of mangrove trees (Lovelock et al., 2010) constrain the recruitment events to the periods 1952—63 (event R1), 1963—77 (R2), 1977—87 (R3), and 1987—96 (R4) (Fig. 1.7). No major recruitment events have occurred since the mid-1990s.

The episodic nature of mangrove forest development in the southern Firth is also preserved in the structure of the present-day forest. Tree architecture varies from open, spreading tree forms typical of tree growth in open settings, to densely packed trees with straight trunks that are typical of growth within light-limited stands. Canopy height, tree density, and tree basal area vary across the mangrove forest upper-intertidal platform, with relict and present-day fringe forests (tree height up to 3.5 m) alternating with scrub forest (<1 m high) on their landward sides (Lovelock et al., 2010; Swales et al., 2015). Direct observations of seedling recruitment success on intertidal mudflats immediately seaward of the mangrove forest during the 2006 summer showed that propagule and seedling numbers

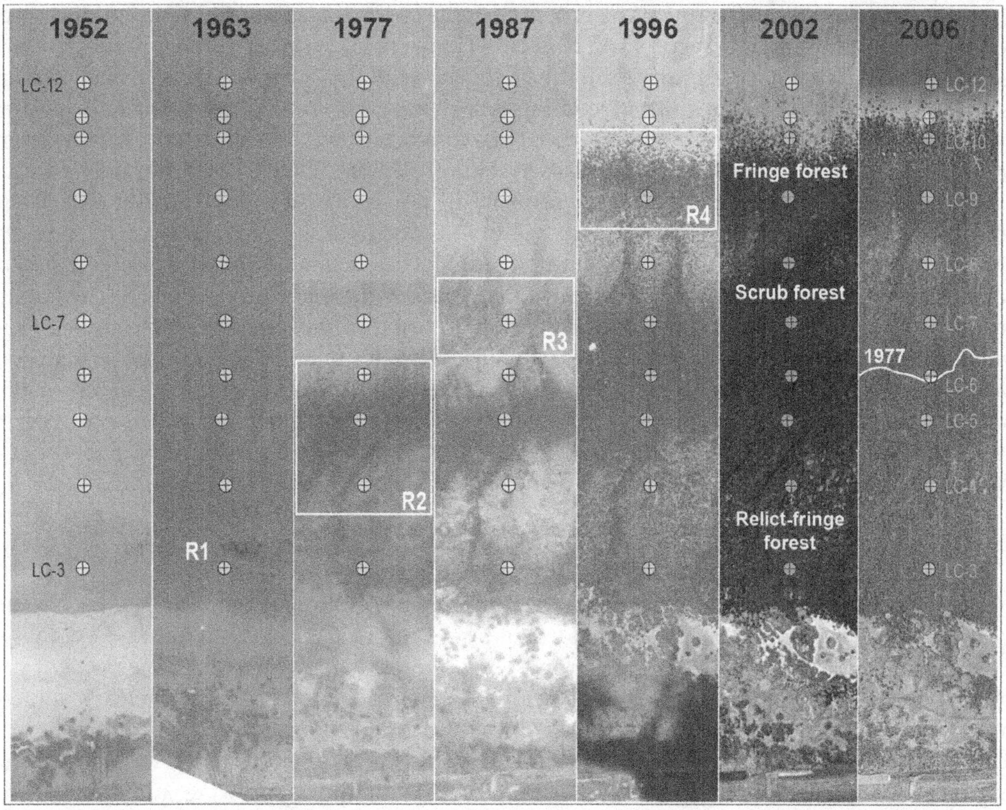

FIGURE 1.7 Historical record of major recruitment events (R1–R4) and resulting mangrove forest development at the sediment-core transect, southern Firth of Thames, 1952–2006. Location of sediment-core sites and seaward edge of fringe forest in 1977 also shown. *Reproduced from Swales, A., Bentley, S.J., Lovelock, C.E., 2015. Mangrove-forest evolution in a sediment-rich estuarine system: opportunists or agents of geomorphic change? Earth Surf. Process. Landforms 40, 1672–1687. https://doi.org/10.1002/esp.3759.*

declined exponentially over time from a peak of about 190 m^{-2} to 10% of their initial abundance within 60 days. Only one seedling remained after 180 days (Swales et al., 2007). Repeated-measures ANOVA analysis showed that propagule and seedling attrition was not significantly influenced by the cross-shore distance from the seaward edge of the mangrove forest. Poor recruitment of seedlings was attributed to frequent reworking of mudflat surficial sediment by waves. This conclusion was based on X-radiographs and penetration depths of the relatively short-lived radioisotope Berrylium-7 (^{7}Be, half-life 53 days) in sediment cores that indicated that surficial sediments consisted of an intensely mixed, low-density mud-layer to about 7-cm depth (Swales et al., 2007).

Manipulative field and laboratory experiments of seedling stability and data-based modeling of recruitment "Windows of Opportunity" were used by Balke et al. (2015) to identify the environmental conditions under which major seedling recruitment events occur in the Firth of Thames. Observations showed that physical disturbance of the substrate by waves prevented successful establishment of propagules that required a 2- to 5-day inundation-free period for root growth and successful anchoring (Balke et al., 2015). Modeling indicated that seedling survival followed a biweekly pattern, in step with the spring-neap tidal cycle so that substrate disturbance and resulting seedling-recruitment failure coincides with spring-tide inundation of the mid-to-upper intertidal flat. Stochastic sea-level and wind-climate variability associated with the Southern Oscillation (La Niña, El Niño) over annual–decadal time scales are also influential factors controlling mangrove-seedling recruitment. These physical controls on seedling recruitment are ubiquitous in Auckland's east-coast estuaries. Mangrove forests in these disturbance-limited systems are unable to fully utilize their potential intertidal habitat (above MSL), with occupancy ratios varying from 0.08 to 0.8. The largest proportional increases in mangrove habitat have occurred in the smallest, fetch-limited/low wave-energy estuaries (i.e., high tide areas $< 2\ km^2$) (Swales et al., 2009). Thus, this coupling of regular and predictable tidal variations in water level with stochastic wave-driven disturbance of the substrate and resulting bed-level dynamics (e.g., Balke et al., 2015) and longer-term sea-level variations appear to be the key drivers of mangrove-seedling recruitment and forest development in these infilled estuarine systems.

5.3 Biogeomorphic evolution

Swales et al. (2015) reconstructed the biogeomorphic evolution of the Firth mangrove forest on the former intertidal sand flat that existed until the 1950s. Two contrasting hypotheses were explored (i.e., Bird, 1986): (1) mangroves are opportunistic, with forest development primarily driven by physical processes; or (2) biophysical feedbacks strongly influence sedimentation and the resulting geomorphology of these vegetated intertidal systems. In this latter case, increases in SAR on unvegetated tidal flats would be expected to occur following seedling recruitment as the mangrove forest developed. These questions were addressed using high-resolution sedimentary records extracted from sediment cores and a detailed chronology of mangrove forest development provide by historical aerial photographs (1944–2006).

Sediment cores collected along a 700 m shore-normal transect sampled the transition over several decades from a muddy sandflat to unvegetated mudflat and finally to mangrove forest (Fig. 1.7). Excess ^{210}Pb ($^{210}Pb_{ex}$) profiles typically extended to 1.7 m depth in the sediment column, providing fine-scale temporal resolution, with apparent ^{210}Pb SAR ranging from ~10 to 100 mm yr^{-1} that were validated using ^{137}Cs dating (Fig. 1.6, Swales et al., 2015). Four major sediment facies were identified (i.e., oldest to most recent): (1) laminated to bedded muddy sands (Ts facies) in basal sediments characterized by downward increasing strata thickness, sand content and grain size with occasional bivalve shells; (2) partially bioturbated muds with sparse silt laminae (Tm facies); (3) bioturbated muds with mangrove roots (Mf facies) with rare partially bioturbated silt laminations and abundant mangrove roots ranging from hair-like to ~1 cm diameter; and (4) a mature mangrove forest facies (Mm) composed of nodular bioturbated muds with abundant roots, root casts, and precipitates of iron and manganese that appear as halos around mangrove roots (Swales et al., 2015).

Analysis of the sediment cores and historical records provided unambiguous evidence that biophysical feedbacks did not measurably enhance sedimentation over annual to decadal time scales following mangrove colonization of the mudflat (Swales et al., 2015). Excess ^{210}Pb profiles displayed prominent changes in log-linear gradients that indicated changes

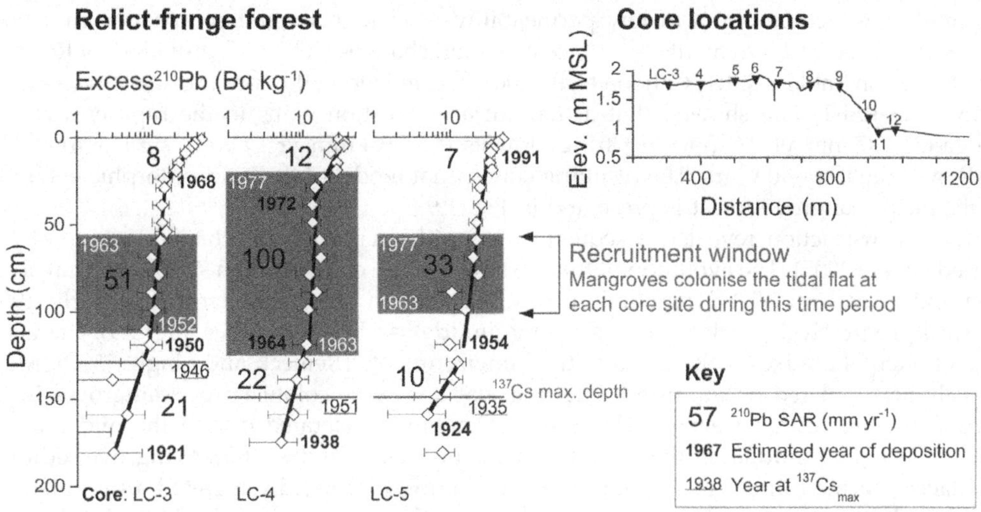

FIGURE 1.8 Examples of excess ^{210}Pb profiles (with 95% confidence intervals) measured in sediment cores collected from the relict-fringe mangrove forest (pre-1980, Sites LC-3—LC-5). Regression fits to the excess ^{210}Pb data are shown along with time-averaged sediment accumulation rates (SAR, mm yr^{-1}). ^{210}Pb-age estimates shown at changes in profile slope, at the bottom of the profile, and at the maximum depth of ^{137}Cs-labelled (i.e., post-1953) sediment (age underlined) are shown. The green (gray in print version)-shaded region identifies the potential recruitment window for mangrove arrival at each core site, based on the aerial-photographic record. Note: the mean sea level (MSL) position for each core is relative to the 2005 substrate surface. *Reproduced and modified from Swales, A., Bentley, S.J., Lovelock, C.E., 2015. Mangrove-forest evolution in a sediment-rich estuarine system: opportunists or agents of geomorphic change? Earth Surf. Process. Landforms 40, 1672–1687. https://doi.org/10.1002/esp.3759.*

in SAR over time (Fig. 1.8). For example, $^{210}Pb_{ex}$ profiles from the relict fringe forest that established prior to 1980 exhibit relatively low SAR near the top (7–12 mm yr^{-1}) and bottom (10–22 mm yr^{-1}) of these cores.

These low-SAR periods are separated by an interval of higher SAR (33–100 mm yr^{-1}). This low–high–low SAR pattern generally corresponded to the depth distributions of sedimentary facies. The low-SAR zone observed in the lowermost section of all the cores from the mangrove forest coincides with the unvegetated, sand and mud tidal–flat facies. Above the basal low–SAR zone, large increases in SAR (to 33–100 mm yr^{-1}) generally coincided with the Mf (colonization phase) facies. The near-surface low-SAR zone coincided with the Mm (mature forest) facies, characterized by abundant mangrove roots, mineral nodules, and root casts (Swales et al., 2015). These 2- to 10-fold increases in $^{210}Pb_{ex}$ SAR were unambiguously large and occurred years to decades before seedling recruitment, demonstrating that the colonization of the mudflat by mangroves did not measurably enhance sedimentation over annual to decadal time scales. Swales et al. (2015) concluded that mangrove forest development in the Firth of Thames was largely dependent on physical processes, with the potential for forests to develop only after the mudflats had reached a suitable elevation in the intertidal zone.

The biogeomorphic evolution of the Firth mangrove forest was reconstructed using the $^{210}Pb_{ex}$ SAR as a proxy for rate of substrate surface-elevation change (ΔE, m/yr) at each core site. Evidence supporting this approach included independent ^{137}Cs dating, the rapid accumulation (i.e., cm yr^{-1}) of low permeability mud with limited compaction in surface layers and direct measurements of surface-elevation changes (2007–15) provided by Rod Surface Elevation Tables (RSET, Cahoon et al., 2002) located along the core transect (Swales et al., 2015). The RSET data showed that actual surface-elevation gains in the rapidly accreting fringe (31–37 mm yr^{-1}) and relict fringe forests (6.7–9.4 mm yr^{-1}) were similar to $^{210}Pb_{ex}$ SAR over the last ~40 years. The resulting conceptual model of the biogeomorphic evolution of the Firth mangrove forest is presented in Fig. 1.9.

The reconstruction revealed a sequence of rapid environmental change over a 60-year period (1944–2005). The evolution of the forest was more complex than simple vertical accretion and lateral mudflat progradation. Mud deposition (Tm facies) from the early 1950s onward progressively buried the preexisting muddy-sand flat (Ts facies, Fig. 1.9), characteristic of energetic mixed-sediment intertidal environments (Reineck and Singh, 1980). Muds initially accumulated on the upper intertidal flat and were colonized by mangroves in the early 1960s (recruitment event R1). Vertical accretion accelerated during the mid-1960s to mid-1970s before mudflat in the midintertidal zone was colonized (R2). Mangrove sediment (Mf facies) formed under this young forest as the mudflat rapidly accreted and a platform developed close to Mean High Water Spring tide elevation (i.e., +1.6m MSL) by the mid-1970s. Concurrently with the formation of this fringe forest (i.e., relict fringe forest), a secondary mud depocentre developed on the mudflat ~300-m seaward (Fig. 1.9). Although not investigated by Swales et al. (2015), the development of this secondary mud depocentre due to rapid vertical accretion is consistent with a horizontal gradient in the suspended-sediment flux (i.e., flux divergence). Sediment-flux gradients are driven by cross-shore wave-energy gradients associated with wave dissipation in shallow water on muddy beds (e.g., Elgar and Raubenheimer, 2008), currents, and upwind sediment availability (e.g., Nittrouer and Wright, 1994; Harris and Wiberg, 2002). Surface-elevation gain at this depocentre was rapid during the 1970s and coincided with relatively lower SAR further landward. To a large extent this depocentre isolated the relict fringe forest from mud supplied by wave reworking of the unvegetated tidal flats. Further seaward expansion of the forest occurred in the mid-1980s (R3) with the mud depocentre migrating ~100m landward. Large storms occasionally perturb the system, eroding mudflat flanking the fringe forest and resulting in tree loss (i.e., during late 1970s).

The last major recruitment event (R4, early–mid-1990s), associated with the secondary mud depocentre, forms the present-day fringe forest and a shallow basin immediately landward. Radiocarbon ages of trees in the fringe forest (1991 ± 1.4 yr to 1995 ± 1.0 yr, Lovelock et al., 2010) are consistent with sediment core dating. The tidal basin rapidly infilled (~50 mm yr^{-1}) and was colonized by mangroves, forming the present-day scrub-forest complex, near the upper limit of the tidal frame (Swales et al., 2015). Radiocarbon ages for scrub-forest trees (1995 ± 1.1 yr to 2000 ± 1.6 yr, Lovelock et al., 2010) are consistent with the biogeomorphic reconstruction. The fringe forest continues to rapidly accumulate muds (~20 mm yr^{-1}) eroded from the mudflat by waves.

The ^{210}Pb geochronology can be summarized in the form of age–surface-elevation curves for the mangrove forest (Fig. 1.10). The negative feedback between surface elevation and hydroperiod is evident, with the rate of surface-elevation gain in the relict fringe forest displaying an asymptotic trajectory. Rates of surface-elevation gain had declined by the 1970s as the mangrove platform reached the upper limit of the tidal frame (Swales et al., 2015).

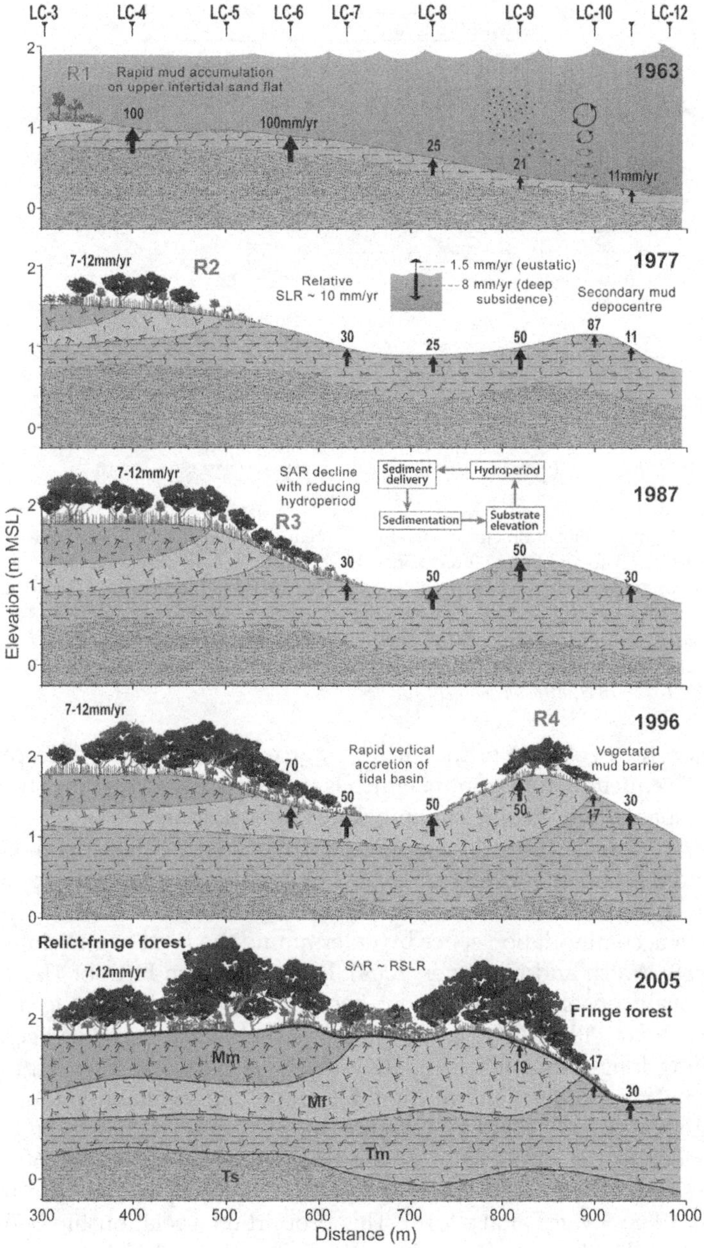

FIGURE 1.9 Biogeomorphic evolution of the Firth of Thames mangrove forest (1963–2005). ^{210}Pb sediment accumulation rates (SAR, mm yr^{-1}) at core locations indicated by upward arrows and sediment facies as defined in the main text. *Reproduced from Swales, A., Bentley, S.J., Lovelock, C.E., 2015. Mangrove-forest evolution in a sediment-rich estuarine system: opportunists or agents of geomorphic change? Earth Surf. Process. Landforms 40, 1672–1687. https://doi.org/10.1002/esp.3759.*

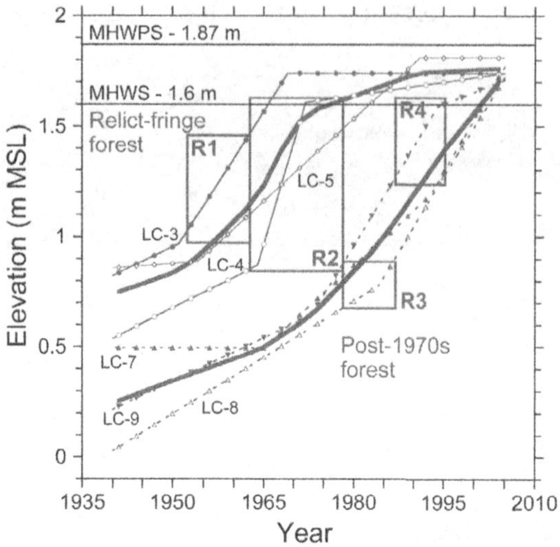

FIGURE 1.10 Age–surface-elevation curves (relative to mean sea level, MSL) for (1) Relict-fringe (LC-3–LC-5) and (2) recent (Post-1970s Scrub and Fringe-forest zones, LC-7–LC-9) Firth of Thames mangrove forest corrected for deep subsidence (~8 mm yr^{-1}). The age–surface-elevation curves are for individual cores (*symbols*) and averages (*bold line*) for the relict-fringe growth and recent forests. Major seedling recruitment windows (R1–R4) identified from the historical aerial photography (Fig. 1.7) are also shown. *Reproduced from Swales, A., Bentley, S.J., Lovelock, C.E., 2015. Mangrove-forest evolution in a sediment-rich estuarine system: opportunists or agents of geomorphic change? Earth Surf. Process. Landforms 40, 1672–1687. https://doi.org/10.1002/esp.3759.*

The biogeomorphic evolution of the Firth mangrove forest can be conceptualized within the framework of sequence stratigraphy, which is primarily based on the concept of sediment accommodation space. Over geological time scales, the combination of sea-level fluctuations and subsidence drive cycles of sedimentation and erosion associated with changes in base level and accommodation space (Jervey, 1988; Posamentier et al., 1988; Schlager, 1993; Coe et al., 2002). In coastal and estuarine environments, sediment resuspension by waves also influences sediment accommodation space by determining the local equilibrium profile or base level (Jervey, 1988; Walsh and Nittrouer, 2004). In the southern Firth of Thames, a 0.7–1 km thick sequence of unconsolidated sediment has accumulated during the quaternary alone (Hochstein et al., 1986; Pillans, 1986). This suggests that subsidence has been the dominant process controlling long-term rates of relative sea-level change. The mangrove forest platform, presently near the upper limit of the tidal frame, represents a limiting case of marine accommodation space. The geomorphic reconstruction showed that the relict fringe forest had aggraded close to the upper limit of the tidal frame by the late 1960s (Swales et al., 2015). In this system, increases in sediment accommodation space are directly proportional to RSLR (Jervey, 1988; Swales et al., 2016). This proportional relationship between SLR, vertical land motion, and sedimentation and creation of accommodation space has been termed a "keep-up" cycle (Soreghan and Dickinson, 1994).

Repeat static-GPS measurements of the RSET benchmarks located across the Firth mangrove forest platform over almost a decade (2007–16, Swales et al., 2016) demonstrate that the mangrove forest is rapidly subsiding, averaging 8.8 ± 0.3 mm yr^{-1} (range: 7.7 ± 0.5 to 9.4 ± 0.3 mm yr^{-1}).

These subsidence rates are five- to six-fold higher than measured at a reference site located on nearby greywacke basement rock (1.6 ± 0.5 mm yr^{-1}). The low-SAR regime observed in the relict fringe forest since the 1960s ($^{210}Pb_{ex}$ SAR = 7–11.6 mm yr^{-1}) is consistent with RSLR largely driven by subsidence of the sedimentary basin by sediment compaction rather than tectonic processes (Swales et al., 2016). The similarity of the vertical trend measured at the three RSET stations suggests that this subsidence is a relatively uniform process in the mangrove forest, at least over spatial scales of hundreds of meters. This rapid subsidence, large sediment supply and elevation of the mangrove forest in the extreme upper intertidal zone, implies that increases in sediment accommodation volume within the forest will be largely achieved by seaward progradation. This progradation has been confirmed by surveys of the cross-shore profile (i.e., 3.8 m yr^{-1}, 2005–2013, Swales et al., 2016). Increases in sedimentation rates since the early 1970s (^{210}Pb SAR (25–31 mm/yr)) and RSET data (32–37 mm yr^{-1}, 2007–19) from the fringe forest (Swales et al., 2015) also suggest that the mangrove forest sedimentary complex had switched from an aggradation- to progradation-dominated regime by the early 1970s.

5.4 Surface-elevation dynamics

The relative contributions of biophysical process controlling surface-elevation changes in mangrove forests vary markedly with local environmental conditions. Surface processes include deposition and erosion of mineral sediment primarily supplied by rivers, and seasonal variations due to sediment wetting and drying and leaf litter and microbial mat production (Woodroffe, 1985; McKee, 2011; Webb et al., 2013; Gladstone-Gallagher et al., 2014; Swales et al., 2015). Subsurface processes include tectonic processes that influence vertical land motion (VLM) (Kaye and Barghoorn, 1964; Cahoon and Reed, 1995; Cahoon, 2015; Swales et al., 2016; Woodroffe et al., 2016), subsidence associated with sediment compaction by dewatering, as well as fluctuations in groundwater storage associated with evapotranspiration, seasonal rainfall, and tidal cycle recharge and discharge (Cahoon et al., 2011; Rogers et al., 2014). In organic-rich mangrove soils, root growth and decay and net organic matter accumulation can substantially influence surface-elevation dynamics (McKee, 2011; Krauss et al., 2014).

Biophysical processes controlling surface-elevation dynamics in the Firth mangrove forest were investigated by Swales et al. (2019) using data from RSET stations located along the sediment core transect (Swales et al., 2015). The RSET measurements included the full range of geomorphic development, from the rapidly accreting fringe forest located in the mid-intertidal zone to the relict fringe forest on the upper-intertidal platform. Key hypotheses tested were (1) local hydroperiod, as a proxy for sediment delivery, is a key predictor of sediment accretion and surface-elevation gain; (2) meteorological drivers such as episodic flood discharges from rivers and storm tides are important mechanisms enhancing sediment delivery to the mangrove forest; (3) distance from proximal sediment sources (i.e., unvegetated mudflats) is an important secondary factor controlling sediment accretion; and (4) mineral sediment accretion and compaction rather than biotic processes are the key factors controlling surface-elevation dynamics in fluvial sediment–rich systems such as the Firth of Thames.

Factors driving sediment accretion (ΔA) and surface-elevation change (ΔE) measured seasonally over a 3-year period (2011–14) were explored using statistical (Generalized

Additive Models, GAM) and machine-learning (Boosted-Regression Trees, BRT) methods. BRT have several advantages over GAM, which include fitting of nonlinear interactions between predictors and identification of response thresholds (Elith et al., 2008). BRT models for ΔA and ΔE were determined by simplification based on the minimization of cross-validated deviance (Swales et al., 2019). Factors considered in the analysis for which data were available included forest zone, horizontal distance from the unvegetated mud flat (i.e., sediment supply), mean onshore wind speed, river discharge, duration of storm surges greater than 0.3 m, hydroperiod (i.e., duration and frequency of inundation), mean water depth, and mean air temperature. Hydroperiod and water depth were measured at RSET stations using two independent methods: (1) water depth inferred from pressure sensors; and (2) direct measurements of water depth using time-lapse images of hydrometric staff gauges (Fig. 1.11). Partial dependency (PD) plots (Friedman, 2001) for the final BRT models provided functional information on the relationships between each retained predictor variable and sediment accretion (ΔA) and surface-elevation change (ΔE) response variables, as well as their relative influence (Fig. 1.11).

Exploratory data analysis was used to identify strong correlations between factors that would otherwise degrade model performance. Functional dependencies between the ΔA and ΔE response variables and predictor variables in the final BRT models were represented in PD plots (Friedman, 2001). The relative influence of each predictor in the final BRT models

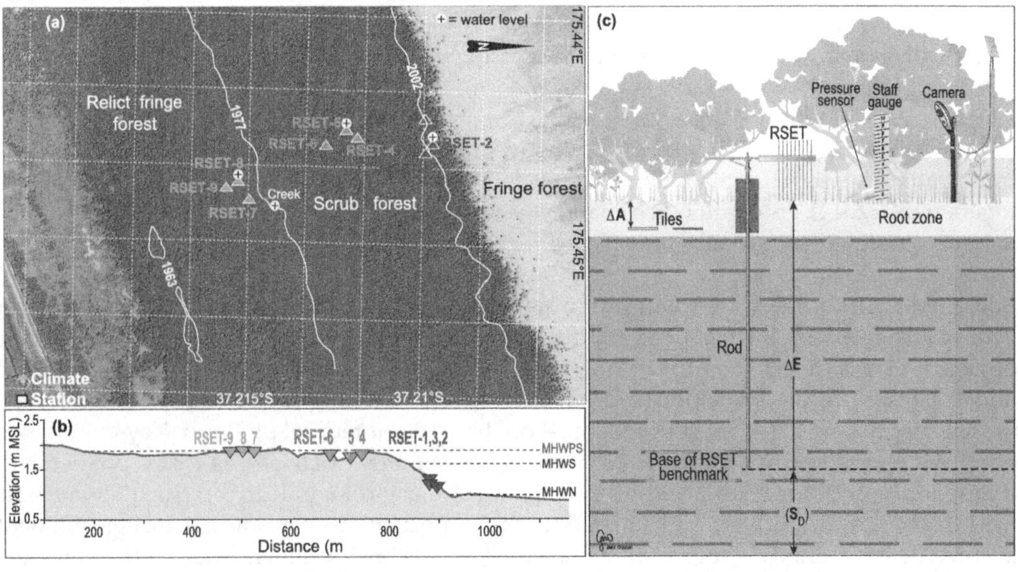

FIGURE 1.11 Location map of (A) Rod Surface Elevation Table (RSET) benchmarks, climate station and water-level recorders; (B) cross-shore transect showing the elevations of the RSET relative to mean sea level. Mean high water neap (MHWN), spring (MHWS), and perigean spring (MHWPS, moon at closest approach to earth) tide levels are also indicated; and (C) schematic of sediment accretion (ΔA), surface-elevation change (ΔE), and hydroperiod measurements made at RSET stations. Deep subsidence (S_D) and ΔE are measured relative to the base of the stainless steel RSET benchmark. *Reproduced from Swales, A., Reeve, G., Cahoon, D.R., Lovelock, C.E., 2019. Landscape evolution of a fluvial sediment-rich Avicennia marina mangrove forest: insights from seasonal and inter-annual surface-elevation dynamics. Ecosystems. https://doi.org/10.1007/s10021-018-0330-5.*

represents the contribution of each predictor to the total reduction in the fitted model's predictive (i.e., cross validation) deviance (Swales et al., 2019).

Surface-elevation trends in the Firth mangrove forest measured over the entire 9.2-year RSET record displayed consistent and linear increases in elevation that varied systematically between the relict fringe, scrub, and present-day fringe-forest zones (Fig. 1.12). Rates of

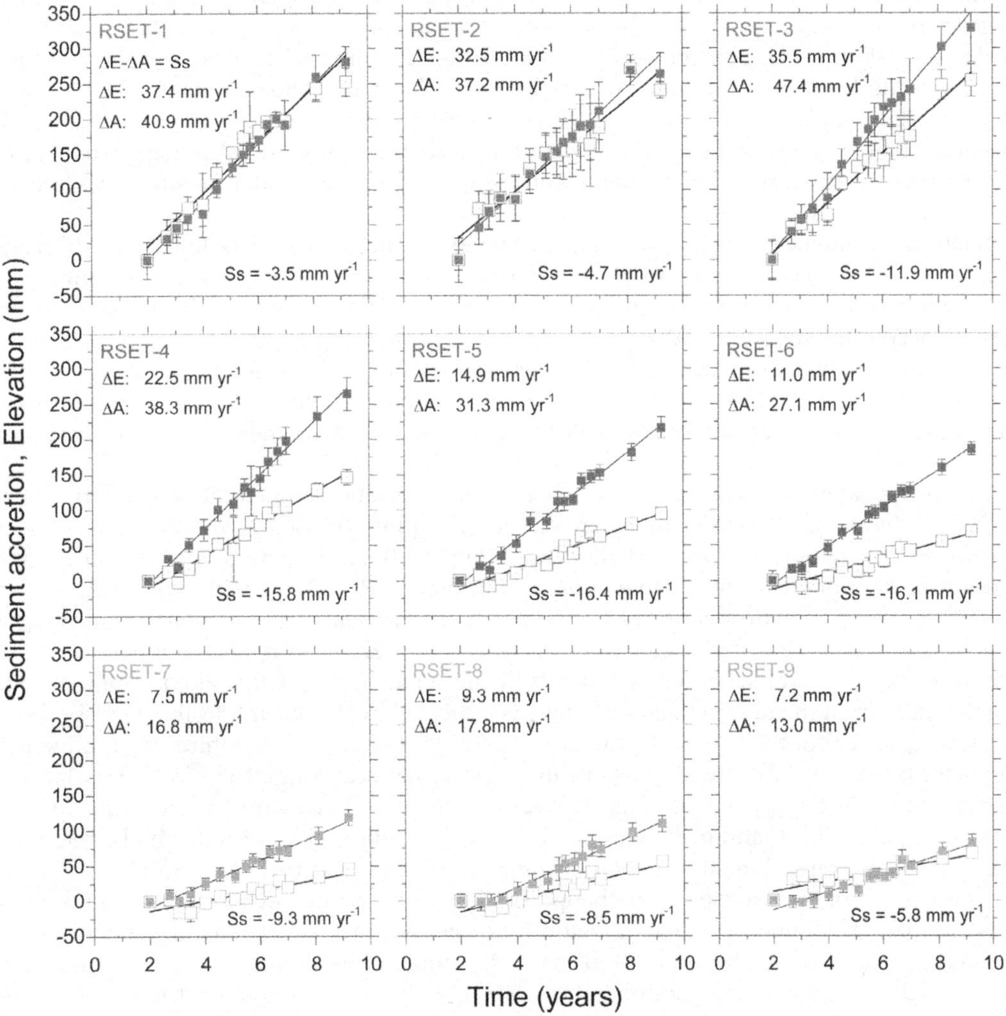

FIGURE 1.12 Cumulative elevation trends and shallow subsidence (SS, mm/yr) inferred from surface-elevation (ΔE) and sediment accretion (ΔA) trends at RSET stations (27 January 2009–19 April 2016) determined from linear regression of mean surface elevations and sediment accretion (40 9 40 cm plates) measured at fringe forest (RSET 1–3); scrub forest (RSET 4–6); and relict fringe forest (RSET 7–9) sites. $P \leq .00001$ (n = 15) at all stations. One standard deviation of mean values plotted. *Reproduced from Swales, A., Reeve, G., Cahoon, D.R., Lovelock, C.E., 2019. Landscape evolution of a fluvial sediment-rich Avicennia marina mangrove forest: insights from seasonal and inter-annual surface-elevation dynamics. Ecosystems. https://doi.org/10.1007/s10021-018-0330-5.*

surface-elevation gain, estimated from linear regression analysis, were highest in the fringe forest (31–35 mm yr^{-1}), close to the unvegetated mudflat, and lowest in the relict fringe forest (8–10 mm yr^{-1}) several hundred meters landward. In the scrub forest, rates of surface-elevation gain (9–20 mm yr^{-1}) show a progressive reduction with distance landward across the upper-intertidal platform Swales et al. (2019). These net surface-elevation tends were dominated by sediment accretion and compaction due to shallow subsidence, with the delivery of sediment to the forest controlled by coupling of mud resuspension on the mudflats by onshore sea breezes with the fortnightly occurrence of spring tides (Swales et al., 2019). The surface-elevation trends are consistent with the longer-term record of surface-elevation trajectory over the last several decades inferred from ^{210}Pb geochronology (Fig. 1.8). Elevation changes also displayed a seasonal cycle (i.e., winter to summer) of cm-scale changes in elevation within the scrub and relict fringe forest zones. This seasonal pattern was related to sediment wetting and drying and resulting in compaction, that was superimposed on the decadal surface-elevation trends.

Shallow subsidence rates (S_S), calculated as the difference between the linear elevation and accretion trends, were highest and most uniform in the scrub forest ($S_S = 15.8–16.4$ mm yr^{-1}) and lowest and most variable in the fringe forest ($S_S = 3.5–11.9$ mm yr^{-1}) where muds are rapidly accumulating (Fig. 1.12).

Rates of shallow subsidence in the relict fringe forest ($S_S = 5.8–9.3$ mm yr^{-1}) were similar to the fringe forest but less variable. These results are contrary to studies of peat-dominated mangrove systems where subsidence rates typically increase with increasing sediment accretion rates (e.g., Cahoon et al., 2006 review).

The BRT analysis enabled key drivers of sediment accretion and surface-elevation change in the Firth mangrove forest to be identified. PD plots for sediment accretion showed nonlinear responses to the retained predictors (Fig. 1.13). Hydroperiod (Hp) was the single most important predictor of ΔA (mean relative influence, RI = 37%), with most sediment accretion occurring at hydroperiods of 10%–15% duration. Mean onshore wind speed (W_{mean}) was the next most influential predictor of ΔA (mean RI = 27%), with most sediment accretion occurring at W_{mean} values of 3.8–4.2 m s^{-1} (Swales et al., 2019). A threshold in rate of sediment supply from the unvegetated mud flats is indicated by the abrupt decline in response of ΔA to W_{mean} at about 4.2 m s^{-1}. Mean air temperature (T_{mean}) was the third most influential predictor (mean RI = 21%) with most sediment accretion occurring at 13–15°C (Fig. 1.13). A derived variable Hz_{resid} (accounting for strong correlation between distance from unvegetated mudflat [MFdist] and mean water depth [h_{mean}]) indicated that MFdist had a relatively minor influence on sediment accretion in the mangrove forest (mean RI = 15%).

Key drivers of surface-elevation change (ΔE) are split almost evenly between the three most influential retained predictors, being hydroperiod, mean water depth, and mean air temperature (Fig. 1.13). The response of ΔE to hydroperiod is similar to ΔA, albeit a noisier signal, with most sediment accretion coinciding with hydroperiod of 10%–17% (mean RI = 36%). Likewise, longer hydroperiods did not substantially influence sediment accretion. Mean water depth was the next most influential predictor of ΔE (mean RI = 33%). The partial response of ΔE indicates that most sediment accretion occurs at mean water depths of 0.2–0.3 m. Some linear interaction between mean water depth and hydroperiod is indicated by their weak correlation (adjusted $r^2 = 0.32$). The similarity of the PD plot for hydroperiod with ΔA suggests that this interaction is minor (Fig. 1.13). The analysis also indicated that

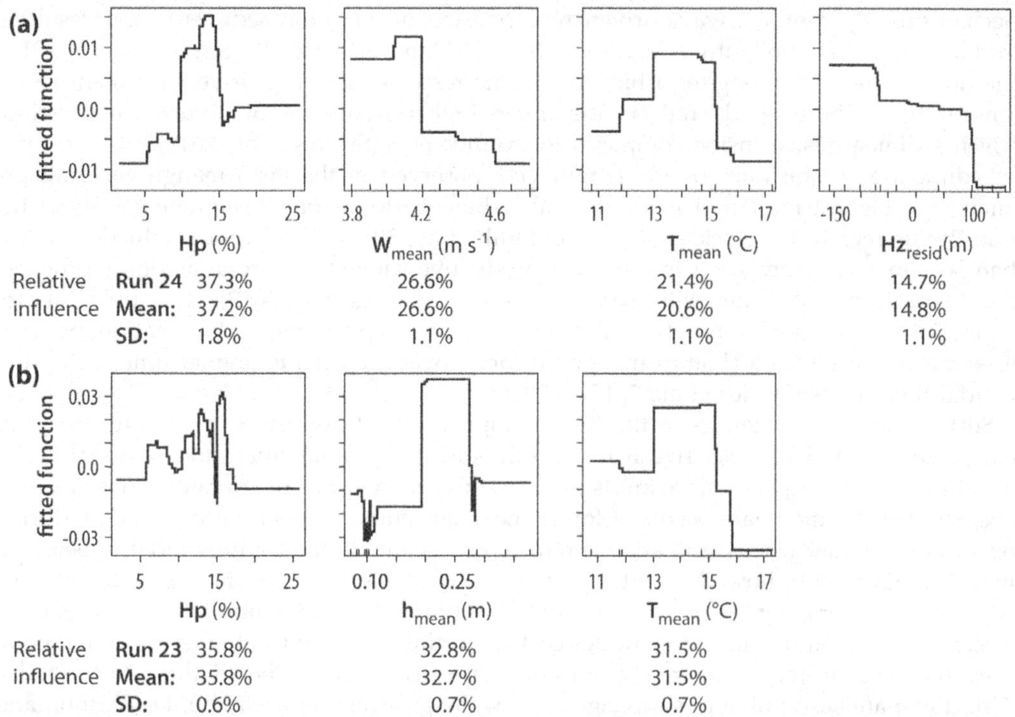

FIGURE 1.13 (A) Sediment accretion (ΔA) and (B) surface-elevation change (ΔE) models: partial dependency plots of retained predictors from analysis of model deviance and their relative influence (RI) on ΔA and ΔE. Shown are results for model runs that are the closest matches to the mean RI values for all model runs. Predictors: hydroperiod (Hp), mean onshore wind speed (W_{mean}), mean air temperature (T_{mean}), residual (Hz_{resid}) of linear-regression between distance of Rod Surface Elevation Table from the unvegetated mud flat with mean water depth (h_{mean}). Final model settings (based on simplification by minimizing cross-validation deviance): tree complexity = 1; learning rates = 0.005 (ΔA) and 0.01 (ΔE); bag fraction = 0.5; data type—Gaussian. Number of trees fitted: 1200–3400 (ΔA) and 3350–9350 (ΔE). Model performance: percentage of cross-validation deviance explained = 34% (ΔA) and 52% (ΔE). *Reproduced from Swales, A., Reeve, G., Cahoon, D.R., Lovelock, C.E., 2019. Landscape evolution of a fluvial sediment-rich Avicennia marina mangrove forest: insights from seasonal and inter-annual surface-elevation dynamics. Ecosystems. https://doi.org/10.1007/s10021-018-0330-5.*

mean air temperature is correlated with a process that influences surface-elevation changes at the RSET stations (mean RI = 32%). In a similar way to ΔA, most surface-elevation response coincides with a mean air temperature of 13–15°C, with no response above 16°C (Fig. 1.13).

The analysis revealed that sediment accretion in the Firth mangrove forest is controlled by the coupling of frequent onshore winds and resulting resuspension of intertidal muds by small estuarine waves, with the fortnightly cycle of spring tide inundation, rather than episodic storm tides or river floods (Swales et al., 2019). These coupled "day-to-day" processes regularly deliver muds sequestered in unvegetated mudflats to the mangrove forest. The relative unimportance of storm surges and river floods to sediment accretion is exacerbated by the position of the mangrove forest platform in the upper-intertidal zone, so that these episodic events must coincide with spring high tides to be effective sediment delivery mechanisms. By contrast, wave-driven resuspension of intertidal sediments and resulting "turbid fringe" is a ubiquitous feature of New Zealand estuaries (Green and Coco, 2014). The dominance of the long-term biogeomorphic response to wave-driven sediment resuspension by onshore winds reflects the much higher frequency of occurrence of wind-driven sediment resuspension compared to storm events (Swales et al., 2019).

Sediment accretion rates of 13–47 mm yr^{-1} observed in the Firth mangrove forest are among the highest measured in the fluvial sediment–dominated mangrove forests of the Indo-Pacific region (Lovelock et al., 2015b; Sidik et al., 2016). Mean root production of less than 50 g m^{-2} yr^{-1} measured in the most productive fringe-forest zone accounted for less than 1% of the post-2009 deposited sediment volume (Swales et al., 2019). Thus, physical processes, in the absence of biophysical feedbacks, can explain the sedimentation patterns observed in the Firth of Thames mangrove forest over seasonal-to-annual time scales and decadal time scales (Swales et al., 2015, 2019).

Surface-elevation dynamics in the Firth mangrove forest are driven by in situ processes and partly by the balance between fortnightly spring-tide inundation and seasonal cycles of wetting and drying of surficial muds that controls shallow subsidence rates. Consequently, most surface-elevation gain occurs at lower mean air temperatures (that is, $< 15°C$), during the autumn–winter when desiccation is reduced compared to the summer months. Seasonal time series (2011–13) showed a winter–summer cycle of elevation loss in the scrub and relict fringe forests, averaging 5–7 mm (maximum—16 mm), where tidal inundation is infrequent. Surface compaction due to sediment desiccation is directly related to higher mean temperatures that occur in the summer. The smectite clay–rich muds of the southern Firth (Naish et al., 1993) are susceptible to shrinkage and swelling (Swales et al., 2019). Desiccation and resulting sediment compaction is a characteristic of water-saturated intertidal mudflats and coastal wetlands that are regularly subaerially exposed. This process is exacerbated by high pore-water salinity (Reineck and Singh, 1980; Allen, 2000), as occurs in the Firth mangrove forest muds (30–55 PSU, Lovelock et al., 2010). Sediment compaction-by-desiccation during the summer months is also likely promoted by forest transpiration, higher air temperatures, and wind speeds (Swales et al., 2019).

A summary and conceptual model of key physical drivers that control sedimentation and surface-elevation dynamics and resultant biogeomorphic evolution of the Firth of Thames mangrove forest is presented in Fig. 1.14. Muds sequestered in intertidal flats are resuspended and delivered to the forest by spring tides coupled with small waves generated by frequent onshore winds. Settling of muds drives a sedimentation gradient across the forest platform, particularly in the scrub forest. Rates of surface-elevation gain (ΔE) in the fringe, scrub, and relict fringe forests reflect the balance between sediment accretion, deep subsidence (S_D), and shallow subsidence (S_S) of surficial sediments. On the forest platform, S_S is partly driven by sediment desiccation and compaction during the summer. Progradation (*P*) of the mangrove forest complex since ~2005 is driven by rapid infilling of available accommodation space created by S_D and SLR.

When corrected for deep subsidence, net surface-elevation trends highlight differences in the relative importance of spatially-varying shallow subsidence and the relatively uniform deep subsidence in each forest zone. In the rapidly accreting fringe forest, S_D accounts for a greater fraction of net elevation change than does shallow subsidence. In the scrub forest, shallow subsidence due to seasonal desiccation and compaction of surficial sediments is an

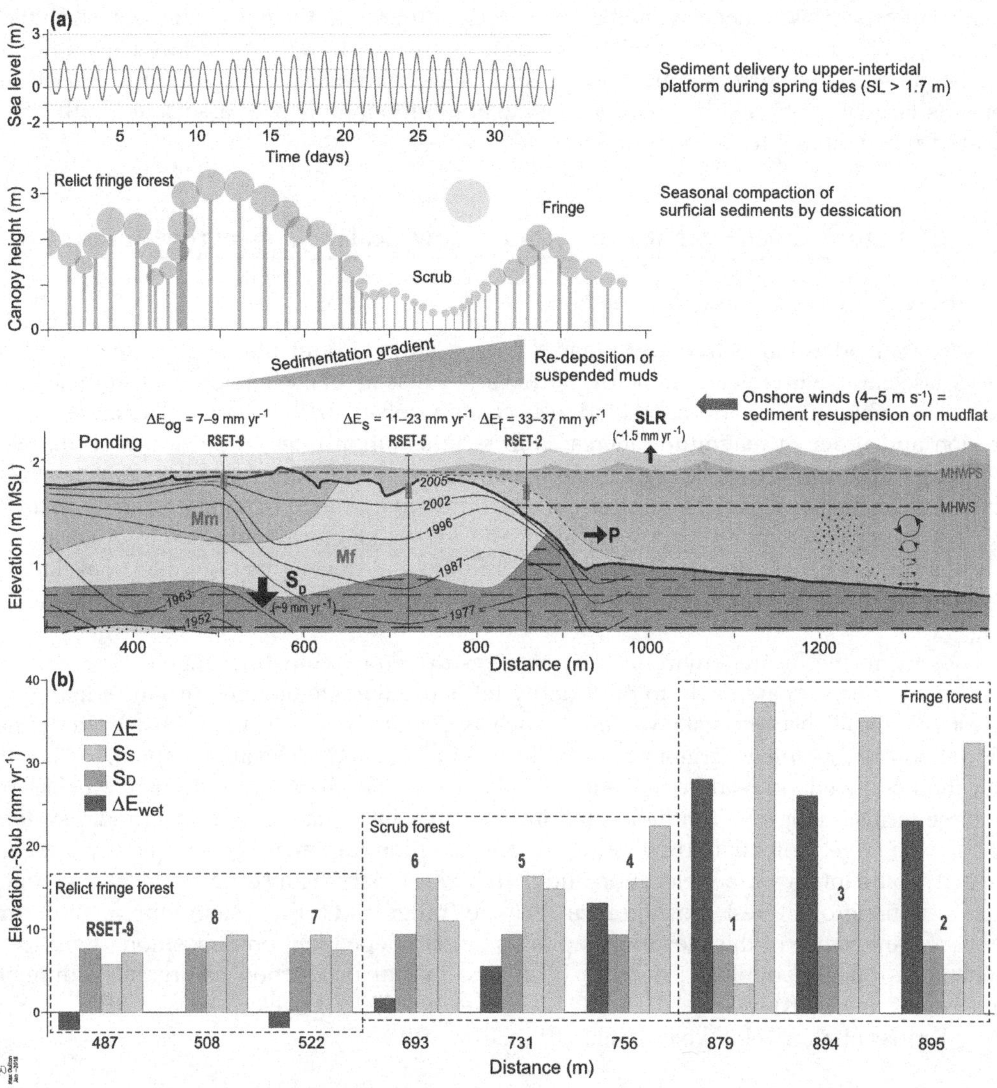

FIGURE 1.14 (A) Conceptual model of surface-elevation dynamics and key drivers in the Firth of Thames mangrove forest. Sedimentary facies: partially bioturbated muds with rare silt laminae (unvegetated mudflats) overlain by bioturbated muds with mangrove roots (Mf); and nodular-bioturbated mud with abundant mangrove roots and precipitates (Mm). Isochrones: substrate surface elevations (1952–2005) based on radioisotope dating of cores and RSET data. (Biogeomorphic reconstruction adapted from Swales et al., 2015, 2016). (B) Trends in surface-elevation dynamics (2009–16, mm/yr) at RSET stations: surface-elevation (ΔE), S_S, S_D, and net rate of change in substrate elevation relative to long-term relative SLR (ΔEwet). Notes: (1) RSLRsed is the weighted-average post-1960s ^{210}Pb SAR in the relict fringe mangrove forest [9.3 mm yr^{-1}]; (2) S_D based on GPS trends for RSET benchmarks in each forest zone (Swales et al., 2016). *Reproduced from Swales, A., Reeve, G., Cahoon, D.R., Lovelock, C.E., 2019. Landscape evolution of a fluvial sediment-rich Avicennia marina mangrove forest: insights from seasonal and inter-annual surface-elevation dynamics. Ecosystems. https://doi.org/10.1007/s10021-018-0330-5.*

important process with net elevation gain of 2–13 mm yr^{-1}. In the relict fringe forest, shallow and deep subsidence rates are similar and near-equivalent to rates of surface-elevation gain so that net elevation increases are close to zero. Net surface-elevation gain in the relict fringe forest is largely controlled by creation of new accommodation space associated with RSLR, most of which is due to deep subsidence.

6. Future shock—what's the fate of New Zealand's mangrove forests?

6.1 Elevation capital and future fate

New Zealand estuaries have experienced large-scale environmental changes since the mid-1800s associated with catchment deforestation and subsequent conversion to pastoral agriculture. This ubiquitous pattern of land use change substantially increased catchment soil erosion and order of magnitude increases in sedimentation rates (e.g., Swales et al., 1997, 2002a,b). Rapid estuary infilling and a shift from sand- to mud-dominated systems has characterized the historical era. This pattern of environmental change in New Zealand's estuaries mirrors patterns described by Roy et al. (2001) for estuaries in southeastern Australia. In Auckland's east-coast estuaries, the extent of infilling and development of intertidal flats appears to be strongly dependent on the rate of sediment supply from catchments to estuaries (Fig. 1.6).

Looking to the future, environmental changes driven by climate change during the 21st century and beyond are likely to profoundly influence the fate of New Zealand's mangrove forests. The resilience of coastal wetlands, such as mangrove forests, to SLR has been defined by Cahoon and Guntenspergen (2010) in terms of a wetlands "elevation capital," being the elevation of a wetland relative to the lowest viable elevation in its growth range. Considered in these terms, mangrove forests located in the upper-intertidal zone in macrotidal systems will have a larger potential growth range, more elevation capital, and consequently a greater capacity to maintain surface elevation and thereby keep pace with rising sea level in comparison to a mangrove forest growing near its lower range and/or a system with a small tidal range. Maintenance of this elevation capital is largely dependent on deposition of sufficient mineral sediments supplied primarily by rivers, in situ production of organic sediments, and root growth (Krauss et al., 2003; McKee et al., 2007; Lovelock et al., 2011; McKee, 2011; Krauss et al., 2014; Swales et al., 2015, 2016, 2019).

6.2 Estuaries—predicting the shape of things to come

Numerical models are increasingly being used to explore the interactions and biophysical feedbacks that drive the maintenance of elevation capital as well as simulate the long-term biogeomorphic evolution of coastal wetlands over decadal to centennial time scales (e.g., Marani et al., 2010; D'Alpaos et al., 2007; Coco et al., 2013; Kakeh et al., 2016).

One modeling approach is the zero-dimensional (0-D) or point model that have relatively simple numerical schemes encapsulating the dynamics of biophysical feedbacks controlling the long-term biogeomorphic development of tide- and wave-dominated sedimentary systems. The 0-D modeling approach focuses on overall system development rather than their "spatially distributed properties (e.g., topography and biological patterns within a tidal platform)" (Marani et al., 2010). Development and validation of these geomorphic models require quantitative data that integrate processes over time scales that are appropriate to the system of interest (Swales et al., 2015). Geomorphic validation of these models has not yet been achieved.

A 0-D model based on the modeling approach of Marani et al. (2010) is used here to consider the likely effect of changes in sediment supply and SLR on the historical and future development of intertidal flats in New Zealand's mangrove estuaries. In particular, the model is used here to explore long-term changes (i.e., decades to century) in the surface elevation of unvegetated intertidal flats relative to sea level. Fig. 1.15 summarizes the key physical processes encapsulated in the model, which include rate of sediment supply, mean rate of SLR, tides, fetch-limited waves, and wetting and drying of intertidal flats. Sediment deposition or erosion on the platform is governed by the sediment particle settling velocity and the critical bed erosion thresholds under waves. The nonlinear equations solved in the model describe the main physical processes that control sedimentation and intertidal platform elevation relative to a sea-level datum. How these processes influence the development and resilience of unvegetated intertidal flats to SLR is the focus, given that tidal flats have historically been the major habitat for mangrove forest expansion in New Zealand estuaries.

Key aspects of the model setup are briefly described here. Scenarios considered include historical (20th century) and future tidal flat evolution during the 21st century for end member SLR scenarios, for a range of sediment supply (i.e., time-average suspended sediment concentrations [SSCs]). We used data from the Auckland study estuaries (Swales et al., 2002a, 2009) to constrain the model for tidal basins of various sizes (i.e., wave fetch [F]), ranging from tidal creeks ($F = 0.1$ km), small ($F = 1$ km), and the largest estuaries ($F = 10$ km) characteristic of the study area. A tidal range of 2 m, that is characteristic of the dominant M_2 constituent for these North Island systems was adopted. Sediment supply is parameterized as time-averaged SSC ranging from 20 to 100 mg L^{-1}. Conceptually, the pool of suspended sediment available for deposition is supplied by rivers and can be augmented by resuspension of tidal flat sediments elsewhere on the intertidal flat (Fig. 1.15).

Physical processes represented in the model include tidal-flat wetting and drying and sediment deposition and erosion controlled by critical bed shear stress for waves ($\tau_{crit.} = 0.13$ N m^{-2}). The behavior of a mud/sand mixture is simulated, with median particle size of 50 μm that are characteristic of the study estuaries (Swales et al., 2002a,b). A wind-speed threshold of 5 m s^{-1} for mud resuspension is based on optical backscatter sensor measurements in Auckland estuaries (Green et al., 2000; Swales et al., 2004; Green and Coco, 2007). These studies show low "background" SSC values (i.e., <20 mg L^{-1}) over submerged intertidal flats at wind speeds below the resuspension threshold. To mimic these observations in the 0-D model, sediment supply to the intertidal platform was limited to 10% of the nominal time-average SSC during below-threshold wind periods.

This approach implicitly recognizes the intermittent nature of sediment transport and sedimentation events and that sediment dynamics elsewhere in the system influences tidal platform development at any given point. Fetch-limited waves are generated by average daily wind speeds based on the formulation of Young and Verhagen (1996a,b). Observed winds from Auckland Airport (1957–2016) are used for hindcast simulations and down-scaled predictions from the UK MetOffice Hadgem Unified Climate Model are used for future scenarios

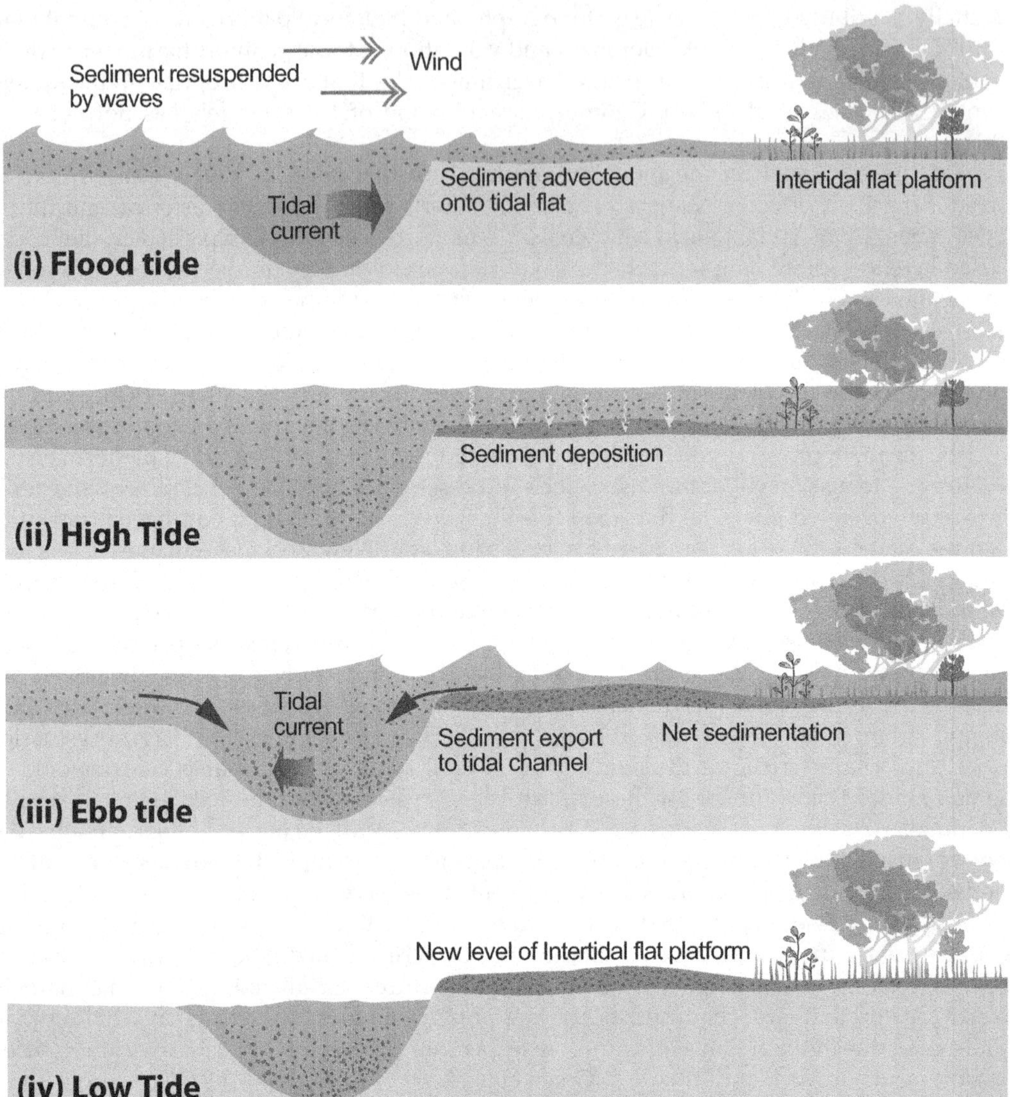

FIGURE 1.15 Conceptual diagram of the zero-dimensional model used to simulate the long-term evolution of estuarine intertidal flats. The diagram shows the key processes occurring over a tidal cycle: (i) fine sediment is resuspended on the intertidal flats by fetch-limited waves as they submerge during flood tides. Tidal- and wave-driven currents advect the suspended sediment onto down-wind intertidal flats; (ii) Sediment deposition occurs at high tide. Physical processes favoring fine-sediment deposition on intertidal flats include tidal asymmetry; advection; shoreward wave-energy dissipation due to shoaling; scour and settling lags and flocculation in shallow water depths.

(2006–2120) (McBride et al., 2016). SLR scenarios for the 21st century were based on Representative Concentration Pathways (RCPs) for greenhouse gas concentration trajectories

(IPCC, 2013). Four RCPs that describe a range of climate futures based on human activities and associated greenhouse gas emissions. These RCPs relate to radiative forcing values that will result by 2100 AD relative to preindustrial values: RCP2.6 (i.e., 2.6 W m^{-2}), RCP4.5, RCP6, and RCP8.5 and resulting SLR scenarios RCP2.6 (SLR = 4.8 mm yr^{-1}), RCP4.5 (SLR = 5.8 mm yr^{-1}), and RCP8.5 (SLR = 9.2 mm yr^{-1}) were included in our model. Results for RCP2.6 and RCP8.5 are presented, as predictions for RCP2.6 and RCP4.5 are very similar. Cumulative probability density functions show that wind speeds exceed the 5 m s^{-1} 20% of the time for the historical record and 40% of the time for the future RCP scenarios. The major difference between the RCP scenarios is the average rates of SLR that are expected to occur during the 21st century.

Model calibration was achieved for the historical/20th century scenario using measurements from the study estuaries to estimate an initial platform height (IPH) (i.e., circa 1900) and tune model parameters to achieve realistic IPH values by the early 2000s. Data included present-day average intertidal flat elevation from RTK-GPS surveys of 87 transects in the study estuaries (Swales et al., 2009) and time-averaged SAR over the 20th century estimated from ^{210}Pb and ^{137}Cs dating of sediment cores (Swales et al., 2002a,b, 2008). The rate of SLR over the 20th century has averaged 1.5 mm yr^{-1} at the Port of Auckland (1899–2008, Hannah and Bell, 2012). An IPH (1900 AD) of −0.3 m MSL (1900 AD) was based on average elevation of +0.3 m MSL (2008 AD) and average SAR of 5 mm yr^{-1} in the study estuaries. Estuary-average intertidal platform elevations ranged from −0.14 to 0.59 m MSL.

The results for the 20th century hindcast indicate that an average sediment supply rate (i.e., SSC) of about 40 mg L^{-1} was sufficient to develop intertidal flat platforms with average elevations of 0.3 m above MSL across all three estuary scales (0.1–10 km fetch) by the early 2000s (Fig. 1.16). These conditions correspond with a sediment supply rate averaging ~120 t km^{-2} yr^{-1} for these largely agricultural lowland catchments (Swales et al., 2009).

Model simulations for the 21st century low- and high-SLR IPCC scenarios (i.e., [4.8 and 9.2 mm yr^{-1}) suggest that intertidal flat habitats in tidal creeks (0.1 km fetch) fringing the upper reaches of estuaries will be able to keep pace with SLR even at relatively low sediment-supply rates (i.e., time-average SSC 20–40 mg/L) characteristic of estuary development during the 20th century (Fig. 1.16). In small estuaries (1 km fetch), the model results suggest that intertidal flats will also be able to maintain their elevation relative to MSL at relatively low sediment supply rates (i.e., time-averaged SSC 40–60 mg L^{-1}). By contrast, intertidal flats in the largest estuaries (10 km fetch) will be more susceptible to inundation by rising sea levels and even at the lower rate of SLR anticipated under the IPCC RCP2.6 scenario (i.e., 4.8 mm yr^{-1}). Loss of intertidal flats could begin as early as the 2020s if relative (local) rates of SLR exceed ~5 mm yr^{-1} and where sediment supply is limited. Model results also indicate that catchment sediment supply to these largest estuaries would need to be twofold higher than the historical rates (~40 mg L^{-1}) that sustained the vertical accretion and development of the intertidal flats in the study estuaries during the 20th century (Fig. 1.16F). Sediment-poor estuarine systems, such as coastal embayment's with relatively small catchment sediment supply (Fig. 1.6), will likely be most susceptible to loss of intertidal flats and fringing mangrove forests.

Our predictions are generally consistent with previous studies employing point models. In general, rapid increases in sea level will increase water depths, thereby changing wave and sediment dynamics in estuaries, particularly in the intertidal zone. Marani et al. (2010) found

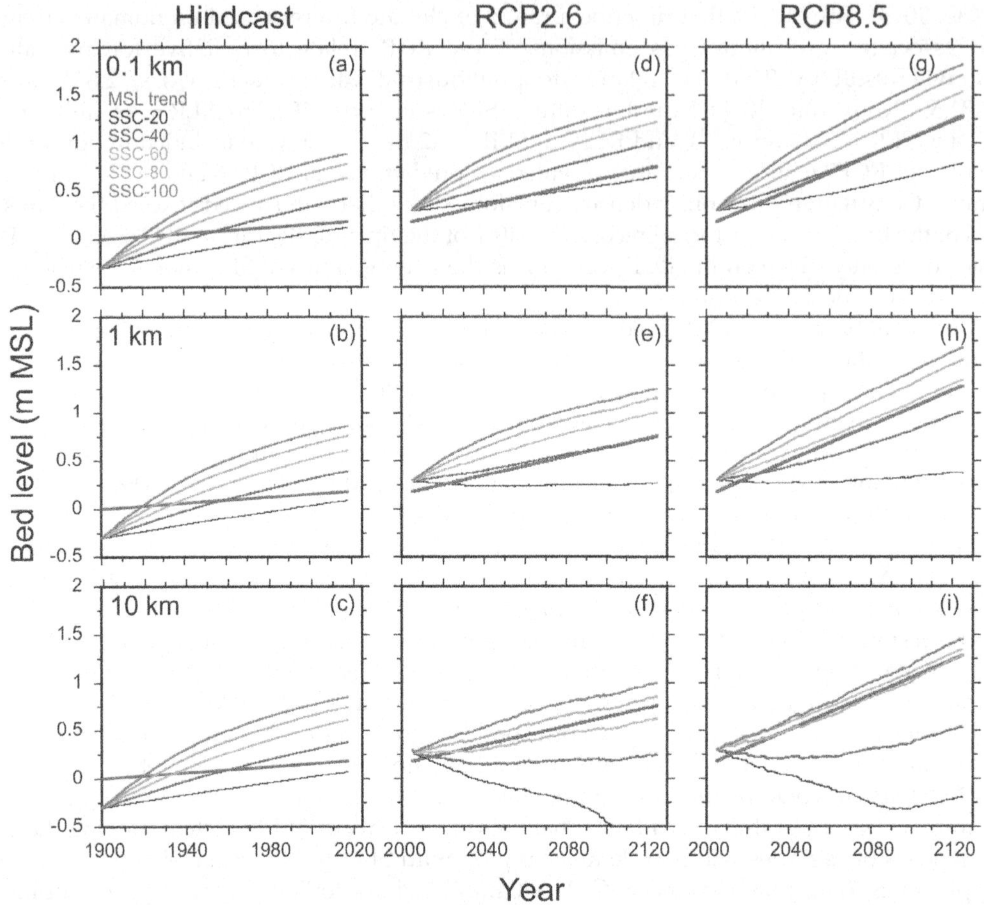

FIGURE 1.16 Zero-dimensional (0-D) model simulations of intertidal flat evolution for Auckland's east-coast estuaries for a range of time-average suspended sediment concentrations (SSC, mg/L) and fetch (0.1–10 km) for the historical period 1900–2018 (A–C) and future climate-change scenarios (2005–2125). Bed level is relative to mean sea level (MSL) Auckland Vertical Datum-1946. Future scenarios are based on IPCC (2013) climate and sea-level rise (SLR) for the Representative Concentration Pathways RCP2.6 (D–F, SLR = 4.8 mm yr^{-1}) and RCP8.5 (G-I, SLR = 9.2 mm yr^{-1}). The MSL trend for the historical period is for the Port of Auckland (1.6 mm yr^{-1}, 1899–2005). **Note**: the minimum bed level for scenario RCP2.6 with 10 km fetch is –0.57 m MSL.

that increases in average SSCs (i.e., from 10 to 20 mg L^{-1}) supported saltmarsh development in the upper intertidal zone. At moderate to high rates of SLR (i.e., 5.4–10.6 mm yr^{-1}), marshes are transformed to tidal flats as wave-forcing and erosion increase. Intertidal flats disappear completely at higher rates of SLR, with a shift to a subtidal geomorphic regime. Kakeh's et al. (2016) study suggest loss of unvegetated intertidal flats in small estuaries (i.e., 2 km fetch) at low sediment concentrations (i.e., 10–20 mg L^{-1}) due to bed erosion by waves. Using a 3D modeling approach, Kirwan and Murray (2007) also found that increasing sea levels may result in loss of unvegetated intertidal flats in settings where deepening and erosion of tidal channels occur. Sediment supply rates were also critical to maintenance of intertidal flats and saltmarshes, with a transition from stable to unstable systems likely to occur at SLR between 5 and 10 mm yr^{-1} at SSC of 20 mg L^{-1}.

6.3 Legacy sediments

Legacy sediments have played a role in the expansion of coastal wetlands in a range of environmental settings over the last several-hundred years. As observed in New Zealand's northern estuaries, mangrove-habitat expansion has followed historical pulses of sediment delivery associated with catchment deforestation, agriculture, and mining activities (Morrisey et al., 2010; James, 2013; Weston, 2014; Lovelock et al., 2015a). Substantial reduction in sediment delivery to estuaries and coasts has occurred globally over the last 50 years or so due to construction of dams (1.4 ± 0.3 billion t yr^{-1}, Syvitski et al., 2005), as well as improved land management (Walling, 1999). This reduction in catchment sediment delivery is a major issue that is likely to contribute to future mangrove forest loss in the Indo-Pacific (Lovelock et al., 2015b). This is less likely to be the case for New Zealand's mangrove estuaries where fluvial fine sediment is delivered from relatively small catchments (i.e., <1000 km^2 Hicks et al., 2011) by rivers that have typically been modified by flood-defense stop banks in their lower reaches, but rarely dammed. These structures have likely increased sediment delivery to estuaries by reducing the frequency of overbank discharges during storms.

The potential role of legacy sediments stored in estuaries, as well as contemporary fluvial inputs from rivers, to enable intertidal habitats to keep pace with accelerated SLR in the future has not been considered in our 0-D model. The simulations suggest that intertidal flats in the largest estuaries (i.e., 10 km fetch) will be the most vulnerable to accelerated SLR during the 21st century due to increased bed erosion by wave action where sediment supply is insufficient. Our case study of the Firth of Thames suggests that legacy sediments delivered over the last ~150 years partly sustained the rapid accretion of intertidal flats and enabled mangrove colonization in the early 1960s (Swales et al., 2015). Our analysis of Auckland east-coast estuaries (Fig. 1.2) suggests that legacy sediment deposits would only likely contribute to maintaining intertidal flat and fringing mangrove forest habitats in the largest estuarine systems with a subtidal "sediment pool" (e.g., Waitemata) or estuaries adjacent to sheltered coastal marine areas (e.g., inner Hauraki Gulf) that have accumulated large quantities of legacy sediments (e.g., Whitford, Wairoa, Swales et al., 2002b). Whether or not intertidal flats and fringing mangrove forest habitats will persist in these estuaries in the future will more likely depend more on the contemporary supply of fluvial sediments from catchments (Fig. 1.6).

7. Summary and conclusions

Mangrove forests composed of monospecific stands of the gray mangrove (*Avicennia marina* var *australasica*) occur in the estuaries of northern New Zealand. Geomorphic settings include barrier-enclosed estuaries, headland-enclosed estuaries, and coastal embayments often with small catchments relative to estuary area. These mangrove estuaries range in size (i.e., high-tide area) from less than 1 km^2 to more than 750 km^2. Contrary to the global pattern of ongoing mangrove forest loss during the 20th century, rapid expansion of mangrove forests has occurred in many New Zealand estuaries, averaging 4.1% yr^{-1} since the 1930s. These rates are twofold higher than for temperate *Avicennia* mangrove forests of southeastern Australia.

Habitat expansion has followed large-scale catchment deforestation and conversion to pastoral agriculture from the mid-1800s onward. The resulting increase in soil erosion has led to a shift from sand-to-mud–dominated systems and accelerated the development of intertidal flats in receiving estuaries that have been colonized by mangroves once they became "ecologically suitable" (Bird, 1986). Estuary infilling and resulting development of mudflats and mangrove forests in the mid- to upper-intertidal over the last century has resulted in the loss of sand flats and subtidal seagrass ecosystems in particular (e.g., Swales et al., 2002a; Inglis, 2003; Thrush et al., 2004; Morrisey et al., 2010; Swales et al., 2015). Data from 18 estuaries located near Auckland City on the east coast of New Zealand's North Island were analyzed to determine relationships between intertidal-flat area above MSL (i.e., potential mangrove habitat), catchment size, and sediment load. In this analysis, the ratio of potential mangrove habitat (i.e., intertidal area above MSL) to high-tide estuary area was compared to the ratio of catchment annual sediment load to estuary spring-tidal prism volume. Potential mangrove habitat above MSL was well predicted by catchment annual sediment load. Coastal embayment–type estuaries are less infilled due to their relatively small catchment sediment supply and consequently have smaller intertidal areas above MSL (Fig. 1.6). By contrast, barrier-enclosed estuaries are substantially more infilled than coastal embayments due to their relatively larger sediment supply. Consequently, barrier-enclosed estuaries have proportionally larger intertidal areas above MSL suitable for mangrove forest development. These estuary types have also historically experienced the largest percentage increases in mangrove-habitat area.

Exploratory modeling approaches are increasingly being used to simulate the future biogeomorphological evolution of estuaries. Questions remain about how to model these systems over meaningful spatial and temporal scales and whether knowledge of mangrove forest sedimentary systems is sufficient to confidently predict their long-term evolution. Quantitative observations of processes and long-term geomorphic response that are required to underpin and validate biogeomorphic models of mangrove systems are rare. Research undertaken over the last decade or so in New Zealand's Firth of Thames (Fig. 1.2) has begun to address these knowledge gaps. Mangroves colonized rapidly accreting mudflats in the southern Firth of Thames from the early 1960s. New insights have been gained into processes controlling mangrove-seedling recruitment, forest succession, biogeomorphic evolution of the mangrove forest and role in estuary sedimentation, influence of large-scale geological processes (e.g., basin subsidence), and processes that drive sedimentation and surface-elevation dynamics in the mangrove forest over seasonal-to-decadal time scales.

Mangrove-seedling recruitment events resulting in major phases of forest development are rare, occurring once per decade on average. During a typical summer, physical disturbance of the substrate by waves prevents successful establishment of propagules that require a 2- to 5-day inundation-free period for root growth and successful anchoring. Seedling survival appears to follow a biweekly pattern, in step with the spring-neap tidal cycle. Stochastic sea-level and wind-climate variability associated with the Southern Oscillation (La Niña, El Niño) over annual–decadal time scales are also influential factors controlling mangrove-seedling recruitment. In Auckland estuaries, this means that mangrove forests are unable to fully utilize their potential intertidal habitat, with occupancy ratios varying from 0.08 to 0.8. The largest historical increases in mangrove habitat have occurred in the smallest, fetch-limited/low wave-energy estuaries (i.e., high tide areas $< 2\ km^2$). This interaction of regular and predictable tidal variations in water level with stochastic wave-driven disturbance of the substrate and bed-level dynamics and longer-term sea-level variations are the key drivers of mangrove-seedling recruitment and forest development in these disturbance-limited systems.

This research has shown that mangrove-seedling recruitment events resulting in major phases of forest development area are rare, occurring once per decade on average. The interaction of regular and predictable tidal variations in water level with stochastic wave-driven disturbance and longer-term sea-level variations are key drivers of mangrove-seedling recruitment and forest development in these disturbance-limited systems. Reconstruction of the biogeomorphic evolution of the Firth mangrove forest since the early 1950s using sedimentary and historical records revealed that evolution of the forest was not only the result of simple vertical accretion and lateral mudflat progradation but rather a complex interplay between wave-driven sediment fluxes and the evolving geomorphology. Age–surface-elevation curves based on ^{210}Pb geochronology highlight the negative feedback between surface elevation and hydroperiod. The rate of surface-elevation gain displaying an asymptotic trajectory with sedimentation rates limited to the creation of new sediment accommodation space by relative sea level rise. This process is primarily driven by subsidence of the sedimentary basin (averaging $8.8 \pm 0.3\ mm\ yr^{-1}$). Surface-elevation dynamics in the mangrove forest are largely controlled by physical processes. In particular, fortnightly spring-tide inundation coupled with frequent onshore winds that resuspend and deliver mud to the mangrove forest. Desiccation of surficial deposits during the summer also results in shallow subsidence.

Ultimately the net trajectory of surface elevation in mangrove forests must be evaluated relative to local sea level. In the Firth of Thames, RSLR of ~ 1 cm/yr is largely driven by compaction and resulting deep subsidence of the sedimentary basin. When corrected for deep subsidence, net surface–elevation trends at RSET stations highlights differences in the relative importance of spatially varying shallow subsidence and the relatively uniform deep subsidence in each forest zone. In the rapidly accreting fringe forest ($23–28\ mm\ yr^{-1}$), deep subsidence accounts for a greater fraction of net elevation change than does shallow subsidence. In the scrub forest, shallow subsidence due to seasonal desiccation and compaction of surficial sediments is an important process with net elevation gain of $2–13\ mm\ yr^{-1}$. In the relict fringe forest, shallow and deep subsidence rates are similar and near-equivalent to rates of surface-elevation gain so that net elevation increases are close to zero. Net surface–elevation gain in the relict fringe forest is largely controlled by creation of new accommodation space associated with RSLR. Thus, physical processes, primarily explain sedimentation patterns and geomorphic evolution of the Firth of Thames mangrove forest over seasonal-to-decadal time scales.

The response of New Zealand's sediment-rich mangrove forests to future SLR remains uncertain. This is primarily because (1) the rate of SLR over the coming century and beyond will likely be determined by human actions or inaction over the next 10–20 years; and (2) future rates of sediment supply to estuaries will depend on factors such as catchment physiography, climate-induced changes in rainfall and runoff, land use, and management practices. Model simulations of Auckland's estuaries for the 21st century low- and high-SLR scenarios suggest that intertidal flat habitats in small tidal creeks (0.1 km fetch) and small estuaries (1 km fetch) will be able to maintain their elevation relative to MSL at relatively low sediment supply rates. By contrast, intertidal flats in the largest estuaries (10 km fetch) will be more susceptible to inundation by rising sea levels, even at the lower rate of SLR anticipated under the IPCC RCP2.6 scenario (i.e., 4.8 mm yr^{-1}). Loss of intertidal flats could begin as early as the 2020s. Model results also indicate that catchment sediment supply to these largest estuaries would need to be twofold higher than the historical rates that sustained vertical accretion and development of intertidal flats during the 20th century.

Legacy sediments may contribute to maintaining intertidal flat and fringing mangrove forest habitats in the largest estuarine systems with a subtidal "sediment pool" or estuaries adjacent to sheltered coastal marine areas that have accumulated large quantities of legacy sediments. Otherwise, maintenance of these intertidal habitats will more likely depend on the contemporary supply of fluvial sediments from catchments (Fig. 1.6). Sediment-poor estuarine systems, such as coastal embayment's with small catchments, will be most susceptible to loss of intertidal flats and fringing mangrove forests.

The critical importance of sediment supply to maintain intertidal habitats in New Zealand's estuaries under rising sea-levels mirrors assessments of mangrove forest vulnerability to SLR in the Indo-Pacific region where the majority of the world's mangrove forests occur. Lovelock et al. 2015a,b analyzed data from RSET stations at more than 150 sites in the Indo-Pacific. They concluded that observed surface-elevation gains at 69% of sites were not sufficient to keep pace with rates of SLR over recent decades. Modeling based on data and SLR predictions also indicated loss of mangrove forests from coasts with low sediment supply by the 2070s under moderate rates of SLR. Mangrove forests and salt marshes may also accommodate rising sea levels by retreating landwards where this is not impeded by coastal infrastructure such as flood-defenses (e.g., stop banks, levee), reclamations, roads, and sea walls. In New Zealand, as elsewhere, the potential for landward retreat will be limited in highly modified urban estuaries and in rural estuaries with low-lying shorelines that are protected by coastal storm–defense infrastructure.

Acknowledgments

The authors thank the reviewers for their constructive comments on the manuscript. Fig. 1.4 (Bridge over the Mangemangeroa Creek near Howick, 1860s. John Kinder photographs, item PH 233) was reproduced with permission of Special Collections, University of Auckland Libraries and Learning Services. Figs. 1.7–1.10 were reproduced and modified from Swales et al. (2015) Earth Surface Processes and Landforms 40: 1672–87 with permission (Copyright © John Wiley and Sons, Ltd.). Aerial photographs in Fig. 1.3 were taken by Alastair Jamieson (Wild Earth Media) and reproduced with permission. Figs. 1.11–1.14 were reproduced from Swales et al. (2019) *Ecosystems* under the terms of the Creative Commons Attribution 4.0 International License. Max Oulton drafted Figs. 1.1, 1.2 and 1.5. This study was supported by the New Zealand Ministry of Business, Innovation and Employment (contracts FWCE1903 and FWCE1911, NIWA's Catchments to Estuaries Programme). We thank Auckland Council for permission to incorporate data from Swales et al. (2009), Auckland Council Technical Report No. 079 to produce Fig. 1.6.

References

Allen, J.R.L., 2000. Morphodynamics of holocene saltmarshes: a review sketch from the Atlantic and Southern North Sea Coasts of Europe. Quat. Sci. Rev. 19, 1155–1231.

Alongi, D.M., 2008. Mangrove forests: resilience, protection from tsunamis, and responses to global climate change. Estuar. Coast Shelf Sci. 76, 1–13.

Alpert, P., 1985. Distribution quantified by microtopography in an assemblage of saxicolous mosses. Vegetatio 64, 141–149.

Balke, T., Swales, A., Lovelock, C.E., Herman, P.M.J., Bouma, T.J., 2015. Limits to seaward expansion of mangroves: translating physical disturbance mechanisms into seedling survival gradients. J. Exp. Mar. Biol. Ecol. 467, 16–25.

Barker, R., Sheaves, M., Johnston, R., 2015. Geographic variation in mangrove flooding and accessibility for fishes and nektonic crustaceans. Hydrobiologia 762, 1–14.

Beaglehole, J.C., 1968. The Journals of Captain James Cook on His Voyages of Discovery. The Voyage of the Endeavour 1768–1771. Cambridge University Press, Cambridge.

Beard, C.M., 2006. Physiological constraints on the latitudinal distribution of the mangrove *Avicennia marina (Forsk.) Vierh. subsp. australasica (Walp.)* J. Everett (*Avicenniaceae*) in New Zealand. PhD thesis. University of Waikato, Hamilton, New Zealand.

Bedin, T., 2001. The progression of a mangrove forest over a newly formed delta in the Umhlatuze Estuary, South Africa. South Afr. J. Bot. 67, 433–438.

Bird, E.C.F., 1986. Mangroves and intertidal morphology in Westernport Bay, Victoria, Australia. Mar. Geol. 69, 251–271.

Bulmer, R.H., Schwendenmann, L., Lohrer, A., Lundquist, C.J., 2017. Sediment carbon and nutrient fluxes from cleared and intact temperate mangrove ecosystems and adjacent sandflats. Sci. Total Environ. 599–600, 1874–1884. https://doi.org/10.1016/j.scitotenv.2017.05.139.

Burns, B.R., Ogden, J., 1985. The demography of the temperate mangrove (*Avicennia marina* (Forsk.) Vierh.) at its southern limit in New Zealand. Aust. J. Ecol. 10, 125–133.

Cahoon, D.R., Hensel, P.F., Spencer, T., Reed, D.J., McKee, K.L., Saintilan, N., 2006. Coastal wetland vulnerability to relative sea-level rise: wetland elevation trends and process controls. In: Verhoeven, J.T.A., et al. (Eds.), Wetlands and Natural Resource Management. Springer-Verlag, Berlin, pp. 271–292.

Cahoon, D.R., 2015. Estimating relative sea-level rise and submergence potential at a coastal wetland. Estuar. Coast 38 (3), 1077–1084. https://doi.org/10.1007/s12237-014-9872-8.

Cahoon, D.R., Guntenspergen, G.R., January–February 2010. Climate change, sea-level rise and coastal wetlands. Natl. Wetl. Newsl. 8–12.

Cahoon, D.R., Reed, D.J., 1995. Relationships among marsh surface topography, hydroperiod and soil accretion in a deteriorating Louisiana Salt Marsh. J. Coast Res. 11 (2), 357–369.

Cahoon, D.R., Lynch, J.C., Perez, B.C., Segura, B., Holland, R.D., Stelley, C., Stephenson, G., Hensel, P., 2002. High-precision measurements of wetland elevation II. The rod surface elevation table. J. Sedim. Res. 72 (5), 734–739.

Cahoon, D.R., Perez, B.C., Segura, B.D., Lynch, J.C., 2011. Elevation trends and shrink–swell response of wetland soils to flooding and drying. Estuar. Coast Shelf Sci. 91, 463–474.

Carvalho do Amaral, P.G., Ledru, M.P., Branco, F.R., Giannin, P.C.F., 2006. Late holocene development of a mangrove ecosystem in Southeastern Brazil (Itanhaem, state of Sao Paulo). Palaegeogr. Palaeoclim. Palaeoecol. 241, 608–620.

Chapman, V.J., Ronaldson, J.W., 1958. The mangrove and saltmarsh flats of the Auckland isthmus. N. Z. Dep. Sci. & Ind. Res. Bull. No. 125 (Wellington, New Zealand).

Chapman, V.J., 1976. Coastal Vegetation. Pergamon Press, New York, 292 pp.

Church, J.A., White, N.J., 2011. Sea-level rise from the late 19th to the early 21st century. Surv. Geophys. 32, 585–602. https://doi.org/10.1007/s10712-011-9119-1.

Church, J.A., Clark, P.U., Cazenave, A., Gregory, J.M., Jevrejeva, S., Levermann, A., Merrifield, M.A., Milne, G.A., Nerem, R.S., Nunn, P.D., Payne, A.J., Pfeffer, W.T., Stammer, D., Unnikrishnan, A.S., 2013. Sea level change. In: Stocker, T.F., Qin, D., Plattner, G.-K., Tignor, M., Allen, S.K., Boschung, J., Nauels, A., Xia, Y., Bex, V., Midgley, P.M. (Eds.), Climate Change 2013: The Physical Science Basis. Contribution of Working Group I to the Fifth Assessment Report of the Intergovernmental Panel on Climate Change. Cambridge University Press, Cambridge, United Kingdom and New York, NY, USA.

Clarke, L.D., Hannon, N.J., 1970. The mangrove swamp and salt marsh communities of the Sydney district. III. Plant growth in relation to salinity and waterlogging. J. Ecol. 58, 351–369.

Clarke, P.J., 1993. Dispersal of grey mangrove (*Avicennia marina*) propagules in Southeastern Australia. Aquat. Bot. 45, 195–204.

Clarke, P.J., Myerscough, P.J., 1993. The intertidal distribution of the grey mangrove (*Avicennia marina*) in Southeastern Australia: the effects of physical conditions, interspecific competition, and predation on propagule establishment and survival. Aust. J. Ecol. 18, 307–315.

Clarke, P.J., Kerrigan, R.A., 2002. The effects of seed predators on the recruitment of mangroves. J. Ecol. 90, 728–736.

Clements, F.E., 1916. Plant Succession: an Analysis of the Development of Vegetation. Carnegie Institutes Publication No. 242, Washington D.C.

Coco, G., Zhou, Z., van Maanen, B., Olabarrieta, M., Tinoco, R., Townend, I., 2013. Morphodynamics of tidal networks: advances and challenges. Mar. Geol. 346, 1–16.

Coe, A., Bosence, D., Church, K., Flint, S., Howell, J., Wilson, C., 2002. The Sedimentary Record of Sea Level Change. Cambridge University Press, 288 pp.

Corenblit, D., Baas, A.C.W., Bornette, G., Darrozes, J., Delmotte, S., Francis, R.A., Gurnell, A.M., Julien, F., Niaman, R.J., Steiger, J., 2011. Feedbacks between geomorphology and biota controlling earth surface processes and landforms: a review of foundation concepts and current understandings. Earth Sci. Rev. 106 (3), 307–331.

Cowles, H.C., 1899. The ecological relations of the vegetation on the sand dunes of Lake Michigan. Bot. Gaz. 27, 95–117, 167–202, 281–308, 361–391.

Cowles, H.C., 1901. The physiographic ecology of Chicago and vicinity. Bot. Gaz. 31, 73–108, 145–182.

Curran, M., Cole, M., Allaway, W.G., 1986. Root aeration and respiration in young mangrove plants (*Avicennia marina* (Forsk.) Vierh.). J. Exp. Bot. 37, 1225–1233.

Davis, J.H., 1940. Ecology and Geologic Role of Mangroves in Florida. Carnegie Institution of Washington Publication No. 517. Carnegie Institute of Washington, Washington, D.C., USA.

D'Alpaos, A.S., Lanzoni, S., Marani, M., Rinaldo, A., 2007. Landscape evolution of tidal embayments: modeling the interplay of erosion, sedimentation and vegetation dynamics. J. Geophys. Res. 112 https://doi.org/10.1029/2006JF000537.

De Lange, W.P., de Lange, P.J., 1994. An appraisal of factors controlling the latitudinal distribution of mangrove (*Avicennia marina* var. *resinifera*) in New Zealand. J. Coast Res. 10, 539–548.

Dietrich, W.E., Perron, J.T., 2006. The search for a topographic signature of life. Nature 439, 411–418.

Duke, N.C., 1991. A systematic revision of the mangrove genus *Avicennia* (*Avicenniaceae*) in Australasia. Aust. Syst. Bot. 4, 299–324.

Elgar, S., Raubenheimer, B., 2008. Wave dissipation by muddy seafloors. Geophys. Res. Lett. 35, LL07611.

Elith, J., Leathwick, J.R., Hastie, T., 2008. A working guide to boosted regression trees. J. Anim. Ecol. 77, 802–813.

Elliot, A.H., Shankar, U., Hicks, D.M., Woods, R.A., Dymond, J.R., 2008. SPARROW regional regression for sediment yields in New Zealand rivers. In: Schmidt, J., Cochrane, T., Phillips, C., Elliot, S., Davies, T., Basher, L. (Eds.), Sediment Dynamics in Changing Environments (Proceedings of a Symposium Held in Christchurch, New Zealand, December 2008, vol. 325. IAHS Publication.

Ellis, J., Nicholls, P., Craggs, R., Hofstra, D., Hewitt, J., 2004. Effects of terrigenous sedimentation on mangrove physiology and associated macrobenthic communities. Mar. Ecol. Prog. Ser. 270, 71–82.

Ellison, J.C., 1993. Mangrove retreat with rising sea-level, Bermuda. Estuar. Coast Shelf Sci. 37, 75–87.

Ferrar, H.T., 1934. The geology of the Dargaville–Rodney Subdivision, Hokianaga and Kaipara Divisions., No. 34. Wellington, New Zealand: Geology Survey Branch, Department of Scientific and Industrial Research.

Friedman, J.H., 2001. Greedy function approximation: a gradient boosting machine. Ann. Stat. 29, 1189–1232.

Galloway, R.W., 1982. Distribution and physiographic patterns of Australian mangroves. In: Clough, B.F. (Ed.), Mangrove Ecosystems in Australia: Structure, Function and Management. Australian Institute of Marine Science, Townsville, Australia, pp. 31–54.

Giri, C., Ochieng, E., Tieszen, L.L., Zhu, Z., Singh, A., Loveland, T., Maesk, J., Duke, N., 2011. Status and distribution of mangrove forests of the world using earth observation satellite data. Global Ecol. Biogeogr. 20, 154–159.

Gladstone-Gallagher, R., Lundquist, C.J., Piltditch, C.A., 2014. Mangrove (*Avicennia marina* subsp. *australasica*) litter production and decomposition in a temperate estuary. N. Z. J. Mar. Freshw. Res. 48 (1), 24–37.

Goodson, J.M., Gurnell, A.M., Angold, P.G., Morrissey, I.P., 2002. Riparian seed banks along the lower River Dove, UK: their structure and ecological implications. Geomorphology 47, 45–60.

Goring, D.G., Bell, R.G., 1999. El Nino and decadal effects on sea-level variability in northern New Zealand: a wavelet analysis. N. Z. J. Mar. Freshw. Res. 33, 587–598.

Green, M.O., Black, K.P., Amos, C.L., 1997. Control of estuarine sediment dynamics by interactions between currents and waves at several scales. Mar. Geol. 144, 97–116.

Green, M.O., Bell, R.G., Dolphin, T.J., Swales, A., 2000. Silt and sand transport in a deep tidal channel of a large estuary (Manukau Harbour, New Zealand). Mar. Geol. 163, 217–240.

Green, M.O., Coco, G., 2007. Sediment transport on an estuarine intertidal flat: measurements and conceptual model of waves, rainfall and exchanges with a tidal creek. Estuar. Coast Shelf Sci. 72, 553–569.

Green, M.O., Coco, G., 2014. Review of wave-driven sediment resuspension and transport in estuaries. Rev. Geophys. 52, 77–117.

Hack, J.T., Goodlett, J.C., 1960. Geomorphology and Forest Ecology of a Mountain Region in the Central Appalachians. U.S. Geological Survey Professional Paper 347, 66 pp.

Hall, T.F., Smith, G.E., 1955. Effects of flooding on woody plants, west sandy dewatering project, Kentucky reservoir. J. For. 53, 281–285.

Hamilton, S.E., Casey, D., 2016. Creation of a high spatio-temporal resolution global database of continuous mangrove forest cover for the 21st century (CGMFC-21). Global Ecol. Biogeogr. 25, 729–738.

Hannah, J., 2004. An updated analysis of long-term sea level change in New Zealand. Geophys. Res. Lett. 31, L03307. https://doi.org/10.01029/02003GL019166, 033.

Hannah, J., Bell, R.G., 2012. Regional sea level trends in New Zealand. J. Geophys. Res. 117, C01004. https://doi.org/10.1029/2011JC007591.

Harris, C.K., Wiberg, P., 2002. Across-shelf sediment transport: interacctions between suspended sediment and bed sediment. J. Geophys. Res. 107 (C1), 3008.

Harty, C., 2009. Mangrove planning in New Zealand and South East Australia – a reflection on approaches. Ocean Coast Manag. 52, 278–286.

Healy, T., 2002. Muddy coasts of mid-latitude oceanic islands on an active plate margin – New Zealand. In: Healy, T., Wang, Y., Healy, J.-A. (Eds.), Muddy Coasts of the World: Processes, Deposits and Function. Elsevier Science, Amsterdam, pp. 347–374.

Hicks, D.M., Shankar, U., McKerchar, A.I., Basher, L., Jessen, M., Lynn, I., Page, M., 2011. Supsended sediment yields from New Zealand rivers. J. Hydrol. (NZ) 50 (1), 81–142.

Hochstein, M.P., Nixon, I.M., 1979. Geophysical study of the Hauraki depression, North Island, New Zealand. N. Z. J. Geol. Geophys. 22, 1–19.

Hochstein, M.P., Tearney, K., Rawson, S., Davey, F.J., Davidge, S., Henreys, S., Backshal, D., 1986. Structure of the Hauraki Rift (New Zealand). Bull. Roy. Soc. N. Z. 24, 333–348.

Horstman, E.M., Dohmen-Janssen, C.M., Bouma, T.J., Hulscher, S.J.M.H., 2015. Tidal-scale flow routing and sedimentation in mangrove forests: combining field data and numerical modelling. Geomorphology 228, 244–262.

Horstman, E.M., Lundquist, C.J., Bryan, K.R., Bulmer, R.H., Mullarney, J.C., Stokes, D.J., 2018. The dynamics of expanding mangroves in New Zealand. Coastal Research Library 25. In: Makowski, C., Finkl, C.W. (Eds.), Threats To Mangrove Forests: Hazards, Vulnerability and Management. Springer International Publishing AG, part of Springer Nature. https://doi.org/10.1007/978-3-319-73016-5_2.

Hovenden, M.J., Curran, M., Cole, M.A., Goulter, P.F.E., Skelton, N.J., Allaway, W.G., 1995. Ventilation and respiration in roots of one-year-old seedlings of grey mangrove *Avicennia marina* (Forsk.) Vierh. Hydrobiologia 295, 23–29.

Hume, T.M., McGlone, M.S., 1986. Sedimentation patterns and catchment use changes recorded in the sediments of a shallow tidal creek, Lucas Creek, Upper Waitemata Harbour, New Zealand. J. Roy. Soc. N. Z. 19, 305–317.

Hume, T.M., Herdendorf, C.E., 1988. A geomorphic classification of estuaries and its application to coastal resource management – a New Zealand example. Ocean Shorel. Manag. 11, 249–274.

Hume, T.M., 2003. Estuaries and tidal inlets. In: Goff, J.R., et al. (Eds.), The New Zealand Coast: Te Tai O Aotearoa. Dunsmore Press, Palmerston North, New Zealand, pp. 191–213.

Hume, T.M., Snelder, T., Weatherhead, M., Liefting, R., 2007. A controlling factor approach to estuary classification. Ocean Coast Manag. 50, 905–929.

Hupp, C.R., Osterkamp, W.R., Howard, A.D., 1995. Preface. Geomorphology 13 (1–4), vii–viii. https://doi.org/10.1016/0169-555X(95)90009-I.

Imeson, A.C., 1976. Some effects of burrowing animals on slope processes in the Luxembourg Ardennes. Geogr. Ann. 58A, 115–125.

Inglis, G.J., 2003. The seagrasses of New Zealand. In: Green, E.P., Short, F.T. (Eds.), World Atlas of Seagrasses. University of California Press, Berkeley, California, pp. 148–157.

IPCC, 2013. Climate change. 2013. The physical science Basis. In: Stocker, T.F., Qin, D., Plattner, G.-K., Tignor, M., Allen, S.K., Boschung, J., Nauels, A., Xia, Y., Bex, V., Midgley, P.M. (Eds.), Contribution of Working Group I to the Fifth Assessment Report of the Intergovernmental Panel on Climate Change. Cambridge University Press, Cambridge, United Kingdom and New York, NY, USA, 1535 pp.

James, L.A., 2013. Legacy sediment: definitions and processes of episodically produced anthropogenic sediment. Anthropocene 2, 16–26.

Jervey, M.T., 1988. Quantitative geological modelling of siliciclastic rock sequences and their seismic expressions. In: Wilgus, C.K., Hastings, B.S., Kendall, C.G.S.C., Posamentier, H.W., Ross, C.A., Van Wagoner, J.C. (Eds.), Sea Level Changes: An Integrated Approach, vol. 42. Society of Economic Palaeontologists and Mineralogists Special Publication, pp. 47–69.

Kakeh, N., Coco, G., Marani, M., 2016. On the morphodynamic stability of intertidal environments and the role of vegetation. Adv. Water Resour. 93, 303–314.

Kaye, C.A., Barghoorn, E.S., 1964. Late quaternary sea-level change and crustal rise at Boston Massachusetts, with notes on the autocompaction of peat. Geol. Soc. Am. Bull. 75, 63–80.

Kirby, M., 1995. Modelling the links between vegetation and landforms. Geomorphology 13, 319–336.

Kirwan, M.L., Murray, A.B., 2007. A coupled geomorphic and ecological model of tidal marsh evolution. Proc. Natl. Acad. Sci. U. S. A. 104 (15), 6118–6122.

Krauss, K., Allen, J.A., Cahoon, D.R., 2003. Differential rates of vertical accretion and elevation change among aerial root types in Micronesian mangrove forests. Estuar. Coast Shelf Sci. 56, 251–259.

Krauss, K.W., McKee, K.L., Lovelock, C.E., Cahoon, D.R., Saintilan, N., Reef, R., Chen, L., 2014. How mangroves adjust to rising sea level. New Phytol. 202, 19–34. https://doi.org/10.1111/nph.12605.

Lovelock, C.E., Adame, M.F., Bennion, V., Hayes, M., Reef, R., Santini, N., Cahoon, D.R., 2015a. Sea level and turbidity controls on mangrove surface elevation change. Estuar. Coast Shelf Sci. 153, 1–9.

Lovelock, C.E., Cahoon, D.R., Friess, D.A., Guntenspergen, G.R., Krauss, K.W., Reef, R., Rogers, K., Saunders, M., Sidik, F., Swales, A., Saintilan, N., Thuyen, L.X., Triet, T., 2015b. The vulnerability of Indo-Pacific mangrove forests to sea level rise. Nature 526, 559–563. https://doi.org/10.1038/nature15538.

Lovelock, C., Feller, I.C., Ellis, J., Schwarz, A.M., Hancock, N., Nichols, P., Sorrell, B., 2007. Mangrove growth in New Zealand estuaries: the role of nutrient enrichment at sites with contrasting rates of sedimentation. Oecologia 153, 633–641.

Lovelock, C., Sorrell, B.K., Hancock, N., Hua, Q., Swales, A., 2010. Mangrove forest and soil development on a rapidly accreting shore in New Zealand. Ecosystems 13, 437–641.

Lovelock, C.E., Bennion, V., Grinham, A., Cahoon, D.R., 2011. The role of surface and subsurface processes in keeping pace with sea level rise in intertidal wetlands of Moreton Bay, Queensland, Australia. Ecosystems 14 (5), 745–757.

Lugo, A.E., Snedaker, S.C., 1974. The ecology of mangroves. Annu. Rev. Ecol. Systemat. 5, 39–64.

Lundquist, C.J., Morrisey, D.J., Gladstone-Gallagher, R.V., Swales, A., 2014. Managing mangrove habitat expansion in New Zealand. In: Faridah-Hanum, et al. (Eds.), Mangrove Ecosystems of Asia. Springer Science and Business Media, New York. https://doi.org/10.1007/978-1-4614-8582-7_19.

Marani, M., D'Alpaos, A., Lanzoni, S., Carniello, L., Rinaldo, A., 2010. The importance of being coupled: stable states and catastrophic shifts in tidal biomorphodynamics. J. Geophys. Res. 115, F04004.

Marani, M., Da Lio, C., D'Alpaos, A., 2013. Vegetation engineers marsh morphology through multiple competing stable states. Proc. Natl. Acad. Sci. U. S. A. 110 (9), 3259–3263.

McBride, G., Reeve, G., Pritchard, M., Lundquist, C., Daigneault, A., Bell, R., Blackett, P., Swales, A., Wadhwa, S., Tait, A., Zammit, C., 2016. The Firth of Thames and Lower Waihou River. Synthesis Report RA2, Coastal Case Study. Climate Changes, Impacts and Implications (CCII) for New Zealand to 2100. Ministry of Business, Innovation and Employment Contract CO1X1225, 50 pp. https://ccii.org.nz/app/uploads/2017/02/RA2-Coastal-Case-Study-Synthesis-report.pdf.

McKee, K.L., Cahoon, D.R., Feller, I.C., 2007. Caribbean mangroves adjust to rising sea level through biotic controls on change in soil elevation. Global Ecol. Biogeogr. 16, 545–556.

McKee, K.L., 2011. Biophysical controls on accretion and elevation change in Caribbean mangrove ecosystems. Estuar. Coast Shelf Sci. 91, 475–483.

Morrisey, D.J., Swales, A., Dittmann, S., Morrison, M.A., Lovelock, C.E., Beard, C.M., 2010. Ecology and management of temperate mangroves. Oceanogr. Mar. Biol. Annu. Rev. 48, 43–160.

Murray, A.B., Lazarus, E., Ashton, A., Baas, A., Coco, G., Coulthard, T., Fonstad, M., Haff, P., McNamara, D., Paola, C., Pelletier, J., Reinhardt, L., 2009. Geomorphology, complexity and the emerging science of the Earth's surface. GeoMorphol. 103 (3), 496–505.

Naish, T.R., Nelson, C.S., Hodder, P.W., 1993. Evolution of Holocene sedimentary bentonite in a shallow marine embayment, Firth of Thames, New Zealand. Mar. Geol. 109, 267–278.

Naylor, L.A., Viles, H.A., Carter, N.E.A., 2002. Biogeomorphology revisited: looking towards the future. Geomorphology 47, 3–14.

Nerem, R.S., van Dam, T.M., Schenewerk, M.S., 1998. Chesapeake Bay subsidence monitored as wetlands loss continues. EOS Trans. 79, 149–157.

Nittrouer, C.A., Wright, L.D., 1994. Transport of particles across continental shelves. Rev. Geophys. 32, 85–113.

Oh, R.R.Y., Friess, D.A., Brown, B.M., 2017. The role of surface elevation in the rehabilitation of abandoned aquaculture ponds to mangrove forests, Sulawesi, Indonesia. Ecol. Eng. 100, 325–334.

Oldman, J.W., Black, K.P., Swales, A., Stroud, M.J., 2009. Prediction of annual average sedimentation rates in an estuary using numerical models with verification against core data - Mahurangi Estuary, New Zealand. Estuar. Coast Shelf Sci. 84, 483–492.

Olsen, J.S., 1958. Lake Michigan dune development 2. Plants as agents and tools in geomorphology. J. Geol. 66 (4), 345–351.

Osterkamp, W.R., Hupp, C.R., 2010. Fluvial processes and vegetation – glimpses of the past, present and futures. Geomorphology 116, 274–285.

Panapitukkul, N., Duarte, C.M., Thampanya, U., Kheowvongsri, P., Srichai, N., Geertz-Hansen, O., Terrados, J., Boromthanarath, S., 1998. Mangrove colonization: mangrove progression over the growing Pak Phanang (SE Thailand) mud flat. Estuar. Coast Shelf Sci. 47, 51–61.

Phillips, J.D., 1995. Biogeomorphology and landscape evolution: the problem of scale. Geomorphology 13, 337–347.

Pillans, B., 1986. A late Quaternary uplift map for North island, New Zealand. Bulletin of the Royal Society of New Zealand. 24, 409–417.

Posamentier, H.W., Jervey, M.T., Vail, P.R., 1988. Eustatic controls on clastic deposition I —conceptual framework. In: Wilgus, C.K., Hastings, B.S., Kendall, C.G.S.C., Posamentier, H.W., Ross, C.A., Van Wagoner, J.C. (Eds.), Sea Level Changes: an Integrated Approach, vol. 42. SEPM Spec. Publ, pp. 110–124.

Pritchard, M., Swales, A., Green, M.O., 2015. Influence of buoyancy- and wind-coupling on sediment dispersal and deposition in the Firth of Thames, New Zealand. In: Australasian Coasts and Ports Conference, Auckland, New Zealand, 15–18 September 2015, 6 pp.

Reineck, H.E., Singh, I.B., 1980. Depositional Sedimentary Environments, Second Revised Edition. Springer-Verlag, Berlin.

Richards, D.R., Friess, D.A., 2016. Rates and drivers of mangrove deforestation in Southeast Asia, 2000–2012. Proc. Natl. Acad. Sci. U. S. A. 113, 344–349.

Rogers, K., Saintilan, N., Woodroffe, C.D., 2014. Surface elevation change and vegetation distribution dynamics in a subtropical coastal wetland: implications for coastal wetland response to climate change. Estuar. Coast Shelf Sci. 149, 49–56.

Roy, P.S., Williams, R.J., Jones, A.R., Yassini, I., Gibbs, P.J., Coates, B., West, R.J., Scanes, P.R., Hudson, J.P., Nichol, S., 2001. Structure and function of south-east Australian estuaries. Estuar. Coast Shelf Sci. 53, 351–384.

Saintilan, N., Rogers, K., McKee, K.L., 2009. Salt marsh-mangrove interactions in Australasia and the Americas. In: Perillo, G.M.E., et al. (Eds.), Coastal Wetlands – An Integrated Ecosystem Approach. Elsevier, Amsterdam, pp. 855–883.

Schlager, W., 1993. Accommodation and supply — a dual control on stratigraphic sequences. Sediment. Geol. 86, 111–136.

Scoffin, T.P., 1970. Trapping and binding of subtidal carbonate sediments by marine vegetation in Bimimi Lagoon, Bahamas. J. Sediment. Petrol. 40, 249–273.

Sheffield, A.T., Healy, T.R., McGlone, M.S., 1995. Infilling rates of a steepland catchment estuary, Whangamata, New Zealand. J. Coast Res. 11 (4), 1294–1308.

Sidik, F., Neil, D., Lovelock, C.E., 2016. Effect of high sedimentation rates on surface sediment dynamics and mangrove growth in the Porong River, Indonesia. Mar. Pollut. Bull. 107 (1), 355–363.

Snedaker, S.C., 1995. Mangroves and climate change in the Florida and Caribbean region: scenarios and hypotheses. Hydrobiologia 295, 43–49.

Soreghan, G.S., Dickinson, W.R., 1994. Generic types of stratigraphic cycles controlled by eustacy. Geology 22, 759–761.

Spalding, M., Kainuma, M., Collins, L., 2010. World Atlas of Mangroves (Version 1.1), A Collaborative Project of ITTO, ISME, FAO, UNEPWCMC, UNESCO-MAB, UNU-INWEH and TNC. Earthscan, London.

Spiers, A.G., 1999. Review of international/ continental wetland resources. In: Finlayson, C.M., Spiers, A.G. (Eds.), Global Review of Wetland Resources and Priorities for Wetland Inventory, pp. 63–104. Supervising Scientist Report 144. Canberra, Australia.

Swales, A., Hume, T.M., McGlone, M.S., Pilvio, R., Ovenden, R., Zviguina, N., Hatton, S., Nicholls, P., Budd, R., Hewitt, J., Pickmere, S., Costley, K., 2002a. Evidence for the Physical Effects of Catchment Sediment Runoff Preserved in Estuarine Sediments: Phase II (Field Study). NIWA Client Report: HAM2002-067.

Swales, A., Hume, T.M., Oldman, J.W., Green, M.O., 1997. Holocene sedimentation and recent human impacts on infilling in a drowned valley estuary. In: Lumsden (Ed.), Pacific Coasts and Ports'97 (Christchurch, New Zealand), Proc. Of 13th Australasian Coastal and Ocean Engineering Conference and the 6th Australasian Port and Harbour Conference, pp. 895–900 (2).

Swales, A., MacDonald, I.T., Green, M.O., 2004. Influence of wave and sediment dynamics on cord grass (*Spartina anglica*) growth and sediment accumulation on a wave-exposed intertidal flat. Estuaries 27 (2), 225–243.

Swales, A., Gibbs, M.M., Ovenden, R., Budd, R., Hermansphan, N., 2008. Sedimentation in the Okura - Weiti - Karepiro Bay System. NIWA Client Report HAM2008–153. NIWA Hamilton, 40 pp.

Swales, A., Gorman, R., Oldman, J.W., Altenberger, A., Hart, C., Bell, R.G., Claydon, L., Wadha, S., Ovenden, R., 2009. Potential Future Changes in Mangrove Habitat in Auckland's East-coast Estuaries. Auckland Council Technical Report No. 079 (June 2009). ISSN 1179-0512 (online), 219 pp.

Swales, A., Ovenden, R., Costley, K., Hermanspahn, N., Budd, R., Rendle, D., Hart, C., Wadhwa, S., 2011. Patterns and Rates of Recent Sedimentation and Intertidal Vegetation Changes in the Kaipara Harbour. NIWA Client Report HAM2011–040 prepared for Auckland Council and Northland Regional Council, 130 pp.

Swales, A., Bentley, S.J., Lovelock, C.E., 2015. Mangrove-forest evolution in a sediment-rich estuarine system: opportunists or agents of geomorphic change? Earth Surf. Process. Landforms 40, 1672–1687. https://doi.org/10.1002/esp.3759.

Swales, A., Bentley, S.J., Lovelock, C., Bell, R.G., 2007. Sediment processes and mangrove-habitat expansion on a rapidly-prograding muddy coast, New Zealand. In: Sixth International Conference on Coastal Engineering and Science of Coastal Sediment Processes, New Orleans, May 2007. American Society of Civil Engineers, Reston, VA, pp. 1441–1454.

Swales, A., Denys, P., Pickett, V.I., Lovelock, C.E., 2016. Evaluating deep subsidence in a rapidly accreting mangrove forest using GPS monitoring of surface-elevation benchmarks and sedimentary records. Mar. Geol. 380, 205–218.

Swales, A., Reeve, G., Cahoon, D.R., Lovelock, C.E., 2019. Landscape evolution of a fluvial sediment-rich *Avicennia marina* mangrove forest: insights from seasonal and inter-annual surface-elevation dynamics. Ecosystems. https://doi.org/10.1007/s10021-018-0330-5.

Swales, A., Williamson, R.B., Van Dam, L.F., Stroud, M.J., McGlone, M.S., 2002b. Reconstruction of urban stormwater contamination of an estuary using catchment history and sediment profile dating. Estuaries 25 (1), 43–56.

Syvitski, J.P.M., Vorosmarty, C.J., Kettner, A.J., Green, P., 2005. Impact of humans on the flux of terrestrial sediment to the global coastal ocean. Science 308, 376–380.

Thom, B.G., 1967. Mangrove ecology and deltaic geomorphology, Tabasco, Mexico. J. Ecol. 55, 301–343.

Thom, B.G., Wright, L.D., Coleman, J.M., 1975. Mangrove ecology and deltaic geomorphology: Cambridge–ord river, Western Australia. J. Ecol. 63, 203–232.

Thrush, S.F., Hewitt, J.E., Cummings, V.J., Ellis, J.I., Hatton, C., Lohrer, A., Norkko, A., 2004. Muddy waters: elevating sediment input to coastal and estaurine habitats. Front. Ecol. Environ. 2, 299–306.

Van Maanen, B., Coco, G., Bryan, K.R., 2015. On the ecogeomorphological feedbacks that control tidal channel network evolution in a sandy mangrove setting. Proc. Math. Phys. Eng. Sci. 471 (2180).

Vann, J.H., 1980. Shoreline changes in mangrove areas (Brazil and Guiana). Zeitschrift für Geomorphologie Supplementband 34, 255–261.

Vaughan, T.W., 1909. Geologic work of mangroves in Southern Florida. Smithsonian Misc. Collect. 52, 461–464.

Viles, H.A., 1988. Biogeomorphology. Basil Blackwell, Oxford, 365 pp.

Walbert, K., 2002. Investigations in the New Zealand mangrove, *Avicennia marina var australasica*. MSc thesis. University of Waikato, Hamilton, New Zealand.

Walling, D.E., 1999. Linking land use, erosion and sediment yields in river basins. Hydrobiologia 410, 223–240.

Walsh, J.P., Nittrouer, C.A., 2004. Mangrove-bank sedimentation in a mesotidal environment with large sediment supply, Gulf of Papua. Mar. Geol. 208, 225–248.

Watson, C.S., White, N.J., Church, J.A., King, M.A., Burgette, R.J., Legresy, B., 2015. Unabated global mean sea level rise over the satellite altimeter era. Nat. Clim. Change 5, 565–569.

Webb, E.L., Friess, D.A., Krauss, K.W., Cahoon, D.R., Guntenspergen, G.R., Phelps, J., 2013. A global standard for monitoring coastal wetland vulnerability to accelerated sea-level rise. Nat. Clim. Change 3, 458–465.

Weston, N.B., 2014. Declining sediments and rising seas: an unfortunate convergence for tidal wetlands. Estuar. Coast 37, 1–23.

Wheaton, J.M., Gibbins, C., Wainwright, J., Larsen, L., McElroy, B., 2011. Preface: multiscale feedbacks in ecogeomorphology. Geomorphology 126, 265–268.

Widdows, J., Blauw, A., Heip, C.H.R., Herman, P.M.J., Lucas, C.H., Middelburg, J.J., Schmidt, S., Brinsley, M.D., Twisk, F., Verbeek, H., 2004. Role of physical and biological processes in sediment dynamics of a tidal flat in Westerschelde Estuary, SW Netherlands. Mar. Ecol. Prog. Ser. 274, 41–56.

Willemsen, P.W.J.M., Horstman, E.M., Borsje, B.W., Friess, D.A., Dohmen-Janssen, C.M., 2016. Sensitivity of the sediment trapping capacity of an estuarine mangrove forest. Geomorphology 273, 189–201.

Wilmshurst, J.M., Anderson, A.J., Higham, T.F.G., Worthy, T.H., 2008. Dating the late prehistoric dispersal of Polynesians to New Zealand using the commensal Pacific rat. Proc. Natl. Acad. Sci. U. S. A. 105, 7676–7680.

Woodroffe, C.D., 1985. Studies of a mangrove basin, Tuff Crater, New Zealand: I. Mangrove biomass and production of detritus. Estuar. Coast. & Shelf Sci. 20, 265–280.

Woodroffe, C.D., 1992. Mangrove sediments and geomorphology. In: Robertson, A.I., Alongi, D.M. (Eds.), Tropical Mangrove Ecosystems. American Geophysical Union, Washington DC, pp. 7–41.

Woodroffe, C.D., Rogers, K., McKee, K.L., Lovelock, C.E., Mendelssohn, I.A., Saintilan, N., 2016. Mangrove sedimentation and response to relative sea level rise. Annu. Rev. Mar. Sci. 8, 243–266.

Young, I.R., Verhagen, L.A., 1996a. The growth of fetch limited waves in water of finite depth. Part 1. Total energy and peak frequency. Coast. Eng. 29 (1–2), 47–78.

Young, I.R., Verhagen, L.A., 1996b. The growth of fetch limited waves in water of finite depth. Part 2. Spectral evolution. Coast. Eng. 29 (1–2), 79–99.

CHAPTER

2

Mangroves: a natural early-warning system of erosion on open muddy coasts in French Guiana

Christophe Proisy[1,2], *Romain Walcker*[3], *Elodie Blanchard*[1], *Antoine Gardel*[4], *Edward J. Anthony*[4,5]

[1]AMAP, University of Montpellier, IRD, CIRAD, CNRS, INRAE, Montpellier, France; [2]French Institute of Pondicherry, Pondicherry, India; [3]Laboratoire Ecologie Fonctionnelle et Environnement, Univ Toulouse, CNRS, Toulouse, France; [4]LEEISA, CNRS, IFREMER, Univ Guyane, Cayenne, French Guiana, France; [5]CEREGE, Univ Aix Marseille, CNRS, Collège de France, INRAE, IRD, Aix-en-Provence, France

OUTLINE

1. Introduction 48
2. Mud dynamics and vulnerability of the wave-exposed mangrove coast of French Guiana 49
3. Status and dynamics of mangroves on a rapidly changing muddy coast 52
4. Mangrove-based early warning of coastal vulnerability to mud erosion 55
4.1 *Results based on cross-shore mangrove shoreline fluctuations* 56
4.2 *Perspectives with inputs from alongshore mud-bank migration rates and oceanic data* 60
5. Concluding remarks 62
Acknowledgments 62
References 63

https://doi.org/10.1016/B978-0-12-816437-2.00011-2

1. Introduction

Interactions between sediments and mangroves have long challenged researchers intrigued by the land-building capacity of mangroves in many places in the world (Carlton, 1974; Davis, 1938; Woodroffe, 1992). Bird (1971) elegantly explained that, "Once established, mangroves influence patterns of accretion, gradually confining the ebb and flow of tides to residual tidal creek systems." In the course of millions of years, mangroves have indeed demonstrated an exceptional ability to adapt to sea-level fluctuations through contraction or extension of their range (Ellison et al., 1999). They found refuge on continental shelves that were most likely subjected to massive sediment supply during lower sea levels, as suggested by Quaternary pollen analysis (Woodroffe and Grindrod, 1991), and can demonstrate a capacity for landward expansion with sea-level rise (Krauss et al., 2014).

This land-building capacity caused by efficient sediment trapping (Horstman et al., 2015; McKee et al., 2007; Van Santen et al., 2007) imparts on mangroves important functions that underpin multiple services (Barbier et al., 2011), such as reduction of water flow velocity and energy (Dasgupta et al., 2019; Marois and Mitsch, 2015; Mazda et al., 2006; Quartel et al., 2007; Tanaka et al., 2007), increase in fishery yields (Aburto-Oropeza et al., 2008), and wastewater treatment (Capdeville et al., 2019). In contrast, when mangroves are degraded by man-made infrastructures, coastal erosion is accelerated and may become persistent on wave-exposed coasts as observed, for example, in French Guiana (Brunier et al., 2019), Guyana (Anthony and Gratiot, 2012), Indonesia (Yulianda et al., 2014), Malaysia (Stanley and Lewis, 2009), Thailand (Thampanya et al., 2006; Winterwerp et al., 2005), and Vietnam (Besset et al., 2019; Mazda et al., 2002; Nguyen et al., 2015). Mangrove loss comes to prominence when assessed on a global scale (Hamilton and Casey, 2016; Thomas et al., 2017; Worthington and Spalding, 2018). This message is often inaccurate when examined regionally, in the case of pristine or naturally preserved mangroves, i.e., mangroves that are remote from direct human influence. Among others, the regions of West Papua in Indonesia, Gulf of Papua in Papua New Guinea (Walsh and Nittrouer, 2004), Andaman and Nicobar Islands in India (Kiruba-Sankar et al., 2018), Makassar Strait in Indonesia (Yulianto et al., 2004), Amapá and Pará in Brazil (Cohen et al., 2012; Souza Filho et al., 2006), and French Guiana all experience not only significant mangrove erosion but also progradation at yearly to decadal time scales. In most of these cases, it appears that the natural back-and-forth alternation of mangroves, at the same place or across a region, is intimately related to morphosedimentary processes induced by oceanic forcing.

Surprisingly, mangrove shoreline fluctuations have rarely been studied from the perspective of early warning of coastal erosion by waves. This latter function would, however, be invaluable for the livelihood and security of millions of inhabitants of mangrove coasts. Furthermore, as shown by Gratiot et al. (2007) and Walcker et al. (2015), oceanic data have proven their value in the interpretation of mud dynamic patterns responsible for mangrove shoreline fluctuations on various spatiotemporal scales in French Guiana. These fluctuations thus deserve, in return, more attention for a better understanding of the role of waves and currents in forcing changes in mangrove shorelines (Proisy et al., 2016). Among platforms issuing oceanic data from the reanalysis of the global atmosphere since 1979, the ERA-INTERIM archive (Berrisford et al., 2011) managed by the European Center for Medium-Range Weather Forecasts (ECMWF) gives access to significant wave height, mean wave period, and direction data, that are updated monthly with a 2-month delay relative to the observation date. In addition, ocean current data can be obtained from the Operational Mercator global ocean analysis and forecast system at 1/12 degrees with 10 days of global ocean forecasts updated daily and issued by the E.U. Copernicus Marine Environment Monitoring Service (CMEMS).

Here we present a synthesis on how rich observational and integrative modeling research on mud–vegetation interactions in French Guiana may support predictions of coastal vulnerability to erosion from the monitoring of the fluctuations of mangrove shorelines over the last seven decades.

2. Mud dynamics and vulnerability of the wave-exposed mangrove coast of French Guiana

The French Guiana coast, situated about 400-km downdrift of the mouth of the Amazon River (Fig. 2.1), is part of the longest and muddiest coast in the world (Anthony et al., 2010). Mangroves can periodically occupy almost all of the 320 km-long French Guiana coastline which is continuously reshaped by the northwestward migration of massive, 10-60 km-long, mud banks (Froidefond et al., 1988), the initial formation of which is observed near the Cabo Orange (Allison et al., 2000), 30 km southeast of the border of French Guiana with Brazil. These shoreface-attached mud banks are up to 2 to 5 m-thick and spaced at intervals of 15–25 km (Fig. 2.1). They migrate alongshore at rates varying from 0.9 to 3.7 $km.y^{-1}$ (Abascal Zorrilla et al., 2018). The resulting shifting of bank and inter-bank phases generates alternations of mud depositional (Allison and Lee, 2004) and erosional processes (Plaziat and Augustinus, 2004), while altering the local hydrodynamics (Chevalier et al., 2008).

The French Guiana coast is exposed to North Atlantic wind-driven waves blowing from E–NE (Gratiot et al., 2007) and reaching the shoreline with oblique incidence angles ranging from 15 degrees to 35 degrees (Gardel and Gratiot, 2005). These waves have mean periods of 6–10 s and offshore heights between 0.5 and 2 m. Longer waves generated by North Atlantic depressions may reach the French Guiana shorelines with a directional range from north to north–northwest (Anthony et al., 2010) and significantly impact seaward mangrove expansion (Walcker et al., 2015). As reviewed by Anthony et al. (2013), a better understanding of the combined action of multiple processes induced by waves and currents, such as cross-shore and alongshore mud advection and budgets, mud-bank consolidation, fluidization and liquefaction, is still required for explaining mud-bank formation and migration. This remains a challenging research question given the complexity of the multiscale measurement and monitoring of intimately interwoven processes.

The ability to predict the recurrent morphosedimentary variability in French Guiana is, however, of crucial economic and ecological importance. Indeed, about 80% of the inhabitants of this territory live on the coast (INSEE, 2017). Industrial activities (Dias, 2018; Rongier, 2013) generated by Europe's Spaceport in Kourou, which hosts Ariane, Soyuz, and Vega rocket launch sites a few kilometers inland of the unstable shoreline, are particularly important to the French Guiana economy, as is agriculture which is mainly practiced along the coast. This instability also calls into question the sustainability of small-scale shrimp fisheries, which represent the third economic sector (Diop et al., 2018).

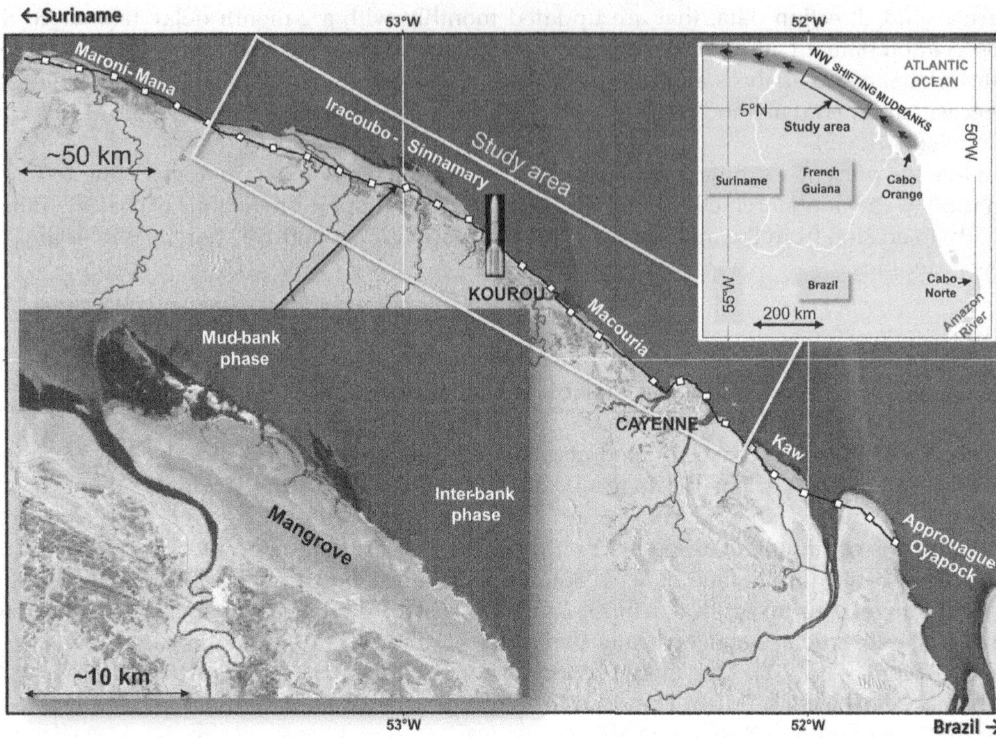

FIGURE 2.1 The open muddy mangrove coast of French Guiana, between Brazil and Venezuela. The background image is a color composite ALOS-PALSAR L-band image acquired in 2007 at low tide (© ALOS Kyoto & Carbon Initiative, Lucas et al. (2007)). The intertidal surface of the mud bank appears very smooth at the L-band and is revealed by specular scattering in the radar image, i.e., black areas (e.g., van der Wal et al. (2005)). The sawtooth pattern typical of coastal areas in erosion is clearly seen in the lower right of the image extract (bottom left). The black line with white square dots corresponds to the reference shoreline fronting settlements within the study area (yellow (gray in print version) rectangle). Refer to Abascal Zorrilla et al. (2018) for identification of alongshore-migrating mud banks, and Brunier et al. (2019) for the shoreline dynamics of the region between the mouths of the Maroni and the Mana Rivers. The Kaw and Approuague-Oyapock river-mouth regions are uninhabited. The rocket image locates Europe's Spaceport of Kourou.

Alternating mud deposition and coastal erosion processes generate contrasting risks for the durability of these coast-located socioeconomic activities. On the one hand, the massive mud supply in the coastal waters hampers easy ship access to commercial ports located a few kilometers upstream of the mouths of the Kourou, Cayenne, and Mahury rivers (Orseau et al., 2017). Millions of euros are spent each year in daily mud dredging in order to maintain navigation channels deep enough to accommodate container ships, ships delivering rocket components, or shrimp trawlers. On the other hand, during inter-bank phases, when mangroves are absent or intensively eroded, the risk for human settlements is high (Fig. 2.2A).

Particularly vulnerable to erosion processes due to the absence of mangroves are the embayed beaches of the bedrock promontory of Cayenne (Fig. 2.2B) which experience a

FIGURE 2.2 Coastal instability in the Cayenne Peninsula. Ongoing mangrove erosion in front of downtown Cayenne during an inter-bank phase (A). Beach erosion (refer to Anthony and Dolique (2004) for the description of beach rotation during erosional and depositional mud phases) and man-made hard-rock sea defense during an erosional phase (B).

sand budget rotation pattern involving complex interactions between mud and sand (Anthony and Dolique, 2004). The persistent erosion of the Mana polders (Brunier et al., 2019) is also illustrative of how anthropogenic modifications of shorelines can aggravate both the coastal erosion risk and the recover capacity of the coast following mangrove degradation.

On a larger scale, mud dynamics impact essential ocean variables that influence coastal biodiversity (e.g., biomass and diversity of plankton, distribution and abundance of fish, birds, and marine turtles) as shown by Miloslavich et al., 2011, on the scale of the North Brazil Shelf. The adaptability of animal species to such severe coastal geomorphological changes is underpinned by questions relating to behavior and preferential sites, as illustrated by the deleterious effects on the nesting and conservation of a number of sea turtle species (Baudouin et al., 2015; Kelle et al., 2007) when sandy beaches (and their access pathways) are alternatively blanketed by mud or eroded by wave action, two processes that jeopardize the success rates of nesting and egg-hatching.

How should alerts on risks to socioeconomic and ecological habitats related to ongoing and alternating mud depositional or erosional processes be raised? How can the impacts of intensive and extensive coastal changes that remain insufficiently understood, monitored, and quantified be confronted? Mangroves have considerable potential for informing on coastal changes (Fromard et al., 2004; Souza Filho et al., 2006). Apart from their recognized role in sustaining coastal biodiversity and fisheries (Isaac and Ferrari, 2017; Miloslavich et al., 2017), healthy and still preserved mangroves in the regions influenced by the Amazon River deserve our attention.

3. Status and dynamics of mangroves on a rapidly changing muddy coast

From the Amazon River mouth to the north of French Guiana, hundreds of square kilometers of preserved mangroves have fascinating dynamics closely linked to mud-bank migration (Fig. 2.3). Their story needs to be told so that their early-warning role in highlighting risks associated with coastal changes induced by mud dynamics may be highlighted.

Coastal instability in French Guiana is thus not tantamount to vulnerability of mangroves (Baltzer et al., 2004) but rather highlights their exceptional ecological adaptability to cope with sedimentary morphodynamics (Figs. 2.3B and 2.4). There, instability determines the extent, structure, and composition of mangrove forests (Fromard et al., 1998) and interdisciplinary research is required for evaluating the propensity of any new mud-bank phase to generate a protective mangrove belt.

First, the region exhibits the highest above-ground mangrove biomass levels (Fig. 2.4A) in the world exceeding 400 tons of dry matter per hectare (Fromard et al., 1998; Olagoke et al., 2016; Proisy et al., 2007). Second, the exceptional rate of colonization exhibited by pioneer *Avicennia germinans* mangroves (Proisy et al., 2009) during a mud-bank phase (Figs. 2.4C and D) may explain the strong mangrove resilience observed at the regional scale. The underlying ecological mechanisms are closely dependent on mud consolidation and mud-crack development in the intertidal zone (Gardel et al., 2009; Proisy et al., 2009) that are preceded by the formation of ephemeral linear mud bars close inshore (Anthony et al., 2008).

No less impressive are the visible effects of erosion processes induced by waves on mangroves (Figs. 2.3A and 2.4B). In French Guiana, 25–40 m high mangrove tree trunks are not directly broken by 1–2 m high waves as is observed elsewhere in the world after storm or tsunami waves have swept across a mangrove region (Tanaka et al., 2007). During the inter-bank phase, the wave erosive power erodes the cohesive structure of the shoreface-welded mud substrate on which mangrove trees develop and thrive (Fig. 2.4B). Mud removal from the substrate destabilizes trees and leads to their fall. Notwithstanding the observed superficial tree anchorage in the muddy substratum (<1.0 m for trees > 30 m high) in the uppermost consolidated mud layers for both *A. germinans* and *Rhizophora* ssp., wave energy will, however, be dissipated consistently by the complex and dense above-ground mangrove vegetation structure and this dissipation is amplified where the mangrove belt is extensive. The presence and renewal of mangroves are thus key elements for keeping waves far from inhabited shorelines. These observations have a corollary in the north of French Guiana, in Surinam, and in Guyana. When mangroves are degraded and natural colonization processes

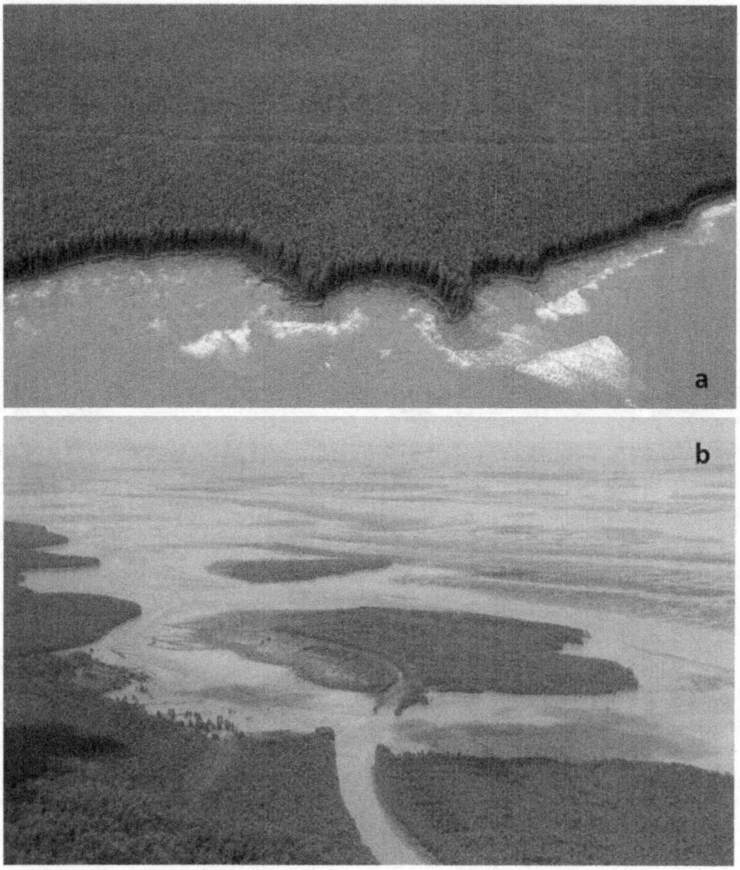

FIGURE 2.3 Illustration of the amplitude of mud erosional and depositional processes from aerial photographs (Fromard et al., 2004; Proisy et al., 2016; Walcker et al., 2015). A typical sawtooth mangrove shoreline plan shape caused by erosion induced by waves, region of Kourou. Hundreds of uprooted trees are visible (A). A complex coastal landscape in eastern French Guiana where mud and mangrove accommodate evolving mud-bank locations and elevation (B).

hindered by man-made sea defenses, persistent and catastrophic erosion rates can become prevalent for years (Anthony and Gratiot, 2012; Brunier et al., 2019).

With these objectives, radar, passive optic, and lidar remote sensing have supported groundbreaking research for showcasing the exceptional mangrove dynamics on the Amazon-influenced coast of the Guianas. Mangrove forest dynamics in terms of forest growth stages or zonations can be mapped based on the finest spatial resolution (pixel size around 1 m), as demonstrated by Proisy et al. (2007), or with radar remote sensing (Proisy et al., 2000, 2002). Remote sensing–based methods have long proven their feasibility and value for monitoring fluctuating mangrove shorelines in French Guiana (e.g., Lucas et al. (2007); Fromard et al. (2004); Proisy et al. (2009)). These fluctuations are discernible and

FIGURE 2.4 Mixed-mangrove stands can reach 40-m high in French Guiana (Olagoke et al., 2016; Proisy et al., 2007) (A). Illustration of the intensity of mud erosion. Total removal of the muddy strata results in uprooting of mangrove trees, which are gradually disintegrated by oceanic forces (B). Illustration of the rapidity and intensity of mud accretion. A new mud deposit phase allows rapid colonization by new mangrove seedlings (Gensac et al., 2011), while dead upright trunks and root systems of fallen trees are visible leftovers of the previous inter-bank phase (C). Rapid colonization by *Avicennia germinans* from shoreward migrating mud bars, aided by mud crack features, may compensate for erosion loss by wave erosion (Anthony et al., 2008; Proisy et al., 2009) (D).

quantifiable, from 1 year to the next, in terms of cross-shore distances (km) or surface area (km^2) using medium-resolution images (pixel size > 5 m), as shown in numerous studies (Brunier et al., 2019; Fromard et al., 2004; Walcker et al., 2015). Photogrammetry based on manned or unmanned aerial vehicles is taking studies of mangrove environments and substrates in French Guiana (Brunier et al., 2020) to unprecedented levels of resolution (mm scale).

In Fig. 2.5, the following effects of mud-bank and inter-bank phases on mangrove shorelines are assessed (refer to the above-mentioned papers for the satellite image illustrations):

1. Mud-bank phase I: Seaward mangrove expansion can be detected by intense near infrared signatures from dense and young mangrove formations (see Proisy et al. (2018) for an illustration of the detection of young seedlings). If the image is acquired at low tide and at high spatial resolution, colonization patterns (Fig. 2.4D) are also discernible and the uppermost layer of the intertidal bare muddy zone can be roughly delineated. In French Guiana, young mangroves can rapidly develop, reach 25 m tall (Fig. 2.4A) in about 15 years, and form forests covering several km^2. In some regions, mangroves can expand over 8 km seaward.

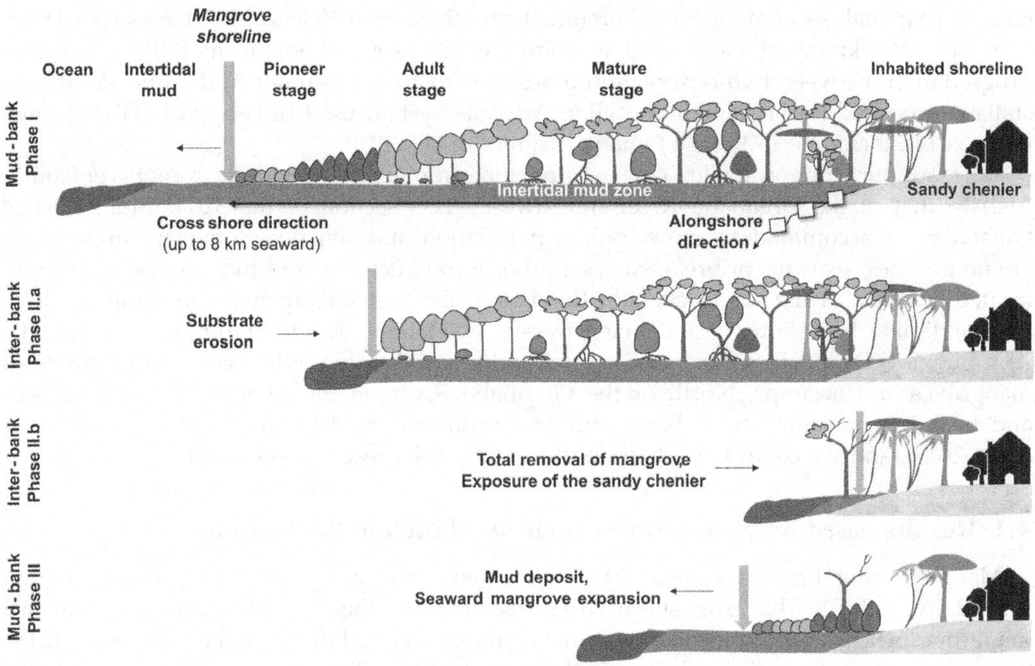

FIGURE 2.5 Cross-shore transects showing mangrove shorelines at different mud-bank and inter-bank phases. Phase I corresponds to a mud deposit phase during which mangroves are expanding seaward. Phase II. (A) illustrates a partial mangrove shoreline retreat due to substrate erosion by waves. Phase II. (B) presents the case of total mangrove shoreline retreat up to the sandy chenier fronting settlements. Phase III shows how new mud deposits followed by mangrove colonization reduce inhabited shoreline vulnerability to wave erosion.

2. Inter-bank phase II.A: Mangrove retreat is observable from one image to the next. Breaking wave patterns and a sawtooth-like coastal morphology are clear indicators (Fig. 2.3A).
3. Inter-bank phase II.B: Total removal of mangroves is easily observable (Fig. 2.4B) since sandy beach (chenier) vegetation and human infrastructures combine to form a clear landscape limit (Fig. 2.2).
4. Mud-bank phase III: The arrival of a new mud-bank phase is not only detectable at low tide through, for example, the increasing distance of the breaking wave zone from the inhabited shoreline but also through rapid mangrove colonization and development (Fig. 2.4C). This is a new colonization phase similar to the first phase but at a different moment, after an inter-bank phase (Fig. 2.4D).

4. Mangrove-based early warning of coastal vulnerability to mud erosion

In this chapter, we have used the database built by Walcker et al. (2015) and have extended it through a complementary analysis over the period 2011–17 in order to provide an up-to-date 68-year analysis of mangrove shoreline fluctuations from 1950 to 2017. Cross-shore transects spaced 1 km-apart were used to compute mangrove shoreline mobility (retreat or progradation) between two observation dates. The method is similar to the End Point Rate statistic proposed in the Digital Shoreline Analysis System used (Jonah et al., 2016; Kaliraj et al., 2014), including in French Guiana (Brunier et al., 2019).

We focus our work on 160 km of coast stretching from the estuary of the Kaw River (southeast) to that of the Organabo River (northwest). This section of the 320 km-long French Guiana coast accommodates most of the population and sensitive infrastructures. There are no groynes, seawalls or breakwaters, harbor infrastructures, and industry or agriculture located seaward of the sandy chenier. Roads and the first settlements were built on these natural limits, backshore of the mangroves or wetlands. South of the Kaw River, the Approuague–Oyapock region remains virtually uninhabited with very large tracts of mangroves and swamps. North of the Organabo River, in the Mana region, mangroves and backshore swamps have been converted into rice polders since the 1980s (Brunier et al., 2019), and we could not integrate this area in the present assessment.

4.1 Results based on cross-shore mangrove shoreline fluctuations

Mangrove shorelines have, since 1950, fluctuated significantly depending on the location (Fig. 2.6A and B). The cross-shore distances between the inhabited and the seaward mangrove margin can attain 8 km and have shifted over 6 km in the last 68 years in the Sinnamary–Iracoubo region, whereas in the south between Sinnamary and Kourou, seaward of Europe's Spaceport, mangrove shorelines have been confined within a band up to 3 km wide. Between Kourou and Cayenne, the mangrove shorelines fluctuate by up to 3 km. The mangrove band on the west side of the Cayenne promontory does not exceed a few hundred meters. South of the Mahury River, the seaward border of the marshy swamps is part of the 95,000 ha of the Nature Reserve of Kaw, which is occupied by mangroves, the variability in extent of which is not presented here (refer to Proisy et al. (2007)). Stable mangrove areas, corresponding to the presence of persistent mangroves over the last 68 years, are located in the Iracoubo–Sinnamary and Macouria regions. They can prograde by up to 4 km seaward in Sinnamary, whereas in Macouria the maximum extension is only about 1 km from the inhabited shoreline. Both Cayenne and Europe's Spaceport regions experienced in 2017 almost total removal of mangroves.

It is worth noting that these mangrove shoreline fluctuations, along with the abundant literature available since the 1980s on the French Guiana coastal dynamics, do not endorse the values given in the global maps 2.1 and 3.1 proposed by Worthington and Spalding (2018): (1) the proportion of mangroves lost since 1996, reported as 5%–9%, has no meaning in the case of rapidly changing mangrove coasts such as those of French Guiana, and (2) indicating tens of km^2 of restorable areas disregards the remarkable resilience of French Guiana mangroves in the course of mud-bank phases.

Progradation and erosion rates are up to $500\,m.yr^{-1}$ and $300\,m.yr^{-1}$, respectively (Fig. 2.6C). These values showcase the exceptional dynamics of the French Guiana coast that can, at one and the same time, lead to devastation of mangroves growing in one place and constitution of a new mangrove ecosystem in another. Average change rates computed for the last three 5-year periods, between 2003 and 2017, highlight an NW shifting of

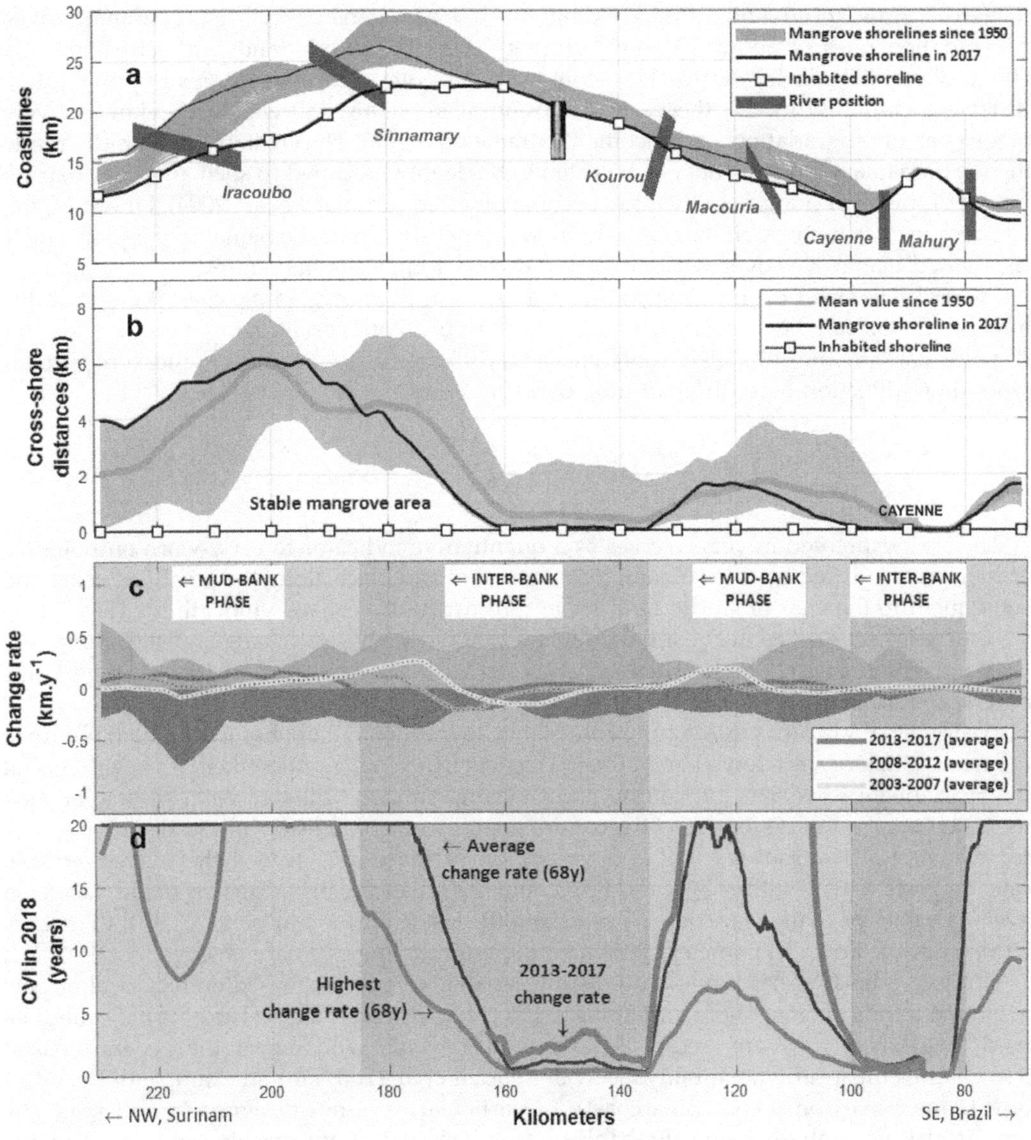

FIGURE 2.6 Fluctuations of mangrove shorelines between 1950 and 2017 along 160 km of inhabited coastline (refer to Walcker et al. (2015) for a complete analysis of these fluctuations). Mangrove shoreline variability along the French Guiana coastline (A) with reference to the inhabited coastline (B), change rates recorded over 68 years and during the 5-year periods 2003–07, 2008–12, and 2013–17 (average values) (C), and coastal vulnerability index (CVI) for the inhabited shoreline, computed and expressed in years, using the highest and average values of erosion rates (i.e., negative change rates) observed over 68 years and during the 2013–17 period (D).

mangrove seaward advance. Around kilometer 180, an extensive vast region of mangroves, progradation rates of about 300 $m.y^{-1}$ during 2003–07 precede landward retreat rates of about 350 $m.y^{-1}$ during 2013–17. Moreover, opposite processes prevail within short distances. During 2013–17, the peak of erosion at kilometer 180 was just 15 km eastward of the peak of progradation approaching the Iracoubo region. The erosion front rapidly reaching the Sinnamary River estuary is expected to be deeply modified to such an extent that the mouth of this river may, as in the past, become diverted (Fromard et al., 2004). However, this last decade does not exhibit extreme variation rates eastward of the Sinnamary region due to the limited mangrove cover which does not exceed 1 km at its maximum.

For each kilometer of shoreline and for a given year of study, simple coastal vulnerability indices (CVIs) can be computed (Eq. 2.1) from the corresponding mangrove shoreline distance dMS to the inhabited shoreline, and different change rates cR measured in the cross-shore direction over different time periods:

$$CVI_{(years)} = \frac{dMS_{(km)}}{cR_{(km.year^{-1})}} \tag{2.1}$$

The CVI, expressed in years, serves as a quantitative indicator to early-warn on potential erosion risks induced by rates of mangrove retreat at any place along the inhabited coast and subsequent to the year of study. In order to forecast future coastal vulnerability (Fig. 2.6D), we thus selected 2017 as the year of study and worked with three change rates composed of (1) the average and (2) highest values observed over 68 years and (3) the average value observed during 2013–17.

We limited CVI values to 20 years, a threshold suggesting either absence of vulnerability to the erosion risk or very low change rates. Whatever the change rate taken, it is evident that the most inhabited regions of Kourou and Cayenne are and will stay vulnerable to erosion in the coming years, as is presently confirmed by a major erosion phase destroying seafront houses in Kourou. Secondly, in the region of Macouria, even if the highest erosion rate suggests a risk within 5 years, this could be balanced by estimates made based on average values of change rate over 68 years and the last (calm) 5-year period. The CVI reaches 20 years in the area of kilometer 120 where persistent mangroves are observed.

However, this CVI has limited forecasting capability since it is based on historical or past evolution trends of cross-shore variations of the mangrove shorelines (Table 2.1). It cannot be used to estimate how any recent or exceptional events will impact the general coastal dynamics or the local vulnerability (see van Ledden et al. (2009) for an example of the effect of extreme waves on the Guyana coast). In particular, it is not designed to anticipate the erosion risk generated by an inter-bank phase unavoidably approaching any point of the inhabited coastline at any moment. For example, the inhabited settlements of the region of Sinnamary and Iracoubo (eastward of kilometer 200) may be impacted in the coming years by the approaching inter-bank phase located westward of Kourou, suggesting that CVI values of 20 years are overestimated. As mud banks migrate alongshore at rates greater than 0.9 km $km.y^{-1}$ (Abascal Zorrilla et al., 2018), the alongshore component of the mangrove retreat rate could indeed exceed the cross-shore component. There is room for enhancing the present CVI through a better consideration of alongshore processes and oceanic forcing data, as described below.

TABLE 2.1 Synthesis of the three mangrove-based approaches to early warning on coastal vulnerability in French Guiana. Data, parameters, equations, potential, and limits of each approach are given. dMS and dME are distances, respectively, in the cross-shore and alongshore directions (see Fig. 2.7).

Coastal vulnerability	Simple index CVI (years)	Compound index CVIr (years)	Simulated index
Data requirements	Time series of remote sensing images		
			Wave $\overrightarrow{W}(WAL, WAC)$ data Current $\overrightarrow{C}$ (C_{AL}, C_{AX}) data
Spatial and temporal resolutions	~1 km, ~ 1 year		10–100 km, daily with a delay of 2 months
Forecasting capacity	<1–2 years	<5 years	From near real time to < 5 years
Methods	Visual delineation of mangrove shorelines at each date of image acquisition		
	Derivation of erosion rates	Derivation of erosion and mud-bank migration rates	Equations of mud-bank migration, mangrove exposure to erosion, and mangrove colonization rate (see Proisy et al. 2016)
Advantages	• Easy dMS-delineation • Early warning based on historical records (hindcast predictions)	• Similar to CVI advantages • Alongshore processes are considered	• Based on forcing signals • *dMS* and *dME* observations for calibration/validation • Multiscale flexible modeling with proven potential (see Proisy et al. (2016)) • Linkable to climate scenarios
Limitations	• Alongshore processes not considered • No real-time forecast • No forcing data	• Similar to CVI limitations • In progress • Accuracy of *dME* when data are missing	• In progress • Complex modeling implementation • Multisource data integration
Perspectives	• Integration of mud-bank location, shape, and extent obtained from maps of suspended sediments • Integration of wave dampening equations • Nearshore intertidal (local) mud consolidation processes • Colonization, acceleration of vertical accretion, and mud substrate stabilization by mangroves • Impact of the interannual variability of Amazon sediment supply and climate in North Atlantic Ocean (Gensac et al., 2016)		

4.2 Perspectives with inputs from alongshore mud-bank migration rates and oceanic data

We discuss here the implementation, potential, and limitations of two ongoing approaches to early-warning coastal erosion from mangrove shoreline monitoring using time series of remote sensing data. As mangroves exposed to wave erosion will retreat, not only in the cross-shore but also in the alongshore direction, forced by morphosedimentary processes operating within the inter-bank phase, the development of a complementary alongshore "eastward-looking" approach may help in refining coastal vulnerability predictions in French Guiana. This requires the consideration of a mud-bank followed by an inter-bank phase as a coupled system (Fig. 2.7) where fluctuations of mangrove shorelines imposed by the alongshore migration of the erosion front are methodically studied.

The first approach aims at enhancing the forecasting capacity of the CVI previously presented with the expected added value, still to be demonstrated, of increased anticipation of coming erosional or depositional phases (Table 2.1). The same dataset of remote sensing images, cross-shore distances, and denoted dMS (Fig. 2.7) between the inhabited and the mangrove shorelines and erosion rates is still to be used. It is to be completed by the integration of a second distance obtained in the alongshore direction from the visual interpretation of seaward mangrove margins which, like vegetation ripples, culminate during a mud-bank phase while dwindling during the following inter-bank phase. This distance, noted *dME*, corresponds to the shortest distance between any place *P* on the inhabited shoreline and the NW shifting erosion front. It could be obtained by searching eastward for the next confluence point, noted *AE* in Fig. 2.7, located between the erosional trailing edge of the mud bank and the mangrove shoreline, which actually corresponds to the mangrove area to be most recently exposed to wave erosion. In most of the remote sensing images, *AE* is distinguishable at low or high tides when the sawtooth-like pattern of mangrove shoreline is detected. At low tide, using very high spatial resolution images, the observation of trees lying on the bottom of a still standing forest (Fig. 2.3A) or on an exposed sandy chenier (Fig. 2.4B) will reveal *AE*. At high tide, using medium-resolution imagery, *AE* is discernible as the area where waves break on the mangrove front (Fig. 2.3A). A refined two component–based CVIr could be computed

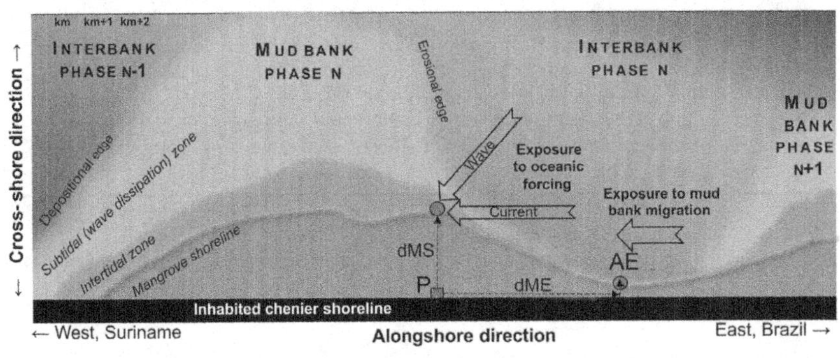

FIGURE 2.7 Schematic illustration of monitoring shoreline vulnerability to erosional processes using time series of remote sensing images and oceanic wave and current data. At a given location along the inhabited shoreline, dMS and dME correspond, respectively, to the distance to the mangrove shoreline (cross-shore direction) and to the lowest southeast cross-shore mangrove expansion in the along-shore direction (point AE). Both distances can be measured and monitored in time series of typical remote sensing images. Oceanic forcing is computed from wave and current data such as those delivered by the ERA-INTERIM archive.

from the minimum value between the cross-shore and alongshore erosion rates at a given moment at a given place as follows:

$$CVIr_{(years)} = \text{minimum}\left(\frac{dMS}{cR}, \frac{dME}{aR}\right) \tag{2.2}$$

where *cER* and *aMR* are the change rates, respectively, in cross-shore and alongshore directions. The migration rate *aMR* of the erosion phase would be derived from the temporal analysis of NW shifting of the *AE* point. However, when data are missing at a given date or over a portion of coastline, *AE* has to be approximately located along an interpolated mangrove shoreline without the support of visual interpretation but according to previous and subsequent observations.

The second approach for coastal vulnerability assessment would be based on a simulation of mangrove shoreline fluctuations observed in remote sensing images as a function of daily ERA-Interim/ECMWF waves and Mercator-Ocean currents/CMEMS currents. The expected added value relies on the consideration of near real-time access to forcing variables of the French Guiana coastal dynamics that can, in return, early-warn on forthcoming risky erosional events induced, for example, by high energy swell waves associated with the North Atlantic Oscillation during winter (Table 2.1). In a first attempt, Proisy et al. (2016) demonstrated that some major interwoven ecological and oceanic processes in French Guiana could be encapsulated as a set of equations describing how ocean, mangrove shoreline, and mud-bank entities interact according to the level of understanding and knowledge available. The model treated the three following processes: (1) mud-bank alongshore migration as a linear function of oceanic wave and current intensity and direction, (2) mangrove shoreline exposure to erosion as a function of its distance from the wave breaking zone, and (3) mangrove expansion rate over a mud bank. An iteration process was used to minimize the divergence between simulated and satellite-based mangrove shorelines at any point on the coast over 15 years and, then, derive the linear equation coefficients. It is worth noting the convergence of this simple modeling approach with the important role of ocean wave and coastal current forcing in explaining mangrove shoreline fluctuations. The impact of waves on mangrove shorelines was estimated at almost twice that induced by coastal currents confirmed by the latest observational results obtained by Walcker et al. (2015) and Gratiot and Anthony (2016).

This simulation work was, however, built on a rough approximation of both shape and extent of mud banks simulated as ellipses of constant size. Furthermore, there is no reason to keep the linear relation coefficients fixed over seasons and years, from one mud bank to the following one: oceanic forcing is not stationary and its impact on mud-bank migration rates is locally dependent on geomorphic features such as bedrock headlands or river mouth size, location, and shape as well as freshwater outflows from the main rivers.

Recent advances in the delineation of the subtidal part of mud banks and monitoring of their migration rates through estimates of suspended particulate matter (SPM) concentrations using medium-scale multispectral images (Abascal Zorrilla et al., 2018) should provide crucial information that has so far been missing in the building of a mangrove-based early-warning system for coastal erosion. The access to SPM also opens up perspectives for the integration of wave dampening equations (Winterwerp et al., 2007) into the description of the general mechanism underlying mangrove shoreline exposure to erosion. Finally, the use of both *dMS* and *dME* distances for the calibration and validation of the ongoing models still remains a crucial step.

5. Concluding remarks

Mangrove retreat is commonly, and intuitively, perceived as an indicator of erosion, a major threat to human infrastructures and local socioeconomic activities (Barbier et al., 2011). Through a synthesis of research undertaken in French Guiana since the 1980s, we provided evidence that the monitoring of mangrove shoreline fluctuations over decades could lead to quantitative and predictive approaches to coastal vulnerability to erosion. The role of remote sensing and oceanic data remains crucial for mapping these fluctuations and acquiring a multiscale perception of the prevailing coastal processes responsible for mangrove retreat or seaward progradation.

As revealed by the exceptional mangrove coast dynamics in French Guiana, the building of a mangrove-based quantitative costal vulnerability index relies on the historical range, the time lapse between observations, and the spatial resolution of the remote sensing data. The use of time series of very high spatial resolution images (including historical aerial photographs) is, of course, to be prioritized for small mangrove regions or annual rates of mangrove retreat of a few meters. The scope of this biogeomorphological index can be easily augmented by the integration of context-dependent ecological and socioeconomic variables (Nguyen et al., 2016; Nicholls et al., 2008; Szlafsztein and Sterr, 2007).

In any mangrove region in the world, the role of mangrove shorelines in the early warning of coastal vulnerability deserves more investigation (Table 2.1). French Guiana provides a unique opportunity to address the question of how alternating and intense morphosedimentary processes, induced by current and wave forcing, interplay with the exceptional adaptability of mangroves to erosional and depositional phases. How much is wave erosion of a mud substrate reduced by the presence of mangroves and what are the physical and ecological factors controlling this protection (Marois and Mitsch, 2015; Osorio-Cano et al., 2019; Phan et al., 2019; Tanaka et al., 2007)? The combined use of remote sensing and oceanic data archives provides significant insights via support from numerical models.

Following the observation of Moir (1972) on natural areas, preserved mangroves serve as bench marks in evaluating mud–vegetation interactions at the coastal interface of subtropical and tropical latitudes. It is thus our strongly held belief that context-dependent experimental studies (Gedan et al., 2011) and situational understanding of mangrove dynamics (Ellison, 2019) are insufficiently represented in global analyses of the future of mangroves. The message about worldwide mangrove loss should not overshadow historical and local evidence of the fascinating mangrove adaptability to coastal change: reasons and solutions for their protection both rely on a better understanding of coastal landscape building capacity and dynamics. Because early warning of coastal changes is invaluable for the sustainability of millions of people in the coming years, pristine and preserved mangroves deserve more consideration and active protection based on groundbreaking modeling research on complex mud–mangrove interactions.

Acknowledgments

This research was conducted in the framework of the MANGWATCH project funded by the French Centre National de la Recherche Scientifique (CNRS) research program "Pépinières Interdisciplinaires de Guyane" and the *Infolittoral-1* project funded by the French "Unique Inter-Ministerial Fund" and certified by the "Aerospace Valley competitiveness cluster." This work was also supported by the Agence Nationale de la Recherche through an Investissement d'Avenir grant (Labex CEBA, ref. ANR-10-LABX-25-01). We acknowledge funding by the ERDF project OYAMAR. This is a contribution from the LIGA French scientific research network. We thank the European Centre for Medium-Range Weather Forecasting (ECMWF) for the ERA-Interim reanalysis data used in this study. We thank Jérôme Le Fol and Damien Galarraga, GIS Pole, Region Guyane for their unwavering support. All photographs were taken by Christophe Proisy.

References

Abascal Zorrilla, N., Vantrepotte, V., Gensac, E., Huybrechts, N., Gardel, A., 2018. The advantages of landsat 8-OLI-derived suspended particulate matter maps for monitoring the subtidal extension of Amazonian coastal mud banks (French Guiana). Rem. Sens. 10, 1733.

Aburto-Oropeza, O., Ezcurra, E., Danemann, G., Valdez, V., Murray, J., Sala, E., 2008. Mangroves in the Gulf of California increase fishery yields. Proc. Natl. Acad. Sci. U. S. A. 105, 10456–10459.

Allison, M.A., Lee, M.T., 2004. Sediment exchange between Amazon mudbanks and shore-fringing mangroves in French Guiana. Mar. Geol. 208, 169–190.

Allison, M.A., Lee, M.T., Ogston, A.S., Aller, R.C., 2000. Origin of Amazon mudbanks along the northeastern coast of South America. Mar. Geol. 163, 241–256.

Anthony, E.J., Dolique, F., 2004. The influence of Amazon-derived mud banks on the morphology of sandy headland-bound beaches in Cayenne, French Guiana: a short- to long-term perspective. Mar. Geol. 208, 249–264.

Anthony, E.J., Dolique, F., Gardel, A., Gratiot, N., Proisy, C., Polidori, L., 2008. Nearshore intertidal topography and topographic-forcing mechanisms of an Amazon-derived mud bank in French Guiana. Continent. Shelf Res. 28, 813–822.

Anthony, E.J., Gardel, A., Gratiot, N., Proisy, C., Allison, M.A., Dolique, F., Fromard, F., 2010. The Amazon-influenced muddy coast of South America: a review of mud-bank-shoreline interactions. Earth Sci. Rev. 103, 99–121.

Anthony, E.J., Gardel, A., Proisy, C., Fromard, F., Gensac, E., Peron, C., Walcker, R., Lesourd, S., 2013. The role of fluvial sediment supply and river-mouth hydrology in the dynamics of the muddy, Amazon-dominated Amapá–Guianas coast, South America: a three-point research agenda. J. S. Am. Earth Sci. 44, 18–24.

Anthony, E.J., Gratiot, N., 2012. Coastal engineering and large-scale mangrove destruction in Guyana, South America: averting an environmental catastrophe in the making. Ecol. Eng. 47, 268–273.

Baltzer, F., Allison, M., Fromard, F., 2004. Material exchange between the continental shelf and mangrove-fringed coasts with special reference to the Amazon-Guianas coast. Mar. Geol. 208, 115–126.

Barbier, E.B., Hacker, S.D., Kennedy, C., Koch, E.W., Stier, A.C., Silliman, B.R., 2011. The value of estuarine and coastal ecosystem services. Ecol. Monogr. 81, 169–193.

Baudouin, M., de Thoisy, B., Chambault, P., Berzins, R., Entraygues, M., Kelle, L., Turny, A., Le Maho, Y., Chevallier, D., 2015. Identification of key marine areas for conservation based on satellite tracking of post-nesting migrating green turtles (*Chelonia mydas*). Biol. Conserv. 184, 36–41.

Berrisford, P., Dee, D.P., Poli, P., Brugge, R., Fielding, K., Fuentes, M., Kållberg, P.W., Kobayashi, S., Uppala, S., Simmons, A., 2011. The ERA-Interim archive version 2.0. In: ERA Report Series. ECMWF, Shinfield Park, Reading, p. 23.

Besset, M., Gratiot, N., Anthony, E.J., Bouchette, F., Goichot, M., Marchesiello, P., 2019. Mangroves and shoreline erosion in the Mekong River delta, Vietnam. Estuar. Coast. Shelf Sci. 226, 106263.

Bird, E.C.F., 1971. Mangroves as land builders. Vic. Nat. 88, 189–197.

Brunier, G., Anthony, E.J., Gratiot, N., Gardel, A., 2019. Exceptional rates and mechanisms of muddy shoreline retreat following mangrove removal. Earth Surf. Process. Landforms 44, 1559–1571.

Brunier, G., Michaud, E., Fleury, J., Anthony, E.J., Morvan, S., Gardel, A., 2020. Assessing the relationship between macro-faunal burrowing activity and mudflat geomorphology from UAV-based Structure-from-motion photogrammetry. Rem. Sens. Environ. 241, 111717.

Capdeville, C., Abdallah, K., Walcker, R., Rols, J.L., Fromard, F., Leflaive, J., 2019. Contrasted resistance and resilience of two mangrove forests after exposure to long-term and short-term anthropic disturbances. Mar. Environ. Res. 146, 12–23.

Carlton, J.M., 1974. Land building and stabilization by mangroves. Environ. Conserv. 1, 285–294.

Chevalier, C., Froidefond, J.-M., Devenon, J.-L., 2008. Numerical analysis of the combined action of littoral current, tide and waves on the suspended mud transport and on turbid plumes around French Guiana mudbanks. Continent. Shelf Res. 28, 545–560.
Cohen, M.C.L., Pessenda, L.C.R., Behling, H., de Fátima Rossetti, D., França, M.C., Guimarães, J.T.F., Friaes, Y., Smith, C.B., 2012. Holocene palaeoenvironmental history of the Amazonian mangrove belt. Quat. Sci. Rev. 55, 50–58.
Dasgupta, S., Islam, M.S., Huq, M., Huque Khan, Z., Hasib, M.R., 2019. Quantifying the protective capacity of mangroves from storm surges in coastal Bangladesh. PloS One 14, e0214079.
Davis, J.H.J., 1938. Mangroves, makers of land. Nat. Mag. 31, 551–553.
Dias, N., 2018. Derivation of the French Space Operation Act requirements in the specifications of the future European launcher Ariane 6. J. Space. Saf. Eng. 5, 121–125.
Diop, B., Sanz, N., Duplan, Y.J.J., Guene, E.H.M., Blanchard, F., Pereau, J.-C., Doyen, L., 2018. Maximum economic yield fishery management in the face of global warming. Ecol. Econ. 154, 52–61.
Ellison, A.M., Farnsworth, E.J., Merkt, R.E., 1999. Origins of mangrove ecosystems and the mangrove biodiversity anomaly. Global Ecol. Biogeogr. 8, 95–115.
Ellison, J.C., 2019. Chapter 20 – Biogeomorphology of mangroves. In: Perillo, G.M.E., Wolanski, E., Cahoon, D.R., Hopkinson, C.S. (Eds.), Coastal Wetlands. Elsevier, pp. 687–715.
Froidefond, J.M., Pujos, M., Andre, X., 1988. Migration of mud banks and changing coastline in French Guiana. Mar. Geol. 84, 19–30.
Fromard, F., Puig, H., Mougin, E., Marty, G., Betoulle, J.L., Cadamuro, L., 1998. Structure, above-ground biomass and dynamics of mangrove ecosystems: new data from French Guiana. Oecologia 115, 39–53.
Fromard, F., Vega, C., Proisy, C., 2004. Half a century of dynamic coastal change affecting mangrove shorelines of French Guiana. A case study based on remote sensing data analyses and field surveys. Mar. Geol. 208, 265–280.
Gardel, A., Gratiot, N., 2005. A satellite image-based method for estimating rates of mud bank migration, French Guiana, South America. J. Coast. Res. 21, 720–728.
Gardel, A., Proisy, C., Lesourd, S., Philippe, S., Caillaud, J., Gontharet, S., Anthony, E.J., Brutier, L., 2009. A better understanding of mud cracking processes gained from in situ measurements on an intertidal mudflat in French Guiana. J. Coast. Res. 1, 424–428.
Gedan, K.B., Kirwan, M.L., Wolanski, E., Barbier, E.B., Silliman, B.R., 2011. The present and future role of coastal wetland vegetation in protecting shorelines: answering recent challenges to the paradigm. Clim. Change 106, 7–29.
Gensac, E., Lesourd, S., Gardel, A., Anthony, E.J., Proisy, C., Loisel, H., 2011. Short-term prediction of the evolution of mangrove surface areas: the example of the mud banks of Kourou and Sinnamary, French Guiana. J. Coast. Res. 64, 388–392.
Gensac, E., Martinez, J.-M., Vantrepotte, V., Anthony, E.J., 2016. Seasonal and inter-annual dynamics of suspended sediment at the mouth of the Amazon river: the role of continental and oceanic forcing, and implications for coastal geomorphology and mud bank formation. Continent. Shelf Res. 118, 49–62.
Gratiot, N., Anthony, E.J., 2016. Role of flocculation and settling processes in development of the mangrove-colonized, Amazon-influenced mud-bank coast of South America. Mar. Geol. 373, 1–10.
Gratiot, N., Gardel, A., Anthony, E.J., 2007. Trade-wind waves and mud dynamics on the French Guiana coast, South America: input from ERA-40 wave data and field investigations. Mar. Geol. 236, 15–26.
Hamilton, S.E., Casey, D., 2016. Creation of a high spatio-temporal resolution global database of continuous mangrove forest cover for the 21st century (CGMFC-21). Global Ecol. Biogeogr. 25, 729–738.
Horstman, E.M., Dohmen-Janssen, C.M., Bouma, T.J., Hulscher, S., 2015. Tidal-scale flow routing and sedimentation in mangrove forests: combining field data and numerical modelling. Geomorphology 228, 244–262.
INSEE, 2017. Recensement de la population en Guyane en 2014, N°56, 2 pp. https://www.insee.fr/fr/statistiques/fichier/version-html/2540209/gy_inf_56.pdf. (Accessed 1 November 2019).
Isaac, V.J., Ferrari, S.F., 2017. Assessment and management of the North Brazil shelf large marine ecosystem. Environ. Dev. 22, 97–110.
Jonah, F.E., Boateng, I., Osman, A., Shimba, M.J., Mensah, E.A., Adu-Boahen, K., Chuku, E.O., Effah, E., 2016. Shoreline change analysis using end point rate and net shoreline movement statistics: an application to Elmina, Cape Coast and Moree section of Ghana's coast. Reg. Stud. Mar. Sci. 7, 19–31.
Kaliraj, S., Chandrasekar, N., Magesh, N.S., 2014. Impacts of wave energy and littoral currents on shoreline erosion/accretion along the south-west coast of Kanyakumari, Tamil Nadu using DSAS and geospatial technology. Environ. Earth Sci. 71, 4523–4542.

Kelle, L., Gratiot, N., Nolibos, I., Therese, J., Wongsopawiro, R., De Thoisy, B., 2007. Monitoring of nesting leatherback turtles (*Dermochelys coriacea*): contribution of remote sensing for real-time assessment of beach coverage in French Guiana. Chelonian Conserv. Biol. 6, 142–147.

Kiruba-Sankar, R., Krishnan, P., Dam Roy, S., Raymond Jani Angel, J., Goutham-Bharathi, M.P., Lohith Kumar, K., Ragavan, P., Kaliyamoorthy, M., Muruganandam, R., Rajakumari, S., Purvaja, R., Ramesh, R., 2018. Structural complexity and tree species composition of mangrove forests of the Andaman Islands, India. J. Coast. Conserv. 22, 217–234.

Krauss, K.W., McKee, K.L., Lovelock, C.E., Cahoon, D.R., Saintilan, N., Reef, R., Chen, L., 2014. How mangrove forests adjust to rising sea level. New Phytol. 202, 19–34.

Lucas, R.M., Mitchell, A.L., Rosenqvist, A., Proisy, C., Melius, A., Ticehurst, C., 2007. The potential of L-band SAR for quantifying mangrove characteristics and change: case studies from the Tropics. Aquat. Conserv. Mar. Freshw. Ecosyst. 17, 245–264.

Marois, D.E., Mitsch, W.J., 2015. Coastal protection from tsunamis and cyclones provided by mangrove wetlands – a review. Int. J. Biodiversity. Sci. Ecosyst. Serv. Manag. 11, 71–83.

Mazda, Y., Magi, M., Ikeda, Y., Kurokawa, T., Asano, T., 2006. Wave reduction in a mangrove forest dominated by *Sonneratia* sp. Wetl. Ecol. Manag. 14, 365–378.

Mazda, Y., Magi, M., Nanao, H., Kogo, M., Miyagi, T., Kanazawa, N., Kobashi, D., 2002. Coastal erosion due to long-term human impact on mangrove forests. Wetl. Ecol. Manag. 10, 1–9.

McKee, K.L., Cahoon, D.R., Feller, I.C., 2007. Caribbean mangroves adjust to rising sea level through biotic controls on change in soil elevation. Global Ecol. Biogeogr. 16, 545–556.

Miloslavich, P., Bax, N.J., Simmons, S.E., Klein, E., Appeltans, W., Aburto-Oropeza, O., Andersen Garcia, M., Batten, S.D., Benedetti-Cecchi, L., Checkley, D.M., Chiba, S., Emmett Duffy, J., Dunn, D.C., Fischer, A., Gunn, J., Kudela, R., Marsac, F., Muller-Karger, F.E., Obura, D., Shin, Y.-J., 2017. Essential ocean variables for global sustained observations of biodiversity and ecosystem changes. Global Change Biol. 24, 2416–2433.

Miloslavich, P., Klein, E., Díaz, J.M., Hernández, C.E., Bigatti, G., Campos, L., Artigas, F., Castillo, J., Penchaszadeh, P.E., Neill, P.E., Carranza, A., Retana, M.V., Díaz de Astarloa, J.M., Lewis, M., Yorio, P., Piriz, M.L., Rodríguez, D., Yoneshigue-Valentin, Y., Gamboa, L., Martín, A., 2011. Marine biodiversity in the Atlantic and Pacific coasts of South America: knowledge and gaps. PloS One 6, e14631.

Moir, W.H., 1972. Natural areas. Science 177, 396–400.

Nguyen, H.H., McAlpine, C., Pullar, D., Leisz, S.J., Galina, G., 2015. Drivers of coastal shoreline change: case study of Hon Dat coast, Kien Giang, Vietnam. Environ. Manag. 55, 1093–1108.

Nguyen, T.T.X., Bonetti, J., Rogers, K., Woodroffe, C.D., 2016. Indicator-based assessment of climate-change impacts on coasts: a review of concepts, methodological approaches and vulnerability indices. Ocean Coast. Manag. 123, 18–43.

Nicholls, R.J., Wong, P.P., Burkett, V., Woodroffe, C.D., Hay, J., 2008. Climate change and coastal vulnerability assessment: scenarios for integrated assessment. Sustain. Sci. 3, 89–102.

Olagoke, A., Proisy, C., Féret, J.-B., Blanchard, E., Fromard, F., Mehlig, U., de Menezes, M., dos Santos, V., Berger, U., 2016. Extended biomass allometric equations for large mangrove trees from terrestrial LiDAR data. Trees 30, 935–947.

Orseau, S., Lesourd, S., Huybrechts, N., Gardel, A., 2017. Hydro-sedimentary processes of a shallow tropical estuary under Amazon influence. The Mahury Estuary, French Guiana. Estuar. Coast. Shelf Sci. 189, 252–266.

Osorio-Cano, J.D., Osorio, A.F., Pelaez-Zapata, D.S., 2019. Ecosystem management tools to study natural habitats as wave damping structures and coastal protection mechanisms. Ecol. Eng. 130, 282–295.

Phan, K.L., Stive, M.J.F., Zijlema, M., Truong, H.S., Aarninkhof, S.G.J., 2019. The effects of wave non-linearity on wave attenuation by vegetation. Coast. Eng. 147, 63–74.

Plaziat, J.-C., Augustinus, P.G.E.F., 2004. Evolution of progradation/erosion along the French Guiana mangrove coast: a comparison of mapped shorelines since the 18th century with Holocene data. Mar. Geol. 208, 127–143.

Proisy, C., Couteron, P., Fromard, F., 2007. Predicting and mapping mangrove biomass from canopy grain analysis using fourier-based textural ordination of IKONOS images. Rem. Sens. Environ. 109, 379–392.

Proisy, C., Degenne, P., Anthony, E.J., Berger, U., Blanchard, E., Fromard, F., Gardel, A., Olagoke, A., Santos, V.F., Walcker, R., Lo Seen, D., 2016. A multiscale simulation approach for linking mangrove dynamics to coastal processes using remote sensing observations. J. Coast. Res. 810–814. Special Issue no. 75.

Proisy, C., Gratiot, N., Anthony, E.J., Gardel, A., Fromard, F., Heuret, P., 2009. Mud bank colonization by opportunistic mangroves: a case study from French Guiana using lidar data. Continent. Shelf Res. 29, 632–641.

Proisy, C., Mougin, E., Fromard, F., Karam, M.A., 2000. Interpretation of polarimetric radar signatures of mangrove forests. Rem. Sens. Environ. 71, 56–66.

Proisy, C., Mougin, E., Fromard, F., Trichon, V., Karam, M.A., 2002. On the influence of canopy structure on the polarimetric radar response from mangrove forest. Int. J. Rem. Sens. 23, 4197–4210.

Proisy, C., Viennois, G., Sidik, F., Andayani, A., Enright, J.A., Guitet, S., Gusmawati, N., Lemonnier, H., Muthusankar, G., Olagoke, A., Prosperi, J., Rahmania, R., Ricout, A., Soulard, B., Suhardjono, 2018. Monitoring mangrove forests after aquaculture abandonment using time series of very high spatial resolution satellite images: a case study from the Perancak estuary, Bali, Indonesia. Mar. Pollut. Bull. 131, 61–71.

Quartel, S., Kroon, A., Augustinus, P.G.E.F., Van Santen, P., Tri, N.H., 2007. Wave attenuation in coastal mangroves in the Red River Delta, Vietnam. J. Asian Earth Sci. 29, 576–584.

Rongier, I., 2013. Chapter 2 – Spaceport design for safety. In: Allahdadi, F.A., Rongier, I., Wilde, P.D. (Eds.), Safety Design for Space Operations. Butterworth-Heinemann, Oxford, pp. 27–63.

Souza Filho, P.W.M., Farias Martins, E.S., da Costa, F.R., 2006. Using mangroves as a geological indicator of coastal changes in the Braganca macrotidal flat, Brazilian Amazon: a remote sensing data approach. Ocean Coast. Manag. 49, 462–475.

Stanley, O.D., Lewis, R.R., 2009. Strategies for mangrove rehabilitation in an eroded coastline of Selangor, Peninsular Malaysia. J. Coast. Dev. 12, 144–156.

Szlafsztein, C., Sterr, H., 2007. A GIS-based vulnerability assessment of coastal natural hazards, state of Pará, Brazil. J. Coast. Conserv. 11, 53–66.

Tanaka, N., Sasaki, Y., Mowjood, M.I.M., Jinadasa, K.B.S.N., Samang Homchuen, S., 2007. Coastal vegetation structures and their functions in tsunami protection: experience of the recent Indian Ocean tsunami. Landsc. Ecol. Eng. 3, 33–45.

Thampanya, U., Vermaat, J.E., Sinsakul, S., Panapitukkul, N., 2006. Coastal erosion and mangrove progradation of Southern Thailand. Estuar. Coast. Shelf Sci. 68, 75–85.

Thomas, N., Lucas, R., Bunting, P., Hardy, A., Rosenqvist, A., Simard, M., 2017. Distribution and drivers of global mangrove forest change, 1996–2010. PloS One 12, e0179302.

van der Wal, D., Herman, P.M.J., Wielemaker-van den Dool, A., 2005. Characterisation of surface roughness and sediment texture of intertidal flats using ERS SAR imagery. Rem. Sens. Environ. 98, 96–109.

van Ledden, M., Vaughn, G., Lansen, J., Wiersma, F., Amsterdam, M., 2009. Extreme wave event along the Guyana coastline in October 2005. Continent. Shelf Res. 29, 352–361.

Van Santen, P., Augustinus, P.G.E.F., Janssen-Stelder, B.M., Quartel, S., Tri, N.H., 2007. Sedimentation in an estuarine mangrove system. J. Asian Earth Sci. 29, 566–575.

Walcker, R., Anthony, E.J., Cassou, C., Aller, R.C., Gardel, A., Proisy, C., Martinez, J.-M., Fromard, F., 2015. Fluctuations in the extent of mangroves driven by multi-decadal changes in North Atlantic waves. J. Biogeogr. 42, 2209–2219.

Walsh, J.P., Nittrouer, C.A., 2004. Mangrove-bank sedimentation in a mesotidal environment with large sediment supply, Gulf of Papua. Mar. Geol. 208, 225–248.

Winterwerp, J.C., Borst, W.G., de Vries, M.B., 2005. Pilot study on the erosion and rehabilitation of a mangrove mud coast. J. Coast. Res. 21, 223–230.

Winterwerp, J.C., Graaff, R.F.d., Groeneweg, J., Luijendijk, A.P., 2007. Modelling of wave damping at Guyana mud coast. Coast. Eng. 54, 249–261.

Woodroffe, C., 1992. Mangrove Sediments and Geomorphology. Tropical Mangrove Ecosystems. American Geophysical Union, pp. 7–41.

Woodroffe, C.D., Grindrod, J., 1991. Mangrove biogeography: the role of quaternary environmental and sea-level change. J. Biogeogr. 18, 479–492.

Worthington, T., Spalding, M., 2018. Mangrove restoration potential. A global map highlighting a critical opportunity. In: IUCN, University of Cambridge & The Nature Conservancy. https://doi.org/10.17863/CAM.39153, 36 pp. (Accessed 1 November 2019).

Yulianda, F., Wardiatno, Y., Nurjaya, I.W., Herison, A., 2014. Coastal conservation strategy using mangrove ecology system approach. Asian J. Sci. Res. 7, 513–524.

Yulianto, E., Sukapti, W.S., Rahardjo, A.T., Noeradi, D., Siregar, D.A., Suparan, P., Hirakawa, K., 2004. Mangrove shoreline responses to Holocene environmental change, Makassar Strait, Indonesia. Rev. Palaeobot. Palynol. 131, 251–268.

CHAPTER

3

Groundwater research in mangrove coastal ecosystems—new prospects

Paul Macklin[1,4], *Judith Rosentreter*[2], *Virni Budi Arifanti*[3], *I Gusti Ngurah Agung Suryaputra*[4]

[1]National Marine Science Centre, Southern Cross University, Coffs Harbour, NSW, Australia; [2]Centre for Coastal Biogeochemistry, School of Environment, Science and Engineering Southern Cross University, Lismore, NSW, Australia; [3]Center for Research and Development of Social, Economy, Policy and Climate Change (P3SEKPI), Ministry of Environment and Forestry, Bogor, Indonesia; [4]Department of Chemical Analysis, Faculty of Mathematics and Natural Sciences, Universitas Pendidikan Ganesha, Singaraja, Bali, Indonesia

OUTLINE

1. Introduction 68
2. Linkage between the hydrogeological and carbon cycle in mangrove ecosystems 68
   2.1 The significance of groundwater and porewater 68
   2.2 Tidal pumping 69
3. Mangrove forest degradation and loss 71
   3.1 Conversion of mangrove forest to aquaculture 71
   3.2 Mangrove deforestation 72
4. Available techniques in SGD research 72
   4.1 Radon as a naturally occurring SGD tracer 74
   4.2 Aerial thermal infrared imaging 75
   4.3 Combining ^{222}Rn and ATI methodologies 76
5. Summary 77
References 78

https://doi.org/10.1016/B978-0-12-816437-2.00010-0

1. Introduction

Mangrove forests are salt tolerant ecosystems that can be found extensively in intertidal zones from temperate to tropical climates around the world (Bunting et al., 2018). As the only forests located at the land–ocean interface, mangrove forests have stronger linkages to their surrounding hydrology than any other forests (Alongi and Brinkman, 2011), meaning they are influenced by both freshwater and seawater exchange. Freshwater inputs can be derived from combinations of recharges from upstream rivers, rainfall, and/or surface water runoff (natural and modified surfaces). Oceanic inputs are controlled predominantly by tides. Because of strong hydrogeological linkages, mangrove forests are known to act as effective natural filtration systems between land and coastal waters, with the ability to improve and regulate water quality by buffering pollutants and nutrients before delivery to adjacent coastal habitats, such as seagrass meadows and coral reefs (Barbier et al., 2011; Hayes et al., 2019). As such mangrove forest degradation and loss is known to increase the export of fine sediment, nutrients, and other contaminants, which also impact the water quality of adjacent habitats (Fabricius, 2005).

Here we discuss the importance of groundwater and porewater (collectively known as submarine groundwater discharge (SGD)) in mangrove forests. We define and explain SGD, highlight tidal pumping mechanisms in relation to SGD in mangrove ecosystems, and resulting hydrological impacts on adjacent coastal ecosystems such as seagrasses and coral reefs. We describe the effects of mangrove forest degradation and loss (e.g., conversion of mangrove forest to aquaculture), urbanization, and mangrove deforestation on SGD. Furthermore, we include suggestions for efficient, widespread monitoring and analysis of mangrove hydrology based on a cross-validation approach of 1) in situ methodologies of two natural tracers (^{222}Rn and surface water temperature) and 2) remote sensing methodologies such as thermal infrared imagery which is from here on described as an ATI (drone-based aerial thermal infrared imagery).

2. Linkage between the hydrogeological and carbon cycle in mangrove ecosystems

2.1 The significance of groundwater and porewater

SGD is defined as any flow of water passing through sediments into coastal oceans regardless of water composition or driving force (Moore, 2010). SGD comprises of both fresh groundwater discharge and recirculated porewater (seawater) with a wide range of salinity and multiple, and sometimes complex, origins. Although groundwater and porewater may have different hydrological dynamics, they are currently described combined as SGD. In mangrove ecosystems, groundwater is usually associated with low salinity (freshwater) and water temperatures with detectable differences compared to adjacent surface water temperatures (Kalbus et al., 2006). Porewater is usually saline and associated with oceanic mixing and tidal exchange, but may not be as well-defined due to differing soil permeability and tidal regimes (e.g., tidal range, tidal cycles, groundwater mixing) that can occur in mangrove forest surrounding waters (Tait et al., 2016). Porewater salinity is also subjected to a wide range of seasonal variation due to hydrological inputs such as freshwater inundation from rivers, monsoon rain and episodic rainfall events, and salinity similar to adjacent coastal oceans in dry (drought) seasons (Komiyama et al., 2019).

Groundwater and porewater have higher concentrations of dissolved carbon constitutes (e.g., DIC, free CO_2, CH_4) compared to coastal surface waters (Chen et al., 2018; Santos et al., 2012). The excess carbon constitutes in SGD can be controlled by terrestrial, riverine, and oceanic processes such as 1) carbonate and silicate rock weathering rates (Hagedorn and Cartwright, 2010), 2) soil organic matter properties, 3) root respiration and accumulation within the soil zone (Abril et al., 2000; Kessler and Harvey, 2001), 4) vertical diffusion of greenhouse gases to the atmosphere, 5) microbial degradation processes in sediments (e.g., anaerobic respiration), and 6) hydrological processes and the relative discharge of surface water runoff (Raymond et al., 2000) and SGD (Kessler and Harvey, 2001; Sulman et al., 2012; Fig. 3.1).

2.2 Tidal pumping

Recent studies have shown that high concentrations of carbon dioxide (CO_2), methane (CH_4), and nutrients in mangrove surrounding surface waters are driven predominantly by groundwater and porewater exchanges (Alongi, 2014; Burnett et al., 2003; Hayes et al., 2019; Santos et al., 2009; Taillardat et al., 2018; Taniguchi et al., 2019). As a result, a typical tidal pattern can be observed with highest concentrations at low (or ebb) tide and lowest

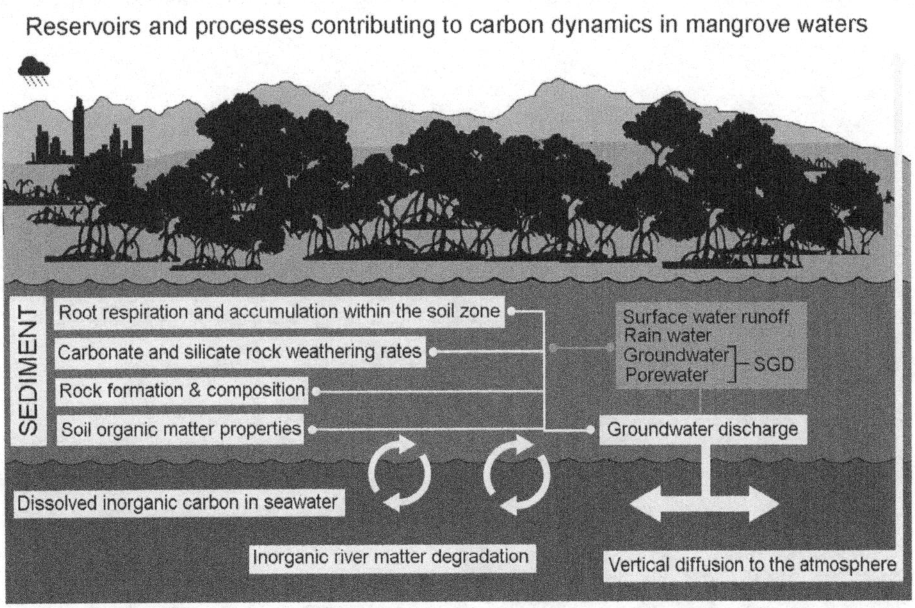

FIGURE 3.1 Urban and mangrove forest principal reservoirs and processes contributing to CO_2 within aquatic systems. Note the land degradation and loss potential to alter the soil/groundwater interfaces (blue [light gray in print version] box).

concentrations at high (or flood) tide (Call et al., 2015: Rosentreter et al., 2018; Santos et al., 2012). This process is referred to as "tidal pumping" (see Geyer and Nepf, 1996; Okubo, 1973; Schijf and Schönfled, 1953; Smith, 1977; Stommel and Farmer, 1952; Uncles et al., 1985; Wang and Elliott, 1978; Zimmerman, 1986). Tidal pumping is defined as a tidally driven exchange of porewater (including groundwater) constituents between sediments and surface waters, controlled by the tidal and lunar cycle (Li et al., 2009; Fig. 3.2). Crab burrows have also been found to be an important mechanism driving SGD exchanges in mangrove forest sediment (Gleeson et al., 2013; Holloway et al., 2016; Mori et al., 2019; Sadat-Noori and Glamore, 2019). For example, Stielglitz (2005) estimated that 10%–40% of surface water infiltrated mangrove sediments (at high tide) via crab burrows in a mangrove forest near the Great Barrier Reef, Australia.

Thus, surface water hydrology and hydrogeology (groundwater hydrology) are closely linked to carbon cycling in mangrove and other "blue carbon" coastal ecosystems (Fig. 3.2). Factors such as population growth and expansion, urbanization, climate change–related shifts in rainfall patterns, and flooding events are likely to impact mangrove forest hydrology in the future.

In addition, SGD transports significant amounts of heavy metals and nutrients, which can impact coastal ecosystems (Adyasari et al., 2018; Garcia-Solsona et al., 2010; Fig. 3.3). Contamination via SGD is recognized as a main pathway for land-derived nutrients and heavy metals to estuaries and coastal waters, more-so in modified regions with rapid urbanization, industry, and aquaculture (Islam et al., 2018; Rashed-Un-Nabi et al., 2011). As an example, a recent study by Luijendijk et al. (2020) estimated that 13% (0.2%–23%) of global coastlines have a significant eutrophication risk via nutrients transported by SGD. Excessive

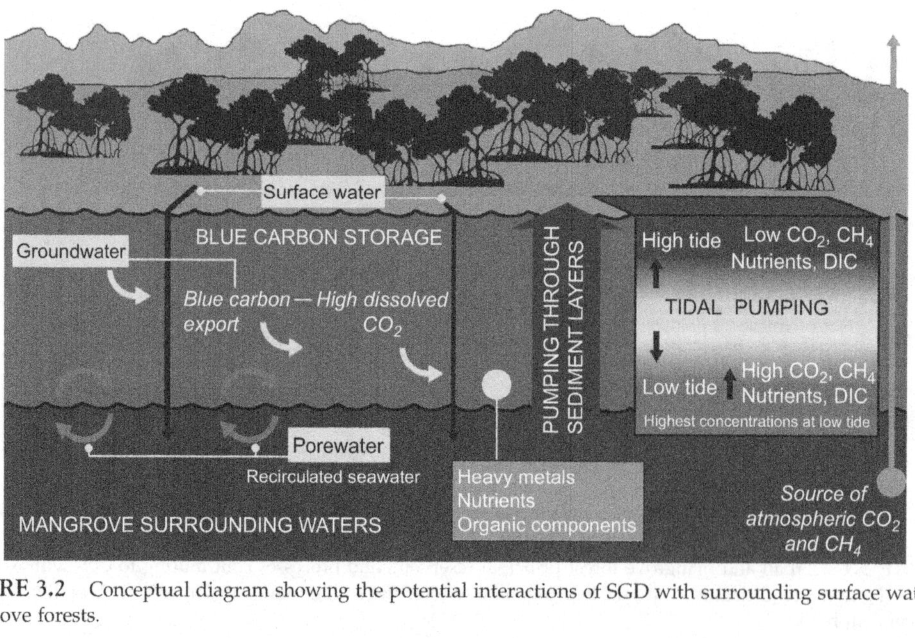

FIGURE 3.2 Conceptual diagram showing the potential interactions of SGD with surrounding surface waters in mangrove forests.

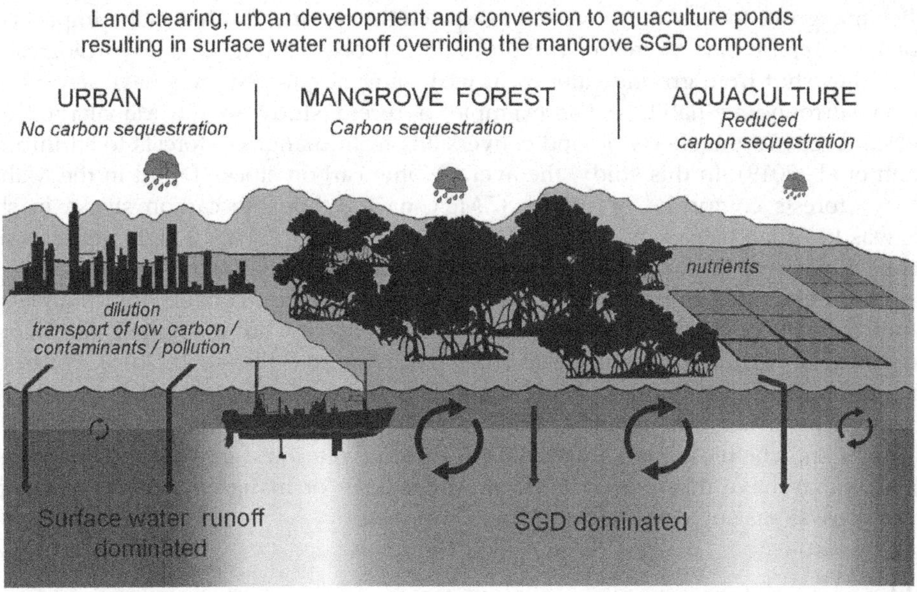

FIGURE 3.3 Conceptual image showing how land conversions have the potential to deplete carbon stocks and alter SGD and surface water regimes. The different sized circular arrows represent potential volumetric changes in SGD which are influenced by surface water runoff associated with impervious surfaces, land clearing, and episodic rainfall.

nutrients (eutrophication) can not only cause ecological impacts such as algal blooms, decrease of water quality, and hypoxia (lack of oxygen), which result in negative environmental and economic impacts in mangrove ecosystems, but also impact biogeochemically connected seagrass habitats, salt marshes, and fringing coral reefs. It is therefore crucial to understand the spatial and temporal variability and magnitude of SGD transport of contaminants and constituents between terrestrial and coastal ecosystems.

3. Mangrove forest degradation and loss

3.1 Conversion of mangrove forest to aquaculture

Land clearing, urban development, and conversion of mangrove forest to aquaculture ponds are often favored strategies for economists that offer short-term economic gains, without accounting for longer-term ecosystem services (Barbier et al., 2011; Rizal et al., 2018). Coastal urbanization, degradation of mangrove forests, and groundwater extraction are also known to increase the delivery of nutrient loads adjacent coastal ecosystems (Adyasari et al., 2018; Paytan et al., 2006). Quantitative assessments of mangrove forest SGD, currently lacking in literature, are vital as to address future degradation of coastal water quality. Recent research has shown how land-use conversions can deplete carbon stocks and alter SGD and surface water regimes, while enhancing further releases of CO_2 to the atmosphere (Sidik and Lovelock, 2013; Fig. 3.3).

When mangrove forests are converted to aquaculture ponds, this has major impacts on the hydrological regime and therefore carbon cycling processes (Fig. 3.3). Hydrological pathways, as they shift from groundwater to surface water dominated, may have significant impacts on surrounding habitats. For example, a recent study in the Mahakam Delta in Indonesia discusses the effects of land conversions from mangrove forests to shrimp ponds (Arifanti et al., 2019). In this study, the average blue carbon stocks found in the Mahakam mangrove forests comprised of 1023 ± 87 Mg C ha^{-1}, while the carbon stocks in shrimp ponds was less than half of those in the forests (499 ± 56 Mg C ha^{-1}), highlighting a significant soil carbon loss similar to 226 years of soil carbon accumulation by the conversion from mangrove forest to shrimp ponds. However, mangrove forest conversions to aquaculture does not only affect the mangrove carbon storage capacity but also alters the mangrove hydrogeological regime (Adyasari et al., 2018; Basyuni and Sulistiyono, 2018). For example, water extracted from rivers (surface and subsurface waters) for irrigation has been shown to alter mangrove surface water salinity (Hogarth, 2015), while introducing effluent (Debenay et al., 2015) and chemical contamination into surface water and groundwaters (Primavera, 2005). This may have major impacts on the hydrology of mangrove forests and adjacent seaward ecosystems such as seagrasses and coral reefs.

3.2 Mangrove deforestation

Mangrove tree loss can also change the hydrology in mangrove ecosystems by subsequently altering the local water supply and water chemistry (Bouillon and Connolly, 2009). Therefore, mangrove forest degradation or loss can further facilitate seaward degradation of seagrass beds and coral reefs (Gilman et al., 2016). Importantly, these processes are becoming more evident in regions where mangrove forest loss is accelerating (Alongi et al., 2008). With rapid global increases in mangrove deforestation, tidal exchanges coupled with episodic rain events and increased surface water runoff during monsoonal seasons or flooding periods may amplify the biogeochemistry and subsequent health of mangrove surrounding waters and adjacent coastal habitats. For example, mangrove carbon export to the coastal ocean may influence the biogeochemistry of seagrass meadows which have been shown to metabolize excess pCO_2 from mangrove tidal exports before subsequent delivery to fringing coral reefs in more pristine mangrove forests (Macklin et al., 2019; Fig. 3.4).

4. Available techniques in SGD research

Currently available techniques to measure SGD include hydrogeological mapping, hydrographic methods, hydrometric analysis, and numerical modeling (Mazda and Wolanski, 2009; Li et al., 2012). Hydrogeological mapping is a method that characterizes groundwater resources by analyzing geological data to create 3D groundwater maps. Hydrographic methods involve the surveying of subsurface physical features and adjacent land areas, mostly using multibeam echo sounders to collect data (e.g., water depth, seafloor and coastline shape, obstructions for navigation). Hydrometric analysis uses the process of grading

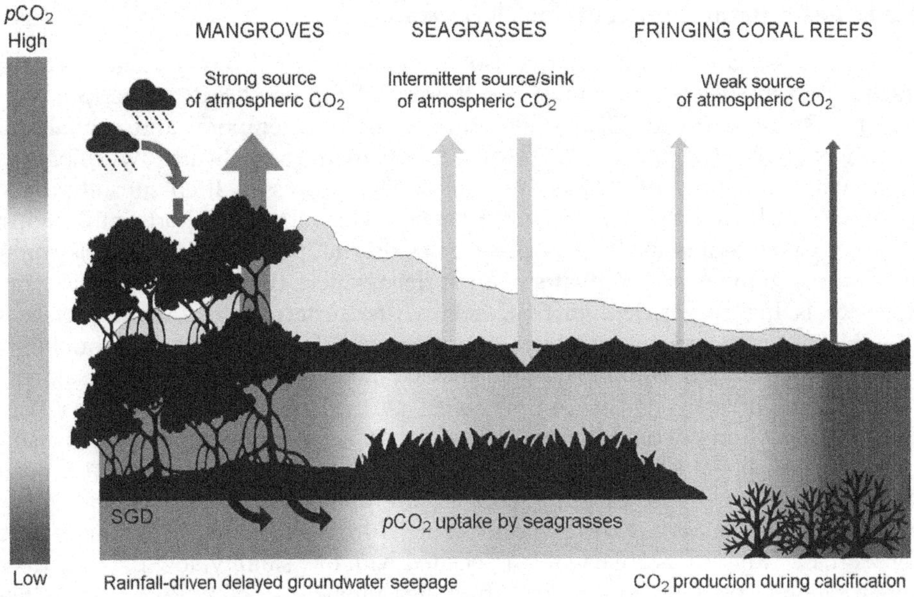

FIGURE 3.4 Mangrove blue carbon export showing how SGD-derived pCO_2 is metabolized by seagrasses before delivery to coastal waters in more pristine mangrove ecosystems. *Source: modified from Macklin, P. A., Suryaputra, I. G. N. A., Maher, D. T., Murdiyarso, D., Santos, I. R., 2019. Drivers of* CO_2 *along a mangrove-seagrass transect in a tropical bay: delayed groundwater seepage and seagrass uptake. Continent. Shelf Res. 172, 57–67.*

fine-grained soils, clays, and silts, which helps to better understand flooding, pollution, and climate risks. Numerical modeling is a technique widely applied to complex geological problems by using simulations scenarios. Numerical modeling has proven to be a very effective method in understanding SGD pathways and to estimate both SGD and seawater intrusion using data to interpolate and extrapolate both spatially and temporally (Fatimah et al., 2008; Akamatsu et al., 2009; Yanagisawa et al., 2010).

Utilizing natural tracers to measure SGD flow is another successful method, most notably in mangrove forests. Natural tracers provide a cost-effective option (e.g., when compared to time intensive and expensive laboratory analysis) and are mostly measured on site (in situ) with automated loggers. A main advantage of this method is the high-resolution, site-specific, and instantaneous assessments of SGD. Several natural tracers can be used for SGD including commonly measured field parameters such as surface water temperature, electrical conductivity, and pH (Rodríguez et al., 2018). Other natural tracers are hydrochemical tracers (both natural major cations and anions), stable isotopes, and other radioactive isotopes (see Burnett et al., 2008; Sappa et al., 2012; Rey et al., 2018). Here, we focus on the benefits of using radon (^{222}Rn) as a naturally occurring SGD tracer which has become a popular tool for larger-scale, high-resolution, automated, and continuous underway SGD investigations in mangrove forest waters (Adyasari et al., 2018; Burnett and Dulaiova, 2003; Burnett et al., 2010; Jeffrey et al., 2018).

4.1 Radon as a naturally occurring SGD tracer

^{222}Rn is a noble gas and radioactive isotope (half-life = 3.8 days), which is produced continually from the decay of uranium and thorium (Uranium-238 decay series). The short half-life of ^{222}Rn makes it an ideal SGD tracer along both temporal and spatial scales in coastal waters (Burnett et al., 2010). ^{222}Rn is popular in groundwater hydrology mainly because it allows for short-term observation of aquatic processes. It enters soil pore spaces and rock fractures, before dissolving in groundwater, which then transports it through spaces in sediments (Gattacceca et al., 2011; McCoy and Corbett, 2009). ^{222}Rn in SGD is commonly measured at magnitudes much greater than surface waters (Burnett et al., 2010). In other words, ^{222}Rn is highly enriched in SGD relative to surface waters, thus surface water recording high ^{222}Rn is usually indicative of a nearby SGD source. Recent technological advances in radon measuring equipment have enabled more efficient and accurate assessments of SGD rates and surface water mixing processes, particularly in mangrove forest waters.

^{222}Rn has been used successfully as a tracer of SGD in a number of recent mangrove studies over large spatial scales. It has been found that radon usually correlates positively with dissolved carbon dioxide, CH_4, and temperature, and negatively with salinity, dissolved oxygen, and pH (Call et al., 2015; Rosentreter et al., 2018; Santos et al., 2012; Tait et al., 2016). In other words, ^{222}Rn signals are usually associated with low salinity, low pH, low dissolved oxygen, and high pCO_2, CH_4, and temperature measurements. The ^{222}Rn tracer method can be used to 1) discriminate SGD from surface water, 2) assess SGD sources in mangrove forest waters, 3) differentiate tidal pumping from terrestrial aquifer groundwater flow, 4) detect seawater intrusion, and 5) locate and identify mixing areas of seawater and freshwater. Fig. 3.5 shows how an automated ^{222}Rn methodology can be deployed underway in a small

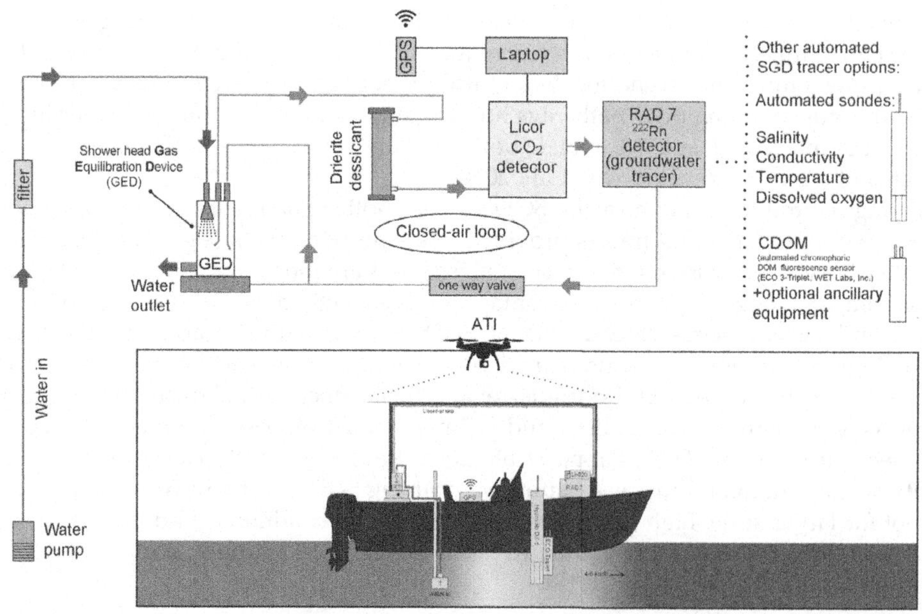

FIGURE 3.5 Schematic of automated instrumentation and underway setup (inset) to determine groundwater discharge (as described by Santos et al., 2012). Note that ATI is included as a supplementary information for Fig. 3.6.

research vessel coupled to ancillary equipment such as CO_2 detectors and other automated/supportive multiparameter water quality sondes (e.g., for ancillary measurements of temperature, conductivity, pH, salinity, dissolved oxygen, turbidity, total dissolved solids, and dissolved organic matter; see Gleeson et al., 2013; Sadat-Noori and Glamore, 2019; Santos et al., 2012; Tait et al., 2016).

An advantage of utilizing radon methodology underway is that the equipment is robust and portable (radon-in-air monitor (Durridge)), relatively easy to assemble and deploy, provides time and date-stamped measurements that can be used simultaneously with supportive ancillary measurements when continuously logged with a GPS (Fig. 3.5). Raw data files can be easily integrated into Geographic Integration System software for spatial analysis and mapping. This method is effective in highly dynamic mangrove waters where biochemical changes may exist over short temporal and small spatial scales and where SGD mixing zones may be identified over large spatial scales (a restriction of laboratory sampling).

In situ surface water temperature measurements have been utilized widely as an important SGD tracers (see Anderson, 2005; Kalbus et al., 2006; Rau et al., 2010; Stoneman and Constantz, 2003; Tristram et al., 2015). Natural tracer methods combined with ATI can now be used to capture and cross-validate differences between surrounding groundwater and surface water temperatures to differentiate SGD sources (Johnson et al., 2008; Kelly et al., 2019) and to investigate SGD residence times (Handcock et al., 2006; Martin and Dean, 1999) over much larger spatial scales.

4.2 Aerial thermal infrared imaging

Remote sensing is a passive acquisitioning of data imagery, which represents physical characteristics of an area without direct contact (e.g., satellites, drones, or aircraft). Remote sensing has been used to extensively gauge mangrove forest extents worldwide (Kuenzer et al., 2011), to map mangrove deforestation in Southeast Asia (Vadrevu et al., 2019), to describe decadal tidal cycling effects on erosion in mangrove forest waters (Fromard et al., 2004), and to investigate climate change effects such as mangrove forest latitudinal expansion (Hu et al., 2018). Aerial thermal infrared imaging (ATI) involves the use of drones mounted with thermal infrared cameras to detect heat (e.g., surface water temperature). In a study by Kumar and Reshmidevi (2013), ATI was used to investigate surface water temperature anomalies (rainfall and surface water runoff) in mangrove-dominated estuaries. Typically, thermal infrared maps indicate color gradients from red (higher temperatures) to blue (lower temperatures). When comparing drone-mounted (e.g., ATI) with in situ surface water temperature monitoring, ATI is usually more time and cost-effective and provides larger spatial coverage and increased precision of surface water assessments (Kumar and Reshmidevi, 2013).

ATI data can be used to quantify spatial water temperature variability in rivers, creeks, and floodplains from <1 m to more than 100 km scales, and to assess thermal landscapes (Handcock et al., 2012). SGD measurements derived from thermal springs, rainfall, surface water runoff, and SGD in tropical regions often display ranges in temperatures ($>1°C$), which is enough to sufficiently differentiate SGD components from surface waters. Also, surface water temperatures in tropical regions may have greater variability, making point source identification more accurate for thermal investigations in tropical latitudes.

4.3 Combining ^{222}Rn and ATI methodologies

Thermal imagery (thermal infrared cameras) can be used to monitor coastal water quality, because radiometric temperature correlates well with salinity, dissolved oxygen, chlorophyll-a, dissolved solids, and turbidity (Gholizadeh et al., 2016). ATI technologies already possess capabilities to enhance ^{222}Rn methodologies, providing wider spatial ranges and cost-effective analysis with the potential to provide new knowledge of biogeochemistry (Ferrara et al., 2017) and CO_2 (Queißer et al., 2018) in mangrove ecosystems. ATI has been paired successfully with ^{222}Rn investigations in a number of studies (Kelly et al., 2013, 2019; Li et al., 2009; Mejías et al., 2012; Mulligan and Charette, 2009; Tamborski et al., 2015; Wilson and Rocha, 2012, 2014). The combination of research combining and cross-validating ATI with statistically validated ^{222}Rn methodologies may be a logical step forward in identifying potential negative hydrological impacts in coastal areas on a regional scale (Burnett et al., 2003), particularly when considering rapid global mangrove deforestation rates (see Figs. 3.5 and 3.6).

Because SGD is a main pathway for nutrient discharges to coastal waters globally (Cho et al., 2018), the combined ^{222}Rn and ATI approach would be particularly helpful for effectively detecting water quality degradation (e.g., excess nutrient discharge) in mangrove waters that are prone to agricultural, industrial, and urban development, and in remote areas affected by extraction industries (see Johnson et al., 2008). A study by Tamborski et al. (2015) combined ^{222}Rn measurements with ATI as a quantitative tool for estimating nonpoint source diffuse SGD. This study revealed that ^{222}Rn positively correlated with lower nearshore

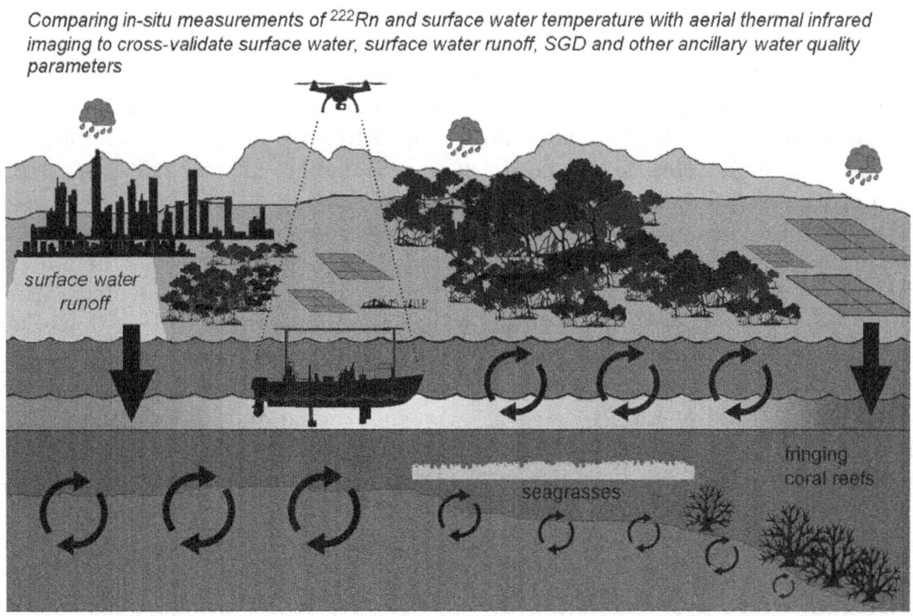

FIGURE 3.6 Potential for comparing ATI data with ^{222}Rn in situ methodology for cross-validation in larger-scale hydrological assessments along mangrove–seagrass–coral reef continuums.

surface water temperature anomalies. At one study site, the surface water temperature anomaly composed of mixed recirculated seawater and SGD with elevated nutrients. However, at a second site, no ATI anomaly was observed, but was composed of circulated seawater SGD only (no freshwater) with low nutrients. This study highlights well how ATI can identify point source mixing between diffuse fresh groundwater and recirculated seawater.

Other studies have demonstrated how ATI data can be used to map water quality parameters such as temperature, turbidity, chlorophyll-a, pH, and total suspended solids (e.g., Gholizadeh et al., 2016; Kumar and Reshmidevi, 2013; Hu et al., 2018). These parameters can be derived (and combined) relatively easily and at low costs from ATI data and potentially used to locate and characterize SGD in mangrove forest waters. Efficient, near-instantaneous, high-resolution assessments of the distribution of SGD locations over large surface water areas can be achieved, and when a specific mangrove forest area has been mapped with a combined ATI/^{222}Rn methodology, upscaling is possible.

Tropical mangrove forest regions would be expected to have large variations in surface water temperature due to potential site-specific discharges from geothermal springs, urban surface water runoff, aquaculture ponds, or industrial pollution. Furthermore tropical countries (e.g., Indonesia with ~22% of global mangrove forests (Giri et al., 2011)) have rapidly expanding human coastal populations, proximate to both modified and more pristine mangrove habitats with mangrove—seagrass—coral reef continuums. This provides potential for assessing SGD in previously understudied mangrove forest regions, with possibilities to make both retrospective and predictive assessments over large scales, and to identify areas of aquatic degradation and provide early mitigation opportunities.

5. Summary

In mangrove and other coastal (blue carbon) ecosystems, SGD, which is groundwater and porewater, is strongly connected with surface water pathways. Mangrove forest degradation and loss has significant implications on the coastal hydrological regime and related carbon cycling and water quality in mangrove waters. Land-use conversions (aquaculture, urbanization) can alter hydrological pathways and water chemistry of mangrove ecosystems, which is becoming more evident in areas where mangrove forest loss is accelerating. Coastal habitats adjacent to mangrove forests, such as salt marshes, seagrasses, and coral reefs, are also negatively impacted. The change and disruption of mangrove hydrogeology can further result in negative social, ecological, and economic impacts. With regard to SGD methodologies, Radon (^{222}Rn) has been successfully used as a natural tracer for larger-scale, high-resolution, automated, and continuous groundwater measurements. By using the combination of ^{222}Rn and ATI tools, we can further validate aerial infrared imaging for more effective and widespread usage, while preparing for emerging remote sensing technologies. As a result, mangrove forests can be monitored over much larger spatial scales. This may lead to a more effective mangrove management system in the coming decades.

References

Abril, G., Etcheber, H., Borges, A.V., Frankignoulle, M., 2000. Excess atmospheric carbon dioxide transported by rivers into the Scheldt estuary. Comptes Rendus Acad. Sci. - Ser. IIA Earth Planet. Sci. 330 (11), 761–768.

Adyasari, D., Oehler, T., Afiati, N., Moosdorf, N., 2018. Groundwater nutrient inputs into an urbanized tropical estuary system in Indonesia. Sci. Total Environ. 627, 1066–1079.

Akamatsu, Y., Ikeda, S., Toda, Y., 2009. Transport of nutrients and organic matter in a mangrove swamp. Estuar. Coast. Shelf Sci. 82 (2), 233–242.

Alongi, D.M., 2014. Carbon cycling and storage in mangrove forests. Annu. Rev. Mar. Sci. 6, 195–219.

Alongi, D.M., Trott, L.A., Tirendi, F., McKinnon, A.D., Undu, M.C., 2008. Growth and development of mangrove forests overlying smothered coral reefs, Sulawesi and Sumatra, Indonesia. Mar. Ecol. Prog. Ser. 370, 97–109.

Alongi, D.M., Brinkman, R., 2011. Hydrology and biogeochemistry of mangrove forests. In: Forest Hydrology and Biogeochemistry. Springer, pp. 203–219.

Anderson, M.P., 2005. Heat as a ground water tracer. Groundwater 43 (6), 951–968.

Arifanti, V.B., Kauffman, J.B., Hadriyanto, D., Murdiyarso, D., Diana, R., 2019. Carbon dynamics and land use carbon footprints in mangrove-converted aquaculture: the case of the Mahakam Delta, Indonesia. For. Ecol. Manag. 433, 17–29. https://doi.org/10.1016/j.foreco.2018.08.047.

Barbier, E.B., Hacker, S.D., Kennedy, C., Koch, E.W., Stier, A.C., Silliman, B.R., 2011. The value of estuarine and coastal ecosystem services. Ecol. Monogr. 81 (2), 169–193.

Basyuni, M., Sulistiyono, N., 2018. Deforestation and reforestation analysis from land-use changes in North Sumatran Mangroves, 1990–2015. In: Paper Presented at the IOP Conference Series: Materials Science and Engineering.

Bouillon, S., Connolly, R.M., 2009. Carbon exchange among tropical coastal ecosystems. In: Ecological Connectivity Among Tropical Coastal Ecosystems. Springer, Dordrecht, pp. 45–70.

Bunting, P., Rosenqvist, A., Lucas, R.M., Rebelo, L.M., Hilarides, L., Thomas, N., Hardy, A., Itoh, T., Shimada, M., Finlayson, C.M., 2018. The global mangrove watch – a new 2010 global baseline of mangrove extent. Rem. Sens. 10 (10) https://doi.org/10.3390/rs10101669.

Burnett, W.C., Bokuniewicz, H., Huettel, M., Moore, W.S., Taniguchi, M., 2003. Groundwater and pore water inputs to the coastal zone. Biogeochemistry 66 (1–2), 3–33.

Burnett, W.C., Dulaiova, H., 2003. Estimating the dynamics of groundwater input into the coastal zone via continuous radon-222 measurements. J. Environ. Radioact. 69 (1), 21–35.

Burnett, W.C., Peterson, R., Moore, W.S., de Oliveira, J., 2008. Radon and radium isotopes as tracers of submarine groundwater discharge—results from the Ubatuba, Brazil SGD assessment intercomparison. Estuar. Coast. Shelf Sci. 76 (3), 501–511.

Burnett, W.C., Peterson, R.N., Santos, I.R., Hicks, R.W., 2010. Use of automated radon measurements for rapid assessment of groundwater flow into Florida streams. J. Hydrol. 380 (3), 298–304.

Call, M., Maher, D.T., Santos, I.R., Ruiz-Halpern, S., Mangion, P., Sanders, C.J., Eyre, B.D., 2015. Spatial and temporal variability of carbon dioxide and methane fluxes over semi-diurnal and spring–neap–spring timescales in a mangrove creek. Geochem. Cosmochim. Acta 150, 211–225.

Chen, X., Zhang, F., Lao, Y., Wang, X., Du, J., Santos, I.R., 2018. Submarine groundwater discharge-derived carbon fluxes in mangroves: an important component of blue carbon budgets? J. Geophys. Res.: Oceans 123 (9), 6962–6979.

Cho, H.M., Kim, G., Kwon, E.Y., Moosdorf, N., Garcia-Orellana, J., Santos, I.R., 2018. Radium tracing nutrient inputs through submarine groundwater discharge in the global ocean. Sci. Rep. 8 (1), 1–7.

Debenay, J.P., Marchand, C., Molnar, N., Aschenbroich, A., Meziane, T., 2015. Foraminiferal assemblages as bioindicators to assess potential pollution in mangroves used as a natural biofilter for shrimp farm effluents (New Caledonia). Mar. Pollut. Bull. 93 (1–2), 103–120.

Fabricius, K.E., 2005. Effects of terrestrial runoff on the ecology of corals and coral reefs: review and synthesis. Mar. Pollut. Bull. 50 (2), 125–146.

Fatimah, E., Wahab, A.K.A., Ismail, H., 2008. Numerical modeling approach of an artificial mangrove root system (ArMS) submerged breakwater as wetland habitat protector. In: Proceeding of the Seventh International Conference on Coastal and Port Engineering, PIANC-COPEDEC VII, vol. 88. United Arab Emirates, pp. 1–20.

Ferrara, C., Lega, M., Fusco, G., Bishop, P., Endreny, T., 2017. Characterization of terrestrial discharges into coastal waters with thermal imagery from a hierarchical monitoring program. Water 9 (7), 500.

Fromard, F., Vega, C., Proisy, C., 2004. Half a century of dynamic coastal change affecting mangrove shorelines of French Guiana. A case study based on remote sensing data analyses and field surveys. Mar. Geol. 208 (2–4), 265–280.

Garcia-Solsona, E., Garcia-Orellana, J., Masqué, P., Rodellas, V., Mejías, M., Ballesteros, B., Domínguez, J.A., 2010. Groundwater and nutrient discharge through karstic coastal springs (Castelló, Spain). Biogeosciences 7 (9), 2625–2638.

Gattacceca, J.C., Mayer, A., Cucco, A., Claude, C., Radakovitch, O., Vallet-Coulomb, C., Hamelin, B., 2011. Submarine groundwater discharge in a subsiding coastal lowland: a ^{226}Ra and ^{222}Rn investigation in the Southern Venice lagoon. Appl. Geochem. 26 (5), 907–920.

Gholizadeh, M., Melesse, A., Reddi, L., 2016. A comprehensive review on water quality parameters estimation using remote sensing techniques. Sensors 16 (8), 1298.

Geyer, W.R., Nepf, H., 1996. Tidal pumping of salt in a moderately stratified estuary. Coast. Estuar. Stud. 213–226.

Gilman, E., Van Lavieren, H., Ellison, J., Jungblut, V., Wilson, L., Areki, F., Henry, M., 2016. Pacific Island Mangroves in a Changing Climate and Rising Sea: UNEP, Secretariat of the Pacific Regional Environment Programme (SPREP) and.

Giri, C., Ochieng, E., Tieszen, L.L., Zhu, Z., Singh, A., Loveland, T., et al., 2011. Status and distribution of mangrove forests of the world using earth observation satellite data. Global Ecol. Biogeogr. 20 (1), 154–159. https://doi.org/10.1111/j.1466-8238.2010.00584.x.

Gleeson, J., Santos, I.R., Maher, D.T., Golsby-Smith, L., 2013. Groundwater–surface water exchange in a mangrove tidal creek: evidence from natural geochemical tracers and implications for nutrient budgets. Mar. Chem. 156, 27–37.

Hagedorn, B., Cartwright, I., 2010. The CO_2 system in rivers of the Australian Victorian Alps: CO_2 evasion in relation to system metabolism and rock weathering on multiannual time scales. Appl. Geochem. 25 (6), 881–899.

Handcock, R.N., Gillespie, A.R., Cherkauer, K.A., Kay, J.E., Burges, S.J., Kampf, S.K., 2006. Accuracy and uncertainty of thermal-infrared remote sensing of stream temperatures at multiple spatial scales. Rem. Sens. Environ. 100 (4), 427–440.

Handcock, R.N., Torgersen, C.E., Cherkauer, K.A., Gillespie, A.R., Tockner, K., Faux, R.N., Carbonneau, P.E., 2012. Thermal infrared remote sensing of water temperature in riverine landscapes. Fluvial Rem. Sens. Sci. Manag. 12, 85–113.

Hayes, M.A., Jesse, A., Welti, N., Tabet, B., Lockington, D., Lovelock, C.E., 2019. Groundwater enhances above-ground growth in mangroves. J. Ecol. 107 (3), 1120–1128.

Hogarth, P.J., 2015. The Biology of Mangroves and Seagrasses. Oxford University Press.

Holloway, C.J., Santos, I.R., Tait, D.R., Sanders, C.J., Rose, A.L., Schnetger, B., Maher, D.T., 2016. Manganese and iron release from mangrove porewaters: a significant component of oceanic budgets? Mar. Chem. 184, 43–52.

Hu, L., Li, W., Xu, B., 2018. The role of remote sensing on studying mangrove forest extent exchange. Int. J. Rem. Sens. 39 (19), 6440–6462.

Islam, M.S., Hossain, M.B., Matin, A., Sarker, M.S.I., 2018. Assessment of heavy metal pollution, distribution and source apportionment in the sediment from Feni River estuary, Bangladesh. Chemosphere 202, 25–32.

Jeffrey, L.C., Santos, I.R., Tait, D.R., Makings, U., Maher, D.T., 2018. Seasonal drivers of carbon dioxide dynamics in a hydrologically modified subtropical tidal river and estuary (Caboolture river, Australia). J. Geophys. Res: Biogeosciences. 123 (6), 1827–1849.

Johnson, A.G., Glenn, C.R., Burnett, W.C., Peterson, R.N., Lucey, P.G., 2008. Aerial infrared imaging reveals large nutrient-rich groundwater inputs to the ocean. Geophys. Res. Lett. 35 (15).

Kalbus, E., Reinstorf, F., Schirmer, M., 2006. Measuring Methods for Groundwater? Surface Water Interactions: A Review.

Kelly, J.L., Dulai, H., Glenn, C.R., Lucey, P.G., 2019. Integration of aerial infrared thermography and in situ radon-222 to investigate submarine groundwater discharge to Pearl Harbor, Hawaii, USA. Limnol. Oceanogr. 64 (1), 238–257.

Kelly, J.L., Glenn, C.R., Lucey, P.G., 2013. High-resolution aerial infrared mapping of groundwater discharge to the coastal ocean. Limnol Oceanogr. Methods 11 (5), 262–277.

Kessler, T.J., Harvey, C.F., 2001. The global flux of carbon dioxide into groundwater. Geophys. Res. Lett. 28 (2), 279–282.

Komiyama, A., Poungparn, S., Umnouysin, S., Rodtassana, C., Pravinvongvuthi, T., Noda, T., Kato, S., 2019. Occurrence of seasonal water replacement in mangrove soil and the trunk growth response of *Avicennia alba* related to salinity changes in a tropical monsoon climate. Ecol. Res. 34 (3), 428–439 (chen).

Kuenzer, C., Bluemel, A., Gebhardt, S., Quoc, T.V., Dech, S., 2011. Remote sensing of mangrove ecosystems: a review. Rem. Sens. 3 (5), 878–928.

Kumar, D.N., Reshmidevi, T.V., 2013. Remote sensing applications in water resources. J. Indian Inst. Sci. 93 (2), 163–188.

Li, X., Hu, B.X., Burnett, W.C., Santos, I.R., Chanton, J.P., 2009. Submarine ground 596 water discharge driven by tidal pumping in a heterogeneous aquifer. Ground Water 47, 558- 597 568.

Li, L., Wang, X.H., Williams, D., Sidhu, H., Song, D., 2012. Numerical study of the effects of mangrove areas and tidal flats on tides: a case study of Darwin Harbour, Australia. J. Geophys. Res.: Oceans 117 (C6).

Luijendijk, E., Gleeson, T., Moosdorf, N., 2020. Fresh groundwater discharge insignificant for the world's oceans but important for coastal ecosystems. Nat. Commun. 11 (1), 1–12.

Macklin, P.A., Suryaputra, I.G.N.A., Maher, D.T., Murdiyarso, D., Santos, I.R., 2019. Drivers of CO_2 along a mangrove-seagrass transect in a tropical bay: delayed groundwater seepage and seagrass uptake. Continent. Shelf Res. 172, 57–67.

Martin, J.B., Dean, R.W., 1999. Temperature as a natural tracer of short residence times for groundwater in karst aquifers. In: Karst Modeling, vol. 5. Karst Waters Institute Special Publication, pp. 236–242.

Mazda, Y., Wolanski, E., 2009. Hydrodynamics and modeling of water flow in mangrove areas. In: Coastal Wetlands. Elsevier, Amsterdam, pp. 69–88.

McCoy, C., Corbett, D.R., 2009. Review of submarine groundwater discharge (SGD) in coastal zones of the Southeast and Gulf Coast regions of the United States with management implications. J. Environ. Manag. 90 (1), 644–651.

Mejías, M., Ballesteros, B.J., Antón-Pacheco, C., Domínguez, J.A., Garcia-Orellana, J., Garcia-Solsona, E., Masqué, P., 2012. Methodological study of submarine groundwater discharge from a karstic aquifer in the Western Mediterranean Sea. J. Hydrol. 464, 27–40.

Moore, W.S., 2010. A reevaluation of submarine groundwater discharge along the southeastern coast of North America. Global Biogeochem. Cycles 24 (4).

Mori, C., Santos, I.R., Brumsack, H.J., Schnetger, B., Dittmar, T., Seidel, M., 2019. Non-conservative behavior of dissolved organic matter and trace metals (Mn, Fe, Ba) driven by porewater exchange in a subtropical mangrove-estuary. Front. Mar. Sci. 6, 481.

Mulligan, A.E., Charette, M.A., 2009. Elements of Physical Oceanography: A Derivative of the Encyclopedia of Ocean Sciences. In: *Groundwater flow to the coastal ocean.* p. 465.

Okubo, A., 1973. Effect of shoreline irregularities on streamwise dispersion in estuaries and other embayments. Neth. J. Sea Res. 6 (1–2), 213–224.

Paytan, A., Shellenbarger, G.G., Street, J.H., Gonneea, M.E., Davis, K., Young, M.B., Moore, W.S., 2006. Submarine groundwater discharge: an important source of new inorganic nitrogen to coral reef ecosystems. Limnol. Oceanogr. 51 (1), 343–348.

Primavera, J.H., 2005. Mangroves and Aquaculture in Southeast Asia. Aquaculture Department, Southeast Asian Fisheries Development Center.

Queißer, M., Burton, M., Kazahaya, R., 2018. Insights into geological processes with CO_2 remote sensing–A review of technology and applications. Earth Sci. Rev.

Rau, G.C., Andersen, M.S., McCallum, A.M., Acworth, R.I., 2010. Analytical methods that use natural heat as a tracer to quantify surface water–groundwater exchange, evaluated using field temperature records. Hydrogeol. J. 18 (5), 1093–1110.

Rashed-Un-Nabi, M., Al-Mamun, M.A., Ullah, M.H., Mustafa, M.G., 2011. Temporal and spatial distribution of fish and shrimp assemblage in the Bakkhali river estuary of Bangladesh in relation to some water quality parameters. Mar. Biol. Res. 7 (5), 436–452.

Raymond, P.A., Bauer, J.E., Cole, J.J., 2000. Atmospheric CO_2 evasion, dissolved inorganic carbon production, and net heterotrophy in the York River estuary. Limnol. Oceanogr. 45 (8), 1707–1717.

Rey, N., Rosa, E., Cloutier, V., Lefebvre, R., 2018. Using water stable isotopes for tracing surface and groundwater flow systems in the Barlow-Ojibway Clay Belt, Quebec, Canada. Can. Water Resour. J./Revue canadienne des ressources hydriques 43 (2), 173–194.

Rizal, A., Sahidin, A., Herawati, H., 2018. Economic value estimation of mangrove ecosystems in Indonesia. Biodiversity Int. J. 2 (3), 00051.

Rodríguez-Rodríguez, M., Fernández-Ayuso, A., Hayashi, M., Moral-Martos, F., 2018. Using water temperature, electrical conductivity, and pH to Characterize surface–groundwater relations in a shallow ponds system (Doñana National Park, SW Spain). Water 10 (10), 1406.

Rosentreter, J.A., Maher, D.T., Erler, D.V., Murray, R., Eyre, B.D., 2018. Seasonal and temporal CO_2 dynamics in three tropical mangrove creeks – a revision of global mangrove CO_2 emissions. Geochem. Cosmochim. Acta 222, 729–745. https://doi.org/10.1016/j.gca.2017.11.026.

Sadat-Noori, M., Glamore, W., 2019. Porewater exchange drives trace metal, dissolved organic carbon and total dissolved nitrogen export from a temperate mangrove wetland. J. Environ. Manag. 248, 109264.

Santos, I.R., Burnett, W.C., Dittmar, T., Suryaputra, I.G., Chanton, J., 2009. Tidal pumping drives nutrient and dissolved organic matter dynamics in a Gulf of Mexico subterranean estuary. Geochem. Cosmochim. Acta 73 (5), 1325–1339.

Santos, I.R., Maher, D.T., Eyre, B.D., 2012. Coupling automated radon and carbon dioxide measurements in coastal waters. Environ. Sci. Technol. 46 (14), 7685–7691.

Sappa, G., Barbieri, M., Ergul, S., Ferranti, F., 2012. Hydrogeological conceptual model of groundwater from carbonate aquifers using environmental isotopes (18O, 2H) and chemical tracers: a case study in Southern Latium Region, Central Italy. J. Water Resour. Protect. 4 (09), 695.

Schijf, J.B., Schönfled, J.C., 1953. Theoretical Considerations on the Motion of Salt and Fresh Water. IAHR.

Sidik, F., Lovelock, C.E., 2013. CO_2 efflux from shrimp ponds in Indonesia. PloS One 8 (6), e66329.

Smith, R., 1977. Long-term dispersion of contaminants in small estuaries. J. Fluid Mech. 82 (1), 129–146.

Stieglitz, T., 2005. Submarine groundwater discharge into the near-shore zone of the Great Barrier Reef, Australia. Mar. Pollut. Bull. 51 (1–4), 51–59.

Stommel, H.M., Farmer, H.G., 1952. On the Nature of Estuarine Circulation, vol. 2. Woods Hole Oceanographic Institution.

Stoneman, D.A., Constantz, J. (Eds.), 2003. Heat as a Tool for Studying the Movement of Ground Water Near Streams, vol. 1260. US Department of the Interior, US Geological Survey.

Sulman, B.N., Desai, A.R., Schroeder, N.M., Ricciuto, D., Barr, A., Richardson, A.D., Chen, G., 2012. Impact of hydrological variations on modeling of peatland CO_2 fluxes: results from the North American carbon program site synthesis. J. Geophys. Res.: Biogeosci. 117 (G1).

Tamborski, J.J., Rogers, A.D., Bokuniewicz, H.J., Cochran, J.K., Young, C.R., 2015. Identification and quantification of diffuse fresh submarine groundwater discharge via airborne thermal infrared remote sensing. Rem. Sens. Environ. 171, 202–217.

Tait, D.R., Maher, D.T., Macklin, P.A., Santos, I.R., 2016. Mangrove pore water exchange across a latitudinal gradient. Geophys. Res. Lett. 43 (7), 3334–3341.

Taillardat, P., Willemsen, P., Marchand, C., Friess, D.A., Widory, D., Baudron, P., Ziegler, A.D., 2018. Assessing the contribution of porewater discharge in carbon export and CO_2 evasion in a mangrove tidal creek (Can Gio, Vietnam). J. Hydrol. 563, 303–318.

Taniguchi, M., Dulai, H., Burnett, K.M., Santos, I.R., Sugimoto, R., Stieglitz, T., Burnett, W., 2019. Submarine groundwater discharge: updates on its measurement techniques, geophysical drivers, magnitudes and effects. Front. Environ. Sci. 7, 141.

Tristram, D.A., Krause, S., Levy, A., Robinson, Z.P., Waller, R.I., Weatherill, J.J., 2015. Identifying spatial and temporal dynamics of proglacial groundwater–surface-water exchange using combined temperature-tracing methods. Freshw. Sci. 34 (1), 99–110.

Uncles, R.J., Elliott, R.C.A., Weston, S.A., 1985. Dispersion of salt and suspended sediment in a partly mixed estuary. Estuaries 8 (3), 256–269.

Vadrevu, K.P., Lasko, K., Giglio, L., Schroeder, W., Biswas, S., Justice, C., 2019. Trends in vegetation fires in south and southeast Asian countries. Sci. Rep. 9 (1), 1–13.

Wang, D.P., Elliott, A.J., 1978. Non-tidal variability in the Chesapeake Bay and Potomac river: evidence for non-local forcing. J. Phys. Oceanogr. 8 (2), 225–232.

Wilson, J., Rocha, C., 2012. Regional scale assessment of submarine groundwater discharge in Ireland combining medium resolution satellite imagery and geochemical tracing techniques. Rem. Sens. Environ. 119, 21–34.

Wilson, J., Rocha, C., 2014. A remote sensing and geochemical tracing approach for localising groundwater discharge to lakes. In: Paper Presented at the Remote Sensing for Agriculture, Ecosystems, and Hydrology XVI.

Yanagisawa, H., Koshimura, S., Miyagi, T., Imamura, F., 2010. Tsunami damage reduction performance of a mangrove forest in Banda Aceh, Indonesia inferred from field data and a numerical model. J. Geophys. Res.: Oceans 115 (C6).

Zimmerman, J.T.F., 1986. The tidal whirlpool: a review of horizontal dispersion by tidal and residual currents. Neth. J. Sea Res. 20 (2–3), 133–154.

CHAPTER

4

Flow and sediment dynamics around structures in mangrove ecosystems—a modeling perspective

Marine Le Minor[1,2], *Martin Zimmer*[3,4,5], *Véronique Helfer*[3], *Lucy Gwen Gillis*[3], *Katrin Huhn*[1]

[1]MARUM - Center for Marine Environmental Sciences, University of Bremen, Bremen, Germany; [2]Géosciences Rennes, Université de Rennes 1, CNRS, UMR 6118, Campus de Beaulieu, Rennes, France; [3]ZMT - Leibniz Centre for Tropical Marine Research, Bremen, Germany; [4]University of Bremen, Faculty of Biology/Chemistry, Bremen, Germany; [5]IUCN-SSC Mangrove Specialist Group

OUTLINE

1. Introduction 84

2. General hydro- and sediment dynamics in mangrove ecosystems 86

3. State-of-the-art numerical modeling on hydro-/sediment dynamics in mangrove ecosystems 89

3.1 Modeled sediment dynamics at the forest scale 89

3.2 Sediment dynamics at the tree scale 94

3.3 Sediment dynamics around idealized cylinders as representation of mangrove structures 98

3.3.1 Single cylinder 98

3.3.2 Arrays of cylinders 99

3.3.3 Sediment properties 101

3.4 Sediment dynamics at the root or seedling scale 102

3.5 Sediment dynamics at the microtopography scale 103

3.5.1 Surface texture 104

3.5.2 Biofouling 105

4. Disparate treatment of spatial scales 107

5. Conclusions and outlook 108

Acknowledgments 110

References 110

Dynamic Sedimentary Environments of Mangrove Coasts
https://doi.org/10.1016/B978-0-12-816437-2.00012-4

1. Introduction

Among the highly diverse coastal vegetated ecosystems, mangroves are one of the most prominent and are globally distributed in tropical and subtropical areas, such as parts of Australia, New Zealand, and South Africa. Their ability to cope with, e.g., high salinity conditions, oscillating current conditions as well as strongly varying sediment supply, allow for their establishment and growth in various types of mostly sheltered intertidal areas at the interface between land and sea, e.g., estuaries, lagoons, and low islands (Ellison, 2019). Different mangrove species can often be found at characteristic tidal elevations within the stretch of a mangrove forest (Cunha-Lignon et al., 2009; Leong et al., 2018), partly driven by inundation regimes and species-specific capacity to handle salinity (Fig. 4.1).

Mangroves are subject to a wide range of hydrodynamic forces such as river flows, waves, and tidal currents depending on their geomorphological location. These different hydrodynamic forces are the main drivers of sediment supply and transport and, thus, play a key role in sedimentation patterns in mangrove ecosystems. The elevation of the forest floor with respect to the tidal water level determines the frequency and depth at which mangroves are inundated (Van Santen et al., 2007). In general, it can be noted that mangroves trap fine sediments by significantly reducing water flow with their above-ground roots, trunks, or even the canopy (branches and leaves). This trapping helps to create mud banks where seedlings can establish (Furukawa and Wolanski, 1996). Furthermore, tidal pumping (i.e., sediment enters with incoming tide and is not transported away at ebb tide due to weak currents that result from mangrove drag) contributes to sustainable sedimentation that is of great importance for shoreline stabilization (Hashim and Catherine, 2013). This continuous sedimentation may help minerogenic mangrove ecosystems to expand (Swales et al., 2002; Hayward et al., 2006; Lovelock et al., 2007; Fricke et al., 2017) and withstand sea-level rise (Ellison, 1993; Huxham et al., 2010; Lovelock et al., 2015; Sasmito et al., 2016). In contrast, extremely large sediment supplies through rivers to mangrove ecosystems may result in

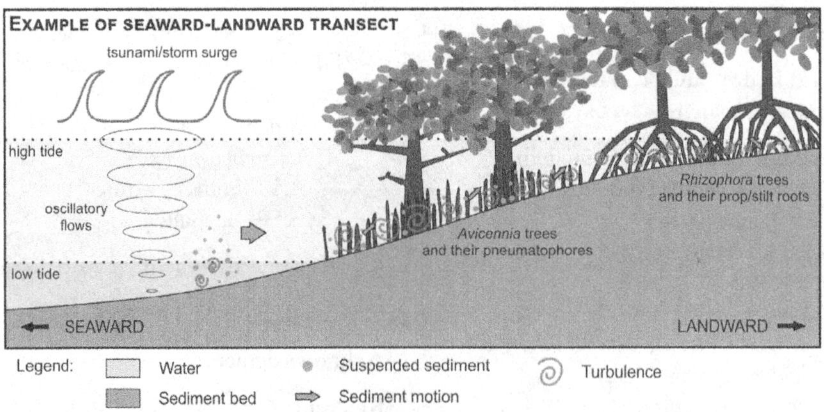

FIGURE 4.1 Example of seaward–landward transect in Southeast Asia. Mangroves on the shoreline are subject to oscillatory flows, tides, or extreme events such as storm surge or tsunami, which control the sediment dynamics. We present one example of a transect with a distribution of mangrove species that in some cases can be reverse.

mangrove burial; for instance, at the mouth of the Porong river in Indonesia (Sidik et al., 2016), where sediment supply rapidly increased due to disturbance from a mud volcano.

A less visible contribution of mangrove ecosystems to shoreline stabilization is the key mechanical role played by below-ground root systems; for instance, rootlets contribute to a greater sediment stability by promoting sediment binding (Spenceley, 1977). This sediment binding capacity is greater for *Rhizophora stylosa* (rootlets occupied half the volume of sediment samples) than *Avicennia* or *Sonneratia* spp. (5% per volume; Spenceley, 1977). The sediment may, therefore, be stabilized by this characteristic, which benefits mangrove trees and other organisms inhabiting mangrove ecosystems, as well as adjacent ecosystems, such as seagrass beds and coral reefs (Guannel et al., 2016).

Many studies have already investigated the interplay between mangroves, hydrodynamics, and sediment dynamics (for more details see Section 2) that, together with the underlying processes, occur on a wide range of spatial scales (Fig. 4.2). However, conducting studies of hydro- and sediment dynamics in mangroves is challenging in most cases due to a wide range of limitations. Firstly, field and laboratory experiments are restricted to a finite number of locations due to logistics (instruments available) and sampling or experimental design (replicated measurements are needed to allow sufficient statistical power). In addition, field measurements are critically affected by the location of the device as well as its performance, especially at small scales. The deployment of instruments inside mangrove forests is difficult due to the presence of mangrove trunks, branches, and roots as well as the soft substrate, typically mud, which is easily disturbed during a field campaign. Interferences with vegetation that lead to bad current measurements may also occur as reported by Horstman et al. (2011). Short periods of

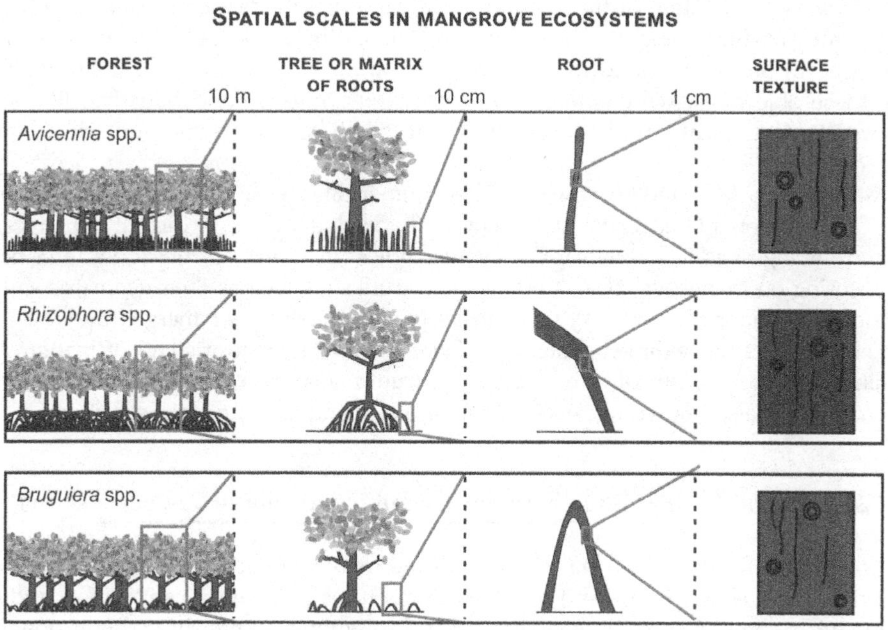

FIGURE 4.2 Spatial scales observed in mangrove ecosystems, using the example of *Avicennia* spp., *Rhizophora* spp., and *Bruguiera* spp.

inundation (tidal submergence) also reduce the measurement opportunities (Zhang et al., 2015a). In the laboratory (e.g., flume experiments), similar issues may be faced, in conjunction with additional problems such as hydrodynamic scaling (Struve et al., 2003; Ismail et al., 2012; Maza et al., 2017; Shan et al., 2019). Finally, in mangrove ecosystems, flow and sediment dynamics exhibit a profusion of spatial and temporal variations that require long-term multiinstrument deployments to be fully covered; however, the costs and efforts necessary to gather an extensive dataset of these variations are inhibitive. In natural conditions, many parameters that cannot be controlled impact the measurements, e.g., wind causing wavelets, and, hence, conducting field experiments implies that a large number of factors might be influencing the data collection. The data interpretation is, therefore, complex, since quantifying the individual contribution of each of these factors to the results is difficult. Furthermore, although a device records high-resolution flow profiles and sediment concentrations in the water column, it is impossible to quantify the flow at every horizontal and vertical point or specific sediment volume over time in a flume or in nature, and thus, details are lacking. As a consequence of all these limitations, process-based numerical simulations are required as complementary tool to field and flume measurements. In turn, experimental data are then needed to assess model performance and quality under conditions mimicking those observed in real mangrove ecosystems. Thus, there is a demand for process-based investigations that could unravel the mechanisms controlling the mangrove system dynamics; to do so, diverse types of studies (field, laboratory, and increasingly numerical modeling) should preferably be used in combination.

This chapter aims to i) introduce the basic knowledge of hydrodynamics and subsequent sediment transport in mangrove ecosystems that is needed to evaluate modeling approaches; ii) give an overview of state-of-the-art numerical experiments dealing with hydro- and sediment dynamics at different spatial scales; and iii) to identify the knowledge gaps that need to be addressed in order to extend the accuracy and application of numerical simulations. To achieve these objectives, we conducted a comprehensive, but nonexhaustive, literature review covering forest-scale models (Section 3.1) to small seedling-scale numerical investigations (Section 3.3).

We used *Google Scholar* and keywords such as "numerical modeling," "numerical simulations," "CFD" (Computational Fluid Dynamics), "flow," "hydrodynamics," "sediment dynamics/transport," and "mangroves" to search for relevant studies. In doing so, we found 163 pertinent papers focusing on mangrove roots, seedlings and trunks (as well as their geometric simplification, a simple cylinder (Section 3.4), that can be partially or fully submerged by water and is, thus, likely to affect the sediment transport). Hereafter, we refer to a stand of mangrove trees or seedlings, the root system of a tree (including trunk), a single root (prop root, knee root, or pneumatophores—also known as pencil roots), or a seedling, as mangrove structures.

2. General hydro- and sediment dynamics in mangrove ecosystems

Hydrodynamic flows are the main drivers of sediment transport in mangrove ecosystems. The interplay between the flow and mangrove structures may significantly modify the sediment dynamics compared to unvegetated intertidal regions, as the three-dimensional structure of mangrove trees alters the hydrodynamics (e.g., drag). Mangrove structures obstruct the flow when partially or fully submerged, which produces flow disturbances resulting in energy absorbance and turbulence generation (Kazemi et al., 2017, 2018a). The magnitude of these two phenomena depends not only on the conditions of currents, tides, and waves inside a mangrove forest but also on the root complexity of different mangrove species and the density of the mangrove forest (Young and Harvey, 1996; Mazda et al., 1997a; Massel et al., 1999; Alongi, 2008; Maza et al., 2015; Mullarney et al., 2017a). The size of mangrove trees is also a factor controlling the magnitude of this drag force; for instance, older mangrove trees are generally bigger (Salmo et al., 2013), which results in a larger drag and a more significant wave dissipation.

The vegetation friction due to mangrove structures is known to cause considerable flow reduction (e.g., Struve et al., 2003; Teh et al., 2009; Horstman et al., 2015). Kathiresan (2003) recorded tidal currents of half the speed in mangrove forests ($9\ cm \cdot s^{-1}$) compared to a nonvegetated bed (up to $20\ cm \cdot s^{-1}$) during a field experiment in the Vellar Estuary in India (Pichavaram mangrove forest). Scoffin (1970) also reported a sharp flow decrease due to *Rhizophora* sp. prop roots, from 40 to $0\ cm \cdot s^{-1}$ over a distance of 1 m. A difference in flow speed of two orders of magnitude was observed between currents in a creek ($1-2\ m \cdot s^{-1}$) and in a dense mangrove forest ($7\ cm \cdot s^{-1}$ at 50 m away from the creek) in Australia (Wolanski et al., 1990; Wolanski, 1992). Waves are also altered due to the presence of mangrove structures that create friction; under normal conditions, the height of swell and wind waves decay considerably between unvegetated areas and the interior of a mangrove forest. For instance, the wave height reduction per meter cross-shore was eight times higher in a mangrove forest than on the fringing bare mudflat when compared to a seaward location (Quartel et al., 2007). Many studies have investigated the impact of mangrove forests on wave height and energy reduction with regards to extreme events that may strike mangrove forests, e.g., storm surges (Montgomery et al., 2018) or tsunamis (Bao, 2011; Husrin et al., 2012; Hashim and Catherine, 2013; Strusińska-Correia et al., 2013). For example, at a distance of 80 m from the mangrove fringe, a tsunami wave height was reduced by 80% by *Rhizophora* trees (Hashim and Catherine, 2013). These wave height reduction and energy dissipation processes are crucial for coastal protection during storm or floods.

Owing to the influence of mangrove structures on the hydrodynamics, a gradient of sediment-trapping and sedimentation occurs along the seaward–landward transect and, consequently, the suspended load that reaches the hinterland is reduced (Kathiresan, 2003; Adame et al., 2010).

Among others, Van Santen et al. (2007) reported sediment accumulation rates reaching almost $6\ g \cdot cm^{-2} \cdot yr^{-1}$ in the front zone of a mangrove forest, while values of less than $1\ g \cdot cm^{-2} \cdot yr^{-1}$ were found in the landward zone. Furthermore, sediment deposition is a selective process (Furukawa et al., 1997; Wolanski et al., 1998). The fine sediment content ($<20\ \mu m$) increases with distance across the mangrove fringe (Yang et al., 2013), since the sediment transported as suspended load mainly consists of fine sediment grains. In contrast, coarse grains are usually found in larger amounts closer to the sea than inside the forest, since the reduced flow speeds are not sufficient to transport such particles further into the mangrove (Yang et al., 2013).

Sediment is transported into mangrove ecosystems through a wide range of processes. While coastal currents are responsible for cross-shore and longshore sediment transport, rivers move sediment from upstream catchments toward the sea, and winds erode beaches and dunes and carry dust into coastal waters. These processes differ not only in terms of spatial but also temporal properties (continuous, periodic – season or tide – or one-time events). Many studies reported that river settings play a key role in sediment availability to coastal zones, in particular to mangrove ecosystems (e.g., Holeman, 1968; Wolanski, 1995). The topography and the soil properties of river catchments affect soil erosion and control the sediment yield which, in turn, regulates sedimentation (Dutta, 2016; Horstman et al., 2017).

Anthropogenic impacts are another factor influencing the sediment supply to mangrove ecosystems. It is well established that human activities such as catchment land-use for agriculture, deforestation, and urbanization increase the sediment runoff to coastal areas (Swales et al., 2002; Hayward et al., 2006; Lovelock et al., 2007; Nayak and Noronha e D'Mello, 2018). The suspended sediment concentration in such zones also rises after disposal of dredged material in the nearshore or owing to waves or ship traffic that resuspends the sediment (Mellor et al., 2018). Conversely, the sediment supply may be reduced due to damming and extraction activities upstream of estuaries (Syvitski et al., 2005), potentially interacting with sediment erosion at the coast (Kondolf, 1997). Furthermore, extreme events such as storms (Cahoon, 2006) and floods (Wolanski, 1994) also contribute greatly to sediment availability (Thrush et al., 2004). In addition, Jennerjahn et al. (2013) observed that the total suspended matter in coastal waters, after the eruption of the mud volcano Lumpur Sidoarjo in Indonesia, was more than twice as much as before. In brief, the sediment availability to mangrove areas is controlled by heterogeneous sediment sources that act over a wide range of time scales.

In general, the suspended sediment concentration within mangrove forests ranges from $10\ mg \cdot L^{-1}$ to $300\ mg \cdot L^{-1}$ (Kathiresan, 2003; Asp et al., 2016). The particle size distribution is mainly driven by salinity, a key factor that controls flocculation, i.e., the formation of aggregates of sediment particles (commonly silt and clay in mangrove ecosystems) and fine particulate organic matter due to electrochemical forces that are enabled by salinity (ions) in brackish waters. Flocs are groups of particles which settle at a higher rate than single sediment particles. The presence of sediment flocs leads to increased near-bed sediment concentrations; Wolanski (1995) reported a suspended sediment concentration of $400\ mg \cdot L^{-1}$ at a mangrove-fringed tidal channel. The settling velocity of the sediment increases with floc size and hence with sediment concentration, reaching a maximum for a suspended sediment concentration of $10\ g \cdot L^{-1}$, then it decreases with rising suspended sediment concentration due to hindered settling (Wolanski, 1995). Mangrove ecosystems are in the range where the settling velocity of flocs increases with sediment load. The organic matter in suspension also plays a role in the flocculation process since small organic particles can attract fine sediment grains and form flocs. Flocs organized around fine sediment grains and sometimes organic particles break apart when they encounter obstacles such as mangrove roots. Furukawa and Wolanski (1996) observed that the interplay between the flow and *Rhizophora* spp. prop root system (Fig. 4.1) generated high microturbulence that prevented the separated sediment particles from settling during flood tide. The main settling inside the forest occurs shortly before high slack tide when the flow speeds are very slow (Orr et al., 2005), and the sediment particles remain there, as ebb currents are too weak to resuspend and/or transport them seaward due to high vegetation friction. Li et al. (1993) observed that the suspended sediment consisted of flocs of clay particles across both high and low tides. Furukawa and Wolanski (1996) observed that the sedimentation rate correlated with flow reduction within a mangrove forest, decreasing with distance from the creek due a reduction of flow velocities and indirectly due to the flocs that settle faster than individual clay and silt particles. Such a spatial sedimentation pattern was reported in many studies (Bird, 1986; Lynch et al., 1989; Cahoon and Lynch, 1997; Smoak and Patchineelam, 1999).

Substantial sediment deposition and erosion events were observed in the energetic zone due to waves by Van Santen et al. (2007). Wave-driven turbulence that reaches the bed keeps the mud layer composed of sediment particles and flocs from consolidating by compaction; conversely, this mud layer can attenuate wave turbulence. The action of waves can also lead to the fluidization of settled mud layers that were formerly consolidated. Deeper in the substrate layer, the deposits get more compacted and, consequently, less likely to be eroded and resuspended.

Many studies have highlighted the sensitivity of mangrove ecosystems to alterations in sediment dynamics (Terrados et al., 1997; Ellison, 1999; Thampanya et al., 2002; Ruiz-Fernández et al., 2018). Mangroves can respond to environmental fluctuations such as change in sediment regime by either growing or dying rapidly (Jiménez et al., 1985). Extreme rates of sediment supply have been demonstrated to affect mangrove establishment, growth, and survival (Jiménez et al., 1985; Terrados et al., 1997; Ellison, 1999; Thampanya et al., 2002; Ellis et al., 2004). Besides these alterations of sediment input, other gradients of stresses related to resources such as oxygen and nutrients can affect mangroves (Horstman et al., 2018a). For example, Yang et al. (2013) observed that in the less-energetic interior zone, where the sediment bed is poorer in oxygen because of finer grain size, mangrove trees had a shorter stature, slimmer stems, and a sparser canopy.

Mangroves are subject to various hydrodynamic forces that cause sediment accretion or erosion and, consequently, alter mangrove ecology. In the following section, we review the numerical studies that have been conducted to unravel the hydro- and sediment dynamics at different spatial scales in mangrove ecosystems.

3. State-of-the-art numerical modeling on hydro-/sediment dynamics in mangrove ecosystems

In this section, we focus on the state of the art in numerical modeling studies simulating the hydro- and sediment dynamics in mangrove ecosystems, at various spatial scales. In doing so, we start with a brief overview of observations, results and open questions with the overarching goal to shed light on knowledge gaps in numerical process simulations. An overview of all the numerical modeling studies found in this chapter is given (Table 4.1).

3.1 Modeled sediment dynamics at the forest scale

Mangrove ecosystems have been mainly studied utilizing hydrodynamic models at large—mangrove forest wide—scales (Fig. 4.2). The motivation for such studies is to gain a deeper insight into flow patterns and current velocities within mangrove forests to quantify their importance for coastline stabilization (wave reduction, sediment retention, etc.). A common approach at this large spatial scale is to consider the mangrove structures as additional (bottom) friction term due to their drag (Table 4.1). However, the major drawback of this

TABLE 4.1 Summary of the numerical experiments reviewed in this study. The scale of the models and the main objectives of the study are indicated. The studies in bold include simulations of sediment dynamics. The model dimensions are written in brackets (0, 1, 2, 3 D). The mangrove species considered for the simulations are listed, and the model representation of mangroves (natural vegetation, idealized as cylinders or additional bottom friction) is specified in brackets.

Scale	Objectives	Literature	Computational suite/model	Species
Mangrove forest	Sea-level rise (climate change)	Oliver et al. (2012)	–	*Avicennia marina* and *Aegiceras corniculatum* (–)
		Di Nitto et al. (2014)	–	*Avicennia marina, Bruguiera gymnorrhiza, Ceriops tagal, Heritiera littoralis, Lumnitzera racemosa, Rhizophora mucronata, Sonneratia alba, Xylocarpus granatum,* and *Pemphis acidula* (–)
		Li et al. (2015)	SLAMM	*Avicennia marina, Aegiceras corniculatum, Kandelia obovata, Rhizophora stylosa,* and *Bruguiera gymnorrhiza* (–)
		Van Maanen et al. (2015)	ELCOM (3D)	*Avicennia marina* (idealized as cylinders)
	Storm surge reduction	Massel et al. (1999)	–	Species not specified (idealized as cylinders)
		Zhang et al. (2012)	CEST (2D)	*Rhizophora mangle, Laguncularia racemosa,* and *Avicennia germinans* (idealized as friction)
		Liu et al. (2013)	FDM with CEST (2D)	Species not specified (idealized as friction)
		Guannel et al. (2015)	InVEST (2D)	*Avicennia germinans* (idealized as cylinders)
		Rahdarian and Niksokhan (2017)	MIKE21 (2D)	Species not specified (idealized as friction)
	Tsunami reduction	Teh et al. (2009)	TUNA-M2 (2D)	*Avicennia officinalis* and *Bruguiera cylindrical* (idealized as cylinders)
		Yanagisawa et al. (2009)	–	*Rhizophora* dominated (idealized as cylinders)
		Yanagisawa et al. (2010)	–	*Rhizophora* dominated (idealized as cylinders)
		Maza et al. (2015)	IHFOAM (3D)	Species not specified (idealized as cylinders)
	Tidal circulation (e.g., in estuary, creek, channel)	Wolanski et al. (1990)	2DSWAMP and HYDRO (2DH)	Species not specified (idealized as friction)
		Wolanski (1992)	–	Species not specified (idealized as friction)

TABLE 4.1 Summary of the numerical experiments reviewed in this study. The scale of the models and the main objectives of the study are indicated. The studies in bold include simulations of sediment dynamics. The model dimensions are written in brackets (0, 1, 2, 3 D). The mangrove species considered for the simulations are listed, and the model representation of mangroves (natural vegetation, idealized as cylinders or additional bottom friction) is specified in brackets.—cont'd

Scale	Objectives	Literature	Computational suite/model	Species
		Furukawa and Wolanski (1996)	(1D)	*Rhizophora* sp., *Avicennia* sp., and *Ceriops* sp. (not included in this simple model)
		Furukawa et al. (1997)	2DSWAMP and LAKE (2D)	*Rhizophora* sp., *Bruguiera gymnorhiza,* and *Ceriops tagal* dominated (idealized as friction)
		Wolanski et al. (1998)	HYDRO (2DH)	Species not specified (idealized as friction)
		Franks and Falconer (1999)	DIVAST (2D) and TRIVAST (3D)	Species not specified (idealized as friction)
		Wu et al. (2001)	(2D)	Species not specified (idealized as cylinders)
		Horstman et al. (2013)	Delft3D-FLOW (3D or 2DH)	Species not specified (idealized as cylinders)
		Li et al. (2012)	FVCOM (3D)	Species not specified (idealized as friction)
		Li et al. (2014)	FVCOM coupled with ESSed (3D)	Species not specified (idealized as friction)
		Horstman et al. (2015)	Delft3D-FLOW (3D and 2DH)	*Rhizophora, Bruguiera, Xylocarpus, Avicennia,* and *Acanthus* (idealized as cylinders)
		Kakeh et al. (2016)	(0D)	*Avicennia marina* (idealized as friction)
		Willemsen et al. (2016)	Delft3D-FLOW (2DH)	*Avicennia alba* and *Sonneratia alba* (idealized as cylinders)
		Bryan et al. (2017)	Delft3D (1D/2DH)	*Sonneratia* (idealized as cylinders)
		Marois and Mitsch (2017)	HEC-RAS (1D)	*Rhizophora* mangle, *Avicennia germinans,* and *Laguncularia racemose* (idealized as friction)
	Water surface elevation (due to roughness)	Struve et al. (2003)	DIVAST (2D)	*Rhizophora* (idealized as cylinders)
		Lee and Shih (2004)	TABS-2 (2D)	*Kandelia candel* (idealized as friction)
	Wave attenuation	Fatimah et al. (2008)	FLOW-3D (2D)	Species not specified but looks like *Rhizophora* (idealized as cylinders)

(Continued)

TABLE 4.1 Summary of the numerical experiments reviewed in this study. The scale of the models and the main objectives of the study are indicated. The studies in bold include simulations of sediment dynamics. The model dimensions are written in brackets (0, 1, 2, 3 D). The mangrove species considered for the simulations are listed, and the model representation of mangroves (natural vegetation, idealized as cylinders or additional bottom friction) is specified in brackets.—cont'd

Scale	Objectives	Literature	Computational suite/model	Species
		Doughty et al. (2017)	InVEST	*Rhizophora mangle*, *Avicennia germinans*, and *Laguncularia racemose* (idealized as cylinders)
Tree	Flow pattern	Furukawa and Wolanski (1996)	VORTEX (2D)	*Rhizophora* root matrix (idealized as cylinders)
		Furukawa et al. (1997)	VORTEX (2D)	*Rhizophora* root matrix (idealized as cylinders)
		Jusoh et al. (2016)	ANSYS Fluent (2D)	*Avicennia marina* pneumatophore matrix and *Rhizophora apiculata* stilt root matrix (idealized as cylinders)
		Kazemi et al. (2018b)	ANSYS Fluent (2D)	*Rhizophora* root matrix (idealized as cylinders)
Stem	Flow pattern	Le Minor et al. (2019)	OpenFOAM (3D)	*Rhizophora mucronata* seedling (natural)

method is that the modeled sediment dynamics are overestimated in comparison to reality, since the increase in bed roughness allows for greater near-bed velocities than when the vertical distribution of the additional drag is considered. More recent approaches consist in representing mangrove structures as rigid vertical cylinders; these cylinders have a known drag coefficient, which allows for the calculation of their contribution to the flow resistance. Numerical studies of hydro- and sediment dynamics in mangrove forests are usually combined with field measurements to calibrate the model and/or to assess the quality of the simulation, i.e., validation (Horstman et al., 2015; Willemsen et al., 2016).

Although sediment dynamics are directly linked to hydrodynamics, and sediment accretion and erosion patterns are essential for the health status of a mangrove forest, sediment dynamic processes are almost neglected in numerical three-dimensional simulations. At the scale of a mangrove forest, the sediment dynamics associated with the flow pattern are simulated mostly in two dimensions (i.e., they are depth averaged). However, since sediment dynamics are clearly related to flow dynamics, numerical models should also take into account three-dimensional coupled flow-sediment approaches. Only 15 out of the 32 reviewed large-scale numerical modeling studies (Table 4.1, Mangrove forests) simulated sediment transport.

Among those 15 large-scale studies, the majority simulated sediment dynamics under tidal circulation conditions (Furukawa and Wolanski, 1996; Wolanski et al., 1998; Franks and Falconer, 1999; Horstman et al., 2013, 2015; Li et al., 2014; Kakeh et al., 2016; Willemsen et al., 2016; Bryan et al., 2017; Doughty et al., 2017). The alteration of the tidal regime by mangroves led to changes in the sediment distribution, for instance, due to tidal pumping (Li et al., 2014). Most of the sediment transported into mangrove forests by the flooding tide remained permanently trapped at the fringe (Wolanski et al., 1998), which resulted in a higher sedimentation rate close to the tidal creek than in the mangrove interior (Furukawa and Wolanski, 1996). Bryan et al. (2017) demonstrated that, although mangroves are well known to reduce tidal flows, water depth within the forest was the main parameter enhancing the tidal flow dissipation. Surface elevation, and thus, water depth rather than mangrove density, controlled sedimentation rates in mangrove forests (Horstman et al., 2015). Sediment supply was also found to be a key parameter that influenced the magnitude of sediment trapping: higher sediment inputs resulted in higher trapping (Horstman et al., 2015; Willemsen et al., 2016). The removal of mangroves caused reduced trapping (Horstman et al., 2015); conversely, extending mangrove area increased sediment trapping capacity (Willemsen et al., 2016). Sediment trapping is of great importance for coastal protection and land building and protects neighboring ecosystems from high turbidity and sedimentation.

Numerical models have practical applications in coastal management: water managers and environmental engineers use numerical simulations to predict flows and sediment transport in mangrove ecosystems (Franks and Falconer, 1999). Modeled results depend significantly on model assumptions and input parameters (Horstman et al., 2013), which make numerical simulations of great use to test the response of mangrove ecosystems to extreme scenarios (e.g., sea-level rise, storm surge, tsunamis) or anthropogenic threats. Only a few studies were conducted to unravel the effect of mangrove forests on sediment dynamics associated with storm surge attenuation (Guannel et al., 2015), sea-level rise (Oliver et al., 2012; Di Nitto et al., 2014; Van Maanen et al., 2015), and water surface elevation (Lee and Shih, 2004). Van Maanen et al. (2015) showed that *Avicennia marina* mangroves affected tidal currents and landscape evolution by changing the bathymetry and the channel network under a uniform sediment input. Due to mangrove flow resistance, flow was concentrated between vegetated areas, and erosion was, thus, enhanced in the nonvegetated areas, leading to incision of tidal channels. In the future, sediment supply could play a significant role, since sediment accretion within forests could help mangroves to keep up with sea-level rise (Di Nitto et al., 2014; Van Maanen et al., 2015). Oliver et al. (2012) predicted changes in mangrove distribution in response to various sea-level rise scenarios associated with a fixed rate of surface-elevation change. Although modeled hydro- and sediment dynamics in mangroves are overall in good agreement with field observations, model calibration is a critical step to predict sediment dynamics in mangrove ecosystems such as sediment concentrations and deposition rates (Guannel et al., 2015).

A positive trend is that with increasing computer capacity, more comprehensive, process-based models, including vegetation dynamics, for longer periods of time and larger areas can be run. For instance, while different types of sediment have been incorporated, ranging from muddy (2 μm, Li et al., 2014, Fig. 4.3A) to sandy (120 μm; Van Maanen et al., 2015), simulating the dynamics of various types of sediment within a same model has been challenging, and thus, more complex models could help tackle this problem.

To date, the numerical approach at the forest scale is used to model sediment transport by estimating the suspended sediment concentration and sediment fluxes through the water column to derive local erosion and deposition rates. Such simulations consider a finite number of grain sizes (or type of sediment) and mainly noncohesive material, thereby considerably

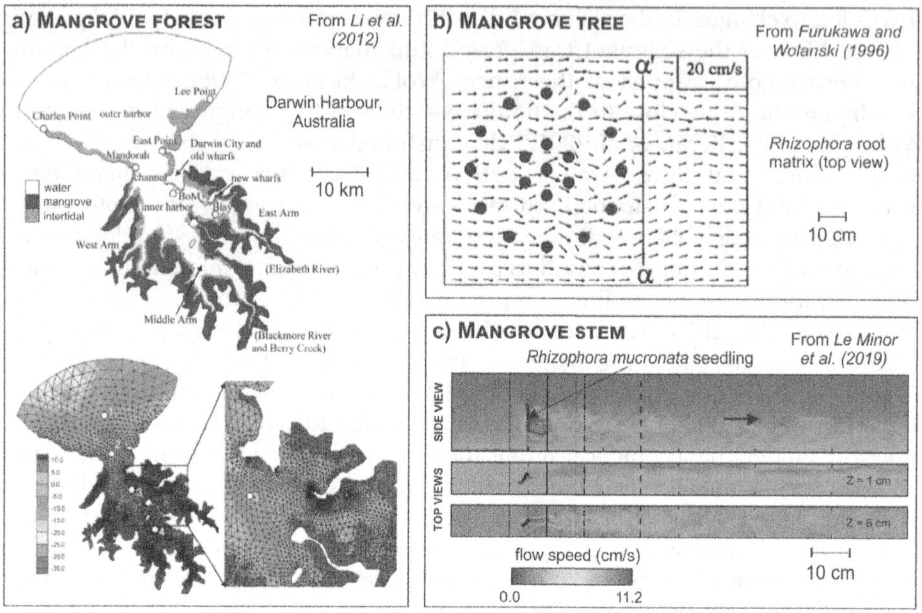

FIGURE 4.3 Use of numerical modeling to study mangrove hydrodynamics at a wide range of spatial scales. (A) Study at the scale of a mangrove forest: estuarine dynamics in Darwin Harbour, Australia. The color scale represents the depth contours of lowest astronomical tides (in meters). (B) Study at the scale of a mangrove tree: flow patterns through an idealized complex *Rhizophora* root matrix. (C) Study at the scale of a mangrove stem: turbulence around a single *Rhizophora mucronata* mangrove seedling. *Credits: (A) Reprinted from Li, L., Wang, X. H., Williams, D., Sidhu, H. & Song, D., 2012. Numerical study of the effects of mangrove areas and tidal flats on tides: a case study of Darwin Harbour, Australia. J. Geophys. Res., 117. http://www.doi.org/10.1029/2011JC007494 with permission from John Wiley and Sons. (B) Adapted by permission from Springer Nature Customer Service Center GmbH: Springer, Furukawa, K., Wolanski, E., 1996. Sedimentation in mangrove forests. Mangroves Salt Marshes, 1, 3–10. https://doi.org/10.1023/A:1025973426404. (C) Reprinted from Le Minor, M., Bartzke, G., Zimmer, M., Gillis, L., Helfer, V., Huhn, K., 2019. Numerical modelling of hydraulics and sediment dynamics around mangrove seedlings: implications for mangrove establishment and reforestation. Estuar. Coast Shelf Sci., 217, 81–95. https://doi.org/10.1016/j.ecss.2018.10.019, with permission from Elsevier.*

simplifying the natural conditions observed in mangrove ecosystems. We advocate that future models/studies incorporate more complex grain size distributions and physical sediment properties. The latter is especially important when investigating the effect of flocculation, which is one main driver of sediment deposition in mangrove ecosystems as outlined above.

3.2 Sediment dynamics at the tree scale

The main types of aerial mangrove roots are prop roots, e.g., *Rhizophora* type, pneumatophores, e.g., *Avicennia/Sonneratia* type, and knee roots, e.g., *Bruguiera* type (Tomlinson, 2016). Above-ground roots play a crucial role for mangrove health since they absorb oxygen, which is then transferred to the entire root system (including the subsurface roots in the waterlogged sediment) and tree, and they provide anchoring in the sediment substrate (Tomlinson, 2016).

Rhizophora trees exhibit very complex structures due to their dense and extensive prop root system, extending from the main trunk and branching off towards the ground. Mazda et al.

(1997a) counted up to 152 prop roots on *R. stylosa* trees, while up to 460 prop roots per *Rhizophora* tree were counted by Horstman et al. (2014). As described by Tomlinson (2016), *Avicennia* and *Sonneratia* mangrove trees consist of a main trunk from which below-ground cable roots extend and aerial roots, pneumatophores, grow upward from the cable roots. As for *Bruguiera* trees, they consist of a main trunk and knee roots, i.e., roots that form a loop above ground. This shows the heterogeneous vertical distributions of the root systems not only in mangrove trees across species but also within individuals; indeed, root systems (matrix of roots) occupy larger volumes close to the sediment bed than away from it (Fig. 4.4). Such vertical variations were reported not only for *Rhizophora* root systems (Mazda et al., 1997a; Zhang et al., 2015a) but also root systems of the genera *Avicennia* (Kamal et al., 2017), *Sonneratia* (Liénard et al., 2016; Norris et al., 2019) and *Bruguiera* (Mazda et al., 2005). Therefore, the general pattern observed is a decrease of the area blocking the water flow, and thereby an increase in porosity, with root elevation and water depth (Horstman et al., 2014). However, the different root types and root structures influence the hydrodynamics differently and, consequently, lead to different sedimentation rates and patterns. Krauss et al. (2003) found that the annual rate of vertical sediment accretion was higher within prop roots of *Rhizophora* spp. than in *Sonneratia alba* pneumatophores, *Bruguiera gymnorhiza* root knees as well as bare soil. (Kumara et al., 2010) and Phillips et al. (2017) showed that, for *Rhizophora mucronata*, sediment accretion rates (measured over 2.4 years) tend to increase with seedling density, ranging from $8.6\ \mathrm{mm \cdot yr^{-1}}$ for the smallest seedling density ($0.95\ \mathrm{seedlings \cdot m^{-2}}$) to $16.2\ \mathrm{mm \cdot yr^{-1}}$ for the highest seedling density (6.96 per $\mathrm{m^2}$), while bare surfaces had sediment accretion rates of $7.4\ \mathrm{mm \cdot yr^{-1}}$. In particular, the presence of seedlings promoted the accumulation of fine sediments (diameter smaller than 63 μm) when the seedling density was higher. Whereas the percent of fine sediment was only 14.6% for the bare surface, it was higher in the presence of seedlings, and it increased up to 19.8% for a density of $6.96\ \mathrm{seedlings \cdot m^{-2}}$.

FIGURE 4.4 *Rhizophora* trees in Colombia (A) and Brazil (B and C). (A) Prop roots. (B) and (C) Zoom on prop roots, with density of roots changing with distance to the sediment. *Credit: (A) Martin Zimmer, (B) and (C) Mirco Wölfelschneider.*

Process-based numerical simulations could provide a better understanding of the flow mechanisms that lead to the observed trapping of fine sediments within arrays of seedlings; this process knowledge could then inform restoration efforts.

The knowledge of flow and sediment transport around mangrove structures at the scale of a tree or a root matrix comes from observational studies (e.g., Phillips et al., 2017) rather than from numerical modeling. We found four papers that report results of numerical simulations at the scale of a mangrove tree (*Tree scale* in Table 4.1). All four were found to unravel effects of *Rhizophora* mangrove structures on flow patterns (Furukawa and Wolanski, 1996; Furukawa et al., 1997; Jusoh et al., 2016; Kazemi et al., 2018b). These numerical studies showed that eddies, jets, and stagnation zones (i.e., areas with extremely low velocity) developed within the mangrove root system under a tidal current. Flows were complex due to the interplay between these hydrodynamic features, and the root density or porosity controlled the velocity reduction and tidal energy dissipation through turbulence. Among those studies, only one (Jusoh et al., 2016) also addressed the influence of pneumatophore root systems (*Avicennia* type), and to our knowledge no numerical experiment has been conducted on matrices of knee roots (*Bruguiera* type) thus far. As the eddies generated when the flow encounters a mangrove structure are likely to be in three dimensions, models should resolve the three-dimensional characteristics of the flow pattern with a suitable turbulence model. In the field of hydrodynamics, various computational suites are available to model flow patterns around structures. Different turbulence models are incorporated therein and can be chosen in accordance with the study requirements regarding resolution of velocity fluctuations and computational time. The most widely used turbulence models are Reynolds-Averaged Navier–Stokes (RANS), Large-Eddy Simulation (LES), and Direct Numerical Simulation (DNS). RANS equations, with parameters that have to be estimated based on field or flume observations prior to simulations, showed good performance in modeling large-scale time-averaged turbulence (i.e., when studies do not aim at predicting time turbulent fluctuations) within vegetation. For example, Fischer-Antze et al. (2001) successfully employed the RANS approach to simulate open-channel flows within submerged vegetation; they found that this method predicted a flow reduction and a velocity distribution in good agreement with flume experiments. However, the downside of using RANS is that the smallest turbulent scales remain unresolved since the turbulent fluctuations are not reproduced by the model. As identified by Furukawa and Wolanski (1996), spatial averaging of the flow would neglect these highly fluctuating flow features that are critical for sediment dynamics in the water column. Indeed, they revealed that the three-dimensional turbulence generated by current-vegetation interactions played a critical role in the trapping of sediment particles by mangrove structures. For instance, under a flow speed of $10-15\ \mathrm{cm \cdot s^{-1}}$, the sediment particles were retained twice as long in their computational domain with mangrove roots than in the absence of vegetation. Hence, the RANS approach should not be used to model the flow around mangrove trees and roots since the flow is usually highly turbulent in their immediate vicinity, where eddies (whirling motion of water) are generated. The application of LES equations for turbulence modeling showed good agreement with field and flume observations in coastal engineering and has started to be used in ecohydraulics quite recently. In comparison with RANS, LES resolves turbulence at scales larger than the mesh resolution (size of grid element) and models the (averaged) turbulence at smaller scales using SubGrid Scale models. This high-resolution simulation of fluctuating flow gives the opportunity to study more flow statistics, such as the impact of vegetation on turbulent dissipation (Stoesser et al., 2009). Combination of RANS and LES equations, i.e., Detached Eddy Simulation, could also be used to save computational time while still providing high-resolution data (Li and Yu, 2010). Such a hybrid approach, with RANS equations resolving the general dynamics and large-scale turbulence field outside the vegetated area and LES equations resolving the small-scale flow fluctuations within and directly around the vegetation field, could predict the microscale turbulence (smaller than mesh resolution, Li and Yu, 2010).

All numerical experiments reviewed at the tree scale consider artificial mangrove structures with cylindrical shapes, which is a common approach in ecohydraulics (cf. section 3.4). However, it has been highlighted by Vargas-Luna et al. (2016) that using mimicked vegetation stems such as cylinders may cause discrepancies from natural observations. They found that representing real plants by wooden sticks with a smaller diameter (2 mm vs. 1.5 cm) led to estimated friction coefficients of similar magnitude, but Reynolds number significantly differed in flume experiments. Using cylinders of larger diameter resulted in a better depiction of both friction coefficients and Reynolds numbers associated with real plants. Besides, Horstman et al. (2018b) clearly showed that using cylinders with variable height to mimic a natural *Avicennia* root system and the associated morphological variability of the pneumatophores (Fig. 4.5) instead of uniform-height wooden dowels renders results that better represent the flow in mangrove ecosystems. They discovered that uniform-height dowels created (spatially averaged) bed shear stresses twice as great as those from real pneumatophores (variable height, natural shape, and roughness) for similar densities. Future numerical simulations of sediment dynamics around mangrove structures should, consequently, incorporate natural mangrove structures, which is made possible by the recent development of photogrammetry. This optical method consists of the reconstruction of three-dimensional structures from photographs and was successfully applied to digitized mangrove root systems in three dimensions (Kamal et al., 2014; Liénard et al., 2016).

FIGURE 4.5 A) *Avicennia marina* trees (here dwarf individuals) and pneumatophores on Rangitoto Island, New Zealand. (B) *Rhizophora mucronata* seedlings in Brazil. (C) and (D) Patch of *Avicennia* pneumatophores of various density (sparse and dense, respectively) observed on Rangitoto Island, New Zealand. *Credits: (A), (C) and (D) Marine Le Minor, (B) Martin Zimmer.*

Only one study examined sediment transport in addition to flow patterns within a mangrove root system. More than 20 years ago already, Furukawa et al. (1997) ran numerical simulations of the flow and estimated the motion of sediment particles to investigate sediment trapping in prop root systems (Fig. 4.3B). The VORTEX model showed that the particles remained within the *Rhizophora* root matrix for a longer duration with an approach velocity of 20 $cm \cdot s^{-1}$ than 5 $cm \cdot s^{-1}$, since they were trapped in areas with extremely low velocity due to stronger turbulence. Only 400 sediment particles were tracked through the computational domain without influencing the flow, which is a simplification of the natural settings. As mentioned in Section 2, the suspended sediment concentrations in mangrove ecosystems can be high, especially near the bed, and the sediment is likely to alter the hydrodynamics and turbulent structures. Hence, the effects of the sediment on the flow should be simulated as a feedback system. Whereas in the Lagrangian approach, particles are followed individually at any time and everywhere they go (trajectory), the properties of a motionless volume of fluid (fixed in space) are observed as a function of time in the Eulerian approach. In this latter approach, the sediment particles are not considered independent from each other but as a fluid phase dispersed in water, with motion determined by concentration, rheology, and settling velocity (Brennan, 2001). This method appears as a good compromise between model computational time and performance to simulate dynamics of rather small sediment particles at high concentration over a few meters, i.e., in a matrix of mangrove roots or around a mangrove tree.

3.3 Sediment dynamics around idealized cylinders as representation of mangrove structures

A common approach in the field of ecohydraulics is to mimic mangrove structures using circular cylinders commonly employed to study the impact of vegetation on the hydraulics and sediment transport. Therefore, 62 papers on flow and sedimentation patterns within idealized vegetation canopies (i.e., cylinders) were also included in this review, although only those referring to mangrove ecosystems are listed in Table 4.1. In flume studies, stems were mimicked by cylinders made of PVC (Tinoco and Coco, 2016), stainless steel (Kothyari et al., 2009a, 2009b), bamboo (Gillis et al. 2014, 2016), or, more often, wood (White and Nepf, 2007; Tanino and Nepf, 2009). These four materials are usually chosen because they exhibit similar mechanical properties (e.g., stiffness) to vegetation structures such as mangrove pneumatophores (Zhang et al., 2015b). Although using idealized structures may over- or underestimate changes in hydrodynamics, such studies provide valuable insights into flow conditions and subsequent sediment transport within mangrove structures.

3.3.1 *Single cylinder*

An extensive body of literature on flume experiments exists for the hydrodynamics around single cylinders, since these are studied in many fields of research. In the field of engineering, a single cylinder can represent a bridge pier (Graf and Istiarto, 2002), an inclined pile (Kitsikoudis et al., 2017), or a vertical pile (Roulund et al., 2005; Palau-Salvador et al., 2010). In ecohydraulics, vegetation is commonly mimicked using single cylindrical structures (e.g., a tree; Kitsikoudis et al., 2016). Flume experiments based on single cylinders provide valuable data to assess the quality of numerical simulations (e.g., Palau-Salvador et al., 2010). The hydrodynamics that develop around a rigid vertical cylinder have been numerically investigated for various flow conditions, covering both laminar and turbulent regimes (Lima E Silva et al., 2003; Mahesh et al., 2004; Shu et al., 2007; Harichandan and Roy, 2010; Zheng et al., 2016; Gao et al., 2017). These flume and numerical engineering studies revealed a flow separation along the height of the pile and vortex shedding on the sides of the pile, and increasing flow speeds led to more intense turbulence in the wake of the pile that resuspended sediment and transported it downstream. In addition, these papers reported a local downward flow, associated with a horseshoe vortex, in front of the pile that caused sediment removal around its foot, i.e., scour. For instance, scour depth, an important parameter for pile engineering that can also be of high relevance in the case of mangrove seedling establishment or mature tree stability, has been investigated as a function of cylinder diameter (Yagci et al., 2017) or submergence ratios (i.e., the height of the structure divided by the water depth; Zhao et al., 2010).

While a large number of flume experiments have been conducted to examine the scour developing at the base of a cylindrical structure (e.g., Zhao et al., 2010; Yagci et al., 2016), only a few numerical studies incorporated a morphological model to examine this scour process (Roulund et al., 2005; Zhao et al., 2010; Baykal et al., 2015; Zhang et al., 2017). These numerical models accounting for sediment transport, initially designed for engineering purposes, could be used to investigate scour patterns in mangrove ecosystems by adapting flow conditions and sediment properties to those found in natural mangrove forests. It would ultimately help better predict and quantify the scour occurring around isolated mangrove structures and provide insights into hydro- and sediment dynamics that might affect the anchoring of a mangrove tree or seedling. Le Minor et al. (2019), indeed, observed such scouring patterns around a seedling using numerical modeling, while Mullarney et al. (2017b) and Fagherazzi et al. (2017) reported observations of scour around pneumatophores or trunks, respectively, during a field campaign in the Mekong Delta.

The impact of an inclined structure on the flow pattern and the scouring process has been previously investigated for engineering purposes by Kitsikoudis et al. (2017). Using CFD, the authors found that the volume and maximum depth of the scour in front of the base of the cylindrical pile increased with greater inclination angle, while in the rear of the pile, the volume of sediment removed was reduced by a factor of almost four when changing the inclination in the direction of the flow from 0° (vertical) to 42 degrees. When the inclination of the pile was greater, sediment mixing was enhanced and the vortex-shedding tended to weaken and disappear. These results suggest that the inclination of a partially submerged structure plays a key role in turbulence dynamics and the consequent sediment transport. While *Rhizophora* prop roots are known to be bended and to exhibit a wide range of inclination angle, studies evaluating the effect of these properties are clearly lacking in current mangrove research. The use of CFD could remediate this in the future: different bending angles and curvatures could be tested under various flow conditions. To assess the impact of a specific bending angle, a single root with no branching should be first considered, to omit the interfering flow features stemming from neighboring roots. Then, more complex structures could be studied to account for branching and vertical variations. Ultimately a full root system could be investigated.

3.3.2 *Arrays of cylinders*

Numerous studies have investigated sedimentation patterns within and around patches of idealized vegetation, i.e., rigid vertical cylinders. Some of these experiments could be of interest to the mangrove research community; for instance, simulated flow patterns around small cylinders (e.g., 1.32 cm in diameter in Chang et al., 2017) could provide useful insights into flow patterns around mangrove seedlings or around pneumatophores.

Horstman et al. (2018b) evaluated the effect of height variations on the flow by comparing hydrodynamic observations in flume experiments with an array of cylinders with uniformal height, an array of cylinders with variable heights, and an array of real pneumatophores (*A. marina* var. *australasica*). The flow through cylinders with variable heights provided results that were more representative of the hydrodynamics, i.e., velocity attenuation and turbulence intensity, within naturally heterogeneous mangrove pneumatophores. While the depth-averaged velocity increased by 17% within uniform-height dowels, it rose by 10% within variable-height dowels in comparison to the one measured within natural pneumatophores. Regarding the turbulence level, the turbulent kinetic energy measured near the top of the structures decreased by 23% within the variable-height dowels but increased by 114% for the uniform-height dowels compared to the real pneumatophores. In numerical simulations, the natural geometrical and mechanical characteristics of mangrove structures are generally simplified by considering them as rigid, vertical, and smooth cylinders. Such approximations could be avoided in future studies, by using a mesh that accounts for the shape, surface structure, and motion of mangrove structures under hydrodynamic forcing.

The hydrodynamic conditions tested in flume and numerical experiments correspond to observations in mangrove ecosystems. The simulated flow speeds are mainly below $30\ \mathrm{cm \cdot s^{-1}}$ (e.g., White and Nepf, 2003; Liu and Nepf, 2016; Horstman et al., 2018b). Fast flow speeds of around $1\ \mathrm{m \cdot s^{-1}}$ and more, representative of extreme scenarios, have also been investigated (Kothyari et al., 2009a, 2009b; Tanino and Nepf, 2009). Numerical models that aim at predicting the spatial and temporal fluctuations in idealized vegetated systems and at providing an accurate and high-resolution simulation are remarkably time-consuming and demanding in terms of computing resources. Therefore, mainly, two-dimensional models have been employed to simulate flow patterns in vegetated systems (Nicolle and Eames, 2011; De Lima et al., 2015; Zheng et al., 2016; Gao et al., 2017). However, with the development of computer capacities, three-dimensional models of flows within vegetation structures at very high resolution may become possible. In both flume and numerical experiments, the level of submergence of arrays of cylinders has been shown to be a critical parameter that controls the flow pattern. Levels of submergence ranging from a water depth covering 90% of the cylinder height to a water depth five times greater than the cylinder height have been examined in the laboratory (Tanaka and Yagisawa, 2010; Tang et al., 2014). Numerical simulations of both emergent (Stoesser et al., 2010; Huai et al., 2015; Etminan et al., 2017) and submerged (Fischer-Antze et al., 2001; Stoesser et al., 2006, 2009; Tang et al., 2014) cylinders have also been completed. Overall, modeled results exhibited good consensus with laboratory results; for instance, time-averaged streamwise velocities and turbulence intensities modeled with the LES code MGLET were in good agreement with flume measurements along vertical profiles (Stoesser et al., 2009, 2010). These studies addressing the impact of vegetation submergence could help understand how tides affect the flow through mangrove structures. Simulating a wide range of water depths would ultimately show a more complete picture of the flow pattern occurring at the scale of trees or roots over a complete tidal cycle. In addition, some flume experiments have included waves with different periods and heights (Pujol et al., 2013; Yao et al., 2018; Tan et al., 2019). Mimicked vegetation at a density of 1280 cylinders·m^{-2} was found to reduce the wave velocities by 20% when submerged and 45% when emerged in comparison to a bare bed (Pujol et al., 2013). Such oscillatory flows were barely considered in numerical simulations thus far and should be included in future models of flow through root systems at the tree scale (including various root systems), since they are characteristic of fringing mangrove ecosystems.

Different shapes of cylinder arrays have also been studied in flume experiments: circular patches (Tanaka and Yagisawa, 2010), hemispheres (Bennett et al., 2002), squares (Ben Meftah and Mossa, 2013), and, more frequently, rectangles (White and Nepf, 2003; Tanino and Nepf, 2008; Hu et al., 2014; Vargas-Luna et al., 2016). Within these arrays, the rigid vertical cylinders have been distributed in three different arrangements: parallel (Tang et al., 2014), staggered (Kothyari et al., 2009a, 2009b), or random patterns (White and Nepf, 2003). With numerical simulations, array configurations similar to the patterns examined in flumes have been studied; modeled results of flow distribution and magnitude were in accordance with flume measurements. In addition, in-line cylinders with different spacing have also been investigated (Liang et al., 2009; Harichandan and Roy, 2010; Awale, 2014; Kim et al., 2014, 2017; Hosseini et al., 2018). These studies could help understand the flow mechanisms that lead to scour around the pneumatophores that are extending laterally from *Avicennia* and *Sonneratia* trees (Bird, 1986; Mullarney et al., 2017b). Very wide ranges of cylinder densities have been investigated; they correspond to solid fractions (i.e., volume of cylinders by volume of fluids) from 0.2% (e.g., Bouma et al., 2007; Ben Meftah and Mossa, 2013) up to 40% (Tanaka and Yagisawa, 2010), the latter is representative of dense mangrove forests. Ricardo et al. (2013) investigated the impact of spatial changes of stem densities by creating an array of alternating dense (1600 stems·m^{-2}) and sparse (400 stems·m^{-2}) patches. They observed stronger velocity fluctuations (i.e., turbulence) in the dense patches, since eddies generated by flow–stem interaction have less space to lose intensity.

3.3.3 *Sediment properties*

Flume experiments considering sediment transport in vegetated environments mainly deployed sand with a density of 2650 kg·m^{-3}: medium sand (280 μm; Tinoco and Coco, 2018), coarse sand (550 μm, Kothyari et al., 2009a; 500 μm, Follett and Nepf, 2012; Yager and Schmeeckle, 2013; 500 and 950 μm, Vargas-Luna et al., 2016; 700 μm, Yagci et al., 2016; 600–850 μm, Yang et al., 2016; 523 μm, Yagci et al., 2017), and gravel (2.4–4.8 mm, James et al., 2004; 1.7–2 mm, Yang et al., 2016; 1.9 –5.9 mm, Kothyari et al., 2009a). Dynamics of finer sediment fractions have also been analyzed by using glass beads (density of 2500 kg·m^{-3}) with median diameters of 5.6 μm (Liu and Nepf, 2016), 7.1 μm (Liu et al., 2017), and 12 μm (Zong and Nepf, 2010, 2011; Chen et al., 2012). These sediment transport studies aimed at unraveling the role of vegetation structures on sediment deposition and resuspension. For example, Yagci et al. (2017) found that there is a density threshold: until the studied structure occupied 32% of the volume of water, the scour volume increases but, in denser arrays, the scour volume is reduced. Except for the glass beads, the applied sediment properties mentioned above are not representative of the natural sediments found in most mangrove ecosystems in terms of grain size, since mangrove sediments are usually very fine-grained. Furthermore, sediments in mangroves are composed of organic and inorganic matter, while sediments used in flume experiment have thus far been merely inorganic, not accounting for this characteristic. Both grain size and organic matter content affect the flocculation of the fine sediments that, in turn, plays an important role in the sediment behavior and dynamics.

Similarly, numerical simulations that incorporated sediment focused on the sand and gravel fractions: coarse sand (850 μm, Kim et al., 2014, 2017; 500 μm, Vargas-Luna et al., 2016; 600–850 μm, Yang et al., 2016; 800 μm, Guan and Liang, 2017) and gravel (1.7 –2 mm, Yang et al., 2016; 5–10 mm, Zhang et al., 2017). These grain size classes are commonly investigated for engineering purposes, since significant scour depths are observed at the base of bridge piers or offshore piles in sandy sediment beds. Kim et al. (2014) ran simulations of flow around two cylinders side-by-side (or tandem) so that changes in flow direction could be studied. Zhang et al. (2017) revealed that the pile configuration controls the scour pattern and that the spacing appears to play a key role on the scour depth. There was no interaction between flow features of adjacent piles when the spacing was large enough, but small spacing led to merging scour patterns and greater scour depth. In general, sediment deposition was observed in the wake. As suggested by Le Minor et al. (2019), these findings may be relevant for predicting the success of seedling establishment as a function of seedling density. The modeling of fine sediment transport through cylinder arrays is, nevertheless, clearly lacking and should be considered in future investigations of sediment dynamics in mangrove ecosystems.

3.4 Sediment dynamics at the root or seedling scale

In this section, we focus on single structures, such as a single prop root (with no further branching), a single pneumatophore, a single knee root, or a single seedling. As described in the previous section, mangrove structures differ significantly from one species to another, especially when it comes to their architecture (root types), in regards of their size and mechanical properties. Using an analogue caliper, Krauss et al. (2003) measured the diameter of a large number of mangrove roots at 4 cm above the sediment surface in the Federated States of Micronesia. At this specific site, they found that the average diameter of prop roots of *Rhizophora* spp. is 2.7 cm, while it is 2.8 cm for pneumatophores of *S. alba* and 2.5 cm for root knees of *B. gymnorrhiza*. Although the mean diameters of these root structures exhibit similar values, individual root diameters show a great variability and change along the stem length (Zhang et al., 2015a). Furthermore, Liénard et al. (2016) reported a wide range of pneumatophore shapes using photogrammetric reconstruction; the height and diameter of *Sonneratia* pneumatophores are very heterogeneous and change with elevation above the sediment bed. Norris et al. (2017) showed, based on high-resolution field observations of flow patterns, that the magnitude of the hydrodynamic features observed in the rear of *Avicennia* pneumatophores depends on their diameter. The variety of structure dimensions was not taken into account in the design of artificial root systems for laboratory studies, but it has been investigated in the field of marine and river engineering (cf. Section 3.4). For example, Zhao et al. (2010) conducted flume and numerical experiments to determine the link between cylinder height-to-diameter aspect ratio (two diameters and seven heights tested) and local scour. In addition, prop roots exhibit varying curvatures or bending angles; these physical properties affect the flow and sediment transport in a way that is not yet well known. Therefore, numerical modeling could be employed to investigate the impact of prop roots with a wide range of dimensions and shapes to quantify sediment-trapping and resuspension. Another point to consider is the flexibility of the roots, such as the pneumatophores of *Avicennia*, that might get bended when exposed to strong currents.

Mangrove seedlings, especially when establishing on mudflats, are subject to harsh and challenging environmental conditions: tidal flats are flooded during high tide and emerged at low tide (Osborne and Berjak, 1997). *Avicennia* spp. are pioneer mangroves able to successfully colonize bare tidal flats (Osborne and Berjak, 1997; Balke et al., 2011 and references therein). Balke et al. (2011) conducted flume and field experiments to study the impact of hydrodynamic forcing and sediment dynamics on the establishment of *Avicennia alba* seedlings. It has been suggested that for successful establishment of these seedlings, a suitable window of opportunity is needed (Balke et al., 2011). Indeed, rooting thresholds have to be passed to overcome buoyancy during tidal flooding, and to resist hydrodynamic forces from waves and currents that could cause dislodgment due to erosion. For *S. alba* seedlings, it has been showed that their establishment may fail when their structure is too weak to cope with sediment mixing causing toppling and uprooting of the seedling (Balke et al., 2013a), resulting from excavation due to erosion. This could, for instance, be the case, when high-nutrient loads result in reduced root development of mangrove seedlings (Gillis et al., 2019). Under certain circumstances, the opposite process, seedling burial due to sediment accretion, has been observed to affect seedling survival as well (Thampanya et al., 2002). *Avicennia* can withstand nonlethal sediment disturbance through compensating root and shoot growth (Balke et al., 2013b).

There is a sheer difference in number of studies using mimicked vegetation as cylinders and real mangrove structures, and observations revealed that cylinders are poor proxies of pneumatophores (cf. Horstman et al., 2018b). Hence, further investigations are required to unravel the effects of the morphology of natural mangrove structures on flow and sediment dynamics. Numerical simulations could help fill this knowledge gap focusing on the scale of a single structure. To our knowledge, only one study has been conducted at the scale of a single structure, using numerical modeling. Le Minor et al. (2019) simulated the three-dimensional flow pattern and the sedimentation in close vicinity of a *R. mucronata* seedling, digitized through a photogrammetric survey, under different flow settings (Fig. 4.3C). These simulations were run for a single seedling only and, therefore, building arrays of seedlings would provide information on interactions between flow features stemming from neighboring mangrove structures. In this multistructure model, the simulation of the turbulence that develops around a single mangrove seedling could be completed using an LES model, since it provides acceptable results at reasonable computational cost (Le Minor et al., 2019). Le Minor et al. (2019) modeled a uniform grain size of fine silt and showed, using LES, that scour was enhanced around the seedling due to the generation of a horseshoe vortex, while enhanced turbulence in the wake of the seedling kept silt in suspension. In the future, larger ranges of sediment grain sizes could be implemented to better represent natural settings. Future numerical simulations of flow and fine sediment dynamics around mangrove seedlings would provide valuable information on interactions between flow features stemming from adjacent seedlings and the consequent sedimentation pattern developing around them, which could inform restoration planning with respect to ideal spacing among seedlings.

3.5 Sediment dynamics at the microtopography scale

Most flume and numerical studies use straight or curved cylinders to mimic vegetation structures such as mangrove roots. Although these mimics have similar diameter and height to the natural mangrove structures, they are usually smooth and, thus, do not account for the surface roughness of the bark or its coverage by epibionts, i.e., organisms attached to the mangrove roots. To our knowledge, the impact of surface roughness of mangrove structures on hydro- and sediment dynamics at the microtopography scale has neither been investigated in the field, in flume experiments or through numerical simulations. However, one flume study addressed the impact at the meter-scale (arrays) of flow features developing at the surface of pneumatophores: the turbulence production was greater at the rough surface of pneumatophores than at the surface of smooth cylinders (Horstman et al., 2018b). The flow above a rough surface has been simulated for engineering purposes only, by Busse et al. (2015). However, this is an aspect that should be studied in the future, and numerical simulations are urgently needed to determine their significance for sediment dynamics and their possible impact on mangrove ecosystems as outlined in more detail in the following subsections.

3.5.1 *Surface texture*

Mangrove root surface textures exhibit different patterns and microtopography, and have, therefore, different roughness associated with their surface texture (Fig. 4.6). Purnobasuki (2011) reported both smooth and rough types of *A. marina* pneumatophores, the smoothest structures being, in general, the youngest. For absorbing oxygen in an environment of mostly anoxic conditions in the sediment, mangroves rely on their breathing cells (called lenticels) located on the surface of their above-ground roots (Tomlinson, 2016). Air is taken into the aerial roots through their opened lenticels during low tide, when the above-ground root system is not fully submerged (Scholander et al., 1955; Tomlinson, 2016). These lenticels or "chimneys" add to the variation in texture among species (Fig. 4.6A) and exhibit a wide range of size and morphology (crater-like to unopened pimples/pustules; Purnobasuki, 2011) as well as different states: close, partially open, and fully open (Ish-Shalom-Gordon and Dubinsky, 1992).

The surface texture is likely to influence the flow in its immediate vicinity by either increasing or reducing the turbulence in the boundary layer. Natural mangrove structure microtopography should, hence, be included in numerical simulations to determine how the texture affects the hydro- and sediment dynamics. Orihuela et al., (1991) reported roots of *Rhizophora* sp. with their surface covered by sediment. A numerical approach would help understand how sediment tends to accumulate in the surface grooves or in the lenticels and would ultimately lead to a better understanding of small-scale sediment transport processes that may negatively affect mangrove respiration. Reduced mangrove respiration could, in turn, lead to reduced growth or even increased mortality of mangrove individuals (Ellison, 1999; Saenger, 2002). The three-dimensional surface texture could be digitized

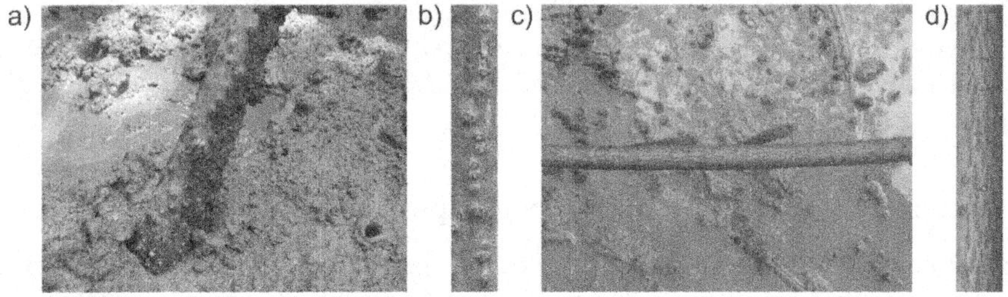

FIGURE 4.6 Surface roughness of *Rhizophora* prop root. The pictures were taken in Brazil. (A) Additional roughness created by algae. (B) Zoom on the lenticels of the prop root shown in (A). (C) Microtopography of root surface. (D) Close view of the surface texture of the prop root shown in (C). *Credits: Mirco Wölfelschneider.*

utilizing a high-resolution optical method such as laser triangulation (Sioma et al., 2018) or photogrammetric measurements (Kamal et al., 2014; Liénard et al., 2016; Le Minor et al., 2019) in order to be incorporated into a model. To work at the scale relevant for revealing the microtopography of mangrove surface texture, the DNS approach appears as a good technique, since, in contrast to LES and RANS, it fully resolves the entire range of turbulence scales occurring in the flow. However, one requirement is that the mesh should be fine enough for all the turbulence scales to be resolved, which implies extremely long computation times. Future developments in computational power may reduce the computational cost of DNS and, thus, may lead to its successful application in ecohydraulic studies of sediment dynamics in mangrove ecosystems, for example.

Furthermore, sediment dynamics could be modeled by a Lagrangian approach that considers sediment grains as individual particles with similar or different grain sizes (heterogeneity) and determines their motion through mechanical equations (settling velocity, contact between grains). A set of equations is solved for every individual particle, and hence, the larger the number of particles, the longer the computational time. The Lagrangian approach has already been applied to simulate flow and sediment dynamics by considering particles with different characteristics such as shape and size and employing the Smoothed Particle Hydrodynamics method (Khanpour et al., 2016). Numerical models that combine both Eulerian and Lagrangian methods have also been implemented; for instance, a coupled CFD-DEM (Discrete Element Method) code was used to simulate turbulence and sand transport by Schmeeckle (2014). In the future, the motion of natural fine sediment grains, i.e., not only with various shapes and sizes but also containing organic matter, could be modeled, if computing capacity were improved.

3.5.2 *Biofouling*

The texture of the surface itself is not the only factor that influences the surface of mangrove structures. Many studies reported biofouling (Schultz, 2004), resulting in a dense cover of algae (Rodriguez and Stoner, 1990; Hayden and Granek, 2015; see also Fig. 4.7A), barnacles or oysters (Linton and Warner, 2003; Liénard et al., 2016; Sahoo et al., 2018a; see also Fig. 4.7B), and even sponges and anemones on aerial mangrove roots. The presence of epibionts from a wide range of species on *Avicennia* pneumatophores (Pongener et al., 2018), or on *Rhizophora* prop roots (Rodriguez and Stoner, 1990), indicates that mangrove

FIGURE 4.7 Biofouling observed on mangrove structures. (A) Algae that contribute to sediment trapping on *Avicennia marina* pneumatophores in Whitianga, New Zealand. (B) Barnacles and mussels covering a root of a *Rhizophora* tree. *Credits: (A) Erik Horstman (B) Martin Zimmer.*

structures are of great importance, as they provide a hard substratum for settlement in an otherwise soft and unstable environment.

Mangrove biofouling patterns vary in space and time. Rodriguez and Stoner (1990) observed a change in the *Rhizophora* biofouling along the seaward–landward gradient in Puerto Rico. A correlation between pneumatophore location and assemblage of epibionts on *Avicennia* was also reported by Gwyther and Fairweather (2005). The diversity of epibionts and their coverage observed on mangrove roots also exhibits seasonal changes (Sahoo and Ansari, 2018). Gwyther and Fairweather (2002) found that the algal coverage of mimicked and natural *Avicennia* pneumatophores varied seasonally: the surface covered by algae was reduced from 40% in summer to 0% in winter. Furthermore, Sahoo and Ansari (2018) observed a vertical heterogeneity of biofouling on *Avicennia officinalis* pneumatophores: the lower part of these mangrove structures was more covered with epibionts, probably due to longer periods of inundation. As pointed out by Hicks (1986), sediment may accumulate on the surface of mangrove structures covered by epiphytes. Consequently, the breathing of mangrove trees may be altered, causing a reduction in mangrove growth or death (Ellison, 1999; Saenger, 2002).

In general, the coverage of a surface by epibionts is characterized by a parameter called Roughness Density (RD), which is calculated as the area covered by epibionts divided by the total surface. Many studies on biofouling revealed that an increase in RD generates an increase in drag (Witman and Suchanek, 1984; Wahl, 1996; Townsin, 2003; Schultz, 2004). It is well established that the force on mangrove structures applied by the flow increases with their increasing submerged area (Mazda et al., 1997b). Therefore, when the biological coverage of mangrove roots expands, the drag force may increase dramatically due to the larger surface exposed to the flow, resulting in a higher risk of mangrove breakage upon wave-hitting. As reviewed by Johnson and Rice (2014), the benthic macrofauna, e.g., barnacles living on mangrove structures, alter hydrodynamics due to their physical presence that leads to zones of high and low flow speeds. The drag force is related to the morphology of the epibionts growing on the mangrove roots, e.g., shape, size, texture (of the shell), as well as their behavior, since these organisms may move over time (Schultz, 2004; Johnson and Rice, 2014).

The presence of epibionts on marine structures has been studied in the field of engineering, since biofouling significantly increases the ship hull drag and, in turn, the fuel consumption for maritime transport (e.g., using numerical simulations as reported by Khor and Xiao, 2011). Flume experiments have been completed to evaluate the impact of biofouling on drag (Holm et al., 2004; Schultz, 2004), using plates covered by small macrofouling organisms (large organisms such as barnacles were removed prior to using the friction disk machine). Flume measurements have also been made to determine the effect, at different densities, of artificial and natural macrozoobenthic features (such as worm tubes, snail shelves, mussels, sand mounds, pits, and cross-stream track furrows) on flow and sediment transport (Friedrichs and Graf, 2009; Friedrichs et al., 2009). To our knowledge, studies on the effect of biofouling on vegetation and especially mangrove structures have not been conducted thus far.

Photogrammetric measurements derived from CT scans have successfully been employed to describe the three-dimensional skeleton of corals (Kaandorp and Garcia-Leiva, 2004). This technique could offer opportunities to digitize the biological coverage of mangrove structures and provide a detailed geometric layout of the structures for representation in a numerical model. Numerical investigations of the contribution of biofouling on mangrove structures to increased drag, flow reduction and turbulence dissipation, and, consequently, sediment trapping are required to better understand small-scale processes as well as their implications for general dynamics in mangrove ecosystems. In the context of ecosystem design (Zimmer, 2018), it would indeed be extremely valuable to understand whether biofouling would reinforce the shoreline-stabilizing role of mangrove trees or not.

4. Disparate treatment of spatial scales

For each spatial scale (forest, tree, pneumatophore, prop root or seedling, and surface texture), the methods used (field, flume, or numerical experiments) as well as the topic investigated (hydro- and/or sediment dynamics) were counted for all papers included in this chapter (Fig. 4.8). These simple statistics highlight the fact that 1) numerical modeling is not the main method used to investigate hydro- and sediment dynamics in mangrove ecosystems; and 2) most of the papers we found focus on large spatial scales. We emphasize the disparate treatment of spatial scales in terms of hydro- and sediment dynamics in mangrove ecosystems. Small-scale physical processes of sediment transport around mangrove structures are greatly understudied in mangrove research, while such processes are likely to affect mangrove ecology.

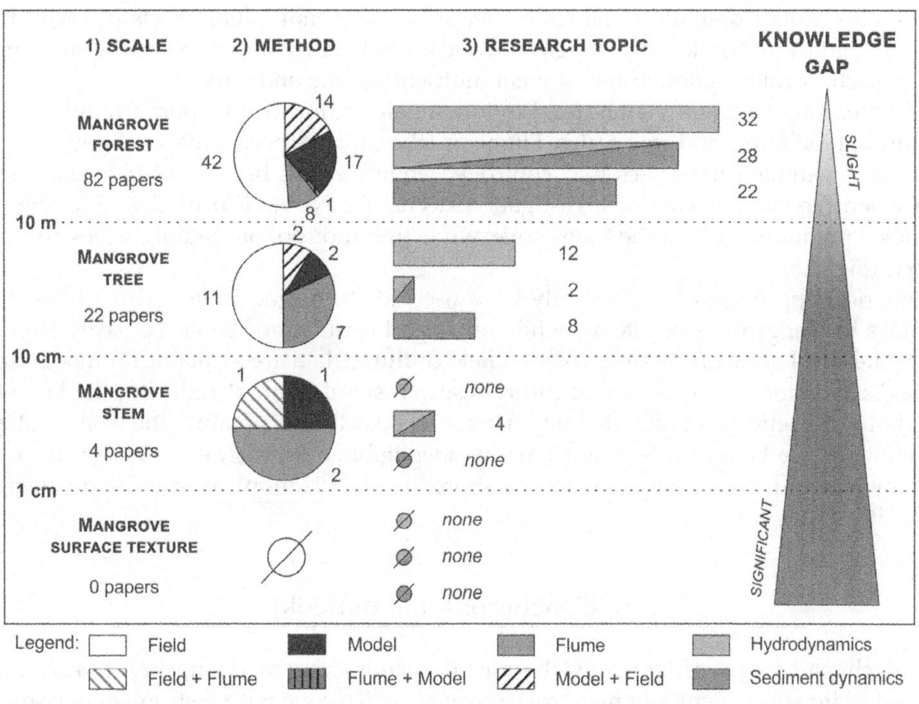

FIGURE 4.8 Classification of papers reviewed in this study. For each spatial scale (forest, tree, stem, and bark), the methods used (field, flume, or numerical experiments) as well as the topic investigated (hydro- and/or sediment dynamics) were counted. Reviews were not incorporated in the paper counts. Gaps of knowledge at smaller scales are highlighted in red (dark gray in print version).

Among the 101 papers on hydrodynamics and sediment transport in mangrove ecosystems reviewed herein, 82 papers focused on the forest scale. Regarding the hydrodynamics in mangrove forests, these studies mainly quantified wave attenuation (Horstman et al., 2014; Henderson et al., 2017), or even tsunami reduction (Harada et al., 2002), through field, flume, and/or numerical experiments. A large number of papers combined hydrodynamics and sediment transport, since currents, tides and oscillatory flows (as well as extreme events such as storm surge and tsunami) are the main drivers of sediment erosion and deposition in mangrove ecosystems (Le Nguyen and Vo Luong, 2019). Studies dedicated to sediment transport aimed principally at linking the above-ground mangrove vegetation to sediment accretion (Krauss et al., 2003; Alongi et al., 2005) and sediment properties (Macintyre et al., 2004).

Regarding the smaller spatial scales, 22 papers analyzed flow patterns and/or sediment transport around root systems or trees, and 4 around single pneumatophores, prop roots or seedlings. To our knowledge, no investigations on the effect of the mangrove bark microtopography on hydrodynamics were conducted so far (Fig. 4.8). This clearly hints at a lack of knowledge on fine-scale processes in mangrove research.

The impact of scale on the physical processes observed was emphasized by Friess et al. (2012): on m^2 scales, flow features and structure properties directly control the local sediment dynamics, whereas on km^2 scales, the factors that dominate sedimentation patterns are erosion, sediment supply and water flows. Although large-scale studies give an overview of the physical processes, the contributions of a wide range of factors tend to be mixed up. Consequently, such studies do not describe small-scale effects. Isolating individual effects of each of these confounding factors would help understand the impacts of sediment dynamics on mangrove ecology such as colonization, establishment, nutrient uptake and survival.

Field measurements are mainly used to investigate sediment transport around mangrove structures at the forest and tree scales. Flume or laboratory experiments, enabling to monitor hydro- and sediment dynamics in a controlled environment, become of increasing importance when the spatial scale of investigation decreases. Numerical modeling studies were conducted predominantly at the forest scale, while tree and root or seedling scales are equally covered (Fig. 4.8).

Numerical approaches have scarcely been used to study fine-scale hydro- and sediment dynamics in mangrove ecosystems, while numerical simulations could be easily applied to investigate small-scale processes, provided field or flume data are available for model calibration and validation. We reckon that future research should aim at reducing the knowledge gap about the effects of small-scale mangrove structures, notably the contribution of biofouling, as we believe this might have nonnegligible effects on ecosystem services provided by mangrove ecosystems (such as shoreline stabilization) as well as on mangrove tree health status.

5. Conclusions and outlook

In this chapter, we have highlighted the potential of numerical modeling for future research on flow and sediment dynamics in mangrove ecosystems. To cope with instrument measurement limitations at very small scales, numerical modeling could be used to contribute to a better mechanistic understanding of small-scale sediment dynamics occurring around mangrove structures.

By isolating individual effects of a wide range of confounding factors, their contribution to the observed dynamics in large-scale studies could thus be unraveled. Although numerical simulations have many advantages, this tool has not been widely used for sediment dynamics in the field of ecohydraulics. The main challenge for numerical modeling is finding a balance between the model resolution and the computational time for small-scale processes.

Small-scale processes are likely to be interconnected and, thus, should be combined in future numerical experiments to assess the sediment dynamics at the ecosystem scale. As already pointed out, studying the effects of small-scale processes separately may lead to a distorted picture of the ecosystem complexity in environmental engineering (Odum, 1996). Therefore, we ultimately recommend increasing the intricacy of numerical simulations once the small-scale processes of sediment transport that take place around mangrove structures have been unraveled. For example, numerical simulations could be run to investigate the hydro- and sediment dynamics in random arrays of mangrove *Rhizophora* prop roots with variable properties such as bending, branching, height, diameter and surface roughness. Conducting such complex numerical experiments, including many factors at a wide range of spatial scales, would help determine the contribution of each one of them to the patterns observed at the scale of the entire mangrove ecosystem.

While we focused on the effect of tree structures, including other organisms attached to those structures (i.e., biofouling), the contribution of other important organisms in mangrove ecosystems, such as crabs (Mokhtari et al., 2016), has not been covered in this chapter. It is well established that crabs substantially affect sediment dynamics through their bioturbating activities: burrow-building and maintenance lead to sediment mixing and reworking (Needham et al., 2010, 2011). At low burrow densities, crab burrows lead to an increase in sediment resuspension and thus promote erosion, whereas at high densities crab burrows enhance sediment deposition (Needham et al., 2013). Crab burrows in mangrove ecosystems affect sediment salinity (Pülmanns et al., 2015; Pestana et al., 2017) and trap particulate organic matter and sediment in large quantities (Iribarne et al., 2000; Le Minor et al., 2020). Through bioturbation, crabs affect not only mangrove seedling establishment, growth and survival but also their spatial distribution within the mangrove forest (Krauss et al., 2008). Implementing models that include both mangrove root systems and crab burrows would provide a deeper insight into sediment dynamics in tropical coastal environments.

Particulate organic matter can affect the formation of flocs, which significantly influences the flow and sediment dynamics in mangrove ecosystems. Future numerical simulations that take into account the organic matter content and its effects on grain size and shape (flocs) would lead to more realistic predictions of sediment dynamics and would also help unravel the mechanisms behind organic matter-trapping in mangrove sediments. Mangroves contribute significantly to organic matter production and sequestration and burial; $23 \cdot 10^{12}$ g C is accumulated per year in mangrove sediments as reported by Jennerjahn and Ittekkot (2002). On one hand, autochthonous particulate organic matter is produced inside the mangrove and stems from both above-ground leaves, branches, trunks and below-ground roots. On the other hand, organic matter is introduced into mangrove ecosystems via aquatic flows that originate from land or sea, i.e., spatial subsidies (see Chapter 15).

Integrating all the different processes (transport, flocculation, turbulence, etc.) discussed in this chapter in a mangrove ecosystem model would be relevant to simulate flow and sediment dynamics over a wide range of spatial and temporal scales. When working toward such integrated models, many challenges are faced not only in terms of computational capacity but also calibration and validation of the models. For instance, researchers should aim for a good balance between computational time and model resolution. Implementing all the processes that take place in a mangrove ecosystem would increase significantly the model complexity and, thus, the computational power required to simulate accurately these processes. To reduce the complexity of numerical simulations, physical processes could be either process based (represented by equations) or in a parameterized form (simplified when too complex or unresolved scale). A hierarchical modeling approach could also be used: for instance, in coastal regions, currents can be simulated with one model at large scales and then other models are nested within such global models to simulate finer resolution spatial patterns. A similar hierarchical approach with the different scales investigated in this chapter could be employed to study mangrove ecosystems from forest scale down to texture effects. In addition, to check whether such models are reliable, large data sets should be collected over a wide range of spatial and temporal scales and the models should be complex enough to include natural processes responsible for dynamics observed during data acquisition.

Acknowledgments

This research was funded by the Deutsche Forschungsgemeinschaft (DFG) through the International Research Training Group INTERCOAST "Integrated Coastal Zone and Shelf-Sea Research" (project no. GRK 1598). The authors are grateful to the reviewers for their valuable comments to improve the quality of the manuscript.

References

Adame, M.F., Neil, D., Wright, S.F., Lovelock, C.E., 2010. Sedimentation within and among mangrove forests along a gradient of geomorphological settings. Estuar. Coast Shelf Sci. 86, 21–30. https://doi.org/10.1016/j.ecss.2009.10.013.

Alongi, D.M., 2008. Mangrove forests: resilience, protection from tsunamis, and responses to global climate change. Estuar. Coast Shelf Sci. 76, 1–13. https://doi.org/10.1016/j.ecss.2007.08.024.

Alongi, D., Pfitzner, J., Trott, L., Tirendi, F., Dixon, P., Klumpp, D., 2005. Rapid sediment accumulation and microbial mineralization in forests of the mangrove *Kandelia candel* in the Jiulongjiang Estuary, China. Estuarine. Coast. Shelf Sci. 63, 605–618. https://doi.org/10.1016/j.ecss.2005.01.004.

Asp, N.E., Gomes, V.J.C., Ogston, A., Borges, J.C.C., Nittrouer, C.A., 2016. Sediment source, turbidity maximum, and implications for mud exchange between channel and mangroves in an Amazonian estuary. Ocean Dynam. 66, 285–297. https://doi.org/10.1007/s10236-015-0910-2.

Awale, S.D., 2014. Flow around row of four circular cylinders. J. Mater. Sci. Mech. Eng. 1, 11–17.

Balke, T., Bouma, T.J., Horstman, E.M., Webb, E.L., Erftemeijer, P.L., Herman, P.M., 2011. Windows of opportunity: thresholds to mangrove seedling establishment on tidal flats. Mar. Ecol. Prog. Ser. 440, 1–9. https://doi.org/10.3354/meps09364.

Balke, T., Bouma, T., Herman, P., Horstman, E., Sudtongkong, C., Webb, E., 2013a. Cross-shore gradients of physical disturbance in mangroves: implications for seedling establishment. Biogeosciences 10, 5411. https://doi.org/10.5194/bg-10-5411-2013.

Balke, T., Webb, E.L., Den Elzen, E., Galli, D., Herman, P.M., Bouma, T.J., 2013b. Seedling establishment in a dynamic sedimentary environment: a conceptual framework using mangroves. J. Appl. Ecol. 50, 740–747. https://doi.org/10.1111/1365-2664.12067.

Bao, T.Q., 2011. Effect of mangrove forest structures on wave attenuation in coastal Vietnam. Oceanologia 53, 807–818. https://doi.org/10.5697/oc.53-3.807.

Baykal, C., Sumer, B.M., Fuhrman, D.R., Jacobsen, N.G., Fredsøe, J., 2015. Numerical investigation of flow and scour around a vertical circular cylinder. Phil. Trans. Math. Phys. Eng. Sci. 373, 20140104. https://doi.org/10.1098/rsta.2014.0104.

Ben Meftah, M., Mossa, M., 2013. Prediction of channel flow characteristics through square arrays of emergent cylinders. Phys. Fluids 25, 045102. https://doi.org/10.1063/1.4802047.

Bennett, S.J., Pirim, T., Barkdoll, B.D., 2002. Using simulated emergent vegetation to alter stream flow direction within a straight experimental channel. Geomorphology 44, 115–126. https://doi.org/10.1016/S0169-555X(01)00148-9.

Bird, E., 1986. Mangroves and intertidal morphology in Westernport Bay, Victoria, Australia. Mar. Geol. 69, 251–271. https://doi.org/10.1016/0025-3227(86)90042-3.

Bouma, T., Van Duren, L., Temmerman, S., Claverie, T., Blanco-Garcia, A., Ysebaert, T., Herman, P., 2007. Spatial flow and sedimentation patterns within patches of epibenthic structures: combining field, flume and modelling experiments. Continent. Shelf Res. 27, 1020–1045. https://doi.org/10.1016/j.csr.2005.12.019.

Brennan, D., (2001). The numerical simulation of two phase flows in settling tanks. Ph.D thesis, Imperial College London (University of London). Available from: https://ethos.bl.uk/OrderDetails.do?uin=uk.bl.ethos.248344.

Bryan, K.R., Nardin, W., Mullarney, J.C., Fagherazzi, S., 2017. The role of cross-shore tidal dynamics in controlling intertidal sediment exchange in mangroves in Cù Lao Dung, Vietnam. Continent. Shelf Res. 147, 128–143. https://doi.org/10.1016/j.csr.2017.06.014.

Busse, A., Lützner, M., Sandham, N.D., 2015. Direct numerical simulation of turbulent flow over a rough surface based on a surface scan. Comput. Fluids 116, 129–147. https://doi.org/10.1016/j.compfluid.2015.04.008.

Cahoon, D.R., Lynch, J.C., 1997. Vertical accretion and shallow subsidence in a mangrove forest of southwestern Florida, USA. Mangroves Salt Marshes 1, 173–186. https://doi.org/10.1023/A:1009904816246.

Cahoon, D.R., 2006. A review of major storm impacts on coastal wetland elevations. Estuar. Coast 29, 889–898. https://doi.org/10.1007/BF02798648.

Chang, W.-Y., Constantinescu, G., Tsai, W.F., 2017. On the flow and coherent structures generated by a circular array of rigid emerged cylinders placed in an open channel with flat and deformed bed. J. Fluid Mech. 831, 1–40. https://doi.org/10.1017/jfm.2017.558.

Chen, Z., Ortiz, A., Zong, L., Nepf, H., 2012. The wake structure behind a porous obstruction and its implications for deposition near a finite patch of emergent vegetation. Water Resour. Res. 48 https://doi.org/10.1029/2012WR012224.

Cunha-Lignon, M., Mahiques, M.M., Schaeffer-Novelli, Y., Rodrigues, M., Klein, D.A., Goya, S.C., Menghini, R.P., Tolentino, C.C., Cintrón-Molero, G., Dahdouh-Guebas, F., 2009. Analysis of mangrove forest succession, using sediment cores: a case study in the Cananéia-Iguape coastal system, São Paulo Brazil. Braz. J. Oceanogr. 57, 161–174. https://doi.org/10.1590/S1679-87592009000300001.

De Lima, P.H., Janzen, J.G., Nepf, H.M., 2015. Flow patterns around two neighboring patches of emergent vegetation and possible implications for deposition and vegetation growth. Environ. Fluid Mech. 15, 881–898. https://doi.org/10.1007/s10652-015-9395-2.

Di Nitto, D., Neukermans, G., Koedam, N., Defever, H., Pattyn, F., Kairo, J.G., Dahdouh-Guebas, F., 2014. Mangroves facing climate change: landward migration potential in response to projected scenarios of sea level rise. Biogeosciences 11, 857–871. https://doi.org/10.5194/bg-11-857-2014.

Doughty, C.L., Cavanaugh, K.C., Hall, C.R., Feller, I.C., Chapman, S.K., 2017. Impacts of mangrove encroachment and mosquito impoundment management on coastal protection services. Hydrobiologia 803, 105–120. https://doi.org/10.1007/s10750-017-3225-0.

Dutta, S., 2016. Soil erosion, sediment yield and sedimentation of reservoir: a review. Model. Earth Syst. Environ. 2, 123. https://doi.org/10.1007/s40808-016-0182-y.

Ellis, J., Nicholls, P., Craggs, R., Hofstra, D., Hewitt, J., 2004. Effects of terrigenous sedimentation on mangrove physiology and associated macrobenthic communities. Mar. Ecol. Prog. Ser. 270, 71–82. https://doi.org/10.3354/meps270071.

Ellison, J.C., 1993. Mangrove retreat with rising sea-level, Bermuda. Estuarine. Coast. Shelf Sci. 37, 75–87. https://doi.org/10.1006/ecss.1993.1042.

Ellison, J.C., 1999. Impacts of sediment burial on mangroves. Mar. Pollut. Bull. 37, 420–426. https://doi.org/10.1016/S0025-326X(98)00122-2.

Ellison, J.C., 2019. Biogeomorphology of Mangroves. Coastal Wetlands. Elsevier.

Etminan, V., Lowe, R.J., Ghisalberti, M., 2017. A new model for predicting the drag exerted by vegetation canopies. Water Resour. Res. 53, 3179–3196. https://doi.org/10.1002/2016WR020090.

Fagherazzi, S., Bryan, K.R., Nardin, W., 2017. Buried alive or washed away: the challenging life of mangroves in the Mekong Delta. Oceanography 30, 48–59. https://doi.org/10.5670/oceanog.2017.313.

Fatimah, E., Wahab, A.K.A., Ismail, H., 2008. Numerical modeling approach of an artificial mangrove root system (ArMS) submerged breakwater as wetland habitat protector. In: Proceeding of the Seventh International Conference on Coastal and Port Engineering, PIANC-COPEDEC Vii, United Arab Emirates, pp. 1–20.

Fischer-Antze, T., Stoesser, T., Bates, P., Olsen, N., 2001. 3D numerical modelling of open-channel flow with submerged vegetation. J. Hydraul. Res. 39, 303–310. https://doi.org/10.1080/00221680109499833.

Follett, E.M., Nepf, H.M., 2012. Sediment patterns near a model patch of reedy emergent vegetation. Geomorphology 179, 141–151. https://doi.org/10.1016/j.geomorph.2012.08.006.

Franks, T., Falconer, R., 1999. Developing procedures for the sustainable use of mangrove systems. Agric. Water Manag. 40, 59–64. https://doi.org/10.1016/S0378-3774(98)00104-8.

Fricke, A., Nittrouer, C., Ogston, A., Vo-Luong, H., 2017. Asymmetric progradation of a coastal mangrove forest controlled by combined fluvial and marine influence, Cù Lao Dung, Vietnam. Continent. Shelf Res. 147, 78–90. https://doi.org/10.1016/j.csr.2017.07.012.

Friedrichs, M., Graf, G., 2009. Characteristic flow patterns generated by macrozoobenthic structures. J. Mar. Syst. 75, 348–359. https://doi.org/10.1016/j.jmarsys.2007.01.015.

Friedrichs, M., Leipe, T., Peine, F., Graf, G., 2009. Impact of macrozoobenthic structures on near-bed sediment fluxes. J. Mar. Syst. 75, 336–347. https://doi.org/10.1016/j.jmarsys.2006.12.006.

Friess, D.A., Krauss, K.W., Horstman, E.M., Balke, T., Bouma, T.J., Galli, D., Webb, E.L., 2012. Are all intertidal wetlands naturally created equal? Bottlenecks, thresholds and knowledge gaps to mangrove and saltmarsh ecosystems. Biol. Rev. 87, 346–366. https://doi.org/10.1111/j.1469-185X.2011.00198.x.

Furukawa, K., Wolanski, E., 1996. Sedimentation in mangrove forests. Mangroves Salt Marshes 1, 3–10. https://doi.org/10.1023/A:1025973426404.

Furukawa, K., Wolanski, E., Mueller, H., 1997. Currents and sediment transport in mangrove forests. Estuar. Coast Shelf Sci. 44, 301–310. https://doi.org/10.1006/ecss.1996.0120.

Gao, Y.-Y., Yin, C.-S., Zhang, H.-Q., Yang, K., Zhao, X.-Z., Sun, Z., 2017. Numerical study on flow around four square-arranged cylinders at low Reynolds numbers. Math. Probl Eng. 2017 https://doi.org/10.1155/2017/6381256.

Gillis, L.G., Hortua, D.A., Zimmer, M., Jennerjahn, T.C., Herbeck, L.S., 2019. Interactive effects of temperature and nutrients on mangrove seedling growth and implications for establishment. Mar. Environ. Res. 104750. https://doi.org/10.1016/j.marenvres.2019.104750.

Gillis, L.G., Zimmer, M., Bouma, T., 2016. Mangrove leaf transportation: do mimic Avicennia and Rhizophora roots retain or donate leaves? Mar. Ecol. Prog. Ser. 551, 107–115. https://doi.org/10.3354/meps11734.

Gillis, L.G., Bouma, T., Kiswara, W., Ziegler, A., Herman, P., 2014. Leaf transport in mimic mangrove forests and seagrass beds. Mar. Ecol. Prog. Ser. 498, 95–102. https://doi.org/10.3354/meps10615.

Graf, W., Istiarto, I., 2002. Flow pattern in the scour hole around a cylinder. J. Hydraul. Res. 40, 13–20. https://doi.org/10.1080/00221680209499869.

Guan, M., Liang, Q., 2017. A two-dimensional hydro-morphological model for river hydraulics and morphology with vegetation. Environ. Model. Software 88, 10–21. https://doi.org/10.1016/j.envsoft.2016.11.008.

Guannel, G., Arkema, K., Ruggiero, P., Verutes, G., 2016. The power of three: coral reefs, seagrasses and mangroves protect coastal regions and increase their resilience. PloS One 11, e0158094. https://doi.org/10.1371/journal.pone.0158094.

Guannel, G., Ruggiero, P., Faries, J., Arkema, K., Pinsky, M., Gelfenbaum, G., Guerry, A., Kim, C.K., 2015. Integrated modeling framework to quantify the coastal protection services supplied by vegetation. J. Geophys. Res.: Oceans 120, 324–345. https://doi.org/10.1002/2014JC009821.

Gwyther, J., Fairweather, P.G., 2002. Colonisation by epibionts and meiofauna of real and mimic pneumatophores in a cool temperate mangrove habitat. Mar. Ecol. Prog. Ser. 229, 137–149. https://doi.org/10.3354/meps229137.

Gwyther, J., Fairweather, P.G., 2005. Meiofaunal recruitment to mimic pneumatophores in a cool-temperate mangrove forest: spatial context and biofilm effects. J. Exp. Mar. Biol. Ecol. 317, 69–85. https://doi.org/10.1016/j.jembe.2004.11.012.

Harada, K., Imamura, F., Hiraishi, T.L., 2002. Experimental study on the effect in reducing tsunami by the coastal permeable structures. In: The Twelfth International Offshore and Polar Engineering Conference. International Society of Offshore and Polar Engineers.

Harichandan, A.B., Roy, A., 2010. Numerical investigation of low Reynolds number flow past two and three circular cylinders using unstructured grid CFR scheme. Int. J. Heat Fluid Flow 31, 154–171. https://doi.org/10.1016/j.ijheatfluidflow.2010.01.007.

Hashim, A.M., Catherine, S.M.P., 2013. A laboratory study on wave reduction by mangrove forests. APCBEE Procedia 5, 27–32. https://doi.org/10.1016/j.apcbee.2013.05.006.

Hayden, H.L., Granek, E.F., 2015. Coastal sediment elevation change following anthropogenic mangrove clearing. Estuar. Coast Shelf Sci. 165, 70–74. https://doi.org/10.1016/j.ecss.2015.09.004.

Hayward, B.W., Grenfell, H.R., Sabaa, A.T., Morley, M.S., Horrocks, M., 2006. Effect and timing of increased freshwater runoff into sheltered harbor environments around Auckland City, New Zeland. Estuar. Coast 29, 165–182. https://doi.org/10.1007/BF02781987.

Henderson, S.M., Norris, B.K., Mullarney, J.C., Bryan, K.R., 2017. Wave-frequency flows within a near-bed vegetation canopy. Continent. Shelf Res. 147, 91–101. https://doi.org/10.1016/j.csr.2017.06.003.

Hicks, G., 1986. Meiofauna associated with rocky shore algae. Ecol. Rocky Coasts 36–56.

Holeman, J.N., 1968. The sediment yield of major rivers of the world. Water Resour. Res. 4, 737–747. https://doi.org/10.1029/WR004i004p00737.

Holm, E., Schultz, M., Haslbeck, E., Talbott, W., Field, A., 2004. Evaluation of hydrodynamic drag on experimental fouling-release surfaces, using rotating disks. Biofouling 20, 219–226. https://doi.org/10.1080/08927010400011245.

Horstman, E.M., Lundquist, C.J., Bryan, K.R., Bulmer, R.H., Mullarney, J.C., Stokes, D.J., 2018a. The dynamics of expanding mangroves in New Zealand. In: Makowski, C., Finkl, C.W. (Eds.), Threats to Mangrove Forests: Hazards, Vulnerability, and Management. Springer International Publishing AG, pp. 23–51. https://doi.org/10.1007/978-3-319-73016-5_2.

Horstman, E.M., Bryan, K.R., Mullarney, J.C., Pilditch, C.A., Eager, C.A., 2018b. Are flow-vegetation interactions well represented by mimics? A case study of mangrove pneumatophores. Adv. Water Resour. 111, 360–371. https://doi.org/10.1016/j.advwatres.2017.11.018.

Horstman, E.M., Dohmen-Janssen, C.M., Bouma, T.J., Hulscher, S.J., 2015. Tidal-scale flow routing and sedimentation in mangrove forests: combining field data and numerical modelling. Geomorphology 228, 244–262. https://doi.org/10.1016/j.geomorph.2014.08.011.

Horstman, E.M., Mullarney, J.C., Bryan, K.R., Sandwell, D.R., 2017. Deposition gradients across mangrove fringes. In: Coastal Dynamics 2017 Conference, pp. 911–922.

Horstman, E., Balke, T., Bouma, T., Dohmen-Janssen, M., Hulscher, S., 2011. Optimizing methods to measure hydrodynamics in coastal wetlands: evaluating the use and positioning of ADV, ADCP AND HR-ADCP. Coast. Eng. Proc. 1, 51. https://doi.org/10.9753/icce.v32.waves.51.

Horstman, E., Dohmen-Janssen, C., Hulscher, S., 2013. Modeling Tidal Dynamics in a Mangrove Creek Catchment in Delft3D.

Horstman, E., Dohmen-Janssen, C.M., Narra, P., Van Den Berg, N., Siemerink, M., Hulscher, S.J., 2014. Wave attenuation in mangroves: a quantitative approach to field observations. Coast. Eng. 94, 47–62. https://doi.org/10.1016/j.coastaleng.2014.08.005.

Hosseini, N., Griffith, M.D., Leontini, J.S., 2018. A Framework for Understanding the Flow Around Groups of Cylinders.

Hu, Z., Suzuki, T., Zitman, T., Uittewaal, W., Stive, M., 2014. Laboratory study on wave dissipation by vegetation in combined current–wave flow. Coast. Eng. 88, 131–142. https://doi.org/10.1016/j.coastaleng.2014.02.009.

Huai, W., Xue, W., Qian, Z., 2015. Large-eddy simulation of turbulent rectangular open-channel flow with an emergent rigid vegetation patch. Adv. Water Resour. 80, 30–42. https://doi.org/10.1016/j.advwatres.2015.03.006.

Husrin, S., Strusińska, A., Oumeraci, H., 2012. Experimental study on tsunami attenuation by mangrove forest. Earth Planets Space 64, 15. https://doi.org/10.5047/eps.2011.11.008.

Huxham, M., Kumara, M.P., Jayatissa, L.P., Krauss, K.W., Kairo, J., Langat, J., Mencuccini, M., Skov, M.W., Kirui, B., 2010. Intra-and interspecific facilitation in mangroves may increase resilience to climate change threats. Phil. Trans. Biol. Sci. 365, 2127–2135. https://doi.org/10.1098/rstb.2010.0094.

Iribarne, O., Botto, F., Martinetto, P., Gutierrez, J.L., 2000. The role of burrows of the SW Atlantic intertidal crab *Chasmagnathus granulata* in trapping debris. Mar. Pollut. Bull. 40, 1057–1062. https://doi.org/10.1016/S0025-326X(00)00058-8.

Ish-Shalom-Gordon, N., Dubinsky, Z., 1992. Ultrastructure of the pneumatophores of the mangrove *Avicennia marina*. South Afr. J. Bot. 58, 358–362. https://doi.org/10.1016/S0254-6299(16)30822-5.

Ismail, H., Wahab, A.A., Alias, N., 2012. Determination of mangrove forest performance in reducing tsunami run-up using physical models. Nat. Hazards 63, 939–963. https://doi.org/10.1007/s11069-012-0200-y.

James, C., Birkhead, A., Jordanova, A., O'sullivan, J., 2004. Flow resistance of emergent vegetation. J. Hydraul. Res. 42, 390–398. https://doi.org/10.1080/00221686.2004.9641206.

Jennerjahn, T.C., Ittekkot, V., 2002. Relevance of mangroves for the production and deposition of organic matter along tropical continental margins. Naturwissenschaften 89, 23–30. https://doi.org/10.1007/s00114-001-0283-x.

Jennerjahn, T.C., Jänen, I., Propp, C., Adi, S., Nugroho, S.P., 2013. Environmental impact of mud volcano inputs on the anthropogenically altered Porong River and Madura Strait coastal waters, Java, Indonesia. Estuar. Coast Shelf Sci. 130, 152–160. https://doi.org/10.1016/j.ecss.2013.04.007.

Jimenez, J.A., Lugo, A.E., Cintron, G., 1985. Tree mortality in mangrove forests. Biotropica 177–185. https://doi.org/10.2307/2388214.

Johnson, M., Rice, S., 2014. Macrozoobenthos, hydraulics and sediment dynamics. In: Users Guide to Ecohydraulic Modelling and Experimentation: Experience of the Ecohydraulic Research Team (PISCES) of the HYDRALAB Network. CRC Press.

Jusoh, M.Z.M., Aziz, N.A., Inayatullah, O., 2016. Computational Fluid Dynamic Simulation of Flow Velocities Dissipation by Mangrove Roots Structure.

Kaandorp, J.A., Leiva, R.A.G., 2004. Morphological analysis of two-and three-dimensional images of branching sponges and corals. Morphometrics 83–96. https://doi.org/10.1007/978-3-662-08865-4_7.

Kakeh, N., Coco, G., Marani, M., 2016. On the morphodynamic stability of intertidal environments and the role of vegetation. Adv. Water Resour. 93, 303–314. https://doi.org/10.1016/j.advwatres.2015.11.003.

Kamal, S., Lee, S.Y., Warnken, J., 2014. Investigating three-dimensional mesoscale habitat complexity and its ecological implications using low-cost RGB-D sensor technology. Methods Ecol. Evol. 5, 845–853. https://doi.org/10.1111/2041-210X.12210.

Kamal, S., Warnken, J., Bakhtiyari, M., Lee, S.Y., 2017. Sediment distribution in shallow estuaries at fine scale: in situ evidence of the effects of three-dimensional structural complexity of mangrove pneumatophores. Hydrobiologia 803, 121–132. https://doi.org/10.1007/s10750-017-3178-3.

Kathiresan, K., 2003. How do mangrove forests induce sedimentation? Rev. Biol. Trop. 51, 355–360.

Kazemi, A., Evans, H.B., Van De Riet, K., Castillo, L., Curet, O.M., 2018b. Hydrodynamics of mangrove roots and its applications in coastal protection. In: OCEANS 2018 MTS/IEEE Charleston. IEEE, pp. 1–8.

Kazemi, A., Van De Riet, K., Curet, O.M., 2017. Hydrodynamics of mangrove-type root models: the effect of porosity, spacing ratio and flexibility. Bioinspiration Biomimetics 12, 056003. https://doi.org/10.1088/1748-3190/aa7ccf.

Kazemi, A., Van De Riet, K., Curet, O.M., 2018a. Drag coefficient and flow structure downstream of mangrove root-type models through PIV and direct force measurements. Phys. Rev. Fluids 3, 073801. https://doi.org/10.1103/PhysRevFluids.3.073801.

Khanpour, M., Zarrati, A., Kolahdoozan, M., Shakibaeinia, A., Amirshahi, S., 2016. Mesh-free SPH modeling of sediment scouring and flushing. Comput. Fluids 129, 67–78. https://doi.org/10.1016/j.compfluid.2016.02.005.

Khor, Y.S., Xiao, Q., 2011. CFD simulations of the effects of fouling and antifouling. Ocean Eng. 38, 1065–1079. https://doi.org/10.1016/j.oceaneng.2011.03.004.

Kim, H.S., Nabi, M., Kimura, I., Shimizu, Y., 2014. Numerical investigation of local scour at two adjacent cylinders. Adv. Water Resour. 70, 131–147. https://doi.org/10.1016/j.advwatres.2014.04.018.

Kim, H., Roh, M., Nabi, M., 2017. Computational modeling of flow and scour around two cylinders in staggered array. Water 9, 654. https://doi.org/10.3390/w9090654.

Kitsikoudis, V., Kirca, V.O., Yagci, O., Celik, M.F., 2017. Clear-water scour and flow field alteration around an inclined pile. Coast. Eng. 129, 59–73. https://doi.org/10.1016/j.coastaleng.2017.09.001.

Kitsikoudis, V., Yagci, O., Kirca, V.O., Kellecioglu, D., 2016. Experimental investigation of channel flow through idealised isolated tree-like vegetation. Environ. Fluid Mech. 16, 1283–1308. https://doi.org/10.1007/s10652-016-9487-7.

Kondolf, G.M., 1997. PROFILE: hungry water: effects of dams and gravel mining on river channels. Environ. Manag. 21, 533–551. https://doi.org/10.1007/s002679900048.

Kothyari, U.C., Hashimoto, H., Hayashi, K., 2009a. Effect of tall vegetation on sediment transport by channel flows. J. Hydraul. Res. 47, 700–710. https://doi.org/10.3826/jhr.2009.3317.

Kothyari, U.C., Hayashi, K., Hashimoto, H., 2009b. Drag coefficient of unsubmerged rigid vegetation stems in open channel flows. J. Hydraul. Res. 47, 691–699. https://doi.org/10.3826/jhr.2009.3283.

Krauss, K.W., Lovelock, C.E., Mckee, K.L., López-Hoffman, L., Ewe, S.M., Sousa, W.P., 2008. Environmental drivers in mangrove establishment and early development: a review. Aquat. Bot. 89, 105–127. https://doi.org/10.1016/j.aquabot.2007.12.014.

Krauss, K., Allen, J.A., Cahoon, D., 2003. Differential rates of vertical accretion and elevation change among aerial root types in Micronesian mangrove forests. Estuar. Coast Shelf Sci. 56, 251–259. https://doi.org/10.1016/S0272-7714(02)00184-1.

Kumara, M., Jayatissa, L., Krauss, K., Phillips, D., Huxham, M., 2010. High mangrove density enhances surface accretion, surface elevation change, and tree survival in coastal areas susceptible to sea-level rise. Oecologia 164, 545–553.

Le Minor, M., Bartzke, G., Zimmer, M., Gillis, L., Helfer, V., Huhn, K., 2019. Numerical modelling of hydraulics and sediment dynamics around mangrove seedlings: implications for mangrove establishment and reforestation. Estuar. Coast Shelf Sci. 217, 81–95. https://doi.org/10.1016/j.ecss.2018.10.019.

Le Minor, M., Mullarney, J.C., Pilditch, C.A., Huhn, K., 2020. Crab burrow aspect ratio influences particle capture rates on intertidal sandflats. Geo Mar. Lett. 40, 197–216. https://doi.org/10.1007/s00367-019-00630-x.

Le Nguyen, H.T., Luong, H.P.V., 2019. Erosion and deposition processes from field experiments of hydrodynamics in the coastal mangrove area of Can Gio, Vietnam. Oceanologia 61, 252–264. https://doi.org/10.1016/j.oceano.2018.11.004.

Lee, H.-Y., Shih, S.-S., 2004. Impacts of vegetation changes on the hydraulic and sediment transport characteristics in Guandu mangrove wetland. Ecol. Eng. 23, 85–94. https://doi.org/10.1016/j.ecoleng.2004.07.003.

Leong, R.C., Friess, D.A., Crase, B., Lee, W.K., Webb, E.L., 2018. High-resolution pattern of mangrove species distribution is controlled by surface elevation. Estuar. Coast Shelf Sci. 202, 185–192. https://doi.org/10.1016/j.ecss.2017.12.015.

Li, C., Yu, L., 2010. Hybrid LES/RANS modelling of free surface flow through vegetation. Comput. Fluids 39, 1722–1732.

Li, S., Meng, X., Ge, Z., Zhang, L., 2015. Evaluation of the threat from sea-level rise to the mangrove ecosystems in Tieshangang Bay, southern China. Ocean Coast Manag. 109, 1–8. https://doi.org/10.1016/j.ocecoaman.2015.02.006.

Li, L., Wang, X.H., Andutta, F., Williams, D., 2014. Effects of mangroves and tidal flats on suspended-sediment dynamics: observational and numerical study of Darwin Harbour, Australia. J. Geophys. Res. 119, 5854–5873. https://doi.org/10.1002/2014JC009987.

Li, L., Wang, X.H., Williams, D., Sidhu, H., Song, D., 2012. Numerical study of the effects of mangrove areas and tidal flats on tides: a case study of Darwin Harbour, Australia. J. Geophys. Res. 117 https://doi.org/10.1029/2011JC007494.

Li, Y., Wolanski, E., Xie, Q., 1993. Coagulation and settling of suspended sediment in the Jiaojiang river estuary, China. J. Coast Res. 390–402.

Liang, C., Papadakis, G., Luo, X., 2009. Effect of tube spacing on the vortex shedding characteristics of laminar flow past an inline tube array: a numerical study. Comput. Fluids 38, 950–964. https://doi.org/10.1016/j.compfluid.2008.10.005.

Liénard, J., Lynn, K., Strigul, N., Norris, B.K., Gatziolis, D., Mullarney, J.C., Karin, R.B., Henderson, S.M., 2016. Efficient three-dimensional reconstruction of aquatic vegetation geometry: estimating morphological parameters influencing hydrodynamic drag. Estuar. Coast Shelf Sci. 178, 77–85. https://doi.org/10.1016/j.ecss.2016.05.011.

Lima E Silva, A., Silveira-Neto, A., Damasceno, J., 2003. Numerical simulation of two-dimensional flows over a circular cylinder using the immersed boundary method. J. Comput. Phys. 189, 351–370. https://doi.org/10.1016/S0021-9991(03)00214-6.

Linton, D.M., Warner, G.F., 2003. Biological indicators in the Caribbean coastal zone and their role in integrated coastal management. Ocean Coast Manag. 46, 261–276. https://doi.org/10.1016/S0964-5691(03)00007-3.

Liu, C., Nepf, H., 2016. Sediment deposition within and around a finite patch of model vegetation over a range of channel velocity. Water Resour. Res. 52, 600–612. https://doi.org/10.1002/2015WR018249.

Liu, C., Hu, Z., Lei, J., Nepf, H., 2017. Vortex structure and sediment deposition in the wake behind a finite patch of model submerged vegetation. J. Hydraul. Eng. 144, 04017065. https://doi.org/10.1061/(ASCE)HY.1943-7900.0001408.

Liu, H., Zhang, K., Li, Y., Xie, L., 2013. Numerical study of the sensitivity of mangroves in reducing storm surge and flooding to hurricane characteristics in southern Florida. Continent. Shelf Res. 64, 51–65. https://doi.org/10.1016/j.csr.2013.05.015.

Lovelock, C.E., Cahoon, D.R., Friess, D.A., Guntenspergen, G.R., Krauss, K.W., Reef, R., Rogers, K., Saunders, M.L., Sidik, F., Swales, A., 2015. The vulnerability of Indo-Pacific mangrove forests to sea-level rise. Nature 526, 559. https://doi.org/10.1038/nature15538.

Lovelock, C.E., Feller, I.C., Ellis, J., Schwarz, A.M., Hancock, N., Nichols, P., Sorrell, B., 2007. Mangrove growth in New Zealand estuaries: the role of nutrient enrichment at sites with contrasting rates of sedimentation. Oecologia 153, 633–641. https://doi.org/10.1007/s00442-007-0750-y.

Lynch, J.C., Meriwether, J.R., Mckee, B.A., Vera-Herrera, F., Twilley, R.R., 1989. Recent accretion in mangrove ecosystems based on 137 Cs and 210 Pb. Estuaries 12, 284–299. https://doi.org/10.2307/1351907.

Macintyre, I.G., Toscano, M.A., Bond, G.B., 2004. Modern sedimentary environments, Twin Cays, Belize, Central America. Atoll Res. Bull. 1–12. https://doi.org/10.5479/si.00775630.509.1.

Mahesh, K., Constantinescu, G., Moin, P., 2004. A numerical method for large-eddy simulation in complex geometries. J. Comput. Phys. 197, 215–240. https://doi.org/10.1016/j.jcp.2003.11.031.

Marois, D.E., Mitsch, W.J., 2017. A mangrove creek restoration plan utilizing hydraulic modeling. Ecol. Eng. 108, 537–546. https://doi.org/10.1016/j.ecoleng.2017.06.063.

Massel, S., Furukawa, K., Brinkman, R., 1999. Surface wave propagation in mangrove forests. Fluid Dynam. Res. 24, 219–249. https://doi.org/10.1016/S0169-5983(98)00024-0.

Maza, M., Adler, K., Ramos, D., Garcia, A.M., Nepf, H., 2017. Velocity and drag evolution from the leading edge of a model mangrove forest. J. Geophys. Res. 122, 9144–9159. https://doi.org/10.1002/2017JC012945.

Maza, M., Lara, J.L., Losada, I.J., 2015. Tsunami wave interaction with mangrove forests: a 3-D numerical approach. Coast. Eng. 98, 33–54. https://doi.org/10.1016/j.coastaleng.2015.01.002.

Mazda, Y., Kobashi, D., Okada, S., 2005. Tidal-scale hydrodynamics within mangrove swamps. Wetl. Ecol. Manag. 13, 647–655. https://doi.org/10.1007/s11273-005-0613-4.

Mazda, Y., Magi, M., Kogo, M., Hong, P.N., 1997b. Mangroves as a coastal protection from waves in the Tong King delta, Vietnam. Mangroves Salt Marshes 1, 127–135. https://doi.org/10.1023/A:1009928003700.

Mazda, Y., Wolanski, E., King, B., Sase, A., Ohtsuka, D., Magi, M., 1997a. Drag force due to vegetation in mangrove swamps. Mangroves Salt Marshes 1, 193–199. https://doi.org/10.1023/A:1009949411068.

Mellor, P., Willson, N., Claydon, N., Paling, E.I., Parker, B., Wylie, N., Mackey, P., Houridis, H., 2018. Using a sedimentation scanner to determine mangrove health responses to sedimentation derived from dredging. An example from northwestern Australia. Environ. Technol. 39, 1497–1504. https://doi.org/10.1080/09593330.2017.1332105.

Mokhtari, M., Abd Ghaffar, M., Usup, G., Che Cob, Z., 2016. Effects of fiddler crab burrows on sediment properties in the mangrove mudflats of Sungai Sepang, Malaysia. Biology 5, 7.

Montgomery, J., Bryan, K., Horstman, E., Mullarney, J., 2018. Attenuation of tides and surges by mangroves: contrasting case studies from New Zealand. Water 10, 1119. https://doi.org/10.3390/w10091119.

Mullarney, J.C., Henderson, S.M., Norris, B.K., Bryan, K.R., Fricke, A.T., Sandwell, D.R., Culling, D.P., 2017b. A question of scale: how turbulence around aerial roots shapes the seabed morphology in mangrove forests of the Mekong Delta. Oceanography 30, 34–47. https://doi.org/10.5670/oceanog.2017.312.

Mullarney, J.C., Henderson, S.M., Reyns, J.A., Norris, B.K., Bryan, K.R., 2017a. Spatially varying drag within a wave-exposed mangrove forest and on the adjacent tidal flat. Continent. Shelf Res. 147, 102–113. https://doi.org/10.1016/j.csr.2017.06.019.

Nayak, G., Noronha-D'mello, C., 2018. Estuarine Mudflat and Mangrove Sedimentary Environments along Central West Coast of India.

Needham, H.R., Pilditch, C.A., Lohrer, A.M., Thrush, S.F., 2011. Context-specific bioturbation mediates changes to ecosystem functioning. Ecosystems 14, 1096–1109. https://doi.org/10.1007/s10021-011-9468-0.

Needham, H.R., Pilditch, C.A., Lohrer, A.M., Thrush, S.F., 2013. Density and habitat dependent effects of crab burrows on sediment erodibility. J. Sea Res. 76, 94–104. https://doi.org/10.1016/j.seares.2012.12.004.

Needham, H., Pilditch, C., Lohrer, A., Thrush, S., 2010. Habitat dependence in the functional traits of *Austrohelice crassa*, a key bioturbating species. Mar. Ecol. Prog. Ser. 414, 179–193. https://doi.org/10.3354/meps08726.

Nicolle, A., Eames, I., 2011. Numerical study of flow through and around a circular array of cylinders. J. Fluid Mech. 679, 1–31. https://doi.org/10.1017/jfm.2011.77.

Norris, B.K., Mullarney, J.C., Bryan, K.R., Henderson, S.M., 2017. The effect of pneumatophore density on turbulence: a field study in a Sonneratia-dominated mangrove forest, Vietnam. Continent. Shelf Res. 147, 114–127. https://doi.org/10.1016/j.csr.2017.06.002.

Norris, B.K., Mullarney, J.C., Bryan, K.R., Henderson, S.M., 2019. Turbulence within natural mangrove pneumatophore canopies. J. Geophys. Res. https://doi.org/10.1029/2018JC014562.

Oliver, T.S., Rogers, K., Chafer, C.J., Woodroffe, C.D., 2012. Measuring, mapping and modelling: an integrated approach to the management of mangrove and saltmarsh in the Minnamurra River estuary, southeast Australia. Wetl. Ecol. Manag. 20, 353–371. https://doi.org/10.1007/s11273-012-9258-2.

Odum, H.T., 1996. Scales of Ecological Engineering. Elsevier.

Orihuela, B., Diaz, H., Conde, J.E., 1991. Mass mortality in a mangrove roots fouling community in a hypersaline tropical lagoon. Biotropica 592–601. https://doi.org/10.2307/2388397.

Orr, M., Zimmer, M., Jelinski, D.E., Mews, M., 2005. Wrack deposition on different beach types: spatial and temporal variation in the pattern of subsidy. Ecology 86, 1496–1507. https://doi.org/10.1890/04-1486.

Osborne, D.J., Berjak, P., 1997. The making of mangroves: the remarkable pioneering role played by seeds of *Avicennia marina*. Endeavour 21, 143–147. https://doi.org/10.1016/S0160-9327(97)01077-6.

Palau-Salvador, G., Stoesser, T., Fröhlich, J., Kappler, M., Rodi, W., 2010. Large eddy simulations and experiments of flow around finite-height cylinders. Flow, Turbul. Combust. 84, 239. https://doi.org/10.1007/s10494-009-9232-0.

Pestana, D.F., Pülmanns, N., Nordhaus, I., Diele, K., Zimmer, M., 2017. The influence of crab burrows on sediment salinity in a *Rhizophora*-dominated mangrove forest in North Brazil during the dry season. Hydrobiologia 803, 295–305. https://doi.org/10.1007/s10750-017-3282-4.

Phillips, D., Kumara, M., Jayatissa, L., Krauss, K.W., Huxham, M., 2017. Impacts of mangrove density on surface sediment accretion, belowground biomass and biogeochemistry in Puttalam Lagoon, Sri Lanka. Wetlands 37, 471–483. https://doi.org/10.1007/s13157-017-0883-7.

Pongener, L., Padmavati, G., Jayabarathi, R., 2018. Meiofauna and microalgae associates on the pneumatophores of *Avicennia marina* from the coastal waters of South Andaman, India. Int. J. Zoo. Stud. 3, 203–212.

Pujol, D., Serra, T., Colomer, J., Casamitjana, X., 2013. Flow structure in canopy models dominated by progressive waves. J. Hydrol. 486, 281–292. https://doi.org/10.1016/j.jhydrol.2013.01.024.

Pülmanns, N., Nordhaus, I., Diele, K., Mehlig, U., 2015. Artificial crab burrows facilitate desalting of rooted mangrove sediment in a microcosm study. J. Mar. Sci. Eng. 3, 539–559. https://doi.org/10.3390/jmse3030539.

Purnobasuki, H., 2011. Structure of lenticels on the pneumatophores of *Avicennia marina*: as aerating device deliver oxygen in mangrove's root. J. Biota 16.

Quartel, S., Kroon, A., Augustinus, P., Van Santen, P., Tri, N., 2007. Wave attenuation in coastal mangroves in the Red River Delta, Vietnam. J. Asian Earth Sci. 29, 576–584. https://doi.org/10.1016/j.jseaes.2006.05.008.

Rahdarian, A., Niksokhan, M.H., 2017. Numerical modeling of storm surge attenuation by mangroves in protected area of mangroves of Qheshm Island. Ocean Eng. 145, 304–315. https://doi.org/10.1016/j.oceaneng.2017.09.026.

Ricardo, A.M., Silva, T., Pêgo, J.P., Maia, R., Franca, M.J., Schleiss, A., Ferreira, R.M., 2013. Turbulent flow within a random array of rigid, emergent stems: laboratory characterization of the drag coefficient. In: Proceedings of 35th IAHR World Congress. TPU, pp. 1–10.

Rodriguez, C., Stoner, A.W., 1990. The epiphyte community of mangrove roots in a tropical estuary: distribution and biomass. Aquat. Bot. 36, 117–126. https://doi.org/10.1016/0304-3770(90)90076-W.

Roulund, A., Sumer, B.M., Fredsøe, J., Michelsen, J., 2005. Numerical and experimental investigation of flow and scour around a circular pile. J. Fluid Mech. 534, 351–401. https://doi.org/10.1017/S0022112005004507.

Ruiz-Fernández, A., Agraz-Hernández, C., Sanchez-Cabeza, J., Díaz-Asencio, M., Pérez-Bernal, L., Keb, C.C., López-Mendoza, P., Y Correa, J.B., Ontiveros-Cuadras, J., Saenz, J.O., 2018. Sediment geochemistry, accumulation rates and forest structure in a large tropical mangrove ecosystem. Wetlands 38, 307–325. https://doi.org/10.1007/s13157-017-0969-2.

Saenger, P., 2002. Mangrove Silviculture and Restoration. Mangrove Ecology, Silviculture and Conservation. Springer.

Sahoo, G., Ansari, Z.A., 2018. Seasonality and vertical distribution of epiphytic diatoms on the pneumatophore surface of mangrove *Avicennia officinalis L.* Mar. Biodivers. 48, 2033–2041. https://doi.org/10.1007/s12526-017-0723-2.

Sahoo, G., Ansari, Z., Shaikh, J.B., Varik, S.U., Gauns, M., 2018. Epibiotic communities (microalgae and meiofauna) on the pneumatophores of *Avicennia officinalis (L.)*. Estuarine. Coast. Shelf Sci. 207, 391–401. https://doi.org/10.1016/j.ecss.2017.08.018.

Salmo, S.G., Lovelock, C., Duke, N.C., 2013. Vegetation and soil characteristics as indicators of restoration trajectories in restored mangroves. Hydrobiologia 720, 1–18. https://doi.org/10.1007/s10750-013-1617-3.

Sasmito, S.D., Murdiyarso, D., Friess, D.A., Kurnianto, S., 2016. Can mangroves keep pace with contemporary sea level rise? A global data review. Wetl. Ecol. Manag. 24, 263–278. https://doi.org/10.1007/s11273-015-9466-7.

Schmeeckle, M.W., 2014. Numerical simulation of turbulence and sediment transport of medium sand. J. Geophys. Res.: Earth Surface 119, 1240–1262. https://doi.org/10.1002/2013JF002911.

Scholander, P., Van Dam, L., Scholander, S.I., 1955. Gas exchange in the roots of mangroves. Am. J. Bot. 42, 92–98. https://doi.org/10.1002/j.1537-2197.1955.tb11097.x.

Schultz, M.P., 2004. Frictional resistance of antifouling coating systems. Trans. ASME-I J. Fluids Eng. 126, 1039–1047. https://doi.org/10.1115/1.1845552.

Scoffin, T.P., 1970. The trapping and binding of subtidal carbonate sediments by marine vegetation in Bimini Lagoon, Bahamas. J. Sediment. Res. 40.

Shan, Y., Liu, C., Nepf, H., 2019. Comparison of drag and velocity in model mangrove forests with random and in-line tree distributions. J. Hydrol. 568, 735–746. https://doi.org/10.1016/j.jhydrol.2018.10.077.

Shu, C., Liu, N., Chew, Y.-T., 2007. A novel immersed boundary velocity correction–lattice Boltzmann method and its application to simulate flow past a circular cylinder. J. Comput. Phys. 226, 1607–1622. https://doi.org/10.1016/j.jcp.2007.06.002.

Sidik, F., Neil, D., Lovelock, C.E., 2016. Effect of high sedimentation rates on surface sediment dynamics and mangrove growth in the Porong River, Indonesia. Mar. Pollut. Bull. 107, 355–363. https://doi.org/10.1016/j.marpolbul.2016.02.048.

Sioma, A., Socha, J., Klamerus-Iwan, A., 2018. A new method for characterizing bark microrelief using 3d vision systems. Forests 9, 30. https://doi.org/10.3390/f9010030.

Smoak, J.M., Patchineelam, S.R., 1999. Sediment mixing and accumulation in a mangrove ecosystem: evidence from 210Pb, 234Th and 7Be. Mangroves Salt Marshes 3, 17–27. https://doi.org/10.1023/A:1009979631884.

Spenceley, A., 1977. The role of pneumatophores in sedimentary processes. Mar. Geol. 24, M31–M37. https://doi.org/10.1016/0025-3227(77)90001-9.

Stoesser, T., Kim, S., Diplas, P., 2010. Turbulent flow through idealised emergent vegetation. J. Hydraul. Eng. 136, 1003–1017. https://doi.org/10.1061/(ASCE)HY.1943-7900.0000153.

Stoesser, T., Liang, C., Rodi, W., Jirka, G., 2006. Large eddy simulation of fully-developed turbulent flow through submerged vegetation. In: International Conference on Fluvial Hydraulics, Sep 06–08. Lisbon, Portugal, pp. 227–234.

Stoesser, T., Salvador, G.P., Rodi, W., Diplas, P., 2009. Large eddy simulation of turbulent flow through submerged vegetation. Transport Porous Media 78, 347–365. https://doi.org/10.1007/s11242-009-9371-8.

Strusińska-Correia, A., Husrin, S., Oumeraci, H., 2013. Tsunami damping by mangrove forest: a laboratory study using parameterized trees. Nat. Hazards Earth Syst. Sci. 13, 483. https://doi.org/10.5194/nhess-13-483-2013.

Struve, J., Falconer, R.A., Wu, Y., 2003. Influence of model mangrove trees on the hydrodynamics in a flume. Estuar. Coast. Shelf Sci. 58, 163–171. https://doi.org/10.1016/S0272-7714(03)00072-6.

Swales, A., Williamson, R.B., Van Dam, L.F., Stroud, M.J., Mcglone, M.S., 2002. Reconstruction of urban stormwater contamination of an estuary using catchment history and sediment profile dating. Estuaries 25, 43–56. https://doi.org/10.1007/BF02696048.

Syvitski, J.P., Vörösmarty, C.J., Kettner, A.J., Green, P., 2005. Impact of humans on the flux of terrestrial sediment to the global coastal ocean. Science 308, 376–380. https://doi.org/10.1126/science.1109454.

Tan, C., Huang, B., Liu, D., Qiu, J., Chen, H., Li, Y., Hu, Z., 2019. Effect of mimic vegetation with different stiffness on regular wave propagation and turbulence. Water 11, 109. https://doi.org/10.3390/w11010109.

Tanaka, N., Yagisawa, J., 2010. Flow structures and sedimentation characteristics around clump-type vegetation. J. Hydro-Environ. Res. 4, 15–25. https://doi.org/10.1016/j.jher.2009.11.002.

Tang, H., Tian, Z., Yan, J., Yuan, S., 2014. Determining drag coefficients and their application in modelling of turbulent flow with submerged vegetation. Adv. Water Resour. 69, 134–145. https://doi.org/10.1016/j.advwatres.2014.04.006.

Tanino, Y., Nepf, H.M., 2008. Lateral dispersion in random cylinder arrays at high Reynolds number. J. Fluid Mech. 600, 339–371. https://doi.org/10.1017/S0022112008000505.

Tanino, Y., Nepf, H.M., 2009. Laboratory investigation of lateral dispersion within dense arrays of randomly distributed cylinders at transitional Reynolds number. Phys. Fluids 21, 046603. https://doi.org/10.1063/1.3119862.

Teh, S.Y., Koh, H.L., Liu, P.L.F., Ismail, A.I.M., Lee, H.L., 2009. Analytical and numerical simulation of tsunami mitigation by mangroves in Penang, Malaysia. J. Asian Earth Sci. 36, 38–46. https://doi.org/10.1016/j.jseaes.2008.09.007.

Terrados, J., Thampanya, U., Srichai, N., Kheowvongsri, P., Geertz-Hansen, O., Boromthanarath, S., Panapitukkul, N., Duarte, C.M., 1997. The effect of increased sediment accretion on the survival and growth of *Rhizophora apiculata* seedlings. Estuar. Coast Shelf Sci. 45, 697–701. https://doi.org/10.1006/ecss.1997.0262.

Thampanya, U., Vermaat, J.E., Terrados, J., 2002. The effect of increasing sediment accretion on the seedlings of three common Thai mangrove species. Aquat. Bot. 74, 315–325. https://doi.org/10.1016/S0304-3770(02)00146-8.

Thrush, S., Hewitt, J., Cummings, V., Ellis, J., Hatton, C., Lohrer, A., Norkko, A., 2004. Muddy waters: elevating sediment input to coastal and estuarine habitats. Front. Ecol. Environ. 2, 299–306. https://doi.org/10.1890/1540-9295(2004)002[0299:MWESIT]2.0.CO;2.

Tinoco, R., Coco, G., 2018. Turbulence as the main driver of resuspension in oscillatory flow through vegetation. J. Geophys. Res. 123, 891–904. https://doi.org/10.1002/2017JF004504.

Tinoco, R.O., Coco, G., 2016. A laboratory study on sediment resuspension within arrays of rigid cylinders. Adv. Water Resour. 92, 1–9. https://doi.org/10.1016/j.advwatres.2016.04.003.

Tomlinson, P.B., 2016. The Botany of Mangroves. Cambridge University Press.

Townsin, R., 2003. The ship hull fouling penalty. Biofouling 19, 9–15. https://doi.org/10.1080/0892701031000088535.

Van Maanen, B., Coco, G., Bryan, K.R., 2015. On the ecogeomorphological feedbacks that control tidal channel network evolution in a sandy mangrove setting. Proc. Math. Phys. Eng. Sci. 471, 20150115. https://doi.org/10.1098/rspa.2015.0115.

Van Santen, P., Augustinus, P., Janssen-Stelder, B., Quartel, S., Tri, N., 2007. Sedimentation in an estuarine mangrove system. J. Asian Earth Sci. 29, 566–575. https://doi.org/10.1016/j.jseaes.2006.05.011.

Vargas-Luna, A., Crosato, A., Calvani, G., Uijttewaal, W.S., 2016. Representing plants as rigid cylinders in experiments and models. Adv. Water Resour. 93, 205–222. https://doi.org/10.1016/j.advwatres.2015.10.004.

Wahl, M., 1996. Fouled snails in flow: potential of epibionts on *Littorina littorea* to increase drag and reduce snail growth rates. Mar. Ecol. Prog. Ser. 138, 157–168. https://doi.org/10.3354/meps138157.

White, B.L., Nepf, H.M., 2003. Scalar transport in random cylinder arrays at moderate Reynolds number. J. Fluid Mech. 487, 43–79. https://doi.org/10.1017/S0022112003004579.

White, B.L., Nepf, H.M., 2007. Shear instability and coherent structures in shallow flow adjacent to a porous layer. J. Fluid Mech. 593, 1–32. https://doi.org/10.1017/S0022112007008415.

Willemsen, P., Horstman, E., Borsje, B., Friess, D.A., Dohmen-Janssen, C., 2016. Sensitivity of the sediment trapping capacity of an estuarine mangrove forest. Geomorphology 273, 189–201. https://doi.org/10.1016/j.geomorph.2016.07.038.

Witman, J.D., Suchanek, T.H., 1984. Mussels in flow: drag and dislodgement by epizoans. Mar. Ecol. Prog. Ser. 16, 259–268. https://doi.org/10.3354/meps016259.

Wolanski, E., 1992. Hydrodynamics of mangrove swamps and their coastal waters. Hydrobiologia 247, 141–161. https://doi.org/10.1007/BF00008214.

Wolanski, E., 1994. Physical Oceanographic Processes of the Great Barrier Reef. CRC Press.

Wolanski, E., 1995. Transport of sediment in mangrove swamps. In: Asia-Pacific Symposium on Mangrove Ecosystems. Springer, pp. 31–42.

Wolanski, E., Mazda, Y., King, B., Gay, S., 1990. Dynamics, flushing and trapping in Hinchinbrook Channel, a giant mangrove swamp, Australia. Estuar. Coast Shelf Sci. 31, 555–579. https://doi.org/10.1016/0272-7714(90)90013-H.

Wolanski, E., Spagnol, S., Ayukai, T., 1998. Field and model studies of the fate of particulate carbon in mangrove-fringed Hinchinbrook Channel, Australia. Mangroves Salt Marshes 2, 205–221. https://doi.org/10.1023/A:1009971326046.

Wu, Y., Falconer, R.A., Struve, J., 2001. Mathematical modelling of tidal currents in mangrove forests. Environ. Model. Software 16, 19–29. https://doi.org/10.1016/S1364-8152(00)00059-1.

Yagci, O., Celik, M.F., Kitsikoudis, V., Kirca, V.O., Hodoglu, C., Valyrakis, M., Duran, Z., Kaya, S., 2016. Scour patterns around isolated vegetation elements. Adv. Water Resour. 97, 251–265. https://doi.org/10.1016/j.advwatres.2016.10.002.

Yagci, O., Yildirim, I., Celik, M.F., Kitsikoudis, V., Duran, Z., Kirca, V.O., 2017. Clear water scour around a finite array of cylinders. Appl. Ocean Res. 68, 114–129. https://doi.org/10.1016/j.apor.2017.08.014.

Yager, E., Schmeeckle, M., 2013. The influence of vegetation on turbulence and bed load transport. J. Geophys. Res.: Earth Surface 118, 1585–1601. https://doi.org/10.1002/jgrf.20085.

Yanagisawa, H., Koshimura, S., Goto, K., Miyagi, T., Imamura, F., Ruangrassamee, A., Tanavud, C., 2009. The reduction effects of mangrove forest on a tsunami based on field surveys at Pakarang Cape, Thailand and numerical analysis. Estuar. Coast Shelf Sci. 81, 27–37. https://doi.org/10.1016/j.ecss.2008.10.001.

Yanagisawa, H., Koshimura, S., Miyagi, T., Imamura, F., 2010. Tsunami damage reduction performance of a mangrove forest in Banda Aceh, Indonesia inferred from field data and a numerical model. J. Geophys. Res.: Oceans 115. https://doi.org/10.1029/2009JC005587.

Yang, J., Chung, H., Nepf, H., 2016. The onset of sediment transport in vegetated channels predicted by turbulent kinetic energy. Geophys. Res. Lett. 43, 11,261–11,268. https://doi.org/10.1002/2016GL071092.

Yang, J., Gao, J., Cheung, A., Liu, B., Schwendenmann, L., Costello, M.J., 2013. Vegetation and sediment characteristics in an expanding mangrove forest in New Zealand. Estuar. Coast. Shelf Sci. 134, 11–18. https://doi.org/10.1016/j.ecss.2013.09.017.

Yao, P., Chen, H., Huang, B., Tan, C., Hu, Z., Ren, L., Yang, Q., 2018. Applying a new force–velocity synchronizing algorithm to derive drag coefficients of rigid vegetation in oscillatory flows. Water 10, 906. https://doi.org/10.3390/w10070906.

Young, B.M., Harvey, E.L., 1996. A spatial analysis of the relationship between mangrove (*Avicennia marina var. australasica*) physiognomy and sediment accretion in the Hauraki Plains, New Zealand. Estuar. Coast. Shelf Sci. 42, 231–246. https://doi.org/10.1006/ecss.1996.0017.

Zhang, K., Liu, H., Li, Y., Xu, H., Shen, J., Rhome, J., Smith Iii, T.J., 2012. The role of mangroves in attenuating storm surges. Estuarine. Coast. Shelf Sci. 102, 11–23. https://doi.org/10.1016/j.ecss.2012.02.021.

Zhang, Q., Zhou, X.-L., Wang, J.-H., 2017. Numerical investigation of local scour around three adjacent piles with different arrangements under current. Ocean Eng. 142, 625–638. https://doi.org/10.1016/j.oceaneng.2017.07.045.

Zhang, X., Chua, V.P., Cheong, H.-F., 2015a. Hydrodynamics in mangrove prop roots and their physical properties. J. Hydro-Environ. Res. 9, 281–294. https://doi.org/10.1016/j.jher.2014.07.010.

Zhang, X., Chua, V.P., Cheong, H.-F., 2015b. Geometrical and material properties of *Sonneratia alba* mangrove roots. Trees 29, 285–297. https://doi.org/10.1007/s00468-014-1117-8.

Zhao, M., Cheng, L., Zang, Z., 2010. Experimental and numerical investigation of local scour around a submerged vertical circular cylinder in steady currents. Coast. Eng. 57, 709–721. https://doi.org/10.1016/j.coastaleng.2010.03.002.

Zheng, S., Zhang, W., Lv, X., 2016. Numerical simulation of cross-flow around three equal diameter cylinders in an equilateral-triangular configuration at low Reynolds numbers. Comput. Fluids 130, 94–108. https://doi.org/10.1016/j.compfluid.2016.02.013.

Zimmer, M., 2018. Ecosystem design: when mangrove ecology meets human needs. In: Makowski, C., Finkl, C.W. (Eds.), Threats to Mangrove Forests: Hazards, Vulnerability, and Management. Springer International Publishing AG, pp. 23–51. https://doi.org/10.1007/978-3-319-73016-5_16.

Zong, L., Nepf, H., 2010. Flow and deposition in and around a finite patch of vegetation. Geomorphology 116, 363–372. https://doi.org/10.1016/j.geomorph.2009.11.020.

Zong, L., Nepf, H., 2011. Spatial distribution of deposition within a patch of vegetation. Water Resour. Res. 47 https://doi.org/10.1029/2010WR009516.

CHAPTER

5

Morphological plasticity and survival thresholds of mangrove plants growing in active sedimentary environments

Alejandra G. Vovides[1], *Uta Berger*[2], *Thorsten Balke*[1]

[1]School of Geographical and Earth Sciences, University of Glasgow, Scotland, United Kingdom; [2]Institute of Forest Growth and Forest Computer Sciences, Technische Universität Dresden, Tharandt, Germany

OUTLINE

1. Introduction 121
2. Colonization and biogeomorphic succession 124
3. Wind resistance and mechanical stability in unstable sediments 127
4. Root responses for mechanical stability 129
   4.1 Plate root systems 130
   4.2 Buttresses 131
   4.3 Flying buttresses and stilt roots 132
5. Plastic responses to sediment dynamics in mature forests 134
6. Concluding remarks 135
References 135

1. Introduction

Morphological plasticity, also called phenotypic plasticity, refers to the potential of organisms to change specific anatomical traits in response to different environments independent

Dynamic Sedimentary Environments of Mangrove Coasts
https://doi.org/10.1016/B978-0-12-816437-2.00025-2

of their genotype (Richards et al., 2006; Sultan, 2003, 2000). Variation in phenotypes can occur as a response to neighboring individuals (e.g., biotic interactions) or changes in abiotic conditions (e.g., nutrients, water, light, temperature, wind, etc.). Plasticity generally allows organisms to optimize their performance, to cope with stress and increase their survival (Callaway et al., 2003; Grime and Mackey, 2002; Miner et al., 2005; Wolf et al., 2011). An example of morphological plasticity in plants is the differential allocation of biomass to roots when nutrients are limited, or to branches and leaves if light is the growth-limiting resource.

Plastic responses that increase individual fitness are considered adaptive (Longuetaud et al., 2013; Sultan, 2003), potentially contributing to species persistence in changing environments (Burggren, 2018). This chapter focuses on our current understanding of mangrove plastic traits that provide an advantage in withstanding abiotic forcing of active coastal sedimentary environments from seedlings to fully grown trees. We will identify knowledge gaps that need to be filled in order to fully appreciate the biophysical interactions along mangrove coasts that determine mangrove resilience and resistance in a changing climate.

Mangroves show not only a range of plastic responses to environmental variability such as salinity and soil anoxia (Ball, 1988; Feller et al., 2010; Tomlinson, 2016) but also in response to coastal physical processes (erosion, sedimentation, hydrodynamic forcing, and wind) and biotic interactions like tree-to-tree competition (Balke et al., 2013; Dahdouh-Guebas et al., 2007, 2004; Vovides et al., 2018). Equipped with mechanisms for salt exclusion or excretion, aerial structures for oxygen exchange, and different root architectures to colonize unstable sediments, the morphological traits of mangrove trees respond to the magnitude of the experienced stress (e.g., pneumatophore height increases with flooding frequency) (Toma et al., 1991; Dahdouh-Guebas et al., 2007) and play critical roles in the survival of individuals throughout their ontogeny.

One of the most crucial life history stages of mangrove trees is the seedling stage. Seedlings not only are most susceptible to stress and disturbance but also show plastic traits that aid rapid anchorage in sedimentary environments (Augustinus, 1996; Balke et al., 2014, 2013, 2011; Krauss et al., 2014). This is especially important on bare tidal flats, where the mangrove canopy cannot provide shelter from wave attack and erosion (Van Santen et al., 2007) to establishing individuals. Thus, the ability of seedlings to modify their morphological traits can determine their survival and fitness (Burggren, 2018; Simpson et al., 2017). In fully established and mature trees, morphological plasticity still occurs and remains a key factor in reducing competition, promoting species coexistence, and in the maintenance of mechanical integrity in unstable sediments (Grueters et al., 2014; Peters et al., 2018; Vovides et al., 2018).

Plant traits can also shape biophysical and biogeomorphic interactions. For example, stilt roots and stems attenuate wave action, while pneumatophores create bottom friction that can further reduce current velocity and promote sedimentation (Horstman et al., 2016; Le Minor et al., 2019; Mazda et al., 2006; Mullarney et al., 2017). Pneumatophore density can also prevent sediment burial in the fringe zones by creating localized scour (Mullarney et al., 2017; Spenceley, 1977). This shows that phenotypic plasticity of mangroves in sedimentary environments is not only an exclusive response to abiotic stressors but is, at least partially, also influenced by the individual species' ecosystem engineering capacity (Jones et al., 1994). The environment can, thus, modulate the response of a plastic trait, whereas effect traits determine the ecosystem engineering capacity of plants to change their abiotic environment (Jones et al., 1994). A given trait can be both a response and an effect trait; pneumatophore height, for example, is a response trait that not only changes with elevation but can also act as an effect trait by increasing sedimentation rates or promoting scouring depending on density and diameter (Dahdouh-Guebas et al., 2007; Mullarney et al., 2017), as shown in Fig. 5.1.

Environmental drivers of mangrove establishment and development are well studied from an ecophysiological perspective (reviewed by Krauss et al., 2008). Changes of anatomical and physiological traits, including dwarfing, can be associated with coping mechanisms for nutrient limitations (Feller, 1995; Feller et al., 2010), salt stress (Ball and Farquhar, 1984; López-Portillo et al., 2014; Méndez-Alonzo et al., 2013), and low temperatures (Cook-Patton et al., 2015; Osland et al., 2014). Such ecophysiological responses are critical as they determine plant architecture and, hence, forest structure and function (Ewel et al., 1998). Reduced growth rates and plant height increase vessel density and reduce vessel size in response to salt stress (Ball, 1988; Robert et al., 2009; Schmitz et al., 2006; Sobrado, 2007), for example, to ensure a safe hydraulic architecture that reduces the risk of embolisms (Robert et al., 2009).

FIGURE 5.1 *Avicennia alba* pneumatophores and a partially excavated seedling at the seaward mangrove fringe illustrating biophysical interactions of hydrodynamics and plant traits (see Balke et al., 2013). *Thorsten Balke.*

Other expressions of plastic responses to salt stress include reduced slenderness and a shift from above- to below-ground biomass allocation (Castañeda-Moya et al., 2011; Peters et al., 2018, 2014). These responses create changes in allometric relationships and also modulate local interactions. With smaller trees, reduced size asymmetry, and increased distance between neighbors, above-ground competition is reduced with salinity increasing up to 60 ppt (Méndez-Alonzo et al., 2012; Vovides et al., 2014, 2018) within tropical coasts. In sites with extreme droughts, where salinity can reach >80 ppt (Vovides et al., 2011), or under nutrient-limiting conditions (Lovelock et al., 2004), trees develop a shrub-like architecture and facilitate each other by changing microclimatic conditions or species associations (Pranchai et al., 2018; Vogt et al., 2014).

2. Colonization and biogeomorphic succession

Not only coastal mangroves and marshes but also dune vegetation and riparian floodplain forests all closely interact with their abiotic environment by stabilizing sediments and reducing hydrodynamic energy and local sediment resuspension. This stabilizing effect allows for further succession after pioneers have established sufficiently and can facilitate further growth of secondary species. The pioneer species hence act as ecosystem engineers (Jones et al., 1994) that start a biogeomorphic succession (Corenblit et al., 2007). This succession can only be initiated when pioneer species are able to colonize a new bare habitat and when growth and survival surpass a biomass density threshold, from which sufficient positive (biogeomorphic) feedbacks to vegetation growth are generated (Fig. 5.2, Balke et al., 2014).

Mangrove forests cannot enter this succession trajectory if the sedimentary environment is not active (i.e., there is no sediment transport and flooding), or if too strong or too

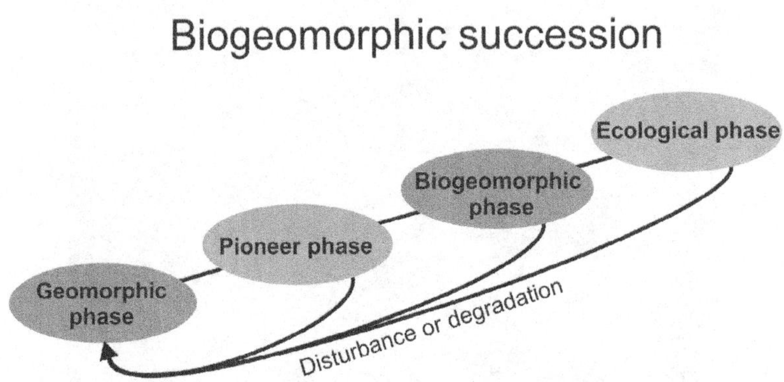

FIGURE 5.2 Biogeomorphic succession adopted after Corenblit et al., (2007). For mangroves specifically, the geomorphic phase represents the unvegetated tidal flat. During the pioneer phase, seedlings establish on the tidal flat, exposed to waves and sediment dynamics, without sufficiently impacting abiotic forcing yet. As the vegetation gets denser and taller, it significantly attenuates hydrodynamic energy, promoting sediment accretion and stabilization during the biogeomorphic phase. Further, it increases plant growth and sedimentation, leading to an ecological state no longer significantly impacted by inundation, as the system gradually accretes out of the tidal frame. *Thorsten Balke, adapted from Corenblit, D., Tabacchi, E., Steiger, J., Gurnell, A.M., 2007. Reciprocal interactions and adjustments between fluvial landforms and vegetation dynamics in river corridors: a review of complementary approaches. Earth Sci. Rev. 84, 56–86.*

frequent disturbances on the tidal flat prevent mangrove establishment. As Balke et al. (2011) demonstrated, mangrove seedlings can be dislodged by waves and currents before the bed shear stress for significant sediment erosion is exceeded (see conceptual diagram of Fig. 5.3). Mangrove propagules are predominantly buoyant directly after they have dispersed, and initial rooting is too shallow to anchor propagules against water movement. Hence, mangrove propagules require a disturbance/inundation-free period of several days (a defined Window of Opportunity, WoO) to securely anchor against further flooding events. The length of this WoO depends mainly on the root growth rates of the pioneer species, which can be up to 1.5 cm/day for *Avicennia alba* (Balke unpublished data) and, secondly, on the forcing experienced upon flooding (i.e., spring or neap tide, calm or windy day).

Once the seedling is sufficiently anchored, root expansion will have to continue at a high rate in order to surpass erosion or burial thresholds. Where dislodgement is no longer possible, sediment erosion around the seedling can lead to toppling and sediment accretion on the tidal flat can lead to lethal burial and seedling smothering (Balke et al., 2013; Thampanya et al., 2002). As long as erosion and burial are below lethal thresholds, seedlings can adjust to burial by increasing above-ground biomass over root biomass and to erosion by increasing root biomass over above-ground biomass. Lethal thresholds strongly depend on the seedling's response time to disturbances and the time between disturbance events. Plastic

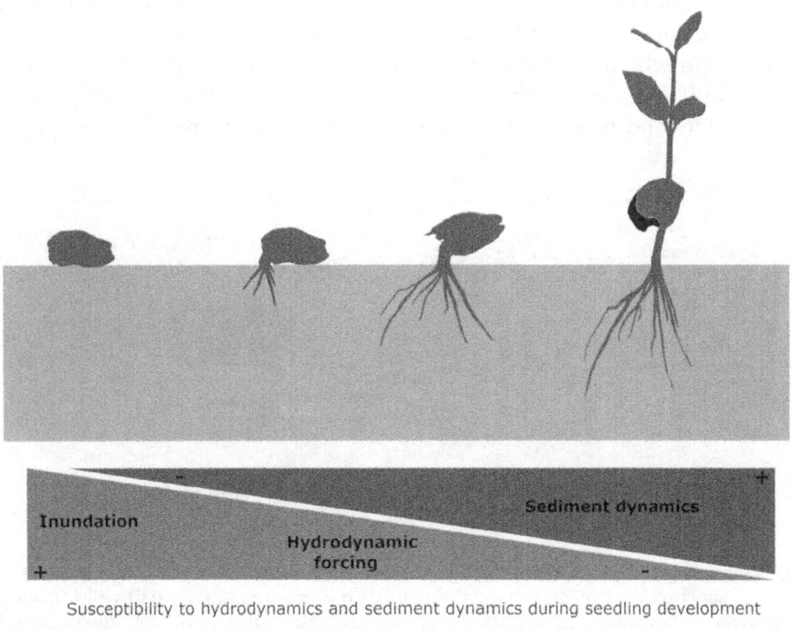

FIGURE 5.3 Stages of mangrove seedling establishment and the variation in susceptibility to hydrodynamics and sediment dynamics. Developmental plasticity allows allocating resources to root biomass, increasing anchorage, thus reducing susceptibility dislodgement by currents. Accretion can trigger shoot elongation, while erosion induces root development. *Adapted from Balke, T., Bouma, T.J., Horstman, E.M., Webb, E.L., Erftemeijer, P.L.A., Herman, P.M.J., 2011. Windows of opportunity: thresholds to mangrove seedling establishment on tidal flats. Mar. Ecol. Prog. Ser. 440, 1–9.*

responses affecting the seedling's architecture in combination with the abiotic disturbance history will then change future failure thresholds. For example, when a period of gradual burial is ended by an abrupt erosion event, the morphological adjustment (root:shoot biomass allocation) achieved in response to burial depth and time will determine how much erosion a seedling can withstand without toppling (Fig. 5.4, Balke et al., 2013).

The ability of species to change the direction of biomass allocation between root and shoot points to morphological plasticity as a key factor for colonization success (Balke et al., 2013; Burggren, 2018; Simpson et al., 2017). Pioneer species with high developmental plasticity (i.e., with fast morphologic responses to environmental variations during ontogeny) will be more successful in colonizing highly dynamic sedimentary environments than species with low developmental plasticity (Burggren, 2018). However, rates of change in sedimentary conditions need to match rates of plant morphological plasticity in order to be meaningful to the individual seedling (Balke et al., 2013).

While the colonization phase of a tidal flat is driven by physical processes, it can only be vegetated by species with pioneer traits (e.g., light demanding and fast growth) and a phenotypic plasticity capable of rapidly responding to sudden environmental changes or strong gradients (Burggren, 2018). Success is determined by a species ability to maintain fitness under harsh conditions, and further take advantage of resources under optimal conditions (Richards et al., 2006). These properties may vary between species and environments. For instance, Thampanya et al. (2002) found that *Sonneratia caseolaris* had higher growth and survival rates to sediment burial than *Avicennia marina* in southern Thailand, while Balke et al. (2013) reported opposite trends for *Sonneratia alba* and *A. alba* in mesocosm experiments. Site-specific conditions might trigger different responses of species traits, such as exposure to abiotic forcing, nutrient use efficiency, or sediment type (Boizard and Mitchell, 2011). A mechanistic understanding of species- and site-specific plastic responses to physical forcing

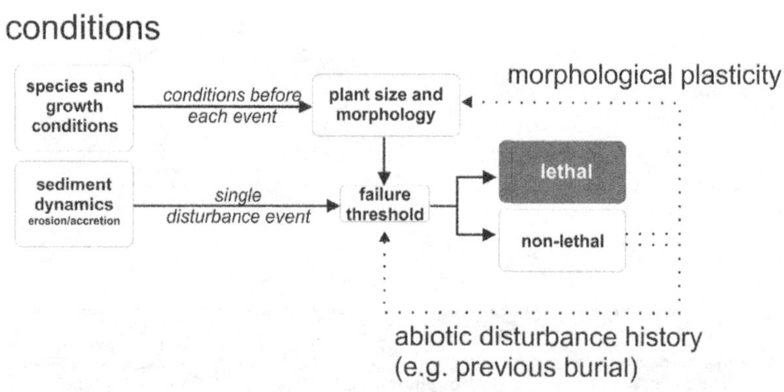

FIGURE 5.4 Conceptual diagram of seedling survival in active sedimentary environments after they have sufficiently anchored against dislodgement. Sediment disturbance events (erosion/burial) will impact plant biomass allocation and therefore future failure thresholds. The abiotic disturbance history of each site (i.e., erosion/burial the seedling has experienced since establishment) also determines seedlings net burial or excavation and hence how much further disturbance they can withstand. The sediment level change rate is crucial, as plants need time to morphologically adjust under rapidly fluctuating conditions (e.g., erosion, accretion). *Adapted from Balke, T., Webb, E.L., van den Elzen, E., Galli, D., Herman, P.M.J., Bouma, T.J., 2013. Seedling establishment in a dynamic sedimentary environment: a conceptual framework using mangroves. J. Appl. Ecol. 50, 740–747.*

during early establishment of mangrove pioneers is important to understand lateral change and for successful mangrove restoration designs.

3. Wind resistance and mechanical stability in unstable sediments

While establishment and stability of pioneer seedlings on tidal flats is mainly dependent on WoO, growth rates, and plastic biomass allocation patterns (Balke et al., 2013; Simpson et al., 2017), large trees are additionally exposed to neighborhood pressure and extreme weather events. Mangrove forests develop on soft unstable sediments and are frequently impacted by tropical storms and hurricanes (Das and Crépin, 2013). The particular morphology of mechanical support structures, such as stilt roots and buttresses, and forest densities are known to mitigate the impact of storm surges, providing valuable coastal protection services (Barbier et al., 2011; Cuc et al., 2015; Das and Crépin, 2013). But the effect that weather events excerpt on supporting structures and their plastic responses to optimize mechanical stability has been much less explored.

Morphological responses of trees to wind dynamic loading, for example, have been well studied in terrestrial forests (MacFarlane and Kane, 2017; Nicoll and Dunn, 2000; Stokes et al., 1997, 1995), but rarely explored in mangrove ecosystems, despite their susceptibility to storm surges and their regulatory ecosystem services. Wind dynamic loading can affect growth rates and survival of trees (MacFarlane and Kane, 2017; Schindler et al., 2012), compromising their mechanical stability. Variations in wood properties, stem to height ratios, and in crown shape and orientation are some of the traits identified that can reduce the negative impact of wind.

In open areas and forest edges, trees possess wide and dense crowns, reduced slenderness, and higher flexural stiffness. These traits are suitable for wind resistance, as dense crowns increase damping, reducing harmonic sway motions and hence increase mechanical stability (Brüchert and Gardiner, 2006; MacFarlane and Kane, 2017; Schindler et al., 2012). While inside the forests, although higher slenderness coefficients are a compromise between light capture and mechanical stability, the higher risk of mechanical failure is compensated by the protection by neighbors, reducing wind forces (Rudnicki et al., 2004).

While the change in allometric relations is a prerequisite for mangrove survival under harsh saline conditions, mechanical stability of mangrove trees interacts with morphological responses to salinity (López-Portillo et al., 2014; Peters et al., 2018, 2014). Higher salinity reduces growth rates, limits tree maximum height, and may prompt wider stems (Peters et al., 2014; Vovides et al., 2014). These responses to salinity are likely to provide a twofold benefit for trees: It can ensure a safe hydraulic architecture by reducing the risk of embolisms in xylem tissues (López-Portillo et al., 2005; Peters et al., 2014; Robert et al., 2009), while also increasing mechanical resistance to wind dynamic loading (Brüchert and Gardiner, 2006; Telewski and Jaffe, 1986).

In terrestrial forests, adaptive growth during tree development allows matching structural properties to environmental variations through responses to light availability (Aakala et al., 2015), crowding by neighbors (Jucker et al., 2015), and wind dynamic loading (MacFarlane and Kane, 2017). While tree height is an adaptive trait that gives light capture advantages to canopy trees, light demanding species also express high crown plasticity (Brisson, 2001;

Longuetaud et al., 2013). Tree crowns are particularly responsive to the biotic and abiotic environment, they can effectively reduce neighborhood competition (Aakala et al., 2015; Brisson, 2001; Longuetaud et al., 2013) and wind exposure when flagged-shaped crowns are positioned parallel to wind direction (Stubbs et al., 2019; Telewski and Jaffe, 1986).

In a study by Vovides et al. (2018), mangrove crowns showed a high degree of crown asymmetry along an increasing salinity gradient, although increasing salinity reduces tree size and above-ground competition for light (Vovides et al., 2014, 2018), which would hypothetically lead to more symmetric crowns. Crown asymmetry and direction of displacement, at increasing salt stress, were correlated to mean wind direction and speed (Vovides et al., 2018), suggesting a shift of the main drivers of crown plasticity from biotic to abiotic as environmental stress increases. In the vertical scale, in benign environments with salinities <35 ppt, at the upper stratum crown displacement is driven by wind, thus crowns are displaced leeward from the mean wind direction (Vovides et al., 2018), whereas at the bottom layers, the upper canopy provides a wind damping effect and below-canopy crown displacement is driven mainly by light availability (Fig. 5.5Aand B). While in more exposed

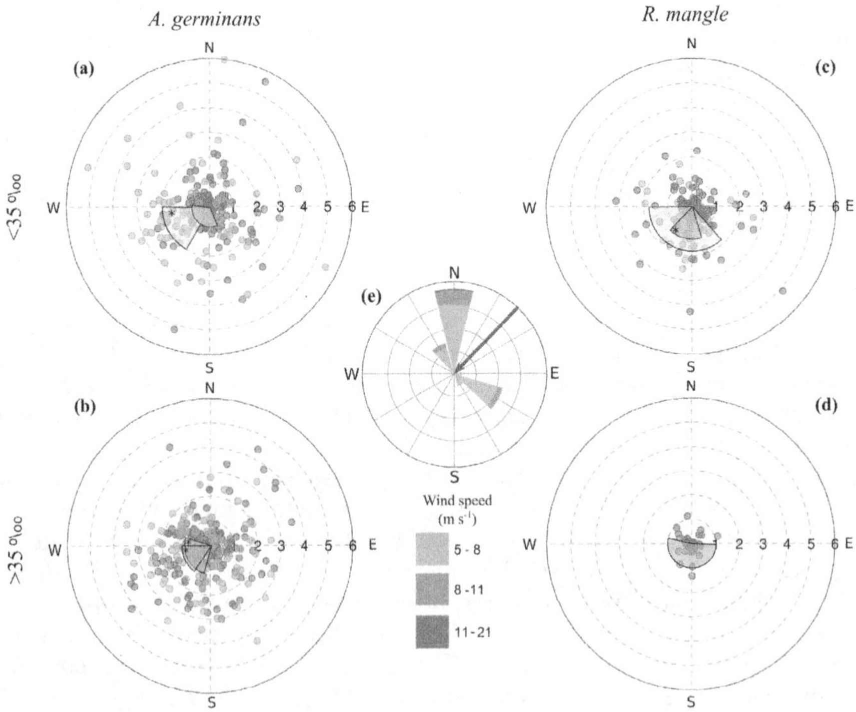

FIGURE 5.5 Directionality and magnitude (meters) of crown displacement in canopy (green (light gray in print version)) and below-canopy trees (blue (gray in print version)) *Avicennia germinans* (A) and (B) and *Rhizophora mangle* (C) and (D) growing at salinities <35 ppt and >35 ppt (upper and lower rows, respectively). The wind rise (E) shows the frequency and intensity of wind directions for the study site between 2011 and 2014, with a circular mean direction (*blue arrow* (dark gray in print version)) at NE. *Adapted from Vovides, A.G., Berger, U., Grueters, U., Guevara, R., Pommerening, A., Lara-Domínguez, A.L., López-Portillo, J., 2018. Change in drivers of mangrove crown displacement along a salinity stress gradient. Funct. Ecol. 32, 2753–2765.*

hypersaline environments (salinity >35 ppt), where trees are smaller, crown displacement at all stratums is mainly driven by wind forcing (Fig. 5.5C and D).

Wind can shape crowns either by increasing crown collisions or through direct branch pruning (Rudnicki et al., 2001; Schindler et al., 2012), and salt stress in mangroves can lead to a reduction of the crown's wind damping capacity, as tree crowns become smaller, increasing intercanopy space (Vovides et al., 2014, 2018). Continuous wind exposure appears to have a greater effect on crown shape than extreme weather events, which despite being highly destructive, are relatively infrequent and hence allow crown recovery (Rudnicki et al., 2001; Rudnicki et al., 2004; Vollsinger et al., 2005). The reduced wind damping capacity of crowns means they receive higher wind dynamic loadings than trees in benign environments, where dense canopy packing provides wind protection (MacFarlane and Kane, 2017). A morphologic response of trees to increasing wind dynamic loading on these exposed locations might be their reduced slenderness. At higher salinity locations trees are shorter and have wider stems, a trait first associated with compensation mechanisms to reduce salt stress (Peters et al., 2014; Vovides et al., 2014). But shorter and wider stems also provide increasing wind resistance (MacFarlane and Kane, 2017; Rudnicki et al., 2004), a topic still relatively unexplored in mangrove forests.

4. Root responses for mechanical stability

Although crown asymmetry is recognized as a plastic adaptive strategy that optimizes light capture and space use (Longuetaud et al., 2013), it can result in an increased risk of mechanical failure due to weight imbalance. A crown displacement of 6 m from the position of the stem (Fig. 5.5) is likely to pose a significant risk of mechanical failure. The stability of a tree is mainly determined by its root architecture and its interactions with the soil, and other tree biomechanical properties, such as stem flexure or the development of tension and compression wood (Stubbs et al., 2019). In order to maintain stability, the soil–root system must provide a restoring moment at least equal to the bending moment caused by external forces (Stubbs et al., 2019), thus, increasing mechanical stability with the interaction between soil and root biomass.

Under waterlogged sediment conditions, the low cohesion of roots and soil is a disadvantage for tree stability, as water reduces particulate cohesion and softens aggregates (Marshall and Holmes, 1988). This, alongside with the reduced depth of root growth in wetland ecosystems, posits a greater challenge for mangrove species to stand erect in an unstable dynamic environment. The risk of mechanical failure in mangrove trees could be reduced by (A) minimizing wind drag forces and (B) increasing the restoring moment at the anchoring root system. Drag forces can be reduced through asymmetric crown growth, reducing the crown area exposed to wind (Rudnicki et al., 2001; Stubbs et al., 2019; Vovides et al., 2018) and minimizing the risk of stem breakage. Whereas the restoring moment increases through plastic responses of the anchoring root system. Such plasticity increases root length (Coutts, 1986; Coutts et al., 1999; Stokes et al., 1995; Stubbs et al., 2019) and compression wood in the leeward side of stem and roots, and tension wood in the windward side (Stubbs et al., 2019; Young and Perkocha, 1994). Overall, within mangroves we recognize three anchoring systems: plate root systems, buttresses, and stilt roots or flying buttresses.

The soil–root plate, the section around the tree stem where woody roots adhere to the soil and get exposed when a tree topples (Coutts, 1983), develops in response to forces acting on the tree aerial biomass. In terrestrial forests, trees take advantage of taproots or vertical deep sinker roots to gain fundamental anchorage (Danjon et al., 2005; Stubbs et al., 2019; Vennetier et al., 2014). In wetlands, however, anoxic conditions limit root development in deeper soil layers, hence, taproots are normally absent and sinker roots are limited to a maximum depth of approximately 30 cm (McKee, 2001; Twilley and Rivera-Monroy, 2009; Vovides et al., 2016). In turn, some wetland tree species develop elongated main lateral roots (*Avicennia* spp.), buttresses (e.g., *Pachira aquatica* in freshwater wetlands or *Laguncularia* spp. in mangrove swamps), or stilt roots (*Rhizophora* spp.).

Buttresses are ridges that project upward to the stem from the main lateral roots, producing narrow undulating extensions of the stem that increase mechanical stability (Crook et al., 1997). Stilt roots are structures that develop from stem and branch cambium (Fisher, 1982), forming an aerial anchoring system that provides stability in highly unstable sediments. Despite commonly thought of as roots (Gill and Tomlinson, 1971, 1977), they share more anatomical properties with stems than with true roots (De Menezes, 2006), but provide the main anchoring function as roots do in other terrestrial plants. Stilt roots are also referred as flying buttresses because they resemble structures of gothic architecture (Fisher, 1982; Méndez-Alonzo et al., 2015) when they develop secondary growth, which is eccentric and gives a plank-like shape near the insertion to the trunk, as gothic arches in the points of connection with the walls. To avoid confusion, in this section, we refer to fully formed, grounded tree anchoring structures that provide mechanical support as "flying buttresses," and as "stilts" to the developing ones, which have emerged from stem or branches, but have not anchored to the ground, thus do not provide mechanical support to the tree.

4.1 Plate root systems

Plate root systems, such as that of the genus *Avicennia*, are formed by large lateral roots radiating horizontally from the stem (Ennos, 2000). While in terrestrial forests thick soil–root plates result from lateral roots growing from a tap root (Stubbs et al., 2019), in wetlands, the lack of oxygen in the sediments limits deep root development (McKee, 2001; Vovides et al., 2016). To compensate for the loss of restoring momentum in shallow soil–root plates (that result from roots developing over shallow groundwater tables), there should be a greater investment in lateral root length, increasing soil–root plate diameter and thus restoring momentum (Ray and Nicoll, 1998; Stubbs et al., 2019).

Because of the challenges in studying buried biomass in waterlogged environments, this topic has not been studied in depth for mangroves, remaining a critical knowledge gap to understand the plastic responses of trees to physical dynamic loads and failure thresholds. First explorations to deal with these challenges suggest the use of the estimated soil–root plate as a viable strategy to elucidate anchoring root responses to the forces received by the aboveground biomass (Vovides et al., 2016). With the aid of a portable ultrasonic Doppler device and a set of steel rods, the length of lateral woody roots can be correlated to the length of exposed soil–root plates, and further, to stem diameter (Cucchi et al., 2004; Vovides et al., 2016), and overturning mechanisms (Cucchi et al., 2004; Fourcaud et al., 2008). Fig. 5.6 depicts

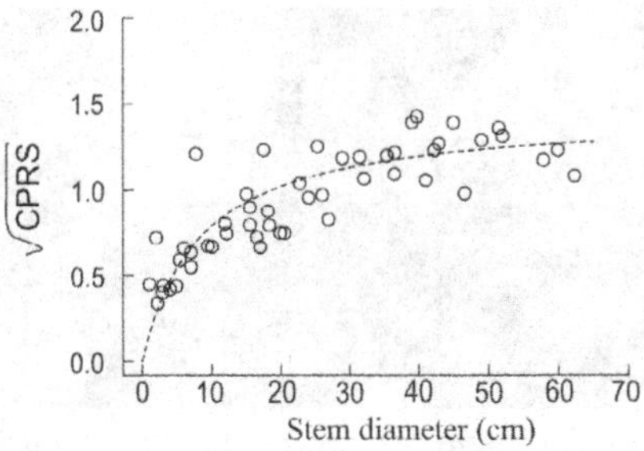

FIGURE 5.6 Log transformed mean soil–root plate radius of *Avicennia germinans* trees along different stem diameters in a mangrove forest of the central coast of the Gulf of Mexico. *Adapted from Vovides, A.G., Marín-Castro, B., Barradas, G., Berger, U., López-Portillo, J., 2016. A simple and cost-effective method for cable root detection and extension measurement in estuary wetland forests. Estuar. Coast Shelf Sci. 183, 117–122.*

a maximum asymptotic soil–root plate diameter, suggesting that at stem diameters >30 cm, the soil–root plate area does not increase, and stability is most likely dependent on differential spatial arrangement of roots, compression and tension wood, depending on the direction of the dynamic loading (Stubbs et al., 2019).

This advance opens the possibility to study the spatial configuration of roots examining the cross-sectional allocation of root biomass (Drexhage et al., 1999) as a response to mechanical forces affecting crowns (Cucchi et al., 2004; Nicoll and Dunn, 2000; Stokes et al., 1997). Such information could improve estimations on the coastal protection services that mangroves provide, and better assess ecosystem resistance to storm disturbances.

4.2 Buttresses

Buttress structures are typical of tropical rain forests and have been associated with shallow soils and unstable sediments in wetlands. Common examples are *P. aquatica*, *Ceiba pentandra*, and several species of *Ficus* (Fisher, 1982; Tomlinson, 2016; Turner, 2001). Within mangroves, well-recognized buttressing species include *Laguncularia racemosa*, *Brugueria gymnorhiza*, *Brugueria parviflora*, *Ceriops tagal*, *Xylocarpus granatum*, *Xylocarpus mekongensis*, *Kandelia* spp., and *Pelliciera rhizophorae* (Calderón-Sáenz, 1984; Clough and Scott, 1989; Fisher, 1982; Kumar et al., 2016), but reports also exist for *Avicennia germinans* and *Heriteria litoralis* (Olagoke et al., 2016; Youssef and Saenger, 1996).

Buttresses optimize mechanical stability by increasing the area of contact between stem and soil (Fig. 5.7), and by providing tension and compression tissues that develop as a function of mechanical loads exerted on the tree, either by crown imbalance, wind, or currents (Christensen-Dalsgaard et al., 2007; Crook et al., 1997; Fisher, 1982; Lewis, 1988; Richter, 1984). These structures represent a natural laboratory in mangrove trees, as many aspects of buttress formation have been positively correlated to directional environmental variables, such as a frequency and arrangement of buttresses around the tree in response to prevalent wind direction (Richter, 1984; Young and Perkocha, 1994). Plants modulate wood stiffness and strength as a function of physical pressures. Vessels become smaller and less abundant in points of higher pressure, increasing mechanical support (Christensen-Dalsgaard et al.,

FIGURE 5.7 Buttressed *Kandelia obovata* (left), *Avicennia germinans* (middle), and *Laguncularia racemosa* (right), with its buttress pinned by *Rhizophora mangle* flying buttresses. *Thorsten Balke (left), Alejandra G. Vovides (middle and right).*

2008, 2007). Also, larger buttresses could develop as a response to greater dynamic loading, similar to root length, enabling greater mechanical stability. This could be a good indicator of thigmomorphogenesis in mangroves (Telewski, 2006; Telewski and Jaffe, 1986), as the radial growth of buttresses and their anatomical traits varies as a function of mechanical loadings (Christensen-Dalsgaard et al., 2008). Although buttressing is recognized in several mangrove species, to date, to our knowledge, no studies have explored their frequency, functional anatomy, plastic responses to directional physical pressures, nor on the role they potentially play in accumulating sediments or preventing asphyxiation due to sediment burial (Poungparn et al., 2002) in different mangrove species.

4.3 Flying buttresses and stilt roots

Flying buttresses show a differential investment in structural tissues as a biomechanical response that increases stability and must be plastic in order to adjust to a changing environment. First-order flying buttresses, those emerging directly from the stem, are eccentric and contain a greater proportion of xylem tissues than bark and pith, suggesting a major support function (Fig. 5.8). When developing from other buttresses (second, third, and so forth orders), they have a centric anatomy and higher proportions of bark and pith, hence, while they add structural support, they might be more important for water transport and gas exchange (Méndez-Alonzo et al., 2015). As tree crowns increase in area, more flying buttresses are produced, increasing the area covered with points of anchorage, optimizing stability. This architecture allows trees to reach the canopy space with far greater slenderness than that of coexisting species such as *Avicennia* spp. (Méndez-Alonzo et al., 2015).

Additionally, root development is also modulated by soil bulk density, at the point of insertion to the ground, flying buttresses are longer at lower soil bulk densities, and shorter at higher soil bulk densites (Ola et al., 2019). While this response has been observed also in terrestrial forests, and soil bulk density is known to limit root growth, it is also possible that it affects the production of profuse roots, that although not deep, can potentially improve anchorage (Goldsmith et al., 2001; Ola et al., 2018; Ryser, 2006). As the density of fine roots is associated with the number of flying buttresses in a single tree, and with decreasing distance from the main stem position (Wilkens et al., 2018), a greater number of flying buttresses lead to increasing fine root biomass, improving sediment binding capacity (Gyssels et al., 2005),

FIGURE 5.8 *Rhizophora mangle* with elliptical thickening of its flying buttresses at the point of insertion with the stem (upper circular close up) and cylindrical shape at the points of insertion to the ground. *Alejandra G. Vovides.*

and potentially increasing restoring momentum at the base of the tree. Support on this assumption can be found in a study by McKee et al. (2007), who related soil elevation in *Rhizophora mangle* forests with low nutrient input, promoting root biomass production and subsequent sediment accumulation, whereas high nutrient availability inhibited root growth leading to shallow subsidence. Still, we lack a mechanistic understanding of the processes and sediment properties that regulate below-ground morphological traits and their functional relationship within the whole root system (Ola et al., 2019; Ryser, 2006).

Other environmental processes interact with flying buttresses that remain less studied but could provide a wider understanding of plastic responses that modulate mechanical stability. One of them is the physical injury to developing stilt roots, the only known biological mechanism to prompt stilt branching (Gill and Tomlinson, 1977) before they make contact with the soil. Injuries can induce the production of up to five lateral stilts that would increase mechanical support when anchoring in the ground. Such injuries occur by physical or biological factors, such as branch felling or the boring action of beetles, scale insects, and isopods (Gill and Tomlinson, 1977; Simberloff et al., 1978). Simberloff et al. (1978) estimated that for each stilt produced by a tree, due to the branching effect of boring insects, 1.4 stilts end up grounding and forming a flying buttress. Although the net benefit of boring insects in terms of the plant performance cost effect remains unexplored, Brooks and Bell (2002) point to the observation that no flying buttresses attached to the ground due to boring damage, suggesting there might be a trade-off on developmental plasticity. The reduced growth rates of the stilt and the cost of reappearing damage and producing laterals could be the price to pay for an increased number of flying buttress providing multiple anchoring points, as first suggested by Ribi (1982). This trade-off is also discussed on the responses of *Rhizophora stylosa* to wood boring teredinids, where an energetically costly increase in tannins and overcompensatory growth to repair wounds can minimize the damage caused by insects and salvage the mechanical support (Brooks and Bell, 2002; Hendy and Cragg, 2017).

5. Plastic responses to sediment dynamics in mature forests

Response traits depend on the magnitude and frequency of abiotic forcing. If too frequent or high intensity events prevail, plants lack time to adjust, and failure thresholds are reached (Fagherazzi et al., 2017). For example, hurricanes can cause catastrophic sedimentation leading to immediate mangrove loss (Smith et al., 2009; Zaldívar-Jiménez et al., 2010). High sedimentation rates can reduce forest productivity and resilience, posing a risk of root smothering (Ellison, 1999; Fagherazzi et al., 2017; Sidik et al., 2016). Sedimentation rates above 10 cm/year reported for northeastern Java between 2010 and 2011 resulted in reduced mangrove tree stem width growth (Sidik et al., 2016), possibly due to anoxic stress (Ellison, 1999), or hydrological impairment, as the gain in elevation reduces frequency and depth of inundation events (Adame et al., 2010). Although sediment burial could trigger an increase in pneumatophore height (Dahdouh-Guebas et al., 2007; Ellison, 1999), processes regulating this response are linked to the magnitude of sediment deposition, site-specific conditions, and species-specific traits that are, to date, not fully understood.

Below failure threshold though, in *Avicennia* trees, for example, new aerial roots continue to develop as a response to changes in flooding frequency, depth, and duration, becoming taller and more dense with more frequent, longer, and deeper inundation events (Dahdouh-Guebas et al., 2007; Toma et al., 1991). In turn, as an effect trait, pneumatophores either increase or decrease sedimentation rates depending on the morphology and density of the above-ground structures which create different magnitudes of hydrodynamic turbulence (Mullarney et al., 2017). Such effect traits, coupled with high rates of fine root biomass production (Castañeda-Moya et al., 2011; McKee et al., 2002; McKee, 1993), contribute to organogenic and minerogenic surface elevation gain, supporting ecosystem resilience and adaptation to sea-level rise (Friess et al., 2012; Krauss et al., 2014; Lovelock et al., 2015; Rogers et al., 2015).

Studies on the effects of disturbances on dynamic sedimentary processes have demonstrated that extreme weather events do not represent only a threat to coastal ecosystems but they can also contribute to their resilience and development over time. Stochastic disturbances related to hurricane events can result in elevation gains that could contribute to the stabilization of inner forests and increase the odds of mangrove persistence (Feher et al., 2019). But resilience is inherently linked to complex interactions between disturbance events, growth promoting (e.g., microbiota, nutrient availability), and growth-limiting factors (e.g., sulfides, salinity, temperature) as well as plant traits, highlighting the urgency to develop interdisciplinary approaches that can efficiently enhance our understanding of biogeomorphic and ecological processes that drive ecosystem evolution and coastal change.

6. Concluding remarks

Mangrove plant morphology responds to changes in biotic and abiotic conditions, for example, through changes in tree height, shoot:root biomass ratio, or canopy asymmetry. Plant traits can not only be responsive (i.e., respond to environmental conditions) and adaptive (a plastic response leading to an increase of fitness) but can also determine how plants interact with each other and affect abiotic processes through time (i.e., effect traits determining attenuation of hydrodynamic energy). Understanding these interactions is key to predict mangrove ecosystem functioning in active sedimentary environments. While some aspects (e.g., responses in the root:shoot ratio and the subsequent changes in carbon sequestration) have been extensively studied, morphological plasticity relating to mechanical stability and thigmomorphogenetic response to physical forcing by waves and currents is understudied. Integrative interdisciplinary approaches are needed to understand the impact of plant morphological plasticity on biogeomorphic processes along mangrove coasts in order to predict mangrove ecosystem functioning and resilience.

References

Aakala, T., Shimatani, K., Abe, T., Kubota, Y., Kuuluvainen, T., 2015. Crown asymmetry in high latitude forests: disentangling the directional effects of tree competition and solar radiation. Oikos 125, 1035–1043.

Adame, M.F., Neil, D., Wright, S.F., Lovelock, C.E., 2010. Sedimentation within and among mangrove forests along a gradient of geomorphological settings. Estuar. Coast Shelf Sci. 86, 21–30.

Augustinus, P.G.E.F., 1996. Geomorphology and sedimentology of mangroves. In: Geomorphology and Sedimentology of Estuaries. Elsevier B.V., Amsterdam, The Netherlands, pp. 333–357.

Balke, T., Bouma, T.J., Horstman, E.M., Webb, E.L., Erftemeijer, P.L.A., Herman, P.M.J., 2011. Windows of opportunity: thresholds to mangrove seedling establishment on tidal flats. Mar. Ecol. Prog. Ser. 440, 1–9.

Balke, T., Herman, P.M.J., Bouma, T.J., 2014. Critical transitions in disturbance-driven ecosystems: identifying windows of opportunity for recovery. J. Ecol. 102, 700–708.

Balke, T., Webb, E.L., van den Elzen, E., Galli, D., Herman, P.M.J., Bouma, T.J., 2013. Seedling establishment in a dynamic sedimentary environment: a conceptual framework using mangroves. J. Appl. Ecol. 50, 740–747.

Ball, M.C., 1988. Ecophysiology of mangroves. Trees Struct. Funct. 2, 129–142.

Ball, M.C., Farquhar, G.D., 1984. Photosynthetic and stomatal responses of the grey mangrove, *Avicennia marina*, to transient salinity consitions. Plant Physiol. 74, 7–11.

Barbier, E.B., Hacker, S.D., Kennedy, C., Koch, E.W., Stier, A.C., Silliman, B.R., 2011. The value of estuarine and coastal ecosystem services. Ecol. Monogr. 81, 169–193.

Boizard, S.D., Mitchell, S.J., 2011. Resistance of red mangrove (*Rhizophora mangle* L.) seedlings to deflection and extraction. Trees (Berl.) 25, 371–381.

Brisson, J., 2001. Neighborhood competition and crown asymmetry in *Acer saccharum*. Can. J. For. Res. 31, 2151–2159.

Brooks, R.A., Bell, S.S., 2002. Mangrove response to attack by a root boring isopod: root repair versus architectural modification. Mar. Ecol. Prog. Ser. 231, 85–90.

Brüchert, F., Gardiner, B., 2006. The effect of wind exposure on the tree aerial architecture and biomechanics of Sitka spruce (*Picea sitchensis*, Pinaceae). Am. J. Bot. 93, 1512–1521.

Burggren, W., 2018. Developmental phenotypic plasticity helps bridge stochastic weather events associated with climate change. J. Exp. Biol. 221, jeb161984.

Calderón-Sáenz, E., 1984. Occurrence of the mangrove *Pelliciera rhizophorae* on the caribbean coast of Colombia with biogeographical notes. Bull. Mar. Sci. 35, 105–110.

Callaway, R.M., Pennings, S.C., Richards, C.L., 2003. Phenotypic plasticity and interactions among plants. Ecology 84, 1115–1128.

Castañeda-Moya, E., Twilley, R.R., Rivera-Monroy, V.H., Marx, B.D., Coronado-Molina, C., Ewe, S.M.L., 2011. Patterns of root dynamics in mangrove forests along environmental gradients in the Florida Coastal Everglades USA. Ecosystems 14, 1178–1195.

Christensen-Dalsgaard, K.K., Ennos, A.R., Fournier, M., 2008. Are radial changes in vascular anatomy mechanically induced or an ageing process? Evidence from observations on buttressed tree root systems. Trees Struct. Funct. 22, 543–550.

Christensen-Dalsgaard, K.K., Fournier, M., Ennos, A.R., Barfod, A.S., 2007. Changes in vessel anatomy in response to mechanical loading in six species of tropical trees. New Phytol. 176, 610–622.

Clough, B.F., Scott, K., 1989. Allometric relationships for estimating above-ground biomass in six mangrove species. For. Ecol. Manage. 27, 117–127.

Cook-Patton, S.C., Lehmann, M., Parker, J.D., 2015. Convergence of three mangrove species towards freeze-tolerant phenotypes at an expanding range edge. Funct. Ecol. 29, 1332–1340.

Corenblit, D., Tabacchi, E., Steiger, J., Gurnell, A.M., 2007. Reciprocal interactions and adjustments between fluvial landforms and vegetation dynamics in river corridors: a review of complementary approaches. Earth Sci. Rev. 84, 56–86.

Coutts, M., 1983. Root architecture and tree stability. Plant Soil 188, 171–188.

Coutts, M.P., 1986. Components of tree stability in sitka spruce on peaty gley soil. Forestry 59, 173–197.

Coutts, M.P., Nielsen, C.C.N., Nicoll, B.C., 1999. The development of symmetry, rigidity and anchorage in the structural root system of conifers. Plant Soil 217, 1–15.

Crook, M.J., Ennos, A.R., Banks, J.R., 1997. The function of buttress roots: a comparative study of the anchorage systems of buttressed (*Aglaia* and *Nephelium ramboutan* species) and non-buttressed (*Mallotus wrayi*) tropical trees. J. Exp. Bot. 48, 1703–1716.

Cuc, N.T.K., Suzuki, T., de Ruyter van Steveninck, E.D., Hai, H., 2015. Modelling the impacts of mangrove vegetation structure on wave dissipation in Ben Tre Province, Vietnam, under different climate change scenarios. J. Coast Res. 31, 340–347.

Cucchi, V., Meredieu, C., Stokes, A., Berthier, S., Bert, D., Najar, M., Denis, A., Lastennet, R., 2004. Root anchorage of inner and edge trees in stands of Maritime pine (*Pinus pinaster* Ait.) growing in different podzolic soil conditions. Trees (Berl.) 18, 460–466.

Dahdouh-Guebas, F., De Bondt, R., Abeysinghe, P.D., Kairo, J.G., Cannicci, S., Triest, L., Koedam, N., 2004. Comparative study of the disjunct zonation pattern of the gray mangrove *Avicennia marina* (Forsk.) Vierh. in Gazi Bay (Kenya). Bull. Mar. Sci. 74, 237–252.

Dahdouh-Guebas, F., Kairo, J.G., De Bond, R., Koedam, N., 2007. Pneumatophore height and density in relation to microtopography in the grey mangrove *Avicennia marina*. Belg. J. Bot. 140 (213), 221.

Danjon, F., Fourcaud, T., Bert, D., 2005. Root architecture and wind- firmness of mature *Pinus pinaster*. New Phytol. 168, 387–400.

Das, S., Crépin, A.-S., 2013. Mangroves can provide protection against wind damage during storms. Estuar. Coast Shelf Sci. 134, 98–107.

De Menezes, N.L., 2006. Rhizophores in *Rhizophora mangle* L: an alternative interpretation of so-called "aerial roots.". An. Acad. Bras. Cienc. 78, 213–226.

Drexhage, M., Chauvière, M., Colin, F., Nielsen, C.N.N., 1999. Development of structural root architecture and allometry of *Quercus petraea*. Can. J. For. Res. 29, 600–608.

Ellison, J.C., 1999. Impacts of sediment burial on mangroves. Mar. Pollut. Bull. 37, 420–426.

Ennos, A.R., 2000. The mechanics of root anchorage. Adv. Bot. Res. 33, 133–157.

Ewel, K., Twilley, R.R., Ong, J., 1998. Different kinds of mangroves provide different goods and services. Global Ecol. Biogeogr. Lett. 7, 83–94.

Fagherazzi, S., Bryan, K., Nardin, W., 2017. Buried alive or washed away: the challenging life of mangroves in the Mekong Delta. Oceanography 30, 48–59.

Feher, L.C., Osland, M.J., Anderson, G.H., Vervaeke, W.C., Krauss, K.W., Whelan, K.R.T., Balentine, K.M., Tiling-Range, G., Smith III, T.J., Cahoon, D.R., 2019. The long-term effects of hurricanes Wilma and Irma on soil elevation change in Everglades mangrove forests. Ecosystems 23, 917–931. htpps://doi.org/10.1007/s10021-019-00446-x.

Feller, I.C., 1995. Effects of nutrient enrichment on growth and herbivory of dwarf red mangrove (*Rhizophora mangle*). Ecol. Monogr. 65, 477–505.

Feller, I.C., Lovelock, C.E., Berger, U., McKee, K.L., Joye, S.B., Ball, M.C., 2010. Biocomplexity in mangrove ecosystems. Ann. Rev. Mar. Sci. 2, 395–417.

Fisher, J., 1982. A survey of buttresses and aerial roots of tropical trees for presence of reaction wood. Biotropica 14, 56–61.

Fourcaud, T., Ji, J.N., Zhang, Z.Q., Stokes, A., 2008. Understanding the impact of root morphology on overturning mechanisms: a modelling approach. Ann. Bot. 101, 1267–1280.

Friess, D.A., Krauss, K., Horstman, E., Balke, T., Bouma, T., Galli, D., Webb, E.L., 2012. Are all intertidal wetlands created equal? Bottlenecks, thresholds and knowledge gaps to saltmarsh and mangrove ecosystems. Biol. Rev. 87, 346–366.

Gill, A.M., Tomlinson, P.B., 1971. Studies on the growth of red mangrove (*Rhizophora mangle* L.) 2. Growth and differentiation of aerial roots. Biotropica 3, 63–77.

Gill, A.M., Tomlinson, P.B., 1977. Studies on the growth of red mangrove (*Rhizophora mangle* L.) 4. The adult root system. Biotropica 9, 145–155.

Goldsmith, W., Silva, M., Fischenich, C., 2001. Determining Optimal Degree of Soil Compaction for Balancing Mechanical Stability and Plant Growth Capacity. ERDC TN-EMRRP-SR-26. U.S. Army Engineer Research & Development Center.

Grime, J.P., Mackey, J.M.L., 2002. The role of plasticity in resource capture by plants. Evol. Ecol. 16, 299–307.

Grueters, U., Seltmann, T., Schmidt, H., Horn, H., Pranchai, A., Vovides, A.G., Peters, R., Vogt, J., Dahdouh-Guebas, F., Berger, U., 2014. The mangrove forest dynamics model mesoFON. Ecol. Model. 291, 28–41.

Gyssels, G., Poesen, J., Bochet, E., Li, Y., 2005. Impact of plant roots on the resistance of soils to erosion by water: a review. Prog. Phys. Geogr. 29, 189–217.

Hendy, I.W., Cragg, S.M., 2017. Rhizophora stylosa prop roots even when damaged prevent wood-boring teredinids from toppling the trees. Hydrobiologia 803, 333–344.

Horstman, E.M., Bryan, K.R., Mullarney, J.C., Pilditch, C.A., 2016. Model versus nature: hydrodynamics in mangrove pmeumatophores. In: 35th Conference of Coastal Engineering. Antalya, Turkey, pp. 1–11.

Jones, C.G., Lawton, J.H., Shachak, M., Jones, C.G., Lawton, J.H., Shachak, M., 1994. Organisms as ecosystem engineers. Oikos 69, 373–386.

Jucker, T., Bouriaud, O., Coomes, D.A., Ecology, F., Faculty, F., Coomes, D.A., Street, D., 2015. Crown plasticity enables trees to optimize canopy packing in mixed-species forests. Funct. Ecol. 29, 1078–1086.

Krauss, K.W., Lovelock, C.E., Mckee, K.L., López-Hoffan, L., Ewe, S.M.L., Sousa, W.P., Lo, L., Ewe, S.M.L., Sousa, W.P., 2008. Environmental drivers in mangrove establishment and early development: a review. Aquat. Bot. 89, 105–127.

Krauss, K.W., McKee, K.L., Lovelock, C.E., Cahoon, D.R., Saintilan, N., Reef, R., Chen, L., 2014. How mangrove forests adjust to rising sea level. New Phytol. 202, 19–34.

Kumar, S., Singh, R., Kathiresan, K., 2016. Multi-gene phylogenetic analysis reveals the multiple origin and evolution of mangrove physiological traits through exaptation. Estuar. Coast Shelf Sci. 183, 41–51.

Le Minor, M., Bartzke, G., Zimmer, M., Gillis, L., Helfer, V., Huhn, K., 2019. Numerical modelling of hydraulics and sediment dynamics around mangrove seedlings: implications for mangrove establishment and reforestation. Estuar. Coast Shelf Sci. 217, 81–95.

Lewis, A.R., 1988. Buttress arrangement in *Pterocarpus officinalis* (Fabaceae): effects of crown asymmetry and wind. Biotropica 20, 280–285.

Longuetaud, F., Piboule, A., Wernsdörfer, H., Collet, C., 2013. Crown plasticity reduces inter-tree competition in a mixed broadleaved forest. Eur. J. For. Res. 132, 621–634.

López-Portillo, J., Ewers, F.W., Méndez-Alonzo, R., Paredes López, C., Angeles, G., Jiménez, A.L.A., Lara-Domínguez, A.L., Barrera, M., del, C.T., 2014. Dynamic control of osmolality and ionic composition of the xylem sap in two mangrove species. Am. J. Bot. 101, 1013–1022.

López-Portillo, J.A., Ewers, F.W., Angeles, G., 2005. Sap salinity effects on xylem conductivity in two mangrove species. Plant Cell Environ. 28, 1285–1292.

Lovelock, C.E., Cahoon, D.R., Friess, D.A., Guntenspergen, G.R., Krauss, K.W., Reef, R., Rogers, K., Saunders, M.L., Sidik, F., Swales, A., Saintilan, N., Thuyen, L.X., Triet, T., 2015. The vulnerability of Indo-Pacific mangrove forests to sea-level rise. Nature 526, 559–563.

Lovelock, C.E., Feller, I.C., McKee, K.L., Engelbrecht, B.M.J., Ball, M.C., 2004. The effect of nutrient enrichment on growth, photosynthesis and hydraulic conductance of dwarf mangroves in Panama. Funct. Ecol. 18, 25–33.

MacFarlane, D.W., Kane, B., 2017. Neighbour effects on tree architecture: functional trade-offs balancing crown competitiveness with wind resistance. Funct. Ecol. 31, 1624–1636.

Marshall, T.J., Holmes, J.W., 1988. Soil Physics, second ed. Cambridge University Press, Cambridge.

Mazda, Y., Magi, M., Ikeda, Y., Kurokawa, T., Asano, T., 2006. Wave reduction in a mangrove forest dominated by *Sonneratia* sp. Wetl. Ecol. Manag. 14, 365–378.

McKee, K., Feller, I., Popp, M., Wanek, W., 2002. Mangrove isotopic (δ15N and δ13C) fractionation across a nitrogen vs phosphorus limitation gardient. Ecology 83, 1065–1075.

McKee, K.L., 1993. Soil physicochemical patterns and mangrove species distribution–reciprocal effects? J. Ecol. 81, 477–487.

McKee, K.L., 2001. Root proliferation in decaying roots and old root channels: a nutrient conservation mechanism in oligotrophic mangrove forests? J. Ecol. 89, 876–887.

McKee, K.L., Cahoon, D.R., Feller, I.C., 2007. Caribbean mangroves adjust to rising sea level through biotic controls on change in soil elevation. Global Ecol. Biogeogr. 16, 545–556.

Méndez-Alonzo, R., Hernández-Trejo, H., López-Portillo, J.A., 2012. Salinity constrains size inequality and allometry in two contrasting mangrove habitats in the Gulf of Mexico. J. Trop. Ecol. 28, 171–179.

Méndez-Alonzo, R., Moctezuma, C., Ordoñez, V.R., Angeles, G., Martínez, A.J., López-Portillo, J., 2015. Root biomechanics in *Rhizophora mangle*: anatomy, morphology and ecology of mangrove's flying buttresses. Ann. Bot. 115, 833–840.

Méndez-Alonzo, R., Pineda-García, F., Paz, H., Rosell, J.A., Olson, M.E., 2013. Leaf phenology is associated with soil water availability and xylem traits in a tropical dry forest. Trees Struct. Funct. 27, 745–754.

Miner, B.G., Sultan, S.E., Morgan, S.G., Padilla, D.K., Relyea, R.A., 2005. Ecological consequences of phenotypic plasticity. Trends Ecol. Evol. 20, 685–692.

Mullarney, J., Henderson, S., Norris, B., Bryan, K., Fricke, A., Sandwell, D., Culling, D., 2017. A question of scale: how turbulence around aerial roots shapes the seabed morphology in mangrove forests of the Mekong Delta. Oceanography 30, 34–47.

Nicoll, B.C., Dunn, A., 2000. The effects of wind speed and direction on radial growth of structural roots. In: Stokes, A. (Ed.), The Supporting Roots of Trees and Woody Plants: Form, Function and Physiology. Kluwer Academic Publishers, Dordrecht, The Netherlands.

Ola, A., Gauthier, A.R.G., Xiong, Y., Lovelock, C.E., 2019. The roots of blue carbon: responses of mangrove stilt roots to variation in soil bulk density. Biol. Lett. 15.

Ola, A., Schmidt, S., Lovelock, C.E., 2018. The effect of heterogeneous soil bulk density on root growth of field-grown mangrove species. Plant Soil 432, 91–105.

Olagoke, A., Proisy, C., Féret, J.-B., Blanchard, E., Fromard, F., Mehlig, U., Machado de Meneses, M., Ferreira dos Santos, V., Berger, U., 2016. Extended biomass allometric equations for large mangrove trees from terrestrial LiDAR data. Trees Struct. Funct. 30, 935–947.

Osland, M.J., Day, R.H., Larriviere, J.C., From, A.S., 2014. Aboveground allometric models for freeze-affected black mangroves (*Avicennia germinans*): equations for a climate sensitive mangrove-marsh ecotone. PLoS One 9, e99604.

Peters, R., Olagoke, A., Berger, U., 2018. A new mechanistic theory of self-thinning: adaptive behaviour of plants explains the shape and slope of self-thinning trajectories. Ecol. Model. 390, 1–9.

Peters, R., Vovides, A.G., Luna, S., Grüters, U., Berger, U., 2014. Changes in allometric relations of mangrove trees due to resource availability – a new mechanistic modelling approach. Ecol. Model. 283, 53–61.

Poungparn, S., Komiyama, A., Jintana, V., Piriyoata, S., Sangtiean, T., Tanapermpool, P., Patanaponpaiboon, P., Kato, S., 2002. A Quantitative analysis on the root system of a mangrove, *Xylocarpus granatum* Koenig. Tropics 12, 35–42.

Pranchai, A., Jenke, M., Vogt, J., Grueters, U., Yue, L., Mehlig, U., Menezes, M.M. de, Wagner, S., Berger, U., 2018. Density-dependent shift from facilitation to competition in a dwarf *Avicennia germinans* forest. Wetl. Ecol. Manag. 26, 139–150.

Ray, D., Nicoll, B.C., 1998. The effect of soil water-table depth on root-plate development and stability of *Sitka spruce*. Forestry 71, 169–182.

Ribi, G., 1982. Differential colonization of roots of *Rhizophora mangle* by the wood boring isopod *Sphaeroma terebrans* as a mechanism to increase root density. Mar. Ecol. 3, 13–19.

Richards, C.L., Bossdorf, O., Muth, N.Z., Gurevitch, J., Pigliucci, M., 2006. Jack of all trades, master of some? On the role of phenotypic plasticity in plant invasions. Ecol. Lett. 9, 981–993.

Richter, W., 1984. A structural approach to the function of buttresses of *Quararibea asterolepis*. Ecology 65, 1429–1435.

Robert, E.M.R., Koedam, N., Beeckman, H., Schmitz, N., 2009. A safe hydraulic architecture as wood anatomical explanation for the difference in distribution of the mangroves *Avicennia* and *Rhizophora*. Funct. Ecol. 23, 649–657.

Rogers, K., Mendelssohn, I.A., Saintilan, N., Lovelock, C.E., Woodroffe, C.D., McKee, K.L., 2015. Mangrove sedimentation and response to relative sea-level rise. Ann. Rev. Mar. Sci. 8, 243–266.

Rudnicki, M., Silins, U., Lieffers, V.J., 2004. Crown cover is correlated with relative density tree slenderness and tree height in Lodgepole pine. For. Sci. 50, 356–363.

Rudnicki, M., Silins, U., Lieffers, V.J., Josi, G., 2001. Measure of simultaneous tree sways and estimation of crown interactions among a group of trees. Trees Struct. Funct. 15, 83–90.

Ryser, P., 2006. The mysterious root length. Plant Soil 286, 1–6.

Schindler, D., Bauhus, J., Mayer, H., 2012. Wind effects on trees. Eur. J. For. Res. 131, 159–163.

Schmitz, N., Verheyden, A., Beeckman, H., Kairo, J.G., Koedam, N., 2006. Influence of a salinity gradient on the vessel characters of the mangrove species *Rhizophora mucronata*. Ann. Bot. 98, 1321–1330.

Sidik, F., Neil, D., Lovelock, C.E., 2016. Effect of high sedimentation rates on surface sediment dynamics and mangrove growth in the Porong River, Indonesia. Mar. Pollut. Bull. 107, 355–363.

Simberloff, D., Brown, B.J., Lowrie, S., 1978. Isopod and insect root borers may benefit Florida mangroves. Science 201, 630–632.

Simpson, L.T., Osborne, T.Z., Feller, I.C., 2017. Establishment and biomass allocation of black and red mangroves: response to propagule flotation duration and seedling light availability. J. Coast Res. 335, 1126–1134.

Smith, T.J., Anderson, G.H., Balentine, K., Tiling, G., Ward, G.A., Whelan, K.R.T., 2009. Cumulative impacts of hurricanes on Florida mangrove ecosystems: sediment deposition storm surges and vegetation. Wetlands 29, 24–34.

Sobrado, M.A., 2007. Relationship of water transport to anatomical features in the mangrove *Laguncularia racemosa* grown under contrasting salinities. New Phytol. 173, 584–591.

Spenceley, A.P., 1977. The role of pneumatophores in sedimentary processes. Mar. Geol. 24, 31–37.

Stokes, A., Fitter, A.H., Courts, M.P., 1995. Responses of young trees to wind and shading: effects on root architecture. J. Exp. Bot. 46, 1139–1146.

Stokes, A., Martin, F., Sacriste, S., Fourcaud, T., 1997. Adaptation of tree roots to wind loading—the relationship between mechanical behaviour and wood formation. In: Jeronimidis, G., Vincent, J.F.V.((Eds.), Proceedings of the 2nd Plant Biomechanics Conference, pp. 339–346.

Stubbs, C.J., Cook, D.D., Niklas, K.J., 2019. A general review of the biomechanics of root anchorage. J. Exp. Bot. 70, 3439–3451.

Sultan, S.E., 2000. Phenotypic plasticity for plant development, function and life history. Trends Plant Sci. 5, 537–542.

Sultan, S.E., 2003. Phenotypic plasticity in plants: a case study in ecological development. Evol. Dev. 5, 25–33.

Telewski, F.W., 2006. A unified hypothesis of mechanoperception in plants. Am. J. Bot. 93, 1466–1476.

Telewski, F.W., Jaffe, M.J., 1986. Thigmomorphogenesis: field and laboratory studies of *Abies fraseri* in response to wind or mechanical perturbation. Physiol. Plantarum 66, 211–218.

Thampanya, U., Vermaat, J.E., Terrados, J., 2002. The effect of increasing sediment accretion on the seedlings of three common Thai mangrove species. Aquat. Bot. 74, 315–325.

Toma, T., Nakamura, K., Patanaponpaiboon, P., Ogino, K., 1991. Effect of flooding water level and plant density on growth of pneumatophore of *Avicennia marina*. Tropics 1, 75–82.

Tomlinson, P.B., 2016. The Botany of Mangroves, second. Cambridge University Press, Cambridge, UK.

Turner, I.M., 2001. The Ecology of Trees in the Tropical Rain Forest. Cambridge University Press, Cambridge, UK.

Twilley, R.R., Rivera-Monroy, V.H., 2009. Ecogeomorphic models of nutrient biogeochemestry for mangrove wetlands. In: Perillo, G.M., Wolanski, E., Cahoon, D.R., Brinson, M.M. (Eds.), Coastal Wetlands: An Integrated Ecosystem Approach. Elsevier, Amsterdam, The Netherlands, pp. 641–684.

Van Santen, P., Augustinus, P.G.E.F., Janssen-stelder, B.M., Quartel, S., Tri, N.H., 2007. Sedimentation in an estuarine mangrove system. J. Asian Earth Sci. 29, 566–575.

Vennetier, M., Zanetti, C., Meriaux, P., Mary, B., 2014. Tree root architecture: new insights from a comprehensive study on dikes. Plant Soil 387, 81–101.

Vogt, J., Lin, Y., Pranchai, A., Frohberg, P., Mehlig, U., Berger, U., 2014. The importance of conspecific facilitation during recruitment and regeneration: a case study in degraded mangroves. Basic Appl. Ecol. 15, 651–660.

Vollsinger, S., Mitchell, S.J., Byrne, K.E., Novak, M.D., Rudnicki, M., 2005. Wind tunnel measurements of crown streamlining and drag relationships for several hardwood species. Can. J. For. Res. 35, 1238–1249.

Vovides, A.G., Bashan, Y., López-Portillo, J.A., Guevara, R., 2011. Nitrogen fixation in preserved, reforested, naturally regenerated and impaired mangroves as an indicator of functional restoration in mangroves in an arid region of Mexico. Restor. Ecol. 19, 236–244.

Vovides, A.G., Berger, U., Grueters, U., Guevara, R., Pommerening, A., Lara-Domínguez, A.L., López-Portillo, J., 2018. Change in drivers of mangrove crown displacement along a salinity stress gradient. Funct. Ecol. 32, 2753–2765.

Vovides, A.G., Marín-Castro, B., Barradas, G., Berger, U., López-Portillo, J., 2016. A simple and cost-effective method for cable root detection and extension measurement in estuary wetland forests. Estuar. Coast Shelf Sci. 183, 117–122.

Vovides, A.G., Vogt, J., Kollert, A., Berger, U., Grueters, U., Peters, R., Lara-Domínguez, A.L., López-Portillo, J., 2014. Morphological plasticity in mangrove trees: salinity-related changes in the allometry of *Avicennia germinans*. Trees Struct. Funct. 28, 1413–1425.

Wilkens, J.F., Huth, F., Berger, U., Devlin, D., Popow, F., Wagner, S., 2018. Spatial explicit distribution of individual fine root biomass of *Rhizophora mangle* L. (Red Mangrove) in South Florida. Wetl. Ecol. Manag. 26, 775–788.

Wolf, C., Field, B., Berry, J.A., 2011. Neighborhood competition and crown asymmetry in *Acer saccharum*. Ecol. Soc. Am. 21, 1546–1556.

Young, T.P., Perkocha, V., 1994. Treefalls, crown asymmetry and buttresses. J. Ecol. 82, 319–324.

Youssef, T., Saenger, P., 1996. Anatomical adaptive strategies to flooding and rhizosphere oxidation in mangrove seedlings. Aust. J. Bot. 44, 297–313.

Zaldívar-Jiménez, M.A., Herrera-Silveira, J.A., Teutli-Hernández, C., Comín, F.A., Andrade, J.L., Molina, C.C., Ceballos, R.P., 2010. Conceptual framework for mangrove restoration in the Yucatán Peninsula. Ecol. Restor. 28, 333–342.

CHAPTER

6

Microbial communities in mangrove sediments

Véronique Helfer, Christiane Hassenrück

ZMT - Leibniz Centre for Tropical Marine Research, Bremen, Germany

OUTLINE

1. Microbial communities in mangrove sediments—setting the scene 142

2. Microbial communities, their drivers, and their role in element cycling 143

2.1 The different members of the microbial community 143

2.2 Drivers of microbial communities 144

2.3 The role of microbial communities in element cycling 145

2.3.1 The C-cycle 146

2.3.2 The N-cycle 147

2.3.3 The S-cycle 148

2.3.4 Drivers of microbial element cycling 150

3. Microbial communities in mangroves 151

3.1 Peculiarities of mangrove ecosystems and their constraints 151

3.2 Actual knowledge on microbial communities in mangrove ecosystems 152

4. Techniques to depict microbial communities and the processes they drive 157

4.1 Traditional methods 158

4.1.1 Cultivation-dependent approaches 158

4.1.2 Clone libraries 158

4.1.3 Community fingerprinting 159

4.2 'Omics and related approaches 159

4.2.1 Metabarcoding 159

4.2.2 Metagenomics 160

4.2.3 Metatranscriptomics 160

4.2.4 Metaproteomics and metabolomics 161

4.3 Challenges of 'omics data analysis 161

4.4 Besides the 'omics approaches 162

4.4.1 Environmental characterization 163

4.4.2 Visualization of microbes at microscale 163

4.4.3 Supplementing 'omics with quantitative information 163

4.5 Approaches used in mangrove research 164

5. Conclusion and perspectives 165

Dynamic Sedimentary Environments of Mangrove Coasts
https://doi.org/10.1016/B978-0-12-816437-2.00003-3

Appendix 1: Excerpts from the R scripts (full script available under: https://gitlableibniz-zmt.de/vhe/microbial-communities-in-mangroves) used for the bibliographic review, showing the different syntaxes used to group the studies by microorganism types, habitat types and applied analytical approaches 167
Acknowledgments 169
References 169

1. Microbial communities in mangrove sediments—setting the scene

Microbes are extremely important organisms, in terms of biomass (c. 50% of organic carbon and 90% of organic nitrogen on earth; Prescott et al., 2010) and in terms of contribution to ecosystem functioning, via their central role in organic matter turnover and element cycling (including nutrient cycling). The term microbes (Wessner et al., 2013) encompasses several taxonomic groups (from all domains of life), such as acellular organisms (e.g., viruses) and prokaryotic (archaea and bacteria) or uni- and pluricellular eukaryotic (protists, fungi) microorganisms (organisms of small size that cannot be directly investigated by the eye). Hereafter, we will use interchangeably the terms microbial communities or microbiome, although strictly speaking, the microbiome refers normally to the resident microbial communities (therefore excluding transient microbial populations). For a given ecosystem, the term microbiome (e.g., the "mangrove microbiome") encompasses several communities, notably the environmental microbiome (e.g., sediment microbiome) and the host-associated microbiomes (such as animal, plant, or human microbiomes). Those distinct communities can be strongly interlinked, especially in case of the rhizosphere or sediment microbial communities.

While referring to microbial communities, research efforts and therefore our scientific knowledge have been biased toward bacterial communities (specifically for genomic and metagenomic studies; Trivedi et al., 2013), one important component of microbial communities, but certainly not the only influential. For instance, among the microbial modulators of soil carbon storage, fungal communities are known to play an important role (Trivedi et al., 2013), especially for the degradation of recalcitrant organic matter.

In this chapter, we shed light on the different members of the microbial community (see Section 2.1), discuss the important drivers of microbial communities (see Section 2.2), and provide a concise overview of their respective role for element cycling, with an emphasis on C-, N-, and S-cycles (see Section 2.3). We review the knowledge available about microbial communities in mangrove ecosystems (Section 3), with a focus on not only sediment-associated but also leaf- or root-associated microbial communities, and provide a thorough overview of available approaches (see Section 4) to study those communities, with respect to taxonomic and functional diversity. Finally, we point to research gaps and priorities (Section 5) to foster research on the mangrove microbiome.

2. Microbial communities, their drivers, and their role in element cycling

Microorganisms can occupy different habitats in mangrove ecosystems. We will concentrate here on the sediment microbial communities (environmental microbiome) and partly the plant microbiome (microbial communities associated to leaves or roots of mangrove plants). While we will briefly describe the effect of faunal organisms on sediment microbial communities (see Section 2.2 and Section 3.2), we will not include their respective microbiomes.

2.1 The different members of the microbial community

Archaea and bacteria (also known as Eubacteria), the two groups of prokaryotic organisms being part of microbial communities, differ from each other notably in their ribosomal RNA sequence, their membrane lipid composition, and the absence (archaea) or presence (bacteria) of a membrane peptidoglycan. They both play a pivotal role for organic matter turnover and element cycling, notably nitrogen cycling (see Section 2.3), as they are the main, if not the sole organisms able to perform nitrogen fixation and nitrification. The relative contribution of bacteria and archaea depends mainly on sediment depth, related to sediment oxygenation levels: the abundance of bacterial communities (based on 16S rRNA gene copies) decreases with sediment depth, while archaeal communities show the opposite trend (Zhou et al., 2017). Stratification patterns are also observed within bacterial (see also Mendes and Tsai, 2014) and archaeal communities, aerobic taxa being plausibly more abundant in surface layers while anaerobic ones dominate in subsurface layers.

The main eukaryotic members of microbial communities are protists and fungi. They are essential for organic matter turnover, as they are able to process recalcitrant fractions of organic matter (Trivedi et al., 2013; Fossier Marchan et al., 2018). Protists encompass unicellular algae, protozoa, myxomycetes, and aquatic molds (Prescott et al., 2010). Unicellular algae are autotrophic organisms, which are, together with the cyanobacteria, producing about 75% of the Earth oxygen (Prescott et al., 2010). Protozoa, heterotrophic unicellular organisms assigned to the animal realm, are the principal grazers of microbial communities, as they rely on organic matter and other microorganisms for their nutrition. Myxomycetes (Nandipati et al., 2012), also known as plasmodial slime molds, undergo several life stages during which they feed on bacteria, unicellular eukaryotes, or dissolved nutrients. Among protists, thraustochytrids, whose taxonomic affiliation is still debated, are epi- or endobiotic unicellular organisms described as chromists, stramenopiles, or heterotrophic microalgae (Fossier Marchan et al., 2018). Found in various marine habitats, ranging from tropical to subarctic or Antarctic coastal waters, they have been reported in many mangrove ecosystems, mostly occurring on leaf litter (see e.g., Wong et al., 2005). Fungi encompass both unicellular and multicellular organisms. In mangrove ecosystems, fungi can be categorized into various groups, notably saprophytic (feeding on decaying organic matter), parasitic, or symbiotic fungi (Thatoi et al., 2013), such as arbuscular mycorrhizal fungi that are obligate symbiotes of the root system. The trophic traits (e.g., autotrophy, heterotrophy, and mixotrophy) of the distinct microorganisms mentioned hereabove will determine their relative contribution to organic matter and element cycling. For instance, in terrestrial ecosystems, autotrophic microorganisms were estimated to be responsible for 4% of CO_2 fixation per year (Yuan et al., 2012).

Hereafter, we will focus on bacteria, archaea, fungi, and thraustochytrids. We will also provide some figures about studies focusing on viruses in mangroves research, as their effect on microbial communities and other constituents (fauna, flora) of mangrove ecosystems might be non-negligible (see e.g., Jin et al., 2019). But we will not describe the processes they are involved in further detail, as this goes beyond the focus of this chapter, and such knowledge is still scarce.

2.2 Drivers of microbial communities

Microbial community composition and activity is influenced by environmental (abiotic) conditions and biotic interactions. Biotic interactions encompass both the interactions within the complex microbial communities (such as between microalgae and other members; Fig. 6.1D exemplifies the potential high densities of microalgae that can be found in mangrove ecosystems) and also between the microbial communities and other members of the mangrove ecosystem, namely, the fauna and the flora. Those factors act simultaneously and it is therefore crucial, when studying microbial communities in a given ecosystem, to record key environmental factors known to influence microbial communities, their interaction, and therefore the processes they will drive (Shang et al., 2017), as well as information about the other organisms present.

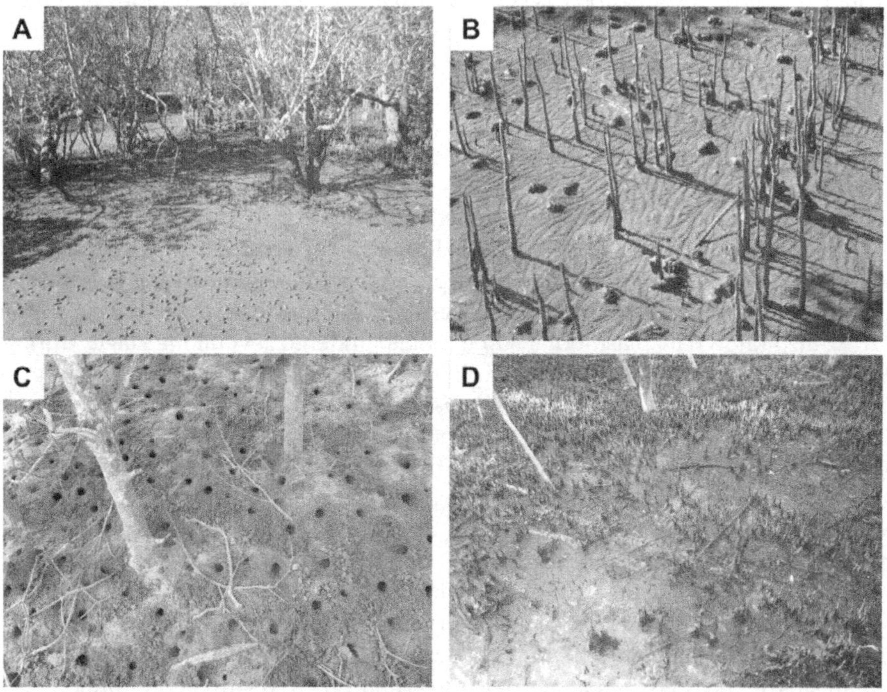

FIGURE 6.1 Mangrove stands with high density of snails (A, B; *Avicennia germinans* stand in South Africa), crabs (C; *Bruguiera gymnorhiza* stand in South Africa), or *Chaetomorpha* sp. algae (D; *Avicennia germinans* stand on San Andrés Island, Colombia). *Credit: Helfer, V., and Zimmer, M.*

Depending on the habitat occupied, relevant environmental parameters will differ. For bulk sediments, key abiotic drivers of microbial community composition are pH (Constancias et al., 2015; Zhou et al., 2017 for archaeal communities), grain size, redox conditions (that will vary over sediment depth), detrital organic matter quality, soil or sediment organic carbon, and soil or sediment water content (see e.g., Shang et al., 2017). For instance, pH and soil C:N ratio (Trivedi et al., 2013 and references herein) are suggested as good predictors of fungal:bacterial dominance (an important metric in soil ecology to assess the functionality of microbial communities). Nonetheless, these associations are mainly based on correlative observations and might be erroneous. Recently developed theoretical approaches to better characterize microbial niche differentiation such as balance trees (Morton et al., 2017) should help further our understanding of microbial responses to environmental parameters. In mangrove ecosystems, subject to fluctuating tidal inundation and therefore variation in water content and salinity conditions over time, the position along the intertidal gradient might be a key driver of microbial community composition and activity. This remains, to our knowledge, scarcely investigated. For host-associated microbial communities, other parameters might be relevant, such as senescence stage for mangrove leaves, or root type (e.g., anchor roots, cable roots, feeding roots, and pneumatophores; Srikanth et al., 2016).

When it comes to biotic interactions, the complexity of possible interactions is immense as it encompasses interactions between different members of the microbial community, as well as direct or indirect interactions with higher organisms, including the plants species and members of the fauna, such as sesarmid or fiddler crabs (known as important ecosystem engineers in mangrove ecosystems; see e.g., Kristensen, 2008), snails, and other, mostly understudied, members of the sediment fauna. The role of the flora will be mediated via the quality of the organic matter and via its effect on sediment conditions (such as sediment oxygenation via their root system). The influence of the fauna will be mainly mediated via trophic interactions (e.g., members of the benthic meiofauna grazing on protists, bacteria, or algae) and physical disturbances (bioturbation; see Fig. 6.1 as example of bioturbation induced by crabs or snails in mangrove ecosystems).

2.3 The role of microbial communities in element cycling

We provide hereafter an overview of the main element cycles (C-, S-, N-cycles) occurring in mangrove sediments, focusing on microbially mediated processes. Besides carbon (C), nitrogen (N), and sulfur (S), manganese (Mn) and iron (Fe) are important elements for the energy metabolism of bacteria and archaea. Describing their dynamics in details is nonetheless beyond the scope of this chapter (see e.g., Ferreira et al., 2007a for iron speciation), but they will be referred to when coupled to other cycles. Indeed, the distinct element cycles are strongly interlinked (for example, iron (Fe) and phosphorous (P) are strongly linked to sulfate-reducing bacteria involved in the S-cycle) and their relative contribution to organic matter dynamics will depend on the ambient environmental conditions. Marine sediments (Canfield and Thamdrup, 2009), and similarly mangrove sediments, are characterized by steep environmental gradients, resulting in a distinct zonation of redox conditions that determine which elements are available for energy metabolism and mineralization processes.

2.3.1 *The C-cycle*

Carbon cycling or carbon dynamics in mangrove ecosystems has received considerable attention in the last two decades, as mangrove ecosystems exhibit the most efficient carbon sequestration ability per unit area (Donato et al., 2011; Mcleod et al., 2011). The carbon cycle in mangrove ecosystems is thus of paramount importance in the context of climate change mitigation. While we will not detail here all processes related to drivers of carbon cycling in mangrove (see e.g., Kristensen et al., 2008 and Alongi, 2014 for reviews), we will focus on sedimentary decomposition pathways involving microbial communities, as documented in Kristensen et al. (2008). The primary step of organic matter decay is the leaching of soluble organic molecules (such as carbohydrates, cyclitols, tannins, and other phenolic compounds). Under oxic conditions, the leached dissolved organic matter (DOM) is efficiently degraded and converted to microbial biomass (with up to 90% conversion efficiency), while in anoxic and nutrient-poor mangrove sediments, the conversion efficiency is much lower (35%; see Kristensen et al., 2008 and references therein). The second phase consists in the decomposition of the remaining particulate organic matter and is mediated by the extracellular enzymatic activity of aerobic or anaerobic prokaryotic microorganisms and marine mycelial decomposers (including notably fungi). Finally, the last steps consist of very slow organic matter decay and the formation of refractory humic compounds and geopolymers (Kristensen, 1997 and references therein).

In surface sediments, aerobic microbial communities have the capacity to perform the complete oxidation of organic carbon to CO_2. In deeper sediment layers, anaerobic degradation processes are conducted in several steps, involving a complex set of microbial communities and interactions, including the hydrolysis of macromolecules via extracellular enzymes (which is the rate-limiting factor of overall degradation rate of organic matter), and the formation of small organic molecules (such as sugars, amino acids, lipids, organic acids) that will undergo complete oxidation (resulting in CO_2 formation) performed by various anaerobic archaeal and bacterial microorganisms, using diverse electron acceptors, such as Mn^{4+}, NO_3^-, Fe^{3+}, and SO_4^{2-}, ranked here in function of their energy yield (Kristensen et al., 2008; Jørgensen et al., 2019; and references therein). Microbial cells that take up small organic molecules conserve energy and grow via multistep fermentation processes, producing various volatile fatty acids together with H_2 and CO_2. Those fermentation products will be used by sulfate-reducing microorganisms, and when the sulfate is depleted in deeper sediments, the terminal degradation of organic matter will be performed by methanogenic archaea that are using mainly H_2 or CO_2.

In mangrove sediments, aerobic respiration and anaerobic sulfate reduction are usually the most important respiration processes (30%–50% each), while denitrification (see N-cycle), manganese, and iron respiration play a minor role (Kristensen et al., 2017 and references therein). In particular cases, the relative contribution of those processes might be different, especially in iron-rich sediments, where iron respiration might be equal if not higher than sulfate reduction, as the latter is hindered when more potent electron acceptors are present. In conditions where all electron acceptors are depleted and electron donors more abundant, fermentative disproportionation of low molecular compounds (such as acetate) or the reduction of CO_2 by hydrogen or alcohols will lead to the production of methane (CH_4). Sulfate reduction can maintain hydrogen and acetate at low levels to hinder methanogens. Even in the presence of active methanogenesis, the simultaneous removal by anaerobic methane oxidizers maintains very low methane concentrations in the porewater of subsurface sediments (Kristensen et al., 2008 and references therein). Recent studies, though, are challenging this statement, as many active methanogenic archaea have been documented in mangrove sediments and the competition between sulfate-reducing and methanogenic bacteria is relaxed when sediment organic matter content is high, or when sulfate reduction is hindered due to sulfate dilution, in areas with a strong freshwater input (see Kristensen et al., 2017 and references therein).

2.3.2 *The N-cycle*

Nitrogen is a key element of many essential biomolecules, such as DNA or proteins, of critical importance for all living organisms. Abundant in the atmosphere as dinitrogen gas (N_2), very few organisms can access it in this form. It becomes available for primary producers as ammonia (NH_3) mainly in the form of ammonium (NH_4^+), although some studies suggest that some plants can assimilate nitrogen as amino acids or even proteins (see Reef et al., 2010 and references herein). Besides dinitrogen gas (N_2) and ammonia (NH_3), nitrogen is mainly found in the environment as nitrate (NO_3^-) or nitrite (NO_2^-). Several microbially mediated processes are involved in the interconversions between these nitrogen species: (1) nitrogen fixation, (2) nitrification s.l. (including also comammox, anammox), (3) denitrification and dissimilatory nitrate reduction to ammonia (DNRA) both involving a nitrate reduction step, and (4) ammonification.

Nitrogen fixation, that consists of transforming dinitrogen gas (N_2) to ammonia (NH_3), is energetically very demanding and only a small selection of prokaryotic microorganisms, referred to as diazotrophs, are able to conduct this process, either as free-living organisms or in symbiosis with plants (e.g., *Rhizobium* species). Diazotrophs include aerobic and anaerobic as well as phototrophic and chemotrophic bacterial or archaeal microorganisms possessing the nitrogenase enzyme complex. In mangrove ecosystems, nitrogen fixation has been documented as free (in e.g., sediments, microbial mats, leaf litter) or associative process in plant roots of various mangrove species (without signs of specific microbes–plant interaction; Reis et al., 2017 and references therein). Associative nitrogen fixation rates reported for mangrove ecosystems, although relying on scarce data, are within the range of global estimates reported for tropical terrestrial forests. Nitrogen fixation rates show variation between the distinct mangrove compartments, the highest rates being observed on leaf litter, roots (associative), cyanobacterial crusts on trunks, and lower rates being reported for senescent leaves, sediments, and pneumatophores (Reis et al., 2017 and references therein). In mangrove sediments, which are often nitrogen limited, nitrogen fixation by members of the Alphaproteobacteria (e.g., *Azotobacter*, *Bradyrhizobium*) as well as sulfate-reducing Deltaproteobacteria is considered a major pathway for the input of bioavailable nitrogen (Zhang et al., 2008; Romero et al., 2015).

Nitrification, the process of converting ammonia (NH_3) to nitrite (NO_2^-) and then nitrite (NO_2^-) to nitrate (NO_3^-), is conducted solely by prokaryotes and occurs in oxic conditions, mainly in the sediment surface layer, in burrow linings, or in deeper layers along oxic root surfaces (Purvaja et al., 2008). Most aerobic ammonia oxidizers are autotrophs and fix CO_2 to produce organic carbon, using ammonia as energy source. Only a restricted number of prokaryotes is able to perform this process, referred to as ammonia-oxidizing archaea (AOA) and ammonia-oxidizing bacteria (AOB). AOA are found in higher abundance than

AOB in many ecosystems; however, in some mangrove sediments, AOB seem to dominate (Li et al., 2011). The second step in the nitrification process, the oxidation of nitrite (NO_2^-) to nitrate (NO_3^-), is undertaken by separate prokaryotic clades, namely nitrite-oxidizing bacteria of the Nitrospirae, Nitrospinae, and some proteobacterial lineages. More recently, comammox bacteria, which are able to perform complete ammonia (NH_3) oxidation to nitrate (NO_3^-), have been identified (Daims et al., 2016). Their presence has been confirmed in mangrove sediments and their abundance was reported to be comparable to AOA and AOB (Liu et al., 2020). Anammox refers to anaerobic ammonium oxidation, in which ammonium combined with nitrite (NO_2^-) is converted to dinitrogen gas (N_2), and is mostly carried out by bacteria belonging to Planctomycetes that are found in various ecosystems, including mangrove sediments.

Denitrification entails the conversion of nitrate (NO_3^-) to dinitrogen gas (N_2), thus rendering nitrogen non-bioavailable. This process, also referred to as nitrate respiration, is anaerobic and occurs mainly in sediments. As intermediate product of this process, nitrous oxide (N_2O) is formed. Denitrification is a process conducted by various microorganisms, including both chemoorganotrophic (requiring organic carbon as energy source) and autotrophic prokaryotes, as well as selected eukaryotic lineages. Denitrification has been shown to be the dominant pathway of nitrogen loss in mangrove ecosystems compared to anammox (Fernandes et al., 2012b; Cao et al., 2017). To avoid the loss of bioavailable nitrogen (N), DNRA converts nitrite (NO_2^-) to ammonia (NH_3) instead of dinitrogen gas (N_2). This pathway can dominate over denitrification in mangrove sediments under nitrogen (N) limitation (Fernandes et al., 2012a).

Finally, ammonification, the transformation of organic compounds into ammonia (NH_3), is a process conducted by various fungi and prokaryotes that results in nutrient availability (mainly in the form of ammonium (NH_4^+)), rendering nitrogen bioavailable to primary production by plants or microorganisms.

Nitrogen transformation in mangrove ecosystems depends strongly on the overall availability of nitrogen species in the ecosystem. Internal recycling of nitrogen plays a major role in meeting the nutrient requirements of mangrove ecosystems. External nitrogen input varies over temporal (tides, seasons, groundwater, and estuarine dynamics) and spatial scales, mangrove type, and physiographic conditions and is strongly influenced by anthropogenic activities. For instance, anthropogenic enrichment in organic and inorganic nitrogen may shift nitrogen transformations to increased ammonification and N_2O production from nitrification and denitrification, while nitrogen fixation rates are expected to decrease (Purvaja et al., 2008; Reis et al., 2017; and references therein). Besides its key contribution for primary production, the relevance of the N-cycle in the context of climate change should not be neglected, as nitrous oxide (N_2O), produced as an intermediate product of denitrification, is an even more powerful climate gas than CO_2.

2.3.3 The S-cycle

Sulfur (S) is estimated to account for 1% of the dry weight of living organisms (Howarth, 1984), being part of important biomolecules such as proteins, sulfo-lipids, and vitamins. Scarce in some ecosystems, sulfur (S) is not a limiting resource in marine and coastal ecosystems, as sulfate (SO_4^{2-}) is found in high concentration in seawater (28 mM at ocean salinity of 35; Jørgensen et al., 2019). The processes involved in the S-cycle include chemical reactions, microbially mediated pathways, and a combination of both. In anoxic marine sediments, microbial dissimilatory sulfate (SO_4^{2-}) reduction (DSR) to sulfide by anaerobic microorganisms is the major process of the sulfur cycle. Combined with fermentative reactions that degrade polymeric organic matter and provide the substrate for sulfate reduction, sulfate reduction is the major process of organic matter degradation in coastal sediments (contributing up to 50% of the total respiration processes in mangrove sediments; Kristensen et al., 2017), which becomes anoxic within a few millimeters of the sediment surface (Howarth, 1984). The reoxidation of most of the sulfide produced by microbially mediated sulfate reduction back to sulfate via various sulfur intermediates (element sulfur (S^0), tetrathionate ($S_4O_6^{2-}$), thiosulfate ($S_2O_3^{2-}$), and sulfite (SO_3^{2-})) takes place then via geochemical or microbial processes involving various potential oxidants, such as oxygen (O_2), nitrate (NO_3^-), manganese (Mn^{4+}, a.k.a Mn (IV)), iron ($Fe3^+$, a.k.a Fe(III)). The remaining sulfide will precipitate with iron or other metals, or via reactions with organic matter, and will be buried in deeper layers (those processes are referred to as pyrite burial and organic sulfur burial, respectively; Jørgensen et al., 2019). We will hereafter describe the major steps of the sulfur cycle (based on Jørgensen et al., 2019 for anoxic marine sediments of the seabed and on Behera et al. (2014) for mangrove sediments): (1) sulfate reduction, (2) sulfide oxidation, and (3) sulfur disproportionation.

There are two major physiological types of sulfate reduction (Behera et al., 2014). During the assimilatory or biosynthetic sulfate reduction, microorganisms reduce only the amount of sulfate needed to meet their nutritional requirements for sulfur. This pathway reduces sulfur compounds for biosynthesis of amino acids and proteins (microbial organic matter). The second pathway is the DSR or respiratory sulfate reduction, during which sulfate (SO_4^{2-}) is reduced to sulfides ($H_2S + HS^- + S^{2-}$), and serves as terminal electron acceptor for anaerobic respiration, in the absence of O_2. Sulfate-reducing microorganisms are composed of a diverse group of anaerobic microorganisms, mainly bacteria, which have the catabolic capacities for a wide range of low molecular weight organic compounds, such as short-chain fatty acids, hydrocarbons, and aromatic compounds. Among them, many belong to the Deltaproteobacteria (including *Desulfovibrionales* and *Desulfobacterales* orders) or to the phyla Firmicutes, Chloroflexi, and Atribacteria in deeper sediments (see Jørgensen et al., 2019 and references herein). Recently, Anantharaman et al. (2018) showed that the potential for sulfate or sulfite reduction is present in many more bacterial and archaeal groups than initially thought. Little is known about autotrophic or mixotrophic sulfate reducers (Hügler and Sievert, 2011) or their role in mangrove ecosystems. However, sulfate-reducing bacteria are known as important diazotrophs that may contribute up to 50% of the total nitrogen fixation in mangrove ecosystems (Shiau and Chiu, 2020).

Sulfide oxidation occurs over the whole sediment column but mainly in the surface zone, where diverse autotrophic and heterotrophic sulfide oxidizers are responsible for dark CO_2 fixation and organic matter degradation, respectively, thus contributing significantly to the organic carbon cycle. Photoautotrophic oxidation of sulfide (also referred to as anoxygenic photosynthesis) is an anaerobic process, carried out by green sulfur bacteria (e.g., *Chlorobium*) or purple sulfur bacteria (e.g., *Allochromatium*), which use H_2S as electron donor for reducing CO_2 into microbial biomass. Colorless sulfur bacteria exhibit a variety of metabolic strategies (Behera et al., 2014) and include (1) obligate chemolithoautotrophs that use inorganic compounds as energy source for CO_2 fixation, (2) facultative chemolithoautotrophic sulfide oxidizers that also grow heterotrophically using complex organic compounds as carbon and energy source, or mixotrophically using both pathways simultaneously, (3) chemolitho-heterotrophs that are unable to fix CO_2 but generate energy from oxidation of reduced sulfur compounds, and (4) chemoorganoheterotrophs that oxidize reduced sulfur compounds without deriving energy from it, using this reaction to detoxify metabolically produced hydrogen peroxide (H_2O_2). Many colorless sulfur bacteria are aerobic, using primarily O_2 as terminal electron acceptor, but some species can grow anaerobically using nitrate or nitrite as terminal electron acceptor. The chemical or microbial oxidation of sulfide via intermediate sulfur species is important to detoxify the sediments, as sulfides can become detrimental for various organisms (inhibiting respiration, oxygen release, nutrient uptake by plant roots, and primary root growth; SamKamaleson and Gonsalves, 2019 and references therein). Sulfide oxidation is mainly driven by specialist oxidizers belonging to the former gammaproteobacterial family Beggiatoaceae (e.g., filamentous: *Beggiatoa* and *Thioploca*; spherical: *Thiomargarita*), now divided into several new *Candidatus* genera, which are found in the surface layer in organic-rich sediments (Jørgensen et al., 2019).

Sulfur disproportionation, a chemolithotrophic process involving the degradation of sulfur intermediates, was discovered in the late 1980s and was compared to the fermentation of organic matter with redistribution of carbon atoms and synthesis of ATP without involving electron transport and proton gradients (Finster, 2008). The first isolated bacterium capable of disproportionating thiosulfate and sulfite belonged to the genus *Desulfovibrio*, a group of sulfate reducers, and was coined *Desulfovibrio desulfodismutans*. Further screening showed that thiosulfate and sulfite disproportionation is rather common within sulfate reducers (particularly in Deltaproteobacteria), but only a few strains could use it for growth. Disproportionation of elemental sulfur has also been demonstrated since then (Finster, 2008 and references therein).

Varon-Lopez et al. (2014) investigated the sulfate-reducing and sulfur-oxidizing microbial communities in Brazilian mangroves. Their results identified Betaproteobacteria, Gammaproteobacteria (for sulfur-oxidizing bacteria), and Deltaproteobacteria (for sulfate-reducing bacteria) as the main groups associated with the sulfur cycle in the studied mangroves. Bioturbation, via benthic macrofauna or plant root systems, will influence oxygen fluxes and consequently the S-cycle and involved microbial communities in mangrove ecosystems. Ferreira et al., 2007b, who observed higher values of pyrite-Fe in vegetated areas, suggested that plants might stimulate the activity of sulfate-reducing bacteria via their root exudates. Furthermore, sulfide can also be oxidized over long distance by "cable bacteria", also found in mangrove sediments (see Jørgensen et al., 2019 and references therein), which form several-centimeter-long chains, spatially separating the individual steps in the sulfide oxidation process by creating an electron current along their length, which results in a unique geochemical zonation.

2.3.4 *Drivers of microbial element cycling*

Generally, in marine sediments, the relative contribution of the different electron acceptors in microbial carbon oxidation and its rate will depend on the quantity and quality of the organic matter, the sediment grain size, and the levels of bioturbation by plants or animals. In mangrove ecosystems, though, other factors might be equally important, such as the forest age (due to higher input of organic carbon and increased amount of fresh organic matter leading to sulfate reduction prevalence), the physiology of the root system of higher plants, and the inundation patterns (Kristensen et al., 2008). Element cycling in mangrove ecosystem is strongly influenced by the redox conditions, which are in turn influenced notably by the length of the inundation period and the sediment grain size (the coarser the sediment, the greater the oxygen penetrability). Therefore, different physiographic positions of mangrove ecosystems will result in distinct processes involved in organic matter decomposition. For example, Ferreira et al., 2007a observed different profiles of iron speciation and degree of pyritization in mangrove sediments from fringe, basin (inland), and estuarine mangroves and attributed such variations to sediment composition (grain size, organic matter content) and the effect of tidal inundation duration and frequency. Consideration and proper documentation of those parameters are of paramount importance when designing field studies, if we want to be able to understand the key drivers of microbial communities at various spatial scales. In addition to the variety of biogeochemical conditions occurring in mangrove ecosystems, the diversity of the microbial communities by itself adds to the complexity of the system, due to the numerous possible microbial interactions.

It is therefore not trivial to decipher the main drivers of organic matter decomposition. Correlative observations linking microbial community composition and/or activity with organic matter characteristics might be misleading, as the direction of the relationship is not known, unless investigated by manipulative experimental approaches. In this context, among the new promising approaches that should help to better decipher the respective role of the distinct members of complex microbial communities in element cycling, stands the so-called SIP-'omics (see Section 4.4) approach, which combines the use of stable isotope probing (SIP) and 'omics to trace the fate of distinct compounds and the microorganisms utilizing them, to thus unravel metabolic pathways (Abram, 2014).

3. Microbial communities in mangroves

3.1 Peculiarities of mangrove ecosystems and their constraints

Alike other coastal ecosystems (e.g., seagrass beds, mudflats, salt marshes), mangrove ecosystems are extremely dynamic ecosystems, especially due to the influence of the tides, inducing drastic fluctuations in oxygenation and salinity. In addition, mangrove ecosystems occur in various environmental conditions, ranging from extreme dry and saline conditions (Oman, Mexico, South India) mainly on coarse sediments with low organic matter content, to more muddy sediments with a higher water holding capacity and high organic matter content. These conditions will drastically influence both the microbial community composition and its activity, which will further lead to different sediments processes.

At the interface between land and sea, and often distributed along estuaries, mangrove ecosystems are under multiple influences (marine, terrestrial, freshwater), and their microbial communities will consequently be influenced by microbes originating from marine, terrestrial, or freshwater communities. In general, mangrove sediment are saline, acidic, and waterlogged and characterized by low inorganic nutrient availability (Reef et al., 2010).

3.2 Actual knowledge on microbial communities in mangrove ecosystems

To get an overview of the current knowledge on microbial communities in mangrove ecosystems, we first searched for mangrove-related publications using the bibliographic database Web of Sciences (http://apps.webofknowledge.com). The following settings were applied: timespan = All years; language = All languages; documents type = "Article," "Book," and "Book Chapter" with the syntax: TS=(*angrov*), resulting in 16,534 entries, on January 18, 2020. The data set was further curated and explored using R version 4.0.2 (R Core Team, 2020) within RStudio v1.3.959 (RStudio Team, 2020). After data curation (removal of certain document types such as "Editorial Material," removal of duplicates, and exclusion of papers not related to mangroves that were included due to the permissive syntax *angrov*), and selecting only the entries that mentioned mangrove or one of the mangrove genera (as listed in Spalding et al., 2010) in either their abstract, title, or keywords (we considered only the Author Keywords (= attribute "DE"), not the Keywords Plus* option (= attribute "ID")), the data set was reduced to 14,678 entries. From those entries, we selected those focusing on microbial communities, by searching for entries mentioning the distinct groups in either the title or the keywords ($N = 1706$; for the syntax, see Appendix 6.1), and further restricted the data set to those entries referring to mangrove (or mangrove genera) in the title or the keywords ($N = 1351$). The R scripts used to perform this bibliographic review are available under https://gitlab.leibniz-zmt.de/vhe/microbial-communities-in-mangroves. We further explored the data set (see Appendix 6.1 for the syntax used for selecting the various groups of papers), screening the abstracts ($N = 1271$; 80 entries had no abstract), titles ($N = 1351$), and Author Keywords ($N = 1094$; 257 entries had no keywords) for the distinct groups of microorganisms, the distinct habitats under focus (sediment, roots, leaves), and the different analytical approaches used in the respective studies.

Starting in the early '60s, the interest in mangrove microbial communities remained low for over four decades and only took off since approximately 2005, with a steep increase over the years since then (Fig. 6.2). This steep increase is mainly observed in studies focusing on sediment microbial communities, while studies on leaves, roots, or combining two or more of the habitats only slightly increased since 2005.

Regarding the distinct groups of microbes, studies focusing on bacteria and fungi increased drastically since 2005. Studies about protists also slightly increased since 2005, but they remained scarce, and studies focusing only on archaea do not show any trend over the years (but as most studies mentioning archaea were also focusing on bacteria (see Fig. 6.3), they are therefore included under the denomination "Combined" studies in Fig. 6.2). In terms of analytical approaches used to study microorganisms in mangroves, the literature is still clearly dominated by studies using traditional approaches (culture-dependent or clone libraries), while 'omics approaches considerably gained in popularity since 2015.

As a result of the yearly trends, the microbial communities linked to the root system of mangrove plants or their leaves or litter have attracted much less attention (197 and 177 studies, respectively) than those linked to the sediments (485 studies), and bacteria and fungi are by far the most investigated microorganisms, followed by protists, archaea, and viruses (see Table 6.1). Regarding the approaches used thus far in mangrove microbiome research, traditional approaches are still used predominantly (784 studies), compared to fingerprinting

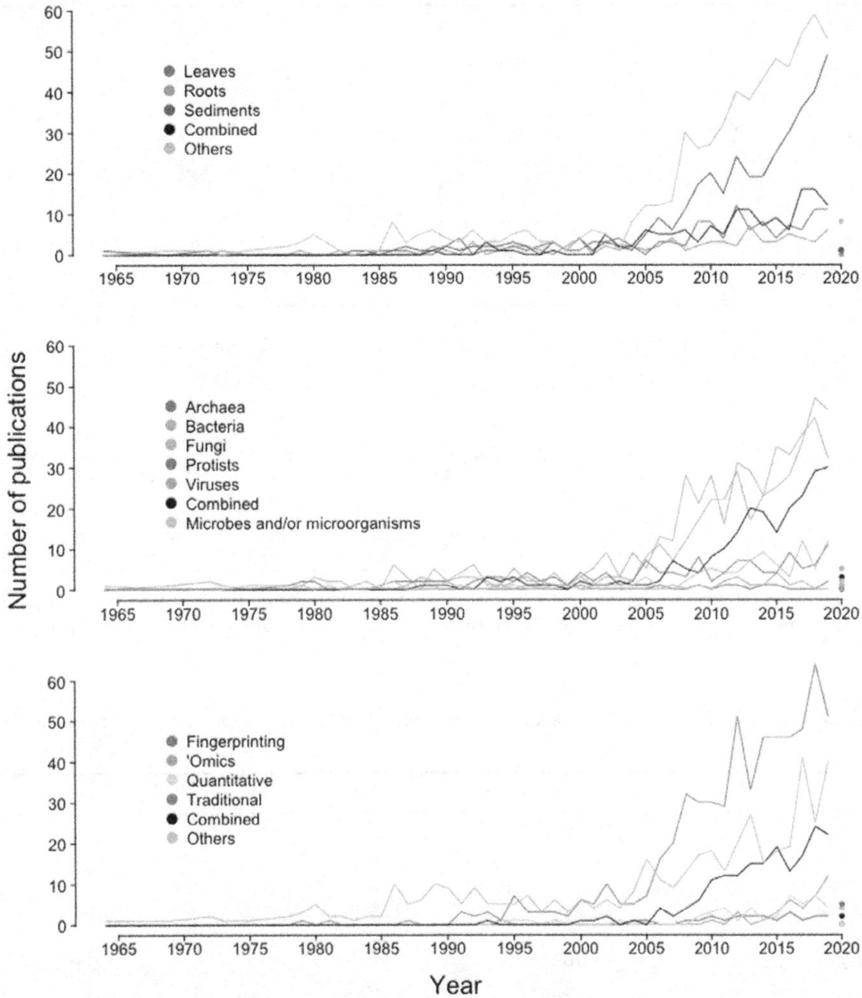

FIGURE 6.2 Number of publications per year for the distinct habitats (top), the distinct groups (middle), and the distinct analytical approaches (bottom; see Section 4 for a description of the approaches). The studies referred to as "Combined" are studies referring to two or more of the classes defined for the distinct habitats, groups, or analytical approaches. We adopted this strategy to avoid having studies represented more than once in the graph, thus biasing the count of publications per year. *Note that publications without information about year of publication were assigned to publications from 2020, as they were belonging to document type DT = "Article; Early Access".*

(62 studies) or 'omics approaches (99). Studies using quantitative approaches were mostly focusing on enzymes or biomarkers (138 studies). The different approaches used in mangrove microbiome research, and microbial ecology in general, are further discussed under Section 4, stressing their respective advantages and limitations.

To visualize to which extent studies conducted so far were integrative in terms of habitat investigated, microorganisms targeted, or analytical approaches used, Fig. 6.3 below displays the figures mentioned above in Venn diagrams.

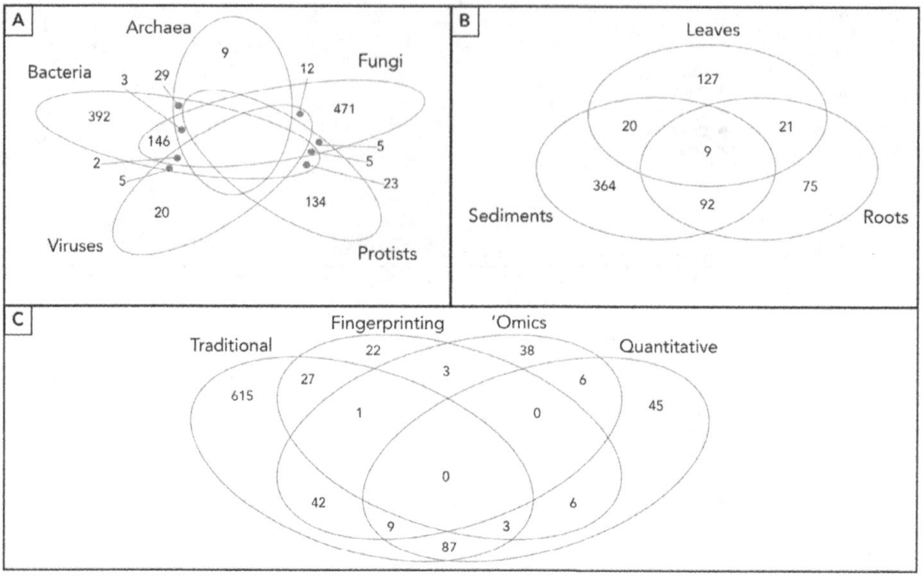

FIGURE 6.3 Venn diagrams showing how many papers were mentioning (in their title, abstract, or keywords) (A) the distinct microorganism groups, (B) the distinct habitats, and (C) the distinct analytical approaches (see Section 4).

TABLE 6.1 Number of publications referring to the distinct groups of microbial organisms (including viruses).

	Abstract, keywords, title	Keywords, title
Archaea	41	24
Bacteria	605	433
Fungi	644	581
Protists	167	153
Viruses	39	33
Others	95	211

Others encompasses publications that were referring to microbes or microbial communities or microorganisms in general, but none of the above groups specifically.

We see that in terms of microorganisms investigated, most studies (1.026) focused on a single group, and only 231 studies were referring to more than one group. The remaining studies (95) were not referring to one of the specific groups but to microbial communities in general and might therefore also be integrative studies (unless the emphasis was on a specific group that we did not target, such as myxomycetes). As most of the sediment processes linked to organic matter dynamics are jointly mediated by various members of the microbial community, integrative studies targeting ideally all the distinct groups of microbes are essential to provide a comprehensive understanding of the biogeochemical processes taking place and their biotic drivers.

Regarding the distinct habitats and their respective microbial communities (the microbial communities linked to the sediments, to plant roots and leaves or leaf litter respectively), most of the studies (566) were targeting only one of the habitats. Only nine studies were mentioning all three habitats (see Fig. 6.3B), but 92 and 20 studies were mentioning jointly the sediments and the roots or the leaves respectively. Nonetheless, the interactions between the microbiome of the rhizosphere and that of the bulk sediment are still poorly understood.

To get a more detailed overview of which communities were studied in which habitat, or what approach was specifically used to study the microbial communities in mangrove ecosystems, we created word clouds based on the abstracts of each publication (when available; $N = 1271$) for the distinct habitats (Fig. 6.4), microorganism groups (Fig. 6.5), and analytical approaches (Fig. 6.7). For each word cloud, only words occurring in at least 10% of the publications (or a minimum of 10 publications) related to that category were retained.

Studies about microbial communities in mangrove sediments relate mainly to bacterial communities ($N = 340$) and much less so to fungal ($N = 102$) or archaeal ($N = 40$; not shown on word cloud as the threshold was set to 49) communities (see Fig. 6.4A). The literature about root-associated microbial communities refers mainly to bacterial and fungal communities (see Fig. 6.4B). Not shown on the word cloud (below threshold), archaeal communities were mentioned only in five studies, while their contribution in terms of abundance and diversity in mangrove rhizosphere is non-negligible and can be higher than in the sediments in some cases (e.g., archaeal diversity in the rhizosphere of *Laguncularia racemosa*, Pires et al., 2012). The determinants of the rhizosphere microbial composition are not yet clearly understood. Whether or not species-specific relationships exist is not well-investigated. Some studies suggest that the rhizosphere is composed of a selection of the environmental microbiome. Nonetheless, further investigations aiming at studying the changes in microbial community composition over the course of the plant growth, from communities associated with mangrove seedlings to the communities associated to the roots and the leaves of trees, might help us decipher the key drivers of the plant-associated microorganisms. Furthermore, mangrove tree species exhibit various root systems, and within each species, the root structure can be further decomposed into several elements, such as anchor roots, cable roots,

FIGURE 6.4 Word clouds computed over all abstracts from studies mentioning the distinct habitats ((A) sediments; (B) roots; (C) leaves of leaf litter) in their abstract, keywords, or title. For the details of the word selection, see https://gitlab.leibniz-zmt.de/vhe/microbial-communities-in-mangroves.

feeding roots, and pneumatophores. Whether or not microbial communities linked to these different structures or elements thereof differ is clearly understudied. The studies focusing on microbial communities associated to the leaves or leaf litter mentioned fungal organisms more often, followed by bacteria (see Fig. 6.4C), while archaea were reported in only one publication (therefore not appearing on the word cloud).

Regarding the distinct groups of microorganisms, the studies focusing on bacterial communities are numerous and relate mainly to sediments (340), while only 100 studies referred to bacterial communities linked to root structures and only 78 to leaves or leaf litter (see Fig. 6.5A). In the studies dedicated to archaea (see Fig. 6.5B), AOA are frequently mentioned in conjunction with AOB. One study in particular focused on the effect of faunal burrows on the diversity and abundance of AOA and AOB, together with anaerobic ammonium oxidizing (anammox) and denitrifying anaerobic methane oxidization (n-damo) bacterial communities (Chen and Gu, 2017). According to the authors, burrowing activity induces higher concentrations of ammonium and lower concentrations of nitrite and nitrate, resulting in an increase in diversity of AOA and AOB. The numerous studies dedicated to fungi ($N = 644$) seem to be mostly related to the bioprospection of compounds as antimicrobial agents (see Fig. 6.5C). Regarding protists, algae (including macroalgae and microalgae) and thraustochytrids seem to be the most studied organisms (see Fig. 6.5D), mainly in the sediments. Thraustochytrids, together with fungi, are important organisms for the

FIGURE 6.5 Word clouds computed over all abstracts from studies mentioning the distinct groups of microorganisms ((A) Bacteria; (B) Archaea; (C) Fungi; (D) Protists; (E) Viruses; (F) Others) in their abstract, keywords, or title. "Others" encompasses publications that were referring to microbes or microbial communities or microorganisms in general, but none of the above groups specifically. For the details of the word selection, see https://gitlab.leibniz-zmt.de/vhe/microbial-communities-in-mangroves.

decomposition of the leaf litter, as they are able to decompose highly recalcitrant organic matter (Raghukumar et al., 1994; Fossier Marchan et al., 2018). They are both endo- and epiphytic (see Raghukumar et al., 1995). They are able to decay cellulose-rich material (notably mangrove leaves but also macroalgae) and are found in higher abundance on leaf litter than on the sediment surface (Wong et al., 2005). Their ability to produce long fatty acids makes them an important food source for detritus feeders. Viruses (see Fig. 6.5E) were mainly mentioned in the context of human health and bioprospection for antiviral properties of mangrove plant species and/or associated microorganisms. One study, though, focused in detail on the diversity and potential biogeochemical impacts of sediment viruses in mangrove ecosystems (Jin et al., 2019), using a virome-specific metagenomics approach and bioinformatic workflow. Their study suggests that mangrove viruses have a direct impact on carbon cycling, as several auxiliary carbohydrate-active enzyme (CAZyme) viral genes, known to participate in the biolysis and therefore the recycling of complex polysaccharides, were identified. Finally, the publications referring to microbial communities in general seem to be related to processes linked to organic matter degradation and to the carbon and nitrogen cycles (see Fig. 6.5F).

Our bibliographic survey shows that bacteria and fungi are the most studied groups in mangrove microbiome research and that many studies were orientated toward the bioprospection of mangrove microbial communities. There is still a lack of integrative studies in mangrove ecological research, with notably little knowledge about archaeal communities.

4. Techniques to depict microbial communities and the processes they drive

Investigating the role of microbial communities requires a comprehensive description of "who is there", i.e., taxonomic diversity, as well as a detailed understanding of "what are they doing", i.e., functional diversity and metabolic activity. However, the information necessary to answer these two questions is difficult to access. The small size (~1 μm), high abundance (10^9 cells mL^{-1} sediment), and extraordinary taxonomic and functional diversity complicate the description of microbial communities in general and individual members of such communities in particular. Genetic information is invaluable to identify and characterize microorganisms. Currently, microbial taxonomy is largely based on the phylogeny of the ribosomal RNA gene (Guillou et al., 2013; Yilmaz et al., 2014; Nilsson et al., 2019). However, given the diversity of metabolic capabilities often found within microbial strains that are closely related based on their ribosomal RNA gene, considering the whole genome in species classification is gaining importance (Parks et al., 2018).

Available methods to study microbial communities often suffer from a trade-off between the detail and the extent of the information they can provide: while cultivation-dependent approaches are suitable to study the morphological, genetic, and metabolic aspects necessary to describe individual organisms in detail, they are often ill-suited to investigate complex communities from environmental samples, such as mangrove sediments. Conversely, high-throughput cultivation-independent methods, such as next-generation sequencing (NGS) of taxonomic or functional marker genes, while ideal for a large-scale screening of microbial diversity, often lack the necessary resolution and detail to elucidate the underlying metabolic mechanisms of microbial interactions and processes to reliably predict responses to a changing environment or to identify candidates for bioprospecting or ecosystem design (Zimmer, 2018). To reconcile this trade-off, recent studies advocate a combination of methods, and advances in methods development in microbial ecology have focused on removing the aforementioned restrictions. In the following section, we will introduce commonly used methods in microbial ecology and discuss their advantages as well as the challenges in their application to microbial communities in mangrove ecosystems. Selecting the most suitable combination of methods from the multitude of approaches available can prove challenging, and there is no universal solution, as this will strongly depend on the specific research questions to be investigated. This section is therefore only intended to aid in the selection progress, enlightening the potential and limitations of methods, but cannot provide one universal recommendation for mangrove microbiome research.

4.1 Traditional methods

4.1.1 Cultivation-dependent approaches

Traditionally, microbiological investigations were based on the isolation and cultivation of microorganisms from different environments. Of primary interest were host-associated microbes due to their role in disease occurrence, and growth media and conditions were therefore optimized mainly for the human microbiome. Far less is known about microbial communities in terrestrial or aquatic, freshwater or marine, pelagic or coastal ecosystems, including mangroves, where conditions differ markedly from human-associated environments, and the microbial community composition is vastly different (Thompson et al., 2017). As a consequence, and despite considerable advances in isolation and cultivation approaches, only a small fraction of the known microbial diversity from those ecosystems is so far culturable (Lloyd et al., 2018). While cultivation approaches are not ideal to depict microbial diversity, they remain an essential tool for studying the functional role and behavior of microorganisms, since a microorganism in culture can be grown, experimented with, and modified without confounding influences of other members of the original microbial community.

4.1.2 Clone libraries

To avoid the bias of cultivation-dependent methods in microbial diversity studies, investigations moved toward working directly with the genetic material of a given sample. Thereby, microbial diversity studies were pioneering the analysis of environmental samples, which have subsequently been extended to include a larger scope of organisms, for instance, the assessment of metazoans communities via eDNA barcoding (Taberlet et al., 2012; Zimmer and Helfer, In prep.). Prior to the onset of the 'omics era and NGS, clone libraries relied on the incorporation of individual DNA fragments, mainly originating from the PCR-based amplification of a taxonomic or functional marker gene, into an easily culturable microorganism, whose colonies were then individually processed for sequencing, yielding one unique DNA sequence per colony (Allen and Banfield, 2005). While clone libraries thereby overcome the cultivation bias, they are limited in resolution and throughput. As such, the number of clones picked for sequencing is often insufficient to cover the full diversity of complex communities, and the approach is not scalable to studies involving a large number of samples and has nowadays been largely replaced by high-throughput NGS approaches.

4.1.3 *Community fingerprinting*

The most commonly used sequence-based community fingerprinting methods are terminal restriction fragment length polymorphisms (TRFLP; Liu et al., 1997), denaturing gradient gel electrophoresis (DGGE; Muyzer and Smalla, 1998), and automated ribosomal intergenic spacer analysis (ARISA; Fisher and Triplett, 1999). By circumventing the need to obtain the exact base sequence directly, fingerprinting approaches solve the issue of high sample throughput at the cost of taxonomic resolution and identification. For instance, ARISA is estimated to approximate community composition and diversity at a taxonomic resolution comparable to family and order level (Gobet et al., 2014). However, to obtain actual taxonomic assignments, additional sequencing is required.

Alternative fingerprinting approaches use differences in the chemical make-up of cells from different taxonomic groups (phospholipid fatty acid profiles; PLFA; Vestal and White, 1989) or functional traits of the whole community (ECOLOG/BIOLOG screening; Vestal and White, 1989) to distinguish microbial communities. Especially the latter is a fast and comparatively inexpensive option if the primary aim is to assess the potential of the bulk microbial community to perform specific processes without the need for taxonomic classification or assignment of metabolic function to specific organisms.

4.2 'Omics and related approaches

The invention of NGS technologies has revolutionized the field of microbial ecology, by enabling the processing of more samples and the generation of more data—the major two shortcoming of clone library and community fingerprinting approaches—at a considerably lower cost and time investment.[1] To date, several NGS platforms are available that differ in the trade-off between the number of sequences they can produce and their length, with the most recent technological developments aiming to maximize both[2] (Jain et al., 2016). NGS technologies have vastly increased our ability to describe microbial communities and the processes they drive. Together with other non—sequencing-based high throughput methods to investigate proteins and metabolites, virtually any aspect of the genetic diversity, transcriptional and translational activity, and metabolic processes of individual microorganisms and whole microbial communities can be studied with this toolbox of so-called 'omics methods. In the following section, the different 'omics approaches, their advantages, and limitations will be further discussed.

4.2.1 *Metabarcoding*

Among the earlier studies employing NGS approaches were metabarcoding projects that rely on the PCR amplification and subsequent sequencing of a marker gene. While metabarcoding uses metagenomic DNA as template, it does not constitute metagenomic sequencing, described in the next section (see also Fig. 6.6; this distinction is not always respected in the literature, where many metabarcoding projects have been published as metagenomic studies). Metabarcoding is preferentially employed for taxonomic surveys to obtain a

[1]https://www.genome.gov/about-genomics/fact-sheets/DNA-Sequencing-Costs-Data.

[2]https://flxlexblog.wordpress.com/2016/07/08/developments-in-high-throughput-sequencing-july-2016-edition/.

detailed overview of the composition of a target community at the desired level of taxonomic resolution, such as bacteria or fungi (Abed et al., 2019), or functional group, such as sulfate-reducing bacteria (Pelikan et al., 2016). As in any PCR-based approach, the PCR step poses both an advantage and a restriction to metabarcoding. The targeted amplification enables the investigation of organisms that only contribute a low amount to the DNA of a given sample without producing a high number of non-target sequences, thereby reducing sequencing costs. Conversely, the primer bias may also limit the taxonomic range of metabarcoding studies and skew the representation of various taxa in the community (Eloe-Fadrosh et al., 2016; Parada et al., 2016). Furthermore, the obtained information is restricted to a comparatively short section of only one marker gene. To extend the scope of metabarcoding studies based on taxonomic markers, such as the rRNA gene, to provide insights into the functional potential of a microbial community, metabolic inference tools have been developed, which estimate functional genes and metabolic pathways based on closely related genomes (Langille et al., 2013; Bowman and Ducklow, 2015; Wemheuer et al., 2018).

4.2.2 *Metagenomics*

As opposed to metabarcoding, in metagenomics all DNA fragments in a sample are sequenced without targeting specific regions. This approach is also referred to as shotgun sequencing and theoretically allows the exploration of the full genetic potential of a microbial community, providing information on taxonomy as well as function. Metagenomic sequencing data are thus highly complex, which poses specific challenges to data analysis and interpretation. For instance, it is often difficult to assign DNA sequences to individual genomes, especially in microbial communities, which host a high diversity, such as those found in mangrove sediments. To overcome these challenges, approaches to reduce the complexity of the data or utilize further structuring information include (1) hybrid short- and long-read sequencing (Wick et al., 2017; Sanders et al., 2019), (2) single-cell genomics, where individual cells from an environmental sample are sequenced independently (Woyke et al., 2009), (3) mini-metagenomics, where cells are sorted prior to sequencing (Alteio et al., 2020), and (4) contact metagenomics, which uses the 3D structure of the chromosome to accurately resolve genomes (Marbouty et al., 2014).

4.2.3 *Metatranscriptomics*

While metagenomics informs about the taxonomic composition of the total microbial community and its functional potential, metatranscriptomics targets the RNA to investigate the active microbial community and study gene expression. As such, metatranscriptomics will provide a more suitable approximation of ecologically relevant microbial processes than metagenomics, enabling the investigation of short-term responses to environmental changes. However, the short half-life of RNA of minutes (mRNA) to hours and days (rRNA) depending on the environmental conditions (Baudrimont et al., 2017; Steiner et al., 2019; Schostag et al., 2020) also poses specific requirements for sample collection and storage. Immediate sample preservation, either by shock-freezing in liquid nitrogen or by adding a preservation solution, is of paramount importance to acquire an accurate snapshot of microbial gene expression at the time of sampling.

In many cases, both metagenomics and metatranscriptomics are being conducted. The combined analysis of microbial communities on DNA and RNA level offers several advantages. While computational tools for standalone metatranscriptomic data analysis are available (Davids et al., 2016), data processing is often simplified if transcripts can be mapped to corresponding metagenomic data. Furthermore, this approach allows the standardization of gene expression by gene copy number, thereby elucidating if changes in gene expression are related to increased transcription rates or increased gene abundance (Salazar et al., 2019). Lastly, the comparison of DNA- and RNA-based community composition can be used to distinguish the contribution of dormant or dead cells, whose DNA may persist in the environment, from that of living cells, which are the ecologically relevant members of the microbial community (Overman and Sikorski, pers. comm.). The latter two applications of combining DNA- and RNA-based approaches are further not restricted to metagenomics and metatranscriptomics, but can also be implemented in metabarcoding studies.

4.2.4 *Metaproteomics and metabolomics*

The 'omics toolbox is not restricted to sequencing technologies based on nucleic acids. Metaproteomics and metabolomics rely on chemical analytical techniques, such as liquid or gas chromatography and high-resolution mass spectrometry (LG/MS or GC/MS), to characterize the proteins, i.e., enzymes, and metabolites present in a sample. While the building blocks of DNA and RNA are limited in number, proteins, which are composed of 21 discrete elements (i.e., amino acids), and other metabolites, such as polysaccharides, which can occur in a virtually infinite number of chemical configurations, are several orders of magnitude more complex. This poses analytical challenges that we have only begun to tackle and solve. Recent developments in the field of analytical organic chemistry have enabled us to identify and resolve the structure of thousands of different compounds in a natural mixture (Kleiner, 2019; Merder et al., 2019). However, our knowledge of the diversity of enzymes and metabolites is still very limited, so that only a small fraction of identified compounds can be annotated and used to reconstruct metabolic pathways and assess microbial processes. In environmental samples, the analysis and interpretation of metaproteomic and metabolomic data is further complicated by the presence of enzymes and metabolites originating from sources other than the microorganisms of interest. Therefore, the combination of several 'omics approaches (integrated multi'omics), such as metaproteomics and/or metabolomics aided by metagenomics and/or metatranscriptomics, constitutes a powerful strategy to assess microbial communities and the processes they drive.

4.3 Challenges of 'omics data analysis

We have reached a state where it is possible to produce more 'omics data than what can be fully analyzed.[3] Often only a fraction of the information content of a data set is analyzed, and with a different research question or in combination with data from other studies (meta-analyses, data mining), new insights might be obtained. Therefore, long-term storage of 'omics data and its associated metadata are of paramount importance. Global initiatives to further that goal advocate the standardized collection and archiving of metadata (Yilmaz et al., 2011) and promote the principle of FAIR (Findable, Accessible, Interoperable, and Reusable; Wilkinson et al., 2016) data. However, to reliably compare data across studies,

[3]https://www.genome.gov/about-genomics/fact-sheets/DNA-Sequencing-Costs-Data.

standardization needs to encompass more than data archiving and include a consistent methodology for sample collection, considering the scale and material of which samples were collected, recorded contextual environmental parameters, and employed 'omics approaches (Allard et al., 2020). Additional measures to ensure data quality include the use of mock communities (samples containing artificial communities of known composition, i.e., diversity and relative abundance) during DNA extraction and sequencing to assess extraction and primer bias, sequencing errors, and the suitability of the bioinformatic sequence processing pipeline.

The rare biosphere plays an important role in ecosystem functioning (Jousset et al., 2017). Our ability to detect rare species in a microbial community is strongly influenced by the available sequencing depth and error rate. To avoid an artificial inflation of diversity, it is recommended to assess the rare biosphere not by rare sequence occurrences but by low sequence proportions in samples that have been exhaustively sequenced so that even rare organisms are represented by a sufficient number of sequences to be clearly distinguished from noise.

Another critical step in the analysis of 'omics data is the annotation of sequences, proteins, or metabolites. In most cases, 'omics data are annotated using curated reference databases. Consequently, it is only possible to classify what is already known, and organisms from understudied environments, including mangrove sediments, may therefore not be well represented in existing reference databases (Karst et al., 2018). Recent large-scale efforts to improve and expand reference databases include the creation of a reference catalog of unknown (metagenomic) sequences (Vanni et al., 2020) or the revision of rRNA gene phylogenies based on rRNA sequences obtained from PCR-free total RNA sequencing (Karst et al., 2018).

4.4 Besides the 'omics approaches

While the 'omics methods described above provide high-resolution data at high throughput, on their own they illuminate only isolated aspects of microbial life (see Fig. 6.6), and without the integration of multiple methods, collection of contextual data, and experimental validation, their ability to support definite conclusions may be limited. Here, we will briefly address other methodological approaches that can be used in

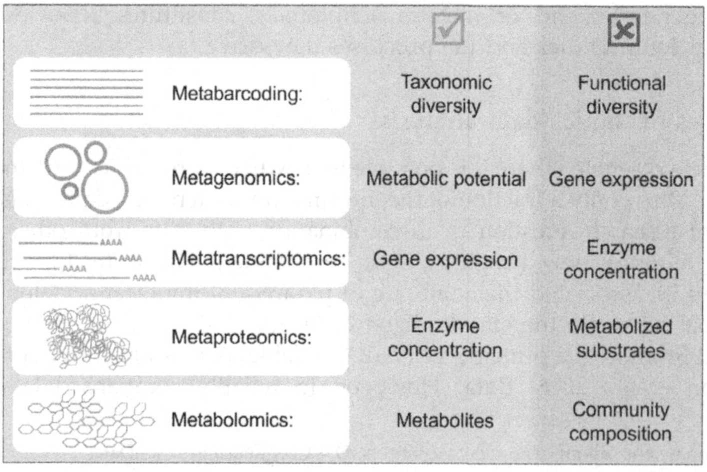

FIGURE 6.6 Overview of 'omics approaches available to investigate microbial communities, summarizing their respective scope and limitations.

combination with 'omics methods and/or traditional microbiological approaches to support and enhance the interpretability of their results.

4.4.1 *Environmental characterization*

To understand the environmental factors that influence microbial community composition and microbial processes, contextual data collected in parallel to the samples for microbial analysis are required. Such parameters may include, but are not limited to, organic matter content, grain size, salinity, inorganic nutrient concentrations, or heavy metal concentrations. As sediments are characterized by steep gradients in physicochemical parameters within the top layers, measurements at microscale by employing microsensors (Meyer et al., 2008) provide a higher resolution of various parameters expected to strongly influence microbial processes, e.g., pH, oxygen concentration, or temperature, than bulk measurements. Oxygen microprofiles can also directly be used to calculate depth-resolved oxygen consumption rates, a key process in organic matter decomposition (Glud, 2008).

We found only one study in our data set that was using microsensors in mangrove ecosystems. Published in 2016, it reported field evidence for the presence of cable bacteria in mangrove ecosystems using microsensor profiling in combination with fluorescence in situ hybridization (see section hereafter). As they are known to have a strong influence on the geochemical cycling in the seafloor, the authors (Burdorf et al., 2016) suggest that those sulfur-oxidizing bacteria could also play a pivotal role in element cycling in mangrove ecosystems.

4.4.2 *Visualization of microbes at microscale*

Microscopy, especially fluorescence in situ hybridization (FISH) techniques, enables the quantification and localization of taxonomic groups (CARD-FISH, DeLong et al., 1989), functional genes (Gene-FISH, Moraru et al., 2010), or proteins (BONCAT-FISH, Hatzenpichler et al., 2014) in a microbial community at microscale. For instance, CARD-FISH has been used to visualize microbial cells on individual sand grains to study biofilm formation (Probandt et al., 2018). The combination of CARD-FISH with mass spectrometry, specifically nanoscale secondary ion mass spectrometry (nanoSIMS), is a powerful analytical method to trace element allocation in microbial systems, as this method provides information on the taxonomy as well as incorporated elements of each cell (Li et al., 2008).

4.4.3 *Supplementing 'omics with quantitative information*

Quantitative information is essential to assess the role of microbial processes in the environment and can vastly increase the scope of 'omics approaches. For instance, multiplying total bacterial cell counts with sequence proportions obtained by 16S metabarcoding is expected to considerably improve network inference in co-occurrence analysis, assuming the PCR bias and differences in 16S copy numbers to be negligible (Peura et al., 2015). Microarrays constitute a potentially quantitative method to measure the abundance of a large but finite selection of genes on DNA or RNA level. As microarrays for environmental samples are mostly based on functional genes, they can be used to directly assess microbial processes biogeochemical cycles or disease occurrence (Lee et al., 2013; Shi et al., 2019). Alternatively, quantitative PCR (qPCR) uses the number of fragments generated during a PCR run to calculate the concentration of template DNA. Enzymatic activity measurements (Boetius and

Lochte, 1994; Zimmer 2020a,b) or isotopic tracer incubations (Reeburgh, 1983) quantify rates of specific microbial processes. The latter approach is especially powerful when combined with molecular techniques in DNA or RNA SIP. By sequencing the fraction of the community enriched in an isotopic signature, the organisms involved in the metabolism and uptake of the labeled substrate can be identified and quantified (DNA-qSIP; Neufeld et al., 2007; Coskun et al., 2018).

4.5 Approaches used in mangrove research

In our bibliographic review (see Section 3.2), studies having applied cultivation-dependent approaches and clone libraries (see Fig. 6.7A) were mainly targeting fungal (467 studies) and, to a lesser extent, bacterial communities (358 studies) and seem to have been mainly

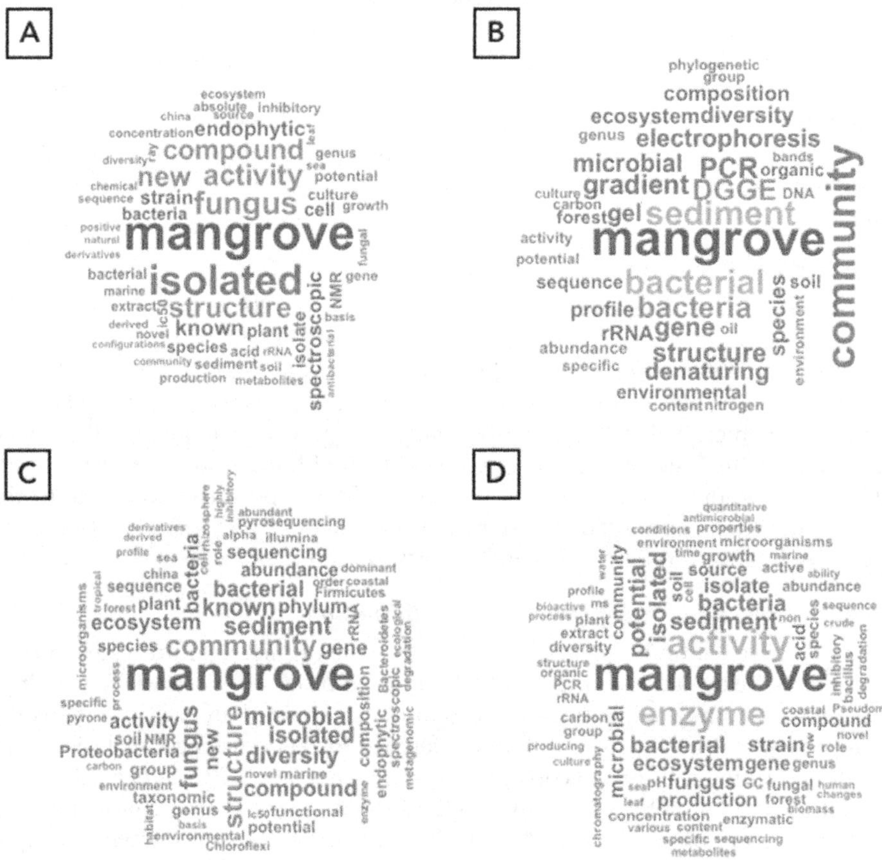

FIGURE 6.7 Word clouds computed over all abstracts from studies mentioning the distinct analytical approaches ((A) cultivation-dependent approaches and clone libraries; (B) community fingerprinting; (C) 'omics approaches; (D) quantitative approaches) in their abstract, keywords, or title. For the details of the word selection, see https://gitlab.leibniz-zmt.de/vhe/microbial-communities-in-mangroves.

conducted with a bioprospection perspective. Notably, such approaches were used to study bacterial communities that are able to degrade polycyclic aromatic hydrocarbons, such as phenanthrene, a common pollutant found in various ecosystems. Mangrove sediments were shown to harbor a variety of populations or strains of phenanthrene-degrading bacteria with the potential for bioremediation applications (Sukhdhane et al., 2019). Studies referring to fingerprinting approaches mainly applied DGGE to describe bacterial diversity in mangrove sediments (see Fig. 6.7B). Nonetheless, the first studies involving fingerprinting used PLFA (e.g., Bouillon et al., 2004) to determine the contribution of various sources of bacterial carbon (local vs. tidal inputs) in different mangrove sites, or TRFLP (Sjolin et al., 2005) to compare bacterial diversity in mangrove sites with distinct levels of deforestation. The first DGGE study evaluated the effect of shrimp farm effluents on bacterial communities in adjacent mangrove ecosystems, revealing characteristic *Vibrio* spp. profiles (Sousa et al., 2006). The latest study in the fingerprinting category combined community-level physiological profiling and 16S rRNA metabarcoding to depict spatiotemporal variability of bacterial communities and their metabolic potential, showing that salinity and organic matter content were the major drivers for microbial community composition (Behera et al., 2019). 'Omics approaches (99 studies; Fig. 6.7C) were mainly applied to sediment microbial communities ($N = 56$; with $N = 16$ and $N = 6$ for roots and leaves, respectively). Despite the potential of 'omics approaches (such as metagenomics) for achieving a broad, if not complete, coverage of microbial communities, studies encompassing several groups of microbes are still very scarce (e.g., only two studies mentioning jointly fungi, bacteria, and archaea and none mentioning all groups targeted in this survey). Quantitative approaches ($N = 156$; Fig. 6.7D) have been mainly used to depict enzymatic activity of not only bacteria (100 studies) or fungi (59 studies) but also of archaea ($N = 12$ is below threshold, therefore not shown) in mangrove ecosystems, often in combination with traditional approaches ($N = 99$).

The increased use of 'omics approaches in mangrove microbiome research is promising, especially to depict multi-taxon microbial community composition. Nonetheless, we advocate the use of various approaches in combination, if we are to understand microbial interactions and the processes they drive in the context of global change.

5. Conclusion and perspectives

This chapter sheds light on the current knowledge on microbial communities in mangrove ecosystems, focusing on sediment microbes and microbial communities associated with roots, leaves, and leaf-litter of mangrove plants, as those are in close contact with the sediment microbiome.

While considerable research efforts have focused on microbial communities in mangrove ecosystems, especially over the last two decades, several important ecological questions remain unanswered. Many studies are rather descriptive and there is a lack of hypothesis-driven research on microbial communities. For instance, increased salinity (salinization), due to the combination of sea-level rise resulting from global warming and human-induced hydrological changes, is one of the major impacts on many mangrove systems. While the effect of salinization on mangrove tree health has been documented to a certain extent (see e.g., Sarker et al., 2016; Banerjee et al., 2017), its effect on the microbial communities living in mangrove sediments has been poorly investigated (but see Das et al., 2015). Tidal inundation is known to strongly affect the conditions in mangrove sediments and is therefore expected to have a strong influence on sediment microbial processes, but the variability of the active microbial community over time or at small spatial scale along the intertidal range has scarcely been investigated. Preliminary results suggest that both community composition and activity fluctuate over time, but the major driver of microbial community composition is the location along the intertidal (Helfer, V., et al., unpublished data). Evaluated in one location, such research questions should be addressed under different settings to be able to generalize the observed patterns. Furthermore, despite the importance of distinct groups of microbial communities in organic matter dynamics in mangrove ecosystems, joint assessments of bacterial, archaeal, and fungal communities have rarely been conducted (see Fig. 6.3), and their response to environmental changes in mangrove ecosystems remains largely unknown.

We strongly advocate global efforts like those applied in other ecosystems (e.g., Earth Microbiome Project[4], Thompson et al., 2017; Ocean Sampling Day, Kopf et al., 2015), in order to understand the distribution and the role of microbial communities in mangrove ecosystems worldwide. As mangrove ecosystems are strongly interlinked with neighboring ecosystems, both on the terrestrial (salt flats, terrestrial vegetation) and the marine margins (e.g., salt marshes, seagrass beds, reefs), approaches should be standardized and align with the standards developed for previous global initiatives, to embed mangrove microbiome research in a broader scope extending beyond ecosystem boundaries. Such global efforts could be supported by the Mangrove Microbiome Initiative (MMI[5]; Allard et al., 2020), an international group of scientists with diverse disciplinary background, aiming at advancing mangrove microbiome research for a better understanding of biogeochemical processes in coastal ecosystems and as support for mangrove ecosystems management. Indeed, the influence and importance of biotic interactions for ecosystem processes call for more integrative hypothesis-driven research studies, encompassing all domains of life, with a multidisciplinary perspective. For instance, while microorganisms play a pivotal role in early and late stages of organic matter degradation, several other groups, such as detritivores, will also have a strong influence on the fate of organic matter. Therefore, to obtain a comprehensive picture of the drivers of organic matter dynamics in mangrove ecosystems, microorganisms should be investigated in conjunction with the higher fauna and primary producers (flora) as those will determine the quality (in terms of chemical composition) of the detritus (see Zimmer and Helfer, This Book) and also interact with microbial communities via sediment oxygenation, root exudates, etc. Furthermore, to be able to document and understand distribution patterns and effects of microbial communities at different spatiotemporal scales, the establishment of additional ILTER (International Long-Term Ecological Research Network) monitoring sites in mangrove ecosystems in strategic locations (for actual status of ILTER for coastal and ocean observation, see Muelbert et al., 2019), and the integration of microbial observations in such ILTER, should be fostered by the mangrove research community, following the example of the microbial observatories in oceanic LTER (Buttigieg et al., 2018).

Microorganisms are pivotal to important processes that govern the well-functioning of mangrove ecosystems and therefore deserve increased scientific attention, especially from

[4]https://www.earthmicrobiome.org.

[5]http://bmmo.microbe.net/mangrove-microbiome-initiative-mmi.

the ecological research perspective. While research efforts have been increasing over the last decades, approximately 50% of the microorganisms detected with 'omics approaches are unknown (pers. obs.). While this can make microbial research challenging, it can also be seen as a window of opportunity to discover new organisms for bioremediation. Although only briefly mentioned here, the role of microbial communities in the context of heavy metals and hydrocarbon pollution should not be neglected, especially since their activity in element cycling is presumably strongly influenced by pollution levels. Some microorganisms (such as the microeukaryotes *Bacillariophyta* sp.: Santos et al., 2010; or fungi: Ghizelini et al., 2019) are indeed used as bioindicators of pollution while others (such as thraustochytrids: Fossier Marchan et al., 2018; several bacterial strains of the genus *Bacillus*: Sukhdhane et al., 2019) could be used for remediation. The idea to use microbial communities for management strategies aiming at capturing and sequestering CO_2 (Trivedi et al., 2013) or to support the rehabilitation of mangrove ecosystems (Holguin et al., 2001) has been formulated approximately two decades ago. More recently, such applications have been envisioned in the concept of ecosystem design (Zimmer, 2018). Ecosystem design aims at designing simple ecosystems composed of a set of organisms known to be able to perform the necessary ecosystem processes that will support the provision of specific ecosystem services (both locally and/or globally), defined according to human needs. Although promising, such an approach nonetheless requires a deep understanding of microbial community interactions and will rely on our ability to detect and define keystone species (Berry and Widder, 2014) for enhancing the different processes (e.g., productivity via plant growth promoting bacteria, such as sulfur-oxidizing plant growth promoting rhizobacteria;see Behera et al., 2014 and reference therein) that would be aimed for in a designed ecosystem.

Appendix 1: Excerpts from the R scripts (full script available under: https://gitlableibniz-zmt.de/vhe/microbial-communities-in-mangroves) used for the bibliographic review, showing the different syntaxes used to group the studies by microorganism types, habitat types and applied analytical approaches

Syntax used to search for publications referring to microbial communities

```
microb <- c("microb", "bacteria", "archal", "archaea", "fungi ", "fungal",
"fungus", "protist", "thraustochytr", "algal", "algae", "virus", "viral", "micro
[ -]*organ")
```

Syntax used to search for publications referring to distinct habitats

```
sed <- c("sediment", "soil")
root <- c("root", "pneumatophor", "rhizosphere")
leaf <- c("litter", "leaf", "leaves")
```

Syntax used to search for publications referring to distinct groups of microorganisms

```
bact <- c("bacteria")
arch <- c("archal", "archaea")
fung <- c("fungi ", "fungal", "fungus")
prot <-c("protist", "thraustochytr", "algal", "algae")
vir <- c("virus", "viral")
```

Syntax used to search for publications referring to distinct approaches used to investigate microbial communities

```
trad <- c("cultivat", "cultur", "isolat", "clon", "librar")
fingerprint <-c("finger[ -]*print",
  "dgge", "denatur[a-z]*[ -]*gradient[ -]*gel[ -]*electrophor",
  "trflp", "terminal[ -]*restriction[ -]*fragment[ -]*length[ -]*polymorph",
  "arisa", "automated[ -]*ribosomal[ -]*intergenic[ -]*spacer[ -]*analysis",
  "plfa", "phospholipid[ -]*deriv[a-z]*[ -]*fatty[ -]*acid",
  "clpp", "community[ -]*level[ -]*physiologic[a-z]*[ -]*profil",
  "ecolog ", "biolog ")
omics <- c("next[ -]*generation", "illumina", "pyro", "454", "pacbio", "nanopore",
"metagenomic", "metabarcoding", "omic", "metabolomic", "proteomic",
"transcriptomic")
qPCR <- c("q[ -]*PCR", "real[ -]*time[ -]*PCR","quantitative[ -]*PCR", "card[ -]
fish", "nanosim",
  "cataly[sz][a-z]*[ -]*reporter[ -]*deposition", "fluorescence[ -]*in[ -]*situ
  [-]*hybridi[sz]ation")
enzym <- c("enzym", "enzym[a-z]*[ -]*activ", "enzym[a-z]*[ -]*rate", "GC[ /-]*MS",
"gas[ /-]*chromatography[ /-]*mass[ /-]*spectromet", "biomarker")
microsens <- c("micro[ -]*sens", "micro[ -]*profil", "micro[ -]*electrode")
```

Note: publications referring to categories "qPCR," "enzyme," and "microsens" were further grouped into "quantitative" approaches.

Acknowledgments

We want to thank Martin Zimmer, Jörg Overman, Olga Jeske, Johannes Sikorski, and members of the MMI (Mangrove Microbiome Initiative) consortium for inspiring discussions and exchanges about the study and the role of microbial communities in mangrove ecosystems. Thanks also to Lucie Büchi for sharing her R scripts to produce some of the figures related to the bibliographic review.

References

Abed, R., Tamm, A., Hassenrück, C., Al-Rawahi, A., Rodrig-uez-Caballero, E., Fiedler, S., Maier, S., Weber, B., 2019. Habitat-dependent composition of bacterial and fungal communities in biological soil crusts from Oman. Sci. Rep. 9, 6468. https://doi.org/10.1038/s41598-019-42911-6.

Abram, F., 2014. Systems-based approaches to unravel multi-species microbial community functioning. Comput. Struct. Biotechnol. J. 13, 24–32. https://doi.org/10.1016/j.csbj.2014.11.009.

Allard, S.M., Costa, M.T., Bulseco, A.N., Helfer, V., Wilkins, L.G.E., Hassenrück, C., Zengler, K., Zimmer, M., Erazo, N., Mazza Rodrigues, J.L., Duke, N., Melo, V.M.M., Vanwonterghem, I., Junca, H., Makonde, H.M., Jiménez, D.J., Tavares, T.C.L., Fusi, M., Daffonchio, D., Duarte, C.M., Peixoto, R.S., Rosado, A.S., Gilbert, J.A., Bowman, J., et al., 2020. Introducing the mangrove microbiome initiative: identifying microbial research priorities and approaches tobetter understand, protect, and rehabilitate mangrove ecosystems. mSystems 5. https://doi.org/10.1128/mSystems.00658-20. e00658–e0020.

Allen, E.E., Banfield, J.F., 2005. Community genomics in microbial ecology and evolution. Nat. Rev. Microbiol. 3, 489–498. https://doi.org/10.1038/nrmicro1157.

Alongi, D.M., 2014. Carbon cycling and storage in mangrove forests. Ann. Rev. Mar. Sci. 6, 195–219. https://doi.org/10.1146/annurev-marine-010213-135020.

Alteio, L.V., Schulz, F., Seshadri, R., Varghese, N., Rodriguez-Reillo, W., Ryan, E., Goudeau, D., Eichorst, S.A., Malmstrom, R.R., Bowers, R.M., Katz, L.A., Blanchard, J.L., Woyke, T., 2020. Complementary metagenomicapproaches improve reconstruction of microbial diversity in a forest soil. mSystems 5. https://doi.org/10.1128/mSystems.00768-19. e00768–e0019.

Anantharaman, K., Hausmann, B., Jungbluth, S.P., Kantor, R.S., Lavy, A., Warren, L.A., et al., 2018. Expanded diversity of microbial groups that shape the dissimilatory sulfur cycle. ISME J. 12, 1715–1728. https://doi.org/10.1038/s41396-018-0078-0.

Banerjee, K., Cazzolla Gatti, R., Mitra, A., 2017. Climate change-induced salinity variation impacts on a stenoecious mangrove species in the Indian Sundarbans. Ambio 46, 492–499.

Baudrimont, A., Voegeli, S., Viloria, E.C., Stritt, F., Lenon, M., Wada, T., Jaquet, V., Becskei, A., 2017. Multiplexed gene control reveals rapid mRNA turnover. Sci. Adv. 3 (7) https://doi.org/10.1126/sciadv.1700006.

Behera, B.C., Mishra, R.R., Dutta, S.K., Thatoi, H.N., 2014. Sulphur oxidsing bacteria in mangrove ecosystem: a review. Afr. J. Biotechnol. 13 (29), 2897–2907. https://doi.org/10.5897/AJB2013.13327.

Behera, P., Mohapatra, M., Kim, J.Y., et al., 2019. Spatial and temporal heterogeneity in the structure and function of sediment bacterial communities of a tropical mangrove forest. Environ. Sci. Pollut. Res. 26, 3893–3908. https://doi.org/10.1007/s11356-018-3927-5.

Berry, D., Widder, S., 2014. Deciphering microbial interactions and detecting keystone species with co-occurrence networks. Front. Microbiol. 5, 219. https://doi.org/10.3389/fmicb.2014.00219.

Boetius, A., Lochte, K., 1994. Regulation of microbial enzymatic degradation of organic matter in deep-sea sediments. Mar. Ecol. Prog. Ser. 104, 299–307. https://doi.org/10.3354/meps104299.

Bouillon, S., Moens, T., Koedam, N., Dahdouh-Guebas, F., Baeyens, W., Dehairs, F., 2004. Variability in the origin of carbon substrates for bacterial communities in mangrove sediments. FEMS Microbiol. Ecol. 49 (2), 171–179. https://doi.org/10.1016/j.femsec.2004.03.004.

Bowman, J.S., Ducklow, H.W., 2015. Microbial communities can be described by metabolic structure: a general framework and application to a seasonally variable, depth-stratified microbial community from the coastal West Antarctic Peninsula. PLoS One 10, 1–18. https://doi.org/10.1371/journal.pone.0135868.

Burdorf, L.D.W., Hidalgo-Martinez, S., Cook, P.L.M., Meysman, F.J.R., 2016. Long-distance electron transport by cable bacteria in mangrove sediments. Mar. Ecol. Prog. Ser. 545, 1–8. https://doi.org/10.3354/meps11635.

Buttigieg, P.L., Fadeev, E., Bienhold, C., Hehemann, L., Offre, P., Boetius, A., 2018. Marine microbes in 4D—using time series observation to assess the dynamics of the ocean microbiome and its links to ocean health. Curr. Opin. Microbiol. 43, 169–185. https://doi.org/10.1016/j.mib.2018.01.015.

Canfield, D.E., Thamdrup, B., 2009. Towards a consistent classification scheme for geochemical environments, or, why we wish the term 'suboxic ' would go away. Geobiology 7, 385–392. https://doi.org/10.1111/j.1472-4669.2009.00214.x.

Cao, W., Guan, Q., Li, Y., Wang, M., Liu, B., 2017. The contribution of denitrification and anaerobic ammonium oxidation to N_2 production in mangrove sediments in Southeast China. J. Soils Sediments 17, 1767–1776. https://doi.org/10.1007/s11368-017-1653-0.

Chen, J., Gu, J.D., 2017. Faunal burrows alter the diversity, abundance, and structure of AOA, AOB, anammox and n-damo communities in coastal mangrove sediments. Microb. Ecol. 74 (1), 140–156. https://doi.org/10.1007/s00248-017-0939-5.

Constancias, F., Saby, N.P., Terrat, S., Dequiedt, S., Horrigue, W., Nowak, V., Guillemin, J.P., Biju-Duval, L., Chemidlin Prévost-Bouré, N., Ranjard, L., 2015. Contrasting spatial patterns and ecological attributes of soil bacterial and archaeal taxa across a landscape. Microbiol. Open 4 (3), 518–531. https://doi.org/10.1002/mbo3.256.

Coskun, Ö.K., Pichler, M., Vargas, S., Gilder, S., Orsi, W.D., 2018. Linking uncultivated microbial populations and benthic. Appl. Environ. Microbiol. 84 https://doi.org/10.1128/AEM.01083-18 e01083-18.

Daims, H., Lücker, S., Wagner, M., 2016. Review A new perspective on microbes formerly known as nitrite-oxidizing bacteria. Trends Microbiol. 24, 699–712. https://doi.org/10.1016/j.tim.2016.05.004.

Das, S., Mukherjee, A., Chakraborty, S., Maiti, T.K., De, M., De, T.K., 2015. Influence of varying degree of salinity-sodicity stress on phytoplankton and cultivable microbes in the estuarine water of the Sundarban Mangrove Forest, India. Int. J. Adv. Life Sci. 8, 424–429.

Davids, M., Hugenholtz, F., Dos Santos, V.M., Smidt, H., Kleerebezem, M., Schaap, P.J., 2016. Functional profiling of unfamiliar microbial communities using a validated de novo assembly metatranscriptome pipeline. PLoS One 11, 1–18. https://doi.org/10.1371/journal.pone.0146423.

DeLong, E.F., Wickham, G.S., Pace, N.R., 1989. Phylogenetic stains: ribosomal RNA-based probes for the identification of single cells. Science 243, 1360–1363.

Donato, D., Kauffman, J., Murdiyarso, D., et al., 2011. Mangroves among the most carbon-rich forests in the tropics. Nat. Geosci. 4, 293–297. https://doi.org/10.1038/ngeo1123.

Eloe-Fadrosh, E.A., Ivanova, N.N., Woyke, T., Kyrpides, N.C., 2016. Metagenomics uncovers gaps in amplicon-based detection of microbial diversity. Nat. Microbiol. 1, 15032. https://doi.org/10.1038/nmicrobiol.2015.32.

Fernandes, S.O., Bonin, P.C., Michotey, V.D., Garcia, N., LokaBharathi, P.A., 2012a. Nitrogen-limited mangrove ecosystems conserve N through dissimilatory nitrate reduction to ammonium. Sci. Rep. 62, 3–7. https://doi.org/10.1038/srep00419.

Fernandes, S.O., Michotey, V.D., Guasco, S., Bonin, P.C., Bharathi, P.A.L., 2012b. Denitrification prevails over anammox in tropical mangrove sediments. Mar. Environ. Res. 74, 9–19. https://doi.org/10.1016/j.marenvres.2011.11.008.

Ferreira, T.O., Otero, X.L., Vidal-Torrado, P., Macías, F., 2007a. Redox processes in mangrove soils under *Rhizophora mangle* in relation to different environmental conditions. Soil Sci. Soc. Am. J. 71 (2), 484–491. https://doi.org/10.2136/sssaj2006.0078.

Ferreira, T.O., Otero, X.L., Vidal-Torrado, P., Macías, F., 2007b. Effects of bioturbation by root and crab activity on iron and sulfur biogeochemistry in mangrove substrate. Geoderma 142 (1), 36–46. https://doi.org/10.1016/j.geoderma.2007.07.010.

Fisher, M.M., Triplett, E.W., 1999. Automated approach for ribosomal intergenic spacer analysis of microbial diversity and its application to freshwater bacterial communities automated approach for ribosomal intergenic spacer analysis of microbial diversity and its application to freshwater. Appl. Environ. Microbiol. 65, 4630–4636.

Finster, K., 2008. Microbiological disproportionation of inorganic sulfur compounds. J. Sulfur Chem. 29 (3–4), 281–292. https://doi.org/10.1080/17415990802105770.

Fossier Marchan, L., Lee Chang, K.J., Nichols, P.D., Mitchell, W.J., Polglase, J.L., Gutierrez, T., 2018. Taxonomy, ecology and biotechnological applications of thraustochytrids: a review. Biotechnol. Adv. 36 (1), 26–46. https://doi.org/10.1016/j.biotechadv.2017.09.003.

Ghizelini, A.M., Martins, K.G., Gießelmann, U.C., Santoro, E., Pasqualette, L., Mendonça-Hagler, L.C.S., Rosado, A.S., Macrae, A., 2019. Fungal communities in oil contaminated mangrove sediments – who is in the mud? Mar. Pollut. Bull. 139, 181–188.

Glud, R.N., 2008. Oxygen dynamics of marine sediments. Mar. Biol. Res. 4 (4), 243–289. https://doi.org/10.1080/17451000801888726.

Gobet, A., Boetius, A., Ramette, A., 2014. Ecological coherence of diversity patterns derived from classical fingerprinting and Next Generation Sequencing techniques. Environ. Microbiol. 16, 2672–2681. https://doi.org/10.1111/1462-2920.12308.

Guillou, L., Bachar, D., Audic, S., Bass, D., Berney, C., Bittner, L., Boutte, C., Burgaud, G., De Vargas, C., Decelle, J., et al., 2013. The Protist Ribosomal reference database (PR2): a catalog of unicellular eukaryote small sub-unit rRNA sequences with curated taxonomy. Nucleic Acids Res. 41, 597–604. https://doi.org/10.1093/nar/gks1160.

Hatzenpichler, R., Scheller, S., Tavormina, P.L., Babin, B.M., Tirrell, D.A., Orphan, V.J., 2014. In situ visualization of newly synthesized proteins in environmental microbes using amino acid tagging and click chemistry. Environ. Microbiol. 16 (8), 2568–2590. https://doi.org/10.1111/1462-2920.12436.

Holguin, G., Vazquez, P., Bashan, Y., 2001. The role of sediment microorganisms in the productivity, conservation, and rehabilitation of mangrove ecosystems: an overview. Biol. Fertil. Soils 33, 265–278. https://doi.org/10.1007/s003740000319.

Howarth, R.W., 1984. The ecological significance of sulfur in the energy dynamics of salt marsh and coastal marine sediments. Biogeochemistry 1, 5–27. https://doi.org/10.1007/BF02181118.

Hügler, M., Sievert, S.M., 2011. Beyond the Calvin cycle: autotrophic carbon fixation in the ocean. Ann. Rev. Mar. Sci. 3, 261–289. https://doi.org/10.1146/annurev-marine-120709-142712.

Jain, M., Olsen, H.E., Paten, B., Akeson, M., Branton, D., Daniel, B., Deamer, D., Andre, M., Hagan, B., Benner, S., et al., 2016. The Oxford Nanopore MinION: delivery of nanopore sequencing to the genomics community. Genome Biol. 17, 239. https://doi.org/10.1186/s13059-016-1103-0.

Jin, M., Guo, X., Zhang, R., et al., 2019. Diversities and potential biogeochemical impacts of mangrove soil viruses. Microbiome 7, 58. https://doi.org/10.1186/s40168-019-0675-9.

Jørgensen, B.B., Findlay, A.J., Pellerin, A., 2019. The biogeochemical sulfur cycle of marine sediments. Front. Microbiol. 10, 849. https://doi.org/10.3389/fmicb.2019.00849.

Jousset, A., Bienhold, C., Chatzinotas, A., Gallien, L., Gobet, A., Kurm, V., Küsel, K., Rillig, M.C., Rivett, D.W., Salles, J.F., et al., 2017. Where less may be more: how the rare biosphere pulls ecosystems strings. ISME J. 11, 853–862. https://doi.org/10.1038/ismej.2016.174.

Karst, S.M., Dueholm, M.S., McIlroy, S.J., Kirkegaard, R.H., Nielsen, P.H., Albertsen, M., 2018. Thousands of primer-free, high-quality, full-length SSU rRNA sequences from all domains of life. Nat. Biotechnol. 10, 070771. https://doi.org/10.1101/070771.

Kleiner, M., 2019. Metaproteomics: much more than measuring gene expression in microbial communities. mSystems 4, 1–6. https://doi.org/10.1128/msystems.00115-19.

Kopf, A., Bicak, M., Kottmann, R., et al., 2015. The ocean sampling day consortium. GigaScience 4, 27. https://doi.org/10.1186/s13742-015-0066-5.

Kristensen, E., Connolly, R.M., Otero, X.L., Marchand, C., Ferreira, T.O., Rivera-Monroy, V.H., 2017. Biogeochemical cycles: global approaches and perspectives. In: Rivera-Monroy, I.V.H., Lee, S.Y., Kristensen, E., Twilley, R.R. (Eds.), Mangrove Ecosystems: A Global Biogeographic Perspective. Structure, Function and Ecosystem Services. Springer, pp. 163–209. https://doi.org/10.1007/978-3-319-62206-4_6.

Kristensen, E., 1997. Carbon, sulfur and nitrogen biogeochemistry of tropical mangrove sediments. In: Haq, B.U., Haq, S.M., Kullenberg, G., Stel, J.H. (Eds.), Coastal Zone Management Imperative for Maritime Developing Nations, Coastal Systems and Continental Margins, vol. 3. Springer, Dordrecht. https://doi.org/10.1007/978-94-017-1066-4_12.

Kristensen, E., 2008. Mangrove crabs as ecosystem engineers; with emphasis on sediment processes. J. Sea Res. 59, 30–43. https://doi.org/10.1016/j.seares.2007.05.004.

Kristensen, E., Bouillon, S., Dittmar, T., Marchand, C., 2008. Organic carbon dynamics in mangrove ecosystems. Aquat. Bot. 89, 201–219. https://doi.org/10.1016/j.aquabot.2007.12.005.

Langille, M.G.I., Zaneveld, J., Caporaso, J.G., McDonald, D., Knights, D., Reyes, J.A., Clemente, J.C., Burkepile, D.E., Vega Thurber, R.L., Knight, R., et al., 2013. Predictive functional profiling of microbial communities using 16S rRNA marker gene sequences. Nat. Biotechnol. 31, 814–821. https://doi.org/10.1038/nbt.2676.

Lee, Y.-J., van Nostrand, J.D., Tu, Q., Lu, Z., Cheng, L., Yuan, T., Deng, Y., Carter, M.Q., He, Z., Wu, L., et al., 2013. The PathoChip, a functional gene array for assessing pathogenic properties of diverse microbial communities. ISME J. 7, 1974–1984. https://doi.org/10.1038/ismej.2013.88.

Li, M., Cao, H., Hong, Y., Gu, J., 2011. Spatial distribution and abundances of ammonia-oxidizing archaea (AOA) and ammonia-oxidizing bacteria (AOB) in mangrove sediments. Appl. Microbiol. Biotechnol. 89, 1243–1254. https://doi.org/10.1007/s00253-010-2929-0.

Li, T., Wu, T.D., Mazéas, L., Toffin, L., Guerquin-Kern, J.L., Leblon, G., Bouchez, T., 2008. Simultaneous analysis of microbial identity and function using NanoSIMS. Environ. Microbiol. 10 (3), 580–588. https://doi.org/10.1111/j.1462-2920.2007.01478.x.

Liu, W.T., Marsh, T.L., Cheng, H., Forney, L.J., 1997. Characterization of microbial diversity by determining terminal restriction fragment length polymorphisms of genes encoding 16S rRNA. Appl. Environ. Microbiol. 63, 4516–4522.

Liu, Z., Zhang, C., Wei, Q., Zhang, S., Quan, Z., Li, M., 2020. Temperature and salinity drive comammox community composition in mangrove ecosystems across southeastern China. Sci. Total Environ. 140456. https://doi.org/10.1016/j.scitotenv.2020.140456.

Lloyd, K.G., Steen, A.D., Ladau, J., Yin, J., Crosby, L., 2018. Phylogenetically novel uncultured microbial cells dominate earth microbiomes. mSystems 3, 1–12. https://doi.org/10.1128/msystems.00055-18.

Marbouty, M., Cournac, A., Flot, J.F., Marie-Nelly, H., Mozziconacci, J., Koszul, R., 2014. Metagenomic chromosome conformation capture (meta3C) unveils the diversity of chromosome organization in microorganisms. Elife 3, e03318. https://doi.org/10.7554/eLife.03318.

Mcleod, E., Chmura, G., Bouillon, S., Salm, R., Björk, M., Duarte, C.M., Lovelock, C.E., Schlesinger, W.H., Silliman, B.R., 2011. A blueprint for blue carbon: toward an improved understanding of the role of of vegetated coastal habitats in sequestering CO_2. Front. Ecol. Environ. 9, 552–560.

Mendes, L.W., Tsai, S.M., 2014. Variations of bacterial community structure and composition in mangrove sediment at different depths in southeastern Brazil. Diversity 6, 827–843.

Merder, J., Freund, J.A., Feudel, U., Niggemann, J., Singer, G., Dittmar, T., 2019. Improved mass accuracy and isotope confirmation through alignment of ultrahigh-resolution mass spectra of complex natural mixtures. Anal. Chem. https://doi.org/10.1021/acs.analchem.9b04234.

Meyer, R.L., Allen, D.E., Schmidt, S., 2008. Nitrification and denitrification as sources of sediment nitrous oxide production: a microsensor approach. Mar. Chem. 110 (1–2), 68–76. https://doi.org/10.1016/j.marchem.2008.02.004.

Moraru, C., Lam, P., Fuchs, B.M., Kuypers, M.M., Amann, R., 2010. GeneFISH–an in situ technique for linking gene presence and cell identity in environmental microorganisms. Environ. Microbiol. 12 (11), 3057–3073. https://doi.org/10.1111/j.1462-2920.2010.02281.x.

Morton, J.T., Sanders, J., Quinn, R.A., McDonald, D., Gonzalez, A., Vázquez-Baeza, Y., Navas-Molina, J.A., Song, S.J., Metcalf, J.L., Hyde, E.R., Lladser, M., Dorrestein, P.C., Knight, R., 2017. Balance trees reveal microbial niche differentiation. mSystems 2 (1). https://doi.org/10.1128/mSystems.00162-16 e00162-16.

Muelbert, J., Nidzieko, N., Acosta, A., Beaulieu, S., Bernardino, A., Boikova, E., Bornman, T., Cataletto, B., Deneudt, K., Eliason, E., Kraberg, A., Nakaoka, M., Pugnetti, A., Ragueneau, O., Scharfe, M., Soltwedel, T., Sosic, H., Stanisci, A., Stefanova, K., Stephan, P., Stier, A., Wikner, J., Zingone, A., 2019. ILTER - the International Long-Term Ecological Research network as a platform for global coastal and ocean observation. Front. Mar. Sci. 6 (527) https://doi.org/10.3389/fmars.2019.00527.

Muyzer, G., Smalla, K., 1998. Application of denaturing gradient gel electrophoresis and temperature gradient gel electrophoresis in microbial molecular ecology. Antonie Leeuwenhoek 73, 127–141. https://doi.org/10.1023/a:1000669317571.

Nandipati, S.C., Haugli, K., Coucheron, D.H., Haskins, E.F., Johansen, S.D., 2012. Polyphyletic origin of the genus Physarum (Physarales, Myxomycetes) revealed by nuclear rDNA mini-chromosome analysis and group I intron synapomorphy. BMC Evol. Biol. 12, 166. https://doi.org/10.1186/1471-2148-12-166.

Neufeld, J.D., Vohra, J., Dumont, M.G., Lueders, T., Manefield, M., Friedrich, M.W., Murrell, C.J., 2007. DNA stable-isotope probing. Nat. Protoc. 2 (4), 860–866. https://doi.org/10.1038/nprot.2007.109.

Nilsson, R.H., Larsson, K.H., Taylor, A.F.S., Bengtsson-Palme, J., Jeppesen, T.S., Schigel, D., Kennedy, P., Picard, K., Glöckner, F.O., Tedersoo, L., et al., 2019. The UNITE database for molecular identification of fungi: handling dark taxa and parallel taxonomic classifications. Nucleic Acids Res. 47, D259–D264. https://doi.org/10.1093/nar/gky1022.

Parada, A.E., Needham, D.M., Fuhrman, J.A., 2016. Every base matters: assessing small subunit rRNA primers for marine microbiomes with mock communities, time series and global field samples. Environ. Microbiol. 18, 1403–1414. https://doi.org/10.1111/1462-2920.13023.

Parks, D.H., Chuvochina, M., Waite, D.W., Rinke, C., Skarshewski, A., Chaumeil, P.-A., Hugenholtz, P., 2018. A proposal for a standardized bacterial taxonomy based on genome phylogeny. bioRxiv 256800. https://doi.org/10.1101/256800.

Pelikan, C., Herbold, C.W., Hausmann, B., Müller, A.L., Pester, M., Loy, A., 2016. Diversity analysis of sulfite- and sulfate-reducing microorganisms by multiplex dsrA and dsrB amplicon sequencing using new primers and mock community-optimized bioinformatics. Environ. Microbiol. 18, 2994–3009. https://doi.org/10.1111/1462-2920.13139.

Peura, S., Bertilsson, S., Jones, R.I., Eiler, A., 2015. Resistant microbial cooccurrence patterns inferred by network topology. Appl. Environ. Microbiol. 81, 2090–2097. https://doi.org/10.1128/AEM.03660-14.

Pires, A.C., Cleary, D.F., Almeida, A., Cunha, A., Dealtry, S., Mendonça-Hagler, L.C., Smalla, K., Gomes, N.C., 2012. Denaturing gradient gel electrophoresis and barcoded pyrosequencing reveal unprecedented archaeal diversity in mangrove sediment and rhizosphere samples. Appl. Environ. Microbiol. 78 (16), 5520–5528. https://doi.org/10.1128/AEM.00386-12.

Prescott, Harley, Klein, Willey, Sherwood, Woolverton (Eds.), 2010. Microbiologie. de boeck.

Probandt, D., Eickhorst, T., Ellrott, A., Amann, R., Knittel, K., 2018. Microbial life on a sand grain: from bulk sediment to single grains. ISME J. 12, 623–633. https://doi.org/10.1038/ismej.2017.197.

Purvaja, R., Ramesh, R., Ray, A.K., Rixen, T., 2008. Nitrogen cycling: a review of the processes, transformations and fluxes in coastal ecosystems. Curr. Sci. 94 (11).

Raghukumar, S., Sharma, S., Raghukumar, C., Sathe-Pathak, V., Chandramohan, D., 1994. Thraustochytrid and fungal component of marine detritus. IV. Laboratory studies on decomposition of leaves of the mangrove *Rhizophora apiculata* Blume. J. Exp. Mar. Biol. Ecol. 183 (1), 113–131. https://doi.org/10.1016/0022-0981(94)90160-0.

Raghukumar, S., Sathe-Pathak, V., Sharma, S., Raghukumar, C., 1995. Thraustochytrid and fungal component of marine detritus. III. Field studies on decomposition of leaves of the Mangrove Laboratory studies on decomposition of leaves of the mangrove *Rhizophora apiculata*. Aquat. Microb. Ecol. 9 (2) https://doi.org/10.3354/ame009117.

R Core Team, 2020. R: A Language and Environment for Statistical Computing. R Foundation for Statistical Computing, Vienna, Austria. https://www.R-project.org/.

Reeburgh, W.S., 1983. Rates of biogeochemical processes in anoxic sediments. Ann. Rev. Earth Planet Sci. 11, 269–298. https://doi.org/10.1146/annurev.ea.11.050183.001413.

Reef, R., Feller, I.C., Lovelock, C.E., 2010. Nutrition of mangroves. Tree Physiol. 30 (9), 1148–1160. https://doi.org/10.1093/treephys/tpq048.

Reis, C.R.G., Nardoto, G.B., Oliveira, R.S., 2017. Global overview on nitrogen dynamics in mangroves and consequences of increasing nitrogen availability for these systems. Plant Soil 410, 1–19. https://doi.org/10.1007/s11104-016-3123-7.

Romero, I., Jacobson-Meyers, M., Fuhrman, J., Capone, D., 2015. Phylogenetic diversity of diazotrophs along an experimental nutrient gradient in mangrove sediments. J. Mar. Sci. Eng. 3, 699–719. https://doi.org/10.3390/jmse3030699.

RStudio Team, 2020. RStudio: Integrated Development Environment for R. RStudio, PBC, Boston, MA. http://www.rstudio.com/.

Salazar, G., Paoli, L., Alberti, A., Huerta-Cepas, J., Ruscheweyh, H.J., Cuenca, M., Field, C.M., Coelho, L.P., Cruaud, C., Engelen, S., Gregory, A.C., Labadie, K., Marec, C., Pelletier, E., Royo-Llonch, M., Roux, S., Sánchez, P., Uehara, H., Zayed, A.A., et al., 2019. Gene expression changes and community turnover differentially shape the global ocean metatranscriptome. Cell 179 (5), 1068–1083.e21. https://doi.org/10.1016/j.cell.2019.10.014.

SamKamaleson, A., Gonsalves, M.-J., 2019. Role of sulfur-oxidizing bacteria on the ecology in tropical mangrove sediments. Reg. Stud. Mar. Sci. 28, 100574. https://doi.org/10.1016/j.rsma.2019.100574.

Sanders, J.G., Nurk, S., Salido, R.A., Minich, J., Xu, Z.Z., Zhu, Q., Martino, C., Fedarko, M., Arthur, T.D., Chen, F., Boland, B.S., Humphrey, G.C., Brennan, C., Sanders, K., Gaffney, J., Jepsen, K., Khosroheidari, M., Green, C., Liyanage, M., et al., 2019. Optimizing sequencing protocols for leaderboard metagenomics by combining long and short reads. Genome Biol. 20 (1), 1–14. https://doi.org/10.1186/s13059-019-1834-9.

Santos, H.F., Cury, J.C., Carmo, F.L., Rosado, A.S., Peixoto, R.S., 2010. 18S rDNA sequences from microeukaryotes reveal oil indicators in mangrove sediment. PLoS One 5, e12437.

Sarker, S.K., Reeve, R., Thompson, J., Paul, N.K., Matthiopoulos, J., 2016. Are we failing to protect threatened mangroves in the Sundarbans world heritage ecosystem? Sci. Rep. 6, 21234. https://doi.org/10.1038/srep21234.

Schostag, M.D., Albers, C.N., Jacobsen, C.S., Priemé, A., May 2020. Low turnover of soil bacterial rRNA at low temperatures. Front. Microbiol. 11, 1–5. https://doi.org/10.3389/fmicb.2020.00962.

Shang, Y., Sikorski, J., Bonkowski, M., Fiore-Donno, A.-M., Kandeler, E., Marhan, S., et al., 2017. Inferring interactions in complex microbial communities from nucleotide sequence data and environmental parameters. PLoS One 12 (3), e0173765. https://doi.org/10.1371/journal.pone.0173765.

Shi, Z., Yin, H., Van Nostrand, J.D., Voordeckers, J.W., Tu, Q., Deng, Y., Yuan, M., Zhou, A., Zhang, P., Xiao, N., et al., 2019. Functional gene array-based ultrasensitive and quantitative detection of microbial populations in complex communities. mSystems 4, 1–16. https://doi.org/10.1128/msystems.00296-19.

Shiau, Y.-J., Chiu, C.-Y., 2020. Biogeochemical processes of C and N in the soil of mangrove forest ecosystems. Forests 11, 492.

Sjöling, S., Mohammed, S.M., Lyimo, T.J., Kyaruzi, J.J., 2005. Benthic bacterial diversity and nutrient processes in mangroves: impact of deforestation. Estuar. Coast Shelf Sci. 63 (3), 397–406. https://doi.org/10.1016/j.ecss.2004.12.002.

Sousa, O.V., Macrae, A., Menezes, F.G.R., Gomes, N.C.M., Vieira, R.H.S.F., Mendonça-Hagler, L.C.S., 2006. The impact of shrimp farming effluent on bacterial communities in mangrove waters, Ceará, Brazil. Mar. Pollut. Bull. 52 (12), 1725–1734. https://doi.org/10.1016/j.marpolbul.2006.07.006.

Spalding, M., Kainuma, M., Collins, L., 2010. World Atlas of Mangroves. Earthscan, London, UK and Washington DC,USA.

Srikanth, S., Lum, S.K.Y., Chen, Z., 2016. Mangrove root: adaptations and ecological importance. Trees 30, 451–465.

Steiner, P.A., De Corte, D., Geijo, J., Mena, C., Yokokawa, T., Rattei, T., Herndl, G.J., Sintes, E., 2019. Highly variable mRNA half-life time within marine bacterial taxa and functional genes. Environ. Microbiol. 21 (10), 3873–3884. https://doi.org/10.1111/1462-2920.14737.

Sukhdhane, K.S., Pandey, P.K., Ajima, M.N.O., Jayakumar, T., Vennila, A., Raut, S.M., 2019. Isolation and characterization of phenanthrene-degrading bacteria from PAHs contaminated mangrove sediment of Thane Creek in Mumbai, India. Polycycl. Aromat. Comp. 39 (1), 73–83. https://doi.org/10.1080/10406638.2016.1261911.

Taberlet, P., Coissac, E., Pompanon, F., Brochmann, C., Willerslev, E., 2012. Towards next-generation biodiversity assessment using DNA metabarcoding. Mol. Ecol. 21, 2045–2050. https://doi.org/10.1111/j.1365-294X.2012.05470.x.

Thatoi, H., Behera, B.C., Mishra, R.R., 2013. Ecological role and biotechnological potential of mangrove fungi: a review. Mycology 4 (1), 54–71. https://doi.org/10.1080/21501203.2013.785448.

Thompson, L.R., Sanders, J.G., McDonald, D., Amir, A., Ladau, J., Locey, K.J., Prill, R.J., Tripathi, A., Gibbons, S.M., Ackermann, G., et al., 2017. A communal catalogue reveals Earth's multiscale microbial diversity. Nature 551, 457–463. https://doi.org/10.1038/nature24621.

Trivedi, P., Anderson, I.C., Singh, B.K., 2013. Microbial modulators of soil carbon storage: integrating genomic and metabolic knowledge for global prediction. Trends Microbiol. 21, 641–651.

Vanni, C., Schechter, M., Acinas, S., Barberán, A., Luigi Buttigieg, P., Delmont, T.O., Duarte, C.M., Murat Eren, A., Finn, R., Mitchell, A., Siren, K., Steinegger, M., Oliver Glöckner, F., Fernandez, A., 2020. Light into the darkness: unifying the known and unknown coding 1 sequence space in microbiome analyses. BioRxiv. https://doi.org/10.1101/2020.06.30.180448, 2020.06.30.180448.

Varon-Lopez, M., Dias, A.C., Fasanella, C.C., et al., 2014. Sulphur-oxidizing and sulphate-reducing communities in Brazilian mangrove sediments. Environ. Microbiol. 16 (3), 845–855. https://doi.org/10.1111/1462-2920.12237.

Vestal, J.R., White, D.C., 1989. Lipid analysis in microbial ecology: quantitative approaches to the study of microbial communities. Bioscience 39, 535–541. https://doi.org/10.2307/1310976.

Wemheuer, F., Taylor, J.A., Daniel, R., Johnston, E., Meinicke, P., Thomas, T., Wemheuer, B., 2018. Tax4Fun2: A R-Based Tool for the Rapid Prediction of Habitat-Specific Functional Profiles and Functional Redundancy Based on 16S rRNA Gene Marker Gene Sequences.

Wessner, D.R., Dupont, C., Charles, T.C. (Eds.), 2013. Microbiology. Wiley.

Wick, R.R., Judd, L.M., Gorrie, C.L., Holt, K.E., 2017. Unicycler: resolving bacterial genome assemblies from short and long sequencing reads. PLoS Comput. Biol. 13 (6), 1–22. https://doi.org/10.1371/journal.pcbi.1005595.

Wilkinson, M.D., Dumontier, M., Aalbersberg, I.J., Appleton, G., Axton, M., Baak, A., Blomberg, N., Boiten, J.W., da Silva Santos, L.B., Bourne, P.E., et al., 2016. The FAIR guiding principles for scientific data management and stewardship. Sci. Data 3, 1–9. https://doi.org/10.1038/sdata.2016.18.

Wong, M., Vrijmoed, L., Au, D., 2005. Abundance of thraustochytrids on fallen decaying leaves of *Kandelia candel* and mangrove sediments in Futian National Nature Reserve, China. Bot. Mar. 48 (5).

Woyke, T., Xie, G., Copeland, A., González, J.M., Han, C., Kiss, H., Saw, J.H., Senin, P., Yang, C., Chatterji, S., et al., 2009. Assembling the marine metagenome, one cell at a time. PLoS One 4. https://doi.org/10.1371/journal.pone.0005299.

Yilmaz, P., Kottmann, R., Field, D., Knight, R., Cole, J.R., Amaral-Zettler, L., Gilbert, J.A., Karsch-Mizrachi, I., Johnston, A., Cochrane, G., et al., 2011. Minimum information about a marker gene sequence (MIMARKS) and minimum information about any (x) sequence (MIxS) specifications. Nat. Biotechnol. 29, 415–420. https://doi.org/10.1038/nbt.1823.

Yilmaz, P., Parfrey, L.W., Yarza, P., Gerken, J., Pruesse, E., Quast, C., Schweer, T., Peplies, J., Ludwig, W., Glöckner, F.O., 2014. The SILVA and "All-species living tree project (LTP)" taxonomic frameworks. Nucleic Acids Res. 42, D643–D648. https://doi.org/10.1093/nar/gkt1209.

Yuan, H., Ge, T., Chen, C., O'Donnell, A.G., Wu, J., 2012. Significant role for microbial autotrophy in the sequestration of soil carbon. Appl. Environ. Microbiol. 78 (7), 2328–2336. https://doi.org/10.1128/AEM.06881-11.

Zhang, Y., Dong, J., Yang, Z., Zhang, S., Wang, Y., 2008. Phylogenetic diversity of nitrogen-fixing bacteria in mangrove sediments assessed by PCR – denaturing gradient gel electrophoresis. Arch. Microbiol. 190, 19–28. https://doi.org/10.1007/s00203-008-0359-5.

Zhou, Z., Meng, H., Liu, Y., Gu, J.-D., Li, M., 2017. Stratified bacterial and archaeal community in mangrove and intertidal Wetland mudflats revealed by high throughput 16S rRNA gene sequencing. Front. Microbiol. 8, 2148. https://doi.org/10.3389/fmicb.2017.02148.

Zimmer, M., 2020a. Cellulases. In: Bärlocher, F., Gessner, M.O., Graça, M.A.S. (Eds.), Methods to Study Litter Decomposition: A Practical Guide, second ed. Springer Nature Switzerland, pp. 397–404.

Zimmer, M., 2020b. Phenol oxidation. In: Bärlocher, F., Gessner, M.O., Graça, M.A.S. (Eds.), Methods to Study Litter Decomposition: A Practical Guide, second ed. Springer Nature Switzerland, pp. 433–438.

Zimmer, M., 2018. Ecosystem Design: when mangrove ecology meets human needs. In: Makowski, C., Finkl, C.W. (Eds.), Threats to Mangrove Forests: Hazards, Vulnerability and Management. Springer, pp. 367–376.

Zimmer, M., Helfer, V. Submitted. Quantity and Quality of Organic Matter in Mangrove Sediments In: This Book.

PART II

Long-term sedimentary processes and sea-level rise

CHAPTER

7

The history of surface-elevation paradigms in mangrove biogeomorphology

Daniel A. Friess[1], *Karen L. McKee*[2]

[1]Department of Geography, National University of Singapore, Singapore; [2]U. S. Geological Survey, Wetland and Aquatic Research Center, Lafayette, LA, United States

OUTLINE

1. Introduction 180

2. Surface elevation determines mangrove species distribution and growth (Paradigm 1) 181

2.1 The importance of elevation in determining mangrove distribution and growth 181

2.2 Historical origins of Paradigm 1 182

2.3 Quantitative descriptions of mangrove-elevation linkages 184

2.4 Are inundation and elevation the only controls on mangrove distribution? 187

2.5 Mangrove species distributions are not static 187

3. Mangroves can modify their surface elevations (Paradigm 2) 188

3.1 The importance of elevation adjustment by mangroves 188

3.2 Historical origins of Paradigm 2 190

3.2.1 Mangroves as "land builders" 190

3.2.2 Mangroves as "land consolidators" 191

3.2.3 Land building or land consolidating depends on biogeomorphic setting 192

3.3 Are mangroves land builders or land consolidators? 193

4. Conclusions 194

Acknowledgments 195

References 195

Dynamic Sedimentary Environments of Mangrove Coasts
https://doi.org/10.1016/B978-0-12-816437-2.00007-0

1. Introduction

Mangrove forests are model ecosystems for understanding biogeomorphology and the physical–ecological linkages that underlie wetland habitat establishment, development, and stability during sea-level changes (e.g., Balke et al., 2013). Studies of physical–ecological linkages have led to two key paradigms in wetland science that apply to mangrove biogeomorphology: (1) the distribution of mangroves across the intertidal zone is controlled primarily by tidal inundation and (2) mangroves can adjust their elevation relative to the tidal frame through various physical and biological processes.

The first paradigm—that the distribution and growth of mangroves are controlled in large part by physical processes such as tidal inundation—is one that is firmly established in the academic literature. Mangroves are controlled by their physiological tolerance to flooding (Ball, 1988; Krauss et al., 2008), alongside edaphic (e.g., Ellison et al., 2000; Lovelock et al., 2005) and biotic factors (e.g., Rabinowitz, 1978a,b; Smith, 1987) that are often cocorrelated with tidal inundation. In many natural settings where water flow is not constrained, the degree of tidal inundation is primarily determined by the elevation of the mangrove surface relative to the tidal frame (Crase et al., 2013).

The second paradigm—that mangrove systems can potentially adjust their surface elevations through a range of physical (minerogenic) and biological (biogenic) processes—has also become well established in the literature, and our contemporary knowledge of this paradigm has been comprehensively reviewed by Krauss et al. (2014); Woodroffe et al. (2016); and others. The combination of minerogenic and biogenic contributions to the soil column allows mangroves to continually adjust their soil surfaces relative to the tidal frame as sea level changes. This vertical adjustment is, in turn, linked to feedback effects of elevation and flooding on mangrove survival and growth (*Paradigm 1*).

A robust understanding of the processes underlying these two paradigms, and their linkages, is key if we are to understand and predict the potential of mangroves to keep pace with accelerated sea-level rise (e.g., Lovelock et al., 2015; Woodroffe et al., 2016) by the end of the 21st century and beyond. Thus, these two linked paradigms have been a key focus of contemporary biogeomorphological research in mangroves (e.g., Lovelock et al., 2015; Swales et al., 2015; McKee and Vervaeke, 2018) and recent data and conceptual reviews (e.g., Krauss et al., 2014; Sasmito et al., 2016; Woodroffe et al., 2016). Much research has investigated the links between elevation and mangrove vegetation the multitude of surface and subsurface processes contributing to surface-elevation change (see Krauss et al., 2014 for a full summary of these processes). However, these two paradigms have a long and largely unacknowledged historical basis, even if they were not as precisely measured in the past as they are today. Such a history can be lost when contemporary scientific articles cite only the most recent publications, and search engines limit the time range of literature searches, effectively erasing older, seminal contributions. Such practices put us in danger of losing touch with our historical scientific roots and, more importantly, obscure the evolution of contemporary scientific views (Graham and Dayton, 2002).

The aim of this chapter is to highlight the historical basis of these interlinked paradigms and show how they are shaping contemporary research agendas and mangrove management today. In particular, we focus on the historical observations and experiments that formed the origins of these biogeomorphological paradigms, and how they have informed our current understanding of elevation dynamics in mangrove forests. While not an empirical study, this investigation relies heavily on a broad literature review of archival sources, particularly regional libraries and online repositories such as www.jstor.org and www.archive.org.

2. Surface elevation determines mangrove species distribution and growth (Paradigm 1)

2.1 The importance of elevation in determining mangrove distribution and growth

The position of the mangrove soil surface relative to the tidal frame determines the depth and duration of flooding during a tidal cycle, which over the long term influences mangrove vegetation in two broad ways: (1) the inundation regime determines where mangrove plants can colonize and (2) the inundation regime influences edaphic conditions controlling subsequent mangrove plant growth.

First, the inundation regime interacts with individual species tolerances to flooding, and thereby their distribution across the intertidal zone (Ball, 1988; Krauss et al., 2008; Spier et al., 2016). Variation in plant tolerance to flooding leads to species grouped around an "inundation optimum" along a tidal gradient (Smith, 1992). In sites with unrestricted tidal flow, the degree of inundation experienced by a species at a particular location, i.e., its inundation optimum, is linked to the surface elevation relative to the tidal frame. In such sites, mangroves can be found growing at elevations and associated inundation regimes at which they can best colonize, establish, and grow (Crase et al., 2013; Spier et al., 2016; Rogers et al., 2017). The relationship may not be found at the largest scales, such as across the entire Sundarbans mangrove complex in South Asia (Ellison et al., 2000). However, the relationship between inundation and mangrove distribution can be found at the estuary scale. Using a modeling approach to look at patterns at a large spatial scale, Crase et al. (2013) showed that inundation regime (linked to underlying surface elevation) was the key driver of species distribution throughout Darwin Harbour, Australia. The relationship is most apparent at the site scale, such as a forest in Singapore where a high-resolution elevation survey of >4300 trees showed that tree species formed distinct distributions across an elevation transect (Leong et al., 2018). In that case, pioneer species such as *Sonneratia alba* grew slightly below mean sea level (msl), while *Avicennia alba* grew at a median 0.6 m above msl, *Bruguiera cylindrica* at 1.3 m above msl, and *Excoecaria agallocha* at 1.4 m above msl (Leong et al., 2018, Fig. 7.1).

Second, variation in elevation across the intertidal zone influences mangrove growth (He et al., 2007; Chen and Ye, 2013), primarily through its influence on edaphic conditions, such as differential flooding effects on root aeration and metabolism (see Krauss et al., 2008 for a review). For example, work in Florida and Belize found that flooding can affect mangrove growth through the blockage of oxygen movement from the atmosphere to subsurface roots (McKee, 1996), directly affecting root respiration (McKee and Mendelssohn, 1987) or by limiting root foraging for nutrients (McKee et al., 2002). Flooding of the soil can also promote the anaerobic formation of sulfide, a phytotoxin that reduces mangrove growth (McKee,

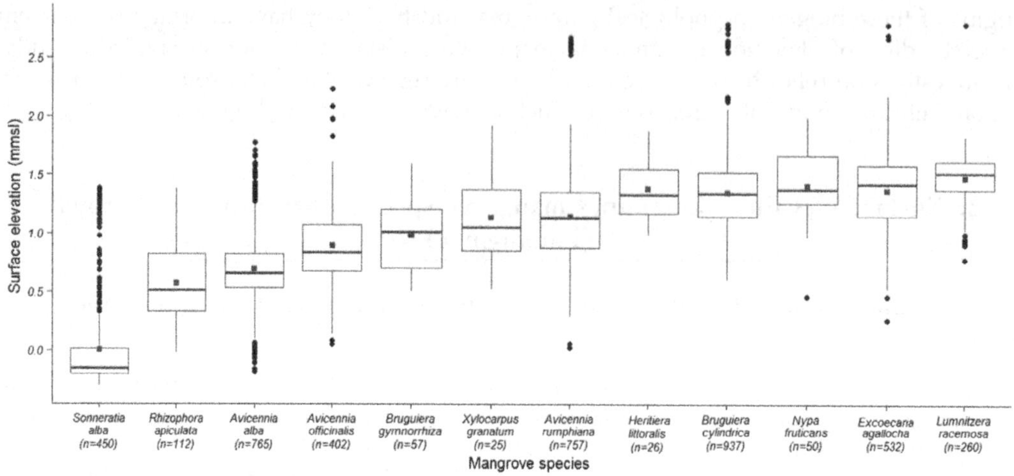

FIGURE 7.1 Elevation ranges for different mangrove species along an elevation gradient in a mangrove forest in northwest Singapore. Median elevation values are shown in bold lines, and mean elevation values are shown in square points. *Leong RC, Friess DA, Crase B, Lee WK, Webb EL. 2018. High-resolution pattern of mangrove species distribution is controlled by surface elevation. Estuar. Coast Shelf Sci. 202, 185–192; reproduced with permission.*

1993). Spatial variation in tree growth across a flooding gradient can ultimately generate different height forms within a single species, a pattern commonly observed in neotropical mangroves (e.g., Lugo and Snedaker, 1974; Lovelock et al., 2005). Growth response to flooding (*Paradigm 1*) has a feedback effect on mangrove capacity to adjust to changes in sea level (*Paradigm 2*, below) through biotic contributions to soil volume and upward expansion of the soil surface, e.g., in peat-forming settings of Florida and the Caribbean Region (McKee et al., 2007a; McKee, 2011).

Information on how elevation influences species distribution and growth and how these interactions affect mangrove capacity to accommodate sea-level rise is crucial for contemporary mangrove management and rehabilitation. Because inundation is a key constraint on seedling establishment and tree growth (Krauss et al., 2008), the surface elevation of newly rehabilitated sites is a primary determinant of mangrove colonization and successful rehabilitation (Lewis, 2005). Many mangrove rehabilitation projects are ineffective with low survival rates, primarily because species are planted at inappropriate locations in the lower intertidal or subtidal zone, outside their tolerance limits (Primavera and Esteban, 2008; Samson and Rollon, 2008; Kodikara et al., 2017). Physical parameters such as inundation and surface elevation can now be readily measured in the field using water level recorders (e.g., van Loon et al., 2016), land surveying techniques, or LiDAR, which, along with knowledge of the appropriate elevations for planting, can help increase mangrove rehabilitation success (Lewis and Brown, 2014).

2.2 Historical origins of Paradigm 1

While our contemporary knowledge of inundation and surface-elevation controls on mangrove distribution is now strong, this concept has a long history of scientific inquiry.

Some of the earliest descriptions of mangrove distribution were made by the Greek scholar Theophrastus in his 10-volume *Historia Plantarum*, written between 350 BC and 287 BC. Theophrastus broadly described the geomorphology of mangroves in the Persian Gulf, noting their variable position in the tidal frame, with some "*growing in the sea*" and other "*trees all washed by the sea up to their middle.*" Theophrastus also noted the role of above-ground roots in anchoring against tidal flows in this dynamic part of the intertidal zone (translated text by Hort, 1916; see also Chapman, 1944a). This treatise may potentially be the first written biogeomorphological description of mangroves.

Written scientific records, which became more common with the increase in geography and ecology journals in the 19th century, are useful resources to understand the historical view of various mangrove processes (Friess, 2016). These early journals contain many observational accounts of mangrove biogeomorphology that describe the importance of surface elevation in controlling mangrove distribution. During surveys of the Indus River in modern-day Pakistan, the engineer Charles William Tremenheere noted that "*extensive mudflats, some of which are covered by mangrove bushes*" were found "*generally at a level of 6 or 8 feet above low water, or from 1 to 3 feet below high water at spring tide*" (Tremenheere, 1867). An early account of mangroves in the Florida Keys by Agassiz (1880) described their establishment and development across the intertidal zone: "*As these trees grow down to the water's edge, and as at high tide the interior of the Mangrove Islands is submerged, the new plants* [i.e., propagules] *are of course dropped into the water, and are swept about and scattered by its movements.*" Others noted that mangroves were found in locations "*not covered by water at low tide*" (Warming, 1909).

Biogeomorphological descriptions continued into the early 20th century and occurred over a variety of geomorphic settings, including muddy estuarine mangroves (Tansley and Fritsch, 1905), reef and atoll mangroves (Vesey-Fitzgerald, 1942; Chapman, 1944a, 1944b), and deltaic mangrove salt marsh ecotones (Penfound and Hathaway, 1938).

The botanists Arthur Tansley and Felix Fritsch provided an in-depth description of estuarine mangrove distribution in South Asia, noting that genera such as *Sonneratia*, *Rhizophora*, *Bruguiera*, *Acanthus*, and *Nypa* grew in different locations in the intertidal zone, often according to tidal inundation. In many instances, they linked these observations with descriptions of various physiological adaptations to tidal inundation and salinity (Tansley and Fritsch, 1905).

Vesey-Fitzgerald (1942) took these biogeomorphological descriptions further by classifying mangroves found on lagoonal islands in the Indian Ocean into four broad geomorphic settings. The first three classes followed an elevation gradient from inland to the shore: *Bruguiera* spp. and *Ceriops* spp. at the head of creeks in thin soils, *Rhizophora*-dominated forests farther down the creek where soils were deeper, and open flats dominated by *Avicennia* spp. that were flooded most frequently and experienced the most sediment deposition. A fourth class comprised a mix of *Lumnitzera racemosa* and mangrove associates found in brackish depressions cut off from regular tidal influence.

20th century observations of mangroves and their relationship to surface elevation also came from the neotropics. In a description of a 1939 expedition to Jamaica, Chapman (1944a) said "*With regard to tidal range it is known that* Rhizophora *will rarely establish itself in water where the average depth of flooding is 2 ft or more, and the small tidal range in Jamaica undoubtedly acts as a limiting factor to seaward spread.*" Davis (1940a) described mangrove zones in Florida, USA, in relation to the tidal frame. The red mangrove, *Rhizophora mangle*, occupied

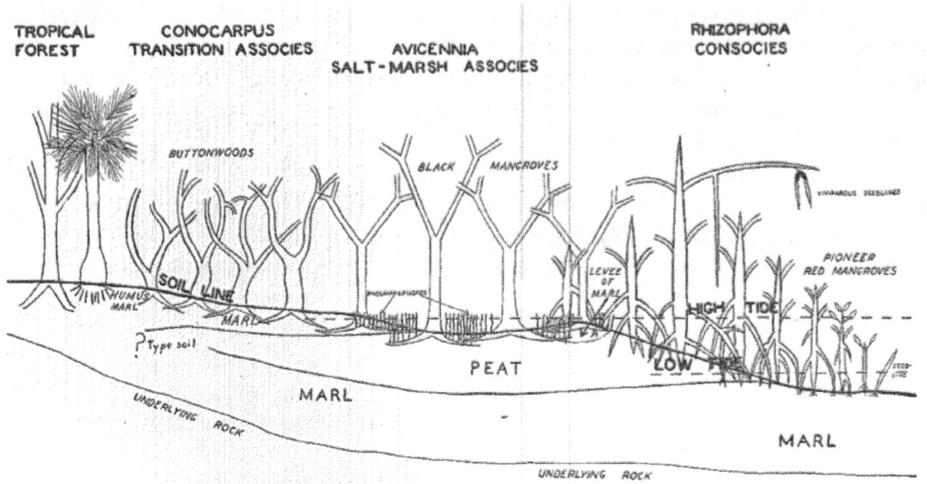

FIGURE 7.2 Spatial variation in mangrove and associated vegetation zones in Florida, USA, in relation to the tidal frame and soil conditions. *From Davis JH. 1940a. The ecology and geologic role of mangroves in Florida. Papers from Tortugas Laboratory 32, 311–384; Courtesy Carnegie Institution for Science.*

lower elevations, and other mangrove and marsh species grew at higher elevations (Fig. 7.2). In addition, Davis described effects of tidal elevation on tree height growth, with "*dwarfed trees*" of three species (*R. mangle*, *Avicennia germinans*, and *Laguncularia racemosa*) found in "*unfavorable*" conditions of two types: "*… in paludal basins where the soil level is low and the swamp flooded by stagnant water; and on elevated soils above the usual tide level…*"

Penfound and Hathaway (1938) described another system in the Americas in which *A. germinans* coexisted at its latitudinal limit with salt marsh in Louisiana, USA. Here, *A. germinans* was only found on riverside levees at an elevation "3 inches" higher than the surrounding salt marsh community. Penfound and Hathaway also noted a relationship between tree height and elevation, with the tallest trees found at the highest elevations where growing conditions were more favorable.

2.3 Quantitative descriptions of mangrove-elevation linkages

Although the earliest accounts of mangroves and their relationships to elevation and geomorphic setting were generally descriptive in nature, some were more quantitative. The British forester James Gilbert Watson and his colleagues at the Forest Department of the Federated Malay States (now Malaysia) presented one of the first rigorous scientific tests of the elevation–vegetation paradigm in the 1920s. These researchers divided various Southeast Asian mangrove species into five classes, depending on the average inundation frequencies in which they grew and reproduced. These classes were defined as between 56–62, 45–59, 20–45, 2–20, and 0–2 flood events per month (assuming a regular semidiurnal tide) and corresponded to areas that were flooded at all high tides, medium high tides, normal high tides, spring high tides only, and equinoctial tides only (Watson, 1928). When grouping species by inundation frequency, Watson was indirectly identifying the thresholds to inundation tolerance by different species. Watson showed that some species such as *Rhizophora mucronata* were cosmopolitan and found across as many as three classes, while other species (particularly mangrove associates) were only found in one class, often the class experiencing the lowest flooding frequency. Watson further mapped out these inundation classes spatially (Fig. 7.3). While Watson's classification may be an oversimplification, it essentially shows how inundation classes often translate into shore-parallel bands of vegetation following the local topography as elevation increases landward.

Over time, Watson's inundation classes have become particularly influential, with several authors expanding upon the classification decades after their introduction (e.g., De Haan, 1931; Chapman, 1976). In the decades after Watson's map of species distribution, elevation

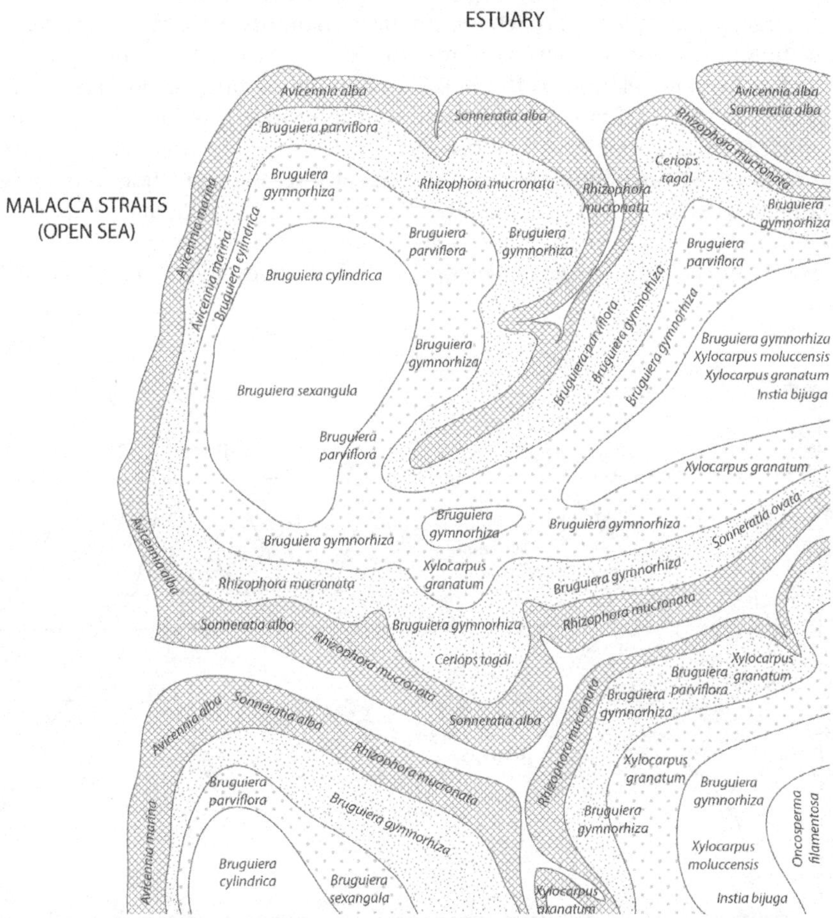

FIGURE 7.3 A diagram to illustrate the typical, but by no means inevitable, distribution of the more important mangrove species. *Redrawn and updated from Watson JG. 1928. Mangrove forests of the Malay Peninsula. Malayan Forest Records 6, 1–275; Friess DA. 2017. J.G. Watson, Inundation Classes, and their influence on paradigms in mangrove forest ecology. Wetlands 37, 603–613; reproduced with permission.*

surveys of mangrove species have become commonplace to show how their distributions vary across gradients of tidal inundation or elevation, using a combination of field-based (Castañeda-Moya et al., 2006; Spier et al., 2016; Leong et al., 2018) and remote sensing/modeling approaches (Crase et al., 2013; Rogers et al., 2017). Watson's book has also been highly cited, receiving over 530 citations on Google Scholar as of July 2018, although this is assumed to be a gross underestimation because many of the citing documents predate modern citation databases. Perhaps most importantly, Watson's concept of inundation classes has been incorporated into several mangrove rehabilitation handbooks, suggesting that it has influenced and been applied to mangrove management across the tropics (Friess, 2017).

Watson's inundation classes were useful in characterizing mangroves in Southeast Asia and similar settings, but were perhaps overly complicated for regions such as the Caribbean with fewer species and a smaller tide range. Chapman, for example, acknowledged Watson's scheme but suggested another approach for Jamaican mangroves that was better suited to the particular setting he observed. Chapman presented the vertical ranges of the three dominant species of mangroves in relation to the heights of the maximum spring and neap tides in Kingston Harbour (Fig. 7.4). His work showed that *R. mangle* was inundated by most tides, Avicennia *nitida* (syn. *A. germinans)* by most spring tides and some neap tides in late summer to fall, and *L. racemosa* only by higher spring tides. By reference to tide data at Galveston, Texas, Chapman (1944a) calculated that the *R. mangle* zone in Jamaica was experiencing 524 to 700+ inundations per year, whereas areas occupied by other species were inundated fewer times (*A. nitida* (syn. *A. germinans)*, 213 to 524 inundations; *L. racemosa*, 4 to 213 inundations).

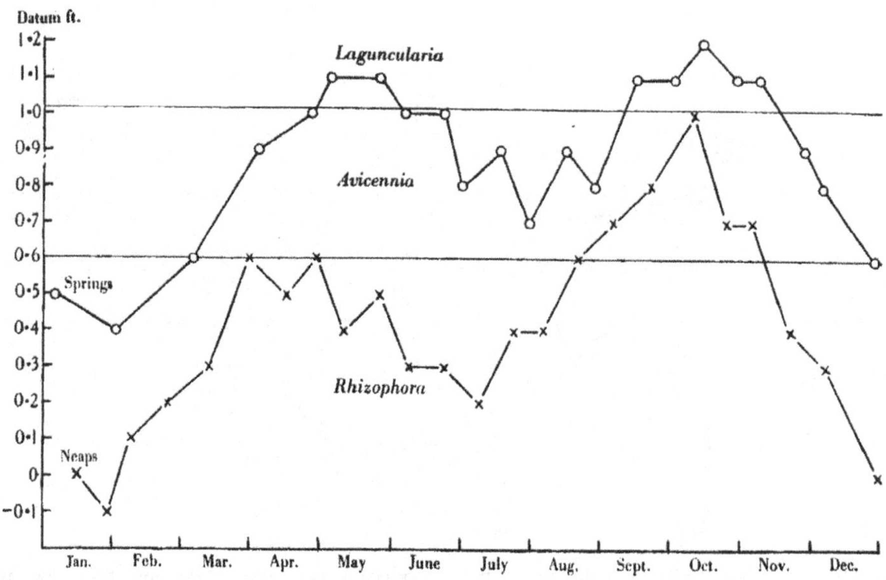

FIGURE 7.4 Relationship between mangroves and tidal heights in Kingston Harbour, Jamaica. *Chapman VJ. 1944. 1939 Cambridge University Expedition to Jamaica.—Part 1. A study of the botanical processes concerned in the development of the Jamaican shore-line. Bot. J. Linn. Soc. 52, 407–447.*

2.4 Are inundation and elevation the only controls on mangrove distribution?

While Watson showed that species distribution was correlated with inundation and surface elevation, he noted that other factors also contributed to mangrove distribution. Indeed, Watson stated that his study illustrated the "*typical, but by no means inevitable distribution of the more important mangrove species*" (Watson, 1928). Other researchers knew this before (e.g., Bowman, 1917) and after Watson's publication. J.H. De Haan added salinity to Watson's Inundation Classes when he investigated mangrove species distribution in Indonesia and found this additional chemical factor to be important (De Haan, 1931). Today, we know that tidal inundation is a dominant driver of species distribution across a site, in conjunction with other abiotic variables such as salinity, light availability, water table, and nutrient status (e.g., Youssef and Saenger, 1999; Satyanarayana et al., 2002; Krauss et al., 2008) and biotic factors such as dispersal, seed predation, herbivory, competition, and facilitation (e.g., Rabinowitz, 1978a,b; Ball, 1980; Smith, 1987; Feller, 2002; McKee et al., 2007b). Physical and chemical variables influence spatial distribution through effects on mangrove physiology and growth, particularly during early development (reviewed by Krauss et al., 2008). The effect of tidal inundation on the early stages of mangrove development was noted by several pioneering ecologists. For example, Chapman (1944b) observed in Jamaican mangrove forests that seedlings were vulnerable to both desiccation (in areas infrequently flooded) and mechanical uprooting (in areas frequently flooded), and suggested that "*future investigators might well study the zonation in mangrove swamps from the point of view of seedling establishment.*" Decades later, Rabinowitz (1978a,b), working in Panamanian mangrove forests, examined the interaction of tidal action with features of dispersing propagules (size and buoyancy) and the influence of "tidal sorting" on initial distribution of species. In addition to physical factors, interactions of propagules and seedlings with other biota (e.g., crabs and insects) in the mangrove habitat can influence establishment and growth (e.g., Smith, 1987; Feller, 1995). However, the dominant driver of many of these physical, chemical, and biological influences is still tidal inundation.

2.5 Mangrove species distributions are not static

Early interpretations of vegetation patterns were based on a static model and did not explicitly consider how ecological processes such as succession could change vegetation distribution over time. However, debates were ongoing in the broader field of ecology in the early 1900s, with successional theory suggesting that species structure of a plant community changed over time in a predictable sequence from pioneering species to a so-called "climax" community (Clements, 1916). This classical theory of plant succession was challenged by Henry A. Gleason (1927, 1939), who argued that species distributions were the result of individualistic reactions to environmental conditions across a gradient.

Mangroves had a role to play in these early debates about succession once researchers began to think about the temporal variability of mangrove species distribution. Davis (1940a,b), for example, conducted an extensive study of mangrove vegetation associations common to Florida, USA, and interpreted zonation patterns as evidence of succession (in agreement with the Clementsian view) and seaward movement of the mangrove community through land building.

However, mangroves provided a cautionary tale to early conceptualizations of succession, which needed to more strongly consider vegetation–environment linkages. Gleason (1927), writing in the journal *Ecology*, used mangroves as an example of why it was important to consider the broader physical conditions that influence vegetation distribution, instead of attributing all spatial patterns to ecological processes. In this early paper, Gleason theorized how mangrove succession would proceed in reaction to the local relative sea-level trajectory and envisioned mangroves as a pioneer stage proceeding toward a stage dominated by other plant species. He also noted that mangroves were "*efficient soil-formers*" and that this influence on the environment had "*the steady effect of tending to enlarge the area of the swamp to seaward.*" Gleason wondered, though, about "*what happens on shores which are sinking or rising?*" If the land was rising (or sea levels were falling), the environment became less favorable for mangroves and more favorable for other species. If the land was sinking, the successional trajectory was toward more flood-tolerant species or open water. Gleason seemed to recognize that successional patterns were complicated by the fact that mangroves could influence their surface elevations (*Paradigm 2, below*), but that the time spans involved in elevation change were much greater than the vegetation reaction, i.e., physical controls could obscure ecological patterns of succession (Gleason, 1927).

Gleason's hypotheses have gone on to influence later debates on mangrove successional theory. Several decades later, Lugo (1980) tackled the ideas of mangrove succession and zonation and came to three important conclusions: (1) mangrove species zonation does not always recapitulate succession, (2) succession in mangroves is cyclic, and (3) mangrove forests are steady-state systems. He further recognized that the mangrove capacity for "*active land accretion*" was "*important to an evaluation of their successional state.*" Thus, mangrove species distribution as both a static and a temporal concept relies heavily on overarching physical factors such as surface elevation.

Thus, the relationship between surface elevation and mangrove distribution and growth (*Paradigm 1*) was recognized by the earliest observers, which set the stage for later quantitative work on elevation and the mechanisms involved in generating zonation and successional patterns, as well as investigations of mangrove response to sea-level rise (reviewed by Alongi, 2008; McKee et al., 2012; Sasmito et al., 2016). Linked to this elevation–mangrove relationship (*Paradigm 1*) is the influence of mangroves on sediment accretion and land building (*Paradigm 2*), which we address in the next section.

3. Mangroves can modify their surface elevations (Paradigm 2)

3.1 The importance of elevation adjustment by mangroves

While elevation relative to the tidal frame is important for the establishment and long-term survival of mangroves, it is not static, but can change through time due to a variety of physical and biological processes. These processes are now relatively well known and can be broadly divided into physical and biological processes, which contribute to surface and subsurface movements (Krauss et al., 2014) (Fig. 7.5). Surface processes include physical processes such as deposition/erosion of mineral sediment or detritus, as well as biological processes such as litterfall (leaves, wood, reproductive matter) and benthic algal mat growth.

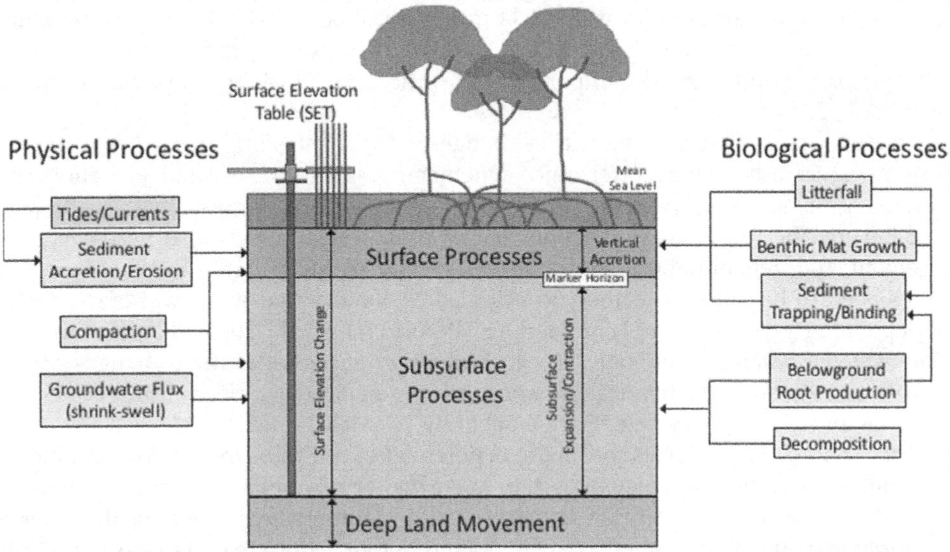

FIGURE 7.5 Physical and biological processes influencing surface and subsurface movements in mangrove forests. Processes are shown in relation to measurements conducted with the Surface Elevation Table-Marker Horizon (SET-MH) system, a common, standardized method to measure surface and subsurface processes (See Krauss et al., 2014 for a full description of the surface elevation processes shown).

Subsurface processes include the physical processes of compaction of deposited material and shrink–swell movements due to groundwater flux. Subsurface biological processes include root production and decomposition. Additions of inorganic or organic matter increase soil volume and cause upward movement of the soil surface, whereas erosion, compaction, decomposition, and desiccation lead to soil shrinkage and a lowering of the soil surface. Both surface and subsurface processes thus contribute to elevation change, which can be measured with a Surface Elevation Table–Marker Horizon system (Cahoon et al., 1995, Fig. 7.5). Accretion above a marker horizon of sand or feldspar clay is measured simultaneously with surface-elevation change relative to a benchmark (SET). The difference between accretion and elevation change rates equals subsurface subsidence or expansion (Cahoon et al., 1995). To avoid submergence, net movement of the soil surface must equal relative sea-level rise, which is the combination of ocean height increase and local land movement.

Direct evidence for elevation control by mangroves can be found in studies in which mangrove growth or density was experimentally manipulated (Kumara et al., 2010; Lang'at et al., 2014). For example, experimental plantings of mangroves at higher densities enhanced sediment accretion and elevation gain in a Sri Lankan lagoon (Kumara et al., 2010). In another study, phosphorus fertilization of stunted mangroves on a peat-forming island in Belize stimulated below-ground root production and altered surface-elevation trajectories (+4.8 mm/yr) relative to unfertilized controls (−3.7 mm/yr) (McKee et al., 2007a). Evidence for elevation control by mangroves can also be seen in the field after natural or human disturbances (Hayden and Granek, 2015). For example, a hurricane caused mortality of peat-forming mangroves in the Bay Islands, Honduras, interrupting biological inputs to surface elevation and causing a loss of surface elevation of 11 mm/yr (Cahoon et al., 2003). Another study at Gazi Bay, Kenya, found that tree girdling followed by cutting led to rapid subsidence (32 mm/yr) of the soil surface compared to elevation gain (+4.2 mm/yr) in control sites (Lang'at et al., 2014).

An understanding of surface-elevation change and its controlling processes is important because elevation adjustment determines whether mangroves can build elevation capital (*sensu* Reed, 2002) and keep pace (or not) with sea-level rise (e.g., Krauss et al., 2010). Knowledge of the specific processes underpinning elevation adjustment is also necessary for the management and rehabilitation of mangrove forests (Krauss et al., 2017). Management strategies need to be tailored to the processes controlling mangrove elevations, which can differ depending on the geomorphic setting (Woodroffe et al., 2016). For example, peat-forming systems reliant on biological contributions to surface-elevation changes are more vulnerable to natural (e.g., hurricanes) and anthropogenic (e.g., clear-cutting) events that cause mortality of mangroves (e.g., Cahoon et al., 2003; McKee and Vervaeke, 2009). A study by Kumara et al. (2010) suggests that there is potential to optimize mangrove planting to increase sediment accretion and elevation gain in created or rehabilitated sites in sediment-rich settings. Mangroves in sediment-rich settings are also more sensitive to actions that alter sediment supply (e.g., dams and other barriers; upland deforestation) (e.g., Lovelock et al., 2010, 2015; Allison et al., 2017). These insights can lead to better strategies for improving resilience to sea-level rise.

3.2 Historical origins of Paradigm 2

3.2.1 Mangroves as "land builders"

The concept that mangroves can influence their position in the tidal frame began with early descriptions of mangroves as "land builders" (Nelson, 1853; Curtiss, 1888; Pollard, 1902; Vaughan, 1909). The term "land builder" refers to the idea that mangroves act as geological agents by actively accreting sediments and/or by autogenic peat formation, thereby creating the conditions suitable for horizontal (seaward) or vertical land expansion and the associated succession of the mangrove community (Davis, 1940b) as tidal inundation changes. One of the earliest descriptions of mangrove land building in the scientific literature was in a memoir by Captain R.J. Nelson communicated by Sir Charles Lyell, a British geologist, to the Geological Society of London in 1852 (Nelson, 1853). In it, we find mention of mangroves as land-building agents: "*The Mangrove (*Rhizophora*) .... is an important agent in the conversion of swamps and littoral tracts into dry land.*" The notion of mangroves as geological agents was attractive and often repeated, e.g., in descriptions of mangroves in Florida (Curtiss, 1888; Pollard, 1902; Vaughan, 1909) and Jamaica (Chapman, 1940). In a letter to a horticultural magazine, Allen Hiram Curtiss, a noted American botanist, observed that *R. mangle* in Florida, USA, was found "*encroaching on the domain of waters and helping build what in future ages will be dry land*" so that "*the present islands will in time be added to the mainland*" (Curtiss, 1888, p.100).

The first attempts to explain the land-building mechanism focused on trapping of sediment and organic debris by mangrove roots (Pollard, 1902; Vaughan, 1909). Expeditions to Africa in the late 1860s noted the role of mangroves in trapping and consolidating sediments;

the British explorer Verney Lovett Cameron found in his surveys along the coast of modern-day Ghana that mangrove roots were able to trap alluvial silts derived from interior soil erosion, with mangroves promoting vertical expansion of the soil surface (Cameron, 1882).

In areas with little sediment, however, land building was thought to be driven by biogenic processes of peat formation. Chapman (in Steers et al., 1940) considered this a slow and *"hitherto undescribed method"* of land building, although previous authors had presumably come to similar conclusions. John H. Davis, working in Florida, conducted the first extensive study of mangroves as geological agents. The first section of his report (Davis, 1940a) described the vegetation associations of South Florida and their successional relationships. In the second section, Davis made some observations about the geologic role of mangroves, which were based in part on soil profiles showing the presence of peat layers up to 4 meters thick and positioned below the modern tide range, indicating postdeposition sea-level rise. Stratigraphic patterns in cores collected across vegetation zones were interpreted to be successional stages and evidence of seaward progradation (Davis, 1940b).

The description of mangroves as land builders, however, was challenged by both ecologists and geologists who argued that they instead colonized land or retained land where it already existed and expanded only where sedimentation was high (e.g., Egler, 1952; Scholl, 1964; Thom, 1967). Perhaps the most vocal critic was Frank Egler, who did not think that mangroves created land by prop root extension or seedling establishment seaward or that mangrove zones represented a successional sequence. Although Egler (1952) and others (Scholl, 1964; Thom, 1967) acknowledged that mangroves did contribute to vertical accretion by peat formation under certain conditions, the consensus was that land building was not a general phenomenon. In fact, the land-building hypothesis faced a chicken-and-the-egg dilemma: what came first, the land or the mangroves? In particular, how could mangroves build land if the starting point was at an elevation low in the tidal frame, beyond the physiological tolerance of mangroves?

The hypothesis that mangroves were land builders was reexamined by Jedfrey Carlton in 1974, with a review article on the subject (Carlton, 1974). This article described the history of the land-building hypothesis up to that point and summarized the two viewpoints: mangroves as land builders versus mangroves as land consolidators. Carlton concluded that the latter viewpoint had prevailed, and the matter seemed to be settled in favor of mangroves as retainers of land, rather than builders.

3.2.2 *Mangroves as "land consolidators"*

Instead of viewing mangroves as geomorphic agents that create their own land, some researchers considered them to be more opportunistic in their role as ecological engineers. In this hypothesis, mangroves colonize intertidal surfaces once they are at an elevation where tidal inundation allows seedling establishment and survival. Once colonized, mangroves then consolidate sediments, create soil through biogenic processes, and encourage further sediment deposition though physical–ecological feedbacks (Swales et al., 2015). This view holds that mangroves are land modifiers and consolidators, and not specifically or solely land builders.

The view of mangroves as land consolidators has been understood for at least 150 years. As early as the 1860s, it was known that mangroves were *"admirably adapted to consolidate and favour the deposit of alluvial matter,"* according to observations made during surveys of the East

African coast (Kirk, 1865). Louis Agassiz (1880), who surveyed the coral reefs in Florida, USA, noted that "*The mangroves are among the most important geological agents in this region; but for them the loose sand and mud would remain an ever-shifting ground of movable particles.*"

Vaughan (1909); p. 461, also working in Florida, observed that mangroves were active agents in land formation, but only when "*trees have grown sufficiently for the development of a tangle of roots they catch and hold sediment and any floating debris.*" This statement suggests an understanding that mangroves must be able to colonize an area before they can modify it, i.e., the surface elevation has to be suitable for mangrove colonization prior to their effect on sediment trapping.

Influenced by his work on Inundation Classes, J.G. Watson also believed that mangroves could not be considered land builders because of the physiological controls on colonization imposed by tidal flooding. Instead, vegetation had to establish first upon suitable intertidal surfaces, such that "*mangroves following silting rather than cause it*" (Watson, 1928, p.5).

The land consolidation hypothesis gained real traction in the academic community when it was framed within the emerging context of rising sea level (Egler, 1952; Scholl, 1964; Thom, 1967). David Scholl (1964, 1965), for example, ascribed the deposition of fibrous peat in coastal swamps of Florida, USA, to mangroves and used radiocarbon-dated material to reconstruct sea-level rise over 4000 years. This vertical deposition differed from the hypothesis that mangroves build new land in a seaward direction by actively accumulating autochthonous and allochthonous material.

3.2.3 Land building or land consolidating depends on biogeomorphic setting

So far in this section, we have discussed the two viewpoints of land formation separately because most historical accounts focused on one or the other hypothesis. However, some seminal studies considered both hypotheses together. These studies highlight the challenges inherent in gaining a true understanding of surface-elevation dynamics.

The report of V.J. Chapman's 1939 expedition to Jamaica highlights the complexity and linkages between the two hypotheses. In their expedition report, Chapman devoted seven pages to a description of mangroves and their relation to land formation (Steers et al., 1940). Chapman described how both land-building and land-consolidating mangroves can exist, based on their environmental and biogeomorphic settings and associated dominance of biogenic versus minerogenic processes (Steers et al., 1940; Chapman, 1944a, Fig. 7.6). Their survey suggested a typology of three different sedimentary habitats: (1) *Sand-dominated mangroves* "*growing on sand cays, shoals, or sand spits*" and, after colonization, begin forming peat; (2) *Mud-dominated mangroves* "*situated at the mouths of rivers or else in protected areas*" where land formation was usually by "*the mechanical trapping of silt,*" except in areas with "*little or no supply of silt in the water*" where peat formation occurred by "*accumulation of the roots of the plants*" and was "*responsible for the development of new land*"; and (3) *Reef-dominated mangroves* "*with their roots in the holes and crevices of the porous coral rock,*" which did not "*play any part in the development of new land*" (Steers et al., 1940). These three categories illustrated how sedimentary setting determined whether mangroves in Jamaica were building or consolidating land and by what process (biogenic, minerogenic, or a combination of the two).

Bruce Thom (1967) studied the role of mangroves as geological agents in the Grijalva–Usumacinta Delta (Tabasco and Campeche states, Mexico) and also concluded that their role as land builders versus land stabilizers depended on the environmental setting. Referring

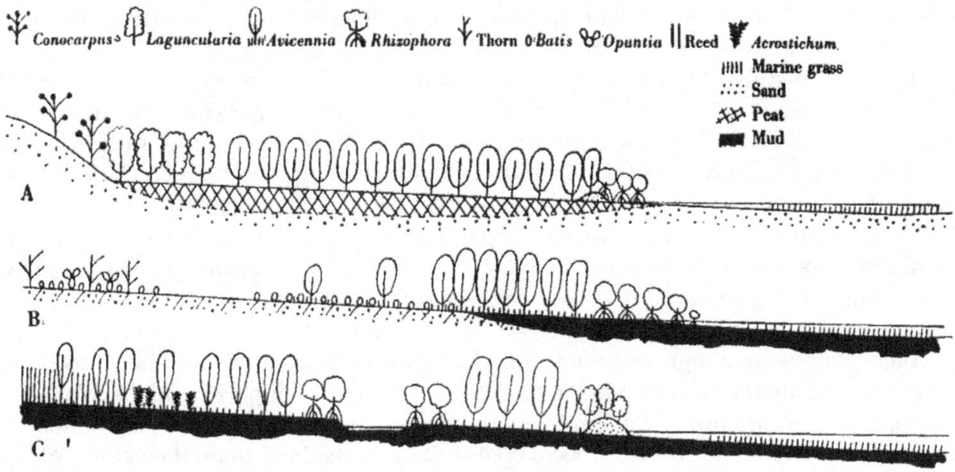

FIGURE 7.6 Examples of vegetation profiles across the intertidal zone in different sedimentary habitats (sand, mud, and reef) in Jamaica. *Chapman VJ. 1944. 1939 Cambridge University Expedition to Jamaica.—Part 1. A study of the botanical processes concerned in the development of the Jamaican shore-line. Bot. J. Linn. Soc. 52, 407–447.*

to the mudflats of Laguna dos Terminos, Campeche and Laguna del Carmen, Tabasco, Thom wrote, "*The gentle increase in slope inland of inactive mudflats and their lack of organic accumulations lead to the conclusion that mangrove plants play a secondary role in the accumulating process of this landform type in Tabasco-Campeche.*" This type of habitat is dominated by inorganic substrate, though Thom also described other habitats in which organic detritus accumulated or fibrous peat formed thick layers. Of these, Thom said, "*But in certain environments, particularly along abandoned river channels, mangroves in Tabasco are showing greater competence in* 'land-building' *than other plants.*" Some researchers at the time thought that only certain mangrove species were land builders (Moldenke, 1960) or observed deposits of fibrous peat mostly associated with one species such as *R. mangle* (Davis, 1940b; Chapman, 1944). However, others questioned whether mangroves promoted accretion or just occupied a part of the intertidal zone where accretion would take place with or without the presence of vegetation (e.g., Egler, 1952; Bird, 1971).

3.3 Are mangroves land builders or land consolidators?

The historical debate about mangroves as land builders or land stabilizers is instructive because it reflects, in part, differences in the biogeomorphic settings in which these respective roles were formulated. Researchers working in sediment-poor environments found thick deposits of mangrove peat (e.g., Davis, 1940b), indicating a biogenic process at work. Others (e.g., Watson, 1928) studying mangroves in sediment-rich settings observed accreting mudflats that were later colonized by mangroves. Still other investigators observed evidence of both land building and land consolidating within the same geomorphic setting (e.g., Steers et al., 1940; Thom, 1967), suggesting that depositional environment (or *sedimentary habitats* as Chapman described them) within a geomorphic type may also be important (see

Woodroffe et al., 2016 for a detailed discussion of sedimentation at different scales in four main types of mangrove environments: river delta, tidal estuary, lagoon, and carbonate reef). However, because few studies experimentally tested the concept of land building, the role of mangroves as land builders or land consolidators was never convincingly settled.

In a review of the land-building concept, Lee et al. (2014) concluded that most researchers accepted the role of mangroves as land stabilizers and that mangroves followed land formation in sediment-rich environments (Woodroffe et al., 1985; Lovelock et al., 2010). Although a few studies have shown that mangroves can expand onto adjacent mudflats or sandy shoals, they did not experimentally test accretionary influences by mangroves in association with this colonization (see Lee et al., 2014 and references therein). Other work has shown greater sediment accretion associated with mangrove aerial roots (Bird, 1986; Krauss et al., 2003), but not in conjunction with mangrove colonization and seaward expansion. Another consideration is that in the formation of a land mass, expansion may occur both horizontally and vertically. For example, the occurrence of thick sequences of peat (up to 10 m thick) in the stratigraphic record is evidence of an autogenic process contributing to the land mass (Lee et al., 2014 and references therein). Thus, although existing evidence fails to show mangroves directly influencing seaward progradation, mangroves may still be considered land builders as they can actively contribute to vertical land development by promoting sedimentation or peat formation.

The disagreement between the land-building and land-consolidating camps may also reflect different definitions of land building. If land building is more narrowly defined as only occurring in the horizontal plane or without any prior physical input, then vertical accretion through peat formation is not land building. On the other hand, if a land platform is considered to be a three-dimensional mass, then biogenic expansion in the vertical direction contributes to land building—even if that vertical expansion initiates after the land platform reaches a mangrove-colonizable elevation through physical processes (e.g., sea-level lowering or inorganic sedimentation). The key point is that the active role played by mangroves, whether termed land building or land consolidating, is important to the persistence of the mangrove habitat as sea level rises.

Our knowledge of mangrove surface-elevation change is now more nuanced and based on a broader appreciation of historic research in both biogenic and minerogenic settings. Whether we ultimately view them as land builders or land consolidators, it is clear that mangroves are active agents or ecological engineers in the landscape. We now know that mangroves can modify surface elevation via two general mechanisms: (1) directly through biological processes, by contributing organic matter to soil volume through litterfall and below-ground root production and (2) indirectly through physical processes by slowing water velocities, promoting sedimentation, and by trapping and binding sediment. However, these processes operate within boundary conditions that are largely physical in nature, such as geomorphic setting, sediment supply, and sea-level history. Thus, our historical and contemporary knowledge of surface-elevation dynamics suggests a combination of land-building and land-consolidating processes is at play.

4. Conclusions

The linked paradigms that (1) surface elevation influences mangrove species distribution and growth and (2) mangroves can modify their position in the tidal frame form a cornerstone of mangrove science. These paradigms are not just academic in nature but are essential to the success of mangrove rehabilitation projects across the tropics. By understanding the elevations and inundation regimes where different species can colonize, we can better promote the resilience of mangrove forests to accelerated sea-level rise over the next century. Our knowledge of physical–ecological linkages in mangroves has real-world impacts and is important for the survival of these habitats and the coastal communities who rely on them.

Historical studies such as those described in this chapter help us understand the origins of key paradigms in mangrove ecology and how they shape our thinking about physical–ecological linkages today. Our contemporary view of surface-elevation paradigms is strongly influenced by quantitative measurements in the early 20th century as well as scientific observations from the 19th century and earlier. Current knowledge is thus shaped by the biases that these earlier researchers had, and while some of these biases may still be present today, we now have a much broader, holistic, and accurate view of biogeomorphological processes and relationships in mangrove forests. At a time when the field of mangrove research is rapidly expanding, it is pertinent to reflect upon and celebrate the rich history of mangrove biogeomorphology and the contributions it has made to our understanding of this important coastal ecosystem.

Acknowledgments

This chapter has benefited from discussions with participants of the 2016 Australian Mangrove and Saltmarsh Network (AMSN) meeting in Darwin (Australia) and members of the Mangrove Lab (National University of Singapore). We also thank Kerrylee Rogers (University of Wollongong), Nicole Cormier (Macquarie University), Ken Krauss, Don Cahoon, Laura Feher (United States Geological Survey), and Irving Mendelssohn (Louisiana State University) for many critical discussions on this topic, and for comments on a previous draft of this manuscript. Any use of trade, firm, or product names is for descriptive purposes only and does not imply endorsement by the US Government. This manuscript is submitted for publication with the understanding that the US Government is authorized to reproduce and distribute reprints for Governmental purposes.

References

Agassiz, L., 1880. Report on the Florida reefs. Mem. Museum Comparat. Zool. 7, 1–182.

Allison, M.A., Nittrouer, C.A., Ogston, A.S., Mullarney, J.C., Nguyen, T.T., 2017. Sedimentation and survival of the Mekong Delta: a case study of decreased sediment supply and accelerating rates of relative sea level rise. Oceanography 30, 98–109.

Alongi, D.M., 2008. Mangrove forests: resilience, protection from tsunamis, and responses to global climate change. Estuar. Coast Shelf Sci. 76, 1–13.

Balke, T., Bouma, T.J., Herman, P.M.J., Horstman, E.M., Sudtongkong, C., Webb, E.L., 2013. Cross-shore gradients of physical disturbance in mangroves: implications for seedling establishment. Biogeosciences 10, 5411–5419.

Ball, M.C., 1980. Patterns of secondary succession in a mangrove forest of southern Florida. Oecologia 44, 226–235.

Ball, M.C., 1988. Ecophysiology of mangroves. Trees Struct. Funct. 2, 129–143.

Bird, E.C.F., 1971. Mangroves as land-builders. Victorian Nat. 88, 189–197.

Bird, E.C.F., 1986. Mangroves and intertidal morphology in westernport Bay, Victoria, Australia. Mar. Geol. 69, 251–271.

Bowman, H.H., 1917. Ecology and physiology of the red mangrove. Proc. Am. Phil. Soc. 56, 589–672.

Cahoon, D.R., Reed, D.J., Day, J.W., 1995. Estimating shallow subsidence in microtidal salt marshes of the southeastern United States: kaye and Barghoorn revisited. Mar. Geol. 128, 1–9.

Cahoon, D.R., Hensel, P., Rybczyk, J., McKee, K.L., Proffitt, C.E., Perez, P.C., 2003. Mass tree mortality leads to mangrove peat collapse at Bay Islands, Honduras after Hurricane Mitch. J. Ecol. 91, 1093–1105.

Cameron, V.L., 1882. Report of the evening meetings, session 1881-2. Proc. Royal Geographic. Soc. Month. Record of Geograph. 4, 500–507.

Carlton, J.M., 1974. Land-building and stabilization by mangroves. Environ. Conserv. 1, 285–294.

Castañeda-Moya, E., Rivera-Monroy, V.H., Twilley, R.R., 2006. Mangrove zonation in the dry life zone of the Gulf of Fonseca, Honduras. Estuar. Coast 29, 751–764.

Chapman, V.J., 1940. A note upon the geological work of mangroves in Jamaica. PGA (Proc. Geol. Assoc.) 51, 346–348.

Chapman, V.J., 1944. 1939 Cambridge University Expedition to Jamaica.—Part 1. A study of the botanical processes concerned in the development of the Jamaican shore-line. Bot. J. Linn. Soc. 52, 407–447.

Chapman, V.J., 1944. 1939 Cambridge University expedition to Jamaica.-Part 2. A study of the environment of *Avicennia nitida* jacq. In Jamaica. Bot. J. Linn. Soc. 52, 448–486.

Chapman, V.J., 1976. Mangrove Vegetation. J. Cramer, Vadus, 426pp.

Chen, Y.P., Ye, Y., 2013. Growth and physiological responses of saplings of two mangrove species to intertidal elevation. Mar. Ecol. Prog. Ser. 482, 107–118.

Clements, F.E., 1916. Plant Succession: An Analysis of the Development of Vegetation. Carnegie Institution of Washington, Washington DC.

Crase, B., Liedloff, A., Vesk, P.A., Burgman, M.A., WIntle, B.A., 2013. Hydroperiod is the main driver of the spatial pattern of dominance in mangrove communities. Global Ecol. Biogeogr. 22, 806–817.

Curtiss, A.H., 1888. How the mangrove forms islands. Garden Forest 1, 100.

Davis, J.H., 1940a. The ecology and geologic role of mangroves in Florida. Paper. Tortugas Lab. 32, 311–384.

Davis, J.H., 1940b. The peat deposits of Flǫrida—their occurrence, development, and uses. Fla. Geol. Surv. Bull. 30, 247 pp.

De Haan, J.H., 1931. Het een en ander over de Tijlatjapsche vloedbosschen. Tectona 24, 39–75.

Egler, F.E., 1952. Southeast saline Everglades vegetation, Florida, and its management. Vegetatio 3 (4–5), 213–265.

Ellison, A.M., Mukherjee, B.B., Karim, A., 2000. Testing patterns of zonation in mangroves: scale dependence and environmental correlates in the Sundarbans of Bangladesh. J. Ecol. 88, 813–824.

Feller, I.C., 1995. Effects of nutrient enrichment on growth and herbivory of dwarf red mangrove (*Rhizophora mangle*). Ecological Monographs 65, 477–505.

Feller, I.C., 2002. The role of herbivory by wood-boring insects in mangrove ecosystems in Belize. Oikos 97, 167–176.

Friess, D.A., 2016. Ecosystem services and disservices of mangrove forests: insights from historical colonial observations. Forests 7, 183.

Friess, D.A., 2017. J.G. Watson, Inundation Classes, and their influence on paradigms in mangrove forest ecology. Wetlands 37, 603–613.

Gleason, H.A., 1927. Further views on the succession concept. Ecology 8, 299–326.

Gleason, H.A., 1939. The individualistic concept of the plant association. Bull. Torrey Bot. Club 44, 463–481.

Graham, M.H., Dayton, P.K., 2002. On the evolution of ecological ideas: paradigms and scientific progress. Ecology 83, 1481–1489.

Hayden, H.L., Granek, E.F., 2015. Coastal sediment elevation change following anthropogenic mangrove clearing. Estuar. Coast Shelf Sci. 165, 70–74.

He, B.Y., Lai, T.H., Fan, H.Q., Wang, W.Q., Zheng, H.L., 2007. Comparison of flooding-tolerance in four mangrove species in a diurnal tidal zone in the Beibu Gulf. Estuar. Coast Shelf Sci. 74, 254–262.

Hort, A., 1916. Theophrastus Enquiry into Plants and Minor Works on Odours and Weather Signs. William Heinemann and G. P. Putnam's Sons, London and New York.

Kirk, J., 1865. Note on two expeditions up the river Rovuma, East Africa. J. Roy. Geogr. Soc. Lond. 35, 154–167.

Kodikara, K.A., Mukherjee, N., Jayatissa, L.P., Dahdouh-Guebas, F., Koedam, N., 2017. Have mangrove restoration projects worked? An in-depth study in Sri Lanka. Restor. Ecol. 25, 705–716.

Krauss, K.W., Allen, J.A., Cahoon, D.R., 2003. Differential rates of vertical accretion and elevation change among aerial root types in Micronesian mangrove forests. Estuar. Coast Shelf Sci. 56, 251–259.

Krauss, K.W., Lovelock, C.E., McKee, K.L., Lopez-Hoffman, L., Ewe, S.M.L., Sousa, W.P., 2008. Environmental drivers in mangrove establishment and early development. Aquat. Bot. 89, 105–127.

Krauss, K.W., Cahoon, D.R., Allen, J.A., Ewel, K.C., Lynch, J.C., Cormier, N., 2010. Surface elevation change and susceptibility of different mangrove zones to sea-level rise on Pacific high islands of Micronesia. Ecosystems 13, 129–143.

Krauss, K.W., McKee, K.L., Lovelock, C.E., Cahoon, D.R., Saintilan, N., Reef, R., Chen, L., 2014. How mangrove forests adjust to rising sea level. New Phytol. 202, 19–24.

Krauss, K.W., Cormier, N., Osland, M.J., Kirwan, M.L., Stagg, C.L., Nestlerode, J.A., Russell, M.J., From, A.S., Spivak, A.C., Dantin, D.D., Harvey, J.E., Almario, A.E., 2017. Created mangrove wetlands store belowground carbon and surface elevation change enables them to adjust to sea-level rise. Sci. Rep. 7, 1030.

Kumara, M.P., Jayatissa, L.P., Krauss, K.W., Phillips, D.H., Huxham, M., 2010. High mangrove density enhances surface accretion, surface elevation change, and tree survival in coastal areas susceptible to sea-level rise. Oecologia 164, 545–553.

Lang'at, J.K.S., Kairo, J.G., Mencuccini, M., Bouillon, S., Skov, M.W., Waldron, S., Huxham, M., 2014. Rapid losses of surface elevation following tree girdling and cutting in tropical mangroves. PloS One 9, 8.

Lee, S.Y., Primavera, J.H., Dahdouh-Guebas, F., McKee, K., Bosire, J.O., Cannicci, S., Diele, K., Fromard, F., Koedam, N., Marchand, C., Mendelssohn, I., Mukherjee, N., Record, S., 2014. Ecological role and services of tropical mangrove ecosystems: a reassessment. Global Ecol. Biogeogr. 23, 726–743.

Leong, R.C., Friess, D.A., Crase, B., Lee, W.K., Webb, E.L., 2018. High-resolution pattern of mangrove species distribution is controlled by surface elevation. Estuar. Coast Shelf Sci. 202, 185–192.

Lewis, R.R., 2005. Ecological engineering for successful management and restoration of mangrove forests. Ecol. Eng. 24, 403–418.

Lewis, R.R., Brown, B., 2014. Ecological Mangrove Rehabilitation: A Field Manual for Practitioners. Restoring Coastal Livelihoods Program, OXFAM, 275pp.

Lovelock, C.E., Feller, I.C., McKee, K.L., Thompson, R., 2005. Variation in mangrove forest structure and sediment characteristics in Bocas del Toro, Panama. Caribb. J. Sci. 41, 456–464.

Lovelock, C.E., Sorrell, B.K., Hancock, N., Hua, Q., Swales, A., 2010. Mangrove forest and soil development on a rapidly accreting shore in New Zealand. Ecosystems 13, 437–451.

Lovelock, C.E., Cahoon, D.R., Friess, D.A., Guntenspergen, G.R., Krauss, K.W., Reef, R., Rogers, K., Saunders, M.L., Sidik, F., Swales, A., Saintilan, N., Thuyen, L.X., Triet, T., 2015. The vulnerability of Indo-Pacific mangrove forests to sea-level rise. Nature 526, 559–563.

Lugo, A.E., 1980. Mangrove ecosystems: successional or steady state? Biotropica 12, 65–72.

Lugo, A.E., Snedaker, S.C., 1974. The ecology of mangroves. Annu. Rev. Ecol. Evol. Syst. 5, 39–64.

McKee, K.L., 1993. Soil physicochemical patterns and mangrove species distribution – reciprocal effects? J. Ecol. 81, 477–487.

McKee, K.L., 1996. Growth and physiological responses of neotropical mangrove seedlings to root zone hypoxia. Tree Physiol. 16, 883–889.

McKee, K.L., 2011. Biophysical controls on accretion and elevation change in Caribbean mangrove ecosystems. Estuar. Coast Shelf Sci. 91, 475–483.

McKee, K.L., Mendelssohn, I.A., 1987. Root metabolism in the black mangrove, *Avicennia germinans* (L.) L.: response to hypoxia. Environ. Exp. Bot. 27, 147–156.

McKee, K.L., Vervaeke, W.C., 2009. Impacts of human disturbance on soil erosion and habitat stability of mangrove-dominated islands in the Pelican Cays and Twin Cays ranges, Belize. Smithsonian Contributions to the Marine Sciences 38, 415–427.

McKee, K.L., Vervaeke, W.C., 2018. Will fluctuations in salt marsh - mangrove dominance alter vulnerability of a subtropical wetland to sea-level rise? Global Change Biol. 24, 1224–1238.

McKee, K.L., Feller, I.C., Popp, M., Wanek, W., 2002. Mangrove isotopic ($d^{15}N$ and d^{13} C) fractionation across a nitrogen vs. phosphorous limitation gradient. Ecology 83, 1065–1075.

McKee, K.L., Cahoon, D.R., Feller, I.C., 2007. Caribbean mangroves adjust to rising sea level through biotic controls on change in soil elevation. Global Ecol. Biogeogr. 16, 545–556.

McKee, K.L., Rooth, J.E., Feller, I.C., 2007. Mangrove recruitment after forest disturbance is facilitated by herbaceous species in the Caribbean. Ecol. Appl. 17, 1678–1693.

McKee, K.L., Rogers, K., Saintilan, N., 2012. *Response of salt marsh and mangrove wetlands to changes in atmospheric CO_2, climate, and sea level*, pp. 63–96. In: Middleton, B. (Ed.), Global Change and the Function and Distribution of Wetlands. Springer, The Netherlands.

Moldenke, H.N., 1960. Materials toward a monograph of the genus *Avicennia*. I. *Phytologia* 7 (3), 123–168.

Nelson, R.J., 1853. On the geology of the Bahamas, and on coral-formations generally. Q. J. Geol. Soc. Lond. 9, 200–214.

Penfound, W.T., Hathaway, E.S., 1938. Plant communities in the marshlands of southeastern Louisiana. Ecol. Monogr. 8, 1–56.

Pollard, C.L., 1902. Plant agencies in the formation of the Florida Keys. Plant World 5 (1), 8–10.

Primavera, J.H., Esteban, J.M., 2008. A review of mangrove rehabilitation in the Philippines: successes, failures and future prospects. Wetl. Ecol. Manag. 16, 345–358.

Rabinowitz, D., 1978. Dispersal properties of mangrove propagules. Biotropica 10, 47–57.

Rabinowitz, D., 1978. Early growth of mangrove seedlings in Panama, and an hypothesis concerning the relationship of dispersal and zonation. J. Biogeogr. 5, 113–133.

Reed, D.J., 2002. Sea-level rise and coastal marsh sustainability: geological and ecological factors in the Mississippi delta plain. Geomorphology 48, 233–243.

Rogers, K., Lymburner, L., Salum, R., Brooke, B.P., Woodroffe, C.D., 2017. Mapping of mangrove extent and zonation using high and low tide composites of Landsat data. Hydrobiologia 803, 49–68.

Samson, M.S., Rollon, R.N., 2008. Growth performance of planted mangroves in the Philippines: revisiting forest management strategies. Ambio 27, 234–240.

Sasmito, S.D., Murdiyarso, D., Friess, D.A., Kurnianto, S., 2016. Can mangroves keep pace with contemporary sea level rise? A global data review. Wetl. Ecol. Manag. 24, 263–278.

Satyanarayana, B., Raman, A.V., Dehairs, F., Kalavati, C., Chandramohan, P., 2002. Mangrove floristic and zonation patterns of Coringa, Kakinada Bay, east coast of India. Wetl. Ecol. Manag. 10, 25–39.

Scholl, D.W., 1964. Recent sedimentary record in mangrove swamps and rise in sea-level over the south-western coast of Florida. I. Mar. Geol. 1, 344–366.

Scholl, D.W., 1965. Recent sedimentary record in mangrove swamps and rise in sea-level over the southwestern coast of Florida. II. Mar. Geol. 2, 343–364.

Smith III., T.J., 1987. Seed predation in relation to tree dominance and distribution in mangrove forests. Ecology 68, 266–273.

Smith III., T.J., 1992. Forest structure. In: Robertson, A.I., Alongi, D.M. (Eds.), Tropical Mangrove Ecosystems. American Geophysical Union, Washington DC, USA, pp. p101–136.

Spier, D., Gerum, H.L.N., Noernberg, M.A., Lana, P.C., 2016. Flood regime as a driver of the distribution of mangrove and salt marsh species in a subtropical estuary. J. Mar. Syst. 161, 11–25.

Steers, J.A., Chapman, V.J., Colman, J., Lofthouse, J.A., 1940. Sand cays and mangroves in Jamaica Cambridge University Jamaica expedition, 1939 meeting of the society, 15 April 1940. Geogr. J. 96, 305–328.

Swales, A., Bentley, S.J., Lovelock, C.E., 2015. Mangrove-forest evolution in a sediment-rich estuarine system: opportunists or agents of geomorphic change? Earth Surf. Process. Landforms 40, 1672–1687.

Tansley, A.G., Fritsch, F.E., 1905. Sketches of vegetation at home and abroad. I. The Flora of the Ceylon littoral (continued). New Phytol. 4, 27–55.

Thom, B.G., 1967. Mangrove ecology and deltaic geomorphology: Tabasco, Mexico. J. Ecol. 55, 301–343.

Tremenheere, C.W., 1867. On the lower portion of the River Indus. J. Roy. Geogr. Soc. Lond. 37, 68–91.

Van Loon, A.F., Brake, B.T., Van Huijgevoort, M.H., Dijksma, R., 2016. Hydrological classification: a practical tool for mangrove restoration. PloS One 11, e0150302.

Vaughan, T.W., 1909. The geological work of mangroves in southern Florida. Smithsonian Misc. Collect. 52, 461–464.

Vesey-Fitzgerald, D., 1942. Further studies of the vegetation on islands in the Indian Ocean. J. Ecol. 30, 1–16.

Warming, E., 1909. Oecology of Plants; an Introduction to the Study of Plant Communities. Clarendon Press, Oxford, 424pp.

Watson, J.G., 1928. Mangrove forests of the Malay Peninsula. Malay. For. Rec. 6, 1–275.

Woodroffe, C.D., Thom, B.G., Chappell, J., 1985. Development of widespread mangrove swamps in mid-Holocene times in northern Australia. Nature 317, 711–713.

Woodroffe, C.D., Rogers, K., McKee, K.L., Lovelock, C.E., Mendelssohn, L.A., Saintilan, N., 2016. Mangrove sedimentation and response to relative sea-level rise. Ann. Rev. Marine Sci. 8, 243–266.

Youssef, T., Saenger, P., 1999. Mangrove zonation in mobbs Bay – Australia. Estuar. Coast Shelf Sci. 49, 43–50.

CHAPTER

8

Radiocarbon dating of mangrove sediments

Juliet Sefton[1], *Sarah Woodroffe*[1], *Philippa Ascough*[2]

[1]Department of Geography, Durham University, Durham, United Kingdom; [2]Scottish Universities Environmental Research Centre, NERC Radiocarbon Facility, East Kilbride, United Kingdom

OUTLINE

1. Introduction 199
1.1 Nomenclature 200
2. Brief introduction to the principles of radiocarbon dating 200
2.1 Radiocarbon age calibration and corrections 201
2.2 Reservoir and post-depositional effects 203
2.3 Age reporting 203
3. Radiocarbon dating of mangrove sediments 204
3.1 Bulk sediment approaches 205
3.2 Mangrove roots 205
3.3 Bioturbators 207
3.4 Carbon import and export 207
3.5 Aboveground plant macrofossils 208
3.6 Organic and pollen concentrates 208
3.7 Compound-specific approaches 209
4. Other quantitative dating approaches 209
5. Conclusions 210
References 210

1. Introduction

Accurate and precise dating of mangrove sediments is important to improve understanding of (1) long-term changes in blue carbon storage and its implications for the global carbon cycle (Ezcurra et al., 2016; Rogers et al., 2019), (2) past coastal geomorphological changes and

Dynamic Sedimentary Environments of Mangrove Coasts
https://doi.org/10.1016/B978-0-12-816437-2.00014-8

how mangroves adapt and contribute to them (Perry et al., 2008; Swales et al., 2015), (3) the timing of past relative sea-level changes, whether driven by melting of ice sheets, tectonic land movements, or other processes (Woodroffe et al. 2015a, 2016), and (4) past changes in mangrove succession and ecology over millennial timescales (McKee et al., 2002; Wooller et al., 2007). In all of these, the accuracy of dating is crucial to provide quantitative measurements of rates of change—ideally in the form of an absolute dating technique to place sedimentary records on an independent, regular timescale. Mangrove sediments are commonly used as paleoenvironmental archives as their root structure traps and stabilizes sediment (Woodroffe, 1992). A variety of absolute dating methods can be applied to mangrove sediments (e.g., tephrochronology, optically stimulated luminescence dating, ^{210}Pb and ^{137}Cs dating techniques); however, one of the most commonly applied is radiocarbon dating. Mangrove sediments contain an abundance of different sources of carbon that can be extracted from sediments for analysis.

This chapter will briefly introduce the key principles of radiocarbon dating and discuss approaches to radiocarbon dating mangrove sediments. We aim to provide a more general overview for mangrove researchers who wish to understand mangrove dynamics beyond the timescales of ^{210}Pb and ^{137}Ce dating techniques (which span, approximately, the last 100 years). We do not intend this chapter to be a complete overview of the radiocarbon dating technique, and we direct readers to comprehensive reviews elsewhere for this information (Olsson, 2009; Taylor, 2000; Törnqvist et al., 2015).

1.1 Nomenclature

- We refer to "^{14}C" when referring to the isotope itself and to "radiocarbon" when discussing the radiocarbon dating technique more generally.
- "Mangrove" refers to the mangrove environment/ecosystem generally. We refer to a mangrove tree/leaf/root, etc., explicitly.
- "Mangrove sediment" refers to any sediment accumulating under a mangrove forest.

2. Brief introduction to the principles of radiocarbon dating

Carbon has three naturally occurring isotopes: ^{12}C, ^{13}C, and ^{14}C, with ^{12}C being the most abundant (~99% of all carbon). ^{14}C is significantly less abundant and the only unstable (radioactive) carbon isotope, accounting for roughly one per 10^{12} carbon atoms (Taylor, 2000). Cosmic rays entering the upper part of Earth's atmosphere produce ^{14}C when they interact with nitrogen (^{14}N) atoms. Newly formed ^{14}C is quickly oxidized to carbon monoxide and then carbon dioxide (CO_2). ^{14}C production rates are greatest at the Earth's poles, although rapid mixing incorporates ^{14}C into the entire atmosphere on timescales of a few years. The transfer of ^{14}C from the atmosphere into the biosphere and oceans is also rapid with respect to the precision of the radiocarbon method. Therefore, we consider the ^{14}C content of the atmosphere equivalent to that of carbon in both the living biosphere and recently fixed inorganic carbon at any point in time.

The uptake of carbon (and therefore ^{14}C) through biological processes occurs while an organism is alive, but ceases abruptly upon death. For inorganic materials (e.g., soil carbonates), the equivalent is the moment of final formation (after which no new ^{14}C is incorporated into the system). The ^{14}C content in a sample after death or final formation drops at a constant rate determined by the law of radioactive decay. The known half-life of ^{14}C (the amount of time required to decay the ^{14}C content of a sample by 50%) is 5730 years (Godwin, 1962), although for historical reasons, a conventional radiocarbon age is calculated using the "Libby Age" of 5568 years (Stuiver and Polach, 1977). Since its introduction, radiocarbon dating has become one of the most widely applied chronological tools for samples ~40,000 years old due to the wide distribution of ^{14}C throughout the biogeochemical cycles, constant production in the Earth's atmosphere through time, and its well-known decay rate.

It is possible to determine the ^{14}C content of a sample in two alternative ways: firstly, by detecting the number of beta particles emitted—as ^{14}C decays spontaneously to ^{14}N—per sample volume per unit time (known as radiometric or conventional radiocarbon dating) and secondly, by measuring the ratio of ^{14}C atoms relative to ^{12}C or ^{13}C in a sample using accelerator mass spectrometry (AMS) (Muller, 1977). AMS techniques are commonly used in earth sciences and archeology because it allows for rapid measurement of smaller sample sizes, compared to what is required for equivalent results from the conventional dating technique (in particular reducing samples size requirements from >1 g to 2 mg of carbon) (Bowman, 1990).

2.1 Radiocarbon age calibration and corrections

To calculate the age of a sample using radiocarbon, the concentration of ^{14}C in the atmosphere at the time of the death of the sampled organism must be known. However, the atmospheric ^{14}C concentration has changed over time (De Vries, 1958; Revelle and Suess, 1957; Stuiver, 1961) due to a number of factors. Production rates of ^{14}C change through time as geophysical processes alter the incidence of cosmic rays to the Earth's atmosphere, such as fluctuations in the Earth's magnetic field. Different climate states change the relative reservoirs of ^{14}C across the globe; the oceans can store more CO_2 if the climate is substantially colder and ocean circulation changes releasing vast stores of old CO_2 with low ^{14}C content (i.e., the reservoir effect). Human activity such as the burning of fossil fuels from the European industrial revolution onwards (~1750 AD) and nuclear weapons testing between 1950 and 1963 have released large amounts of ^{14}C-dead CO_2 and generated excess ^{14}C, respectively, altering the ^{14}C content of the Earth's atmosphere.

The research community calculates a radiocarbon age by comparing the ^{14}C content (via the $^{14}C/^{12}C$ or $^{14}C/^{13}C$ ratio if using AMS) of a sample with that of a single value (an international standard, currently oxalic acid II), which is taken to represent a constant, preindustrial atmospheric ^{14}C content. To calculate the time in years for radioactive decay of sample ^{14}C content to result in the measured value, the ^{14}C half-life is used. However, as the atmospheric ^{14}C content varies through time, a single ^{14}C measurement of a sample can correspond to a range of possible calendar ages. Therefore, researchers calibrate measured ^{14}C ages to the calendar age ranges they represent using an international radiocarbon calibration curve. These are currently IntCal20, SHCal20, and Marine20 (Heaton et al., 2020; Hogg et al., 2020; Reimer et al., 2020) for northern hemisphere terrestrial biosphere/atmosphere, the southern hemisphere terrestrial biosphere/atmosphere, and marine samples, respectively. A radiocarbon calibration curve is constructed by obtaining ^{14}C measurements of a sequential series of samples of known calendar ages, such as annually resolved tree rings, where the first such attempt was the "curve of knowns" by Libby (Bowman, 1990). Additionally, the atmospheric ^{14}C content of the northern and southern hemispheres can be significantly different, which is largely due to the difference in percent land and ocean cover and because atmospheric mixing between hemispheres is slow, hence the need for a separate southern hemisphere calibration curve (Hogg et al., 2020).

^{12}C, ^{13}C, and ^{14}C behave the same chemically, but some isotopes react faster than others do during chemical processes—termed isotopic fractionation. For example, ^{12}C has a lower atomic weight than other carbon isotopes and is preferentially incorporated into plants during photosynthesis as it has a faster rate of diffusion than the heavier isotopes. Because of this differential fractionation, the concentration of ^{14}C in plants is usually much lower than the atmosphere at a single point in time, but the exact size of this offset modulates with factors such as local environmental conditions (e.g., temperature, humidity) and species type.

Plants and the animals that consume them in different parts of the biosphere therefore have characteristic carbon isotope compositions (Table 8.1), which if left uncorrected would result in apparent ^{14}C age differences. To correct for this, researchers measure the ratio of $^{13}C-^{12}C$ in each sample along with its ^{14}C content and the $^{13}C/^{12}C$ ratio normalized to give a $\delta^{13}C$ value of $-25‰$ relative to an international standard (the Pee Dee Belemnite, a Cretaceous belemnite sample). This process provides a "fractionation factor" that is applied to the sample $^{14}C/^{12}C$ or $^{14}C/^{13}C$ ratio to provide an age correction that amounts to ~16 years for every 1‰ difference in the original measured sample $\delta^{13}C$ from $-25‰$.

TABLE 8.1 Examples of approximate mean $\delta^{13}C$ values for different samples from mangrove environments taken from the literature.

Material	$\delta^{13}C$ value
Mangrove peat (Jacotot et al., 2018; Kemp et al., 2019; McKee et al., 2002; Sen and Bhadury, 2017; Tue et al., 2011)	−22‰ to −26‰
Avicennia marina fine roots (Kelleway et al., 2018; Khan et al., 2019)	−26‰ to −27‰
Mangrove benthic invertebrates (Bouillon et al., 2002)	−16‰ to −23‰
Dwarf *Rhizophora mangle* green leaf (Wooller et al., 2003)	−24.2‰
Avicennia germinans leaves (Khan et al., 2019)	−28.5‰ to −31.6‰
Mangrove sediment porewater (Dittmar et al., 2006)	−31.4‰
Mangrove microalgae (Bouillon et al., 2008; Khan et al., 2019)	−17‰ to −20‰
Seagrass (Gonneea et al., 2004; Khan et al., 2019)	−11‰ to −16‰
Freshwater sediment (Kemp et al., 2019)	−26‰ to −29‰
Atmospheric CO_2	−8‰
Marine carbonates	0‰

2.2 Reservoir and post-depositional effects

The ^{14}C content of the local environment in which a sample forms (i.e., from where its carbon originates) can also affect radiocarbon ages if this differs from the ^{14}C content of the atmosphere at the same point in time. This is known as a "reservoir effect" (Deevey et al., 1954; Stuiver et al., 1986; Reimer and Reimer, 2001; Waelbroeck et al., 2001). Carbon on Earth is contained within a number of different "reservoirs," such as the marine reservoir, and the ^{14}C content of these can differ from the atmospheric content (and hence the terrestrial biosphere). This differs as a function of atmosphere reservoir exchange rates, internal reservoir carbon residence times, and the input of carbon from nonatmospheric sources. If an organism forms its tissues from a reservoir other than the atmosphere/terrestrial biosphere, then its measured ^{14}C content (and therefore radiocarbon age) will require correcting before it is representative of a "true" calendar age. Reservoir effects particularly influence marine and freshwater systems. Carbon ions dissolved in water can come from a large variety of sources, such as dissolution of ancient carbon from carbonate geology (Olsson, 1991). The rate that a reservoir mixes with the atmosphere, its size, and the residence time of carbon are also important. For example, a reservoir with slow rates of mixing and high residence times can result in carbon atoms spending a longer period of time separated from atmospheric contact, during which time the ^{14}C content of the reservoir reduces by radioactive decay (Ascough et al., 2005; Alves et al., 2018). While researchers have made great efforts to characterize local and regional reservoir effects, the processes influencing ^{14}C content can vary greatly over very small distances. It is therefore important to identify the potential for any reservoir effects to influence the ^{14}C age of any organism used for paleoenvironmental reconstruction and, if needed, establish and apply the appropriate correction.

While radiocarbon dating assumes no process other than radioactive decay influences ^{14}C content, there are some situations where this is not the case. Dissolution and reprecipitation of carbonate material can alter the ^{14}C content, meaning it no longer reflects the age of death (Bowman, 1990). Instead, it is an integration of the original sample age and the age of (in this case, more recent) contamination, weighted by mass. Humic and fulvic acids can be mobile within the sediment column and will contaminate organic material intended for radiocarbon dating, similarly altering ^{14}C content (Turetsky et al., 2004). Sediment reworking through bioturbation or through transport of carbon material in suspended sediments also influences radiocarbon measurements by mixing carbon of different ages. Therefore, once collected, samples are usually refrigerated to minimize growth of bacteria or other flora and fauna. Samples are then "pretreated" before radiocarbon measurements to remove any post-depositional carbon contamination that could skew measurements (which is likely to be a different age to the actual sample/event of interest). One must carefully consider and control sample context, collection, and storage prior to any radiocarbon measurements in order to minimize the possibility for addition of exogenous carbon (i.e., any carbon not synthesized by the organism when it was living).

2.3 Age reporting

The radiocarbon dating community has many established protocols and conventions for laboratory procedures, reporting, and rigorous laboratory intercomparison studies (Stuiver and Polach, 1977; Scott et al., 2010). This ensures consistent quality assurance for measurements in participating laboratories worldwide. Laboratories commonly express an uncertainty of 1σ on AMS measurements of sample ^{14}C content, which researchers translate into a radiocarbon age (i.e., a central value ± uncertainty) (see Stuiver and Polach, 1977 for full details). Samples are measured along with a mix of blanks, secondary standards, and process standards, to assess measurement reproducibility, background contamination, and laboratory processing.

The reported error on a radiocarbon age incorporates the uncertainty on measured sample ^{14}C content (usually by AMS) due to Poisson counting statistics, along with other factors such as the uncertainty on blank measurements. It is important to note that the uncertainty on a measurement is as important as the central value and that no radiocarbon age can ever be a point estimate, as it is generated according to Poisson counting statistics, whether beta decay events or measured atoms of ^{14}C in an AMS instrument. Instead, the uncertainty gives a range within which there is a specific level of statistical confidence that the "true" ages lies. Researchers should always consider radiocarbon ages as age ranges, particularly when comparing individual ages to assess whether they are reliably different. A 95% confidence interval or 2σ uncertainty is usually recommended when considering radiocarbon ages in a paleoenvironmental reconstruction context (e.g., Hijma et al., 2015). For calibrated age ranges, researchers should always use the 95% or 2σ confidence interval in interpretation, as statistically this is the range within which there is a good level of confidence that the "true" age lies. It is not correct to give the impression that a particular calendar date or group of dates within this entire calibrated range are more or less likely than any other, unless further Bayesian statistical analysis is applied (Bronk Ramsey, 2009).

3. Radiocarbon dating of mangrove sediments

The radiocarbon dating method is the most widely used dating technique to understand past changes in mangrove systems because the method is applicable over the past ~40,000 years and has relatively small measurement uncertainties. A large proportion of carbon in mangroves is produced in situ (above and below ground), which provides a good basis for obtaining accurate radiocarbon dates (Alongi, 2014; Bouillon et al., 2008; McKee et al., 2007; Woodroffe et al., 2015b; Wooller et al., 2007). Mangrove trees use carbon dioxide from the atmosphere during photosynthesis and therefore do not typically require reservoir corrections. However, there are still some challenges to obtaining reliable chronologies. There is uncertainty about the production, transportation, consumption, and decomposition of mangrove carbon (particularly below ground) and about how this environmental context can affect radiocarbon ages (Alongi, 2014; Bouillon et al., 2008). Thus, the complexity of the mangrove carbon cycling and depositional system means radiocarbon chronologies require careful evaluation.

Observing stratigraphic relationships and sedimentary characteristics are important first steps in constructing a mangrove radiocarbon chronology. For example, whether an organic deposit in a sediment sequence is mangrove- or freshwater-derived is difficult to establish by visual observation alone (Bloom, 1970; Scholl, 1964a,b). Often before sampling for radiocarbon, further analyses are required to assign sedimentary units to a mangrove depositional environment confidently, for example, palynology or micropaleontology (i.e., pollen, diatom, or foraminifera assemblages). Plant macrofossils can also distinguish between mangrove- and freshwater-derived organic deposits, where these are preserved in mangrove sediment. Sedimentary deposition and reworking processes within mangrove environments are highly localized, and detailed modern observations are necessary to interpret paleo records more accurately.

3.1 Bulk sediment approaches

Bulk sediment dating has been a common approach for obtaining chronologies for past mangrove environments. These are samples of mangrove sediments and peats, where we cannot extract identifiable organic remains. Bulk sediment samples can be composed of many individual components of carbon-based material. These are mostly humified organics, but also macrofossils such as leaves or roots, bacteria, algae, pollen, and any dissolved organic carbon (DOC) in porewaters—which can itself be a mix of allochthonous or autochthonous carbon (Beaman et al., 1994). A number of previous studies demonstrate out-of-sequence ages when sampling mangrove bulk sediment and also large individual sample age ranges (Geyh et al., 1979; Grindrod and Rhodes, 1984; Larcombe et al., 1995; Punwong et al., 2013a,b; Tamura et al., 2009; Woodroffe et al., 2015b). This is due to the large range of possible carbon sources contributing to mangrove bulk sediment. It is difficult to determine the success of bulk sediment–derived mangrove chronologies from the literature, where removal of "outliers" is common or where there is only one radiocarbon age in a sediment sequence. Thus, this approach is generally not preferred for constructing radiocarbon chronologies in mangrove deposits. A sample which contains carbon from an unknown number of sources (i.e., a bulk sediment sample) may have a high precision value (i.e., a small level of uncertainty on the measurement), but it always reflects the combined mean age of multiple sources. Because of this, it is difficult to assign bulk sediment ages to sedimentation or organism death events of interest.

3.2 Mangrove roots

While mangroves have relatively shallow roots compared with other terrestrial trees due to sediment anoxia, mangrove roots can penetrate up to 2 m into the underlying sediment (Grindrod, 1988; Kristensen, 2008; Tomlinson, 2016). Roots can make up a large proportion of total biomass in mangroves, but this varies with sedimentary settings and geography. In biogenic settings with low minerogenic input such as carbonate islands in the Caribbean (McKee et al., 2007; McKee and Faulkner, 2000), root production is high relative to minerogenic settings in estuaries and deltas which can produce shallow roots (Lovelock et al., 2007) or tall canopies (Krauss et al., 2010). Large mangrove roots introduce younger carbon material to the underlying sedimentary sequence, which will not be the same as the "true" age of the time of deposition at a given stratigraphic depth. Some researchers suggest that avoiding large mangrove roots in a sedimentary sequence might result in more reliable radiocarbon chronologies, or that any bulk sediment radiocarbon age should be considered as a minimum age for the sediment depth (Larcombe et al., 1995; Woodroffe, 1990, 1981). Young fine roots, which branch off these larger roots, also introduce younger carbon into the sedimentary sequence but decompose more readily than large roots (Fig. 8.1). These are difficult to identify by macro- or microscopic examination. A bulk sediment sample that does not contain any visible large roots may still be younger than its equivalent stratigraphic depth as it may contain undetected material from decomposed fine roots (Gischler, 2006).

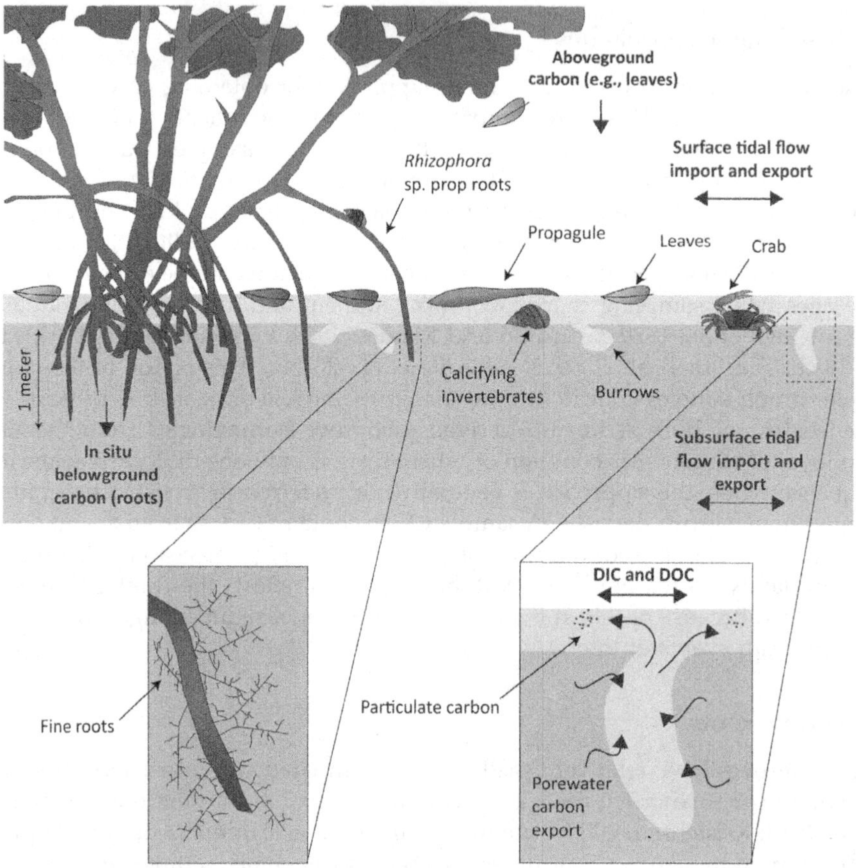

FIGURE 8.1 *Rhizophora* sp. root structure and relationship with underlying sediments. Major carbon sources and transport processes are in bold purple text. Large roots penetrate through the stratigraphy and fine roots branch off these. Aboveground macrofossils (leaves, seeds, fruits, propagules, etc.) are assumed to be incorporated into sediments in situ, and their preservation in the stratigraphy indicates rapid burial. The largest carbon sources are surface and subsurface in situ production, followed by allochthonous carbon transport via water flow. Carbon production via macrofauna such as crabs is minimal relative to other sources (Alongi, 2014). *DIC*, dissolved inorganic carbon; *DOC*, dissolved organic carbon. *Symbols courtesy of the Integration and Application Network (ian.umces.edu/symbols/), University of Maryland Center for Environmental Science.*

3.3 Bioturbators

Crabs, mollusks, shrimp, and other microfauna (e.g., foraminifera) can burrow up to 1 m deep into the sediment column (Fig. 8.1) (Koch et al., 2005; Stieglitz et al., 2000). This can rework young and old carbon, but particularly younger carbon through the process of crabs burying leaves in their burrows (Middleton and McKee, 2001; Skov and Hartnoll, 2002). These processes potentially reduce the resolution of paleoenvironmental reconstructions and radiocarbon chronologies (Grindrod and Rhodes, 1984; Debenay et al., 2004; Deng et al., 2006; Perry et al., 2008). Bioturbation can also drive oxygenation of the sediment column and therefore modify the preservation (or decomposition) of carbon material measured for radiocarbon dating. Oxygenated sediments are more likely to provide habitats for a larger variety of organisms. In Northern Australia, upper mangrove sediments can contain living foraminifera to depths of 50 cm (Berkeley et al., 2007) and up to 80 cm in Malaysia (Culver et al., 2015). The severity of biological mixing of the sediment column is highly dependent on locality and does not necessarily impede paleoenvironmental reconstructions and chronologies to a great extent depending on sedimentation rates and habitat optimums for burrowing species (Behling et al., 2001; Culver et al., 2015). The problem of bioturbation is also not unique to mangrove settings (Berkeley et al., 2007). Bacterial, fungal, algal, insect, and invertebrate carbon contributions to mangrove sediment composition are not well-documented, but are likely to be minimal with respect to total mangrove biomass (Alongi, 2014; Bouillon et al., 2008; Bouillon and Boschker, 2006; Huxham et al., 2010; Koch et al., 2005; Pan et al., 2019; Vane et al., 2013).

3.4 Carbon import and export

Although large-scale reworking of a deposit is usually easy to identify on the basis of whether material is horizontally bedded or roots are penetrating vertically (Cohen, 1970; Woodroffe, 1981), finer scale reworking, such as particulate transport during a single tidal cycle, could play a significant role in the taphonomy of carbon within the sample analyzed. Import and export of particulate and dissolved inorganic carbon and DOC into and from mangrove sediments occurs during porewater transport, driven by tidal cycles (Maher et al., 2013; Taillardat et al., 2018). The rates at which these processes occur are highly variable and localized, but can be significant in terms of both carbon export from mangroves and import of allochthonous carbon from adjacent marine environments (Adame and Lovelock, 2011; Bouillon et al., 2008; Dittmar et al., 2006). Currents (tidal, river, or wave) and storm activity all work to redistribute carbon within and proximal to mangroves. The $\delta^{13}C$ composition of a sediment sample should reflect any mix of carbon sources present (see Table 8.1). Many researchers use bulk sediment $\delta^{13}C$ to interpret increasing marine or terrestrial influences (and therefore relative sea-level changes) in mangroves and other intertidal settings such as saltmarshes (Kemp et al., 2019, 2010; Khan et al., 2015; Lamb et al., 2006; Mackie et al., 2007; Wilson, 2017). Therefore, analyzing $\delta^{13}C$ composition in combination with ^{14}C content can be a useful approach to interpret the degree of autochthonous and allochthonous carbon (or terrestrial vs. marine contribution) in mangrove sedimentary records.

3.5 Aboveground plant macrofossils

Aboveground mangrove plant macrofossils (e.g., leaves, seeds, flowers, propagules, and fruit) can provide reliable radiocarbon ages where they are present and identifiable (Wooller et al., 2007). Aboveground material is likely to have entered the deposit very shortly after death of the plant tissue and therefore should record the age of sediment deposition (i.e., the event of interest). Macrofossils like leaves represent up to 18 months of mangrove tree growth or as little as a single flowering season (Reef et al., 2010; Reich et al., 1992; Suárez and Medina, 2005; Tomlinson, 2016). Mangrove leaf litter is one of the largest contributors (along with root biomass) to the organic matter pool in mangrove environments (Fiala et al., 2017; Koch et al., 2005). Leaves, seeds, flowers, and fruit are therefore optimal materials for obtaining radiocarbon ages. Sometimes other aboveground plant material like timber can also be present in mangrove sediment. Timber is longer-lived than leaves, and different parts of a wood sample may have different ages depending on species morphology, species growth rate, and affinity for decomposition (the "old-wood" effect, Schiffer, 1986). As mangroves grow, new wood forms from reserves of carbohydrates from the previous year (Worbes and Junk, 1989), resulting in a gradual age offset between newer (outer) and older (inner) parts of the tree. It is very rare to collect whole pieces of timber for radiocarbon dating in mangroves, particularly when retrieving dating material from sediment cores. In addition, it is very difficult to identify whether wood found in mangrove sediments is from above ground (timber) or below ground (large roots). Therefore, this kind of material is very rarely dated.

Aboveground plant macrofossils are most often found where sedimentation and/or peat accumulation rates are high, for example, where Holocene relative sea-level rise has been continuous and drives mangrove peat accretion, such as in the Caribbean region (McKee et al., 2007; Monacci et al., 2009; Rogers et al., 2019; Wooller et al., 2007). In other parts of the world where late Holocene relative sea-level has fallen (such as Seychelles and Tanzania), sediment accommodation space is reduced, sedimentation and peat accumulation rates are lower, and surface plant material totally decomposes or is consumed at the sediment surface before burial (Punwong et al., 2013a,b; Woodroffe et al., 2015a,b).

3.6 Organic and pollen concentrates

If aboveground plant macrofossils are not readily available for dating in mangrove sediments, it is advisable to use other carbon fractions and sources. Specific organic concentrates from mangrove sediments are one such approach. A sequence of bulk sediment radiocarbon ages from the same cores produced out-of-sequence ages (Punwong et al., 2013b). Targeting the 10–63 μm size fraction, Woodroffe et al. (2015b) produced in-sequence Holocene ages with depth from cores collected in a mangrove in Tanzania. Woodroffe et al. (2015b) chose this size fraction to avoid large fragments like roots (which may be a source of younger carbon) and to better concentrate in situ pollen. The results of this study show that targeting specific carbon sources in this way could improve mangrove sediment chronologies over Holocene timescales. However, there has only been one attempt at this approach in one mangrove sediment sequence with widely spaced ages. Future research following similar methods is required to substantiate these results further to determine if this is a reliable method.

Although the Woodroffe et al. (2015b) study identifies the 10–63 μm size fraction as being the most likely to date in situ sedimentation, this sediment fraction may include fine organic residue of unknown origin along with mangrove pollen. To include predominantly above-ground material in the dated sample (i.e., the pollen and not unidentifiable humified matter), methods that are routinely used in other paleoenvironments need to be adopted to concentrate mangrove pollen for dating (Howarth et al., 2013; Li et al., 2014; Neulieb et al., 2013; Newnham et al., 2007; Vandergoes and Prior, 2003; Zhang et al., 2017). However, it requires complex laboratory processes with potential for contamination at each stage and often results in very small samples for dating (and therefore requiring a lot of material for processing) (Howarth et al., 2013). Pollen concentrates can still result in problematic radiocarbon ages where significant quantities of pollen are reworked from surrounding environments, which may not be the same age as the deposition of the sediment (Howarth et al., 2013; Kilian et al., 2002; Li et al., 2014).

3.7 Compound-specific approaches

Compound-specific radiocarbon dating is another targeted approach that has been applied to other environments where it is also difficult to distinguish between different carbon sources, e.g., shallow marine environments (Mollenhauer et al., 2007; Smittenberg et al., 2004), high-latitude environments (Ohkouchi and Eglinton, 2008), and terrestrial peats (Pancost et al., 2000). We can attribute some organic compounds (also called biomarkers) to individual species (or a small range of species), which is useful for identifying material in mangroves for radiocarbon dating (Eglinton et al., 1997). While there is established literature on mangrove biomarkers (Basyuni et al., 2007; Ghosh et al., 1985; He et al., 2014; Koch et al., 2011; Ladd and Sachs, 2015), no study (to our knowledge) has yet attempted compound-specific radiocarbon dating of mangrove sediments. Radiocarbon ages of this type are more time-consuming to obtain than those previously discussed because the sample extraction procedure requires specialist techniques and equipment (such as solvent extraction and gas chromatography mass spectrometry). The relationship between the age of sediment deposition and a biomarker age is also difficult to evaluate, as biomarkers can be stored in the environment for a period of time before deposition or can be reworked (Grossi et al., 2003; Jaffé et al., 2001). Additionally, the usefulness of biomarkers in radiocarbon dating is highly dependent on how well we can characterize the original source and formation of the organic compound. For example, the lipid taraxerol is found in considerably higher concentrations in *Rhizophora* sp. leaves than in other higher plants and is therefore considered a species-specific biomarker (Versteegh et al., 2004). However, other biomarkers found in mangroves are not necessarily unique to a particular component of a mangrove tree (i.e., leaf or root) or to a single species of mangrove.

4. Other quantitative dating approaches

Further research should focus on obtaining multi-proxy chronologies to validate radiocarbon approaches in mangroves. This could be by utilizing other techniques such as tephrochronology, optically stimulated luminescence dating, and ^{210}Pb and ^{137}Cs dating alongside radiocarbon methods. Radioisotope techniques such as ^{210}Pb and ^{137}Cs are commonly applied to mangrove sediments; however, they are only suitable for sediments <200 years old. ^{210}Pb and ^{137}Cs dating is not applicable in all settings, as the approach requires minerogenic material (as these isotopes do not adsorb to organic particulates) and high enough concentrations of the isotopes (which varies geographically due to different atmospheric production and/or fallout). A tephrochronological approach requires presence of volcanic ash (tephra) that is readily identifiable in a mangrove sedimentary sequence and that can be assigned to a known volcanic eruption event. So far, only one study has used a tephra layer to provide chronological constraints in a mangrove sequence (Adame et al., 2015), and no known studies have applied optically stimulated luminescence dating to mangrove sediments.

5. Conclusions

Establishing reliable chronologies is vital for understanding past mangrove environment dynamics. Radiocarbon dating is the most widely used technique for producing chronologies for sediment sequences, but it is challenging due to complex carbon cycling processes in mangroves. Targeted approaches that sample specific organic fractions such as pollen or biomarkers will likely yield more reliable chronologies than bulk sediment ages; however, further testing in multiple mangrove settings and with independent dating approaches is required.

References

Adame, M.F., Lovelock, C.E., 2011. Carbon and nutrient exchange of mangrove forests with the coastal ocean. Hydrobiologia 663, 23–50. https://doi.org/10.1007/s10750-010-0554-7.

Adame, M.F., Santini, N.S., Tovilla, C., Vázquez-Lule, A., Castro, L., Guevara, M., 2015. Carbon stocks and soil sequestration rates of tropical riverine wetlands. Biogeosciences 12, 3805–3818. https://doi.org/10.5194/bg-12-3805-2015.

Alongi, D.M., 2014. Carbon cycling and storage in mangrove forests. Ann. Rev. Mar. Sci. 6, 195–219. https://doi.org/10.1146/annurev-marine-010213-135020.

Alves, E.Q., Macario, K., Ascough, P., Bronk Ramsey, C., 2018. The Worldwide marine radiocarbon reservoir effect: definitions, mechanisms, and prospects. Rev. Geophys. 56, 278–305. https://doi.org/10.1002/2017RG000588.

Ascough, P., Cook, G., Dugmore, A., 2005. Methodological approaches to determining the marine radiocarbon reservoir effect. Prog. Phys. Geogr. Earth Environ. 29, 532–547. https://doi.org/10.1191/0309133305pp461ra.

Basyuni, M., Oku, H., Baba, S., Takara, K., Iwasaki, H., 2007. Isoprenoids of Okinawan mangroves as lipid input into estuarine ecosystem. J. Oceanogr. 63, 601–608. https://doi.org/10.1007/s10872-007-0053-2.

Beaman, R., Larcombe, P., Carter, R.M., 1994. New evidence for the Holocene sea-level high from the inner shelf, central Great Barrier Reef, Australia. J. Sediment. Res. 64, 881–885. https://doi.org/10.1306/D4267EF1-2B26-11D7-8648000102C1865D.

Behling, H., Cohen, M.C.L., Lara, R.J., 2001. Studies on holocene mangrove ecosystem dynamics of the Bragança Peninsula in north-eastern Pará, Brazil. Palaeogeogr. Palaeoclimatol. Palaeoecol. 167, 225–242. https://doi.org/10.1016/S0031-0182(00)00239-X.

Berkeley, A., Perry, C.T., Smithers, S.G., Horton, B.P., Taylor, K.G., 2007. A review of the ecological and taphonomic controls on foraminiferal assemblage development in intertidal environments. Earth Sci. Rev. 83, 205–230. https://doi.org/10.1016/j.earscirev.2007.04.003.

Bloom, A.L., 1970. Paludal stratigraphy of Truk, Ponape, and Kusaie, eastern Caroline islands. GSA Bull. 81, 1895–1904. https://doi.org/10.1130/0016-7606(1970)81[1895:PSOTPA]2.0.CO;2.

Bouillon, S., Boschker, H.T.S., 2006. Bacterial carbon sources in coastal sediments: a cross-system analysis based on stable isotope data of biomarkers. Biogeosciences 3, 175–185.

Bouillon, S., Connolly, R.M., Lee, S.Y., 2008. Organic matter exchange and cycling in mangrove ecosystems: recent insights from stable isotope studies. J. Sea Res. 59, 44–58. https://doi.org/10.1016/j.seares.2007.05.001. Mangrove Macrobenthos Special Issue Proceedings of the Mangrove Macrobenthos Meeting II.

Bouillon, S., Raman, A.V., Dauby, P., Dehairs, F., 2002. Carbon and nitrogen stable isotope ratios of subtidal Benthic invertebrates in an estuarine mangrove ecosystem (Andhra Pradesh, India). Estuar. Coast Shelf Sci. 54, 901–913. https://doi.org/10.1006/ecss.2001.0864.

Bowman, S., 1990. Radiocarbon Dating. University of California Press.

Bronk Ramsey, C., 2009. Bayesian analysis of radiocarbon dates. Radiocarbon 51, 337–360. https://doi.org/10.1017/S0033822200033865.

Cohen, A.D., 1970. An allochthonous peat deposit from southern Florida. GSA Bull. 81, 2477–2482. https://doi.org/10.1130/0016-7606(1970)81[2477:AAPDFS]2.0.CO;2.

Culver, S.J., Leorri, E., Mallinson, D.J., Corbett, D.R., Shazili, N.A.M., 2015. Recent coastal evolution and sea-level rise, Setiu Wetland, Peninsular Malaysia. Palaeogeogr. Palaeoclimatol. Palaeoecol. 417, 406–421. https://doi.org/10.1016/j.palaeo.2014.10.001.

De Vries, H., 1958. Variation in Concentration of Radiocarbon with Time and Location on Earth, vol. 61. Koninklijke Nederlandse Akademie van Wetenschappen B, pp. 1–9.

Debenay, J., Guiral, D., Parra, M., 2004. Behaviour and taphonomic loss in foraminiferal assemblages of mangrove swamps of French Guiana. Marine Geol. 208, 295–314. https://doi.org/10.1016/j.margeo.2004.04.013. Material Exchange Between the Upper Continental Shelf and Mangrove Fringed Coasts with Special Reference to the N. Amazon-Guianas Coast.

Deevey, E.S., Gross, M.S., Hutchinson, G.E., Kraybill, H.L., 1954. The natural C^{14} contents of materials from hard-water lakes. Proc. Natl. Acad. Sci. U.S.A. 40, 285–288. https://doi.org/10.1073/pnas.40.5.285.

Deng, Y., Horrocks, M., Ogden, J., Anderson, S., 2006. Modern pollen-vegetation relationships along transects on the Whangapoua estuary, great barrier island, northern New Zealand. J. Biogeogr. 33, 592–608.

Dittmar, T., Hertkorn, N., Kattner, G., Lara, R.J., 2006. Mangroves, a major source of dissolved organic carbon to the oceans. Global Biogeochem. Cycles 20. https://doi.org/10.1029/2005GB002570.

Eglinton, T.I., Benitez-Nelson, B.C., Pearson, A., McNichol, A.P., Bauer, J.E., Druffel, E.R.M., 1997. Variability in radiocarbon ages of individual organic compounds from marine sediments. Science 277, 796–799. https://doi.org/10.1126/science.277.5327.796.

Ezcurra, P., Ezcurra, E., Garcillán, P.P., Costa, M.T., Aburto-Oropeza, O., 2016. Coastal landforms and accumulation of mangrove peat increase carbon sequestration and storage. Proc. Natl. Acad. Sci. U.S.A. 113, 4404–4409. https://doi.org/10.1073/pnas.1519774113.

Fiala, K., Hernández, L., Holub, P., 2017. Comparison of vertical distribution of live and dead fine root biomass in six types of Cuban forests. J. Trop. For. Sci. 29, 275–281.

Geyh, M.A., Streif, H., Kudrass, H.-R., 1979. sea-level changes during the late Pleistocene and holocene in the strait of Malacca. Nature 278, 441–443. https://doi.org/10.1038/278441a0.

Ghosh, A., Misra, S., Dutta, A.K., Choudhury, A., 1985. Pentacyclic triterpenoids and sterols from seven species of mangrove. Phytochemistry 24, 1725–1727. https://doi.org/10.1016/S00319422(00)82541-8.

Gischler, E., 2006. Comment on "Corrected western Atlantic sea-level curve for the last 11,000 years based on calibrated ^{14}C dates from *Acropora palmata* framework and intertidal mangrove peat" by Toscano and Macintyre. Coral Reefs 22:257–270 (2003), and their response in Coral Reefs 24:187–190 (2005). Coral Reefs 25, 273–279. https://doi.org/10.1007/s00338-006-0101-1.

Godwin, H., 1962. Half-life of radiocarbon. Nature 195, 984. https://doi.org/10.1038/195984a0.

Gonneea, M.E., Paytan, A., Herrera-Silveira, J.A., 2004. Tracing organic matter sources and carbon burial in mangrove sediments over the past 160 years. Estuar. Coast Shelf Sci. 61, 211–227. https://doi.org/10.1016/j.ecss.2004.04.015.

Grindrod, J., 1988. The palynology of holocene mangrove and saltmarsh sediments, particularly in Northern Australia. Rev. Palaeobot. Palynol. 55, 229–245. https://doi.org/10.1016/0034-6667(88)90088-7. Quaternary Palynology of Tropical Areas.

Grindrod, J., Rhodes, E.G., 1984. Holocene sea level history of a tropical estuary: Missionary Bay, north Queensland. In: Thom, B.G. (Ed.), Coastal Geomorphology in Australia. Academic Press, Australia, pp. 151–178.

Grossi, V., Caradec, S., Gilbert, F., 2003. Burial and reactivity of sedimentary microalgal lipids in bioturbated Mediterranean coastal sediments. Mar. Chem. 81, 57–69. https://doi.org/10.1016/S0304-4203(02)00139-1.

He, D., Mead, R.N., Belicka, L., Pisani, O., Jaffé, R., 2014. Assessing source contributions to particulate organic matter in a subtropical estuary: a biomarker approach. Org. Geochem. 75, 129–139. https://doi.org/10.1016/j.orggeochem.2014.06.012.

Heaton, T.J., Köhler, P., Butzin, M., Bard, E., Reimer, R.W., Austin, W.E.N., Ramsey, C.B., Grootes, P.M., Hughen, K.A, Kromer, B., Reimer, P.J, Adkins, J., Burke, A., Cook, M.S., Olsen, J., Skinner, L.C., et al., 2020. Marine20 — The Marine Radiocarbon Age Calibration Curve (0–55,000 cal BP). Radiocarbon 62 (4), 779–820. https://doi.org/10.1017/RDC.2020.68.

Hijma, M.P., Engelhart, S.E., Törnqvist, T.E., Horton, B.P., Hu, P., Hill, D.F., 2015. A protocol for a geological sea-level database. In: Handbook of Sea-Level Research. Wiley, pp. 536–553.

Hogg, A.G., Heaton, T.J., Hua, Q., Palmer, J.G., Turney, C.S.M., Southon, J., Bayliss, A., Blackwell, P.G., Boswijk, G., Ramsey, C.B., Pearson, C., Petchey, F., Reimer, P., Reimer, R., Wacker, L., et al., 2020. SHCal20 Southern Hemisphere Calibration, 0–55,000 years cal BP. Radiocarbon 62 (4), 759–778. https://doi.org/10.1017/RDC.2020.59.

Howarth, J.D., Fitzsimons, S.J., Jacobsen, G.E., Vandergoes, M.J., Norris, R.J., 2013. Identifying a reliable target fraction for radiocarbon dating sedimentary records from lakes. Quat. Geochronol. 17, 68–80. https://doi.org/10.1016/j.quageo.2013.02.001.

Huxham, M., Langat, J., Tamooh, F., Kennedy, H., Mencuccini, M., Skov, M.W., Kairo, J., 2010. Decomposition of mangrove roots: effects of location, nutrients, species identity and mix in a Kenyan forest. Estuar. Coast. Shelf Sci. 88, 135–142. https://doi.org/10.1016/j.ecss.2010.03.021.

Jacotot, A., Marchand, C., Rosenheim, B.E., Domack, E.W., Allenbach, M., 2018. Mangrove sediment carbon stocks along an elevation gradient: influence of the late Holocene marine regression (New Caledonia). Mar. Geol. 404, 60–70. https://doi.org/10.1016/j.margeo.2018.07.005.

Jaffé, R., Mead, R., Hernandez, M.E., Peralba, M.C., DiGuida, O.A., 2001. Origin and transport of sedimentary organic matter in two subtropical estuaries: a comparative, biomarker-based study. Org. Geochem. 32, 507–526. https://doi.org/10.1016/S0146-6380(00)00192-3.

Kelleway, J.J., Mazumder, D., Baldock, J.A., Saintilan, N., 2018. Carbon isotope fractionation in the mangrove *Avicennia marina* has implications for food web and blue carbon research. Estuar. Coast Shelf Sci. 205, 68–74. https://doi.org/10.1016/j.ecss.2018.03.011.

Kemp, A.C., Vane, C.H., Horton, B.P., Culver, S.J., 2010. Stable carbon isotopes as potential sea-level indicators in salt marshes, North Carolina, USA. Holocene 20, 623–636. https://doi.org/10.1177/0959683609354302.

Kemp, A.C., Vane, C.H., Khan, N.S., Ellison, J.C., Engelhart, S.E., Horton, B.P., Nikitina, D., Smith, S.R., Rodrigues, L.J., Moyer, R.P., 2019. Testing the utility of geochemical proxies to reconstruct holocene coastal environments and relative sea level: a case study from Hungry Bay, Bermuda. Open Quat. 5, 1. https://doi.org/10.5334/oq.49.

Khan, N.S., Vane, C.H., Engelhart, S.E., Kendrick, C.P., Horton, B.P., 2019. The application of δ13C, TOC and C/N geochemistry of mangrove sediments to reconstruct Holocene paleoenvironments and relative sea levels, Puerto Rico. Mar. Geol. 415, 105963. https://doi.org/10.1016/j.margeo.2019.105963.

Khan, N.S., Vane, C.H., Horton, B.P., Hillier, C., Riding, J.B., Kendrick, C.P., 2015. The application of δ13C, TOC and C/N geochemistry to reconstruct Holocene relative sea levels and paleoenvironments in the Thames Estuary, UK. J. Quat. Sci. 30, 417–433. https://doi.org/10.1002/jqs.2784.

Kilian, M.R., van der Plicht, J., van Geel, B., Goslar, T., 2002. Problematic 14C-AMS dates of pollen concentrates from lake Gosciaz (Poland). Quat. Int. 88, 21–26. https://doi.org/10.1016/S1040-6182(01)00070-2. The value of annually Laminated sediments in palaeoenvironment Re constructions: Dedicated to Bjorn E. Berglund.

Koch, B.P., Harder, J., Lara, R.J., Kattner, G., 2005. The effect of selective microbial degradation on the composition of mangrove derived pentacyclic triterpenols in surface sediments. Org. Geochem. 36, 273–285. https://doi.org/10.1016/j.orggeochem.2004.07.019.

Koch, B.P., Souza Filho, P.W.M., Behling, H., Cohen, M.C.L., Kattner, G., Rullkötter, J., ScholzBöttcher, B., Lara, R.J., 2011. Triterpenols in mangrove sediments as a proxy for organic matter derived from the red mangrove (*Rhizophora mangle*). Org. Geochem. 42, 62–73. https://doi.org/10.1016/j.orggeochem.2010.10.007.

Krauss, K.W., Cahoon, D.R., Allen, J.A., Ewel, K.C., Lynch, J.C., Cormier, N., 2010. Surface elevation change and susceptibility of different mangrove zones to sea-level rise on Pacific high islands of Micronesia. Ecosystems 13, 129–143. https://doi.org/10.1007/s10021-009-9307-8.

Kristensen, E., 2008. Mangrove crabs as ecosystem engineers; with emphasis on sediment processes. J. Sea Res. 59, 30–43. https://doi.org/10.1016/j.seares.2007.05.004. Mangrove Macrobenthos Special IssueProceedings of the Mangrove Macrobenthos Meeting II.

Ladd, S.N., Sachs, J.P., 2015. Influence of salinity on hydrogen isotope fractionation in *Rhizophora mangroves* from Micronesia. Geochem. Cosmochim. Acta 168, 206–221. https://doi.org/10.1016/j.gca.2015.07.004.

Lamb, A.L., Wilson, G.P., Leng, M.J., 2006. A review of coastal palaeoclimate and relative sea-level reconstructions using δ13C and C/N ratios in organic material. Earth-Sci. Rev. 75, 29–57. https://doi.org/10.1016/j.earscirev.2005.10.003. ISOtopes in PALaeoenvironmental reconstruction (ISOPAL).

Larcombe, P., Carter, R.M., Dye, J., Gagan, M.K., Johnson, D.P., 1995. New evidence for episodic post-glacial sea-level rise, central Great Barrier Reef, Australia. Mar. Geol. 127, 1–44. https://doi.org/10.1016/0025-3227(95)00059-8.

Li, C., Li, Y., Burr, G.S., 2014. Testing the accuracy of ^{14}C age data from pollen concentrates in the Yangtze delta, China. Radiocarbon 56, 181–187. https://doi.org/10.2458/56.16920.

Lovelock, C.E., Feller, I.C., Ellis, J., Schwarz, A.M., Hancock, N., Nichols, P., Sorrell, B., 2007. Mangrove growth in New Zealand estuaries: the role of nutrient enrichment at sites with contrasting rates of sedimentation. Oecologia 153, 633–641. https://doi.org/10.1007/s00442-007-0750y.

Mackie, E.A.V., Lloyd, J.M., Leng, M.J., Bentley, M.J., Arrowsmith, C., 2007. Assessment of δ13C and C/N ratios in bulk organic matter as palaeosalinity indicators in Holocene and Lateglacial isolation basin sediments, northwest Scotland. J. Quat. Sci. 22, 579–591. https://doi.org/10.1002/jqs.1081.

Maher, D.T., Santos, I.R., Golsby-Smith, L., Gleeson, J., Eyre, B.D., 2013. Groundwater-derived dissolved inorganic and organic carbon exports from a mangrove tidal creek: the missing mangrove carbon sink? Limnol. Oceanogr. 58, 475–488. https://doi.org/10.4319/lo.2013.58.2.0475.

McKee, K.L., Cahoon, D.R., Feller, I., 2007. Caribbean mangroves adjust to rising sea level through biotic controls on soil elevation change. Global Ecol. Biogeogr. 16, 545–556. https://doi.org/10.1111/j.1466-8238.2007.00317.x.

McKee, K.L., Faulkner, P.L., 2000. Mangrove peat analysis and reconstruction of vegetation history at the Pelican Cays, Belize. Atoll Res. Bull. 12.

McKee, K.L., Feller, I.C., Popp, M., Wanek, W., 2002. Mangrove isotopic (δ15N AND δ13C) fractionation across a nitrogen vs. phosphorus limitation gradient. Ecology 83, 1065–1075. https://doi.org/10.1890/0012-9658(2002)083[1065:MINACF]2.0.CO;2.

Middleton, B.A., McKee, K.L., 2001. Degradation of mangrove tissues and implications for peat formation in Belizean island forests. J. Ecol. 89, 818–828. https://doi.org/10.1046/j.0022-0477.2001.00602.x.

Mollenhauer, G., Inthorn, M., Vogt, T., Zabel, M., Damsté, J.S.S., Eglinton, T.I., 2007. Aging of marine organic matter during cross-shelf lateral transport in the Benguela upwelling system revealed by compound-specific radiocarbon dating. Geochem. Geophys. Geosyst. 8 https://doi.org/10.1029/2007GC001603.

Monacci, N.M., Meier-Grünhagen, U., Finney, B.P., Behling, H., Wooller, M.J., 2009. Mangrove ecosystem changes during the holocene at Spanish lookout Cay, Belize. Palaeogeogr. Palaeoclimatol. Palaeoecol. 280, 37–46. https://doi.org/10.1016/j.palaeo.2009.05.013.

Muller, R.A., 1977. Radioisotope dating with a cyclotron. Science 196, 489–494.

Neulieb, T., Levac, E., Southon, J., Lewis, M., Pendea, I.F., Chmura, G.L., 2013. Potential pitfalls of pollen dating. Radiocarbon 55, 1142–1155. https://doi.org/10.1017/S0033822200048050.

Newnham, R.M., Vandergoes, M.J., Garnett, M.H., Lowe, D.J., Prior, C., Almond, P.C., 2007. Test of AMS ^{14}C dating of pollen concentrates using tephrochronology. J. Quat. Sci. 22, 37–51. https://doi.org/10.1002/jqs.1016.

Ohkouchi, N., Eglinton, T.I., 2008. Compound-specific radiocarbon dating of Ross Sea sediments: a prospect for constructing chronologies in high-latitude oceanic sediments. Quat. Geochronol. 3, 235–243. https://doi.org/10.1016/j.quageo.2007.11.001. Prospects for the New Frontiers of earth and Environmental Sciences.

Olsson, I.U., 2009. Radiocarbon dating history: early days, questions, and problems met. Radiocarbon 51, 1–43. https://doi.org/10.1017/S0033822200033695.

Olsson, I.U., 1991. Accuracy and precision in sediment chronology. Hydrobiologia 214, 25–34. https://doi.org/10.1007/BF00050928.

Pan, J., Chen, Y., Wang, Y., Zhou, Z., Li, M., 2019. Vertical distribution of Bathyarchaeota communities in mangrove wetlands suggests distinct Niche preference of Bathyarchaeota subgroup 6. Microb. Ecol. 77, 417–428. https://doi.org/10.1007/s00248-018-1309-7.

Pancost, R.D., Geel, B., Baas, M., Damsté, J.S.S., 2000. δ13C values and radiocarbon dates of microbial biomarkers as tracers for carbon recycling in peat deposits. Geology 28, 663–666. https://doi.org/10.1130/0091-7613(2000)28<663:CVARDO>2.0.CO;2.

Perry, C.T., Berkeley, A., Smithers, S.G., 2008. Microfacies characteristics of a tropical, Mangrove Fringed Shoreline, Cleveland Bay, Queensland, Australia: sedimentary and taphonomic controls on mangrove facies development. J. Sediment. Res. 78, 77–97. https://doi.org/10.2110/jsr.2008.015.

Punwong, P., Marchant, R., Selby, K., 2013a. Holocene mangrove dynamics from Unguja Ukuu, Zanzibar. Quat. Int. 298, 4–19. https://doi.org/10.1016/j.quaint.2013.02.007. Zanzibar to the Yellow Sea: A transect of Quaternary Studies from 6°S, 39°E to 35°N, 127°E.

Punwong, P., Marchant, R., Selby, K., 2013b. Holocene mangrove dynamics in Makoba Bay, Zanzibar. Palaeogeogr. Palaeoclimatol. Palaeoecol. 379–380, 54–67. https://doi.org/10.1016/j.palaeo.2013.04.004.

Reef, R., Feller, I.C., Lovelock, C.E., 2010. Nutrition of mangroves. Tree Physiol. 30, 1148–1160. https://doi.org/10.1093/treephys/tpq048.

Reich, P.B., Walters, M.B., Ellsworth, D.S., 1992. Leaf life-span in relation to leaf, plant, and stand characteristics among diverse ecosystems. Ecol. Monogr. 62, 365–392. https://doi.org/10.2307/2937116.

Reimer, P.J., Austin, W.E.N., Bard, E., Bayliss, A., Blackwell, P.G., Ramsey, C.B., Butzin, M., Cheng, H., Edwards, R.L., Friedrich, M., Grootes, P.M., Guilderson, T.P., Hajdas, I., Heaton, T.J., Hogg, A.G., Hughen, K.A., Kromer, B., Manning, S.W., Muscheler, R., Palmer, J.G., Pearson, C.G., van der Plicht, J., Reimer, R.W., Richards, D.A., Scott, E.M., Southon, J.R., Turney, C.S.M., Wacker, L., Adolphi, F., Büntgen, U., Capano, M., Fahrni M, S., Fogtmann-Schulz, A., Friedrich, R., Köhler, P., Kudsk, S., Miyake, F., Olsen, J., Reinig, F., Sakamoto, M., Sookdeo, A., Talamo, S., et al., 2020. The IntCal20 Northern Hemisphere Radiocabon Age Calibration Curve (0–55 cal kBP). Radiocabon 62 (4), 725–757. https://doi.org/10.1017/RDC.2020.41.

Reimer, P.J., Reimer, R.W., 2001. A marine reservoir correction database and on-line interface. Radiocarbon 43, 461–463. https://doi.org/10.1017/S0033822200038339.

Revelle, R., Suess, H.E., 1957. Carbon dioxide exchange between atmosphere and ocean and the question of an increase of atmospheric CO_2 during the past decades. Tellus 9, 18–27. https://doi.org/10.1111/j.2153-3490.1957.tb01849.x.

Rogers, K., Kelleway, J.J., Saintilan, N., Megonigal, J.P., Adams, J.B., Holmquist, J.R., Lu, M., SchileBeers, L., Zawadzki, A., Mazumder, D., Woodroffe, C.D., 2019. Wetland carbon storage controlled by millennial-scale variation in relative sea-level rise. Nature 567, 91. https://doi.org/10.1038/s41586-019-0951-7.

Schiffer, M.B., 1986. Radiocarbon dating and the "old wood" problem: the case of the Hohokam chronology. J. Archaeol. Sci. 13, 13–30. https://doi.org/10.1016/03054403(86)90024-5.

Scholl, D.W., 1964a. Recent sedimentary record in mangrove swamps and rise in sea level over the southwestern coast of Florida: Part 1. Mar. Geol. 1, 344–366. https://doi.org/10.1016/0025-3227(64)90020-9.

Scholl, D.W., 1964b. Recent sedimentary record in mangrove swamps and rise in sea level over the southwestern coast of Florida: part 2. Mar. Geol. 2, 343–364. https://doi.org/10.1016/0025-3227(64)90047-7.

Scott, E.M., Cook, G.T., Naysmith, P., 2010. The fifth international radiocarbon intercomparison (VIRI): an assessment of laboratory performance in stage 3. Radiocarbon 52, 859–865.

Sen, A., Bhadury, P., 2017. Intertidal foraminifera and stable isotope geochemistry from world's largest mangrove, Sundarbans: assessing a multiproxy approach for studying changes in sealevel. Estuar. Coast. Shelf Sci. 192, 128–136. https://doi.org/10.1016/j.ecss.2017.05.010.

Skov, M.W., Hartnoll, R.G., 2002. Paradoxical selective feeding on a low-nutrient diet: why do mangrove crabs eat leaves? Oecologia 131, 1–7. https://doi.org/10.1007/s00442-001-0847-7.

Smittenberg, R.H., Hopmans, E.C., Schouten, S., Hayes, J.M., Eglinton, T.I., Sinninghe Damsté, J.S., 2004. Compound-specific radiocarbon dating of the varved Holocene sedimentary record of Saanich Inlet, Canada. Paleoceanography 19. https://doi.org/10.1029/2003PA000927.

Stieglitz, T., Ridd, P., Müller, P., 2000. Passive irrigation and functional morphology of crustacean burrows in a tropical mangrove swamp. Hydrobiologia 421, 69–76. https://doi.org/10.1023/A:1003925502665.

Stuiver, M., 1961. Variations in radiocarbon concentration and sunspot activity. J. Geophys. Res. 66, 273–276. https://doi.org/10.1029/JZ066i001p00273.

Stuiver, M., Pearson, G.W., Braziunas, T., 1986. Radiocarbon age calibration of marine samples back to 9000 cal yr BP. Radiocarbon 28, 980–1021. https://doi.org/10.1017/S0033822200060264.

Stuiver, M., Polach, H.A., 1977. Discussion reporting of ^{14}C data. Radiocarbon 19, 355–363. https://doi.org/10.1017/S0033822200003672.

Suárez, N., Medina, E., 2005. Salinity effect on plant growth and leaf demography of the mangrove, *Avicennia germinans* L. Trees 19, 722. https://doi.org/10.1007/s00468-005-0001-y.

Swales, A., Bentley, S.J., Lovelock, C.E., 2015. Mangrove-forest evolution in a sediment-rich estuarine system: opportunists or agents of geomorphic change? Earth Surf. Process. Landforms 40, 1672–1687. https://doi.org/10.1002/esp.3759.

Taillardat, P., Ziegler, A.D., Friess, D.A., Widory, D., Truong Van, V., David, F., Thành-Nho, N., Marchand, C., 2018. Carbon dynamics and inconstant porewater input in a mangrove tidal creek over contrasting seasons and tidal amplitudes. Geochem. Cosmochim. Acta 237, 32–48. https://doi.org/10.1016/j.gca.2018.06.012.

Tamura, T., Saito, Y., Sieng, S., Ben, B., Kong, M., Sim, I., Choup, S., Akiba, F., 2009. Initiation of the Mekong River delta at 8 ka: evidence from the sedimentary succession in the Cambodian lowland. Special Theme: Modern Analogues in Quaternary Palaeoglaciological Reconstruction Quat. Sci. Rev. 28, 181–260. https://doi.org/10.1016/j.quascirev.2008.10.010, 327–344.

Taylor, R.E., 2000. Fifty years of radiocarbon dating: this widely applied technique has made major strides since its introduction a half-century ago at the University of Chicago. Am. Sci. 88, 60–67.

Tomlinson, P.B., 2016. The Botany of Mangroves. Cambridge University Press.

Törnqvist, T.E., Rosenheim, B.E., Hu, P., Fernandez, A.B., 2015. Radiocarbon dating and calibration. In: Shennan, I., Long, A.J., Horton, B.P. (Eds.), Handbook of Sea-Level Research. Wiley, pp. 349–360.

Tue, N.T., Hamaoka, H., Sogabe, A., Quy, T.D., Nhuan, M.T., Omori, K., 2011. The application of δ13C and C/N ratios as indicators of organic carbon sources and paleoenvironmental change of the mangrove ecosystem from Ba Lat Estuary, Red River, Vietnam. Environ. Earth Sci. 64, 1475–1486. https://doi.org/10.1007/s12665-011-0970-7.

Turetsky, M.R., Manning, S.W., Wieder, K., 2004. Dating recent peat deposits. Wetlands 24, 324–356. https://doi.org/10.1672/0277-5212(2004)024[0324:DRPD]2.0.CO;2.

Vandergoes, M.J., Prior, C.A., 2003. AMS dating of pollen concentrates—a methodological study of late quaternary sediments from South Westland, New Zealand. Radiocarbon 45, 479–491. https://doi.org/10.1017/S0033822200032823.

Vane, C.H., Kim, A.W., Moss-Hayes, V., Snape, C.E., Diaz, M.C., Khan, N.S., Engelhart, S.E., Horton, B.P., 2013. Degradation of mangrove tissues by arboreal termites (*Nasutitermes acajutlae*) and their role in the mangrove C cycle (Puerto Rico): chemical characterization and organic matter provenance using bulk δ13C, C/N, alkaline CuO oxidation-GC/MS, and solid-state ^{13}C NMR. G-cubed 14, 3176–3191. https://doi.org/10.1002/ggge.20194.

Versteegh, G.J.M., Schefuß, E., Dupont, L., Marret, F., Sinninghe Damsté, J.S., Jansen, J.H.F., 2004. Taraxerol and Rhizophora pollen as proxies for tracking past mangrove ecosystems 1. Geochem. Cosmochim. Acta 68, 411–422. https://doi.org/10.1016/S00167037(03)00456-3.

Waelbroeck, C., Duplessy, J.C., Michel, E., Labeyrie, L., Paillard, D., Duprat, J., 2001. The timing of the last deglaciation in North Atlantic climate records. Nature 412, 724–727. https://doi.org/10.1038/35089060.

Wilson, G.P., 2017. On the application of contemporary bulk sediment organic carbon isotope and geochemical datasets for Holocene sea-level reconstruction in NW Europe. Geochem. Cosmochim. Acta 214, 191–208. https://doi.org/10.1016/j.gca.2017.07.038.

Woodroffe, C.D., 1992. Mangrove sediments and geomorphology. In: Robertson, A.I., Alongi, D.M. (Eds.), Tropical Mangrove Ecosystems. American Geophysical Union, pp. 7–41. https://doi.org/10.1029/CE041p0007.

Woodroffe, C.D., 1990. The impact of sea-level rise on mangrove shorelines. Prog. Phys. Geogr. 14, 483–520. https://doi.org/10.1177/030913339001400404.

Woodroffe, C.D., 1981. Mangrove swamp stratigraphy and holocene transgression, Grand Cayman island, west Indies. Mar. Geol. 41, 271–294. https://doi.org/10.1016/00253227(81)90085-2.

Woodroffe, C.D., Rogers, K., McKee, K.L., Lovelock, C.E., Mendelssohn, I.A., Saintilan, N., 2016. Mangrove sedimentation and response to relative sea-level rise. Ann. Rev. Mar. Sci. 8, 243–266. https://doi.org/10.1146/annurev-marine-122414-034025.

Woodroffe, S.A., Long, A.J., Milne, G.A., Bryant, C.L., Thomas, A.L., 2015a. New constraints on late Holocene eustatic sea-level changes from Mahé, Seychelles. Quat. Sci. Rev. 115, 1–16. https://doi.org/10.1016/j.quascirev.2015.02.011.

Woodroffe, S.A., Long, A.J., Punwong, P., Selby, K., Bryant, C.L., Marchant, R., 2015b. Radiocarbon dating of mangrove sediments to constrain Holocene relative sea-level change on Zanzibar in the southwest Indian Ocean. Holocene 25, 820–831. https://doi.org/10.1177/0959683615571422.

Wooller, M.J., Morgan, R., Fowell, S., Behling, H., Fogel, M., 2007. A multiproxy peat record of Holocene mangrove palaeoecology from Twin Cays, Belize. Holocene 17, 1129–1139. https://doi.org/10.1177/0959683607082553.

Wooller, M.J., Smallwood, B.J., Scharler, U., Jacobson, M., Fogel, M., 2003. A taphonomic study of δ13C and δ15N values in *Rhizophora mangle* leaves for a multi-proxy approach to mangrove palaeoecology. Org. Geochem. 34, 1259–1275. https://doi.org/10.1016/S01466380(03)00116-5.

Worbes, M., Junk, W.J., 1989. Dating tropical trees by means of 14C from bomb tests. Ecology 70, 503–507. https://doi.org/10.2307/1937554.

Zhang, J.F., Xu, B., Turner, F., Zhou, L., Gao, P., Lü, X., Nesje, A., 2017. Long-term glacier melt fluctuations over the past 2500 yr in monsoonal high Asia revealed by radiocarbon-dated lacustrine pollen concentrates. Geology. G38690.1. https://doi.org/10.1130/G38690.1.

CHAPTER

9

Australian mangroves through the Holocene: interactions between sea level, mangrove extent, and carbon sequestration

Neil Saintilan[1], *Jeff Kelleway*[2], *Kerrylee Rogers*[2]

[1]Department of Environmental Sciences, Macquarie University, NSW, Australia; [2]School of Earth, Atmospheric and Life Sciences, University of Wollongong, NSW, Australia

OUTLINE

1. Introduction 218
2. Early phases of the postglacial marine transgression: 13,000–9000 BP 219
3. Late phases of the postglacial marine transgression: 9000–6000 BP 221
4. Estuary infill in the mid-late Holocene 224
5. Contemporary patterns of mangrove–saltmarsh interaction 226
6. 21st century sea-level rise acceleration and the challenges of coastal planning 227
7. Conclusions 230

References 230

Dynamic Sedimentary Environments of Mangrove Coasts
https://doi.org/10.1016/B978-0-12-816437-2.00009-4

1. Introduction

The Australian continent supports one-third of the world's mangroves (Spalding, 2010). Lying close to the centers of mangrove biodiversity in Southeast (SE) Asia, species diversity in the tropical north is high, with up to 28 species recorded in a single estuary, the Daintree (16° 16′S; 145° 24′E). Diversity and extent decrease along latitudinal and aridity gradients, with sporadic representation in the estuaries of south-west Western Australia and sympatric association on the SE coast with increasingly diverse saltmarsh south to Corner Inlet in Victoria (38° 54′S; 146° 18′E). Rhizophoraceae extend into northern New South Wales (NSW) to a latitudinal limit of *Rhizophora stylosa* at South West Rocks Creek (30° 53′S; 153° 02′E: (Wilson and Saintilan, 2012). *Aegiceras corniculatum* extends south to Lake Merimbula (36.88°S, 149.90°E), though is not as common SE Australia as *Avicennia marina*. Mangroves are not found in Tasmania (Rogers and Krauss, 2018).

Australian mangroves are largely protected from development by state government policy and legislation (Rogers et al., 2016) in place in most jurisdictions for three decades or more, though early losses to industrial, agricultural, and transport development were locally significant (Harty, 2004). Since the 1970s, legislative and policy direction has been influenced by research identifying links between mangrove habitat and fisheries productivity, including mud crab across the tropical north (Manson et al., 2005), wild prawn fisheries in the Gulf of Carpentaria (Staples et al., 1986) and SE Queensland (Skilleter et al., 2005), and crustacean and fin fish in NSW (Saintilan, 2004). The nursery role of mangrove and saltmarsh and their significance in providing a nutritional subsidy to estuarine fish has been supported by studies of trophic ecology (Guest and Connolly, 2005; Mazumder et al., 2006; Saintilan and Mazumder, 2017). More recently, the carbon sequestration capacity of tidal wetlands (termed "blue carbon") has generated interest in ecosystem service provision by mangrove and saltmarsh, while opening up new avenues for restoration and management (Kelleway et al., 2017a).

Sea-level rise is now recognized as the most significant emerging threat to Australian mangroves. Projections of sea-level rises of up to 80 cm by 2100 have prompted studies exploring physiological and landscape-scale responses of mangroves to sea-level rise. Rates of sedimentation and vertical accretion in mangroves have been informed by Surface Elevation Tables (SETs) installed in SE Australia from the early 2000s. These have allowed an assessment of relationships between sedimentation, autocompaction, and surface elevation gain in mangroves across a range of geomorphological settings, including estuarine embayments in Victoria (Rogers et al., 2005), drowned river valleys and barrier estuaries in NSW (Rogers et al. 2006, 2014), and back-barrier marine and coastal terrigenous settings in SE Queensland (Lovelock et al., 2014). Australian data have also contributed to regional assessments of drivers of mangrove vertical accretion and surface elevation gain in relation to sea-level rise (Lovelock et al., 2015; Rogers et al., 2006).

Data on rates of vertical accretion derived from the SET network have been important in informing models of mangrove response to sea-level rise projected to 2100 under a range of IPCC scenarios. Rogers et al. (2012) developed a dynamic elevation model relating intertidal wetland extent in the Hunter estuary to high- and low-range IPCC sea-level scenarios, based on SET data for the estuary. The model demonstrated the opportunities for mangrove expansion under low rates of relative sea-level rise, as well as the likely severe reductions in extent under the higher end of sea-level rise to 2100 projected by the IPCC. Similarly, Traill et al. (2011) used the Sea Level Affecting Marshes Model (SLAMM) to explore the fate of wetlands under IPCC B1 and A1 scenarios, finding extensive opportunities for landward mangrove expansion. SET data have also been incorporated into efforts seeking to compare and validate models (Oliver et al., 2012; Mogensen and Rogers, 2018), and this work is likely to be ongoing, given the need to understand and project the response of mangrove to anticipated sea level rise.

Dynamic elevation models that incorporate feedback between accretion, surface elevation adjustment, and sea-level rise are becoming an important coastal planning tool as they can more effectively project the response of mangrove to sea-level rise than simpler bath fill approaches (Rogers et al., 2012). However, projections concerning the fate of mangroves under the higher range of projected sea-level rise are constrained by our understanding of the upper limits of mangrove vertical accretion that can be sustained under the pressure of rapid sea-level rise. Over the previous century, the rate of sea-level rise in eastern Australia has accelerated from ~2.0 mm yr^{-1} (averaged from 1966 to 2009) to ~3.4 mm yr^{-1} (averaged from 1990 to 2009: White et al., 2014). In most Australian settings for which SET data are available, mangrove vertical accretion has been equivalent to local sea-level trends, though falling short of projected rates of change over the coming century (e.g., Rogers et al., 2012). Higher rates of vertical accretion in mangrove have been recorded across SE Asia (Lovelock et al., 2015), but the limited duration of the record (often <5 years) gives limited insight into whether these rates can be sustained over decades. With high-emission IPCC scenarios suggesting rates of sea-level rise climbing to 10 mm yr^{-1} by the end of the century, an understanding of mangrove responses to sustained higher rates of sea-level rise are required.

In this chapter we explore mangrove responses to sea level on the Australian continent since the last glacial maxima (~22,000 calibrated years BP, henceforth BP). This period incorporates phases of very rapid sea-level rise in which mangroves over vast areas were unable to survive. In other periods, mangroves were able to sustain their position in situ for centuries while rates of sea-level rise were substantially higher than those encountered by the contemporary SET record. We review these data with a view to determine upper thresholds of mangrove survival and the possible biological and sedimentological controls influencing these observations. In particular, the role of mangrove root system production in mediating the response to sea-level rise is discussed in the context of carbon sequestration and the potential of negative feedback on radiative forcing.

2. Early phases of the postglacial marine transgression: 13,000–9000 BP

Glacial low sea-level deposits (14,000–21,000 calibrated years BP) on the Queensland east coast continental shelf (~17° S; 146° E) show very little mangrove development, and Amaranthaceae/Chenopodiaceae has the highest representation in a 60,000-year pollen record (Moss et al., 2017). The prevalence of saltmarsh on a coastline now dominated by tropical mangrove forest may suggest that the cooler, drier climatic conditions of the last glacial favored saltmarsh development over mangrove. The declining sea level prior to 17,000 years BP may also have restricted the habitat available to mangrove, increasing the proportion of estuarine habitat in the upper intertidal favored by saltmarsh (Fig. 9.1).

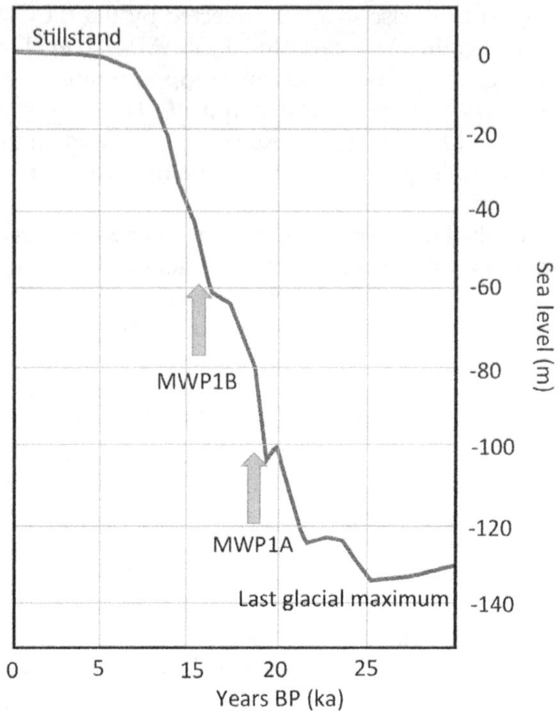

FIGURE 9.1 Postglacial eustatic sea-level reconstruction, indicating the timing of meltwater pulses 1A (MWP1A) and meltwater pulse 1B (MWP1B). *After Lambeck, K., Rouby, H., Purcell, A., Sun, Y., Sambridge, M., 2014. sea Level and global ice volumes from the last glacial maximum to the Holocene. Proc. Natl. Acad. Sci. U.S.A., 111 (43), 15296–15303.*

Following the end of the glacial maxima, mangroves began to expand with transgressing sea levels. Late Pleistocene mangrove sediments are preserved across large areas of the Timor Sea (Sahul Shelf) adjacent to the cost of North West Australia, dating to ~15,000–16,000 BP (Nicholas et al., 2014). Sea-level rise associated with the transition from glacial to interglacial conditions was nonlinear, punctuated with periods of rapid sea-level rise associated with ice-sheet failure. The first of these, meltwater pulse 1A (MWP1A), started no earlier than 14,650 BP and ended before 14,310 BP, rising most probably 14–18 m, a rate of 40 mm yr^{-1} on average, though this rate was probably temporally variable, with associated pulses of higher rates. This effectively drowned all mangroves in situ on the continental shelf. The paleochannels and floodplains associated with this early transgressive "big swamp" phase are largely preserved, the very rapid sea-level rise leaving their morphology in place (Fig. 9.2).

The flooding of the Queensland coastline during this early, rapid phase of marine transgression led ultimately to a sharp increase in the representation of mangrove pollen delivered to the shelf (Moss et al., 2017), particularly *Rhizophora* and *Ceriops/Bruguiera*, as these retreated into former Cyperaceae (freshwater) swamps. There is no evidence of rainfall having increased at this stage, and saltmarsh is still well represented to 13,000 BP.

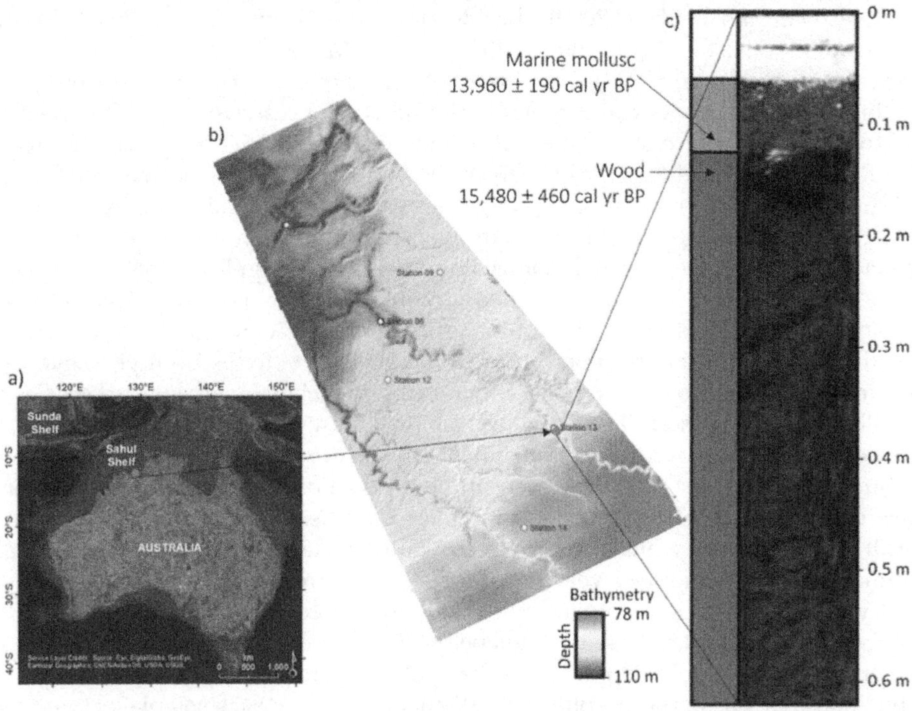

FIGURE 9.2 Abandoned estuarine habitats on the (A) Sahul Shelf, NW Australia, following marine transgression (B) now at bathymetric depths of 78–110 m, and (C) associated late-Pleistocene mangrove peats. *Adapted from Nicholas, W.A., Nichol, S.L., Howard, F.J.F., Picard, K., Dulfer, H., Radke, L.C., et al., 2014. Pockmark development in the Petrel sub-basin, Timor Sea, northern Australia: seabed habitat mapping in support of CO2 storage assessments. Continent. Shelf Res., 83, 129–142.*

The Arafura Sill was breached ~12,000 BP with full marine conditions attained in the former Lake Carpentaria 10,500 yr BP (Harris et al., 2008), and mangrove pollen dating to this time has been identified from cores extracted from the current Gulf of Carpentaria (Chivas et al., 2001). Sea level in the Gulf of Carpentaria coast reached current position around 8000 yr BP though continued to rise and may have reached a maxima of up to 2 m above current sea levels by 7400 BP, though there is regional variability and conflicting evidence regarding this highstand (Lewis et al., 2013). The stabilization of sea levels mid-Holocene set the scene for the development of the most extensive mangrove forests of the Holocene.

3. Late phases of the postglacial marine transgression: 9000–6000 BP

Holocene mangrove peat deposits have been dated to occur in many tropical localities between 10,000 and 7800 BP and their extent is pantropical (Saintilan et al., 2020). Radiocarbon basal dates from this period have been returned from thick mangrove organic soils far separated as the Ivory Coast of Africa (Tossou et al., 2008), Zanzibar (Punwong et al., 2018), the

Ganges delta (Hait and Behling, 2009), the Mekong delta (Tamura et al., 2009), Makassar Strait (Yulianto et al., 2004; Papua Ellison 2005), Belize (Monacci et al., 2011), and the delta of the Amazon (Cohen et al., 2012). The timing of the commencement of mangrove organic soil-building across the tropics corresponds to the slowing of sea-level rise at the concluding stages of the marine transgression. Mangroves had not been locally extirpated during rapid phases of Relative Sea Level Rise (RSLR), given their widespread expansion as sea levels stabilized. During very rapid rates of rise, mangroves can translate landwards by recruiting at high elevations in the intertidal and maintain their higher position with respect to sea level over sufficient time to grow to sexual maturity and deliver propagules to the new higher position of the tidal frame, before being drowned. Accordingly, there is, under this model, a successive role over of the next generation of mangrove to higher elevation.

By the time sea-level rise decelerated from 10 to 20 mm yr^{-1} to 5–7 mm yr^{-1}, mangroves were able to sustain their position in relation to mean sea level by increasing surface elevation in situ. Pollen from the genera *Rhizophora* and *Sonneratia* are particularly prevalent in transgressive sequences. Mangrove in situ adjustment was achieved by the trapping of sediment and the building of root mass, which, even in the absence of terrigenous input, can sustain mangrove elevation gain of up to 5 mm yr^{-1} over centuries (McKee et al., 2007; Saintilan et al., 2020) on coral reefs in which mangroves grow thick peats subsidized by inputs of calcareous sediments. Decelerating sea-level rise occurred in most tropical localities circa 9000–8000 calibrated years BP, and extensive pantropical mangrove forest development coincides with this time, as the rate of sea-level rise declined below 7 mm yr^{-1} (Saintilan et al., 2020). In tropical northern Australia, black soil plains are drained by macrotidal rivers, flooded seasonally during the monsoon season (Woodroffe et al., 1989). Beneath the black soil plains lie extensive mangrove sediments several meters thick. These have been located in King Sound, the Fitzroy River (Jennings, 1975), the Ord River (Thom et al., 1975), and the Daly and Alligator Rivers (Woodroffe et al., 1985). Basal deposits date from ~8100 BP in the Mary River, Northern Territory (Woodroffe, 2018), 8000–7500 BP in the South Alligator River, Northern Territory (Woodroffe et al., 1987), 7500–8000 BP in the Fitzroy River (Jennings, 1975), and 7500 BP in the Ord River, Western Australia (Thom et al., 1975).

By 5000 yr BP these extensive mangrove forests had largely disappeared and the wide upper intertidal and supratidal flats were bare in lower rainfall areas or occupied by sparse grasses, sedges, and, in wetter areas, *Melaleuca* forest (Woodroffe et al., 1989). Mangroves are currently confined to active depositional environments, including channel banks, the edges of mid-channel islands, and prograding coastlines (Woodroffe et al., 1985). Jennings (1975) suggested rainfall as an explanation of the rise and fall of the great early-Holocene mangrove forests, with the Holocene climate optimum associated with higher rainfall across the region. Certainly, the proportion of mangrove and saltmarsh/hypersaline flats in northern Australian estuaries is associated with rainfall; more arid estuaries have lower proportional mangrove cover compared to humid estuaries (Bucher and Saenger, 1991). However, the confinement of mangroves to fringing environments is more likely to be related to geomorphological evolution of the estuaries through the Holocene, as proposed by Woodroffe et al. (1985). With the stabilization of sea level shortly after 8000 yr BP, sedimentation continued and led to the progressive shrinkage of accommodation space, the vertical space into which mangroves can accrete in the intertidal. As estuaries infilled, the intertidal zone was replaced by supratidal flats inundated by seasonal monsoons and cyclonic storm surges. In dryer areas, mangroves gave way to hypersaline flats, while in wetter areas, upper intertidal and supratidal environments were colonized by salt-tolerant grasses and sedges.

On the east coast of tropical Queensland, extensive mangrove sediments dating to ~8000 yr BP have been found in Missionary Bay and Princess Charlotte Bay (Grindrod 1985, 1988), on the Mulgrave River near Innisfail (17° S; 146° E; Crowley et al., 1990), and in the nearby Russell River (Crowley and Gagan, 1995). These forests show a remarkable rate of vertical accretion (approximately 5–6 mm yr^{-1}) sustained for 800–900 years between 7900 and 7000 years BP. These forests were dominated by *Rhizophora*, though there is a noticeable shift to *Ceriops* and *Bruguiera* toward the end of the big swamp phase, indicating a transition toward conditions higher in the tidal frame prior to their replacement by terrestrial and freshwater swamp vegetation by the mid-Holocene.

The southernmost early-Holocene mangrove "big swamp" is situated several kilometers upstream of the current intertidal wetlands on the Richmond River, Northern NSW (29°S; 150.5°E: Hashimoto et al., 2006, Fig. 9.2). These organic silts extend to 3 m depth beneath the current floodplain, which is now used for agriculture. Between ~7800 and 7000 BP, the mangroves sustained a rate of in situ elevation gain of ~3 mm yr^{-1}, before being replaced by supratidal forests of *Casuarina* and *Melaleuca*. Pollen analysis suggests that the mangrove was dominated by Rhizophoraceae (*Rhizophora* and *Bruguiera*) in spite of the rarity of this family on the Richmond River today. Only a few individuals of either species can be found at the mouth of the estuary, among mangrove forests dominated by *Avicennia marina* and *Aegiceras corniculatum*.

It is interesting to speculate why the mid-Holocene mangrove assemblage on the Richmond River was dominated by Rhizophoraceae. Deep mangrove sediments across the globe are almost always associated with *Rhizophora*, suggesting that this genus may have significant capacity to increase elevation in situ and thereby dominate large transgressive forests under conditions of high relative sea-level rise. However, *Rhizophora* is a more prolific pollen producer than *Avicennia* (Crowley et al., 1994), and *Avicennia* pollen is not noted for its preservation, as wet–dry cycles in the upper intertidal can lead to oxidation and humic acids promote degradation, which may partly explain the rarity of *Avicennia* representation in transgressive mangrove facies. Temperatures may have been higher in the region during 8200–8000 BP, a time known as the "Holocene climatic optimum," facilitating the expansion of two species nearing their poleward limit. There is evidence of range expansion and strong population growth of *R. stylosa* over the past few decades in the northern rivers of NSW (Wilson, 2009) associated with higher temperatures.

In spite of extensive stratigraphic investigations across the temperate estuaries of SE Australia, no transgressive mangrove deposits have been discovered. Forests of *Avicennia marina* are well developed in tidal estuaries throughout NSW and Victoria, but estuarine facies dating to 7000–8000 BP are primarily subtidal (e.g., Umitsu et al., 2001), in common with the North Island of New Zealand (Woods, 2012). Several possibilities exist, including a late-Holocene expansion in the population of *Avicennia*, the presence of *Avicennia* though limited pollen and/or organic preservation, or a limited capacity of *Avicennia* to sustain position in situ under long-term transgressive phases.

4. Estuary infill in the mid-late Holocene

Sea-level stabilization has facilitated the infill of estuaries and recession of mangroves toward the present coastline. The extent to which this has occurred is a function of sediment yield, the size of the estuary accommodation space, and the retention of sediment within the estuary. In the relatively small catchments east of the Great Dividing Range, sediment yield varies in relation to catchment size and the resistance of soils to erosion and catchment geology to weathering. In catchments with low sediment yield, estuaries may still be in an "immature" phase of sediment infill to the present day (Roy et al., 2001), while others may be infilled, channelized to the mouth, and delivering sediment to the nearshore. Compared to the estuaries of West Coast USA, east coast Australian estuaries are skewed toward the immature stage of development, though estuaries with larger catchments are more likely to have evolved to maturity (Saintilan et al., 2016). Most estuaries have a predominance of open water habitat compared to lower intertidal unvegetated and upper intertidal vegetated habitat, which is a function of low sediment yield. In these estuaries, mangrove and saltmarsh is less represented, and extensive seagrass beds occur in shallow estuarine waters (Saintilan, 2004). Where estuary entrances close periodically, mangrove may be excluded. Mangroves are best represented in estuaries of intermediate to mature stages of infill and with open entrances (Saintilan, 2004).

The extent of mangrove development in mature estuaries is also a function of the relative importance of tides, river flow, and ocean waves in the formation of the estuary entrance conditions. In macrotidal estuaries, strong tidal currents create wide funnel-shaped estuarine mouths supporting extensive intertidal environments. In the high-wave energy coastlines characteristic of the southern half of the continent, sand barriers often restrict tidal exchange and truncate the development of flood-tidal estuarine deposits that create suitable conditions for intertidal vegetation, and where river flows are small, entrances may periodically close. As outlined above, the channelization of estuaries and gradual reduction in accommodation space has occurred to varying degrees.

These patterns of estuarine infill explain the historic and current development of mangroves. As suggested in Fig. 9.3, the progradational phase of estuary development leaves behind mangrove organic facies dating from the mid- to late Holocene, overlain by fluvial deltaic silts. Working in the middle reaches of the Hawkesbury River estuary in NSW (Fig. 9.4), Saintilan and Hashimoto (1999) found mangrove organic sediments 20–30 cm beneath the present-day wetland surface dating to 1200–1700 yr BP. These are approximately at the same elevation as contemporary mangrove root systems, implying that sea levels had stabilized by this time. The contemporary saltmarsh consisted of an assemblage of *Sporobolus virginicus*, *Juncus kraussii*, and *Casuarina glauca*. The stratigraphy suggests that infill has led to the gradual replacement of mangrove with saltmarsh as the river prograded within the sandstone bedrock valley (Saintilan, 1997). Further south, a similar pattern of mangrove replacement by saltmarsh over the late Holocene was identified for Currambene Creek and Cararma Inlet, small tributary creeks draining into Jervis Bay (Saintilan and Wilton, 2001). Within Currambene Creek, mangrove sediments dated to 1300 yr BP, while those in Carama Inlet dated to 1900 yr BP. Both were overlying subtidal sands. A similar scenario has also been observed further north in the drowned river valley of Port Stephens, where mangrove sediments below

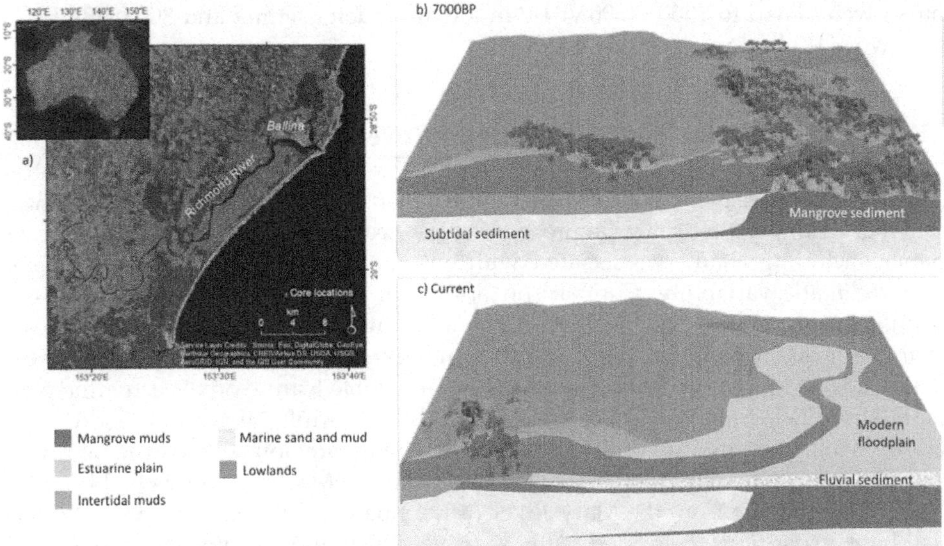

FIGURE 9.3 Widespread Rhizophoraceae mangrove cover identified on the (A) Richmond River at (B) ~7000 yr BP before (C) estuarine infill limited mangrove cover to the intertidal zone near the mouth of the estuary.

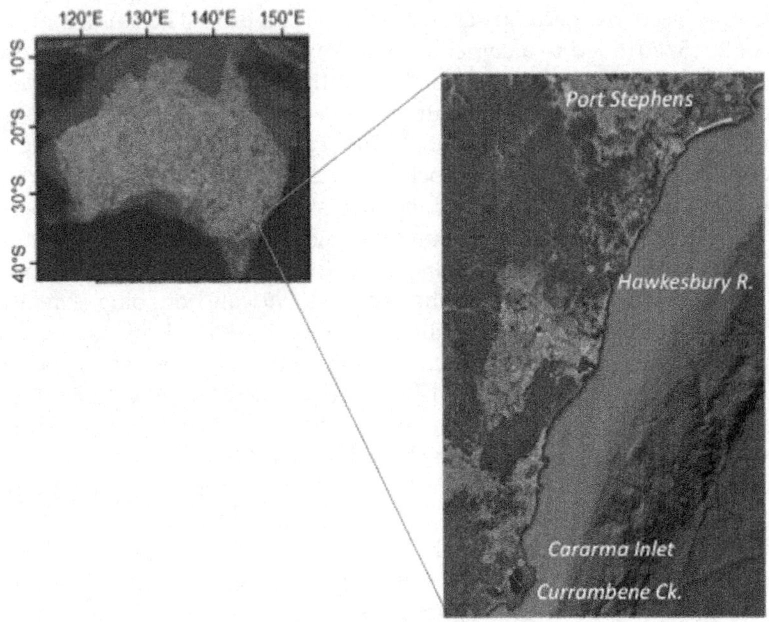

FIGURE 9.4 Location of SE Australian estuaries with mid- to late-Holocene mangrove facies overlain by saltmarsh.

saltmarsh were dated to 1800–1500 yr BP in a marine delta setting and 800 yr BP in a tidal creek network (Kelleway et al., 2017b).

5. Contemporary patterns of mangrove–saltmarsh interaction

Analyses of changes in mangrove and saltmarsh distribution over observational timescales indicate that a range of hydrological processes influence interactions between mangrove and saltmarsh, including changes associated with eustatic sea-level rise and sea-level change driven by climatic variability. Comparison of current mangrove distribution with earliest aerial photography (dating to the 1940s in some locations) has demonstrated widespread landward expansion of mangroves in most settings across the continent. Tidal creek expansion and landward mangrove expansion has been a notable feature on the estuarine plains of northern Australia (Winn et al., 2006; Knighton et al., 1991; Mulrennan and Woodroffe, 1998), spreading mangroves into upper intertidal environments previously free of mangrove since the mid-Holocene. On the East Alligator River, 64% of the *Melaleuca* forests had been lost by 2000 and 17,000 ha lost from the Mary River (Mulrennan and Woodroffe, 1998; Winn et al., 2006). These expansions coincided with sea-level fluctuations with a predominant rise (8–10 mm yr^{-1} 1990–2010: White et al., 2014; Asbridge and Lucas, 2016). Introduced buffalo and their effect on runnelling and tidal barrages may also be contributing to tidal reactivation (Knighton et al., 1991).

Expansion of mangrove across the tropical North occurred in association with relative sea-level rise enhanced by decadal-scale climate variability. The failure of monsoons in the summer of 2015/2016 led to a coincidence of lower relative sea level and lower rainfall, increasing salinity particularly in landward, recently colonized mangrove. The defoliation and dieback of mangrove observed in relation to this event affected an estimated 6% of mangroves in the Gulf of Carpentaria (Duke et al., 2017) and extended into Kakadu (Lucas et al., 2018) and NW Western Australia (Lovelock et al., 2017). Most affected were the landward forests (often dominated by *Avicennia marina*) many of which recruited over the past three decades (Asbridge et al., 2016). Recent estimates suggest that the extent of mangroves post-defoliation/dieback was still greater than a previous minimum (in 1991) because of the longer-term pattern of landward encroachment post 1991 and ongoing seaward expansion in some areas associated with estuary infill and sediment redistributions in a dynamic environment. Phases of contraction and subsequent expansion in extent and canopy cover have occurred previously (Lovelock et al., 2017). Based on 27-year time series (1987–2014) of mangrove dynamics and projected changes in inundation with sea-level rise to 2100, continued expansion of mangrove onto the broad lowland floodplains of northern Australia is probable, continuing the decline of freshwater wetland in the region (Asbridge et al., 2016).

The expansion of mangroves into the saltmarshes of SE Australia has been well-documented (Saintilan and Williams, 1999) and coincides with a pattern of mangrove proliferation at mangrove poleward limits across several continents (Saintilan et al., 2014). Estuarine foreshores have seen a transition from saltmarsh to mangrove dominance, as evidenced by early photography, historical drawings, and ground surveys (see Fig. 9.5 for one example). Saltmarsh loss has been documented in the range of geomorphological settings

FIGURE 9.5 The shoreline of the Cooks River, Sydney, at Tempe, mid-1940s (A) and 2020 (B). Note the occupation of *Avicennia marina* in previously saltmarsh foreshore vegetation.

occupied by subtropical and temperate mangroves in Australia, including back-barrier embayments in southern Queensland, large coastal rivers and drowned river valleys in northern and central NSW, smaller barrier estuaries on the NSW south coast, marine embayments of the Victorian Bass Strait coast, and the inverse estuaries of South Australia (reviewed in Saintilan et al., 2019).

In the estuaries of SE Australia, mangroves are delimited in their landward extent by desiccation of propagules (Clarke and Allaway, 1993; Saintilan, 2003). The landward encroachment of mangrove may therefore be attributed to changes to inundation frequency and water availability in the upper intertidal zone. It is difficult to disentangle the possible effects of rainfall and sea-level trends in driving these changes given their close correlation (Rogers et al., 2014a,b). In tropical and subtropical climates, the proportion of mangrove and saltmarsh in estuaries is correlated to annual rainfall (Bucher and Saenger, 1991; Duke et al., 2019). Where saltmarsh occurs, it often occupies a position in the tidal frame between seaward and landward mangrove zones. Increased rainfall in recent decades has been used to explain the encroachment of tall mangrove forest into saltpan in Hinchinbrook Channel (Duke, 1995). However, the correlation between rainfall and the proportion of mangrove and saltmarsh is not found in NSW (Saintilan, 2003), where the upslope limit of mangrove is more consistently related to frequency of tidal inundation, and there is no landward mangrove fringe. Saltmarsh in most settings has shown little increase in elevation in spite of sea-level rise of ~3.2 mm yr^{-1} over the 20-year measurement period of the SE Australian SETs network (Rogers et al., 2006, 2014), suggesting a vulnerability to encroachment. Saltmarsh survival is therefore contingent on opportunities for lateral expansion.

6. 21st century sea-level rise acceleration and the challenges of coastal planning

Global eustatic sea-level rise has accelerated through the past century, increasing from an average rise of 2.0 ± 0.3 mm yr^{-1} in the period 1966–2009 to 3.4 ± 0.4 mm yr^{-1} in the period 1990–2009 (White et al., 2014). This rate is likely to double by 2100 (Nerem et al., 2018). Based upon the SET and stratigraphic record, sea-level rise acceleration will have profound impacts on the distribution, structure, and carbon burial capacity of Australia's tidal wetlands.

The configuration of coastal lowlands is different today from that encountered by mangroves in the early Holocene. Nearly 7000 years of sea-level stability meant coastal valleys infilled and broad deltas could develop. Substrate compaction over time has provided a relatively stable platform for mangrove colonization, where permitted. Even in the most heavily developed of estuaries, such as the Cooks River in Sydney, opportunities for mangrove colonization under sea-level rise exist where there is a willingness to allow parkland to convert to coastal wetland (Rogers et al., 2019b). Modeling of coastal wetland responses to sea-level rise in the Hunter Estuary NSW has suggested a 20% increase in extent by 2100 under low sea-level rise scenario and 56% under a high sea-level rise scenario, if floodgates are opened (Rogers et al., 2014b). Factoring in both vertical elevation gain and lateral expansion Rogers et al. (2014a,b) estimated 775,000 tonnes of additional carbon burial by 2100. In other settings, such as the deeply incised Triassic sandstone drowned river valleys (for example, the Hawkesbury River and Clyde River in NSW), mangrove lateral expansion will be restricted by topographic impediments.

The expansion of mangrove into saltmarsh has been shown to increase soil carbon in the region. Unlike Gulf of Mexico settings where studies show an increase in soil carbon only in arid zone settings (Yando et al., 2016), mangrove root development in encroachment zones can exceed 600 Mg ha^{-1} (Lamont et al., 2020), among the highest ever recorded. Radiometric dating of a recently encroached wetland in the Sydney region shows an increase in the surface accumulation rate from <3 mm yr^{-1} during the prior saltmarsh phase, increasing up to 5–6 mm yr^{-1} during peak surface soil development by encroaching *Avicennia marina* (Fig. 9.6) in a time of relative sea-level rise. *Avicennia* is capable of keeping pace with current rates of sea-level rise (Rogers et al., 2013, 2014a), doing so partly through the accumulation of root mass (Rogers et al., 2019b).

There is an emerging understanding of the links between sea-level rise and blue carbon at local through to global spatial scales (Rogers et al., 2019a). As sea-level rise accelerates, it increases the vertical and horizontal accommodation space in which carbon-rich mangrove sediments can accumulate. For Australia, like much of the southern hemisphere which has experienced relatively stable sea level over recent millennia, an acceleration in sea-level rise is likely to shift tidal wetlands toward greater proportion of organic material within substrates as accommodation space infills, as seen in the North Atlantic (Rogers et al., 2019a,b) and observed already in Australian encroachment settings (Kelleway et al., 2016). This model of increasing organic carbon accumulation as relative sea-level rise increases available accommodation space has been confirmed at an intertidal wetland where rapid relative sea-level rise was imposed by collapse of an underlying coal mine that resulted in wetland subsidence (Rogers et al., 2019a).

The capacity of mangroves to build elevation while enhancing carbon burial under sea-level rise presents an opportunity to leverage the carbon sequestration benefits of tidal wetlands, while also promoting adaptation. As mangroves also buffer wave energy, bind sediments, and store floodwater, they are able to provide the dual benefits of climate mitigation through carbon sequestration and coastal adaptation to sea-level rise. Management interventions which (1) support the retention of mangrove carbon stocks and accumulation capacity in situ; (2) support the migration and development of new mangrove forests that would not otherwise occur (e.g., due to land use constraints or building of barriers), or (3) support biochemical changes avoiding greenhouse gas emission might be eligible to generate credits

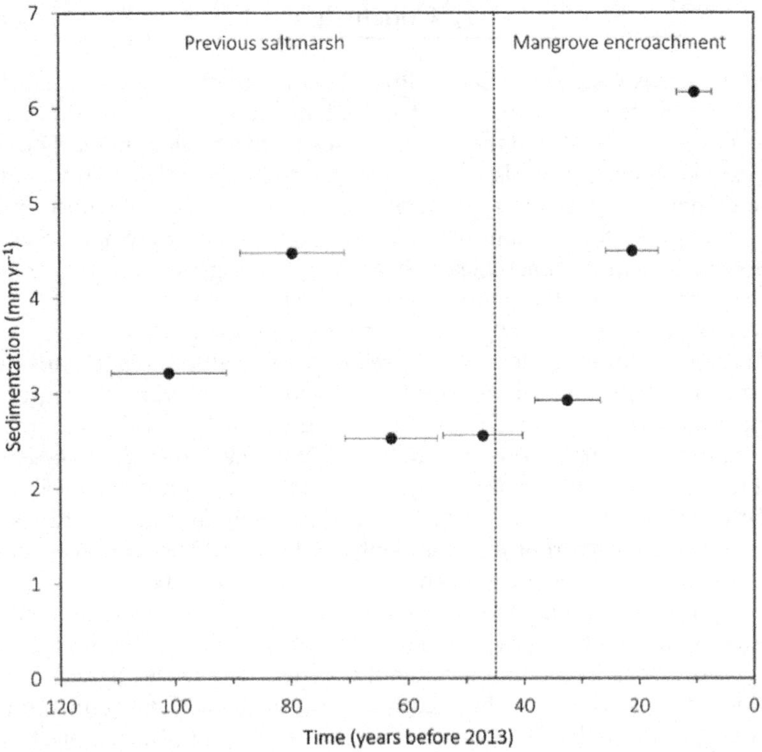

FIGURE 9.6 Sedimentation rate derived from ^{210}Pb dating of sediments that had accumulated in an intertidal wetland following mangrove encroachment of a previous saltmarsh in the Sydney region. Sedimentation rate accelerated after mangrove encroachment (onset indicated by dashed line) driven largely by root production (see Kelleway et al., 2016 for carbon data), reflecting a feedback between vertical accretion and increasing accommodation space that occurred synchronously with mangrove encroachment. The high sedimentation rate during the saltmarsh phase (at ~80 years) coincides with a period of catchment modification Kelleway et al. (2016).

under carbon mitigation schemes (e.g., Kelleway et al., 2017a). An additional benefit of tidal reinstatement in previously isolated wetlands is the effect of saline water in reducing methane gas emissions (Kroeger et al., 2017), though the extent to which these benefits are realized will depend on local groundwater and climatic conditions (Negandhi et al., 2019).

While the opportunities for market incentives in mangrove conservation and planning are exciting, the potential of carbon credits should be seen as one of the many reasons to conserve these important ecosystems, given their demonstrated importance for fisheries productivity, shoreline protection, and habitat provision (Kelleway et al., 2017a). However, current conservation mechanisms largely rely on fixed boundaries, either as habitat maps under State government planning policies or within the boundaries of protected areas. The speed with which climate change is altering the distribution of wetlands poses a challenge to planning instruments, and calls for the introduction of ambulatory boundaries, wider buffers, and other flexible planning provisions have been made (Rogers et al., 2016).

7. Conclusions

The extent of mangrove development within Australian estuaries has been strongly associated with sea-level trends since the last glacial maxima. Phases of rapid sea-level rise (>20 mm yr^{-1}) during at least two meltwater pulses were sufficient not only to drown mangroves over extensive areas but also to create opportunities for mangrove colonization of coastal lowlands newly exposed to the tide. As the rate of sea-level rise decelerated to below 5–7 mm yr^{-1}, mangrove forests dominated by Rhizophoraceae were able to keep pace with sea-level rise over millennial timescales and extensive mangrove forests formed across the tropical north, particularly in the period 8000–6000 BP.

There is growing consensus that biological processes in mangrove forests play a significant role in maintaining mangrove elevation in relation to sea-level rise (Krauss et al., 2014). Mangrove aerial roots promote the deposition of sediment to varying degrees depending on root architecture (Krauss et al., 2003). Fine mangrove roots on the surface of the forest assist in the binding and retention of sediment within the forest (McKee, 2011). Litterfall may contribute to the soil volume (Sukardjo et al., 2013), though this contribution is minor compared with the growth of root material. In Caribbean carbonate settings deprived of terrigenous input, mangrove root development has been sufficient to keep pace with sea-level rise of 3–5 mm yr^{-1} over several centuries (McKee et al., 2007).

The refractory nature of mangrove roots and the anaerobic, saline soil conditions of the mangrove provide an excellent environment for root preservation over extended time periods. These properties have not only facilitated their use in paleostratigraphic investigations but also promoted their value as "blue carbon" ecosystems. Their contribution to carbon sequestration and burial is likely to increase as sea levels rise (Kirwan et al., 2016; Rogers et al., 2019a,b), given the positive feedback between the rate of sea-level rise and carbon burial. An examination of accretion associated with the postglacial marine transgression would suggest that an upper threshold of mangrove vertical accretion might occur at ~5 mm yr^{-1} in carbonate settings (McKee et al., 2007) and 6–7 mm yr^{-1} in terrigenous settings (Crowley and Gagan, 1995). If this rate is exceeded, mangroves will seek to survive by retreating into new intertidal areas, as occurred following MWP1A and MWP1C. Their ability to do so will be subject to local planning decisions (Rogers et al., 2016).

References

Asbridge, E., Lucas, R., Ticehurst, C., Bunting, P., 2016. Mangrove response to environmental change in Australia's Gulf of Carpentaria. Ecol. Evol. 6 (11), 3523–3539.

Asbridge, E., Lucas, R., 2016. Mangrove response to environmental change in Kakadu National Park. IEEE J. Sel. Top. Appl. Earth Obs. Remote Sens. 9 (12), 5612–5620.

Bucher, D., Saenger, P., 1991. An inventory of Australian estuaries and enclosed marine waters: an overview of results. Aust. Geogr Stud. 29, 370–381.

Chivas, A.R., Garcıa, A., van der Kaars, S., Couapel, M.J., Holt, S., Reeves, J.M., Wheeler, D.J., Switzer, A.D., Murray-Wallace, C.V., Banerjee, D., Price, D.M., Wang, S.W., Pearson, G., Edgar, N.T., Beaufort, L., De Deckker, P., Lawson, E., Cecil, C.B., 2001. Sea-level and environmental changes since the last interglacial in the Gulf of Carpentaria, Australia: an overview. Quat. Int. 83, 19–46.

Clarke, P.J., Allaway, W., 1993. The regeneration niche of the grey mangrove *Avicennia marina* – effects of salinity, light and sediment factors on establishment, growth and survival in the field. Oecologia 93, 548–556.

Cohen, M.C.L., Pessenda, L.C.R., Behling, H., de Fátima Rossetti, D., França, M.C., Guimarães, J.T.F., et al., 2012. Holocene palaeoenvironmental history of the Amazonian mangrove belt. Quat. Sci. Rev. 55, 50–58.

Crowley, G.M., Gagan, M.K., 1995. Holocene evolution of coastal wetlands in wet-tropical northeastern Australia. Holocene 5 (4), 385–399.

Crowley, G.M., Grindrod, J., Kershaw, A.P., 1994. Modern pollen deposition in the tropical lowlands of northeast Queensland, Australia. Rev. Palaeobot. Palynol. 83, 299–327.

Crowley, G.M., Anderson, P., Kershaw, A.P., Grindrod, J., 1990. Palynology of a Holocene marine transgressive sequence, lower Mulgrave River valley, north-east Queensland. Aust. J. Ecol. 15 (2), 231–240.

Duke, N.C., 1995. Genetic diversity, distributional barriers and rafting continents – more thoughts on the evolution of mangroves. Hydrobiologia 295, 167–181.

Duke, N.C., Field, C., Mackenzie, J.R., Meynecke, J.O., Wood, A.L., 2019. Rainfall and its possible hysteresis effect on the proportional cover of tropical tidal-wetland mangroves and saltmarsh–saltpans. Mar. Freshw. Res. 70 (8), 1047–1055.

Duke, N.C., Kovacs, J.M., Griffiths, A.D., Preece, L., Hill, D.J., Van Oosterzee, P., et al., 2017. Large-scale dieback of mangroves in Australia's Gulf of Carpentaria: a severe ecosystem response, coincidental with an unusually extreme weather event. Mar. Freshw. Res. 68 (10), 1816–1829.

Ellison, J., 2005. Holocene palynology and sea-level change in two estuaries in Southern Irian Jaya. Palaeogeogr. Palaeoclimatol. Palaeoecol. 220 (3–4), 291–309.

Grindrod, J., 1985. The palynology of mangroves on a prograded shore, Princess Charlotte Bay, North Queensland, Australia. J. Biogeogr. 323–348.

Grindrod, J., 1988. The palynology of Holocene mangrove and saltmarsh sediments, particularly in northern Australia. Rev. Palaeobot. Palynol. 55 (1–3), 229–245.

Guest, M.A., Connolly, R.M., 2005. Fine-scale movement and assimilation of carbon in saltmarsh and mangrove habitat by resident animals. Aquat. Ecol. 38 (4), 599–609.

Hait, A.K., Behling, H., 2009. Holocene mangrove and coastal environmental changes in the western Ganga–Brahmaputra Delta, India. Veg. Hist. Archaeobotany 18 (2), 159–169.

Hashimoto, T.R., Saintilan, N., Haberle, S.G., 2006. Mid-holocene development of mangrove communities featuring Rhizophoraceae and geomorphic change in the Richmond River estuary, New South Wales, Australia. Geogr. Res. 44 (1), 63–76.

Harris, P.T., Heap, A.D., Marshall, J.F., McCulloch, M., 2008. A new coral reef province in the Gulf of Carpentaria, Australia: colonisation, growth and submergence during the early Holocene. Mar. Geol. 251 (1–2), 85–97.

Harty, C., 2004. Planning strategies for mangrove and saltmarsh changes in Southeast Australia. Coast. Manag. 32 (4), 405–415.

Jennings, J.N., 1975. Desert dunes and estuarine fill in the Fitzroy estuary, north-western Australia. Catena 2, 215–262.

Kelleway, J.J., Saintilan, N., Macreadie, P.I., Skilbeck, C.G., Zawadzki, A., Ralph, P.J., 2016. Seventy years of continuous encroachment substantially increases 'blue carbon' capacity as mangroves replace intertidal salt marshes. Global Change Biol. 22 (3), 1097–1109.

Kelleway, J.J., Serrano, O., Baldock, J., Cannard, T., Lavery, P., Lovelock, C.E., Macreadie, P., Masqué, P., Saintilan, N., Steven, A.D., 2017a. Technical Review of Opportunities for Including Blue Carbon in the Australian Government's Emissions Reduction Fund. ACT: CSIRO, Canberra.

Kelleway, J.J., Saintilan, N., Macreadie, P.I., Baldock, J.A., Heijnis, H., Zawadzki, A., Gadd, P., Jacobsen, G., Ralph, P.J., 2017b. Geochemical analyses reveal the importance of environmental history for blue carbon sequestration. J. Geophys. Res. Biogeosci. 122 (7), 1789–1805.

Kirwan, M.L., Temmerman, S., Skeehan, E.E., Guntenspergen, G.R., Fagherazzi, S., 2016. Overestimation of marsh vulnerability to sea level rise. Nat. Clim. Change 6 (3), 253.

Knighton, A.D., Mills, K., Woodroffe, C.D., 1991. Tidal-creek extension and saltwater intrusion in Northern Australia. Geology 19, 831–834.

Krauss, K.W., Allen, J.A., Cahoon, D.R., 2003. Differential rates of vertical accretion and elevation change among aerial root types in Micronesian mangrove forests. Estuar. Coast Shelf Sci. 56 (2), 251–259.

Krauss, K.W., McKee, K.L., Lovelock, C.E., Cahoon, D.R., Saintilan, N., Reef, R., Chen, L., 2014. How mangrove forests adjust to rising sea level. New Phytol. 202 (1), 19–34.

Kroeger, K.D., Crooks, S., Moseman-Valtierra, S., Tang, J., 2017. Restoring tides to reduce methane emission in impounded wetlands: A new and potent Blue Carbon climate change intervention. Sci. Rep. 7 (1), 1–12.

Lambeck, K., Rouby, H., Purcell, A., Sun, Y., Sambridge, M., 2014. sea Level and global ice volumes from the last glacial maximum to the Holocene. Proc. Natl. Acad. Sci. U.S.A. 111 (43), 15296–15303.

Lamont, K., Saintilan, N., Kelleway, J.J., Mazumder, D., Zawadzki, A., 2020. Thirty-year measures of mangrove above- and below-ground biomass reveals unexpectedly high carbon sequestration. Ecosystems 23 (2), 370–382 (in press).

Lewis, S.E., Sloss, C.R., Murray-Wallace, C.V., Woodroffe, C.D., Smithers, S.G., 2013. Post-glacial sea-level changes around the Australian margin: a review. Quat. Sci. Rev. 74, 115–138.

Lovelock, C.E., Adame, M.F., Bennion, V., Hayes, M., O'Mara, J., Reef, R., Santini, N.S., 2014. Contemporary rates of carbon sequestration through vertical accretion of sediments in mangrove forests and saltmarshes of South East Queensland, Australia. Estuar. Coast 37 (3), 763–771.

Lovelock, C.E., Cahoon, D.R., Friess, D.A., Guntenspergen, G.R., Krauss, K.W., Reef, R., et al., 2015. The vulnerability of Indo-Pacific mangrove forests to sea-level rise. Nature 526 (7574), 559.

Lovelock, C.E., Feller, I.C., Reef, R., Hickey, S., Ball, M.C., 2017. Mangrove dieback during fluctuating sea levels. Sci. Rep. 7 (1), 1680.

Lucas, R., Finlayson, C.M., Bartolo, R., Mitchell, A., Woodfroffe, C.D., Asbridge, E., Ens, E., 2018. Historical perspectives on the mangroves of Kakadu National Park. Mar. Freshw. Res. 69 (7), 1047–1063.

Manson, F.J., Loneragan, N.R., Harch, B.D., Skilleter, G.A., Williams, L., 2005. A broad-scale analysis of links between coastal fisheries production and mangrove extent: a case-study for northeastern Australia. Fish. Res. 74 (1–3), 69–85.

Mazumder, D., Saintilan, N., Williams, R.J., 2006. Trophic relationships between itinerant fish and crab larvae in a temperate Australian saltmarsh. Mar. Freshw. Res. 57 (2), 193–199.

McKee, K.L., 2011. Biophysical controls on accretion and elevation change in Caribbean mangrove ecosystems. Estuar. Coast Shelf Sci. 91 (4), 475–483.

McKee, K.L., Cahoon, D.R., Feller, I.C., 2007. Caribbean mangroves adjust to rising sea level through biotic controls on soil elevation change. Global Ecol. Biogeogr. 16, 545–556.

Mogensen, L.A, Rogers, K, 2018. Validation and comparison of a model of the effect of sea-level rise on coastal wetlands. Sci. Rep. 8 (1), 1–14.

Monacci, N.M., Meier-Grünhagen, U., Finney, B.P., Behling, H., Wooller, M.J., 2011. Paleoecology of mangroves along the Sibun river, Belize. Quat. Res. 76 (2), 220–228.

Moss, P.T., Dunbar, G.B., Thomas, Z., Turney, C., Kershaw, A.P., Jacobsen, G.E., 2017. A 60 000-year record of environmental change for the Wet Tropics of north-eastern Australia based on the ODP 820 marine core. J. Quat. Sci. 32 (6), 704–716.

Mulrennan, M.E., Woodroffe, C.D., 1998. Saltwater intrusion into the coastal plains of the lower Mary River, northern Territory, Australia. J. Environ. Manag. 54, 169–188.

Negandhi, K., Edwards, G., Kelleway, J.J., Howard, D., Safari, D., Saintilan, N., 2019. Blue carbon potential of coastal wetland restoration varies with inundation and rainfall. Sci. Rep. 9, 4368.

Nerem, R.S., Beckley, B.D., Fasullo, J.T., Hamlington, B.D., Masters, D., Mitchum, G.T., 2018. Climate-change-driven accelerated sea-level rise detected in the altimeter era. Proc. Natl. Acad. Sci. U.S.A. 115 (9), 2022–2025.

Nicholas, W.A., Nichol, S.L., Howard, F.J.F., Picard, K., Dulfer, H., Radke, L.C., et al., 2014. Pockmark development in the Petrel sub-basin, Timor Sea, northern Australia: seabed habitat mapping in support of CO_2 storage assessments. Continent. Shelf Res. 83, 129–142.

Oliver, T.S., Rogers, K, Chafer, C.J., Woodroffe, C.D., 2012. Measuring, mapping and modelling: an integrated approach to the management of mangrove and saltmarsh in the Minnamurra RIver estuary, southeast Australia. Wetl. Ecol .Manag. 20 (4), 353–371.

Punwong, P., Selby, K., Marchant, R., 2018. Holocene mangrove dynamics and relative sea-level changes along the Tanzanian coast, East Africa. Estuar. Coast Shelf Sci. 212, 105–117.

Rogers, K., Saintilan, N., Heijnis, H., 2005. Mangrove encroachment of salt marsh in Western Port Bay, Victoria: the role of sedimentation, subsidence, and sea level rise. Estuaries 28 (4), 551–559.

Rogers, K., Wilton, K.M., Saintilan, N., 2006. Vegetation change and surface elevation dynamics in estuarine wetlands of southeast Australia. Estuar. Coast Shelf Sci. 66 (3–4), 559–569.

Rogers, K., Saintilan, N., Copeland, C., 2012. Modelling wetland surface elevation dynamics and its application to forecasting the effects of sea-level rise on estuarine wetlands. Ecol. Model. 244, 148–157.

Rogers, K., Saintilan, N., Howe, A.J., Rodríguez, J.F., 2013. Sedimentation, elevation and marsh evolution in a southeastern Australian estuary during changing climatic conditions. Estuar. Coast Shelf Sci. 133, 172–181.

Rogers, K., Saintilan, N., Woodroffe, C.D., 2014a. Surface elevation change and vegetation distribution dynamics in a subtropical coastal wetland: implications for coastal wetland response to climate change. Estuar. Coast Shelf Sci. 149, 46–56.

Rogers, K., Saintilan, N., Copeland, C., 2014b. Managed retreat of saline coastal wetlands: challenges and opportunities identified from the Hunter River Estuary, Australia. Estuar. Coast 37 (1), 67–78.

Rogers, K., Boon, P.I., Branigan, S., Duke, N.C., Field, C.D., Fitzsimons, J.A., et al., 2016. The state of legislation and policy protecting Australia's mangrove and salt marsh and their ecosystem services. Mar. Pol. 72, 139–155.

Rogers, K., Krauss, K.W., 2018. Moving form generalisations to specificity about mangrove-saltmarsh dynamics. Wetlands. https://doi.org/10.1007/s13157-018-1067-9.

Rogers, K., Kelleway, J., Saintilan, N., Megonigal, P., Adams, J., Holmquist, J.R., Meng, L., Schille-Beers, L., Zawadski, A., Mazumder, D., Woodroffe, C., 2019a. Wetland carbon storage controlled by millennial-scale variation in relative sea-level rise. Nature 567 (7746), 91.

Rogers, K., Saintilan, N., Mazumder, D., Kelleway, J.J., 2019b. Mangrove dynamics and blue carbon sequestration. Biol. Lett. 15 (3), 20180471.

Roy, P.S., Williams, R.J., Jones, A.R., Yassini, I., Gibbs, P.J., Coates, B., et al., 2001. Structure and function of south-east Australian estuaries. Estuar. Coast Shelf Sci. 53 (3), 351–384.

Saintilan, N., 1997. Mangroves as successional stages on the Hawkesbury River estuary, New South Wales. Wetlands 16, 99–107.

Saintilan, N., 2003. The influence of nutrient enrichment upon mangrove seedling establishment and growth in the Hawkesbury River estuary, New South Wales, Australia. Wetlands 21, 29–35.

Saintilan, N., 2004. Relationships between estuarine geomorphology, wetland extent and fish landings in New South Wales estuaries. Estuar. Coast Shelf Sci. 61 (4), 591–601.

Saintilan, N., Hashimoto, R., 1999. Mangrove-saltmarsh dynamics on a prograding bayhead delta in the Hawkesbury River estuary, New South Wales, Australia. Hydrobiologia 413, 95–102.

Saintilan, N., Williams, R.J., 1999. Mangrove transgression into saltmarsh environments in south-east Australia. Global Ecol. Biogeogr. Lett. 8, 117–124.

Saintilan, N., Wilton, K., 2001. Changes in the distribution of mangroves and saltmarshes in Jervis Bay, Australia. Wetl. Ecol. Manag. 9, 409–420.

Saintilan, N., Wilson, N.C., Rogers, K., Rajkaran, A., Krauss, K.W., 2014. Mangrove expansion and salt marsh decline at mangrove poleward limits. Global Change Biol. 20 (1), 147–157.

Saintilan, N., Rogers, K., Toms, C., Stein, E.D., Jacobs, D., 2016. Intermittent estuaries: linking hydro-geomorphic context to climate change resilience. J. Coast Res. 75 (sp1), 133–137.

Saintilan, N., Mazumder, D., 2017. Mass spawning of crabs: ecological implications in subtropical Australia. Hydrobiologia 803 (1), 239–250.

Saintilan, N., Rogers, K., McKee, K.L., 2019. The shifting saltmarsh-mangrove ecotone in Australasia and the Americas. In: Coastal Wetlands. Elsevier, pp. 915–945.

Saintilan, N., Khan, N.S., Ashe, E., Kelleway, J.J., Rogers, K., Woodroffe, C.D., Horton, B.P., 2020. Thresholds of mangrove survival under rapid sea level rise. Science 368 (6495), 1118–1121.

Skilleter, G.A., Olds, A., Loneragan, N.R., Zharikov, Y., 2005. The value of patches of intertidal seagrass to prawns depends on their proximity to mangroves. Mar. Biol. 147 (2), 353–365.

Spalding, M.D., Blasco, F., Field, C.D., 2010. World Atlas of Mangroves. Routledge.

Staples, D.J., Vance, D.J., 1986. Emigration of juvenile banana prawns Penaeus merguiensis from a mangrove estuary and recruitment to offshore areas in the wet-dry tropics of the Gulf of Carpentaria, Australia. Mar. Ecol. Prog. Ser. 27 (239), 52.

Sukardjo, S., Alongi, D.M., Kusmana, C., 2013. Rapid litter production and accumulation in Bornean mangrove forests. Ecosphere 4 (7), 1–7.

Tamura, T., Saito, Y., Sieng, S., Ben, B., Kong, M., Sim, I., et al., 2009. Initiation of the Mekong River delta at 8 ka: evidence from the sedimentary succession in the Cambodian lowland. Quat. Sci. Rev. 28 (3–4), 327–344.

Thom, B.G., Wright, L.D., Coleman, J.M., 1975. Mangrove ecology and deltaic-estuarine geomorphology, Cambridge Gulf-Ord River, western Australia. J. Ecol. 63, 203–222.

Tossou, M.G., Akoègninou, A., Ballouche, A., Sowunmi, M.A., Akpagana, K., 2008. The history of the mangrove vegetation in Bénin during the Holocene: a palynological study. J. Afr. Earth Sci. 52 (4–5), 167–174.

Traill, L.W., Perhans, K., Lovelock, C.E., Prohaska, A., McFallan, S., Rhodes, J.R., Wilson, K.A., 2011. Managing for change: wetland transitions under sea-level rise and outcomes for threatened species. Divers. Distrib. 17 (6), 1225–1233.

Umitsu, M., Buman, M., Kawase, K., Woodroffe, C.D., 2001. Holocene palaeoecology and formation of the Shoalhaven River deltaic-estuarine plains, southeast Australia. Holocene 11 (4), 407–418.

Woodroffe, C.D., Thom, B.G., Chappell, J., 1985. Development of widespread mangrove swamps in mid-Holocene times in northern Australia. Nature 317, 711–713.

Woodroffe, C.D., 2018. Mangrovbe response to sea level rise: palaeoecological insights from macrotidal systems in northern Australia. Mar. Freshw. Res. 69 (6), 917–932.

Woodroffe, C.D., Chappell, J., Thom, B.G., Wallensky, E., 1989. Depositional model of a macrotidal estuary and floodplain, South Alligator River, Northern Australia. Sedimentology 36, 737–756.

White, N.J., Haigh, I.D., Church, J.A., Koen, T., Watson, C.S., Pritchard, T.R., et al., 2014. Australian sea levels—trends, regional variability and influencing factors. Earth Sci. Rev. 136, 155–174.

Wilson, N.C., 2009. The Distribution, Growth, Reproduction and Population Genetics of a Mangrove Species, Rhizophora Stylosa Griff. Near its Southern Limits in New South Wales, Australia. Sch. Arts Sci. Australian Catholic University, Sydney.

Wilson, N.C., Saintilan, N., 2012. Growth of the mangrove species Rhizophora stylosa Griff. at its southern latitudinal limit in eawstern Australia. Aquat. Bot. 101, 8–17.

Winn, K.O., Saynor, M.J., Eliot, M.J., Eliot, I., 2006. Saltwater intrusion and morphologicvla change at the mouth of the east Alligator river, Northern Territory. J. Coast Res. 22, 137–149.

Woodroffe, C.D., Thom, B.G., Chappell, J., Wallensky, E., Grindrod, J., 1987. Relative sea level in the South Alligator River region, north Australia, during the Holocene. Search 4, 198–200.

Woods, J.L.D., 2012. The evolution of a Holocene estuarine barrier on the Coromandel Coast, New Zealand. Geogr. Res. 50 (1), 89–101.

Yando, E.S., Osland, M.J., Willis, J.M., Day, R.H., Krauss, K.W., Hester, M.W., 2016. Salt marsh-mangrove ecotones: using structural gradients to investigate the effects of woody plant encroachment on plant–soil interactions and ecosystem carbon pools. J. Ecol. 104 (4), 1020–1031.

Yulianto, E., Sukapti, W.S., Rahardjo, A.T., Noeradi, D., Siregar, D.A., Suparan, P., Hirakawa, K., 2004. Mangrove shoreline responses to Holocene environmental change, Makassar Strait, Indonesia. Rev. Palaeobot. Palynol. 131 (3–4), 251–268.

CHAPTER

10

Responses of mangrove ecosystems to sea level change

Raymond D. Ward[1,2], *Luiz Drude de Lacerda*[3]

[1]Centre for Aquatic Environments, School of the Environment and Technology, University of Brighton, Brighton, United Kingdom; [2]Institute of Agriculture and Environmental Sciences, Estonian University of Life Sciences, Tartu, Estonia; [3]Instituto de Ciências do Mar, Universidade Federal do Ceará, Fortaleza, Ceará, Brazil

OUTLINE

1. Introduction 236
2. Factors contributing to mangrove resilience to sea level rise 237
3. Landward migration and sediment availability in arid and semiarid areas 238
4. New biogeochemical setting following mangrove migration 242
5. Plant productivity 243
6. Measuring mangrove resilience to sea level rise 244
7. Differential responses of mangrove types (estuarine/deltaic, fringing) to sea level rise 245
8. Regions with mangroves under threat from sea level rise 246
   8.1 North America 246
   8.2 Indo-Pacific 247
   8.3 South America 247
   8.4 Africa 248
9. Conclusions 248

References 249

Dynamic Sedimentary Environments of Mangrove Coasts
https://doi.org/10.1016/B978-0-12-816437-2.00002-1

1. Introduction

Mangrove forests are expected to be impacted by climate change throughout their distribution, and between 1990 and 2018, over 1200 publications refer to "Mangroves and Climate Change." Most, however, discuss eventual impacts and scenarios that might occur in the future, based on projections and modeling. Unfortunately, however, this approach inherits two major constraints that make them extremely difficult to support local and regional scale governance aimed at adaptation and mitigation of climate change impacts to coastal regions. First, these scenarios and modeling exercises are based on scarce empirical data, with data supporting them often generated from mangroves located near their latitudinal limits or obtained under greenhouse-controlled conditions. Second, these scenarios consider impacts that may occur in the future, whereas they are actually happening today in many significant mangrove-dominated areas. Less than 5% of those 1200 publications actually discuss observed, quantified impacts of sea level rise, even considering among them the nearly half that deals with controlled laboratory or greenhouse experiments. This present discussion deals with these 5%. For a general appraisal of the mangrove versus climate change issues, one is referred to a number of recent reviews summarizing future scenarios and aspects of climate change impacts on mangroves (Alongi, 2015; Ward et al., 2016b; Jennerjahn et al., 2017a,b; Makowski and Finkl, 2018; Saintilan et al., 2018).

Of the principal impacts of climate change, sea level rise is the factor that is most likely to affect mangrove ecosystems (Ward et al., 2016b; Jennerjahn et al., 2017a,b). Global sea levels have been rising since at least the 1880s, with an acceleration recorded globally since the 1990s to a current recorded rate of 3.2 ± 0.4 mm/yr (Church and White, 2011); this rate is likely to increase in the future to cause a rise in sea level of between 0.28 and 0.98 m by 2100 (IPCC, 2013). Regionally, sea level rise is highly variable as a result of ocean currents, water mass redistribution (gravitational responses), addition of freshwater from the cryosphere, isostatic uplift/crustal subsidence, and air–sea interactions (atmospheric pressure changes, wind regime) (Ward et al., 2016a; Cazenave et al., 2018; Zemp et al., 2019). Rates of local sea level rise are predicted to be greatest at mid- to low latitudes (IPCC, 2013), particularly the Philippines and Indonesia, although some tropical areas including the central and eastern Pacific have lower rates of sea level rise (Rietbroek et al., 2016). Regional variations in current sea level rise (SLR) rates are highly variable, from 1.9 mm/yr in the Caribbean to 7.5 mm/yr in parts of Indonesia (Nerem et al., 2010) and up to 9 mm/yr in the lower Mississippi River Delta (NOAA, 2015). Sea level rise impacts mangroves as the species that make up this ecosystem are sensitive to alterations to the hydroperiod (Ellison, 2015; Sasmito et al., 2016) as increasing flooding duration and frequency pushes species beyond their species-specific physiological thresholds of tolerance to inundation (Friess et al., 2012; Lovelock et al., 2017). Increases in inundation times can result in plant death at the seaward mangrove margins, particularly in fringing mangroves (He et al., 2007) as well as shifts in species composition (Gilman et al., 2008), ultimately leading to a reduction in productivity (Castañeda-Moya et al., 2010) and ecosystem service provision. Thus, sea level rise can have a range of interrelated influences on mangrove ecosystems, including plant productivity, plant health, sediment supply, and resilience to other global change influences.

From a global perspective, some projections suggest losses of coastal wetlands of between 20% and 90% as a result of sea level rise; however, a recent study by Schuerch et al. (2018) suggests that coastal wetlands could potentially increase in extent by up to 60% of the current area taking into account current rates of sediment accumulation, should sufficient landward space be made available with losses of between 0% and 30% should no further landward space be made available. The aim of this chapter is to provide an overview of the response of mangroves to sea level rise and identify the factors that confer resilience or vulnerability with regard to their survival, particularly concerning regional variability and how these may be assessed.

2. Factors contributing to mangrove resilience to sea level rise

Mangroves are able to actively modify their environment through trapping sediments or creating biogenic inputs into the soil, thereby increasing surface elevation relative to sea level. Mangroves are also able to migrate inland to higher elevations, where landward space is available (Krauss et al., 2014). This provides them with considerable resilience to fluctuations in sea level (Woodroffe and Grindrod, 1991). During the Holocene marine transgression, characterized by rapid rates of sea level rise, mangroves were able to relocate further inland throughout their range (Woodroffe et al., 1985; Ellison and Stoddart, 1993; Woodroffe, 1995; Friess et al., 2019). In some areas, rapid mangrove colonization and expansion occurred as a result of rapid early sea level rise (e.g., western India, and SE and N Brazil), with later decreases in extent as sea level continued to rise, as a result of an unfavorable geomorphic setting (Limaye et al., 2014; França et al., 2015). In many areas, mangroves rapidly colonized freshwater wetland systems such as floodplains, sawgrass marshes, and wet forest (*Alnus, Quercus, Ulmaceae*), particularly in the case of those located in deltaic or estuarine settings (Li et al., 2012; Cordero-Oviedo et al., 2019; Jones et al., 2019), as well as expanded into higher latitude areas as a result of climate warming (França et al., 2019). Over the Holocene, mangroves have adapted to rises and falls in sea level of 120 m at an average rate of 1 m per century for 10,000 years (Alongi, 2015). However, this has not occurred without the local/global extinction of some species as a result of rates of sea level rise beyond their tolerance thresholds (Alongi, 2015). Guo et al. (2018) found that local genetic diversity of *Sonneratia alba, Ceriops tagal, Avicennia marina*, and *Rhizophora apiculata* was diminished as a result of high rates of historical and present local sea level rise.

Surface elevation change can be both positive and negative and is influenced by the availability of allochthonous and autochthonous organic matter, subsurface compaction, and the ability of the mangrove root matrix to trap and retain minerogenic sediments (Krauss et al., 2014). Local relative sea level rise is therefore influenced by the interplay between all these factors, combined with increases in eustatic sea level and land uplift/subsidence (Krauss et al., 2014). In addition to trapping of sediments through reducing water velocities in the root network (prop roots, pneumatophores) (Furukawa and Wolanski, 1996), positive surface elevation change is promoted through the trapping and binding of sediments and organic material in microbial filamentous algal mats and through the accumulation of litter and woody debris (Krauss et al., 2005). This can result in vertical accretion rates between 0.7

and 20.8 mm/yr, although there can be substantial inter- and intrasite variation (McKee et al., 2007; Krauss et al., 2010).

3. Landward migration and sediment availability in arid and semiarid areas

Irrespective of the regional geomorphology, mangroves will respond to sea level rise by landward migration where possible. With rising sea level, no landward obstruction, and where sedimentation rates are moderate to high, mangrove fringes will move inland and in estuarine and deltaic systems upriver. However, where any form of natural or artificial landward obstruction exists and/or sedimentation rates are smaller than sea level rise, mangrove migration is impeded and erosion and loss of fringe forests occur (Gilman et al., 2008). Thus, mangrove migration inland will depend on the interaction of three major drivers: sediment supply (a function of basin morphology and regional climate and anthropogenic factors, e.g., dams, dredging, etc.), rate of sea level rise, and the presence of physical or chemical obstructions (e.g., extreme soil pH, salinity, or pollution).

Migration inland is more pronounced under extreme conditions such as on continental semiarid coastlines. Exacerbated arid conditions, caused by a generalized decrease in annual rainfall and longer extended drought periods and/or severity of drought (more than three consecutive years with annual rainfall below the historical mean), are associated with climatic change (Marengo et al., 2018). Mafi-Gholami et al. (2018) have also found that intense drought events can force a change point in mangrove ecosystem functioning and result in a decrease in productivity and ecosystem health, which can continue to impact on mangrove systems even after the drought events. This is likely to impact on a mangroves ability to respond to sea level rise. Moreover, abstraction of freshwater as a result of increased human activity and population growth in semiarid regions is intensifying. This increasing water demand requires more dams and larger reservoirs and a greater exploitation of groundwater resources. Lower rainfall significantly reduces continental runoff facilitating salinization of river lower basins due to the increase in sea level, whereas higher groundwater abstraction facilitates saline intrusion inland.

Along the NE Brazilian coast bordering semiarid hinterlands, for example, no anthropogenic obstruction to mangrove migration inland exists, and remote sense mapping consistently indicates a generalized mangrove area expansion through inland migration on the order of 20%–30% (Maia et al., 2006), one of the largest worldwide. Sea level rise is not particularly high in the region, but local drivers intensify its effects (Fig. 10.1). A study undertaken in the Jaguaribe River, the largest river totally constrained within the semiarid north east of Brazil (a total basin area of 76,000 km^2) using similar satellite image sources from the period 1992–2003, showed an increase of 24.15 ha in the area of islands within the estuary. These islands are quickly colonized by mangrove vegetation. The increase in the rate of sedimentation in the estuary is caused not only by the river flow regulation reducing water flow and sediment flushing but also by the decrease in precipitation over the basin caused by global climate change (Godoy and Lacerda, 2014). In addition, the residence time of materials (sediment and associated trace elements) and water in the estuary is increasing due to strengthening ocean forcing on the continental shelf because of excess heat accumulating in the South Atlantic as a result of climate change (Dias et al., 2013; Lacerda et al., 2013).

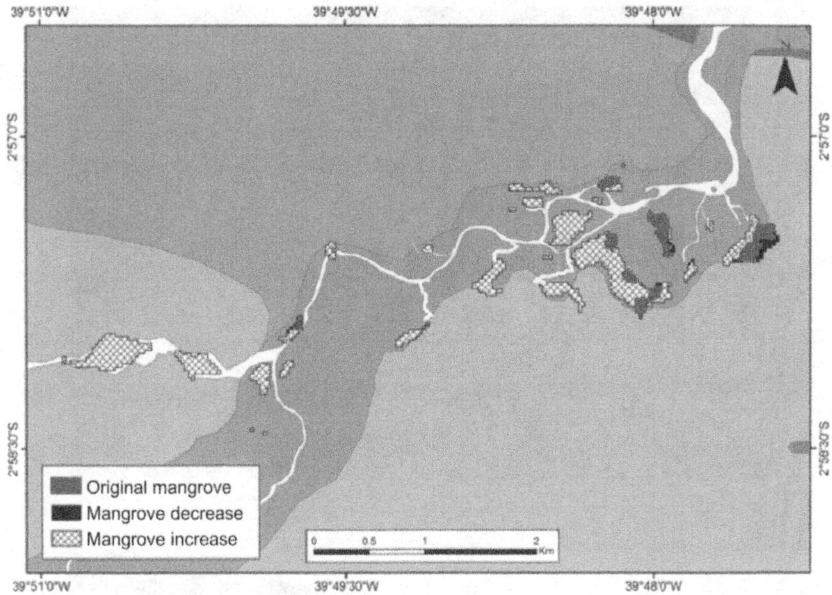

FIGURE 10.1 Mangrove expansion in the Aracatim Mirim estuary, Ceará State, NE Brazil. Expansion reaches about 400%. Since there is no damming in the river watershed and no geomorphologic alteration due to human intervention, so sea level rise is strongly suggested as the major driver of mangrove expansion in this estuary.

In this same area, however, fringe mangroves at the land–sea interface are being eroded (Fig. 10.2). However, where progradation is unlikely to take place (e.g., as a result of low sedimentation rates), migration inland may be limited by dune encroachment as has been seen in several areas of Northeast Brazil (Fig. 10.3).

Results from different sites along the arid Arabian Gulf coast suggest that mangroves respond measurably to changes in land use and global climate change, in particular mangroves along coasts and lagoons in association with corals and submerged seagrass meadows in the shallow continental shelf. Sites with multiple stressors (e.g., drought, reduced surface runoff of upstream catchments, strong winds, extreme temperatures, and g activities) in arid areas have been assessed to be under a very high threat level (Mafi-Gholami and Ward, 2019). Through the interpretation of remote sensing imagery of the Arabian Gulf coast, Moore et al. (2015) found increases in mangrove areas along the coast of Qatar and Abu Dhabi and related these mostly to replanting experiments. However, some changes in mangrove area have been reported to areas without replanting efforts (Howari et al., 2009) and may be due to regional and/or global environmental changes, following sea level rise and imbalances between continental and marine contribution, which results in alterations in coastal sedimentation and the extension of saline intrusion allowing expansion inland without losses at the seaward fringes.

The impacts on mangroves and their respective responses to environmental changes in the Arabian Gulf have been increasingly well-studied in recent years. Consistent SLR rate measurements in the Arabian Gulf, a key step to predict mangrove adaptation, have been estimated at 2.2 mm/yr, taking into account a subsidence rate of 0.7 mm/yr (Alothman

FIGURE 10.2 Erosion of fringe mangroves at the Jaguaribe estuary mouth as a result of river damming, sea level change, and extended drought (A); consequent sediment accumulation and colonization inside the lower estuary due to reduced river transport capacity and augmenting ocean forcing due to sea level rise (B). *Photos by Luiz Drude de Lacerda.*

et al., 2014). Also, modeling of annual rainfall for the Gulf forecasts a 10-day per year increase in the number of wet days (Lelieveld et al., 2012). Present rates of sediment accretion (SAR: sediment build up by trapping continental runoff or carbonate production) in the Arabian Gulf mangrove environments reach 0.21 ± 0.09 mm/yr (Saderne et al., 2018). In relation to regional SLR, this SAR value suggests that a mangrove, free of other constraints, will eventually adapt to SLR by migrating inland, as observed in other arid and semiarid coastlines, with SLR rates higher or similar to SAR. Wetter conditions may decrease the gap between SLR and SAR in Middle Eastern mangroves favoring migration, but this will depend on

FIGURE 10.3 Advancing dunes over mangroves at the Parnaíba River Delta (A) and the Pacoti River (B) in NE Brazil, probably a response to extended drought. Encroachment by moving sand dunes may hamper mangrove expansion following sea level rise. *Photos by Luiz Drude de Lacerda (A) and Raymond Ward (B).*

site characteristics (e.g., wadi mouths, lagoons, khors). However, these may be over or under estimations as they do not account for all the factors relating to surface elevation change.

Notwithstanding the potential capacity for adaptation, there may be constraints to the extent of mangrove adaptation to climate change. Along most of the south-eastern Brazilian coast, mangroves are restricted to small coastal plains or along margins of coastal lagoons limited by the granitic outcrops of the Serra do Mar. Here low sedimentation rates and natural (steep coastal geomorphic constraints)/anthropogenic (hard infrastructure) impediments physically constrain mangrove migration and/or expansion inland (Godoy and Lacerda, 2015). Even in wide coastal plains, such as in NE Brazil and the Arabian Gulf, constraints of a geomorphological or geochemical nature may impede mangrove migration inland following sea level rise. Intensification of sand dune encroachment displacing mangroves, a phenomenon frequently recorded in arid and semiarid environments, is intensified by climate changes (Maia et al., 2005; Lacerda, 2018) and may pose a threat to mangrove resilience. It is also still uncertain if SAR/carbonate production rates measured in Gulf mangroves, for example, apply to all local subenvironments relative to continental runoff (wadi mouths, lagoons, khors) or to other arid/semiarid areas such as the Caribbean coast. Sabkha environments stretching along most of the coast of the Arabian Gulf (Barth and Böer, 2002) and many other arid and semiarid littorals may act as a barrier to mangrove migration inland. Sabkha environments are coastal plains, frequently below mean sea level, occupied by a typical guild of cyanophycean microorganisms that create a characteristic geochemical environment. Sediments develop anoxic conditions and a sulfate-reduction based metabolism, resulting in the accumulation of toxins to higher plants, such as trace metals and sulfur compounds. If this sedimentary environment is well developed, it will pose a barrier to mangrove migration in these areas (Lacerda, 1994).

Mangrove migration as a major response to sea level rise is not, however, restricted to semiarid coastlines (Jennerjahn et al., 2017a,b). Along the northern humid coast of Brazil, harboring the second largest continuous mangrove forest in the world, there is a reduction in mangrove coverage along the coastline, but a simultaneous active growth of mangroves in the topographically higher plains previously occupied by grasses and forbs. The herbaceous vegetation located inland of these mangroves occupied an area of 8.8 km^2 in 1972, which was reduced to 5.6 km^2 in 1997 as a result of encroachment by mangroves. A linear extrapolation of the process showed that the whole of the current herbaceous vegetation would be displaced by mangroves by 2035 (Rubén et al., 2002). Remotely sensed mapping has shown changes in mangrove extent along this mesotidal Amazon coast over a period of 12 years (1996–2008). The total mangrove area in that region increased by 718.6 km^2 (from 6705 to 7423.60 km^2) and this seems typical of the entire mesotidal north coast of Brazil (Nascimento Jr. et al., 2013).

4. New biogeochemical setting following mangrove migration

An extremely important concept relating to the response of mangroves to sea level change is that any change in forest distribution or extension is not a simple case of moving vegetation cover and forest area. At least as significant is the changing distribution of a unique biogeochemical realm, since mangrove environments are one of the few sulfate reduction metabolism-based ecosystems on the Earth surface, and these biogeochemical changes can be of substantial environmental concern. Many chemical elements of environmental significance, including key nutrients, such as phosphorus, and pollutants, mostly calcophilic trace metals, display completely different cycles under the redox conditions of mangrove sediments and porewaters, altering these ecosystems as sources or sinks of these chemicals, with direct impacts on the adjacent habitats and on mangrove functioning (Agraz-Hernández et al., 2018; Lacerda, 1998; Lacerda et al., 1993).

Sulfate reduction produces large amounts of sulfides and dissolved organic matter. Both compounds interact strongly with nutrients and pollutants. Mangrove sediments are so efficient in accumulating calcophilic heavy metals in a nonbioavailable form, that this property has led the widespread use of mangroves to mitigate contamination from landfill areas (Lacerda et al., 2000), industrial and mining sites under pollution (Oyuang and Guo, 2016), as a phytoremediation technique. Similarly, mangroves have been also suggested as efficient filters of sewage and shrimp farm effluents in many tropical areas (Robertson and Phillips, 1995). However, this ability depends on the maintenance of the redox conditions, which depends on sediment stability and accumulation rates, and mangrove productivity, which are strongly affected by sea level rise.

As a consequence of sea level rise, erosion of the mangrove littoral fringe has been most typically observed. This increases the suspended sediment load, which may accumulate in upper estuarine areas, favoring mangrove expansion upriver (Godoy and Lacerda, 2014); oxidation of reduced minerals (mostly sulfides), and consequent acidification and mobilization of deposited pollutants from sediments to pore and surface waters, in particular trace metals (Mounier et al., 2001). This remobilization also increases trace metal bioavailability (Lacerda et al., 2013), which is particularly serious where mangroves have been accumulating large amounts of metal pollutants from anthropogenic sources, such as in the highly polluted Guanabara Bay, in Rio de Janeiro, where mangroves efficiently trap and accumulate trace metals leaching from nearby solid wastes disposal sites (Machado et al., 2002).

Landward mangrove migration increases the trapping of continental sediments, including pollutants; moving mangrove redox conditions upriver; and enlarging the mangrove redox realm (Godoy et al., 2018). Sulfate reduction, the typical metabolism within mangrove sediments, is unable to fully degrade deposited organic matter and thus exports large quantities of dissolved organic complexes to adjacent waters increasing metal mobilization through organic complexation (Lacerda et al., 2013; Lei et al., 2019) and favoring metal bioavailability (Moura and Lacerda, 2018), resulting in more efficient and rapid bioaccumulation rates and increasing human exposure to pollutants (Costa and Lacerda, 2014).

This described sequence of biogeochemical events primarily triggered by a response of mangroves to sea level rise may, unfortunately, be occurring worldwide where landward migration is possible, particularly along broad coastal plains, but has been virtually neglected in studies. Some questions considering mangrove migration and expansion inland of its biogeochemical characteristics remain unanswered, and those require urgent assessment, in order to achieve the necessary conservation and sustainable utilization of mangrove forests in the Anthropocene.

5. Plant productivity

There are a range of factors that influence the vulnerability of mangroves to sea level rise, including tidal range. Lovelock et al. (2015a) suggest that mangroves located in microtidal areas are likely to be at greater risk from sea level rise than those situated in mesotidal- and macrotidal-dominated regions due to their lower elevation capital, as seems the case of the Northern Brazilian coast mentioned previously in this chapter.

The geomorphic setting can also influence mangrove resilience to sea level rise. Sasmito et al. (2016) undertook a global metaanalysis of mangrove surface elevation change studies, with results suggesting that back-barrier mangroves were more resilient than fringing mangroves to sea level rise, as a result of the greater sediment deposition rates and the species that occur there. Their results also showed that the management regime can have a substantial influence on mangrove resilience to sea level rise. Disturbance, either natural or anthropogenic, can result in peat loss or negative surface elevation change, exacerbating the impacts of sea level rise, whereas positive management interventions, e.g., removals of dams increasing sediment supply, bamboo wave dampeners reducing erosion, can cause positive surface elevation gains limiting the impact of sea level rise.

The frequency and severity of storms is likely to influence the response of mangroves to sea level rise, potentially providing additional resilience or vulnerability dependent on a range of factors including direction of incoming wind and waves, historical occurrence of storms in the area, and storm severity. Storms can provide a sudden influx of sediment and allochthonous organic matter, raising surface elevation, decreasing inundation times, and conferring a level of resilience to sea level rise (Ward et al., 2016). Furthermore, storms can provide an important influx of nutrients particularly for mangroves in arid and semiarid zones, which can lead to increased productivity and resultant increases in below- and aboveground biomass (Castañeda-Moya et al., 2020). Increases in belowground biomass increase surface elevation, and increases in aboveground biomass could potentially decrease water velocities through increased drag through roots. Conversely, extreme wind and wave activity can also strip leaves from trees, remove branches, and erode surface sediments, leading to mangrove degradation and tree mortality, resulting in organic matter decomposition, peat collapse, and losses to mangrove area. In areas where tropical cyclones are rare (e.g., the South Atlantic and South East Pacific), mangroves are unlikely to have developed in response to high-frequency, low-energy inundation (tidal) and are thus likely to be altered in terms of structure and geomorphology when exposed to strong wind and wave activity (Krauss and Osland, 2020).

6. Measuring mangrove resilience to sea level rise

Globally, there has been an increase in coastal development including around mangrove margins (both urban and agricultural), reducing available landward migration space. As a result of this limitation, it is important to evaluate rates of surface elevation change within mangroves with a high degree of accuracy in order to evaluate historical impacts of sea level rise and to predict future responses at a site, local regional, and global scale (Ward et al., 2016a). Sediment trapping and resultant vertical accretion is the principle mechanism by which minerogenic mangroves can mitigate the effects of increases in inundation as a result of sea level rise (Lovelock et al., 2015). Thus, it is important to accurately measure SARs, together with geological and biological drivers of surface elevation change, to determine survivability of coastal wetlands to sea level rise at the site scale (Ward et al., 2014). Sediment accretion (defined as the total addition of organic and inorganic material to the surface) can be estimated utilizing artificial radioisotopic markers (^{137}Cs) and natural radioisotope signatures (Pb from point source pollution) or to measure the levels of accumulation (i.e., ^{210}Pb activity, ^{14}C) along the soil profile (Appleby and Oldfield, 1992), depending on the resolution required.

^{137}Cs is an artificial radionuclide produced during nuclear fission reactions with the first significant amounts being released into the environment in the 1950s culminating in a global peak of atmospheric deposition in 1963 as a result of aboveground nuclear weapons testing and the test ban treaty. This has provided an excellent marker horizon for dating coastal wetland sediments and evaluating sediment accretion against this boundary horizon. However, due to its relatively short half-life (~30 years), it is rapidly becoming obsolete in southern hemisphere sites and those located away from later inputs (e.g., Chernobyl 1986 and Fukushima 2007) such as the Americas (Drexler et al., 2018). ^{210}Pb dating does not suffer from the same restrictions as ^{137}Cs and can be used to date sediments up to ~150 years and, dependent on the method used, can also be used to identify periods with elevated SARs (Ward et al., 2014). However, there are a range of assumptions that must be met with these methods including constant sedimentation rates (CF:CS method), constant rates of supply of ^{210}Pb (not always the case in arid or semiarid areas), and no physical or biological (bioturbation) sediment mixing (Andersen, 2017), which can be particularly problematic in sites with large numbers of burrowing crabs (e.g., Americas). ^{14}C dating is a tried and tested method for dating sediments between ~5700 and 50,000 years old, providing a good measure of Holocene sediment accumulation rates. However, there are some specific problems associated with this method including intrusion of new organic material to deeper horizons (mangrove roots can penetrate a meter deep) (Lougheed et al., 2017).

These radioisotope dating methods provide an important insight into sediment accretion relative to dateable horizons in the soil profile. However, the USGS has developed a widely adopted methodology to evaluate very short-term (decadal to event) surface elevation change processes, which also incorporates shallow subsidence and expansion processes, Rod Surface Elevation Table (RSET) (Cahoon et al., 2002; Webb et al., 2013). This method utilizes a steel rod inserted to refusal (e.g., parent material), providing a stable datum to measure expansion or contraction of the soil column. This provides an established method for monitoring net surface elevation change (deposition and erosion) from time of installation and responses to sea level change related to local sea level rise as well as episodic events (e.g., storm surges, tsunamis, high precipitation events, floods, etc.).

7. Differential responses of mangrove types (estuarine/deltaic, fringing) to sea level rise

As mentioned previously, there are a wide range of factors that influence the response of mangroves to sea level rise, including sediment availability, inundation times, velocity of water, and resuspension and erosion as a result of wave activity. It is likely that mangroves situated in estuarine/deltaic settings are likely to be protected from oceanic influences and resultant sediment loss from being situated in a higher energy environment than those located in fringing/atoll settings. Furthermore, mangroves located in estuarine/deltaic settings are more likely to receive high terrestrial sediment loads as a result of fluvial deposition, although this is obviously dependent on catchment geology and geomorphology as well as anthropogenic factors such as sand mining in the catchment and damming upstream. In the Mekong, and to a lesser extent Red River, deltas damming has been suggested to decrease sedimentation by 40% (Veetil et al., 2018). There are currently >100 dams along the Amazon River and tributaries and a further 80 dams at the proposal stage; the effects that these dams have on sediment transport are likely to have impacts on mangroves beyond the immediate Amazon estuary itself and extend along the muddy coastlines of the Guyana and on up to Venezuela, with resultant impacts on responses of mangroves to sea level rise (Latrubesse et al., 2017).

In a global data review of the impacts of sea level rise of mangroves utilizing surface accretion rate data, it was found that basin mangroves were likely to be more resilient than fringing mangroves at lower rates of sea level rise (Sasmito et al., 2016). This was linked to greater suspended sediment concentrations and resultant surface accretion rates found in basin mangroves compared to fringing mangroves. They also found that degraded mangroves (those experiencing multiple stressors) were also less likely to be able to keep pace with sea level rise, most likely related to root mortality, organic matter oxidation, and peat collapse, combined with decreases in aboveground biomass and resultant increases in water flow velocities and erosion and/or decreases in sediment deposition. Conversely, restored sites presented the highest positive surface elevation change as a result of root expansion and increased sediment trapping, together with sites impacted by storms as a result of episodic sediment dumping (e.g., Hurricane Wilma, Whelan et al. 2009). Different species are able to influence surface accretion rates as a result of different aboveground biomass structures such as densities and size of prop roots or pneumatophores such as *Avicennia* spp., *Sonneratia* spp., and *Rhizophora* spp., as a result of a reduction in water velocity as a result of the drag force (Krauss et al., 2003). This factor should be taken into account when considering species selection for restoration sites.

8. Regions with mangroves under threat from sea level rise

8.1 North America

Decreases in terrestrial vegetation cover as a result of tree mortality is occurring in low elevation areas with partial replacement by *Avicennia germinans* in Sugarloaf Key, Florida, and throughout the US coast of the Gulf of Mexico (Ross et al., 2009). In the Mississippi River Delta, USA, a combination of drought and low river outflow has led to widespread dieback of the saltmarsh-dominant *Spartina alterniflora*. More drought-tolerant *A. germinans* has been found to be unaffected, and the absence of killing freezes during this period was found to have resulted in a rapid expansion of *A. germinans* cover in former saltmarsh-dominated areas (McKee et al., 2004). On the other hand, increasing storminess, although potentially threatening, has not been faced vis-à-vis mangrove resilience. Utilizing the Sea Level Affecting Marshes Model (SLAMM), Geselbrecht et al. (2015) suggest that there will be a net gain in mangrove forest cover in the Gulf Coast of Florida (40% increase in area from their study areas), principally as a result of expansion in areas of coastal forest and adjacent undeveloped dryland. However, the results of this study suggested that in some areas there would be a net loss of mangrove area with losses to open water. The United States has a broad network of RSETs (180 in the Gulf of Mexico region), as well as high-resolution airborne LiDAR data covering many mangrove forest areas in the country (Osland et al., 2017), and thus, a lot of work on the impacts of sea level rise on mangrove ecosystems has been focused here. Osland et al. (2017) have proposed the formation of a coordinated and structured network of RSETs to evaluate the impacts of sea level rise over a broader region and identify factors that may confer resilience as well as predict and prepare for the response of mangroves to global change. While sea level rise is likely to be one of the principal factors influencing the ability of mangroves to cope with climate change, it should be noted that other climatic drivers should be taken into account. Osland et al. (2016) suggest that mangroves in the Gulf of Mexico are also particularly prone to the influence of macroclimatic changes (including sea level rise and temperature) and, in the west of this region and the Gulf of California, also to changes in precipitation (drought).

8.2 Indo-Pacific

Lovelock et al. (2015) undertook an extensive study of the vulnerability mangroves in the Indo-Pacific region to sea level rise utilizing a network of RSETs distributed across SE Asia, Australia, New Zealand, and the western Pacific Ocean. Factors that conferred substantial resilience were high SARs and subsurface organic matter accumulation, which resulted in high rates of positive surface elevation change. Other factors studied that had less influence on resilience were geomorphology (river delta, tidal, lagoon, or carbonate island), rate of sea level rise, and ecological classification (fringe, scrub, hammock, basin, overwash, or riverine), the latter through the availability of sediment and ecogeomorphic feedback.

The impacts of sea level rise on mangroves in Asia have been relatively well studied in some countries particularly Vietnam and Indonesia. In Vietnam, the majority of studies have focused on the Mekong and Red River deltas in the South and North of the country, respectively (Veetil et al., 2018). Tide gauge data suggest that recent (last 30 years) rates of sea level rise for the region are ~3 mm/yr (MONRE, 2012), slightly less than observed global rates of sea level rise (3.2 mm/yr) derived from satellite data (IPCC, 2013). However, recent studies have recorded decreases in sediment discharge in both of these deltas as a result of damming in the upper reaches for hydropower and sediment mining (Van et al., 2015a,b; Schmitt et al., 2017). Both the Red River and Mekong deltas are experiencing relatively rapid rates of land subsidence (Red River Delta 0.9 mm/yr, Neumann et al. (2015); Mekong Delta 10–40 mm/yr, Erban et al. (2014)). Losses in mangrove area in these deltas are likely to be geographically variable; Nguyen and Woodroffe (2016) have proposed that the western part of the Mekong Delta is likely to be at a greater threat from sea level rise due to its distal location from the principal distributaries and resultant lower sediment accumulation rates. However, there are suggestions that mangroves in Asia may be more resilient to sea level rise compared with other regions due to high plant species diversity (Ward et al., 2016b) that causes greater interspecific facilitation (Huxham et al., 2010) and increased belowground root production (Lang'at et al., 2014).

8.3 South America

The landward expansion of mangroves observed in most estuaries of the NE coast in Brazil represents the most important adaptation to global warming and regional environmental changes depicted by these ecosystems. However, this necessitates the existence of a large buffer zone along the mangrove-land border (Ferreira and Lacerda, 2016; Godoy and Lacerda, 2015; Godoy et al., 2018). Apart from geological and/or geomorphological restrictions discussed previously, anthropogenic activities in these areas, such as legal and illegal urban and tourism expansion, agriculture, and aquaculture, may pose strong constraints to mangrove adaptation and even survival.

In many countries, although true mangrove forests are protected natural areas, legislation is often feeble or nonexistent regarding salt pans bordering mangroves on their landward margin. In Brazil, which harbors the second largest mangrove area in the world, the Forest Code established in 1965 not only considered mangroves as fully protected areas but also included broad expanses of salt flats, locally known as "apicuns," as an integral part of the mangrove ecosystem. These flats stretch to the maximum influence of the tides or the extreme reaches of saline intrusion (Junk et al., 2014). The interpretation by developers and local authorities to acknowledge these "apicuns" as permanently dry habitats, not considering the natural, broad temporal variation in the water level as an inherent attribute of the system, has, unfortunately, led to a revision of the original 1965 Forest Code. This review released most of these salt pans to aquaculture and infrastructure development. In the northeastern region of Brazil alone, this "new opened area" to development may reach over 600,000 ha. In addition, the review of the old Forest Code has also allowed forest conversion in alleged "public utility" or "social interest" projects, even in permanent protection areas, such as mangroves (Oliveira-Filho et al., 2016). In the Latin America and the Caribbean region, occupation of these buffer zones by shrimp aquaculture is increasing to over 20% per year, mostly as a result of this weakening of legislation (Lacerda et al., 2019). This change of legislation character may drastically reduce mangrove resilience to sea level change.

8.4 Africa

Compared with other global regions, there is a paucity of data concerning the impacts of sea level rise on African mangrove systems. However, from data available, many sites are minerogenic in nature with lower fluvial sediment supply than in comparable systems in Asia, Oceania, or the Americas (Milliman and Meade, 1983). Development around mangrove margins in Africa is typically lower than in the Americas and Asia, suggesting that migration inland as a result of the greater available accommodation space is a possible scenario. In West Africa, a study from Cameroon suggested that long-term low sediment supply combined with recent sea level rise together with degradation (trampling and removal of trees for firewood) resulted in losses of fringing mangroves (Ellison and Zouh, 2012). In the Niger Delta, sea level rise is exacerbated by subsidence (7 mm/yr) as a result of the extraction of hydrocarbons; this is combined with widespread degradation of mangroves in the area as a result of oil spills (Ayanlade et al., 2016), which are likely to result in losses of mangroves (Musa et al., 2016). Only isolated studies have been undertaken in South Africa, where it has been suggested that mangroves in small estuaries such as Mngazana are likely to lose up to 50% of the current area (Yang et al., 2014), although this study used SARs from atoll mangroves in the Caribbean alone, so this may not be an accurate representation of the response of these mangroves to sea level rise as it fails to account for local SARs and biological/geological influences on surface elevation change. Research undertaken in Kenya suggests that SARs in Gazi Bay mangroves are as high as 4.7 mm/yr, far in excess of local sea level rise for the region (1.1 mm/yr).

9. Conclusions

The impacts of sea level rise are strongly influenced by local factors with resilience conferred by positive surface elevation change linked to uplift high suspended sediment loads and high productivity (above- and belowground biomass). However, multiple stressors, e.g., extreme weather events (drought, wind and wave damage, high temperatures exceeding maximum tolerance), plus global change factors including pollution, deforestation, conversion to aquaculture, urban expansion, lack of accommodation space, and subsidence are likely to exacerbate the impacts of sea level rise, resulting in localized losses. Other factors that are likely to influence mangrove resilience to sea level rise are species diversity and type, geochemical barriers, and well-implemented management and protection legislation.

There is also a substantial variation in data concerning the current status of mangroves and their responses to sea level rise, and it should be noted that these are substantial knowledge gaps to conservation and maintenance of the ecosystem services that mangroves provide, with Africa being a notably understudied area. As has been suggested elsewhere, mounting a global campaign to undertake an evaluation of the impacts of sea level rise using RSETs, radionuclide dating as well as methods described by the Blue Carbon Initiative (Howard et al. 2014), combined with better communication with policy makers and land managers, is suggested to secure a more stable future for mangrove ecosystems in the face of sea level rise and combined global change stressors.

References

Alongi, D.M., 2015. The impact of climate change on mangrove forests. Curr. Clim. Chang. Rep. 1, 30–39.

Agraz-Hernández, C.M., Río-Rodríguez, R.E., Chan-Keb, C.A., Osti-Saenz, J., Muñiz-Salazar, R., 2018. Nutrient removal efficiency of *Rhizophora mangle* (L.) seedlings exposed to experimental dumping of municipal waters. Diversity 10 (16), 2–15.

Alothman, A., Bos, M.S., Fernades, R.M.S., Ayham, M.E., 2014. Sea level rise in the north-western part of the Arabian Gulf. J. Geodyn. 81, 105–110.

Andersen, T.J., 2017. Some practical considerations regarding the application of ^{210}Pb and ^{137}Cs dating to estuarine sediments. Dev. Paleoenviron. Res. 121–140. https://doi.org/10.1007/978-94-024-0990-1_6.

Appleby, P., Oldfield, F., 1992. Application of lead-210 to sedimentation studies. In: Harmon, S. (Ed.), Uranium Series Disequilibrium: Application to Earth, Marine and Environmental Science. Oxford Scientific Publications, Oxford, UK, pp. 731–783.

Ayanlade, A., Howard, M., 2016. Environmental impacts of oil production in the Niger Delta: remote sensing and social survey examination. Afr. Geogr. Rev. 35, 272–293.

Barth, H.-J., Böer, B., 2002. Sabkha Ecosystems. Volume I. The Arabian and Peninsula and Adjacent Countries. Kluwer Academic Publishers, Dordrecht, pp. 1–5.

Cahoon, D.R., Lynch, J.C., Hensel, P., Boumans, R., Perez, B., Segura, B., Day Jr., J.W., 2002. A device for high precision measurement of wetland sediment elevation: I. Recent improvements to the sedimentation-erosion table. J. Sediment. Res. 72, 730–733.

Castaneda-Moya, E., Rivera-Monroy, V., Chambers, R.M., Zhao, X., Lamb-Wotton, L., Gorsky, A., Gaiser, E.E., Troxler, T.G., Kominoski, J.S., Hiatt, M., 2020. Hurricanes fertilize mangrove forests in the Gulf of Mexico (Florida Everglades, USA). Proc. Natl. Acad. Sci. U S A 117 (9), 4831–4841.

Castañeda-Moya, E., Twilley, R., Rivera-Monroy, V., Zhang, K., Davis III, S., Ross, M., 2010. Sediment and nutrient deposition associated with hurricane Wilma in mangroves of the Florida Coastal Everglades. Estuar. Coast 33, 45–58.

Cazenave, A., Palanisamy, H., Ablain, M., 2018. Contemporary sea level changes from satellite altimetry: what have we learned? What are the new challenges? Adv. Space Res. 62, 1639–1653.

Church, J.A., White, N.J., 2011. Sea-level rise from the late 19th to the early 21st century. Surv. Geophys. 32 (4–5), 585–602.

Cordero-Oviedo, C., Correa-Metrio, A., Urrego, L., Vázquez, G., Blaauw, M., Escobar, J., Curtis, J., 2019. Holocene establishment of mangrove forests in the western coast of the Gulf of Mexico. Catena 180, 212–223.

Costa, B.G.B., Lacerda, L.D., 2014. Mercury (Hg) in fish consumed by the local population of the Jaguaribe River lower basin, Northeast Brazil. Environ. Sci. Pollut. Control Ser. 21, 13335–13341.

Dias, F.J.S., Castro, B.M., Lacerda, L.D., 2013. Continental shelf water masses off Jaguaribe River (4°S) – Northeastern, Brazil. Continent. Shelf Res. 66, 123–135.

Drexler, J., Fuller, C., Archfield, S., 2018. The approaching obsolescence of ^{137}Cs dating of wetland soils in North America. Quat. Sci. Rev. 199, 83–96.

Ellison, J., 2015. Vulnerability assessment of mangroves to climate change and sea-level rise impacts. Wetl. Ecol. Manag. 23, 115–137.

Ellison, J., Stoddart, D., 1993. Mangrove ecosystem collapse during predicted sea-level rise: Holocene analogues and implications. J. Coast Res. 7, 151–165.

Ellison, J., Zouh, I., 2012. Vulnerability to climate change of mangroves: assessment from Cameroon, Central Africa. Biology 1 (3), 617–638.

Erban, L.E., Gorelick, S.M., Zebker, H.A., 2014. Groundwater extraction, land subsidence, and sea-level rise in the Mekong Delta, Vietnam. Environ. Res. Lett. 9 (8), 084010.

Ferreira, A.C., Lacerda, L.D., 2016. Degradation and conservation of Brazilian mangroves, status and perspectives. Ocean Coast Manag. 125, 38–46.

França, M., Alves, I., Castro, D., Cohen, M., Rossetti, D., Pessenda, L., Lorente, F., AraújoFontes, N., Buso Junior, A., Fonseca Giannini, P., IzarFrancisquini, M., 2015. A multi-proxy evidence for the transition from estuarine mangroves to deltaic freshwater marshes, Southeastern Brazil, due to climatic and sea-level changes during the late Holocene. Catena 128, 155–166.

França, M.C., Pessenda, L.C., Cohen, M.C., de Azevedo, A.Q., Fontes, N.A., Silva, F.B., Macario, K., 2019. Late-Holocene subtropical mangrove dynamics in response to climate change during the last millennium. Holocene 29 (3), 445–456.

Friess, D.A., Krauss, K.W., Horstman, E.M., Balke, T., Bouma, T.J., Galli, D., Webb, E.L., 2012. Are all intertidal wetlands naturally created equal? Bottlenecks, thresholds and knowledge gaps to mangrove and saltmarsh ecosystems. Biol. Rev. 87, 346–366.

Friess, D.A., Rogers, K., Lovelock, C., Krauss, K., Hamilton, S., Lee, S., Lucas, R., Primavera, J., Rajkaran, A., Shi, S., 2019. State of the world's mangrove forests: past, present and future. Annu. Rev. Environ. Resour. 44, 89–115.

Furukawa, K., Wolanski, E., 1996. Sedimentation in mangrove forests. Mangroves Salt Marshes 1, 3–10.

Geselbracht, L.L., Freeman, K., Birch, A.P., Brenner, J., Gordon, D.R., 2015. Modeled sea level rise impacts on coastal ecosystems at six major estuaries on Florida's Gulf coast: implications for adaptation planning. PLoS One 10 (7), e0132079.

Gilman, E.L., Ellison, J., Duke, N.C., Field, C.D., 2008. Threats to mangroves from climate change and adaptation options: a review. Aquat. Bot. 89, 237–250.

Godoy, M.D.P., Lacerda, L.D., 2014. River-island response to land-use change within the Jaguaribe River, Brazil. J. Coast Res. 30, 399–410.

Godoy, M.D.P., Lacerda, L.D., 2015. Mangroves response to climate change: a review of recent findings on mangrove extension and distribution. Anais da Academia Brasileira de Ciências 87 (2), 651–667.

Godoy, M.D.P., Meirelles, J.A., Lacerda, L.D., 2018. Mangrove response to land use change in estuaries along the semiarid coast of Ceará, Brazil. J. Coast Res. 34 (3), 524–533.

Guo, Z., Li, X., He, Z., Yang, Y., Wang, W., Zhong, C., Greenberg, A., Wu, C., Duke, N., Shi, S., 2018. Extremely low genetic diversity across mangrove taxa reflects past sea level changes and hints at poor future responses. Glob. Chang. Biol. 24, 1741–1748.

He, B., Lai, T., Fan, H., Wang, W., Zheng, H., 2007. Comparison of flooding-tolerance in four mangrove species in a diurnal tidal zone in the Beibu Gulf. Estuarine. Coast. Shelf Sci. 74, 254–262.

Howard, J., Hoyt, S., Isensee, K., Pidgeon, E., Telszewski, M. (Eds.), 2014. Coastal Blue Carbon: Methods for Assessing Carbon Stocks and Emissions Factors in Mangroves, Tidal Salt Marshes, and Seagrass Meadows. Conservation International, Intergovernmental Oceanographic Commission of UNESCO, International Union for Conservation of Nature, Arlington, Virginia, USA.

Howari, F.M., Jordan, B.R., Bouchouche, N., Wyllie-Echeverria, S., 2009. Field and remote sensing assessment of mangrove forests and seagrass beds in the northwest part of the United Arab Emirates. J. Coast Res. 25 (1), 48–56.

Huxham, M., Kumara, M.P., Jayatissa, L.P., Krauss, K.W., Kairo, J., Langat, J., Mencuccini, M., Skov, M.W., Kirui, B., 2010. Intra-and interspecific facilitation in mangroves may increase resilience to climate change threats. Philos. Trans. R. Soc. 365, 2127–2135.

IPCC, 2013. Climate change 2013: the physical science basis. In: Contribution of Working Group I to the Fifth Assessment Report of the Intergovernmental Panel on Climate. Cambridge University Press, Cambridge, United Kingdom.

Jennerjahn, T.C., Gilman, E., Krauss, K.W., Lacerda, L.D., Nordhaus, I., Wolanski, E., 2017a. Climate change. In: Rivera-Monroy, V.H., Lee, S.Y., Kristensen, E., Twilley, R.R. (Eds.), Mangrove Ecosystems: A Global Biogeographic Perspective. Springer Verlag, pp. 211–244 (Chapter 7).

Jennerjahn, T.C., Gilman, E., Krauss, K., Lacerda, L.D., Nordhaus, I., Wolanski, E., 2017b. Mangrove ecosystems under climate change. Mangrove Ecosystems: A Global Biogeographic Perspective. In: Rivera-Monroy, V.H., Lee, S.Y., Kristensen, E., Twilley, R.R. (Eds.), Mangrove Ecosystems: A Global Biogeographic Perspective. Springer International Publishing, Switzerland.

Jones, M.C., Wingard, G.L., Stackhouse, B., Keller, K., Willard, D., Marot, M., Landacre, B., Bernhardt, C., 2019. Rapid inundation of southern Florida coastline despite low relative sea-level rise rates during the late-Holocene. Nat. Commun. 10, 3231.

Junk, W.J., Piedade, M.T.F., Lourival, R., Wittmann, F., Kandus, P., Lacerda, L.D., Agostinho, A., 2014. Brazilian wetlands: definition, delineation and classification for sustainable management and protection. Aquat. Conserv. Mar. Freshw. Ecosyst. 24, 5–22.

Krauss, K., Allen, J., Cahoon, D., 2003. Differential rates of vertical accretion and elevation change among aerial root types in Micronesian mangrove forests. Estuar. Coast Shelf Sci. 56 (2), 251–259.

Krauss, K., Cahoon, D., Allen, J., Ewel, K., Lynch, J., Cormier, N., 2010. Surface elevation change and susceptibility of different mangrove zones to sea-level rise on Pacific High Islands of Micronesia. Ecosystems 13, 129–143.

Krauss, K., McKee, K., Lovelock, C., Cahoon, D., Saintilan, N., Reef, R., Chen, L., 2014. How mangrove forests adjust to rising sea level. New Phytol. 202, 19–34.

Krauss, K., Osland, M., 2020. Tropical cyclones and the organization of mangrove forests: a review. Ann. Bot. 125 (2), 213–234.

Krauss, K., Twilley, R., Smith III, T.J., Whelan, K., 2005. Woody debris in the mangrove forests of south Florida. Biotropica 37, 9–15.

Lacerda, L.D., 1994. Biogeochemistry of heavy metals in coastal lagoons. In: Kjerfve, B. (Ed.), Coastal Lagoon Processes. Elsevier, Amsterdam, pp. 221–242.

Lacerda, L.D., 2018. Burial of mangroves by mobile dunes: a climate change threat in semiarid coasts. ISME/GLOMIS Electron. J. 16 (2), 6–10.

Lacerda, L.D., Machado, W., Moscatelli, M., 2000. Use of mangroves in landfill management. GLOMIS Electron. J. 1 (1), 1–2.

Lacerda, L.D., Dias, F.J.S., Marins, R.V., Soares, T.M., Godoy, J.M., Godoy, M.L.D.P., 2013. Pluriannual watershed discharges of Hg into tropical semi-arid estuary of the Jaguaribe River, NE Brazil. J. Braz. Chem. Soc. 24, 1719–1731.

Lacerda, L.D., Borges, R., Ferreira, A.C., 2019. Neotropical mangroves: conservation and sustainable use in a scenario of global climate change. Aquat. Conserv. Mar. Freshw. Ecosyst. 29, 1347–1364.

Lacerda, L.D., 1998. Biogeochemistry of Trace Metals and Diffuse Pollution in Mangrove Ecosystems. Okinawa: International Society for Mangrove Ecosystems, Japan, 65pp.

Lacerda, L.D., Carvalho, C.E., Tanizaki, K.F., Ovalle, A.R.C., Rezende, C.E., 1993. The biogeochemistry and trace metals distribution of mangrove rhizospheres. Biotropica 25, 252–257.

Lang'at, J.K.S., Kairo, J.G., Mencuccini, M., Bouillon, S., Skov, M.W., Waldron, S., Huxham, M., 2014. Rapid losses of surface elevation following tree girdling and cutting in tropical mangroves. PLoS One 10 (3), e0118334.

Latrubesse, E., Arima, E., Dunne, T., Park, E., Baker, V., d'Horta, F., Wight, C., Wittmann, F., Zuanon, J., Baker, P., Ribas, C., Norgaard, R., Filizola, N., Ansar, A., Flyvbjerg, B., Stevaux, J., 2017. Damming the rivers of the Amazon basin. Nature 546, 363–369.

Lei, P., Zhong, H., Duan, D., Pan, K., 2019. A review on mercury biogeochemistry in mangrove sediments: hotspots of methylmercury production? Sci. Total Environ. 680, 140–150.

Lelieveld, J.P., Hadjinicolaou, E., Kostopoulou, J., Chenowet, h M., El Maayar, C., Giannakopoulos, C., Hannides, M.A., Lange, M., Tanarhte, E., Tyrlis, E., Xoplaki, E., 2012. Climate change and impacts in the eastern Mediterranean and the Middle East. Clim. Chang. 114, 667–687.

Li, Z., Saito, Y., Mao, L., Tamura, T., 2012. Mid-Holocene mangrove succession and its response to sea-level change in the upper Mekong River delta, Cambodia. Quat. Res. 78 (2), 386–399.

Limaye, R., Kumaran, K., Padmalal, D., 2014. Mangrove habitat dynamics in response to Holocene sea level and climate changes along southwest coast of India. Quat. Int. 325, 116–125.

Lougheed, B., Obrochta, S., Lenz, C., Mellström, A., Metcalfe, B., Muscheler, R., Reinholdsson, M., Snowball, I., Zillén, L., 2017. Bulk sediment ^{14}C dating in an estuarine environment: how accurate can it be? Paleoceanogr. Paleoclimatol. 32 (2), 123–131.

Lovelock, C., Cahoon, D.R., Friess, D.A., Guntenspergen, G.R., Krauss, K.W., Reef, R., Rogers, K., Saunders, M., Sidik, F., Swales, A., Saintilan, N., Thuyen, L., Triet, T., 2015. The vulnerability of Indo-Pacific mangrove forests to sea-level rise. Nature 526, 559–563.

Lovelock, C., Feller, I.C., Reef, R., Hickey, S., Ball, M., 2017. Mangrove dieback during fluctuating sea levels. Sci. Rep. 7, 1680.

Machado, M., Moscatelli, M., Rezende, L.G., Lacerda, L.D., 2002. Mercury, zinc, and copper accumulation in mangrove sediments surrounding a large landfill in southeast Brazil. Environ. Pollut. 120, 455–461.

Mafi-Gholami, D., Ward, R.D., 2019. Assessment of the probability of occurrence of multiple Environmental hazards in mangrove habitats using remote sensing and geographic information system. J. Environ. Stud. 44 (3), 425–443.

Mafi-Gholami, D., Zenner, E., Jaafari, A., Ward, R.D., 2018. Modeling multi-decadal mangrove leaf area index in response to drought along the semi-arid southern coasts of Iran. Sci. Total Environ. 656, 1326–1336.

Maia, L.P., Freire, G.S.S., Lacerda, L.D., 2005. Accelerated dune migration and sand transport during El Niño events along the NE Brazilian coast. J. Coast Res. 21, 1121–1126.

Maia, L.P., Lacerda, L.D., Monteiro, L.H.U., Souza, G.M., 2006. Atlas dos Manguezais do Nordeste do Brasil: Avaliação das Áreas de Manguezais dos Estados do Piauí, Ceará, Rio Grande do Norte, Paraíba e Pernambuco. Fortaleza: Secretaria do Meio Ambiente do Estado do Ceará, Brazil.

Makowski, C., Finkl, C., 2018. Threats to Mangrove Forests: Hazards, Vulnerability, and Management. Springer Verlag, Berlin.

Marengo, J.A., Alves, L.M., Alvala, R.C.S., Cunha, A.P., Brito, S., Moraes, O.L.L., 2018. Climatic characteristics of the 2010–2016 droughts in the semiarid Northeast Brazil region. Academia Brasileira de Ciências 90, 1973–1985.

McKee, K.L., Mendelssohn, I., Materne, M., 2004. Acute salt marsh dieback in the Mississippi River deltaic plain: a drought-induced phenomenon? Glob. Ecol. Biogeogr. 13, 65–67.

McKee, K.L., Cahoon, D.R., Feller, I.C., 2007. Caribbean mangroves adjust to rising sea level through biotic controls on change in soil elevation. Glob. Ecol. Biogeogr. 16, 545–556.

Milliman, J., Meade, R., 1983. World-wide delivery of river sediment to the oceans. J. Geol. 91, 1–21.

MONRE, 2012. National Strategy on Climate Change. Ministry of Natural Resources and Environment, Hanoi.

Moore, G.E., Grizzle, R.E., Ward, K.M., Alshihi, R.M., 2015. Distribution, pore-water chemistry, and stand characteristics of the mangroves of the United Arab Emirates. J. Coast Res. 31 (4), 957–963.

Mounier, S., Lacerda, L.D., Marins, R.V., Benaim, J., 2001. Copper and mercury complexing capacity of organic matter from a mangrove mud flat environment (Sepetiba Bay, Brazil). Bull. Environ. Contam. Toxicol. 67, 519–525.

Moura, V.L., Lacerda, L.D., 2018. Contrasting mercury bioavailability in the marine and fluvial dominated areas of the Jaguaribe River, Ceará, Brazil. Bull. Environ. Contam. Toxicol. 101, 49–54.

Musa, Z., Popescu, I., Mynett, A., 2016. Assessing the sustainability of local resilience practices against sea level rise impacts on the lower Niger delta. Ocean Coast Manag. 130, 221–228.

Nascimento Jr., W.R., Souza-Filho, P.W., Proisy, C., Lucas, R.M., Rosenqvist, A., 2013. Mapping changes in the largest continuous Amazonian mangrove belt using object-based classification of multisensor satellite imagery. Estuar. Coast Shelf Sci. 117, 83–93.

Nerem, R., Chamber, D., Choe, C., Mitchum, G., 2010. Estimating mean sea level change from the TOPEX and Jason Altimeter Missions. Mar. Geodes. 33, 435–446.

Neumann, J., Emanuel, K., Ravela, S., Ludwig, L., Verly, C., 2015. Risks of coastal storm surge and the effect of sea level rise in the Red River delta, Vietnam. Sustainability 7 (6), 6553–6572.

Nguyen, T.T.X., Woodroffe, C.D., 2016. Assessing relative vulnerability to sea-level rise in the western part of the Mekong River Delta in Vietnam. Sustain. Sci. 11 (4), 645–659.

NOAA, 2015. National Oceanic and Atmospheric Administration. National Ocean Service, Center for Operational Oceanographic Products and Services, Tides and Currents. http://tidesandcurrents.noaa.gov/.

Oliveira-Filho, R.R., Rovai, A.S., Menghini, R.P., Coelho-Júnior, C., Novelli, Y.S., Cintrón, G., 2016. On the impact of the Brazilian Forest Code on mangroves: a comment to Ferreira and Lacerda (2016). Ocean Coast Manag. 132, 36–37.

Osland, M.J., Enwright, N.M., Day, R.H., Gabler, C.A., Stagg, C.L., Grace, J.B., 2016. Beyond just sea-level rise: considering macroclimatic drivers within coastal wetland vulnerability assessments to climate change. Glob. Chang. Biol. 22, 1–11.

Osland, M.J., Griffith, K.T., Larriviere, J.C., Feher, L.C., Cahoon, D.R., Enwright, N.M., 2017. Assessing coastal wetland vulnerability to sea-level rise along the northern Gulf of Mexico coast: gaps and opportunities for developing a coordinated regional sampling network. PLoS One 12 (9), e0183431.

Ouyang, X., Guo, F., 2016. Paradigms of mangroves in treatment of anthropogenic wastewater pollution. Sci. Total Environ. 544, 971–979.

Rietbroek, R., Brunnabend, S., Kusche, J., Schröter, J., Dahle, C., 2016. Revisiting the sea-level budget. Proc. Natl. Acad. Sci. U S A 113 (6), 1504–1509.

Robertson, A.I., Phillips, M.J., 1995. Mangroves as filters of shrimp pond effluent: predictions and biogeochemical research needs. Hydrobiologia 295 (1–3), 311–321.

Ross, M.S., O'Brien, J.J., Ford, R.G., Zhang, K., Morkill, A., 2009. Disturbance and the rising tide: the challenge of biodiversity management on low-island ecosystems. Front. Ecol. Environ. 7, 471–478.

Rubén, L., Szlafsztein, C., Cohen, M., Berger, U., Glaser, M., 2002. Implication of mangroves dynamics for private land use in Bragança, North Brazil: a case study. J. Coast. Coast Conserv. 8, 97–102.

Saderne, V., Cusack, M., Almahasheer, H., Serrano, O., Masqué, P., Arias-Ortiz, A., Kadinjappalli, P., Lotfi, K., Mohammad, R., Qurban, A., Duarte, C., 2018. Accumulation of carbonates contributes to coastal vegetated ecosystems keeping pace with sea level rise in an arid region (Arabian Peninsula). J. Geophys. Res. Biogeosci. 123, 1498–1510.

Saintilan, N., Rogers, K., Kelleway, J., Ens, E., Sloane, D.R., 2018. Climate change impacts on the coastal wetlands of Australia. Wetlands 39, 1145–1154.

Sasmito, S.D., Murdiyarso, D., Friess, D.A., Kurnianto, S., 2016. Can mangroves keep pace with contemporary sea level rise? A global data review. Wetl. Ecol. Manag. 1–16.

Schmitt, R.J.P., Rubin, Z., Kondolf, G.M., 2017. Losing ground - scenarios of land loss as consequence of shifting sediment budgets in the Mekong Delta. Geomorphology 294, 58–69.

Schuerch, M., Spencer, T., Temmerman, S., Kirwan, M., Wolff, C., Lincke, D., McOwen, C., Pickering, M., Reef, R., Vafeidis, A., Hinkel, J., Nicholls, R., Brown, S., 2018. Future response of global coastal wetlands to sea-level rise. Nature 561, 231–234.

Van, M.N., Dung, N.V., Hung, N.N., Kummu, M., Merz, B., Apel, H., 2015a. Future sediment dynamics in the Mekong Delta floodplains: impacts of hydropower development, climate change and sea level rise. Glob. Planet. Chang. 127, 22–33.

Van, T.T., Wilson, N., Tung, H.T., Quisthoudt, K., Minh, V.Q., Tuan, L.X., Guebas, F.D., Koedam, N., 2015b. Changes in mangrove vegetation area and character in a war and land use change affected region of Vietnam (Mui Ca Mau) over six decades. Acta Oecol. 63, 71–81.

Veetil, B., Ward, R.D., Quang, N., Trang, N., Giang, T., 2018. Mangroves of Vietnam: historical development, current state of research and future threats. Estuarine. Coast. Shelf Sci. 218, 212–236.

Ward, R.D., Burnside, N., Joyce, C., Sepp, K., Teasdale, P.A., 2014. Recent rates of sedimentation on irregularly flooded Boreal Baltic coastal wetlands: responses to recent changes in sea level. Geomorphology 217, 61–72.

Ward, R.D., Burnside, N., Joyce, C., Sepp, K., Teasdale, P.A., 2016a. Improved modelling of the impacts of sea level rise on coastal wetland plant communities. Hydrobiol. Wetl. Biodivers. Process. 1–14.

Ward, R.D., Friess, D.A., Day, R., Mackenzie, R., 2016b. Impacts of climate change on mangrove ecosystems: a region by region overview. Ecosys. Health Sustain. 2 (4), e01211.

Webb, E., Friess, D.A., Krauss, K., Cahoon, D., Guntenspergen, G., Phelps, J., 2013. A global standard for monitoring coastal wetland vulnerability to accelerated sea level rise. Nat. Clim. Chang. 3, 458–465.

Woodroffe, C., Thom, B.G., Chappell, J., 1985. Development of widespread mangrove swamps in mid-Holocene times in northern Australia. Nature 317, 711–713.

Woodroffe, C., 1995. Response of tide-dominated mangrove shorelines in Northern Australia to anticipated sea-level rise. Earth Surf. Process. Landforms 20, 65–85.

Woodroffe, C., Grindrod, J., 1991. Mangrove biogeography: the role of Quaternary environmental and sea level change. J. Biogeogr. 18, 479–492.

Yang, S., Riddin, T., Adams, J., Shih, S., 2014. Predicting the spatial distribution of mangroves in a South African estuary in response to sea level rise, substrate elevation change and a sea storm event. J. Coast Conserv. 18, 459–469.

Zemp, M., Huss, M., Thibert, E., Eckert, N., McNabb, R., Huber, J., Barandun, M., Machguth, H., Nussbaumer, S., Gärtner-Roer, I., Thomson, J., Paul, F., Maussion, F., Kutuzov, S., Cogley, J., 2019. Global glacier mass changes and their contributions to sea-level rise from 1961 to 2016. Nature 568, 382–386.

CHAPTER

11

Does geomorphology determine vulnerability of mangrove coasts to sea-level rise?

Karen L. McKee[1], *Ken W. Krauss*[1], *Donald R. Cahoon*[2]

[1]**U.S. Geological Survey, Wetland and Aquatic Research Center, Lafayette, LA, United States;** [2]**U.S. Geological Survey, Patuxent Wildlife Research Center, Beltsville, MD, United States**

OUTLINE

1. Introduction 256

2. Methods 260

2.1 *Review of published elevation change and accretion rates* 260

2.2 *Surface elevation change and component variables* 261

2.3 *Calculation of relative sea-level rise, vertical land development, and deficits* 262

2.4 *Site-specific comparison (Belize vs. Micronesia)* 262

2.5 *Statistical analyses* 263

3. Results and discussion 263

3.1 *Global dataset* 263

3.2 *Biogenic versus minerogenic settings* 266

3.3 *Management considerations* 268

3.4 *Surmountable challenges in determining vulnerability to sea-level rise* 269

4. Conclusions 269

Acknowledgments 270

References 270

https://doi.org/10.1016/B978-0-12-816437-2.00005-7

1. Introduction

Low-lying coastal wetlands are vulnerable to relative sea-level rise (RSLR, combined ocean and land surface movements) and must build their soil surfaces vertically at an equivalent rate or become submerged. In spite of this simple relationship, the processes involved are complex and interconnected by feedback linkages, and their relative influences vary across sedimentary settings (Cahoon et al., 2006; McKee, 2011; Kirwan and Guntenspergen, 2012). Soil surface elevation at any given time is a function of processes such as (1) sedimentation or erosion of mineral matter and/or accretion of organic detritus (fallen leaves, woody debris) on the soil surface; (2) shallow subsurface movement of two strata (root zone, sub-root zone) resulting from expansion of growing roots, decomposition of soil organic matter, shrink-swell due to water fluxes, and physical compaction of deposits; and (3) deeper, geologic movements (tectonics, isostasy) (Krauss et al., 2014; Woodroffe et al., 2016). Depending on the geomorphological setting, however, rates and drivers of elevation change in coastal wetlands may differ greatly among geographic locations. In addition, differences in ocean warming can generate spatial and temporal variation in the rate at which the ocean surface height changes (Meyssignac and Cazenave, 2012).

Mangrove-dominated wetlands, which serve as nursery grounds for marine life (Mumby et al., 2004) and buffers against tropical cyclones and tsunamis (Barbier, 2006; Marois and Mitsch, 2015; Cochard et al., 2008), occur across a broad spectrum of sedimentary settings and thereby provide an opportunity to examine controls on soil elevation and relative vulnerability of tropical coastlines to RSLR. Mangroves occupy 137,600 km^2 of global land area (Bunting et al., 2018), but can account for a proportionally large percentage of land area on oceanic islands where they provide critical goods and services to support subsistence, economic development, and coastal protection (Ewel et al., 1998a; Himes-Cornell et al., 2018). These intertidal ecosystems develop on different substrates and in different geomorphic settings, ranging from alluvial sediment in riverine settings to peat deposits on carbonate platforms (Woodroffe et al., 2016). Stability of mangrove coastlines depends partly upon inorganic sediment deposition for vertical accretion, but accumulation of organic matter can be of equal or greater importance (McKee, 2011). Conceptual models of mangrove habitat response to sea-level rise often recognize subsurface processes but emphasize allochthonous sedimentary processes as a primary mechanism for how mangroves respond to sea-level rise (Soares, 2009). However, even where coastal topography allows landward migration of mangroves, vertical land development must balance RSLR to avoid submergence, and sediments are not available in all settings (Ellison and Stoddart, 1991).

We tested the paradigm that mangrove-dominated coasts where autochthonous processes prevail, i.e., peat formation and/or calcareous sedimentation, are less likely to keep pace with RSLR than minerogenic settings receiving large inputs of allochthonous sediment (McKee, 2011; Woodroffe, 1990; Ellison, 1993; Parkinson et al., 1994; Middleton and McKee, 2001; Gonzalez et al., 2010). We analyzed data from 11 published studies (see Table 11.1) reporting elevation change, accretion, and subsidence rates based on the surface elevation table-marker horizon (SET-MH) system, which allows high-resolution measurements of surface and subsurface land movements (Cahoon et al., 2002a, 2002b; Callaway et al., 2013). To better understand differences between biogenic and minerogenic sedimentary settings, we also

TABLE 11.1 Summary of published data on mangrove elevation dynamics in different geomorphic settings (carbonate platform, tide-dominated, riverine), including local sea-level trend based on tide gauge records; elevation (elev.) change, surface accretion, and shallow subsidence (or expansion) based on the Surface Elevation Table-Marker Horizon system; relative sea-level rise (RSLR) (tide gauge trend + shallow subsidence); and elevation or accretion deficit (tide gauge trend minus elev. change or accretion). Negative values for subsidence or deficits indicate expansion or surplus, respectively. Tide gauge data are as reported by the original reference, except for Source 11, which was acquired from http://www.psmsl.org.

ID	Location	Site	Geomorphic setting	Tide gauge	Elev. change	Surface accretion	Shallow subsidence	RSLR	Elev. deficit	Accr. deficit	Source
1	Rookery Bay, Florida, USA	Fringe	Carbonate	2.3	3.5	7.8	4.3	6.6	−1.2	−5.5	1
2	Rookery Bay, Florida, USA	Basin	Carbonate	2.3	3.7	6.0	2.3	4.6	−1.4	−3.7	1
3	Rookery Bay, Florida, USA	Island shoreline	Carbonate	2.3	2.5	6.3	3.8	6.1	−0.2	−4	1
4	Rookery Bay, Florida, USA	Island interior	Carbonate	2.3	0.6	4.4	3.8	6.1	1.7	−2.1	1
5	Gulf of Fonseca, Honduras	Interior	Riverine	1.0	4.6	2.9	−1.7	−0.7	−3.6	−1.9	2
6	Gulf of Fonseca, Honduras	Shoreline	Riverine	1.0	4.6	2.9	−1.7	−0.7	−3.6	−1.9	2
7	Roatan, Honduras	Interior	Carbonate	2.0	4.8	2.0	−2.8	−0.8	−2.8	0.0	3
8	Roatan, Honduras	Shoreline	Carbonate	2.0	4.8	2.0	−2.8	−0.8	−2.8	0.0	3
9	Roatan, Honduras	Shoreline	Carbonate	2.0	4.8	2.0	−2.8	−0.8	−2.8	0.0	3
10	New South Wales, Australia	Cararma Inlet	Tide	4.1	−0.8	3.0	3.8	8.0	4.9	1.1	4
11	New South Wales, Australia	Currambene Creek	Tide	4.1	0.3	0.7	0.4	4.5	3.8	3.4	4
12	Victoria, Australia	French Island	Tide	2.7	−2.1	9.5	11.6	14.3	4.8	−6.8	4
13	New South Wales, Australia	Homebush Bay	Riverine	0.9	5.6	4.6	−1.1	−0.2	−4.7	−3.7	4

(Continued)

TABLE 11.1 Summary of published data on mangrove elevation dynamics in different geomorphic settings (carbonate platform, tide-dominated, riverine), including local sea-level trend based on tide gauge records; elevation (elev.) change, surface accretion, and shallow subsidence (or expansion) based on the Surface Elevation Table-Marker Horizon system; relative sea-level rise (RSLR) (tide gauge trend + shallow subsidence); and elevation or accretion deficit (tide gauge trend minus elev. change or accretion). Negative values for subsidence or deficits indicate expansion or surplus, respectively. Tide gauge data are as reported by the original reference, except for Source 11, which was acquired from http://www.psmsl.org.—cont'd

ID	Location	Site	Geomorphic setting	Tide gauge	Elev. change	Surface accretion	Shallow subsidence	RSLR	Elev. deficit	Accr. deficit	Source
14	New South Wales, Australia	Kooragang Island	Riverine	0.3	1.9	4.7	2.7	3.1	−1.7	−4.4	4
15	Victoria, Australia	Kooweerup	Tide	2.7	0.0	7.2	7.2	9.9	2.7	−4.5	4
16	Victoria, Australia	Quail Island	Tide	2.7	−2.6	6.8	9.4	12.0	5.3	−4.1	4
17	Victoria, Australia	Rhyll/Phillip Island	Tide	2.66	0.9	5.1	4.2	6.8	1.7	−2.44	4
18	New South Wales, Australia	Ukerebagh Island	Tide	−0.4	2.2	2.2	0.0	−0.4	−2.6	−2.6	4
19	New South Wales, Australia	Minnamurra River	Tide	−0.5	0.6	6.6	6.0	5.6	−1.1	−7.1	4
20	Twin Cays, Belize	Transition	Carbonate	2.3	−1.1	2.0	3.1	5.4	3.4	0.3	5
21	Twin Cays, Belize	Dwarf	Carbonate	2.3	−3.7	0.7	4.4	6.7	6.0	1.6	5
22	Twin Cays, Belize	Fringe	Carbonate	2.3	4.1	1.6	−2.5	−0.2	−1.8	0.7	5
23	Everglades, Florida, USA	Shark River	Riverine	1.9	1.4	6.5	5.1	7.0	0.5	−4.6	6
24	Kosrae, Micronesia	Yela shoreline	Tide	1.8	−3.0	11.6	14.6	16.4	4.8	−9.8	7
25	Kosrae, Micronesia	Yela riverine	Riverine	1.8	−2.7	12.9	15.6	17.4	4.5	−11.1	7
26	Kosrae, Micronesia	Yela interior	Tide	1.8	1.3	12.0	10.7	12.5	0.5	−10.2	7
27	Kosrae, Micronesia	Utwe shoreline	Tide	1.8	1.2	11.9	10.7	12.5	0.6	−10.1	7

TABLE 11.1 Summary of published data on mangrove elevation dynamics in different geomorphic settings (carbonate platform, tide-dominated, riverine), including local sea-level trend based on tide gauge records; elevation (elev.) change, surface accretion, and shallow subsidence (or expansion) based on the Surface Elevation Table-Marker Horizon system; relative sea-level rise (RSLR) (tide gauge trend + shallow subsidence); and elevation or accretion deficit (tide gauge trend minus elev. change or accretion). Negative values for subsidence or deficits indicate expansion or surplus, respectively. Tide gauge data are as reported by the original reference, except for Source 11, which was acquired from http://www.psmsl.org.—cont'd

ID	Location	Site	Geomorphic setting	Tide gauge	Elev. change	Surface accretion	Shallow subsidence	RSLR	Elev. deficit	Accr. deficit	Source
28	Kosrae, Micronesia	Utwe riverine	Riverine	1.8	6.3	18.7	12.4	14.2	−4.5	−16.9	7
29	Kosrae, Micronesia	Utwe interior	Tide	1.8	1.3	12.9	11.6	13.4	0.5	−11.1	7
30	Pohnpei, Micronesia	Sapwalap riverine	Riverine	1.8	−0.6	14.1	14.7	16.5	2.4	−12.3	7
31	Pohnpei, Micronesia	Sapwalap shoreline	Tide	1.8	−2.3	4.1	6.4	8.2	4.1	−2.3	7
32	Pohnpei, Micronesia	Sapwalap interior	Tide	1.8	0.9	8.2	7.3	9.1	0.9	−6.4	7
33	Pohnpei, Micronesia	Enipoas shoreline	Tide	1.8	−5.8	6.6	12.4	14.2	7.6	−4.8	7
34	Pohnpei, Micronesia	Enipoas riverine	Riverine	1.8	−1.4	6.3	7.7	9.5	3.2	−4.5	7
35	Pohnpei, Micronesia	Enipoas interior	Tide	1.8	−2.8	2.9	5.7	7.5	4.6	−1.1	7
36	Rookery Bay, Florida, USA	Fringe	Carbonate	2.1	0.6	5.7	5.1	7.2	1.5	−3.6	8
37	Rookery Bay, Florida, USA	Basin	Carbonate	2.1	3.9	2.0	−1.9	0.2	−1.8	0.1	8
38	Rookery Bay, Florida, USA	Basin	Carbonate	2.1	1.1	7.6	6.5	8.6	1.0	−5.5	8
39	Queensland, Australia	Moreton Bay east	Tide	1.9	5.8	8.0	2.2	4.1	−3.9	−6.1	9
40	Queensland, Australia	Moreton Bay west	Tide	1.9	1.7	9.2	7.5	9.4	0.2	−7.3	9
41	Louisiana, USA	Port Fourchon	Riverine	9.0	3.8	9.8	6.0	15.0	5.2	−0.8	10
42	Hainan, China	Sanjiang	Tide	6.3	8.6	13.2	4.6	10.9	−2.3	−6.9	11
43	Hainan, China	Houpai	Tide	6.3	2.2	2.7	0.5	6.8	4.1	3.6	11
44	Hainan, China	Houpai	Tide	6.3	7.8	7.4	−0.4	5.9	−1.5	−1.1	11

(Continued)

TABLE 11.1 Summary of published data on mangrove elevation dynamics in different geomorphic settings (carbonate platform, tide-dominated, riverine), including local sea-level trend based on tide gauge records; elevation (elev.) change, surface accretion, and shallow subsidence (or expansion) based on the Surface Elevation Table-Marker Horizon system; relative sea-level rise (RSLR) (tide gauge trend + shallow subsidence); and elevation or accretion deficit (tide gauge trend minus elev. change or accretion). Negative values for subsidence or deficits indicate expansion or surplus, respectively. Tide gauge data are as reported by the original reference, except for Source 11, which was acquired from http://www.psmsl.org.—cont'd

ID	Location	Site	Geomorphic setting	Tide gauge	Elev. change	Surface accretion	Shallow subsidence	RSLR	Elev. deficit	Accr. deficit	Source
Summary statistics											
		Mean		2.2	1.6	6.5	4.9	7.3	0.8	−4.1	
		SE		0.2	0.5	0.6	0.8	0.8	0.5	0.7	
		Maximum		9.0	8.6	18.7	15.6	17.4	7.6	3.6	
		Minimum		−0.5	−5.8	0.7	−2.8	−0.8	−4.7	−16.9	

Data sources as follows: (1) Cahoon and Lynch (1997); (2) Cahoon et al. (2002); (3) Cahoon et al. (2003); (4) Rogers et al. (2006); (5) McKee et al. (2007); (6) Whelan et al. (2009); (7) Krauss et al. (2010); (8) McKee (2011); (9) Lovelock et al. (2015); (10) McKee and Vervaeke (2018); (11) Fu et al. (2018).

compared elevation dynamics on oceanic islands representing a sedimentation gradient: peat-forming low islands in the Caribbean (Belize) versus Pacific high islands (Micronesia), which receive moderate to high inputs of mineral sediment.

2. Methods

2.1 Review of published elevation change and accretion rates

Data from 11 published studies (McKee, 2011; McKee and Vervaeke, 2018; Cahoon and Lynch, 1997; Cahoon et al., 2002c, 2003; Rogers et al., 2006; McKee et al., 2007; Whelan et al., 2009; Krauss et al., 2010; Lovelock et al., 2015a; Fu et al., 2018) conducted in mangrove-dominated sites and representing a range of Holocene sea level histories (Rogers et al., 2019) were used for this analysis (summarized in Table 11.1). Inclusion required that both accretion and elevation change rates were reported and that the methods used to acquire the data were marker horizons and SETs, respectively. We considered data from both the pipe-style SET technique (Boumans and Day, 1993), having a maximum depth of insertion of up to 6 m, and the new Rod SET (Cahoon et al., 2002a), having a maximum depth of insertion of up to 28 m. Subsidence was calculated as the difference between elevation change and accretion rates for all sites for consistency (some studies reported subsidence rates derived from linear trends). Furthermore, we did not include datasets that were influenced significantly by anthropogenic activity or that were associated with restoration recovery. For sites with multiple reports, we used the most recent data representing the longest observation period. Sasmito et al. (2016) provided a listing of studies from which to choose; only a few failed to meet our data selection requirements. The study locations that we selected included Florida, USA

(Rookery Bay, The Everglades); Louisiana, USA (Port Fourchon); Belize (Twin Cays); Honduras (Bay Islands, Gulf of Fonseca); Federated States of Micronesia (Kosrae, Pohnpei); Hainan, China (Sanjiang, Houpai), and Australia (New South Wales, Victoria, Queensland). Although techniques were the same, the period of observation varied among studies (1–8 years). Tide range (0.2–3 m) and local sea-level trend from tide gauges (−0.5 to 9.0 mm yr^{-1}) also varied across study sites. Because not all possible geomorphic settings were represented in this dataset, the mangrove sites were assigned to three broad categories following the classification scheme described by Woodroffe (Woodroffe, 1992), including river-dominated (riverine) and tide-dominated, as well as mangroves that occur on carbonate platforms (Table 11.1), which can also occur in a mix of tide- and wave-dominated environments.

2.2 Surface elevation change and component variables

The following description, illustrated in Fig. 11.1, summarizes the basic protocols used to measure elevation dynamics in the reviewed studies. Soil surface elevations were tracked

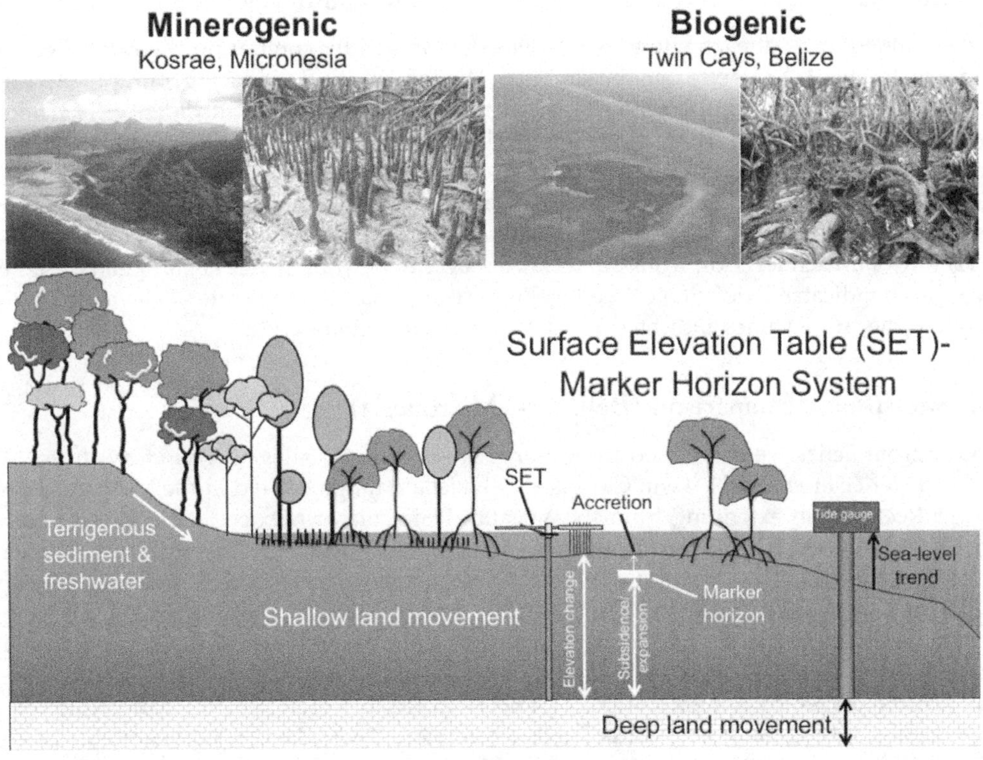

FIGURE 11.1 Top Panel: Aerial and ground views of a minerogenic (Kosrae, Federated States of Micronesia) and a biogenic (Twin Cays, Belize) mangrove system. Bottom Panel: Diagram illustrating the surface elevation table-marker horizon system used to measure elevation change, surface accretion, and shallow subsurface movement in mangrove forests. Sea-level trends acquired from nearby tide gauges reflect change in ocean height plus deep land movement. *Photos courtesy of USDA Forest Service/The Nature Conservancy, Nicole Cormier (U.S. Geological Survey), Karen McKee (U.S. Geological Survey), and Candy Feller (Smithsonian Institution).*

in situ using surface elevation tables (SETs) (Cahoon et al., 2002a, 2002b), which at all study sites consisted of a benchmark rod (or pipe) driven to the point of refusal. This technology allowed for high-precision, repetitive measurements of soil surface elevations relative to the benchmark base. Benchmark rods were installed permanently, and a portable table was fit into the exact same location on the benchmark rod for each measurement series. On each measurement date, nine pins were extended from the table to the soil surface in four directions for a total of 36 measurements for each SET. The height of the pins above the table was recorded to the nearest millimeter.

Simultaneously, the build-up of mineral sediment and organic matter (e.g., leaf litter fall) on the soil surface (surface accretion) was independently tracked by laying marker horizons of artificial soil (e.g., feldspar clay or sand) near each SET (Cahoon and Lynch, 1997). Shallow subsidence was calculated as the difference between surface accretion and elevation change rates (for each SET-MH station). However, some sites exhibited subsurface expansion (i.e., elevation change exceeded accretion), which resulted in a negative value for subsidence.

2.3 Calculation of relative sea-level rise, vertical land development, and deficits

The vulnerability of each site to RSLR was determined by comparing the rate of vertical land development to the rate of submergence. The rate of RSLR for each geographic location was estimated by adding the sea-level trend to the rate of shallow subsidence determined with the SET-MH system (Fig. 11.1). Sea-level data based on long-term tide gauge records were as reported in the original studies, except for one record that was acquired from the Permanent Service for Mean Sea Level (http://www.psmsl.org) (Table 11.1). Elevation or accretion deficits for all sites were calculated by subtracting the elevation change (SET) or accretion (MH) rates for each location from the local sea-level trend (tide gauge) (Cahoon, 2015). Positive values indicated a deficit, i.e., sea level was rising faster than the rate of elevation gain or accretion; negative values indicated an elevation or accretion surplus.

2.4 Site-specific comparison (Belize vs. Micronesia)

Data from Belize were collected from mangrove-dominated sites on Twin Cays from 2001 to 2005 (McKee et al., 2007). Twin Cays is a 75-ha island range located in the Meso-American Barrier Reef system extending from the Yucatan Peninsula of Mexico to the Bay Islands of Honduras. This site is ca. 17 km from the mainland, has no source of terrigenous sediment, and is underlain by 10–12 m of peat. Data from Micronesia were collected from two watersheds each on Kosrae (Yela and Utwe) and Pohnpei (Enipoas and Sapwalap), Federated States of Micronesia, from 1997 to 2004 (Krauss et al., 2010). Kosrae and Pohnpei contain 1600 and 5300 ha, respectively, of mangroves and have high rates of sedimentation from upland erosion driven by annual rainfall of 4.8–5.4 m (Krauss et al., 2010). Islands differed in age (Kosrae, 1.2–2.6 million; Pohnpei, 3.0–8.6 million; Twin Cays, 8000 years old) as well as forest structure and dominant species. Mean tide range was greater in Micronesia (0.7 m) than in Belize (0.2 m); surface elevations (relative to a local datum) varied from shoreline to interior as follows: 0.17–0.92 m (Micronesia) and −0.09–0.002 m (Belize). Average forest heights varied in Belize from the shoreline (5–7 m) to interior (1–2.5 m) (McKee et al.,

2007) and in Micronesia from the shoreline (17–21 m) to interior (24–30 m) (Ewel et al., 1998b). The dominant mangrove tree species in Belize was *Rhizophora mangle*, while Micronesian forests were composed of mixed communities of *Rhizophora apiculata*, *Sonneratia alba*, and *Bruguiera gymnorrhiza*.

SETs were installed in Belize in three forest zones (shoreline, transition, interior) in each of three replicate sites (total = 9); three replicate SETs were installed in Micronesia in each of three zones (shoreline, interior, riverine) in two river basins on two islands (total = 36). Benchmarks were inserted to refusal at depths of 9–12 m in Belize and 2–5 m in Micronesia. Marker horizons were established with carbonate sand, feldspar clay, or a mixture. Due to bioturbation, horizons in Micronesia provided comparative accretion data for only 1.5 years; however, reapplication of marker horizons expanded the overall accretion record to 2.7–2.9 years. For Micronesia, the records from two tide gauges on Pohnpei (Pohnpei B: 1976–2009, Pohnpei C: 2002–10) were used. For Belize, the tide gauge record at Key West (1913–2010) was used as this was the nearest long-term record. The respective sea-level trends for Micronesia and Belize were 1.8 and 2.3 mm yr^{-1}, which reflected sea level changes plus deep subsidence (below the tide gauge benchmark).

2.5 Statistical analyses

Data from published studies were analyzed with a fixed effects model ANOVA using geomorphic setting (riverine, tide-dominated, carbonate) as a grouping factor (JMP, Version 14.0.0) (SAS, 2018). Bivariate relationships between elevation components were assessed with correlation analysis. Site-specific data were analyzed with ANOVA, using site (Twin Cays, Kosrae, Pohnpei) and position nested within site (shoreline, interior, and transition or riverine) as grouping factors. Since site and spatial position were preselected, a fixed effects model was used. Post-ANOVA multiple comparisons were conducted with Tukey's HSD. Some data were log transformed to equalize variance prior to analysis; when variance homogeneity could not be achieved, a Welch's ANOVA (testing equal means, given standard deviations not equal) was used.

3. Results and discussion

3.1 Global dataset

Previous reviews (Cahoon et al., 2006; McKee, 2011; Krauss et al., 2010; Sasmito et al., 2016) report a broad range in rates of surface elevation change, from a loss of 9.5–20 mm yr^{-1} for mangrove coastlines undergoing peat collapse after a hurricane (Cahoon et al., 2003) to a gain of 11.3 mm yr^{-1} due to storm-driven sedimentation (Whelan et al., 2009). Values for sites included in our study also showed substantial variation in rate and direction of elevation change (from a loss of 5.8 to a gain of 8.6 mm yr^{-1}; Table 11.1). Surface accretion rates varied from 0.6 to 18.7 mm yr^{-1} (Table 11.1, Fig. 11.2) and differed significantly among geomorphic types (log-transformed data, $F_{2,41} = 4.68$, $P = .015$), suggesting a greater potential for vertical land development with greater mineral input. Sites receiving river sediment had the highest accretion rates (8.3 ± 1.7 mm yr^{-1}); those in carbonate settings had the lowest

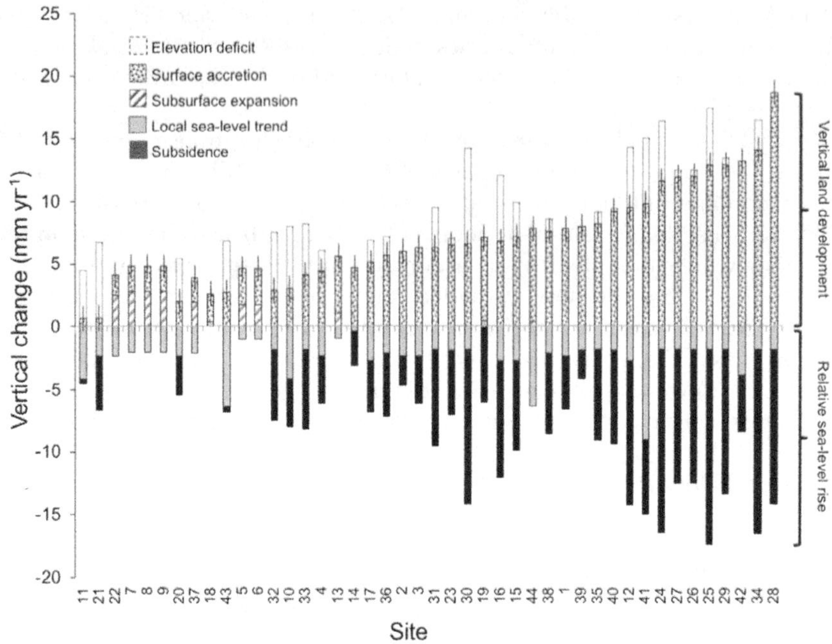

FIGURE 11.2 Data from published studies (see Table 11.1), ranked by increasing rates of accretion, illustrate the balance between rates of relative sea-level rise (RSLR) and vertical land development; the difference equals the elevation deficit. Local sea-level trends (from tide gauge records) and shallow subsidence (from Surface Elevation Table-Marker Horizon measurements) are plotted as negative values to facilitate visual comparison with positive land building rates (accretion plus subsurface expansion). Elevation deficits are also plotted to show how much additional accretion or subsurface expansion would be required to balance RSLR. Values (n = 44) are single observations for each location or the mean ± 1 standard error in cases of replicate plots (accretion only). *x*-axis labels indicate the site identification number (see Table 11.1).

(3.9 ± 0.7 mm yr^{-1}), and tide-dominated sites were intermediate (7.2 ± 0.8 mm yr^{-1}) and not significantly different from the riverine sites (Tukey's HSD). Despite differences in accretion rates, there were no significant differences in elevation change rates among geomorphic types. Elevation change rates on carbonate sites (2.3 ± 0.7 mm yr^{-1}) were similar to riverine sites (2.4 ± 1.0 mm yr^{-1}) and tide-dominated sites (0.7 ± 0.8 mm yr^{-1}) ($F_{2,41} = 1.35$, $P = .27$). The lack of correspondence between accretion and elevation change patterns was due to the additional movement occurring below the soil surface (i.e., subsidence or subsurface expansion) (Table 11.1, Fig. 11.2). In this dataset, higher subsidence rates occurred in tide-dominated (6.5 ± 1.0 mm yr^{-1}) and riverine (6.0 ± 2.1 mm yr^{-1}) settings compared to carbonate (1.6 ± 1.0 mm yr^{-1}) settings (Welch's ANOVA: $F_{2,20.4} = 6.49$, $P = .0066$).

If surface accretion is compared to long-term sea-level trends (tide gauge records), all three geomorphic types exhibit an average accretion surplus (sea-level trend − accretion = negative value), with tide-dominated sites (−4.6 ± 0.9 mm yr^{-1}) intermediate to carbonate sites (−1.7 ± 0.7 mm yr^{-1}) and riverine sites (−6.2 ± 1.7 mm yr^{-1}) ($F_{2,41} = 3.74$, $P = .032$), which differed significantly (Tukey's HSD). The lack of an accretion deficit suggests that most of the sites have the capacity to keep pace with historical rates of RSLR, and that those with the greatest surface accretion (riverine and tide-dominated) are best buffered against future RSLR. However, such a comparison does not fully assess relative vulnerability because it fails to incorporate shallow subsidence, which often exceeds local sea-level trends derived from tide gauge records (Table 11.1). Long-term trends in sea level (tide gauge), which reflect changes in ocean height plus deep land movement, were thus added to shallow subsidence (derived from SET-MH measurements) to calculate a rate of RSLR for each location (Rybczyk and Cahoon, 2002). This calculation showed that elevation gains due to high sediment accretion were often offset by high rates of subsidence, and sites with low sediment accretion compensated through shallow subsurface expansion (Table 11.1). When vertical land development (accretion plus subsurface expansion) was slower than the RSLR, an elevation deficit occurred (Table 11.1, Fig. 11.2). The mean elevation deficit at tide-dominated sites (1.9 ± 0.7 mm yr^{-1}) was higher than at riverine or carbonate sites, which showed a slight elevation surplus (-0.2 ± 1.2, -0.1 ± 0.8 mm yr^{-1}), but differences were not significant ($F_{2,41} = 2.28$, $P = .115$).

Relative vulnerability of individual sites can be depicted by plotting surface accretion in relation to the RSLR rate; sites with values falling below a 1:1 line were not keeping pace (Fig. 11.3). Viewed this way, there is not a consistent pattern showing that geomorphic settings with greater sediment accretion (e.g., riverine) are less vulnerable than those with minimal accretion (e.g., carbonate platform); not quite half of the sites examined in this analysis were keeping pace with RSLR (Fig. 11.3). Along with the lack of a correlation between accretion and elevation change ($r = 0.001$, $P > .05$) or elevation deficit ($r = 0.011$, $P > .05$), these results show that mineral accretion alone is not predictive of the capacity to keep up with sea-level rise and that consideration of all surface and subsurface movements in relation to

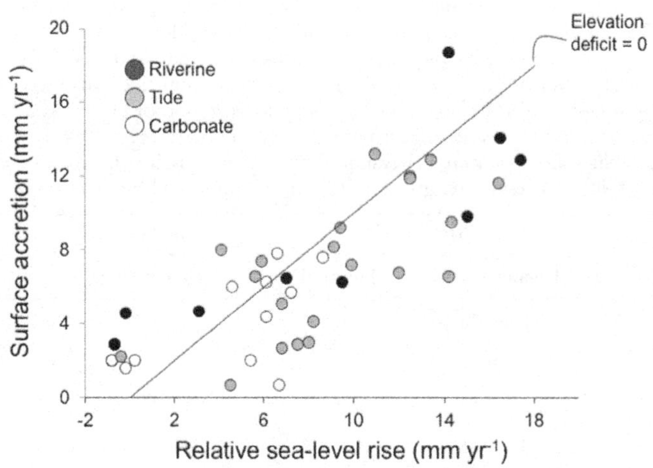

FIGURE 11.3 Vulnerability of mangrove sites (Table 11.1) to sea-level rise. The scatterplot shows rates of surface accretion in relation to the relative sea-level rise (RSLR) rates across geomorphic types (riverine, tide-dominated, carbonate platform) represented in the dataset. The line indicates where the elevation deficit is zero, i.e., accretion is equal to the rate that sea level is rising. Mangrove sites below the line are not keeping pace with RSLR (sea-level trend from tide gauges plus shallow subsidence from SET-MH).

local sea-level trends is necessary to assess the potential for submergence (Fig. 11.3). In fact, there was a general pattern of increasing subsidence rate with increasing accretion rate (with some exceptions) ($r = 0.78$, $P < .0001$), which may reflect one or more correlated processes: (1) subsidence creates accommodation space for sediment deposition and thereby promotes accretion (Woodroffe et al., 2016); (2) accretion influences subsidence by contributing to overburden and compression of underlying strata (Cahoon et al., 2000; van Asselen, 2011); (3) sediment particle density (and related pore space) varies with rate and type of accreting material, which influences the potential for compaction (Toscano et al., 2018); and/or (4) an indirect association exists due to an unknown causal factor or autocorrelation. Although we cannot distinguish among these possibilities based on these data, they do suggest future areas of investigation.

3.2 Biogenic versus minerogenic settings

To better understand controls on elevation dynamics in biogenic versus minerogenic systems, we compared oceanic mangrove islands representing the accretion gradient depicted in Fig. 11.2. Peat-forming low islands in the Caribbean kept pace with RSLR as well as Micronesian islands with higher rates of accretion (Table 11.2). The Pacific high island of Kosrae displayed some of the highest sustained rates of accretion recorded in the literature (12–19 mm yr^{-1}) (Krauss et al., 2010). At the other extreme were Caribbean low islands (Twin Cays, Belize) without inputs of terrigenous sediment where accretion rates were low (0–4.5 mm yr^{-1}). On Pohnpei, intermediate rates of accretion were found (3–14 mm yr^{-1})

TABLE 11.2 Comparison of elevation dynamics in minerogenic (Kosrae), intermediate (Pohnpei), and biogenic (Twin Cays) mangrove systems. The observation period was 1.4–3 yr for Micronesia and 3 yr for Belize (McKee et al., 2007). Shallow subsidence was calculated as the difference between accretion and elevation change (a negative value indicates subsurface expansion); relative sea-level rise (RSLR) equals the local trend from tide gauge records (1.8, 2.3 mm yr^{-1} for Micronesia, Belize) plus shallow subsidence; accretion or elevation deficit equals the local tide gauge trend minus surface accretion or elevation change (negative value indicates a surplus). Values are the mean ± 1 standard error (n = 9, Belize; 6, Micronesia); analysis of variance (ANOVA) results are given and P value is indicated by superscript.

	Micronesia		Belize	ANOVA (F-ratio)	
Variable	Kosrae	Pohnpei	Twin Cays	Site	Position
Elev. Change	0.7 ± 1.4a	−2.0 ± 0.9a	−0.2 ± 1.4a	1.3	2.1
Surface accretion	13.3 ± 1.1a	7.0 ± 1.6b	1.5 ± 0.5c	39.3^{b}	1.3
Subsidence/Expansion	12.6 ± 0.8a	9.0 ± 1.5a	1.7 ± 1.6b	18.0^{a}	1.4
RSLR	14.4 ± 0.8a	10.8 ± 1.5a	4.0 ± 1.6b	16.3^{a}	1.4
Accr. Deficit/Surplus	−11.5 ± 1.1a	−5.2 ± 1.6b	0.8 ± 0.5c	42.9^{b}	1.3
Elev. Deficit/Surplus	1.1 ± 1.4a	3.8 ± 0.9a	2.5 ± 1.4a	1.2	2.1

[a](P ≤ .001) *or.*
[b](P ≤ .0001); *differences among means are indicated by letters (Tukey's HSD).*

(Krauss et al., 2010). Despite higher accretion rates, however, Micronesian sites exhibited only slight gains or a loss in elevation, which were not significantly different overall from those in Belize (Table 11.2). The rate of RSLR was three times higher for Micronesian sites compared to Belize, which indicated a greater submergence potential and, hence, greater requirement for vertical land building to keep pace (Table 11.2). This requirement was met at only one riverine site on Kosrae where accretion exceeded the RSLR rate (i.e., a surplus: −4.5 mm yr^{-1}), but all other sites showed elevation deficits (0.5–4.8 mm yr^{-1}). Pohnpei, which had significantly lower accretion rates compared to Kosrae, had similar rates of subsidence and elevation deficits (Table 11.2). Although spatial position (riverine, interior, fringe) was not a significant effect for any variable, indicating no consistent spatial pattern, there was considerable within-site variation (Table 11.2), which provides clues as to site-specific controls on resilience to RSLR.

A key difference between biogenic and minerogenic mangrove coastlines resides in the nature of the processes contributing to soil volume and, hence, vertical movement of the soil surface. Shoreline stands in Micronesia, which had the greatest elevation deficits (2.7 mm yr^{-1} Kosrae, 5.9 mm yr^{-1} Pohnpei) (site #24, 31, 33), were highly erosional, of low elevation relative to mean sea level, and with slower growing trees than other mangrove forests on the same island (Krauss et al., 2007). Interior and riverine plots on both islands varied, however, depending on the site-specific balance between accretion and subsidence, and showed no consistent pattern (site # 25–26, 28–30, 32, 34–35). In Belize, greatest elevation deficits (3.4, 6.0 mm yr^{-1}) occurred in transitional and interior stands (site # 20, 21) with stunted trees of low productivity, whereas an elevation surplus (−1.8 mm yr^{-1}) occurred in more productive shoreline stands (site #22) where subsurface expansion, rather than subsidence, prevailed. The latter pattern is mainly attributable to spatial variation in root production rates relative to decomposition, which determines whether shallow subsurface movement is positive or negative in these peat-forming systems (Middleton and McKee, 2001; McKee et al., 2007). The peat substrate contains living and dead mangrove roots and other organic components (McKee, 2011; McKee et al., 2007). The uppermost stratum (~30 cm) is essentially a living mat composed of actively growing roots and filamentous algae (McKee, 2011). Because of their refractory nature and the anaerobic and nutrient-poor conditions of the soil environment, roots decay slowly and tend to accumulate (Middleton and McKee, 2001). The result is a tough, tight matrix of fibrous material with high shear strength that not only resists erosion (McKee, 2011) but also increases in volume as more roots are produced (McKee et al., 2007). Even though peat has a higher compaction potential than clastic materials (van Asselen et al., 2009), its contribution to subsidence may not be fully realized due to continual expansion of the living mat (McKee, 2011) and the relative resistance of dead roots and water voids to compaction with soil depth (Toscano et al., 2018). If mangrove productivity declines or mortality occurs, however, decomposition and compaction processes continue, leading to a cessation of soil expansion and eventual elevation deficit (Cahoon et al., 2003; McKee et al., 2007). By comparison, unconsolidated mineral sediment can undergo progressive compaction and has a lower shear strength and greater potential for re-suspension and erosion (Whelan et al., 2009). The bulk density of mineral sediment also imparts a greater loading stress (overburden) on underlying deposits compared to an equivalent volume of peat (van Asselen, 2011). Compaction of peat layers interspersed with denser mineral layers can increase subsidence in deltaic sequences (van Asselen, 2011; Tornqvist et al., 2008) and might be a factor causing faster subsidence in mangrove settings receiving periodic pulses of mineral sediment (tide-dominated and riverine settings) or sand or marl (carbonate settings). Some compaction is necessary, however, to create accommodation space for continued clastic sedimentation or peat formation (Woodroffe et al., 2016; Tornqvist et al., 2008). Shallow subsidence occurring beneath the expanding, live root mat in Belize was estimated to be 5.5–6.5 mm yr^{-1} (McKee et al., 2007). This subsidence rate was counterbalanced by the expanding root mat along shorelines but not in interior stands where conditions limited root growth (McKee et al., 2007). The fact that the Belize system has built vertically for 7000 to 8000 years, accumulating over 10 meters of peat, is evidence of its capacity to keep up with past rates of RSLR via peat formation (McKee et al., 2007). None of these mangrove systems, however, may survive a rise in sea level that exceeds its capacity to keep pace; our analysis does not preclude the possibility that these systems with different geomorphologies may differ with respect to the upper limit of RSLR that could lead to permanent submergence.

3.3 Management considerations

Our analysis suggests that although biogenic and minerogenic systems may not differ in overall vulnerability to historical rates of RSLR, they are not necessarily equally sensitive to the same external and internal forces controlling soil elevations. Both biological and physical processes influence elevation dynamics in mangrove systems, but their relative contributions differ with sedimentary setting and have different consequences for vertical movement.

Minerogenic mangrove settings, such as in river deltas, rely on high sedimentation rates to retain their vertical positions in the intertidal plane. Mineral sedimentation is influenced by various internal and external physical factors, not all related to changing sea level. Spatial variation in sediment supply and erosive forces (as well as trapping by aerial roots) determine whether accretion rates balance subsidence in minerogenic settings such as Micronesia (Krauss et al., 2010). Human activities that affect sediment supplies (dams, levees, deforestation) would alter elevation dynamics in such settings (Lovelock et al., 2015b). In addition, spatial and temporal variation in sea-level trends and geophysical phenomena could further influence the relative vulnerability of these sites, e.g., drought years superimposed on years with temporary sea-level reductions (Lovelock et al., 2017). In contrast, the regional variation in mean sea-level trend (based on satellite altimetry) suggests hotspots with rates up to three times higher than the global average, e.g., in the western Pacific (Cazenave and Llovel, 2010), and particularly in relation to tropical Pacific islands (Becker et al., 2012). The combination of high rates of sea-level rise locally and high subsidence rates might also substantially increase vulnerability of mangrove coastlines in such regions.

Biogenic mangroves, at the other extreme, may be more vulnerable to changes directly affecting organic matter accumulation, and may need greater protections from forest clearing or eutrophication events that may, even temporarily, affect surface elevations (Lang'at et al., 2014). Vulnerability of such systems also may vary with regional rates of sea-level rise, and peat formation may be especially responsive to changing sea level due to feedback effects of flooding and salinity on root growth (Kirwan and Guntenspergen, 2012; McKee, 1996) and decomposition (Middleton and McKee, 2001), which can be further influenced by other drivers of climate change (Ward et al., 2016; Alongi, 2015).

3.4 Surmountable challenges in determining vulnerability to sea-level rise

Processes that control vulnerability of mangroves to sea-level rise vary spatially; however, time-scale bias and choice of assessment method can also influence the interpretation of how mangroves respond to RSLR and other submergence factors along coastlines (e.g., impoundments that isolate mangroves hydrologically). In our analysis, the SET-MH system in conjunction with tide gauge trends provided a relatively short-term estimate of elevation change rates. Other methods that integrate over longer time periods, such as dating of sediment cores using ^{137}Cs, ^{210}Pb, or ^{14}C, may estimate lower rates of vertical land development (Breithaupt et al., 2018) potentially indicating greater vulnerability (Parkinson et al., 2017). Radiometric methods, however, also have limitations and sources of error. For example, radiometric dating of cores can produce estimates of sediment accumulation rates, but cannot distinguish between surface and subsurface contributions to elevation gain, as the SET-MH system can. A combination of methods may overcome some of these limitations, e.g., radiocarbon dating of peat cores in conjunction with SET-MH measurements to compare past rates with modern rates of vertical land development (McKee et al., 2007).

Spatial replication of standardized SET-MH stations in different types of mangrove systems (Webb et al., 2013) can aid in identifying general patterns of elevation change associated with different sedimentary settings. In addition, calculations should be standardized globally but interpreted regionally. For example, the standard way of determining vulnerability of a mangrove forest to rising sea level is to compare elevation data (from SETs) with tide gauge trends, assuming that both are installed relative to similar survey data, are close enough in space, are aligned temporally, and are subjected to similar rates of deep subsidence (Cahoon, 2015). Alternatively, approaches that make use of vertical accretion of sediments in comparison with RSLR (i.e., including direct measures of shallow subsidence from SETs, deep subsidence from repetitive GPS surveys, and regional sea-level trends) offer another option (Swales et al., 2016; Jankowski et al., 2017). These methods of evaluating vulnerability may produce different results but can be informative. For example, McKee and Vervaeke (McKee and Vervaeke, 2018) calculated vulnerability of a salt marsh-mangrove community in coastal Louisiana using different time intervals and different ways of calculating RSLR. That approach showed that the community was not keeping pace with either recent or long-term rates of RSLR and was experiencing a minimum elevation deficit of 3 mm yr^{-1}. This approach of using different methods to measure and calculate elevation change can aid in disentangling the processes controlling elevation trajectories over different time scales, thereby improving assessments of vulnerability to sea-level rise.

4. Conclusions

Data were not consistent with the paradigm that tropical coastlines characterized by peat-forming mangroves are generally more vulnerable to RSLR compared to minerogenic systems. Instead, our findings demonstrate that high rates of mineral sedimentation do not necessarily ensure accommodation of RSLR; nor do low rates of mineral accretion automatically mean a greater risk of submergence. Our results point to the importance of assessing an elevation deficit to evaluate vulnerability to RSLR, rather than relying on a simple accretion deficit (Rybczyk and Cahoon, 2002). Also, vertical land development must be evaluated within the context of local factors determining submergence rates, especially shallow subsurface movement (subsidence or expansion), which may vary substantially among and within sedimentary settings (Table 11.1). Recognition of such differences is essential to proper management of tropical coastlines to ensure their resilience in the face of future sea-level rise.

Acknowledgments

We thank James A. Allen, Nicole Cormier, James C. Lynch, Jill A. Rooth, Richard A. MacKenzie, and William C. Vervaeke for assistance with SET installation, data collection, and collaboration, and Tom Cronin, Neil Saintilan, and Edward Webb for constructive comments on previous versions of this manuscript. We acknowledge the USGS Climate R&D Program and USGS LandCarbon Program, which provided critical research support to the authors. Any use of trade, product, or firm names is for descriptive purposes only and does not imply endorsement by the U.S. Government.

References

Alongi, D.M., 2015. The impact of climate change on mangrove forests. Curr. Clim. Change 1 (1), 30–39.

Barbier, E.B., 2006. Natural barriers to natural disasters: replanting mangroves after the tsunami. Front. Ecol. Environ. 4 (3), 124–131.

Becker, M., Meyssignac, B., Letetrel, C., Llovel, W., Cazenave, A., Delcroix, T., 2012. Sea level variations at tropical Pacific islands since 1950. Global Planet. Change 80–81, 85–98.

Boumans, R.M., Day, J.W.J., 1993. High precision measurements of sediment elevation in shallow coastal areas using a sedimentation-erosion table. Estuaries 16 (2), 375–380.

Breithaupt, J.L., Smoak, J.M., Byrne, R.H., Waters, M.N., Moyer, R.P., Sanders, C.J., 2018. Avoiding timescale bias in assessments of coastal wetland vertical change. Limnol. Oceanogr. 63, S477–S495.

Bunting, P., Rosenqvist, A., Lucas, R.M., 2018. The global mangrove watch—a new 2010 global baseline of mangrove extent. Rem. Sens. 10 (10), 1669.

Cahoon, D.R., Lynch, J.C., 1997. Vertical accretion and shallow subsidence in a mangrove forest of southwestern Florida, U.S.A. Mangroves Salt Marshes 1, 173–186.

Cahoon, D.R., Marin, P.E., Black, B.K., Lynch, J.C., 2000. A method for measuring vertical accretion, elevation, and compaction of soft shallow water sediments. J. Sediment. Res. 70 (5), 1250–1253.

Cahoon, D.R., Lynch, J.C., Hensel, P., et al., 2002a. High-precision measurements of wetland sediment elevation: I. recent improvements to the sedimentation-erosion table. J. Sediment. Res. 72 (5), 730–731.

Cahoon, D.R., Lynch, J.C., Perez, B.C., et al., 2002b. High-precision measurements of wetland sediment elevation: II. the rod surface elevation table. J. Sediment. Res. 72 (5), 734–739.

Cahoon, D.R., Hensel, P., Rybczyk, J., Perez, B.C., 2002c. Hurricane Mitch: Impacts on Mangrove Sediment Elevation Dynamics and Long-Term Mangrove Sustainability. U.S. Geological Survey. USGS Open File Report OFR 03-184.

Cahoon, D.R., Hensel, P., Rybczyk, J., McKee, K.L., Proffitt, C.E., Perez, B.C., 2003. Mass tree mortality leads to mangrove peat collapse at Bay Islands, Honduras after Hurricane Mitch. J. Ecol. 91 (6), 1093–1105.

Cahoon, D.R., Hensel, P.F., Spencer, T., Reed, D.J., McKee, K.L., Saintilan, N., 2006. Coastal wetland vulnerability to relative sea-level rise: wetland elevation trends and process controls. In: Verhoeven, J.T.A., Beltman, B., Bobbink, R., Whigham, D.F. (Eds.), Wetlands and Natural Resource Management. Springer-Verlag, Berlin Heidelberg, pp. 271–292.

Cahoon, D.R., 2015. Estimating relative sea-level rise and submergence potential at a coastal wetland. Estuar. Coast 38, 1077–1084.

Callaway, J.C., Cahoon, D.R., Lynch, J.C., 2013. The surface elevation table-marker horizon method for measuring wetland accretion and elevation dynamics. In: DeLaune, R.D., Reddy, K.R., Richardson, C.J., Megonigal, J.P. (Eds.), Methods in Biogeochemistry of Wetlands. Vol SSSA Book Series no. 10. Soil Science Society of America, Madison, WI.

Cazenave, A., Llovel, W., 2010. Contemporary sea level rise. Annu. Rev. Mar. Sci. 2, 145–173.

Cochard, R., Ranamukhaarachchi, S.L., Shivakoti, G.P., Shipin, O.V., Edwards, P.J., Seeland, K.T., 2008. The 2004 tsunami in Aceh and Southern Thailand: a review on coastal ecosystems, wave hazards and vulnerability. Perspect. Plant Ecol. Evol. Systemat. 10 (1), 3–40.

Ellison, J.C., Stoddart, D.R., 1991. Mangrove ecosystem collapse during predicted sea-level rise: Holocene analogues and implcations. J. Coast Res. 7 (1), 151–165.

Ellison, J.C., 1993. Mangrove retreat with rising sea-level, Bermuda. Estuar. Coast Shelf Sci. 37, 75–87.

Ewel, K.C., Twilley, R.R., Ong, J.E., 1998a. Different kinds of mangrove forests provide different goods and services. Global Ecol. Biogeogr. 7 (1), 83–94.

Ewel, K.C., Bourgeois, J.A., Cole, T.G., Zheng, S., 1998b. Variation in environmental characteristics and vegetation in high-rainfall mangrove forests, Kosrae, Micronesia. Global Ecol. Biogeogr. Lett. 7, 49–56.

Fu, H.F., Wang, W.Q., Ma, W., Wang, M., 2018. Differential in surface elevation change across mangrove forests in the intertidal zone. Estuar. Coast Shelf Sci. 207, 203–208.

Gonzalez, C., Urrego, J., Martinez, J.I., Polania, J., Yokoyama, Y., 2010. Mangrove dynamics in the southwestern Caribbean since the 'Little Ice Age': A history of human and natural disturbances. Holocene 20 (6), 849–861.

Himes-Cornell, A., Pendleton, L., Atiyah, P., 2018. Valuing ecosystem services from blue forests: a systematic review of the valuation of salt marshes, sea grass beds, and mangrove forests. Ecosyst. Serv. 30 (Part A), 36–48.

Jankowski, K.L., Tornqvist, T.E., Fernandes, A.M., 2017. Vulnerability of Louisiana's coastal wetlands to present-day rates of relative sea-level rise. Nat. Commun. 8, 14792.

Kirwan, M.L., Guntenspergen, G.R., 2012. Feedbacks between inundation, root production, and shoot growth in a rapidly submerging brackish marsh. J. Ecol. 100 (3), 764–770.

Krauss, K.W., Keeland, B.D., Allen, J.A., Ewel, K.C., Johnson, D.J., 2007. Effects of season, rainfall, and hydrogeomorphic setting on mangrove tree growth in Micronesia. Biotropica 39 (2), 161–170.

Krauss, K.W., Cahoon, D.R., Allen, J.A., Ewel, K.C., Lynch, J.C., Cormier, N., 2010. Surface elevation change and susceptibility of different mangrove zones to sea-level rise on Pacific high islands of Micronesia. Ecosystems 13, 129–143.

Krauss, K.W., McKee, K.L., Lovelock, C.E., et al., 2014. How mangrove forests adjust to rising sea level. New Phytol. 202, 19–24.

Lang'at, J.K.S., Kairo, J.G., Mencuccini, M., et al., 2014. Rapid losses of surface elevation following tree girdling and cutting in tropical mangroves. PloS One 9 (9), 8.

Lovelock, C.E., Adame, M.F., Bennion, V., et al., 2015a. Sea level and turbidity controls on mangrove soil surface elevation change. Estuar. Coast Shelf Sci. 153, 1–9.

Lovelock, C.E., Cahoon, D.R., Friess, D.A., et al., 2015b. The vulnerability of Indo-Pacific mangrove forests to sea-level rise. Nature 526, 559–563.

Lovelock, C.E., Feller, I.C., Reef, R., Hickey, S., Ball, M.C., 2017. Mangrove dieback during fluctuating sea levels. Sci. Rep. 7, 8.

Marois, D.E., Mitsch, W.J., 2015. Coastal protection from tsunamis and cyclones provided by mangrove wetlands—a review. Int. J. Biodiversity Sci. Ecosyst. Serv. Manag. 11 (1), 71–83.

McKee, K.L., Vervaeke, W.C., 2018. Will fluctuations in salt marsh - mangrove dominance alter vulnerability of a subtropical wetland to sea-level rise? Global Change Biol. 24 (3), 1224–1238.

McKee, K.L., Cahoon, D.R., Feller, I.C., 2007. Caribbean mangroves adjust to rising sea level through biotic controls on change in soil elevation. Global Ecol. Biogeogr. 16, 545–556.

McKee, K.L., 1996. Growth and physiological responses of neotropical mangrove seedlings to root zone hypoxia. Tree Physiol. 16, 883–889.

McKee, K.L., 2011. Biophysical controls on accretion and elevation change in Caribbean mangrove ecosystems. Estuar. Coast Shelf Sci. 91, 475–483.

Meyssignac, B., Cazenave, A., 2012. Sea level: a review of present-day and recent-past changes and variability. J. Geodyn. 58, 96–109.

Middleton, B.A., McKee, K.L., 2001. Degradation of mangrove tissues and implications for peat formation in Belizean island forests. J. Ecol. 89, 818–828.

Mumby, P.J., Edwards, A.J., Arias-Gonzalez, J.E., et al., 2004. Mangroves enhance the biomass of coral reef fish communities in the Caribbean. Nature 427, 533–536.

Parkinson, R.W., DeLaune, R.D., White, J.R., 1994. Holocene sea-level rise and the fate of mangrove forests within the wider Caribbean region. J. Coast Res. 10 (4), 1077–1086.

Parkinson, R.W., Craft, C., DeLaune, R.D., et al., 2017. Marsh vulnerability to sea-level rise. Nat. Clim. Change 7 (11), 756.

Rogers, K., Wilton, K.M., Saintilan, N., 2006. Vegetation change and surface elevation dynamics in estuarine wetlands of southeast Australia. Estuar. Coast Shelf Sci. 66 (3–4), 559–569.

Rogers, K., Kelleway, J.J., Saintilan, N., et al., 2019. Wetland carbon storage controlled by millennial-scale variation in relative sea-level rise. Nature 567 (7746), 91–95.

Rybczyk, J., Cahoon, D.R., 2002. Estimating the potential for submergence for two wetlands in the Mississippi River delta. Estuaries 25 (5), 985–998.

SAS, 2018. Discovering JMP 14. SAS Institute Inc., Cary, NC, USA.

Sasmito, S.D., Murdiyarso, D., Friess, D.A., Kurnianto, S., 2016. Can mangroves keep pace with contemporary sea level rise? A global data review. Wetl. Ecol. Manag. 24 (2), 263–278.

Soares, M.L.G., 2009. A conceptual model for the responses of mangrove forests to sea level rise. J. Coast Res. SI56, 267–271.

Swales, A., Denys, P., Pickett, V.I., Lovelock, C.E., 2016. Evaluating deep subsidence in a rapidly-accreting mangrove forest using GPS monitoring of surface-elevation benchmarks and sedimentary records. Mar. Geol. 380, 205–218.

Tornqvist, T.E., Wallace, D.J., Storms, J.E.A., et al., 2008. Mississippi delta subsidence primarily caused by compaction of Holocene strata. Nat. Geosci. 1 (3), 173–176.

Toscano, M.A., Gonzalez, J.L., Whelan, K.R.T., 2018. Calibrated density profiles of Caribbean mangrove peat sequences from computed tomography for assessment of peat preservation, compaction, and impacts on sea-level reconstructions. Quat. Res. 89 (1), 201–222.

van Asselen, S., Stouthamer, E., van Asch, T.W.J., 2009. Effects of peat compaction on delta evolution: a review on processes, responses, measuring and modeling. Earth Sci. Rev. 92 (1–2), 35–51.

van Asselen, S., 2011. The contribution of peat compaction to total basin subsidence: implications for the provision of accommodation space in organic-rich deltas. Basin Res. 23 (2), 239–255.

Ward, R.D., Friess, D.A., Day, R.H., MacKenzie, R.A., 2016. Impacts of climate change on mangrove ecosystems: a region by region overview. Ecosys. Health Sustain. 2 (4), e01211.

Webb, E.L., Friess, D.A., Krauss, K.W., Cahoon, D.R., Guntenspergen, G.R., Phelps, J., 2013. A global standard for monitoring coastal wetland vulnerability to accelerated sea-level rise. Nat. Clim. Change 3 (5), 458–465.

Whelan, K.R.T., Smith, T.J.I., Anderson, G.H., Ouellette, M.L., 2009. Hurricane Wilma's impact on overall soil elevation and zones within the soil profile in a mangrove forest. Wetlands 29 (1), 16–23.

Woodroffe, C.D., Rogers, K., McKee, K.L., Lovelock, C.E., Mendelssohn, I.A., Saintilan, N., 2016. Mangrove sedimentation and response to relative sea-level rise. Annu. Rev. Mar. Sci. 8 (1), 243–266.

Woodroffe, C.D., 1990. The impact of sea-level rise on mangrove shorelines. Prog. Phys. Geogr. 14 (4), 483–520.

Woodroffe, C.D., 1992. Mangrove sediments and geomorphology. In: Alongi, D., Robertson, A. (Eds.), Tropical Mangrove Ecosystems. American Geophysical Union, Washington, D.C., pp. 7–41

PART III

Blue carbon

CHAPTER

12

Environmental drivers of blue carbon burial and soil carbon stocks in mangrove forests

Richard MacKenzie[1], *Sahadev Sharma*[2], *Andre R. Rovai*[3,4]

[1]Institute of Pacific Islands Forestry, Pacific Southwest Research Station, USDA Forest Service, Hilo, HI, United States; [2]Institute of Ocean and Earth Sciences, University of Malaya, Kuala Lumpur, Selangor, Malaysia; [3]Department of Oceanography and Coastal Sciences, College of Coast and Environment, Louisiana State University, Baton Rouge, LA, United States; [4]Departamento de Oceanografia, Universidade Federal de Santa Catarina, Florianópolis, Santa Catarina, Brazil

OUTLINE

1. Introduction 276

2. Abiotic factors 279

2.1 Climatic drivers: temperature and precipitation 279

2.2 Storms 280

2.3 Geomorphology and nutrients 281

2.4 Salinity 282

3. Biotic factors 283

3.1 Tree diversity 283

3.2 Forest age 284

3.3 Algae 286

3.4 Crabs 286

4. Conclusion 287

References 288

Dynamic Sedimentary Environments of Mangrove Coasts
https://doi.org/10.1016/B978-0-12-816437-2.00006-9

1. Introduction

There is growing evidence that the conservation or restoration of blue carbon ecosystems such as mangroves, saltmarshes, or seagrass meadows offers effective nature-based solutions to climate change impacts (Alongi et al., 2015). These highly productive ecosystems can help offset greenhouse gas emissions by sequestering and storing large amounts of atmospheric derived CO_2 in the form of aboveground and belowground carbon (C). Belowground C pools are often stored under waterlogged and anoxic conditions that can inhibit or slow down microbial breakdown of organic matter. As a result, C buried in sediments of blue carbon habitats can remain there for millennia if left undisturbed, making them critical long-term C sinks.

Blue carbon was first introduced in 2009 in response to the increased importance of terrestrial or "green" carbon in climate change adaptation and mitigation. Organisms living in the seas can capture more than half the amount of C captured in forested lands. Blue carbon is therefore equally important for climate change adaptation and mitigation (Nellemann and Corcoran, 2009). Mangroves, saltmarshes, and seagrass beds are the main blue carbon ecosystems, storing up to 70% of all C in ocean sediments despite only covering <0.5% of the ocean floor (Chmura et al., 2003; Duarte et al., 2004; Bouillon et al., 2008; Lo Iacono et al., 2008). Long-term C sequestration per unit area is also much greater in blue carbon ecosystems compared with terrestrial forests despite the global area of the latter being one to two orders of magnitude smaller (Mcleod et al., 2011). A study conducted by Donato et al. in 2011 further supported these arguments. They found that C stocks from Indo-Pacific mangroves were three times greater than any other major forest domain (e.g., boreal, temperate, tropical upland; Fig. 12.1). This was due to the greater belowground C stores measured in mangrove forests and that represented 71%–98% of total mangrove C stocks (Donato et al., 2011). Since then, the role mangrove forests play in climate change and mitigation strategies has significantly grown (Duarte et al., 2013b; Lovelock and Duarte, 2019; Macreadie et al., 2019) as have the number of studies that have documented mangrove C stocks and the impacts of land use change on those stocks (Atwood et al., 2017; Cameron et al., 2019; Kauffman et al., 2020; Sharma et al., 2020).

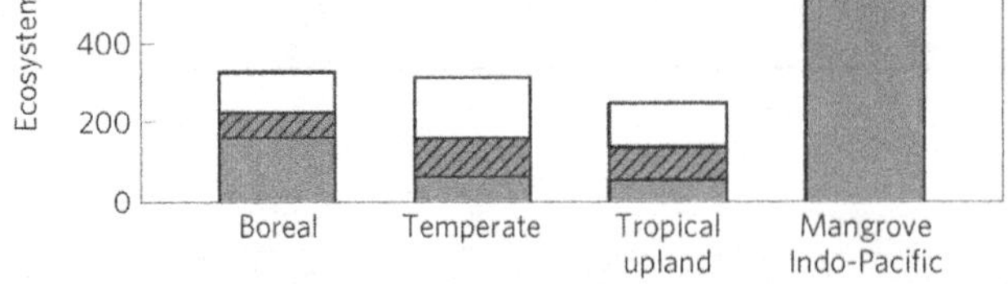

FIGURE 12.1 Comparison of mean Indo-Pacific mangrove C stocks (±95% confidence interval) with that of major global forest domains. *Reprinted from Donato, D.C., Kauffman, J.B., Murdiyarso, D., Kurnianto, S., Stidham, M., Kanninen, M., 2011. Mangroves among the most carbon-rich forests in the tropics. Nat. Geosci. 4, 293–297 with permission.*

Most C stored in mangrove sediments is largely derived from mangrove roots, though this varies across different coastal landforms and stage of peat development (e.g., Ezcurra et al., 2016). Compared with terrestrial forests, mangrove trees allocate more sequestered C to belowground roots to replace sloughed root hairs or fine roots (Alongi, 2014). Thus, the bulk of belowground C is stored in dead roots that represent 75%–95% of root biomass (Alongi et al. 2003, 2004b; Cormier et al., 2015). Other autochthonous C sources include macro- and microalgae growing on or within aboveground roots and sediments, respectively (Alongi, 2014). External sources of C also contribute to mangrove soil C stores, though this can vary within a mangrove. Mangrove areas adjacent to or near riverine or upland systems can receive C inputs from upland forests, while areas adjacent to the mangrove-ocean ecotone can receive inputs from seagrass ecosystems (Saavedra-Hortua et al., 2020; Sasmito et al., 2020). Seagrass beds and coral reefs directly adjacent to mangrove forests can also enhance mangrove soil C stocks by protecting mangroves through reduced wave action (Gillis et al., 2014; Huxham et al., 2018). The combined high inputs of autochthonous and allochthonous C result in C burial rates in mangroves being 20–70x greater than upland forests (Mcleod et al., 2011). High C burial rates also result in 10%–15% of coastal sediment C being stored in mangrove soils and, along with saltmarshes and seagrass systems, account for 50%–70% of annual C burial in oceans despite only representing ~0.5% of global coastal area (Chmura et al., 2003; Duarte et al., 2004; Bouillon et al., 2008; Lo Iacono et al., 2008).

The amount of belowground C stored in mangrove sediments can significantly vary among countries, regions, and land uses. For example, Jardine and Siikamäki (2014) used a climate-based model to estimate global mangrove soil organic C stocks. They found soil C stocks to 1 m vary nearly threefold (272 ± 49 to 703 ± 38 Mg C/ha), while (Atwood et al., 2017), based on country-level soil C density reference values, found a larger and more than tenfold difference in theirs (80.7–935.67 Mg C/ha). Later, Rovai et al. (2018) found global mangrove soil organic C stocks range from 33 to 465 Mg C/ha and suggested that differences in soil C stocks across these studies are partly explained by the omission of key geophysical drivers (e.g., river/sediment discharge, tidal influence) and partly by the inclusion criteria established by former studies to extrapolate soil C stocks from site to global level. More recently, Kauffman et al. (2020) reported a greater range of 46–2076 Mg C/ha that is due to the inclusion of soil C stocks to 3 m depth.

Soil carbons stocks in mangroves are largely influenced by C burial rates (Kristensen et al., 2008), though evidence suggests this may be an inverse relationship (Breithaupt et al., 2012). Regardless, burial rates, like soil C stocks, exhibit high levels of variability among countries, regions, and land uses. Alongi (2012) reported a global mean (±1 SD) of 174 ± 184 g C/m^2 yr with a range of 10–920 g C/m^2 yr. Breithaupt et al. (2012) argued that a geometric mean of 163 (+40; −31) g OC/m^2 yr (95% CI) may be a more reasonable global mean estimate due to the extreme low and high (5–1129 g C/m^2 yr) values in their review. For this review, we report a new global mean in C burial rates of 239 g C/m^2 yr, with a range of 5–1722 g C/m^2 yr (Fig. 12.2).

While plausible relationships between mangrove soil organic C stocks and climate and geophysical drivers seem to have been reasonably determined (Kristensen et al., 2008; Atwood et al., 2017; Rovai et al., 2018), it is still unclear what abiotic or biotic drivers, either independently or combined, and across different spatial scales, are responsible for the variability observed in C burial rates. However, because soil C stocks represent a "snapshot"

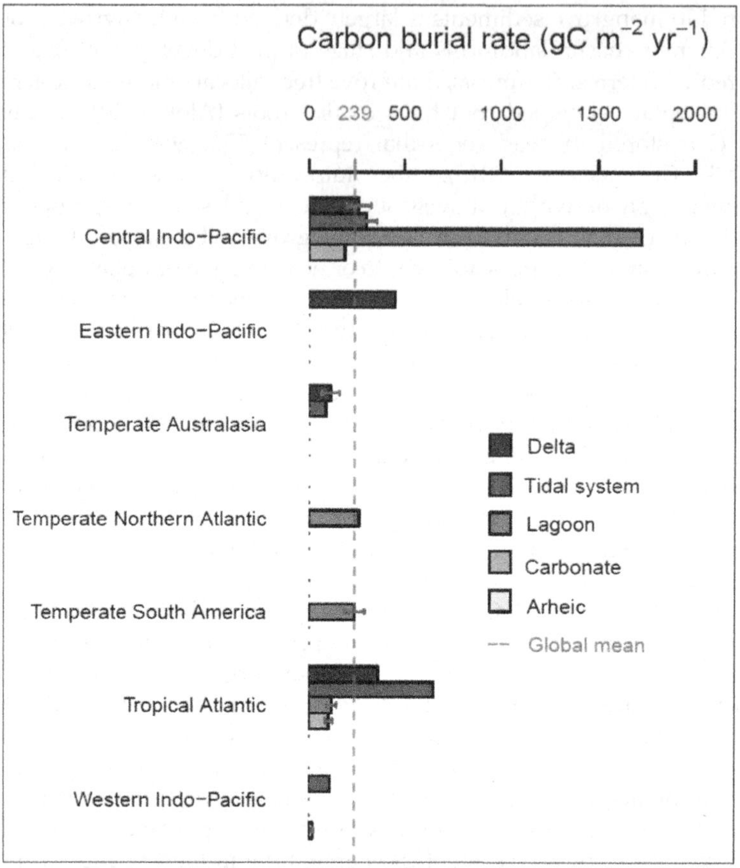

FIGURE 12.2 Average carbon burial rates from regions estimated from various studies (Lynch et al., 1989; Furukawa and Wolanski, 1996; Callaway et al., 1997; Fujimoto et al., 1999; Alongi and Dixon, 2000; Alongi et al. 2001, 2004a, 2005; Brunskill et al. 2002, 2004; Chmura et al., 2003; Duarte et al. 2004, 2013a; Gonneea et al., 2004; Tateda et al., 2005; Sanders et al. 2008, 2010a, 2010b, 2010c; Xiaonan et al., 2008; Alongi 2009, 2011; Cerón-Bretón et al., 2011; Ray et al., 2011; Smoak et al., 2013; Yang et al., 2014; Kusumaningtyas et al., 2019; MacKenzie et al., 2016).

of the process of soil formation, it is reasonable to assume that C burial in these ecosystems is controlled by similar mechanisms that explain soil C stock distributions (e.g., Atwood et al., 2017; Rovai et al., 2018; Twilley et al., 2018). Identifying these factors and understanding the different roles they play in C burial is critical if we wish to implement management efforts to conserve or restore mangrove forests for climate change mitigation and adaptation or to increase the resilience of mangrove forests to climate change impacts, such as sea level rise. This chapter will attempt to identify some of the major abiotic and biotic factors across different spatial scales (e.g., site level to global level) that have been identified in the literature to influence C burial rates and soil C stocks in mangrove forests.

2. Abiotic factors

2.1 Climatic drivers: temperature and precipitation

Temperature and precipitation are two important drivers that can influence mangrove soil C stocks. Global models have predicted that soil C content varies with air temperature, though the strength of this relationship depends on the model used (Chmura et al., 2003; Jardine and Siikamäki, 2014; Rovai et al., 2018). Sanders et al. (2016) reported that 86% of the observed variability in soil C stocks was associated with annual rainfall, which proved to be an effective predictor of mangrove soil C in Indo-Pacific and Australasia mangroves. Latitude, which incorporates both air temperature and rainfall, proved to be a significant predictor of soil C in Cambodia (Sharma et al., 2020) and across the Asia Pacific region (Bukoski et al., 2017), though Atwood et al. (2017), Twilley et al. (2018), and Kauffman et al. (2020) all found a poor relationship between latitude and soil C stocks. Indeed, the usefulness of absolute degrees of latitude is of very limited value as a predictor of species distribution and ecosystem ecological attributes as it does not capture real geographical gradients (Hawkins and Felizola Diniz-Filho, 2004). Furthermore, water temperature, which can also influence productivity (Ellison, 1994; Lovelock and Ellison, 2007), is not controlled by latitude but by other factors such as currents. The independent and interactive influence of temperature and precipitation on soil C stocks clearly varies with scale (e.g., global, regional, local) (Saavedra-Hortua et al., 2020).

As reviewed earlier in this chapter, soil C stocks in mangroves are formed by allochthonous and autochthonous sources. Temperature and precipitation control both the amount of terrestrial C delivered to mangroves from allochthonous upstream sources (Šímová and Storch, 2017; Cragg et al., 2020) and in situ C accumulation via net primary productivity (NPP) (Pool et al., 1975; Twilley, 1995; Day Jr. et al., 1996; Feher et al., 2017). Overall, temperature affects NPP by influencing rates of growth, photosynthesis, and respiration and thereby C storage (Duke, 1990; Lovelock, 2008). Similarly, rainfall reduces stress caused by soil salinity and increases nutrient input via upland runoff, thus stimulating mangrove NPP (Day Jr. et al., 1996; Twilley et al., 1997; Agraz-Hernández et al., 2015). NPP includes aboveground (litterfall, wood) and belowground (root) production that differs in respect to their relative contribution to C sequestration in mangrove soils. The contribution of mangrove aboveground production (e.g., litterfall and wood) to soil organic matter formation is generally negligible (e.g., less than 10%) (Chen and Twilley, 1999) as nearly 50% of the litterfall is exported by tidal action (Jennerjahn and Ittekkot, 2002; Bouillon et al., 2008), though litter retention rates can be greater in basin mangroves with little tidal export (Twilley et al. 1986, 1997). Litterfall export, however, is influenced by rainfall and temperature, with mangroves in regions with lower rainfall and higher temperatures exporting more litter than mangroves in regions with higher rainfall and lower temperatures (Adame and Lovelock, 2011). The remaining amount of leaf litter is either directly buried in soils or is consumed by crabs, other invertebrates, or microorganisms (Robertson and Daniel, 1989; MacKenzie et al., 2020) that then release C to be incorporated into mangrove sediments (Andreetta et al., 2014) or to be lost as dissolved organic carbon to the water column (Lee, 1995; Kristensen et al., 2008). Belowground (or root) production, on the other hand, can account for more than 80% of carbon sequestered and stored in mangrove soils (Saintilan et al., 2013; Kusumaningtyas et al., 2019). Thus, root production is an important biotic mechanism to C sequestration in mangrove soils (McKee et al., 2007; Krauss et al., 2014; Woodroffe et al., 2016).

Temperature and precipitation also influence C sequestration and thereby soil C stocks in mangrove soils by controlling the balance between organic matter production and decomposition rates. For example, on a global scale, temperatures that occur above a certain threshold have been suggested to decrease C stocks in mangrove soils due to increased microbial decomposition (Chmura et al., 2003). However, increases in minimum annual temperatures can also increase soil organic C stocks as it stimulates mangrove root production and thereby C flux to the soil (Rovai et al., 2018). While mangrove root decomposition rates can also increase with elevated temperatures, root C inputs appear to exceed mineralization rates (Ouyang et al., 2017), resulting in organic matter accumulation and a positive C flux to mangrove soils (Pregitzer et al., 2000). Precipitation is also believed to influence C sequestration in mangrove soils as demonstrated for other functionally analogous tidal saline wetlands where decomposition was accelerated by increased precipitation and slowed by drought (Charles and Dukes, 2009), though longer periods of soil inundation that results from increased rainfall can also decrease organic matter decomposition in mangroves (Alongi, 2005).

Rainfall events and thus runoff from mangroves may also offset C storage in mangrove soils. Monthly exports of total organic C from Florida mangroves increased 2–10x during large rain events that increased runoff by 3–20x (Twilley, 1985). Similarly, concentrations of dissolved organic C exported from Okinawa mangroves in Japan nearly doubled after a large rainfall event that resulted from a category 4 typhoon (Goni). Episodic increases in freshwater from large storms likely enhance leaching of dissolved organic C from mangrove litter as well as from soils where they are otherwise retained by salinity-induced aggregation (Kida et al., 2019). However, in Brazil, only export of particulate organic C, not dissolved organic C, increased with rainfall (Dittmar and Lara, 2001).

While relationships between climate variables and soil C sequestration have been demonstrated for terrestrial ecosystems, results from mangrove forests are not always consistent in the literature (Breithaupt et al., 2012; Feher et al., 2017). The interaction of these multiscalar climatic factors and stochastic events, coupled with scarcity of data on C burial rates, may explain the low correlation ($R^2 = 0.05$) between mean annual temperature and rates of soil C burial and suggests that regional or local factors must be the dominant controls on rates of C burial in mangrove soils (Chmura et al., 2003).

2.2 Storms

Typhoons, hurricanes, and tropical cyclones can also influence productivity and thus soil C stocks. The lack of these disturbance results in less damage to mangroves, higher levels of productivity, and enhanced C storage (Kauffman and Cole, 2010; Twilley et al., 2017; Thorhaug et al., 2019). Storms can also influence soil C stocks through increased precipitation that can promote mangrove growth and/or release trees from salinity stress (see discussion above). Another mechanism is through the deposition of C-rich and nutrient-rich sediments onto the mangrove forest floor contributing to organic carbon accretion and NPP (Castañeda-Moya et al., 2010; Breithaupt et al., 2017; Castañeda-Moya et al., 2020). Carbon burial rates within storm deposits from Hurricane Wilma (260–393 g/m^2 yr) were nearly double the overall burial rates in the mangrove forest sampled (151–168 g/m^2 yr) (Smoak et al., 2013).

2.3 Geomorphology and nutrients

Coastal morphology influenced by rivers, tides, wave power, and climate has been proposed as a potential framework to explain global variations in mangrove C storage and to an extent C burial (Twilley et al., 2018). These forcings influence the rate and source of sediment and nutrient inputs that result in clastic coastal environmental settings that include riverine/deltas, tidal systems, lagoons, carbonate, and arheic environments (Thom, 1982; Dürr et al., 2011). Models in Twilley et al. (2018) and Rovai et al. (2018) predicted that soil C densities were greatest in carbonate and arheic settings and lowest in riverine/deltaic mangroves. A recent global assessment of field data found opposite patterns. Soil C stocks were lowest in an arheic mangrove and in carbonate-dominated mangroves in the Yucatan peninsula, Mexico. Deltaic mangroves, on the other hand, had some of the greatest (Mexico, Indonesia) and lowest (Gabon, Brazil) soil C stocks in their analysis (Kauffman et al., 2020). These differences are likely due to differences in scale as well as geomorphic classifications.

The patterns reported by Twilley et al. (2018) and Rovai et al. (2018) are likely influenced by the conspicuous nitrogen (N) and phosphorous (P) loads and limitations associated with each coastal environmental setting. In riverine/deltaic mangroves, soil C sequestration is primarily driven by burial of riverine delivered allochthonous material (Hayes et al., 2017; Kusumaningtyas et al., 2019) resulting in elevated burial rates (Figs. 12.2 and 12.3) but lower soil C density due to the high proportion of mineral alluvial material relative to organic matter inputs (Rovai and Twilley this book). Lower soil C stocks are also driven by large P loads from upland catchments (Harrison et al., 2005) that release mangrove trees from nutrient limitations in mangrove soils (Rovai et al., 2018). Trees can then allocate more biomass to aboveground structures since optimum nutrient uptake efficiency can be attained with less complex root systems (Chapin III et al., 1986; Castañeda-Moya et al., 2011; Cormier et al., 2015). Tide- (estuaries) and wave-dominated (lagoons) coastlines have intermediate levels

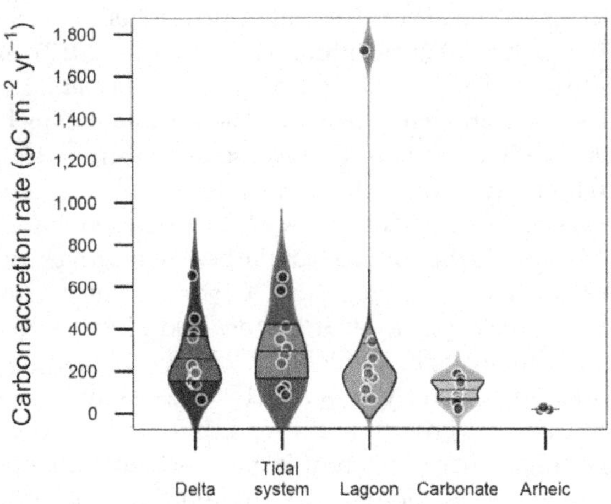

FIGURE 12.3 Average C burial rates across coastal environmental settings as defined by Twilley et al. (2018). See Fig. 12.2 for references used to produce this figure.

of soil C stocks but high levels of C burial rates (Figs. 12.2 and 12.3). High C burial rates reflect the relative contribution of river-borne organic particles and in situ root production to soil formation. Tides also play a fundamental role in C burial in mangrove soils by promoting nutrient exchange and soil oxygenation and reducing sulfide levels, which altogether regulate root production and organic matter decomposition rates in mangrove soils (Chapman et al., 2019). Tides also carry abundant sediment, promoting trapping of C-rich mud due to reduced water velocity within mangrove forests (Woodroffe et al., 2016). Conversely, the structural development of mangrove forests in carbonate settings (e.g., South Florida and Caribbean region) has been linked to P limitation and hydrologic stress. Phosphorous supply is often limited due to carbonate chemisorption reactions and the sequestering of P into recalcitrant organic pools, which reduces the availability of exchangeable P to plants (Koch et al., 2001). As a result, plants allocate more C to root systems than trunks and branches to increase access to limited P, resulting in well-developed root systems and elevated high soil C stocks (Twilley et al., 2018), but reduced aboveground biomass (Feller et al., 2003; Castañeda-Moya et al., 2013; Adame, 2014). Arheic environments are arid coastlines where mangrove distribution is restricted to tidal range, which regulates soil properties (e.g., salinity, sulfides) and nutrient availability (Castañeda-Moya et al., 2013; Lovelock et al., 2017; Schile et al., 2017). In this coastline type soil C burial is lowest due to negligible freshwater inputs (e.g., rainfall, river discharge) and low NPP rates due to aridity conditions.

2.4 Salinity

Salinity is an important driver of C storage and burial rates, though its impact is not straightforward. Decreased porewater salinities that result from increased precipitation or location (e.g., riverine) can result in increased soil C stocks. Soil C stocks were negatively correlated to porewater salinities in a wall-to-wall C assessment of mangroves across Cambodia (Sharma et al., 2020). Similar results have been reported in Mexico (Adame et al., 2013) and to a degree from West-Central Africa, where correlations between soil C stocks and salinity were much weaker (Kauffman and Bhomia, 2017). Saintilan (1997b) found that root biomass of both *Avicennia marina* and *Aegiceras corniculatum* significantly decreased with salinity in a brackish mangrove in Australia. These relationships between porewater salinities and soil C stocks are likely due to the release of mangrove trees from osmotic stress that increases their productivity and soil C stocks (Sharma et al., 2020).

Alternatively, higher soil C stocks have also been reported under elevated porewater salinities (Ball, 1988; Matsui, 1998). In the Sundarban mangroves of Bangladesh, salinity enhanced soil C stocks relative to aboveground C stocks. Soil C stocks represented only 57.2% of total ecosystem C stocks in freshwater zones and 71.9% of total ecosystem C stocks in higher salinity zones (Rahman et al., 2015). Ball (1988) found that the root:shoot ratios of *A. corniculatum* and *A. marina* increased by 50% and 15%, respectively, when salinity increased from ~1 to 30 ppt. Increased root growth and thus C burial allowed shoots to increase their water demands, though this is more apparent in more salt-tolerant species (e.g., *A. marina*). Higher salinities can also decrease microbial respiration that would also result in increased soil C stocks. Ouyang (2017) found that decay rates declined exponentially with porewater salinity ($R^2 = 0.16$, $P < .001$), similar to other findings (Matsui, 1998).

3. Biotic factors

3.1 Tree diversity

It has been suggested that as mangrove tree species diversity increases, so do the ecosystem services provided by those mangroves (Twilley et al., 1996; Farnsworth, 1998; Field et al., 1998). This would include increased productivity that could potentially increase C burial. Lang'at et al. (2013) found that species mixing boosts root yield in 3–4-year-old planted mangrove saplings, especially when plots included *A. marina*. Belowground production significantly increased within mixed plots of four mangrove species compared with plots of individual species. A global soils analysis found that mangrove stands that had five genera had 70%–90% greater soil C stocks than all other richness levels (Atwood et al., 2017). Similarly, C burial rates increased with each additional species until reaching an average maximum (±1 SD) of 280.5 ± 168.0 gC/m^2 yr at four species, after which C burial decreased (MacKenzie et al., 2016). Belowground root growth and thus soil C stocks and burial rates may be greater in less diverse mangroves forests due to less competition. Increased diversity of an upland forest increased belowground interspecific competition and thus root growth of some species, but not others (Leuschner et al., 2001).

Individual tree species may also influence root production, soil C stocks, and C burial rates. However, it is not clear if this is simply due to the abiotic factors discussed above or if it's a true biotic response. For example, does one species produce more root biomass than another or are there differences in interspecific responses to abiotic factors? McKee (1995) found that *Laguncularia racemosa* produced more roots (1.10 ± 0.11 g) than either *Rhizophora mangle* (0.9 ± 0.08 g) or *Avicennia germinans* (0.84 ± 0.12 g) in a Belize mangrove (0.84 ± 0.12 g). Different monotypic stands of low salinity mangroves along the upper portion of Mary River in Australia exhibited stark differences in total root biomass. Total root biomass from stands of *A. marina* (72.6 Mg/ha) was greater than stands of *A. corniculatum* (13.2 Mg/ha), *Excoecaria agallocha* (45.0 Mg/ha), or *Ceriops tagal* (32.4 Mg/ha). Monotypic stands of mangroves that experienced higher salinities at the mouth of Mary River also exhibited stark differences in total root biomass. Total biomass from stands of *Rhizophora stylosa* (102.0 Mg/ha) were greater than *A. marina* (74.4 Mg/ha), *A. corniculatum* (50.4 Mg/ha), or *E. agallocha* (14.4 Mg/ha) (Saintilan, 1997a). Similarly in Mexico, *A. germinans* and *R. mangle* dominated mangroves had higher root biomass (11.5–30.5 Mg/ha) and root production (1.0–1.9 g/m^2 d) than a forest dominated by *L. racemosa* (9.6 Mg/ha and 0.5 g/m^2 d, respectively) (Adame, 2014). In Kenya, *Sonneratia alba* had greater root biomass (48.4–53.4 Mg/ha) than *A. marina* (39.1–43.7 Mg/ha^{-1}) or *Rhizophora mucronata* (7.5–35.8 Mg/ha) (Tamooh et al., 2008), though values for *A. marina* and *R. mucronata* were similar to those for *Avicennia* spp. and *Rhizophora* spp. from Australia and Mexico (Tamooh et al., 2008). In Micronesia, root ingrowth rates of *S. alba* were also nearly 10x greater than *Bruguiera gymnorrhiza* or *Rhizophora* sp. (Gleason and Ewel, 2002). When data were analyzed at a global scale, soil C stocks ranged fourfold across monotypic stands of different genera. *Laguncularia* sp. (424 ± 262 MgC/ha) and *Rhizophora* sp. (388 ± 227 MgC/ha) had the highest averaged soil C stocks (±1 SE) than any other genera (Atwood et al., 2017). Kauffman et al. (2020) reported no significant differences in soil C stocks among genera (779.1 ± 112.7 to 976.2 ± 92.0 Mg/ha) with the exception of significantly lower soil C stocks for *Avicennia* spp. (361.6 ± 57.2 Mg/ha).

This was attributed to higher porewater salinity. The mean porewater salinity across mangrove forest stands dominated by *Avicennia* spp. was 44 ± 4 ppt, while the mean porewater salinity of mangrove forests dominated by *Rhizophora* spp. was 28 ± 1 ppt ($P \geq .05$) and that of all other stands was <21 ppt.

Increased genera diversity could also result in increased structural complexity on the mangrove forest floor from different aerial root structures (e.g., prop roots vs. pneumatophores). This could affect C burial as different root structures can potentially be more effective at trapping sediments than others. For example, sedimentation can be greater in mangroves with complex trunk and root structures (e.g., *Rhizophora* spp. prop roots) compared to simpler forms, such as single trunked species (e.g., *Ceriops* sp.) or species with pneumatophores (e.g., *S. alba*) or knee roots (e.g., *Bruguiera gymnorrhiza*) as the more complex roots are more effective at slowing the flow of water and allowing C-rich suspended sediments or organic matter to settle out onto the mangrove forest floor (Furukawa and Wolanski, 1996; Furukawa et al., 1997). Average vertical accretion rates (±1 SE) via sediment deposition were enhanced in areas with *Rhizophora* spp. prop roots (11.0 ± 2.9 mm/yr) compared with *S. alba* pneumatophores (7.2 ± 1.1 mm/yr), *B. gymnorrhiza* knee roots (9.3 ± 1.2 mm/yr), and bare soil controls (9.4 ± 2.6 mm/yr) (Krauss et al., 2003). Aside from tertiary structure, the density of aerial roots, stems, and trunks can also increase vertical accretion. This has been demonstrated using natural *A. marina* roots as well as artificial structures (Spenceley, 1977; Bird, 1986; Young and Harvey, 1996) and mangrove seedlings (Huxham et al., 2010; Kumara et al., 2010). In addition to root complexity and density, the influence of pneumatophores of *A. marina* and *S. alba* on sediment deposition may be due to their ability to bind and retain of sediments (Spenceley, 1977). Mangrove species also differ in terms of their specific gravity of wood (e.g., g/cm^3) (see review by Rovai et al., 2016) and thereby C content in belowground woody biomass (Rodrigues et al., 2015; Gillerot et al., 2018). In addition, mangrove species differ as to the quality of the litter which ultimately influence organic matter turnover and C remineralization (Twilley et al., 1986).

3.2 Forest age

Forest age can also influence soil C stocks. Several studies using restored mangroves as chronosequences have found that soil C stocks increase with age. In Florida, C stocks in the upper 10 cm of soils from mangroves that had been naturally restored with *R. mangle, L. racemosa*, and *A. germinans* increased linearly over a 20-year period at a rate of 218 g C/m^2 yr. Restored mangroves reached soil C stock levels (52.5 ± 24.4 MgC/ha) similar to reference sites (28.1–50.7 MgC/ha) after 20 years (Osland et al., 2012). In Indonesia, naturally recolonized tidal mangroves in highly productive areas had soil C values that were within 80% of reference values after only 10 years. In contrast, similar aged naturally recolonized mangrove stands in low productive areas on a coral atoll had lower soil C values that were only 20% of reference values (Cameron et al., 2019). Similar results have been seen in planted sites. In Thailand, root biomass of *Rhizophora apiculata* replanted mangrove forests significantly increased from 3 to 25 years until reaching a maximum value of 35.6 Mg/ha (Alongi and Dixon, 2000) and in Brisbane, Australia, 25-year-old replanted stands of *A. marina* (118 Mg/ha) had similar amounts of root biomass as natural stands (121 Mg/ha) (Mackey, 1993). In China, Chen et al. (2018)

reported that average soil C stocks (±1 SE) in the top meter of soil increased from 130 ± 5 to 151 ± 7 to 170 ± 9 MgC/m^2 in 12-, 24-, and 48-year-old planted mangrove forests, respectively. Similar results were reported for *R. mucronata* planted forests in Kenya, where dry root biomass significantly increased from 7.5 ± 0.4 to 24.9 ± 1.6 to 35.8 ± 1.1 Mg/ha in 6-year, 12-year, and natural mangrove forests. However, while dry root biomass also increased with age in *S. alba* and *A. marina* planted forests in Kenya, dry root biomass was highest in the 12-year-old stands (75.5 ± 2.0 and 43.7 ± 1.7 Mg/ha, respectively) compared with natural stands (48.4 ± 0.7 and 39.1 ± 0.7 Mg/ha, respectively) (Tamooh et al., 2008).

In natural mangroves of French Guiana, Marchand (2017) also found that mangrove forest soil C stocks increased linearly from 4 to 107 Mg OC/ha with forest age (Fig. 12.4). However, while C burial rates also increased with forest age, they peaked after ~10 years at 480 gC/m^2 yr, after which they decreased to 135 gC/m^2 yr in 40-year-old forests and then increased again to 486 gC/m^2 yr in ~50-year-old forests. The increase in the older forest was attributed to these sites being isolated from the sea and developing a community of mangrove associates (fern, epiphytes, etc.) that would have contributed to the NPP of the ecosystem. MacKenzie (unpublished data) also found that C burial increased with stand age until reaching a maximum of (±1 SD) 384.3 ± 176.0 gC/m^2 yr at 10 years, after which C burial decreased. In restored sites in Malaysia, C burial rates were greatest in the first 5 years of recovery (290–610 gC/m^2 yr), which were 2x greater than burial rates in mature forests (>50 years; ~120 gC/m^2 yr) (Adame et al., 2018). Alongi et al. (2004a) reported that while sedimentation rates decreased from 5 to 18 to 85-year-old mangrove forests, average C burial rates increased from 274.4 to 296.2 to 352.2 g/m^2 yr with forest age. However, the greatest C burial observed was from a 5-year-old site (403.7 g/m^2 yr).

It is not entirely clear why the above patterns in C stocks and burial rates were observed. Greater C stocks and C burial rates in younger stands were attributed to greater root growth and thus greater belowground production rates of younger trees compared with older trees (Osland et al., 2012; Krauss et al., 2017; Sharma et al., 2020). However, Alongi et al. (2004a)

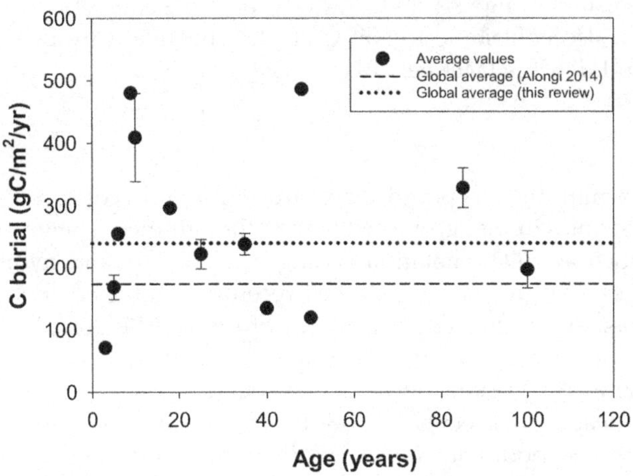

FIGURE 12.4 Average C burial rates (±1 SE) versus stand age (Alongi et al., 2004a; Marchand, 2017; MacKenzie et al., unpublished data; Adame et al., 2018). *Dotted line* represents the global average (Alongi, 2014).

argues the opposite, that the increase in C burial rates observed in Malaysia were due to increase rates of primary production with age. While aboveground NPP did increase with age across the Malaysian sites studied, it is unclear if belowground production also increased. Additional studies are needed to confirm this especially as countries, agencies, and nongovernmental organizations continue to invest in mangrove restoration to offset greenhouse gas emissions or to participate in voluntary C markets.

The lower soil C stocks and burial rates observed in mangroves less than 10 years old likely reflect initial lower values reported from degraded sites or bare landscapes (Kauffman et al., 2017; Sasmito et al., 2020) prior to reforestation or afforestation, respectively. The lower organic matter content from these sites could also result in oxic–suboxic sediment conditions that would allow organic matter decomposition to initially occur until a threshold is reached where sediments become anoxic and decomposition is reduced (Marchand, 2017).

3.3 Algae

Primary production from seagrass litter, macroalgae (e.g., *Ulva pertusa*, *Enteromorpha intestinalis*), filamentous algae, microphytobenthic algae, or diatoms growing on aerial roots or on/within sediments can also be a source of allochthonous or autochthonous C to mangrove sediments (Kristensen 2007, 2008; Alongi, 2014; Sasmito et al., 2020; MacKenzie et al., 2020). A recent review found that gross primary production of microphytobenthic algae can be comparable to the amount of C delivered as litterfall from trees, though standing stocks may be less due to consumption by crabs (Kristensen, 2008; MacKenzie et al., 2020). Contributions of algae to sediment C likely varies with season (Mfilinge et al., 2005) as well as forest age. Marchand et al. (2003) reported that algae growing on the surface of sediments in younger stands of *A. germinans* in French Guiana significantly contributed to organic C content in the upper portions of the sediment. However, he concluded algal contribution to soil C stocks/burial rates likely decreased over time as older stands developed closed canopies (Marchand, 2017). Turnover of algae is also likely to be much greater than leaf litter as algae can decompose quite rapidly before being buried in the sedimentary column. Algae may also contribute to soil C stocks and burial by helping trap and retain C-rich sediments and detritus (McKee, 2011).

3.4 Crabs

Bioturbating sesarmid and ocypodid crabs are important ecological engineers that may also influence C dynamics in mangrove sediments, though this depends on various environmental condition such as soil, vegetation (Wang et al., 2010), and hydrological conditions (Sharma et al., 2014). One of the few studies recently conducted in Kenya found higher soil C stocks in forests with greater abundances of sesarmid crabs (Andreetta et al., 2014). While this might not necessarily imply causation, soil C stocks can be affected via the complex burrow networks created by crabs. Sesarmid tree crabs typically create burrows that can be meters deeper than ocypodid fiddler crabs (Robertson and Daniel, 1989; Smith III et al., 1991), though ocypodid burrows typically occur at much greater densities sesarmids (Cannicci et al., 2008). Burrows can increase the amount of sediment surface area that is exposed to the atmosphere at low tide. Increased surface area is greater in sesarmid burrows

(150%–380%) than ocypodid burrows (~1%), though the greater densities of ocypodids result in comparable surface area increase for both types of crabs. As a result, oxygen can diffuse into anaerobic sediments at low tides when sediments are exposed, which oxidizes elements toxic to plants (i.e., iron and sulfur) (Mitsch and Gosselink, 1993), and can increase tree productivity (Smith III et al., 1991; Kristensen and Alongi, 2006) and potentially belowground production and C burial. During high tides, many crabs plug their holes and thus provide an effective barrier at shielding organic matter from flushing at high tides (Kristensen, 2008).

Crabs also likely influence soil C stocks and burial through feeding. Mechanical fragmentation of decayed and fresh leaves by sesarmid feeding coupled with enzymatic activity within their guts reduces leaf litter particle size (Mouton and Felder, 1996; Botto and Iribarne, 2000; Werry and Lee, 2005) and increases the surface area available for microbial colonization litter (Lee 1997, 2005). Fragmentation, enzymatic activity, and increased surface area all increase leaf litter decomposition and release of C that is either incorporated into mangrove sediments (Andreetta et al., 2014) or lost through outwelling to nearshore waters (Lee, 1995). Crabs also directly transfer C to sediments through the direct transport of leaf litter, propagules, algae, feces, and exuviae into their burrows (Lee, 1997; Alongi, 2002; Huang et al., 2007). The importance of vertical transport into sediments by crabs was demonstrated via radiocarbon dating of sediment cores from a Kenyan mangrove. Modern C was found as deep as 115 cm, confirming that sesarmids are supplying fresh organic matter to deeper sediments (Andreetta et al., 2014). The relationship between macrofauna (e.g., burrowing crabs, polychaete worms, etc.) and soil C stocks and burial rates deserved increased attention in the future. It appears that these macrofaunal diversity could be a key driver of C dynamics in these systems and these organisms are currently threatened by human activities such as pollution, eutrophication, and sea level rise (Andreetta et al., 2014).

4. Conclusion

Many studies have attempted to identify key parameters to develop models that can be used to predict soil C stocks and burial rates as this is where the majority of C is stored in mangrove systems. While we have attempted to present these factors, what is clear is that there are a range of values that differ at the local, regional, and global scale. Some factors that are important at one scale may not be as important at another scale (e.g., porewater salinity). While global or regional models are valuable and play an important role in predicting soil C dynamics, site-level assessments, upon which the quality of these larger-scale efforts rely on, are still scarce, thus preventing the development of robust extrapolation of C burial rates across spatial scales. Complications also arise in comparing results of different studies or in creating global models, as methods used to collect soil C data often vary, which can significantly influence reporting of soil C stocks or burial rates, particularly timescale measurement bias (Breithaupt et al., 2018). It is also clear that, while the number of studies that report soil C burial has significantly increased, gaps remain and additional studies are needed at different scales to increase our understanding of various abiotic and biotic factors. For example, there appear to be several studies that have focused on abiotic factors (e.g., temperature, precipitation, salinity), but fewer studies have examined biotic influence, specifically the role of burrowing crabs or other macrofauna. This information can be used to guide more successful conservation and restoration efforts globally.

References

Adame, M., Zakaria, R., Fry, B., Chong, V., Then, Y., Brown, C., Lee, S., 2018. Loss and recovery of carbon and nitrogen after mangrove clearing. Ocean Coast Manag. 161, 117–126.

Adame, M.F., Kauffman, J.B., Medina, I., Gamboa, J.N., Torres, O., Caamal, J.P., Reza, M., Herrera-Silveira, J.A., 2013. Carbon stocks of tropical coastal wetlands within the karstic landscape of the Mexican Caribbean. PLoS One 8.

Adame, M.F., Lovelock, C.E., 2011. Carbon and nutrient exchange of mangrove forests with the coastal ocean. Hydrobiologia 663, 23–50.

Adame, M.F., Teutli, C., Santini, N.S., Caamal, J.P., Zaldívar-Jiménez, A., Hernández, R., Herrera-Silveira, J.A., 2014. Root biomass and production of mangroves surrounding a karstic oligotrophic coastal lagoon. Wetlands 34, 479–488.

Agraz-Hernández, C.M., Keb, C.A.C., Iriarte-Vivar, S., Venegas, G.P., Serratos, B.V., Sáenz, J.O., 2015. Phenological variation of *Rhizophora mangle* and ground water chemistry associated to changes of the precipitation. Hidrobiológica 25, 49–61.

Alongi, D., Murdiyarso, D., Fourqurean, J., Kauffman, J., Hutahaean, A., Crooks, S., Lovelock, C., Howard, J., Herr, D., Fortes, M., 2015. Indonesia's blue carbon: a globally significant and vulnerable sink for seagrass and mangrove carbon. Wetl. Ecol. Manag. 1–11.

Alongi, D., Pfitzner, J., Trott, L., Tirendi, F., Dixon, P., Klumpp, D., 2005. Rapid sediment accumulation and microbial mineralization in forests of the mangrove *Kandelia candel* in the Jiulongjiang Estuary, China. Estuar. Coast Shelf Sci. 63, 605–618.

Alongi, D., Sasekumar, A., Chong, V., Pfitzner, J., Trott, L., Tirendi, F., Dixon, P., Brunskill, G., 2004a. Sediment accumulation and organic material flux in a managed mangrove ecosystem: estimates of land–ocean–atmosphere exchange in peninsular Malaysia. Mar. Geol. 208, 383–402.

Alongi, D., Wattayakorn, G., Pfitzner, J., Tirendi, F., Zagorskis, I., Brunskill, G., Davidson, A., Clough, B., 2001. Organic carbon accumulation and metabolic pathways in sediments of mangrove forests in southern Thailand. Mar. Geol. 179, 85–103.

Alongi, D.M., 2002. Present state and future of the world's mangrove forests. Environ. Conserv. 29, 331–349.

Alongi, D.M., 2005. Mangrove-microbe-soil relations. In: Interactions between Macro-And Microorganisms in Marine Sediments, pp. 85–103.

Alongi, D.M., 2009. The Energetics of Mangrove Forests. Springer Dordrecht, The Netherlands.

Alongi, D.M., 2011. Carbon payments for mangrove conservation: ecosystem constraints and uncertainties of sequestration potential. Environ. Sci. Pol. 14, 462–470.

Alongi, D.M., 2012. Carbon sequestration in mangrove forests. Carbon Manag. 3, 313–322.

Alongi, D.M., 2014. Carbon cycling and storage in mangrove forests. Annu. Rev. Mar. Sci. 6, 195–219.

Alongi, D.M., Clough, B.F., Dixon, P., Tirendi, F., 2003. Nutrient partitioning and storage in arid-zone forests of the mangroves *Rhizophora stylosa* and *Avicennia marina*. Trees (Berl.) 17, 51–60.

Alongi, D.M., Dixon, P., 2000. Mangrove Primary Production and Above-And Below-Ground Biomass in Sawi Bay, Southern Thailand. Phuket Marine Biological Center Special Publication 22.

Alongi, D.M., Wattayakorn, G., Tirendi, F., Dixon, P., 2004b. Nutrient capital in different aged forests of the mangrove *Rhizophora apiculata*. Bot. Mar. 47, 116–124.

Andreetta, A., Fusi, M., Cameldi, I., Cimò, F., Carnicelli, S., Cannicci, S., 2014. Mangrove carbon sink. Do burrowing crabs contribute to sediment carbon storage? Evidence from a Kenyan mangrove system. J. Sea Res. 85, 524–533.

Atwood, T.B., Connolly, R.M., Almahasheer, H., Carnell, P.E., Duarte, C.M., Lewis, C.J.E., Irigoien, X., Kelleway, J.J., Lavery, P.S., Macreadie, P.I., 2017. Global patterns in mangrove soil carbon stocks and losses. Nat. Clim. Chang. 7 (7), 523–528.

Ball, M.C., 1988. Salinity tolerance in the mangroves *Aegiceras corniculatum* and *Avicennia marina*. I. Water use in relation to growth, carbon partitioning, and salt balance. Funct. Plant Biol. 15, 447–464.

Bird, E., 1986. Mangroves and intertidal morphology in Westernport Bay, Victoria, Australia. Mar. Geol. 69, 251–271.

Botto, F., Iribarne, O., 2000. Contrasting effects of two burrowing crabs (*Chasmagnathus granulata* and *Uca uruguayensis*) on sediment composition and transport in estuarine environments. Estuar. Coast Shelf Sci. 51, 141–151.

Bouillon, S., Borges, A.V., Castañeda-Moya, E., Diele, K., Dittmar, T., Duke, N.C., Kristensen, E., Lee, S.Y., Marchand, C., Middelburg, J.J., Rivera-Monroy, V.H., Smith III, T.J., Twilley, R.R., 2008. Mangrove production and carbon sinks: a revision of global budget estimates. Glob. Biogeochem. Cycles 22.

Breithaupt, J.L., Smoak, J.M., Byrne, R.H., Waters, M.N., Moyer, R.P., Sanders, C.J., 2018. Avoiding timescale bias in assessments of coastal wetland vertical change. Limnology and Oceanography S477–S495.

Breithaupt, J.L., Smoak, J.M., Rivera-Monroy, V.H., Castañeda-Moya, E., Moyer, R.P., Simard, M., Sanders, C.J., 2017. Partitioning the relative contributions of organic matter and mineral sediment to accretion rates in carbonate platform mangrove soils. Mar. Geol. 390, 170–180.

Breithaupt, J.L., Smoak, J.M., Smith III, T.J., Sanders, C.J., Hoare, A., 2012. Organic carbon burial rates in mangrove sediments: strengthening the global budget. Glob. Biogeochem. Cycles 26.

Brunskill, G., Zagorskis, I., Pfitzner, J., 2002. Carbon burial rates in sediments and a carbon mass balance for the Herbert River region of the Great Barrier Reef continental shelf, North Queensland, Australia. Estuar. Coast Shelf Sci. 54, 677–700.

Brunskill, G.J., Zagorskis, I., Pfitzner, J., Ellison, J., 2004. Sediment and trace element depositional history from the Ajkwa River estuarine mangroves of Irian Jaya (West Papua), Indonesia. Continent. Shelf Res. 24, 2535–2551.

Bukoski, J.J., Broadhead, J.S., Donato, D.C., Murdiyarso, D., Gregoire, T.G., 2017. The use of mixed effects models for obtaining low-cost ecosystem carbon stock estimates in mangroves of the Asia-Pacific. PLoS One 12, e0169096.

Callaway, J., DeLaune, R., Patrick Jr., W., 1997. Sediment accretion rates from four coastal wetlands along the Gulf of Mexico. J. Coast Res. 181–191.

Cameron, C., Hutley, L.B., Friess, D.A., Brown, B., 2019. Community structure dynamics and carbon stock change of rehabilitated mangrove forests in Sulawesi, Indonesia. Ecol. Appl. 29, e01810.

Cannicci, S., Burrows, D., Fratini, S., Smith III, T.J., Offenberg, J., Dahdouh-Guebas, F., 2008. Faunal impact on vegetation structure and ecosystem function in mangrove forests: a review. Aquat. Bot. 89, 186–200.

Castañeda-Moya, E., Rivera-Monroy, V.H., Chambers, R.M., Zhao, X., Lamb-Wotton, L., Gorsky, A., Gaiser, E.E., Troxler, T.G., Kominoski, J.S., Hiatt, M., 2020. Hurricanes fertilize mangrove forests in the Gulf of Mexico (Florida Everglades, USA). Proc. Natl. Acad. Sci. 117 (9), 4831–4841.

Castañeda-Moya, E., Twilley, R.R., Rivera-Monroy, V.H., 2013. Allocation of biomass and net primary productivity of mangrove forests along environmental gradients in the Florida Coastal Everglades, USA. For. Ecol. Manag. 307, 226–241.

Castañeda-Moya, E., Twilley, R.R., Rivera-Monroy, V.H., Marx, B.D., Coronado-Molina, C., Ewe, S.M., 2011. Patterns of root dynamics in mangrove forests along environmental gradients in the Florida Coastal Everglades, USA. Ecosystems 14, 1178–1195.

Castañeda-Moya, E., Twilley, R.R., Rivera-Monroy, V.H., Zhang, K., Davis, S.E., Ross, M., 2010. Sediment and nutrient deposition associated with hurricane Wilma in mangroves of the Florida Coastal Everglades. Estuar. Coast 33, 45–58.

Cerón-Bretón, J., Cerón-Bretón, R., Rangel-Marrón, M., Murielgarcia, M., Cordova-Quiroz, A., Estrella-Cahuich, A., 2011. Determination of carbon sequestration rate in soil of a mangrove forest in Campeche, Mexico. WSEAS Trans. Environ. Dev. 7, 55–64.

Chapin III, F.S., Vitousek, P.M., Van Cleve, K., 1986. The nature of nutrient limitation in plant communities. Am. Nat. 127, 48–58.

Chapman, S.K., Hayes, M.A., Kelly, B., Langley, J.A., 2019. Exploring the oxygen sensitivity of wetland soil carbon mineralization. Biol. Lett. 15, 20180407.

Charles, H., Dukes, J.S., 2009. Effects of warming and altered precipitation on plant and nutrient dynamics of a New England salt marsh. Ecol. Appl. 19, 1758–1773.

Chen, G., Gao, M., Pang, B., Chen, S., Ye, Y., 2018. Top-meter soil organic carbon stocks and sources in restored mangrove forests of different ages. For. Ecol. Manag. 422, 87–94.

Chen, R., Twilley, R.R., 1999. A simulation model of organic matter and nutrient accumulation in mangrove wetland soils. Biogeochemistry 44, 93–118.

Chmura, G.L., Anisfeld, S.C., Cahoon, D.R., Lynch, J.C., 2003. Global carbon sequestration in tidal, saline wetland soils. Glob. Biogeochem. Cycles 17.

Cormier, N., Twilley, R.R., Ewel, K.C., Krauss, K.W., 2015. Fine root productivity varies along nitrogen and phosphorus gradients in high-rainfall mangrove forests of Micronesia. Hydrobiologia 750, 69–87.

Cragg, S.M., Friess, D.A., Gillis, L.G., Trevathan-Tackett, S.M., Terrett, O.M., Watts, J.E., Distel, D.L., Dupree, P., 2020. Vascular Plants Are Globally Significant Contributors to Marine Carbon Fluxes and Sinks.

Day Jr., J., Coronado-Molina, C., Vera-Herrera, F., Twilley, R., Rivera-Monroy, V., Alvarez-Guillen, H., Day, R., Conner, W., 1996. A 7 year record of above-ground net primary production in a southeastern Mexican mangrove forest. Aquat. Bot. 55, 39–60.

Dittmar, T., Lara, R., 2001. Driving forces behind nutrient and organic matter dynamics in a mangrove tidal creek in North Brazil. Estuarine. Coast. Shelf Sci. 52, 249–259.

Donato, D.C., Kauffman, J.B., Murdiyarso, D., Kurnianto, S., Stidham, M., Kanninen, M., 2011. Mangroves among the most carbon-rich forests in the tropics. Nat. Geosci. 4, 293–297.

Duarte, C.M., Kennedy, H., Marbà, N., Hendriks, I., 2013a. Assessing the capacity of seagrass meadows for carbon burial: current limitations and future strategies. Ocean Coast Manag. 83, 32–38.

Duarte, C.M., Losada, I.J., Hendriks, I.E., Mazarrasa, I., Marbà, N., 2013b. The role of coastal plant communities for climate change mitigation and adaptation. Nat. Clim. Chang. 3, 961–968.

Duarte, C.M., Middelburg, J.J., Caraco, N., 2004. Major role of marine vegetation on the oceanic carbon cycle. Biogeosci. Discuss. Eur. Geosci. Union 1, 659–679.

Duke, N., 1990. Phenological trends with latitude in the mangrove tree *Avicennia marina*. J. Ecol. 113–133.

Dürr, H.H., Laruelle, G.G., van Kempen, C.M., Slomp, C.P., Meybeck, M., Middelkoop, H., 2011. Worldwide typology of nearshore coastal systems: defining the estuarine filter of river inputs to the oceans. Estuar. Coast 34, 441–458.

Ellison, J.C., 1994. Climate change and sea level rise impacts on mangrove ecosystems. In: Impacts of Climate Change on Ecosystems and Species. A Marine Conservation and Development Report. IUCN.

Ezcurra, P., Ezcurra, E., Garcillán, P.P., Costa, M.T., Aburto-Oropeza, O., 2016. Coastal landforms and accumulation of mangrove peat increase carbon sequestration and storage. Proc. Natl. Acad. Sci. U States A 113, 4404–4409.

Farnsworth, E.J., 1998. Issues of spatial, taxonomic and temporal scale in delineating links between mangrove diversity and ecosystem function. Glob. Ecol. Biogeogr. Lett. 15–25.

Feher, L.C., Osland, M.J., Griffith, K.T., Grace, J.B., Howard, R.J., Stagg, C.L., Enwright, N.M., Krauss, K.W., Gabler, C.A., Day, R.H., 2017. Linear and nonlinear effects of temperature and precipitation on ecosystem properties in tidal saline wetlands. Ecosphere 8, e01956.

Feller, I.C., Whigham, D.F., McKee, K.L., Lovelock, C.E., 2003. Nitrogen limitation of growth and nutrient dynamics in a disturbed mangrove forest, Indian River Lagoon, Florida. Oecologia 134, 405–414.

Field, C.B., Osborn, J.G., Hoffman, L.L., Polsenberg, J.F., Ackerly, D.D., Berry, J.A., Bjorkman, O., Held, A., Matson, P.A., Mooney, H.A., 1998. Mangrove biodiversity and ecosystem function. Glob. Ecol. Biogeogr. 7, 3–14.

Fujimoto, K., Imaya, A., Tabuchi, R., Kuramoto, S., Utsugi, H., Murofushi, T., 1999. Belowground carbon storage of Micronesian mangrove forests. Ecol. Res. 14, 409–413.

Furukawa, K., Wolanski, E., 1996. Sedimentation in mangrove forests. Mangroves Salt Marshes 1, 3–10.

Furukawa, K., Wolanski, E., Mueller, H., 1997. Currents and sediment transport in mangrove forests. Estuar. Coast Shelf Sci. 44, 301–310.

Gillerot, L., Vlaminck, E., De Ryck, D.J., Mwasaru, D.M., Beeckman, H., Koedam, N., 2018. Inter-and intraspecific variation in mangrove carbon fraction and wood specific gravity in Gazi Bay, Kenya. Ecosphere 9, e02306.

Gillis, L., Bouma, T., Jones, C., Van Katwijk, M., Nagelkerken, I., Jeuken, C., Herman, P., Ziegler, A., 2014. Potential for landscape-scale positive interactions among tropical marine ecosystems. Mar. Ecol. Prog. Ser. 503, 289–303.

Gleason, S.M., Ewel, K.C., 2002. Organic matter dynamics on the forest floor of a Micronesian mangrove forest: an investigation of species composition shifts. Biotropica 34, 190–198.

Gonneea, M.E., Paytan, A., Herrera-Silveira, J.A., 2004. Tracing organic matter sources and carbon burial in mangrove sediments over the past 160 years. Estuar. Coast Shelf Sci. 61, 211–227.

Harrison, J.A., Caraco, N., Seitzinger, S.P., 2005. Global patterns and sources of dissolved organic matter export to the coastal zone: results from a spatially explicit, global model. Glob. Biogeochem. Cycles 19.

Hawkins, B.A., Felizola Diniz-Filho, J.A., 2004. 'Latitude' and geographic patterns in species richness. Ecography 27, 268–272.

Hayes, M.A., Jesse, A., Hawke, B., Baldock, J., Tabet, B., Lockington, D., Lovelock, C.E., 2017. Dynamics of sediment carbon stocks across intertidal wetland habitats of Moreton Bay, Australia. Glob. Chang. Biol. 23, 4222–4234.

Huang, K., Boudreau, B.P., Reed, D.C., 2007. Simulated fiddler-crab sediment mixing. J. Mar. Res. 65, 491–522.

Huxham, M., Kumara, M.P., Jayatissa, L.P., Krauss, K.W., Kairo, J., Langat, J., Mencuccini, M., Skov, M.W., Kirui, B., 2010. Intra-and interspecific facilitation in mangroves may increase resilience to climate change threats. Phil. Trans. Biol. Sci. 365, 2127–2135.

Huxham, M., Whitlock, D., Githaiga, M., Dencer-Brown, A., 2018. Carbon in the coastal seascape: how interactions between mangrove forests, seagrass meadows and tidal marshes influence carbon storage. Curr. For. Rep. 4, 101–110.

Jardine, S.L., Siikamäki, J.V., 2014. A global predictive model of carbon in mangrove soils. Environ. Res. Lett. 9, 104013.

Jennerjahn, T.C., Ittekkot, V., 2002. Relevance of mangroves for the production and deposition of organic matter along tropical continental margins. Naturwissenschaften 89, 23–30.

Kauffman, J.B., Adame, M.F., Arifanti, V.B., Schile-Beers, L.M., Bernardino, A.F., Bhomia, R.K., Donato, D.C., Feller, I., Ferreira, T.O., Garcia, M.C.J., MacKenzie, R.A., Megonigal, J.P., Murdiyarso, D., Simpson, L., Trejo, H.H., 2020. Total ecosystem carbon stocks of mangroves across broad global environmental and physical gradients. Ecol. Monogr.

Kauffman, J.B., Arifanti, V.B., Trejo, H.H., García, M.C.J., Norfolk, J., Cifuentes, M., Hadriyanto, D., Murdiyarso, D., 2017. The jumbo carbon footprint of a shrimp: carbon losses from mangrove deforestation. Front. Ecol. Environ. 15, 183–188.

Kauffman, J.B., Bhomia, R.K., 2017. Ecosystem carbon stocks of mangroves across broad environmental gradients in West-Central Africa: global and regional comparisons. PLoS One 12, e0187749.

Kauffman, J.B., Cole, T.G., 2010. Micronesian mangrove forest structure and tree responses to a severe typhoon. Wetlands 30, 1077–1084.

Kida, M., Tanabe, M., Tomotsune, M., Yoshitake, S., Kinjo, K., Ohtsuka, T., Fujitake, N., 2019. Changes in dissolved organic matter composition and dynamics in a subtropical mangrove river driven by rainfall. Estuar. Coast Shelf Sci. 223, 6–17.

Koch, M., Benz, R., Rudnick, D., 2001. Solid-phase phosphorus pools in highly organic carbonate sediments of northeastern Florida Bay. Estuarine. Coast. Shelf Sci. 52, 279–291.

Krauss, K.W., Allen, J.A., Cahoon, D.R., 2003. Differential rates of vertical accretion and elevation change among aerial root types in Micronesian mangrove forests. Estuar. Coast Mar. Sci. 56, 251–259.

Krauss, K.W., Cormier, N., Osland, M.J., Kirwan, M.L., Stagg, C.L., Nestlerode, J.A., Russell, M.J., From, A.S., Spivak, A.C., Dantin, D.D., 2017. Created mangrove wetlands store belowground carbon and surface elevation change enables them to adjust to sea-level rise. Sci. Rep. 7, 1030.

Krauss, K.W., McKee, K.L., Lovelock, C.E., Cahoon, D.R., Saintilan, N., Reef, R., Chen, L., 2014. How mangrove forests adjust to rising sea level. New Phytol. 202, 19–34.

Kristensen, E., 2007. Carbon balance in mangrove sediments: the driving processes and their controls. Greenh. Gas Carbon Balances Mangrove Coast. Ecosyst. 61–78. Gendai Tosho.

Kristensen, E., 2008. Mangrove crabs as ecosystem engineers; with emphasis on sediment processes. J. Sea Res. 59, 30–43.

Kristensen, E., Alongi, D.M., 2006. Control by fiddler crabs (*Uca vocans*) and plant roots (*Avicennia marina*) on carbon, iron, and sulfur biogeochemistry in mangrove sediments. Limnol. Oceanogr. 51, 1557–1571.

Kristensen, E., Bouillon, S., Dittmar, T., Marchand, C., 2008. Organic carbon dynamics in mangrove ecosystems: a review. Aquat. Bot. 89, 201–219.

Kumara, M., Jayatissa, L., Krauss, K., Phillips, D., Huxham, M., 2010. High mangrove density enhances surface accretion, surface elevation change, and tree survival in coastal areas susceptible to sea-level rise. Oecologia 164, 545–553.

Kusumaningtyas, M.A., Hutahaean, A.A., Fischer, H.W., Pérez-Mayo, M., Ransby, D., Jennerjahn, T.C., 2019. Variability in the organic carbon stocks, sources, and accumulation rates of Indonesian mangrove ecosystems. Estuar. Coast Shelf Sci. 218, 310–323.

Lang'at, J.K.S., Kirui, B.K.Y., Skov, M.W., Kairo, J.G., Mencuccini, M., Huxham, M., 2013. Species mixing boosts root yield in mangrove trees. Oecologia 172, 271–278.

Lee, S.Y., 1995. Mangrove outwelling: a review. Hydrobiologia 295, 203–212.

Lee, S.Y., 1997. Potential trophic importance of the faecal material of the mangrove sesarmine crab Sesarma masse. Mar. Ecol. Prog. Ser. 159, 275–284.

Lee, S.Y., 2005. Exchange of organic matter and nutrients between mangroves and estuaries: myths, methodological issues and missing links. Int. J. Ecol. Environ. Sci. 31, 163–176.

Leuschner, C., Hertel, D., Coners, H., Büttner, V., 2001. Root competition between beech and oak: a hypothesis. Oecologia 126, 276–284.

Lo Iacono, C., Mateo, M.A., Gracia, E., Guasch, L., Carbonell, R., Serrano, L., Serrano, O., Danobeitia, J., 2008. Very high-resolution seismo-acoustic imaging of seagrass meadows (Mediterranean Sea): implications for carbon sink estimates. Geophys. Res. Lett. 35.

Lovelock, C.E., 2008. Soil respiration and belowground carbon allocation in mangrove forests. Ecosystems 11, 342–354.

Lovelock, C.E., Duarte, C.M., 2019. Dimensions of blue carbon and emerging perspectives. Biol. Lett. 15, 20180781.

Lovelock, C.E., Ellison, J., 2007. Vulnerability of Mangroves and Tidal Wetlands of the Great Barrier Reef to Climate Change.

Lovelock, C.E., Feller, I.C., Reef, R., Hickey, S., Ball, M.C., 2017. Mangrove dieback during fluctuating sea levels. Sci. Rep. 7, 1–8.

Lynch, J.C., Meriwether, J.R., McKee, B.A., Vera-Herrera, F., Twilley, R.R., 1989. Recent accretion in mangrove ecosystems based on137Cs and210Pb. Estuaries 12, 284–299.

MacKenzie, R.A., Cormier, N., Demopoulos, A.W., 2020. Estimating the value of mangrove leaf litter in sesarmid crab diets: the importance of fractionation factors. Bull. Mar. Sci.

MacKenzie, R.A., Foulk, P.B., Klump, J.V., Weckerly, K., Purbospito, J., Murdiyarso, D., Donato, D.C., Nam, V.N., 2016. Sedimentation and belowground carbon accumulation rates in mangrove forests that differ in diversity and land use: a tale of two mangroves. Wetl. Ecol. Manag. 24, 245–261.

Mackey, A., 1993. Biomass of the mangrove *Avicennia marina* (Forsk.) Vierh. Near Brisbane, south-eastern Queensland. Mar. Freshw. Res. 44, 721–725.

Macreadie, P.I., Anton, A., Raven, J.A., Beaumont, N., Connolly, R.M., Friess, D.A., Kelleway, J.J., Kennedy, H., Kuwae, T., Lavery, P.S., 2019. The future of Blue Carbon science. Nat. Commun. 10, 1–13.

Marchand, C., 2017. Soil carbon stocks and burial rates along a mangrove forest chronosequence (French Guiana). For. Ecol. Manag. 384, 92–99.

Marchand, C., Lallier-Vergès, E., Baltzer, F., 2003. The composition of sedimentary organic matter in relation to the dynamic features of a mangrove-fringed coast in French Guiana. Estuar. Coast Shelf Sci. 56, 119–130.

Matsui, N., 1998. Estimated stocks of organic carbon in mangrove roots and sediments in Hinchinbrook Channel, Australia. Mangroves Salt Marshes 2, 199–204.

McKee, K.L., 1995. Interspecific variation in growth, biomass partitioning, and defensive characterisitcics of neotropical mangrove seedlings: response to light and nutrient availability. Am. J. Bot. 82, 299–307.

McKee, K.L., 2011. Biophysical controls on accretion and elevation change in Caribbean mangrove ecosystems. Estuar. Coast Shelf Sci. 91, 475–483.

McKee, K.L., Cahoon, D.R., Feller, I.C., 2007. Caribbean mangroves adjust to rising sea level through biotic controls on change in soil elevation. Glob. Ecol. Biogeogr. 16, 545–556.

Mcleod, E., Chmura, G.L., Bouillon, S., Salm, R., Björk, M., Duarte, C.M., Lovelock, C.E., Schlesinger, W.H., Silliman, B.R., 2011. A blueprint for blue carbon: toward an improved understanding of the role of vegetated coastal habitats in sequestering CO_2. Front. Ecol. Environ. 9, 552–560.

Mfilinge, P.L., Meziane, T., Bachok, Z., Tsuchiya, M., 2005. Litter dynamics and particulate organic matter outwelling from a subtropical mangrove in Okinawa Island, South Japan. Estuarine. Coast. Shelf Sci. 63, 301–313.

Mitsch, W.J., Gosselink, J.G., 1993. Wetlands, second ed. John Wiley and Sons, New York.

Mouton, E.C., Felder, D.L., 1996. Burrow distributions and population estimates for the fiddler crabs *Uca spinicarpa* and *Uca longisignalis* in a Gulf of Mexico salt marsh. Estuaries 19, 51–61.

Nellemann, C., Corcoran, E., 2009. Blue Carbon: The Role of Healthy Oceans in Binding Carbon: a Rapid Response Assessment. UNEP/Earthprint.

Osland, M.J., Spivak, A.C., Nestlerode, J.A., Lessmann, J.M., Almario, A.E., Heitmuller, P.T., Russell, M.J., Krauss, K.W., Alvarez, F., Dantin, D.D., 2012. Ecosystem development after mangrove wetland creation: plant–soil change across a 20-year chronosequence. Ecosystems 15, 848–866.

Ouyang, X., Lee, S.Y., Connolly, R.M., 2017. The role of root decomposition in global mangrove and saltmarsh carbon budgets. Earth Sci. Rev. 166, 53–63.

Pool, D.J., Lugo, A.E., Snedaker, S.C., 1975. Litter production in mangrove forests of southern Florida and Puerto Rico. In: International Symposium on Biology and Management of Mangroves, Honolulu, pp. 213–237.

Pregitzer, K.S., King, J.S., Burton, A.J., Brown, S.E., 2000. Responses of tree fine roots to temperature. New Phytol. 147, 105–115.

Rahman, M.M., Khan, M.N.I., Hoque, A.F., Ahmed, I., 2015. Carbon stock in the Sundarbans mangrove forest: spatial variations in vegetation types and salinity zones. Wetl. Ecol. Manag. 23, 269–283.

Ray, R., Ganguly, D., Chowdhury, C., Dey, M., Das, S., Dutta, M., Mandal, S.K., Majumder, N., De, T., Mukhopadhyay, S., 2011. Carbon sequestration and annual increase of carbon stock in a mangrove forest. Atmos. Environ. 45, 5016–5024.

Robertson, A., Daniel, P., 1989. The influence of crabs on litter processing in high intertidal mangrove forests in tropical Australia. Oecologia 78, 191–198.

Rodrigues, D.P., Hamacher, C., Estrada, G.C.D., Soares, M.L.G., 2015. Variability of carbon content in mangrove species: effect of species, compartments and tidal frequency. Aquat. Bot. 120, 346–351.

Rovai, A., Riul, P., Twilley, R., Castañeda-Moya, E., Rivera-Monroy, V., Williams, A., Simard, M., Cifuentes-Jara, M., Lewis, R., Crooks, S., 2016. Scaling mangrove aboveground biomass from site-level to continental-scale. Glob. Ecol. Biogeogr. 25, 286–298.

Rovai, A.S., Twilley, R.R., Castañeda-Moya, E., Riul, P., Cifuentes-Jara, M., Manrow-Villalobos, M., Horta, P.A., Simonassi, J.C., Fonseca, A.L., Pagliosa, P.R., 2018. Global controls on carbon storage in mangrove soils. Nat. Clim. Chang. 8, 534–538.

Saavedra-Hortua, D.A., Friess, D.A., Zimmer, M., Gillis, L.G., 2020. Sources of particulate organic matter across mangrove forests and adjacent ecosystems in different geomorphic settings. Wetlands 1–13.

Saintilan, N., 1997a. Above-and below-ground biomass of mangroves in a sub-tropical estuary. Mar. Freshw. Res. 48, 601–604.

Saintilan, N., 1997b. Above-and below-ground biomasses of two species of mangrove on the Hawkesbury River estuary, New South Wales. Mar. Freshw. Res. 48, 147–152.

Saintilan, N., Rogers, K., Mazumder, D., Woodroffe, C., 2013. Allochthonous and autochthonous contributions to carbon accumulation and carbon store in southeastern Australian coastal wetlands. Estuar. Coast Shelf Sci. 128, 84–92.

Sanders, C.J., Maher, D.T., Tait, D.R., Williams, D., Holloway, C., Sippo, J.Z., Santos, I.R., 2016. Are global mangrove carbon stocks driven by rainfall? J. Geophys. Res.: Biogeosciences 121, 2600–2609.

Sanders, C.J., Smoak, J.M., Naidu, A.S., Araripe, D.R., Sanders, L.M., Patchineelam, S.R., 2010a. Mangrove forest sedimentation and its reference to sea level rise, Cananeia, Brazil. Environ. Earth Sci. 60, 1291–1301.

Sanders, C.J., Smoak, J.M., Naidu, A.S., Patchineelam, S.R., 2008. Recent sediment accumulation in a mangrove forest and its relevance to local sea-level rise (Ilha Grande, Brazil). J. Coast Res. 533–536.

Sanders, C.J., Smoak, J.M., Naidu, A.S., Sanders, L.M., Patchineelam, S.R., 2010b. Organic carbon burial in a mangrove forest, margin and intertidal mud flat. Estuar. Coast Shelf Sci. 90, 168–172.

Sanders, C.J., Smoak, J.M., Sanders, L.M., Naidu, A.S., Patchineelam, S.R., 2010c. Organic carbon accumulation in Brazilian mangal sediments. J. S. Am. Earth Sci. 30, 189–192.

Sasmito, S.D., Kuzyakov, Y., Lubis, A.A., Murdiyarso, D., Hutley, L.B., Bachri, S., Friess, D.A., Martius, C., Borchard, N., 2020. Organic carbon burial and sources in soils of coastal mudflat and mangrove ecosystems. Catena 187, 104414.

Schile, L.M., Kauffman, J.B., Crooks, S., Fourqurean, J.W., Glavan, J., Megonigal, J.P., 2017. Limits on carbon sequestration in arid blue carbon ecosystems. Ecol. Appl. 27, 859–874.

Sharma, S., Hoque, A.R., Analuddin, K., Hagihara, A., 2014. A model of seasonal foliage dynamics of the subtropical mangrove species *Rhizophora stylosa* Griff. growing at the northern limit of its distribution. For. Ecosyst. 1, 15.

Sharma, S., MacKenzie, R.A., Tieng, T., Soben, K., Tulyasuwan, N., Resanond, A., Blate, G., Litton, C.M., 2020. The impacts of degradation, deforestation and restoration on mangrove ecosystem carbon stocks across Cambodia. Sci. Total Environ. 706, 135416.

Šímová, I., Storch, D., 2017. The enigma of terrestrial primary productivity: measurements, models, scales and the diversity–productivity relationship. Ecography 40, 239–252.

Smith III, T.J., Boto, K.G., Frusher, S.D., Giddins, R.L., 1991. Keystone species and mangrove forest dynamics: the influence of burrowing by crabs on soil nutrient status and forest productivity. Estuar. Coast Mar. Sci. 33, 419–432.

Smoak, J.M., Breithaupt, J.L., Smith III, T.J., Sanders, C.J., 2013. Sediment accretion and organic carbon burial relative to sea-level rise and storm events in two mangrove forests in Everglades National Park. Catena 104, 58–66.

Spenceley, A., 1977. The role of pneumatophores in sedimentary processes. Mar. Geol. 24, M31–M37.

Tamooh, F., Huxham, M., Karachi, M., Mencuccini, M., Kairo, J., Kirui, B., 2008. Below-ground root yield and distribution in natural and replanted mangrove forests at Gazi bay, Kenya. For. Ecol. Manag. 256, 1290–1297.

Tateda, Y., Nhan, D., Wattayakorn, G., Toriumi, H., 2005. Preliminary evaluation of organic carbon sedimentation rates in Asian mangrove coastal ecosystems estimated by {210} Pb chronology. Radioprotection 40, S527.

Thom, B.G., 1982. Mangrove ecology – a geomorphological perspective. In: Clough, B.F. (Ed.), Mangrove Ecosystems in Australia: Structure, Function and Management. Australian National University Press, Canberra, Australia, pp. 3–17.

Thorhaug, A.L., Poulos, H.M., López-Portillo, J., Barr, J., Lara-Domínguez, A.L., Ku, T.C., Berlyn, G.P., 2019. Gulf of Mexico estuarine blue carbon stock, extent and flux: mangroves, marshes, and seagrasses: a North American hotspot. Sci. Total Environ. 653, 1253–1261.

Twilley, R., Snedaker, S., Yáñez-Arancibia, A., Medina, E., 1996. Biodiversity and ecosystem processes in tropical estuaries: perspectives of mangrove ecosystems. In: Mooney, H.A., Cushman, J.H., Medina, E., Sala, O.E., Schulze, E.D. (Eds.), Functional Roles of Biodiversity: A Global Perspective. John Wiley and Sons Ltd., New York.

Twilley, R.R., 1985. The exchange of organic carbon in basin mangrove forests in a southwest Florida estuary. Estuar. Coast Shelf Sci. 20, 543–557.

Twilley, R.R., 1995. Properties of mangrove ecosystems related to the energy signature of coastal environments. Maximum Power: The ideas and applications of HT Odum, 43–62.

Twilley, R.R., Castañeda-Moya, E., Rivera-Monroy, V.H., Rovai, A., 2017. Productivity and carbon dynamics in mangrove wetlands. In: Mangrove Ecosystems: A Global Biogeographic Perspective. Springer, pp. 113–162.

Twilley, R.R., Pozo, M., Garcia, V.H., Rivera-Monroy, V.H., Zambrano, R., Bodero, A., 1997. Litter dynamics in riverine mangrove forests in the Guayas River estuary, Ecuador. Oecologia 111, 109–122.

Twilley, R.R., Rovai, A.S., Riul, P., 2018. Coastal morphology explains global blue carbon distributions. Front. Ecol. Environ. 16, 503–508.

Twilley, R.W., Lugo, A.E., Patterson-Zucca, C., 1986. Litter production and turnover in basin mangrove forests in southwest Florida. Ecology 670–683.

Wang, J.Q., Zhang, X.D., Jiang, L.F., Bertness, M.D., Fang, C.M., Chen, J.K., Hara, T., Li, B., 2010. Bioturbation of burrowing crabs promotes sediment turnover and carbon and nitrogen movements in an estuarine salt marsh. Ecosystems 13, 586–599.

Werry, J., Lee, S.Y., 2005. Grapsid crabs mediate link between mangrove litter production and estuarine planktonic food chains. Mar. Ecol. Prog. Ser. 293, 165–176.

Woodroffe, C.D., Rogers, K., McKee, K.L., Lovelock, C.E., Mendelssohn, I., Saintilan, N., 2016. Mangrove sedimentation and response to relative sea-level rise. Annu. Rev. Mar. Sci. 8, 243–266.

Xiaonan, D., Xiaoke, W., Lu, F., Zhiyun, O., 2008. Primary evaluation of carbon sequestration potential of wetlands in China. Acta Ecol. Sin. 28, 463–469.

Yang, G., Chen, H., Wu, N., Tian, J., Peng, C., Zhu, Q., Zhu, D., He, Y., Zheng, Q., Zhang, C., 2014. Effects of soil warming, rainfall reduction and water table level on CH_4 emissions from the Zoige peatland in China. Soil Biol. Biochem. 78, 83–89.

Young, B.M., Harvey, E.L., 1996. A spatial analysis of the relationship between mangrove (*Avicennia marinavar.* australasica) physiognomy and sediment accretion in the Hauraki Plains, New Zealand. Estuar. Coast Shelf Sci. 42, 231–246.

CHAPTER

13

Gaps, challenges, and opportunities in mangrove blue carbon research: a biogeographic perspective

Andre S. Rovai[1,2], *Robert R. Twilley*[1]

[1]Department of Oceanography and Coastal Sciences, College of Coast and Environment, Louisiana State University, Baton Rouge, LA, United States; [2]Departamento de Oceanografia, Universidade Federal de Santa Catarina, Florianópolis, Santa Catarina, Brazil

OUTLINE

1. **Introduction** 295
2. **Mangroves as blue carbon sinks: stocks versus sequestration** 298
3. **Biogeography of blue carbon research in mangroves** 307
   - *3.1 Above-ground biomass* 307
   - *3.2 Below-ground biomass* 307
   - *3.3 Soil organic carbon* 310
4. **Global trends in blue C in mangroves** 315
   - *4.1 Biogeographic and regional controls of C storage in mangrove biomass and soils* 315
   - *4.2 Biogeographic and regional controls of C sequestration in mangrove biomass and soils* 319
5. **Challenges and opportunities to improve mangrove blue C inventories and mapping** 325

References 329

1. Introduction

Mangroves fringe the border of all continents, except Europe and Antarctica, spanning from equatorial to subtropical coastlines (Fig. 13.1A). However, in certain regions they extend all the

https://doi.org/10.1016/B978-0-12-816437-2.00001-X

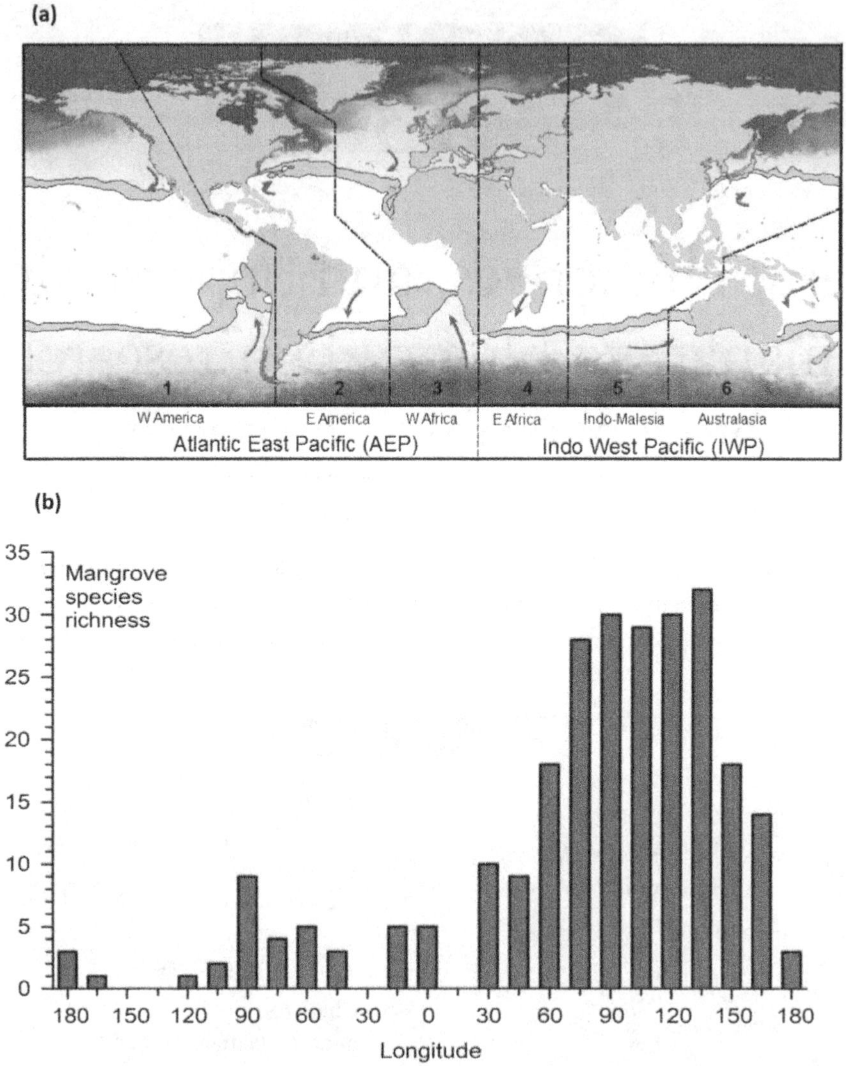

FIGURE 13.1 Global distribution of mangrove forests (light green (light gray in print version)) across biogeographic regions as defined by Duke et al. (1998) (A). Polygons (pink (dark gray in print version)) over the oceans represent updated 20°C isotherms for summer and winter along with the influential currents affecting those isotherms (*arrows*). Mangrove species richness by longitude depicting the greater species richness in Indo West Pacific (IWP) versus Atlantic East Pacific (AEP) biogeographic regions (B). *(A) From Ellison, A.M., Farnsworth, E.J., Merkt, R.E., 1999. Origins of mangrove ecosystems and the mangrove biodiversity anomaly. Global Ecol. Biogeogr. 8, 95–115; (B) From Twilley et al., (in press).*

way into warm temperate zones where the synergetic effect of climatic (e.g., absence of frost) and oceanographic factors provide suitable environments for these plant communities to thrive (Chapman, 1975, but see Bardou et al., 2020, Cavanaugh et al., 2018, 2014 and Osland et al., 2017 for updated discussion on climate drivers). In addition to this latitudinal range, mangroves are widespread across both "old world" and "new world" (Fig. 13.1B). These latitudinal and longitudinal continuums are reflected in much of the conspicuous tree traits and adaptations we see in mangrove species today, but are chiefly expressed in the marked decrease in biodiversity (species richness) from the Indo West Pacific (IWP) to the Atlantic East Pacific (AEP) regions reflecting the old-world origin of mangroves (Duke et al., 1998; Ellison et al., 1999). A total of 36 true species have been described from the IWP area, but fewer than 10 species are found in the AEP (Tomlinson, 2016). In addition, there is a strong effect of latitude on species richness in the eastern hemisphere with maximum richness along the equator and reduction by half at 20° north and south latitude (Ellison, 2002).

While there are clear patterns in tree species richness between eastern and western hemispheres, and north and south of the equator, the total mangrove area in the IWP and AEP regions is very similar (Giri et al., 2011; Hamilton and Casey, 2016; Spalding et al., 2010). For both IWP and AEP regions about 75% of total mangrove area is found between 0 and 10° north and south latitude of the equator. Also, species richness is highest near the equator for both regions likely reflecting broader patterns of species–area relationship observed for other ecosystems and taxa (as reviewed in Drakare et al., 2006 and Würtz and Annila, 2008).

This east–west gradient in species diversity, as well as latitudinal range, has always intrigued coastal wetland scientists in regards to whether these biogeographic provinces provide grounds to explain mangrove epicenter, genetic diversity, and global distribution (Duke et al., 2002; Ellison et al., 1999; Lo et al., 2014; Yan et al., 2016). However, mangrove ecosystem services such as carbon (C) storage in response to biogeographic filters have not yet been fully accounted for in the literature. Moreover, variations in plant traits raise the question whether the diversity in tree life forms (e.g., forest structure, tree architecture) between IWP and AEP species could influence C storage in mangrove biomass or soils.

In addition to biogeographic filters, regional factors are paramount in shaping ecological attributes and conveying ecosystem services in coastal ecosystems (Woodroffe, 2002; Woodroffe et al., 2016). The morphological diversity of coastal environments (or coastline types) reflects distinct energy (or environmental) signatures composed by a set of geophysical (e.g., tidal amplitude, river discharge) and climatic (temperature, precipitation, evapotranspiration) variables that altogether drive nutrient and C dynamics in coastal wetlands (Thom, 1982; Woodroffe, 1992). This set of physical variables explain the formation of distinct coastal environmental settings such as river-dominated (e.g., large rivers and small deltas), tidal systems, lagoons, carbonate, and arheic environments (Boyd et al., 1992; Dürr et al., 2011; Rovai et al., 2018; Thom, 1982; Woodroffe, 1992). This diversity of coastal environmental settings where mangroves thrive is represented along both IWP and AEP regions (Twilley et al., 2018), reflecting the fractal nature of shorelines (Dürr et al., 2011; Mandelbrot, 1967) despite discrete geographical boundaries. Distinct coastal environmental settings have conspicuous soil nutrient (nitrogen and phosphorus) density and stoichiometry (Rovai et al., 2018; Twilley et al., 2018), to which mangrove species respond to by shifting C allocation between above-ground (for example, trunk, branches, leaves) and below-ground (roots) compartments (Castañeda-Moya et al., 2013; Chapin et al., 1986; Hayes et al., 2019, 2017; McGowan, 2006;

Vitousek and Farrington, 1997). For instance, in karstic settings in the Florida Coastal Everglades, scrub mangroves allocate higher biomass to roots compared to above-ground compartments due to fertility and hydrological stress conditions in soils (Castañeda-Moya et al., 2013). Thus, species-specific traits coupled to stress conditions driven by geophysical and climate drivers within a coastal environmental setting determine the degree of plasticity and C allocation patterns in mangrove wetlands (Schlichting, 1986).

In this chapter, we review mangrove blue C research at the global scale and discuss conceptual and methodological aspects when scaling up these ecosystem attributes from site- to global-level. We show how biogeographic and regional (coastal morphology) filters may constrain global mangrove C stocks and sequestration rates in soils and biomass. To advance this topic, we reviewed the literature on mangrove C stocks and sequestration rates in biomass and soils to provide an overview of the spatial distribution of these data sets and account for differences in these ecosystem attributes between biogeographic regions (e.g., AEP vs. IWP) and across diverse coastal morphologies. Here, we focus on biomass and soils as these are the largest and most stable C pools in mangroves, persisting from century to millennia time scales, respectively. We use a worldwide classification of coastline typologies to account for the mangrove area within distinct types of coastal environmental settings, including deltas, estuaries, lagoons, carbonate, and arheic settings (but see also Worthington et al., 2020 for an updated global biophysical typology of mangroves) Lastly, we discuss how issues with data availability and quality pose challenges to mangrove ecologists to account for the spatial variability of these ecosystem properties and show how incorporating coastal environmental settings into modeling frameworks can improve C storage and sequestration estimates in mangroves globally. Our goal is to provide coastal wetlands scientists with a viable framework that will help us understand the spatial variability of mangrove C stocks and sequestration rates in response to key environmental drivers.

2. Mangroves as blue carbon sinks: stocks versus sequestration

Blue C ecosystems (mangroves, tidal marshes, and seagrasses) are known for sequestering C at annual rates per unit area greater than other terrestrial biomes, resulting in proportionally high C stocks (Howard et al., 2017), mainly in the soil where it remains stable for millennia (Donato et al., 2011; McLeod et al., 2011). Among blue C ecosystems, mangrove forests are the most C dense due to the relative high net ecosystem production which tends to decompose slowly in oxygen-depleted soils (McKee et al., 2007; Twilley et al., 1986). Moreover, C accumulation rates in mangrove woody tissues are estimated to be equivalent to C burial rates in soils (Twilley et al., 1992, 2017), and C stored in mangrove biomass on a per unit area is orders of magnitude higher than other blue C ecosystems (McLeod et al., 2011).

Carbon stocks and sequestration are often misinterpreted by the general public. However, these terms refer to distinct ecosystem properties and their conceptualization is important to understand the contribution of mangroves to the C cycle, as well as to guide science-based policymaking. As mangrove trees grow, they store C from the atmosphere in their wood (trunks, branches, roots). Trunks and branches are the main above-ground C pool. Roots slowly decompose and over time form the autochthonous (that is, *in situ*) soil organic C pool (Duarte et al., 1996; Saintilan et al., 2013). In addition to root-derived C inputs, more C builds up in soils with the accumulation of allochthonous (ex situ) organic matter brought in mainly by river and tides (Cragg et al., 2020; Kusumaningtyas et al., 2019; Sasmito et al., 2020). To a minor extent, dead leaves and branches are also incorporated into the soil organic C matrix (Chen and Twilley, 1999a; Jennerjahn and Ittekkot, 2002). At any point in time, the amount of C present in mangrove biomass or soils represents the ecosystem C stocks in that compartment. Together, biomass and soils constitute the two largest and more stable C pools in mangrove ecosystems, here referred to as ecosystem-level C stocks. To facilitate cross-site comparisons, C stocks are usually expressed on a per unit area basis, that is, MgC/ha.

While C stocks are the amount of C stored per unit area, C sequestration is the flow of C that accumulates in that same area per year, usually expressed in MgC ha^{-1} yr^{-1}. In a systems perspective, C stocks and flows are fundamental to understanding the outcome of continuous and dynamic ecogeomorphic processes that govern net primary productivity and sediment accretion rates at the land–ocean interface (Twilley et al., 2018). C sequestration is the C flow in mangroves that results in some permanent storage such as soils or wood, representing a loss of C from the atmosphere. For example, C sequestration is enhanced in deltaic environments because rivers continuously deposit sediment in mangrove soils, leading to higher C burial rates (Breithaupt et al., 2012; Cragg et al., 2020). However, river-born sediments are predominantly of mineral nature, so deltaic mangrove soils have a relatively low C content by volume. In contrast, in carbonate settings such as in Florida Coastal Everglades and Caribbean islands, most of the soil volume is made up of plant roots, so it has a higher C content per volume (Breithaupt et al., 2020). But soils form in deltas more quickly, since rivers are always depositing new sediment (Kusumaningtyas et al., 2019).

C stocks and sequestration are often tightly coupled but understanding the distinction between these ecosystem properties is key for blue C conservation strategies (Jennerjahn, 2020). For example, one can think that for the purpose of curbing climate change, the most important question is how much C an ecosystem sequesters each year, mitigating anthropogenic C emissions. However, destroying mangrove forests to build shrimp ponds, roads, or other commercial projects can release significant quantities of C stocks to the atmosphere, particularly in carbonate settings where much of the soil consists of organic material. By improving our understanding of how mangroves function under different environmental conditions, we can safeguard and increase these valuable blue C stocks and flows. The following sections describe how the coastal environmental setting framework is a powerful tool to account for both C stocks and sequestration rates across diverse types of coastlines from regional to global scales in spite of biogeographic divisions.

To date, few studies have attempted to map and quantify the global variability in mangrove C stock estimates in biomass (Hutchison et al., 2014; Simard et al., 2019) and soil (Jardine and Siikamäki, 2014; Rovai et al., 2018; Sanderman et al., 2018). These studies differ with regards to conceptual and methodological aspects but converge in that they are all based on site-specific observations used to scale up C stocks to the global mangrove area. The geographic distribution of data points and associated C values for each of these studies can be found at their respective sources. We strongly recommend a thorough examination of these studies as they offer complementary insights to accounting for the global variability of C stock in either mangrove biomass or soil. In this review, however, we used a literature-based composite data set for mangrove biomass (Table 13.1) and published studies that have consistently used the coastal environmental setting framework to generate global soil organic C estimates in mangrove ecosystems (Rovai et al., 2018; Twilley et al., 2018), respectively. We demonstrate that the approach used in these studies, namely the coastal environmental setting framework (Thom, 1982; Twilley, 1995; Woodroffe, 1992), allow for a more robust determination of ecosystem-level C stocks, that is, biomass plus soil combined.

TABLE 13.1 Summary statistics of global field–based mangrove above-ground biomass (AGB).

Country	CES	Mean AGB (Mg/ha ±1SE)	AGB range (Mg/ha; N)	References
Australia	II. Small deltas	197.62 ± 27.53	45.8–413.33; 15	Clough, 1998, Hickey et al., 2018, Mackey, 1993, Robertson and Alongi, 2016, Alongi et al., 2005a; Alongi et al., 2003; Alongi, 2000, Ellison and Simmonds, 2003
	III. Tidal systems	68.35 ± 10	2.34–292; 48	Briggs, 1977, Clough, 1975; Clough and Attiwill, 1979, Goulter and Allaway, 1979, van der Valk and Attiwill, 1984; Attiwill and Clough, 1978, Kelleway et al., 2016, Melville and Burchett, 2002, Accad et al., 2019, Lamont et al., 2019, Owers et al., 2018, Woodroffe et al., 1988
Bangladesh	III. Tidal systems	116.98 ± 25.57	14.36–305.56; 11	Rahman et al., 2015, Hossain et al., 2012
Belize	IV. Lagoons	24.57 ± 14.71	6.06–53.63; 3	McKee et al., (2007); Koltes et al., 1998; Feller et al., (2003); Woodrofe, 1995, McKee et al., (2007); Koltes et al., 1998; Feller et al., (2003)
Benin	IV. Lagoons	35.07		Ajonina et al., 2014c
Brazil	I. Large rivers	128.2 ± 14.08	14.26–504.48; 65	Gardunho, 2009, Menezes, 2010, Santos, 2013, Santos et al., 2012, Schaeffer-Novelli et al., 1984, Kauffman et al., 2018a
	II. Small deltas	71.78 ± 17.31	6.15–215.4; 16	Costa, 2010, Maia and Coutinho, 2012, Souza, 2016, Kauffman et al., 2018b
	III. Tidal systems	81.66 ± 3.49	0–481.74; 505	Abreu et al., 2006; Nascimento, 2006, Reise, 1999, Menezes et al., 2003, Ramos-Silva et al., 2007, Cunha et al., 2006, Abreu et al., 2006, Carmo et al., 2000, Kilca et al., 2011, Medeiros and Sampaio, 2008, Mendoza et al., 2012, Schories et al., 2003, Alves and Sassi, 2003, Araujo and Santos, 1987, Araújo, 2002, Bastos and Lobato, 1996, Bernini and Rezende, 2010, Bernini and Rezende, 2011, Bernini and Rezende, 2004, Braga et al., 1989, Carmo et al., 1995, Carmo et al. 1998a, 1998b, Cidreira, 2014,

TABLE 13.1 Summary statistics of global field–based mangrove above-ground biomass (AGB).—cont'd

Country	CES	Mean AGB (Mg/ha ±1SE)	AGB range (Mg/ha; N)	References
				Cintrón and Schaeffer-Novelli, 1981, Coutinho, 1999, Cruz et al., 2013, Cunha, 2001, Motta Portillo et al., 2017, Deus, 2000, Freitas et al., 2002, Kilca et al., 2010, Carvalho, 2016, Lima, 2011, Lopes, 2014, Lugli, 2008, Maia and Coutinho, 2012, Martins et al., 2011, Menezes, 2006, Mochel, 1999, Nascimento, 1999, Nascimento-Filho, 2007, Pellegrini, 2010, Pereira et al., 2009, Petry, 2009, Rocha-Filho, 2009, Rovai et al., 2012, Sales et al., 2009, Schaeffer-Novelli, 1994, Schaeffer-Novelli et al., 1984, Schaeffer-Novelli et al., 1988, Seixas et al., 2006, Silva, 2001, Silva, 2004, Silva et al., 2005, Soares et al., 2008, Soares et al., 2012, Souza and Sampaio, 2001, Souza, 1996, Souza, 2016, Souza et al., 1994, Tognella-de-Rosa et al., 2006, Vale, 1999, Weinberg, 1987, Weller, 2010, Zamprogno et al., 2016, Mehling et al., 2010
	IV. Lagoons	59.8 ± 2.48	0.13–325.51; 317	Silva, 1988; Silva et al., 1998, Araújo e Maciel, 1979, Comelli, 1994, Lima, 2009, Ponte, 1987, Soares and Schaeffer-Novelli, 2005, Adaime, 1985, Calegario, 2012, Castanheira, 2004, CETESB, 1983, Coelho-Jr, 1998, Colpo et al., 2011, Couto, 1996, Cunha-Lignon, 2001, Duque-Estrada, 2009, Henriques, 2012, Menghini, 2004, Menghini, 2008, Neto, 2000, Peria, 1990, Pellegrini, 2000, Pereira, 2011, Portugal, 2002, Reis et al., 2017, Soares, 1999, Lamparelli, 1995, Soares et al., 2003; Chaves, 2007, Schmiegelow, 2009, Sessegolo, 1997
Cameroon	III. Tidal systems	423.4 ± 57.94	238.18–570.8; 5	Ajonina and Usongo, 2001, Ntyam, 2014, Ajonina et al., 2014b, Feka et al., 2011
China	II. Small deltas	164.8 ± 83.7	81.1–248.5; 2	Lin et al., 1990, Wang et al., 2019
	III. Tidal systems	103.88 ± 19.33	14.7–282.75; 17	Lin et al., 1985, Chen et al., 2008; Ye et al., 2013, Li et al., 1999; Li, 1997, Luo et al., 2010, Wang et al., 2019, Alongi et al., 2005b, Wang et al., 2013, Chen et al., 2012
Colombia	I. Large rivers	61.25 ± 7.05	47.19–69.12; 3	Blanco and Castaño, 2012
	II. Small deltas	350.17 ± 66.74	26.8–1090.4; 18	Yepes et al., 2016, Meyer et al., 2019, Agudelo et al., 2015, Sierra-Correa et al., 2007, Velez and Polania, 2007
	IV. Lagoons	48.83 ± 14.33	6.61–271.17; 18	Calderón-Sáenz, 1984, Cardona and Botero, 1998, Orjuela-Rojas et al., 2011, Rodríguez-Ramírez et al., 2004, Valle et al., 2011
	V. Carbonate	178.93 ± 47.44	62.03–388.66; 6	García-Hansen et al., 2002, Urrego et al., 2009

(*Continued*)

TABLE 13.1 Summary statistics of global field–based mangrove above-ground biomass (AGB).—cont'd

Country	CES	Mean AGB (Mg/ha ±1SE)	AGB range (Mg/ha; N)	References
Costa Rica	II. Small deltas	180.81 ± 26.83	49.68–412.01; 16	Fonseca et al., 2007, Pool et al., 1977, BIOMARCC-SINAC-GIZ 2012, Manrow-Villalobos, 2012; Cortés, 1998, Manrow-Villalobos, 2012, Samper-Villarreal, 2012, Kauffman et al., 2017
Curacao	V. Carbonate	52.74 ± 13.53	28.8–102; 5	Pors and Nagelkerken, 1998
Democratic Republic of the Congo	I. Large rivers	409		Ajonina et al., 2014b
Dominican Republic	V. Carbonate	272.4 ± 77	195.4–349.4; 2	Sherman et al., 2000; Sherman et al., 2003
East Timor	II. Small deltas	155.8 ± 52.91	51.1–221.5; 3	Alongi and Carvalho, 2008
Ecuador	II. Small deltas	267.53 ± 46.95	175–327.59; 3	Hamilton et al., 2017; Twilley et al., 1997
Fiji	III. Tidal systems	213.7 ± 17.09	192.4–247.5; 3	Duke, 2013
French Guiana	III. Tidal systems	192.63 ± 28.37	77.6–431.9; 12	Fromard et al., 1998, Fromard et al., 2004
Gabon	III. Tidal systems	341		Ajonina et al., 2014b
Ghana	II. Small deltas	245.22 ± 115.6	16.01–386.03; 3	Ntyam, 2014
	IV. Lagoons	254.82 ± 129.58	36.46–626.03; 4	Ajonina, 2011; Ajonina et al., 2014a
Guadeloupe, France	V. Carbonate	272.59 ± 112.88	47–685.61; 6	Imbert and Rollet, 1989, Migeot and Imbert, 2012
Guinea-Bissau	III. Tidal systems	82.9		Vasconcelos et al., 2015
Hawaii, United States	II. Small deltas	279		Cox and Allan, 1999
Honduras	II. Small deltas	77.84 ± 35.9	36.1–149.3; 3	Castañeda-Moya et al., (2006), Kauffman et al., 2017
Hong Kong	III. Tidal systems	19.82 ± 4.35	1.58–87.14; 21	Tam et al., 1995, Chen et al., 2009, Duke and Khan, 1999
India	I. Large rivers	129.72 ± 26.02	103.7–155.75; 2	Bhomia et al., 2016
	II. Small deltas	235.3 ± 29.41	72.06–379.64; 13	Hebbalalu et al., 2014, Upadhyay and Mishra, 2014, Pandey et al., 2019, Sahu et al., 2016

TABLE 13.1 Summary statistics of global field–based mangrove above-ground biomass (AGB).—cont'd

Country	CES	Mean AGB (Mg/ha ±1SE)	AGB range (Mg/ha; N)	References
	III. Tidal systems	73.78 ± 14.22	0.3–465.9; 45	Thivakaran et al., 2018, Joshi and Ghose, 2014, 2003, Devi and Pathak, 2016, Banerjee and Mitra, 2015, Mitra et al., 2011, Ray et al., 2011
	IV. Lagoons	24.04 ± 13.34	1.98–157.03; 11	Saravanan, 2004, Prasanna et al., 2017
Indonesia	II. Small deltas	242.53 ± 13.01	3.29–907.91; 159	Komiyama et al., 1988, Murdiyarso et al., 2015, Arifanti et al., 2019, Chen et al., 2018, Cameron et al., 2019, Kangkuso et al., 2018, Rumengan et al., 2018, Kusumaningtyas et al., 2019
	III. Tidal systems	331.11 ± 18.68	12.5–1050.09; 109	Aslan, unpublished, Sillanpää et al., 2017, Murdiyarso et al., 2015
	IV. Lagoons	262.2		Kusumaningtyas et al., 2019
	V. Carbonate	161.4		Kusumaningtyas et al., 2019
Iran	VI. Arheic	19.88 ± 6.78	9.21–32.46; 3	Eisini et al., 2006, Ghazemi et al., 2013
Jamaica	V. Carbonate	16.22		Gayle and Woodley, 1998
Japan	II. Small deltas	126.36 ± 18.12	64.9–162.8; 5	Ohtsuka et al., 2019, Kamruzzaman et al., 2017, 2019
Kenya	II. Small deltas	66.69 ± 20.72	33.44–104.73; 3	Kairo, 2001, Dahdouh-Guebas et al., 2002
	IV. Lagoons	115.61 ± 28.15	38.89–249; 7	Kairo et al., 2008, Bosire et al., 2006, Githaiga, 2013, Slim et al., 1996
	VI. Arheic	347.24 ± 149.86	197.38–497.1; 2	Kairo, 2001
Madagascar	III. Tidal systems	125.87 ± 20.87	46.92–222.43; 10	Scales and Friess, 2019, Rakotomavo, 2018, Jones et al., 2015, Jones et al., 2014, Rakotomavo and Fromard, 2010, Benson et al., 2017
	IV. Lagoons	206.23 ± 33.66	114.33–349.67; 6	Alavaisha and Mangora, 2016
Malaysia	II. Small deltas	98.4		Faridah-Hanum, 2012
	III. Tidal systems	291.39 ± 27.73	16–576; 24	Ong et al., 1982; Putz et al., 1986, Ong et al., 1981, Hossain, 2014, Adame et al., 2018, Ong et al., 1979a; Ong et al., 1982b, Zakaria et al., 2018, Goessens et al., 2014, Alongi et al., 2004, Zhila et al., 2014, Tanouchi et al., 2000, Eong, 1993
Martinique, France	V. Carbonate	124.87 ± 47.07	77.8–171.94; 2	Imbert and Ménard, 1997
Mauritius	II. Small deltas	21.8 ± 5.17	16.63–26.96; 2	Abib and Appadoo, 2012

(Continued)

TABLE 13.1 Summary statistics of global field–based mangrove above-ground biomass (AGB).—cont'd

Country	CES	Mean AGB (Mg/ha ±1SE)	AGB range (Mg/ha; N)	References
Mayotte, France	III. Tidal systems	179.26 ± 68.43	110.83–247.69; 2	Capdeville et al., 2018
Mexico	I. Large rivers	381.63 ± 97.41	200.46–534.28; 3	Barreiro-Güemes, 1999, Guerra-Santos et al., 2014
	II. Small deltas	69.58 ± 24.28	4.95–144.24; 5	Téllez-García and Valdez-Hernández 2012, Agraz-Hernández 1999
	III. Tidal systems	82.59 ± 25.82	2.02–304.64; 12	Utrera-López and Moreno-Casasola, 2008, Agraz-Hernández et al., 2011
	IV. Lagoons	89.95 ± 13.67	0.06–557.11; 67	Flores-Verdugo et al., 1987, Agraz-Hernández et al., 2015, Muñoz et al., 2006, Pool et al., 1977, Adame et al., 2013a, Agraz-Hernández 1999, Barreiro-Güemes 1999, Basañez-Muñoz et al., 2008, Cerón et al., 2011, Flores-Cárdenas et al., 2017, Flores-Verdugo et al., 1990, Guerra-Santos et al., 2014, Lara-Dominguez et al., 2005, Day et al., 1987, Day et al., 1996, Coronado-Molina et al., 2012, Torres et al., 2017, Adame et al., 2013b
	V. Carbonate	14.89 ± 6.47	3.74–45.43; 6	Agraz-Hernández, 2006, Adame et al., 2013a, Lara-Dominguez et al., 2005
	VI. Arheic	32.54 ± 9.24	1.2–113.5; 14	Arreola-Lizárraga et al., 2004, Félix-Pico et al., 2006, Domínguez-Cadena et al., 2016, Chávez-Rosales, 2006, Sanchez-Andrés et al., 2010, Ochoa-Gómez et al., 2018
Micronesia	II. Small deltas	338.49 ± 47.85	113.72–624.76; 12	Lang'at, 2008, Steinke et al., 1995, Allen et al., 2000, Kauffman et al., 2011
Mozambique	I. Large rivers	163.66 ± 33.9	75.4–268.5; 5	Stringer et al., 2015
	II. Small deltas	21.51 ± 4.7	4.47–42.33; 10	Bandeira et al., 2009, Macamo et al., 2016
	IV. Lagoons	43.53 ± 32.2	2.09–106.95; 3	Boer, 2000, Bandeira et al., 2009
New Zealand	II. Small deltas	37.32 ± 23.2	4.91–104.1; 4	Woodroffe, 1985, Yang et al., 2013
	IV. Lagoons	43		Bulmer et al., 2015
Nicaragua	IV. Lagoons	200.51		Roth, 1992
Nigeria	III. Tidal systems	65.13		Ukpong, 1997
Oman	VI. Arheic	12.26 ± 3.25	7.56–18.49; 3	Fouda and Al-Muharrami, 1995

TABLE 13.1 Summary statistics of global field–based mangrove above-ground biomass (AGB).—cont'd

Country	CES	Mean AGB (Mg/ha ±1SE)	AGB range (Mg/ha; N)	References
Pakistan	VI. Arheic	26.48 ± 11.07	2.62–173.62; 15	Saifullah and Rasool, 2000, Saifullah et al., 1994, Saifullah and Rasool, 2002
Palau	II. Small deltas	225 ± 33.83	174–289; 3	Kauffman et al., 2011
Panama	II. Small deltas	114.89 ± 20.63	0.65–335.3; 26	Golley et al., 1975, Golley et al., 1969, Gross et al., 2014, McGowan (2006)
	IV. Lagoons	62.33 ± 34.04	8.3–194.3; 5	Guzmán et al., 2005, Lovelock et al., 2005
Philippines	II. Small deltas	88.29 ± 11.42	45.94–135.83; 7	de la Cruz et al., 1967, Castillo et al., 2017, Duncan et al., 2016, Lillo and Fernando 2017, Castillo et al., 2018
Puerto Rico, United States	V. Carbonate	70.07 ± 16.34	3.31–282.98; 20	Golley et al., 1962, Rogers and Cintrón, 1974, Pool et al., 1977, Pool et al., 1975, Cintrón et al., (1978), García et al., 1998, NHSPR, 1972
Republic of Congo	IV. Lagoons	251		Ajonina et al., 2014b
Samoa	III. Tidal systems	174.45 ± 95.45	79–269.9; 2	Duke, 2013
Saudi Arabia	VI. Arheic	14.48 ± 3.37	2.21–62.97; 17	El-Juhany, 2009, Abohassan et al., 2012
Singapore	III. Tidal systems	297.05 ± 46.51	145.04–605.17; 10	Phang et al., 2015
Solomon Islands	II. Small deltas	416.83 ± 74.83	296.1–553.8; 3	Duke, 2013
South Africa	II. Small deltas	25.68 ± 0.7	24.98–26.38; 2	Rajkaran et al., 2009
	IV. Lagoons	36.21		Rajkaran et al., 2009
Sri Lanka	II. Small deltas	164.65 ± 28.15	19–521.25; 26	Perera and Amarasinghe, 2016, Jayakody et al., 2008, Perera and Amarasinghe, 2018, Amarasinghe and Balasubramaniam, 1992, Jonsson and Hedman, 2019, Dayarathne and Kumara, 2013, Dahdouh-Guebas et al., 2002, Satyanarayana et al., 2011, Perera et al., 2013
	IV. Lagoons	10.07		Krishnanantham et al., 2015
Tanzania	II. Small deltas	31.46 ± 11.9	1.67–90; 7	Lupembe, 2014, Msangameno et al., 2017, Bandeira et al., 2009
Thailand	II. Small deltas	124.4 ± 73.44	42–344.2; 4	Alongi and Dixon, 2000; Alongi et al., 2002

(*Continued*)

TABLE 13.1 Summary statistics of global field–based mangrove above-ground biomass (AGB).—cont'd

Country	CES	Mean AGB (Mg/ha ±1SE)	AGB range (Mg/ha; N)	References
	III. Tidal systems	241.12 ± 14.62	1.91–627.59; 66	Okimoto et al., 2013, Komiyama et al., (2000), Komiyama et al., (1987), Tamai et al., 1986, Jachowski et al., 2013, Poungparn et al., 2016, Imai et al., 2006, Umnouysin et al., 2017, Komiyama et al., (2005), Meepol, 2002
	V. Carbonate	136.25 ± 22.75	113.49–159; 2	Christensen, 1978, Komiyama et al., (2005)
Tonga	III. Tidal systems	98.85 ± 20.72	39.1–128.3; 4	Duke, 2013
Torres Strait, Australia	III. Tidal systems	307.61 ± 35.83	91.4–494.4; 14	Duke et al., 2015
Trinidad and Tobago	V. Carbonate	141		Juman, 2005
Turneffe Atoll, Belize	IV. Lagoons	84.88 ± 18.28	26.3–130.5; 5	Blanco and Cho-Ricketts, 2015
United Arab Emirates	VI. Arheic	3.83 ± 2.39	0.39–8.41; 3	Habshi et al., 2007
United States	IV. Lagoons	94.11 ± 7.91	9–194.17; 39	Heald et al., 1979, McCoy et al., 1996, Pool et al., 1975, 1977; Snedaker et al., 1973, Ellis and Bell, 2004, Twilley et al., (1986), Lugo and Snedaker (1974), Parkinson et al., 1999, Shafer and Roberts, 2008, Steyer, 1988, Yando et al., 2016, Simpson et al., 2019, 2017, Doughty et al., 2015, McKee and Faulkner, 2000
	V. Carbonate	83.69 ± 11.38	1.55–461.82; 55	Coronado-Molina et al., 2004, Lahmann, 1988, Pool et al., 1977, Ball, 1980, Chen and Twilley, 1999, Dawes et al., 1999, Ellis et al., 2006, Lugo and Snedaker (1974), Shafer and Roberts, 2008, Snedaker et al., 1992, 1981, Coronado-Molina et al., 2012, Castañeda-Moya et al., (2013), Ross et al., 2001
Vanuatu	III. Tidal systems	214.8 ± 65.33	79.6–463; 5	Duke, 2013
Venezuela	IV. Lagoons	47.38 ± 20.81	8.89–115.67; 5	López et al., 2011
Vietnam	III. Tidal systems	348.26 ± 101.37	130.5–643.6; 5	Dung et al., 2016, Nam et al., 2016

Data are organized by country and coastal environmental settings (CES; see main text and methods for details). Mean values are provided along with standard error (±1SE), range, and number of observations (N).

3. Biogeography of blue carbon research in mangroves

3.1 Above-ground biomass

A recent systematic review of the literature shows that until 2019 approximately 363 independent studies reporting on mangrove forest structure and above-ground biomass (AGB) have been published worldwide (Table 13.1). Together, these studies account for 2041 plots distributed over 62 countries (out of 105 with mangroves), encompassing the diversity of coastal environmental settings (e.g., deltas, estuaries, lagoons, carbonate, arheic) where mangroves thrive in both IWP and AEP regions (Fig. 13.2A, Table 13.1).

Despite the large number of sites with data on mangrove forest structure parameters, there are still many nations underrepresented in terms of data availability, particularly in Central America, Caribbean islands, West Africa, and Southeast Asia (Fig. 13.2A). To date, the large concentration of plots is in the neotropics, with nearly 60% of the studies. It should be noted, however, that this difference in terms of data availability between AEP and IWP regions could be related to search bias; in the neotropics over 50% of the studies consist of thesis and dissertations carried out in Brazil (e.g., "gray literature") (Rovai et al., 2016), in addition to published material (e.g., peer-reviewed articles and book chapters) from the remaining study area. It should also be noted that the substantial increase in number of sites relative to earlier reviews (e.g., Hutchison et al., 2014; Twilley et al., 1992; Kauffman et al., 2020; Simard et al., 2019) was due to the use of a universal allometric model that converts forest structure parameters (that is, basal area in m^2/ha and mean tree height in meters) into AGB (see Rovai et al., 2016 for details). Here we expand the data set collated by Rovai et al. (2016) beyond the AEP region, doubling the number of sites inventoried worldwide in that study (Table 13.1). This increase, however, was largely based on the incorporation of published material due to both the difficulty in finding gray literature via commonplace search engines (i.e., Google Scholar). We hope this review encourages other scientists to assemble country-level inventories with a focus on unpublished literature in order to expand this current data set, reducing the associated search bias. A very nice way of systematizing and making publicly available important gray literature material was recently proposed by Blanco-Libreros and Álvarez-León (2019), who converted novel unpublished mangrove forest structure inventories across Pacific Colombia into an open-source digital database (Blanco-Libreros and Álvarez-León, 2018).

3.2 Below-ground biomass

Mangrove below-ground biomass (BGB) corresponds to woody (axial roots) and nonwoody (fine roots) tissues present in tree's root systems. However, differently from AGB, BGB sampling is a much more laborious process, and large differences in estimates can result depending on the sampling method used. Here we expanded the data set collated by Adame et al. (2017) which considers only observed BGB values (that is, obtained using either trench or coring methods) (Table 13.2). Data on observed mangrove BGB are scarce in the literature, restricted to 37 studies from 19 countries (Fig. 13.2B; Table 13.2). Nearly 76% of these studies are in the IWP region. Australia and China represent most of the studies in the IWP, whereas in the AEP region the Brazil, United States, and Mexico account for most

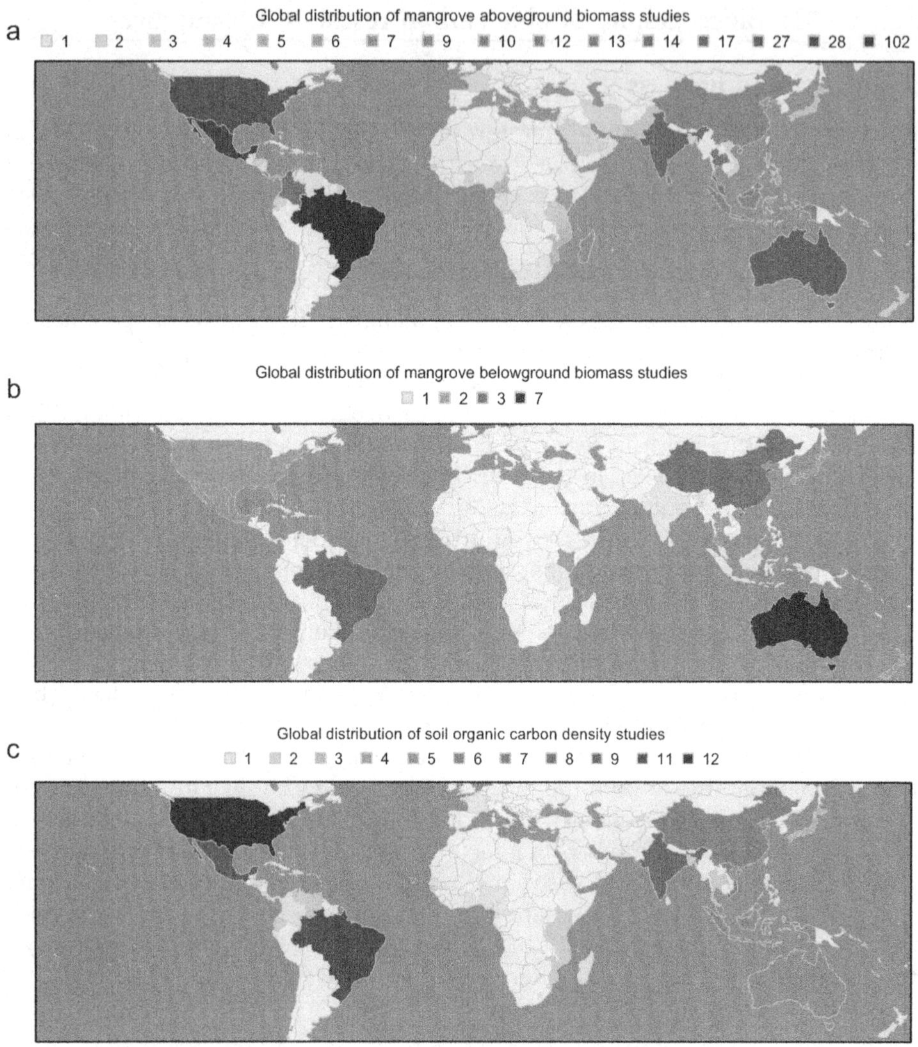

FIGURE 13.2 Number of studies per country on mangrove above- (A) and below-ground (B) biomass, and on soil organic carbon density (C).

of the sites in this region. The reduced and geographically biased number of observations coupled with the lack of sites representing the diversity of coastal environmental settings where mangrove thrive, in addition to methodological inconsistencies (e.g., trench vs. coring) poses a challenge for coastal wetland scientists to map and quantify the global variability of C stocks in mangrove BGB. Moreover, it is not always clear if reported biomass values include necromass. Thus, the common practice is to derive BGB from AGB using a reference ratio, and then converting that biomass to C density to estimate total C stocks in root system. A conversion factor of 0.5 has been consistently used across recent global C stock estimates in mangroves to derive BGB directly from AGB (Hamilton and Friess, 2018; Simard et al.,

TABLE 13.2 Summary statistics of global field–based mangrove below-ground biomass (BGB).

Country	CES	Mean BGB (Mg ha^{-1} ± 1SE)	BGB range (Mg ha^{-1}; N)	References
Australia	II. Small delta	70.83 ± 11.29	14.4–205; 26	Mackey, 1993, Alongi et al., 2000, Robertson and Alongi, 2016, Saintilan, 1997b
	III. Tidal system	248.24 ± 68.87	94.6–506.38; 6	Saintilan, 1997a, Briggs, 1977, Lamont et al., 2019
Brazil	III. Tidal system	23.47 ± 7.05	16.42–30.52; 2	Madi et al., 2017
	IV. Lagoon	130.88 ± 10.3	105.22 –173.36; 6	Santos et al., 2017, Checon et al., 2017; Schaeffer-Novelli et al., 2017
China	II. Small delta	13.77 ± 4.68	2.89–26.78; 5	Xiong et al., 2017
	III. Tidal system	29.93 ± 11.81	7.65–47.85; 3	Tam et al., 1995, He et al., 2018
Dominican Republic	V. Carbonate	110		Sherman et al., 2003
India	III. Tidal system	24.11 ± 2.55	16.94–44.04; 11	Ray et al., 2011
Japan	V. Carbonate	31.97		Khan et al., 2009
Kenya	IV. Lagoon	12.3 ± 1.95	2.08–21; 9	Kairo et al., 2008, Tamooh et al., 2008
Malaysia	II. Small delta	26.01		Muhammad-Nor et al., 2019
Mexico	IV. Lagoon	32.94 ± 4.24	16–48.66; 9	Adame et al., (2014), Torres et al., 2019
Micronesia	V. Carbonate	16.1 ± 2.71	6.87–35.67; 9	Cormier et al., (2015), Gleason and Ewel, 2002
New Zealand	II. Small delta	54.89 ± 4.74	32.97–78.78; 11	Tran, 2013, Bulmer et al., 2016
Puerto Rico, United States	V. Carbonate	31.23		Golley et al., 1962
Sri Lanka	II. Small delta	0.7 ± 0.22	0.35–1.31; 4	Philips et al., 2017
Tanzania	II. Small delta	143.58		Njana et al., 2016
Thailand	III. Tidal system	115.45 ± 37.05	42.4–254.75; 5	Komiyama et al., (1987), Komiyama et al., (2000)

(*Continued*)

TABLE 13.2 Summary statistics of global field–based mangrove below-ground biomass (BGB).—cont'd

Country	CES	Mean BGB (Mg ha^{-1} ± 1SE)	BGB range (Mg ha^{-1}; N)	References
United States	I. Large river	7.01 ± 4.59	2.41–11.59; 2	Yando et al., 2018
	IV. Lagoon	8.21 ± 3.46	4.75–11.66; 2	Yando et al., 2018
	V. Carbonate	22.49 ± 2.53	14.67–28.72; 6	Castañeda-Moya et al., (2011), Castañeda-Moya, 2010

Data are organized by country and coastal environmental settings. Mean values are provided along with standard error (±1 SE), range, and number of observations (N).
Modified from Adame et al. (2017).

2019). Then, C stock in BGB is estimated using conversion factors ranging between 0.39 (Howard et al., 2014) and 0.475 (Hamilton and Friess, 2018). This approach is a plausible alternative given the current lack of measured BGB data, but it is paramount to address the sampling bias discussed above to improve the accuracy of the mangrove global C stock.

3.3 Soil organic carbon

Organic matter accumulation in mangrove soils results from both in situ (autochthonous) and ex situ (allochthonous) sources (Breithaupt et al., 2020, 2012; Kusumaningtyas et al., 2019; Saintilan et al., 2013). Autochthonous inputs of organic matter to soil result mostly from mangrove below-ground (root) productivity and, to a lesser extent, via litterfall (Chen and Twilley, 1999a). Mangrove crabs also play an important role on the soil organic matter budget, removing significant amounts of leaves from the forest floor and burying them into their galleries (Robertson, 1986; Robertson and Daniel, 1989; Smith et al., 1991). However, most of the litterfall is exported as POC (particulate organic carbon) and DOC (dissolved organic carbon) as tides move water in and out of mangrove forests (Chen and Twilley, 1999a; Jennerjahn and Ittekkot, 2002; Kristensen et al., 2008). Allochthonous organic matter comes from upstream areas as well as from incoming tides in the form of both POC and DOC (Kusumaningtyas et al., 2019; Sasmito et al., 2020). As estuarine waters flood mangrove forests part of this material settles and is buried by sediment particles (Adame et al., 2010). The relative contribution of autochthonous and allochthonous material to soil C stocks varies across distinct coastal environmental settings, with allochthonous material accounting for most of C accumulation and stocks in river-dominated coastlines and autochthonous material in peat-dominated (carbonate) systems (Cragg et al., 2020; Saintilan et al., 2013; Jennerjahn, 2020).

Mangrove SOC stocks have been reasonably mapped globally (Jardine and Siikamäki, 2014; Rovai et al., 2018; Sanderman et al., 2018). Currently, about 40% of countries with mangroves have data on SOC density (Fig. 13.2C and Table 13.3; but see CCRCN, 2020, data library for a more recent comprehensive compilation). In addition to data availability, data quality is a constraint particularly for IWP mangroves, where very-poor-quality data have recently been noted (Atwood et al., 2017). Data quality issues refer mainly to methods used to determine SOC content (e.g., elemental analyses vs. conversion factors), soil depth sampled (e.g., shallow cores) and pseudoreplication (e.g., depth intervals along soil profiles

TABLE 13.3 Summary statistics of global field–based mangrove soil organic carbon (SOC) density.

Country	CES	Mean SOC density (mg/cm^3 ±1SE)	SOC range (mg/cm^3; N)	References
Australia	II. Small deltas	29.12 ± 1.26	12.39–40.31; 31	Sanders et al., 2016, Alongi et al., 2000, Brunskill et al., 2002, Matsui, 1998
	III. Tidal systems	14.3 ± 0.23	13.86–14.64; 3	Livesly and Andrusiak, 2012
Bangladesh	III. Tidal systems	12.06 ± 0.82	7.22–16.01; 11	Rahman et al., 2015, Donato et al., (2011)
Brazil	II. Small deltas	17.29 ± 1.98	11.67–21.06; 5	Nóbrega et al., 2016, Mehling et al., 2010
	III. Tidal systems	26.97 ± 3.11	11.67–68.13; 20	Rovai et al., (2018), Sanders et al., 2010, Nóbrega et al., 2016, Mehling et al., 2010, Nobrega et al., 2013
	IV. Lagoons	37.44 ± 3.91	18.53–64.07; 15	Sanders et al., 2008, 2010b, Rovai et al., (2018), Otero et al., 2014, Ferreira et al., 2010, Wasserman et al., 2001
Cameroon	III. Tidal systems	28.9 ± 4.01	22.66–36.39; 3	Ntyam, 2014
China	II. Small deltas	9.99 ± NA	9.99–9.99; 1	Bao et al., 2013
	III. Tidal systems	12.34 ± 2.05	1.4–26.68; 17	Wang et al., 2013, Chen et al., 2012, Lunstrum and Chen, 2014, Alongi et al., 2005, Lu et al., 2014
Colombia	II. Small deltas	21.16 ± NA	21.16–21.16; 1	Ruiz-Fernandez et al., 2011
Costa Rica	II. Small deltas	36.46 ± 3.7	24.46–46.98; 6	Rovai et al. (2018)
Dominican Republic	V. Carbonate	41.91 ± 2.94	26.22–53.37; 9	Kauffman et al., 2014
East Timor	II. Small deltas	22.9 ± 4	15.01–37.74; 5	Alongi et al., 2012
Ecuador	II. Small deltas	49.39 ± 9.05	26.29–78.75; 6	Twilley et al., 1997, DelVecchia et al., 2014
El Salvador	II. Small deltas	21.39 ± 1.99	18.07–24.95; 3	Rovai et al. (2018)
Micronesia	III. Tidal systems	17.47 ± 1.72	14.04–19.22; 3	Cormier, 2003
	V. Carbonate	55.32 ± 2.84	29.19–107.76; 33	Donato et al., (2011), Cormier, 2003, Donato et al., 2012, MacKenzie et al., 2016, Kauffman et al., 2011, Fujimoto et al., 1999, Ono et al., 2015

(*Continued*)

TABLE 13.3 Summary statistics of global field–based mangrove soil organic carbon (SOC) density.—cont'd

Country	CES	Mean SOC density (mg/cm^3 ±1SE)	SOC range (mg/cm^3; N)	References
French Guiana	III. Tidal systems	15.79 ± 1	5.9–32.91; 31	Marchand et al., 2004, 2008, 2003, 2005, Marchand, 2017
Ghana	II. Small deltas	24.54 ± 2.3	20.21–28.03; 3	Ntyam, 2014
Guadeloupe, France	V. Carbonate	62.35 ± 3.4	58.95–65.75; 2	Lallier-Verges et al., 1998
Guinea-Bissau	III. Tidal systems	17.07 ± 5.46	8.58–27.25; 3	Andreetta et al., 2016, Vasconcelos et al., 2015
Honduras	II. Small deltas	25.84 ± 4.62	12.5–33.63; 4	Bhomia et al., 2016
	IV. Lagoons	38.1 ± 5.27	19.07–51.45; 5	Bhomia et al., 2016
	V. Carbonate	46.82 ± 4.37	23.32–70.04; 9	Bhomia et al., 2016
India	II. Small deltas	15.55 ± 0.45	12.85–17.35; 11	Ranjan et al., 2010, 2011, Fernandes et al., 2014, Ray et al., 2011
	III. Tidal systems	10.77 ± 1.09	6.7–12.94; 5	Fernandes et al., 2011, Chatterjee et al., 2009
	I. Large rivers	13.54 ± 2.28	11.22–18.1; 3	Bhomia et al., 2016, Sahu et al., 2016
Indonesia	II. Small deltas	25.94 ± 3.16	7.38–49.07; 15	Alongi et al., 2012, Weiss et al., 2016, Alongi et al., (2008), Donato et al., (2011)
	III. Tidal systems	25.94 ± 3.72	20.73–36.87; 4	Brunskill et al., 2004
	IV. Lagoons	20.86 ± 8.79	10.25–38.3; 3	Weiss et al., 2016
	V. Carbonate	58.73 ± 4.89	32.79–75.74; 10	Weiss et al., 2016, Donato et al., (2011)
Japan	III. Tidal systems	5.73 ± NA	5.73–5.73; 1	Khan et al., 2007
	V. Carbonate	42.65 ± 3.93	31.93–50.79; 4	Higashi and Shinagawa, 1985, Tateda et al., 2005
Kenya	IV. Lagoons	23.84 ± 2.41	7.87–45.31; 21	Andreetta et al., 2014, Middleburg et al., 1996
Madagascar	III. Tidal systems	23.43 ± 3.8	16.52–29.64; 3	Jones et al., 2014

TABLE 13.3 Summary statistics of global field–based mangrove soil organic carbon (SOC) density.—cont'd

Country	CES	Mean SOC density (mg/cm^3 ±1SE)	SOC range (mg/cm^3; N)	References
Malaysia	II. Small deltas	20.63 ± 1.12	16.24–30.31; 14	Arianto et al., 2015, Pazi et al., 2016
	III. Tidal systems	89.94 ± 15.64	22.94–152.69; 10	Hossain, 2014, Hemati et al., 2015, Alongi et al., 2004
Mexico	II. Small deltas	30.98 ± NA	30.98–30.98; 1	Ezcurra et al., 2016
	IV. Lagoons	49.78 ± 3.57	27.96–87.62; 21	Kauffman et al., 2016, Coronado-Molina et al., 2012, Adame et al., (2013), Gonneea et al., 2004, Gutierrez et al., 2016, Day et al., 1996
	I. Large rivers	12.5 ± NA	12.5–12.5; 1	Kauffman et al., 2016
	V. Carbonate	65.87 ± 5.81	20.7–125.55; 22	Adame et al., 2015, Adame et al., (2013)
	VI. Arheic	60.06 ± 11.29	7.52–140; 11	Ezcurra et al., 2016, Giani et al., 1996
Mozambique	II. Small deltas	16 ± NA	16–16; 1	Sitoe et al., 2014
	I. Large rivers	15.53 ± 0.28	14.84–16.5; 5	Stringer et al., 2015
New Caledonia	II. Small deltas	22.02 ± NA	22.02–22.02; 1	Sanders et al., 2016
	III. Tidal systems	36.09 ± 6.69	29.4–42.79; 2	Deborde et al., 2015
	V. Carbonate	41.65 ± 3.24	30.08–47.49; 5	Marchland et al., 2011, Leopold et al., 2015
New Zealand	II. Small deltas	22.34 ± 0.86	8.2–30.47; 37	Yang et al., 2013
Nigeria	III. Tidal systems	55.02 ± 7.65	34.2–82.8; 6	Ukpong, 1995, 2000
Palau	V. Carbonate	47.83 ± 3.04	34.3–60.48; 9	Donato et al., (2011), Donato et al., 2012, Kauffman et al., 2011, MacKenzie et al., 2016
Panama	II. Small deltas	29.1 ± 3.09	23.82–34.52; 3	Rovai et al. (2018)
Philippines	II. Small deltas	14.61 ± 2.94	9.01–24.06; 6	Thompson et al., 2014
Senegal	III. Tidal systems	13.89 ± 5	8.9–18.89; 2	Sakho et al., 2015

(Continued)

TABLE 13.3 Summary statistics of global field–based mangrove soil organic carbon (SOC) density.—cont'd

Country	CES	Mean SOC density (mg/cm^3 ±1SE)	SOC range (mg/cm^3; N)	References
Singapore	III. Tidal systems	32.97 ± NA	32.97–32.97; 1	Phang et al., 2015
Sri Lanka	II. Small deltas	83.61 ± NA	83.61–83.61; 1	Perera and Amarasinghe, 2016
Taiwan	II. Small deltas	15.68 ± NA	15.68–15.68; 1	Fan et al., 2012
Tanzania	II. Small deltas	24.98 ± 3.28	16.43–38.46; 6	Machiwa, 1998, Lupembe, 2014
Thailand	II. Small deltas	21.19 ± 1.56	17.55–24.97; 4	Alongi et al., 2001
	III. Tidal systems	9.75 ± NA	9.75–9.75; 1	Tateda et al., 2005
United States	IV. Lagoons	29.52 ± 5.63	8.04–81.47; 17	Yando et al., 2016, Bianchi et al., 2013, Rovai et al., (2018), Marchio et al., 2016, Doughty et al., 2015
	I. Large rivers	18.85 ± 2.73	15.93–24.3; 3	Henry and Twilley, 2013, Yando et al., 2016
	V. Carbonate	52.25 ± 3.25	25.6–96.2; 25	Rovai et al., (2018), Steyer, 1988, Twilley et al., (1986), Coronado-Molina et al., 2012, Castañeda-Moya et al., (2013), Breithaupt et al., 2014, Osland et al., 2012
Venezuela	II. Small deltas	35.11 ± 4.75	8.19–44; 7	Barreto et al., 2016, Vegas-Vilarrubia et al., 2010
Vietnam	III. Tidal systems	25.25 ± 2.63	15.75–61.5; 17	Tue et al., 2014, MacKenzie et al., 2016, Nam et al., 2016, Dung et al., 2016
	I. Large rivers	12.53 ± 1.14	10.72–14.64; 3	Tateda et al., 2005, Grellier et al., 2017

Data are organized by country and coastal environmental settings. Mean values are provided along with standard error (±1 SE), range, and number of observations (N).
Modified from Rovai et al. (2018).

treated as independent samples). Moreover, it is unclear if roots are removed from soil samples prior to SOC determination. This methodological detail is not transparent in the vast majority of studies reporting on mangrove SOC, but it raises the issue if a portion of C stored in BGB is included in SOC estimates, leading to an overestimation of C stocks in mangrove soils. Here we use the data set collated by Rovai et al. (2018) where much of these issues have been addressed where possible. Briefly, in contrast with other global mangrove SOC stock estimates, this data set only included data from soil profiles that were at least 0.3 m deep, SOC density values (in mg/cm^3) obtained via elemental or wet oxidation analysis (e.g., Walkley-Black method), and SOC stock values (in Mg/ha) estimated from independent depth-integrated soil profiles (but see Rovai et al., 2018 for details).

The quality criteria applied by Rovai et al. (2018) resulted in a final data set with 107 studies totaling 552 sites from 40 countries (Fig. 13.2C; Table 13.3) against previous studies that included over 900 soil C measurements from 28 countries (Jardine and Siikamäki, 2014), 1230 sampling locations from 48 countries (Atwood et al., 2017), and 1613 sites from 48 countries (Sanderman et al., 2018). The marked drop in number of sites in Rovai et al. (2018) relative to the latter two studies resulted from the application of data quality criteria discussed above. However, the number of nations remained nearly the same across these studies. In terms of spatial distribution, the number of sites reporting on SOC density values is similar between biogeographic provinces, with IWP and AEP regions accounting for 56% (N = 308) and 44% (N = 244) of the available data, respectively (Fig. 13.2C; Table 13.3). Also, the diversity of coastal environmental settings where mangrove thrive in both IWP and AEP regions, coupled with the selection of significant environmental drivers of SOC storage, allowed for the development of robust geospatial analyses in spite of the reduction in the total number of sampling sites relative to previous studies as demonstrated by Rovai et al. (2018). But again, we recommend consultation of original sources for points of complementarity among these different approaches (that is, Atwood et al., 2017; Jardine and Siikamäki, 2014; Rovai et al., 2018; Sanderman et al., 2018). Furthermore, the reader is referred to the CCRCN coastal carbon atlas for a comprehensive core-level data inventory which presently includes more than 11,000 cores from blue carbon ecosystems (CCRCN, 2020).

4. Global trends in blue C in mangroves

4.1 Biogeographic and regional controls of C storage in mangrove biomass and soils

Mangrove AGB was twice as large in the IWP region (mean ± 1SE: 196.16 ± 6.16 Mg/ha) relative to AEP mangroves (90.12 ± 2.78 Mg/ha) (Fig. 13.3A). This difference could be partly reflecting the distribution of mangrove area in the IWP region mostly along the equator (e.g., 21,294 km^2 between 5°S and 5°N relative to 13,074 km^2 for the same latitudinal zone in the AEP, based on Hamilton and Casey, 2016 mapping) where environmental conditions (e.g., higher precipitation and temperature, low seasonality) allow for higher forest stand development.

There were no differences in BGB between AEP (49.50 ± 8.86 Mg/ha, in the top 0.5 m of soil) and IWP (58.65 ± 8.30 Mg/ha) mangroves (Fig. 13.3B). However, the limited number of studies (Fig. 13.2B, Table 13.2) and associated uncertainties related to sampling methods (e.g., coring vs. trench; see review by Adame et al., 2017) and environmental heterogeneity of sites preclude robust inferences to be drawn about the spatial variability of mangrove BGB.

SOC was higher in the AEP (385.05 ± 14.81 Mg/ha, in the top 1 m of soil) region relative to IWP mangroves (303.54 ± 12.72 Mg/ha) (Fig. 13.3C). This difference, however, could likely be attributed to the fact that the AEP region shelters 57% of the global tropical carbonate coastlines (Dürr et al., 2011), where mangrove soils are predominantly peat-dominated (Thom, 1982; Woodroffe, 1992).

Within both AEP and IWP regions mangroves develop on diverse coastal morphologies subjected to conspicuous environmental settings (Twilley et al., 2018). Among distinct coastal morphologies, geophysical (tides, river discharge) and climatic (temperature, precipitation,

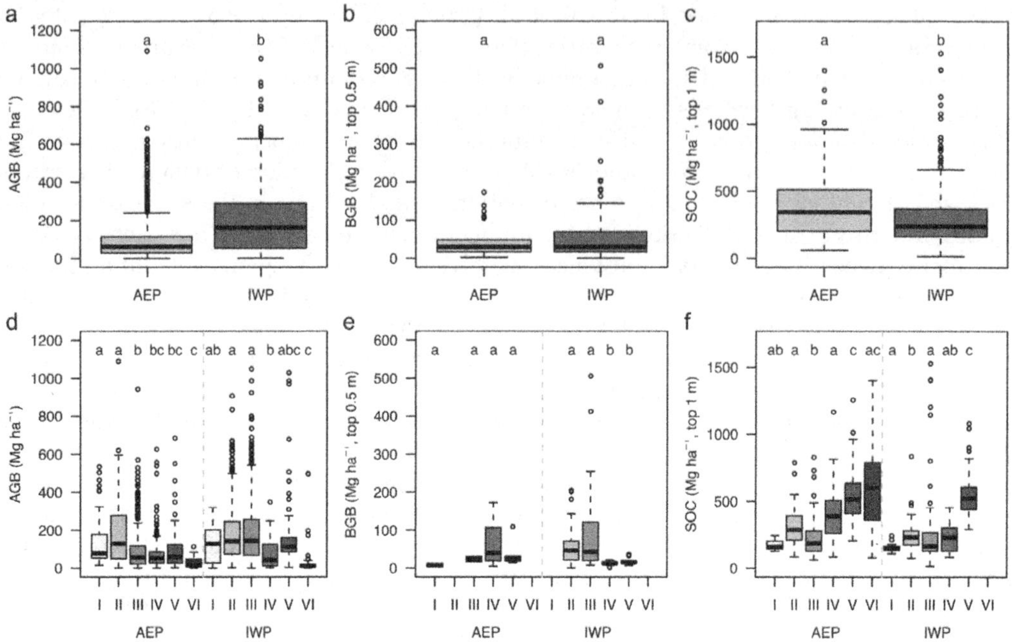

FIGURE 13.3 Data on above- (AGB) and below-ground (BGB) biomass and soil organic carbon (SOC) stocks between Atlantic East Pacific (AEP) and Indo West Pacific (IWP) biogeographic regions and across distinct coastal environmental settings (CES). AGB is higher in IWP mangroves (A), while there were no differences in BGB between biogeographic regions (B). SOC stocks are higher in AEP mangroves (C). AGB decreases from river- and tide-dominate systems to arheic coastlines (D). BGB varied widely across CES (E). SOC density increased from river-dominated coastlines to carbonate and arheic systems (F). Different lowercase letters indicate statistical difference ($P < .05$). I—Large rivers, II—Small deltas, III—Tidal systems, IV—Lagoons, V—Carbonate, VI—Arheic.

evapotranspiration) forcings have long been shown to influence mangrove functional attributes and resulting C stocks (Saenger and Snedaker, 1993; Schaeffer-Novelli et al., 1990a; Twilley, 1995). The analyses presented in this chapter suggest that coastal environmental settings control to a large degree the variability of mangrove attributes across broad geographical gradients and has potential to explain and reconcile spatial and temporal scales in mangrove ecological studies.

Overall, mangrove AGB decreased from river-dominated to arheic coastal environmental settings in both biogeographic regions (Fig. 13.3D). In the AEP, AGB is decreased from 156.4 Mg/ha (mean for types I and II) to 32.5 Mg/ha in arheic (type VI) coastal environmental settings. IWP mangroves had a similar pattern with AGB decreasing from 181.4 and 208.8 Mg/ha in deltaic and tidal systems (type III) to 33.6 Mg/ha in arheic (type VI) coastlines. Riverine mangroves are characterized by optimal structural growth resulting from high nutrient availability and reduced salinity levels in soil (Castañeda-Moya et al., 2006; Cintrón et al., 1978). These results also highlight the high variability in carbonate settings, with relatively high mean AGB values of 96.3 and 136.3 Mg/ha for AEP and IWP mangroves, respectively. This finding is interesting as it underscore the importance of carbonate settings (for example, Shark River estuary, in southwestern Florida Coastal Everglades, and on Holocene reef tops fringing oceanic high islands) (for example, Federate States of Micronesia), as significant hotspots for C storage in AGB (Alongi et al., 2008; Ewel et al., 2003; Woodroffe, 1987). The structural development of mangrove forests developing in carbonate settings (e.g., south Florida and Caribbean region) has been linked to nutrient limitation (e.g., P) and hydrologic stress conditions, with plants allocating more C to the root system than to trunks and branches, resulting in well-developed root systems but reduced AGB, particularly in scrub mangroves (Adame et al., 2014; Castañeda-Moya et al., 2013; Feller et al., 2003). Also, some carbonate settings in arid regions, such as in the Red Sea, may not be similarly limited by nutrient availability, but by the overriding influence of aridity. However, tall well-developed (e.g., basal area = 40 m^2/ha; max height = 20–25 m) mangrove stands can develop over thick peat deposits on carbonate settings, where hurricane-driven allochthonous mineral inputs (i.e., Ca-bound P) enhance soil nutrient concentrations (Castañeda-Moya et al., 2020, 2010; Chen and Twilley, 1999b; Lovelock et al., 2011) resulting in higher AGB relative to prestorm conditions due to increases in stand basal area and stem density (Imbert, 2018; Rivera-Monroy et al., 2019).

This is in contrast with and offers an alternative explanation for some of the unaccounted variability reported in earlier works where a negative relationship between mangrove AGB (using canopy height as a proxy) and hurricanes frequency has been suggested (Simard et al., 2019). Furthermore, while hurricanes may constrain canopy heights in regions subjected to this cyclical disturbance regime, the direct influence of tropical cyclones on mangrove forest structure and function is rather localized, and not a global phenomenon occurring over the totality of mangrove area. Tall mangrove forests also develop around "cenotes" (natural sinkholes) which provide surface connection to subterranean nutrient-rich groundwater (for example, Yucatán peninsula) (Adame et al., 2013). In addition, mangroves colonizing quaternary reef tops surrounding oceanic high islands with high precipitation (>3 m/yr) can receive considerable phosphorus input from weathering of phosphate rock beds (Fosberg, 1957) or via phreatic groundwater from geothermal endo-upwelling (Rougerie et al., 1997), allowing for higher above-ground forest development such as in Micronesia (Cormier et al., 2015) and possibly in other highly structurally developed relict oceanic mangrove stands (Woodroffe, 1988).

Mangrove BGB varies as a function of hydroperiod, resources, and regulators (Castañeda-Moya et al., 2011; Cormier et al., 2015). For example, BGB responds inversely to soil fertility (Castañeda-Moya et al., 2011), which vary significantly across and within distinct coastal environmental settings (Rovai et al., 2018). Therefore, it is reasonable to expect trends in mangrove BGB across distinct coastal environmental settings (that is, deltas, estuaries, lagoons, carbonate, arheic) as well as among hydrogeomorphic zones within these coastal settings (for example, fringe and interior sites). However, the reduced number of studies (Fig. 13.2B; Table 13.2) and different methods used to estimate mangrove BGB (Adame et al., 2017), coupled with high environmental heterogeneity, obscures patterns in biomass allocation to root systems across diverse coastal morphologies (Fig. 13.3E). For example, while the trench method has been shown to quantify more appropriately BGB, the coring method is more preferred yet underestimates BGB (Adame et al., 2017; Komiyama et al., 2000, 1987). Mangrove BGB can also be estimated using allometric equations. Indeed, the majority of mangrove BGB estimates reported in the literature are based on allometric relationships. Here, our rationale to consider only observed mangrove BGB values was based on the fact that site- and species-specific allometric equations are often applied beyond the range of environmental conditions under which these equations were developed, yielding inaccurate estimates (Adame et al., 2017). For example, BGB for northeastern semiarid Brazilian mangroves, where precipitation averages 1000–1200 mm/yr, was estimated using a BGB ~ DBH relationship developed for mangroves subjected to high rainfall (2263–4810 mm/yr), macrotidal regimes and completely different species compositions (Kauffman et al., 2018). Plants in arid environments allocate more biomass into below-ground tissues to optimize nutrient and water absorption and storage capacity (Chapin et al., 1986; Vitousek and Farrington, 1997). Indeed, mangroves in arid regions develop higher root biomass in response to infrequent flooding and high salinity (Adame et al., 2017), such as in northeastern semiarid Brazilian mangroves where soil salinity averages 40‰ (Kauffman et al., 2018). Thus, while the allometric relationship found for mangroves under a macrotidal and high rainfall regime in the IWP region (Komiyama et al., 2005) may hold when applied to similar coastal environmental settings in the AEP, such as the Amazon mangroves, its use to predict BGB in arid coastlines should be interpreted with caution. Moreover, biomass apportioning between plants' above- and below-ground compartments responds differently to the same macroecological drivers. For example, nutrients stimulate AGB growth relative to BGB production (Mooney, 1972; Song et al., 2019); tidal flushing increases AGB but reduces SOC stocks (Rovai et al., 2018, 2016), thus must be accounted for separately to understand how each factor controls partitioning of ecosystem-level resources.

Contrary to AGB, SOC stock consistently increased from river-to peat-dominate and arheic coastlines in both AEP and IWP mangroves (Fig. 13.3F). However, data on arheic mangroves (e.g., Arabian Peninsula and other arid locations in Africa and Indian Ocean) are still limited in the literature and more studies are needed to clarify C partition between above- and below-ground compartments in these environments (Adame et al., 2020). Mean SOC stocks were 172.64 and 157.14 Mg/ha in river-dominated (type I) AEP and IWP mangroves, respectively, while nearly four times this amount in carbonate and arheic coastal environmental settings.

River discharge and tidal amplitude are significant predictors of mangrove SOC stocks (Rovai et al., 2018). Riverine mangroves are characterized by optimal structural growth resulting from high nutrient availability and reduced soil salinity levels (Castañeda-Moya et al., 2006; Cintrón et al., 1978). Rivers are particularly important in arheic coastlines where evapotranspiration far exceeds precipitation and inputs from upstream catchment areas become the only source of freshwater to maintain soil salinity within tolerable thresholds for mangrove development. For instance, dry coastlines with minor river discharge such as in western and northern Australia have experienced massive mangrove diebacks due to fluctuating sea levels, reductions in rainfall, and abnormally high air temperatures (Duke et al., 2017; Lovelock et al., 2017). During these events, only mangroves fringing estuary channels and upstream riverine stands remained healthy and mostly intact (Duke et al., 2017).

Tidal inundation promotes nutrient exchange, soil aeration, and reduction in the levels of sulfides, allowing higher mangrove structural development (Castañeda-Moya et al., 2013; Lugo and Snedaker, 1974). Moreover, tides have been shown to be a major driver of C allocation between above- and below-ground compartments in mangrove forests. For instance, macrotidal systems have been linked to well-developed mangrove forest stands (Rovai et al., 2016). Conversely, mangrove root biomass was found to be higher in sites subjected to infrequent inundation (Adame et al., 2017; Castañeda-Moya et al., 2011). Similarly, higher SOC stocks have been negatively correlated with tides (Rovai et al., 2018). The mechanism by which tides regulate SOC stocks in mangroves is related to soil aeration which stimulates microbial activity thereby reducing soil organic C density (Chapman et al., 2019). In addition to tidal amplitude, tidal cycle duration is also important for mangrove soil biogeochemical processes. Longer tidal cycles allow for higher loads of nutrient-rich particles to settle on the forest soil surface increasing nutrient transformation and availability (Adame et al., 2010) and increasing C sequestration in coastal wetlands (Morris and Bowden, 1986). Tides influence mangrove species zonation (Castañeda-Moya et al., 2006; Crase et al., 2013) as well as the vertical range of suitable environment for mangrove establishment (Hutchings and Saenger, 1987). Varying with fluctuating sea level along the few past millennia, tides promote lateral transport of marine and river-borne sediments filling in the realized accommodation space occupied by coastal wetlands (Rogers et al., 2019). Ultimately, rivers deliver massive amounts of sediments to the coastal oceans (Syvitski et al., 2005), and along with tidally controlled hydroperiod maintain (e.g., relative to sea-level rise) much of the global intertidal substrate at an adequate elevation for mangroves persistence (Saintilan et al., 2020; Rogers et al., 2019; Swales et al., 2019).

The role of temperature and precipitation in constraining mangrove properties has been extensively covered in the literature, thus will not be discussed in depth here. Briefly, temperature and precipitation control mangrove range limits (Cavanaugh et al., 2018; Osland et al., 2017), above- and below-ground C allocation (Hutchison et al., 2014; Rovai et al., 2018, 2016; Simard et al., 2019; Twilley et al., 2018), primary productivity (Bouillon et al., 2008; Twilley et al., 2017; Ribeiro et al., 2019), and decay rates (Ouyang et al., 2017), all important spatial and biogeochemical processes that determine global C stocks. Potential evapotranspiration (PET) is an important regulator water balance in terrestrial forests, representing the amount of water that could potentially be used by plants, but it is transferred back to the atmosphere through evaporation (Holdridge, 1967). PET also plays a major role on continental-scale variability of mangrove AGB (Rovai et al., 2016) and global soil organic C stocks (Rovai et al., 2018). For instance, in equatorial climates, where the ratio between PET and precipitation is generally <1, mangroves can allocate more C into AGB; however, where PET exceeds rainfall, increased water deficit leads to reduced forest structure development (Schaeffer-Novelli et al., 1990a).

4.2 Biogeographic and regional controls of C sequestration in mangrove biomass and soils

We also examined C sequestration rates in mangrove biomass and soils globally. For C sequestration in woody biomass, we use a comprehensive review recently published by Xiong et al. (2019) (Table 13.4). For C sequestration in mangrove soils, we expanded the data set collated by Breithaupt et al. (2012) (Table 13.5) and used it to assess how biogeographic and coastal morphology may affect this ecogeomorphic process. It should be noted that this updated data set only includes organic C burial rates in mangrove obtained using long-term dating stable isotope techniques (e.g., ^{210}Pb, ^{137}Cs) in order to avoid time scale bias relative to other methods utilized to assess vertical change in coastal wetlands soils

TABLE 13.4 Summary statistics of global field–based mangrove net above-ground woody biomass productivity (NPPw).

Country	CES	Mean NPPw (Mg ha^{-1} yr^{-1} ± 1SE)	NPPw range (Mg ha^{-1} yr^{-1}; N)	References
Australia	II. Small deltas	3.92 ± 0.54	0.8–9.44; 23	Alongi, 2011, Clough, 1998
	III. Tidal systems	25.93 ± 6.13	6.3–45.4; 6	Clough, 1992
Bangladesh	III. Tidal systems	7.54 ± 0.45	7.1–7.99; 2	Kamruzzaman et al., 2017, Ray et al., 2011
Brazil	IV. Lagoons	5.13 ± 0.48	4.22–5.87; 3	Estrada et al., 2015
Cameroon	III. Tidal systems	9.99 ± NA	9.99–9.99; 1	Ajonina, 2008
China	III. Tidal systems	1.19 ± NA	1.19–1.19; 1	Peng et al., 2016
Dominican Republic	V. Carbonate	11.57 ± 2.8	6.6–16.3; 3	Sherman et al., 2003
Guadeloupe, France	V. Carbonate	3.5 ± 2.15	0.3–7.6; 3	Imbert and Rollet, 1989 in Alongi (2011)
Indonesia	II. Small deltas	3.5 ± NA	3.5–3.5; 1	Versteegh, 1951 in Devoe and Cole (1998)
Japan	II. Small deltas	12.75 ± 1.95	6.4–21.5; 8	Khan et al., 2009 in Alongi (2011), Suwa and Analuddin, 2008
Malaysia	III. Tidal systems	8.96 ± 2.36	6.51–13.67; 3	Hossain et al., 2008, Noakes, 1955 in Devoe and Cole (1998), Putz and Chan, 1986
Mexico	IV. Lagoons	4.97 ± 2.14	1.08–12.1; 5	Day et al., 1987, Day et al., 1996
Micronesia	II. Small deltas	3.13 ± 1.11	0.28–5.04; 4	Devoe and Cole, 1998
Philippines	II. Small deltas	0.55 ± 0	0.55–0.55; 2	Salmo et al., 2013
Puerto Rico, United States	V. Carbonate	3.1 ± NA	3.1–3.1; 1	Golley et al., 1962
Sri Lanka	II. Small deltas	4.53 ± 1.16	1.4–6.76; 4	Amarasinghe and Balasubramaniam, 1992
Thailand	III. Tidal systems	11.71 ± 2.82	7.39–20; 4	Christensen, 1978, Poungparn et al., 2012

TABLE 13.4 Summary statistics of global field–based mangrove net above-ground woody biomass productivity (NPPw).—cont'd

Country	CES	Mean NPPw (Mg ha^{-1} yr^{-1} ± 1SE)	NPPw range (Mg ha^{-1} yr^{-1}; N)	References
United States	II. Small deltas	15.07 ± 7.79	2.2–29.1; 3	Cox et al., 1999, Ross et al., 2001
	IV. Lagoons	24.41 ± 4.94	10–37.99; 5	Lahmann, 1988, Sell, 1977
	V. Carbonate	3.24 ± 1.14	0.47–10; 8	Castañeda-Moya et al., (2013), Sell, 1977

Data are organized by country and coastal environmental settings. Mean values are provided along with standard error (±1 SE), range, and number of observations (N).
Modified from Xiong et al. (2019).

(e.g., surface elevation tables, marker horizons, event horizons) (Breithaupt et al., 2018). This approach generates estimates compatible with the time scale of anthropogenic CO_2, which are more readily associable to the potential of mangroves to absorb the excess of this greenhouse gas from the atmosphere. However, the use of stable isotope dating techniques has limitations that can also introduce bias in soil geochronology or historical reconstruction since mangrove environments develop on top of heterogeneous sediments with varying inputs of organic material, that are disturbed by natural and anthropogenic processes resulting in sediment mixing and changes in sedimentation rates or erosion (see Arias-Ortiz et al., 2018 for a comprehensive review).

We also constrained our analysis to C sequestration in woody biomass and soils since these are the two largest, long-term C pools in mangrove ecosystems, and mass balance approach shows that these compartments account for the portion of the net ecosystem productivity that is retained in mangrove biomass and soils over time (Twilley et al., 2017). For example, soil organic C sequestration rates based on long-term core dating techniques embed ecogeomorphic feedbacks that control C sequestration in mangrove soils thus representing the net amount of C accumulated over time from diffuse sources including litterfall and roots in addition to allochthonous inputs (Twilley et al., 2017). Therefore, C sequestration estimates based on long-term core dating techniques are rather conservative as they prevent double counting of ecosystem's processes (e.g., adding up C inputs from litterfall and roots to soil organic C sequestration).

The number of studies on C sequestration in mangrove woody biomass available in the literature (see Xiong et al., 2019 for a comprehensive review) is evenly spread across biogeographic regions, with 17 and 12 studies in the IWP and AEP region, respectively (Fig. 13.4A; Table 13.4). There were no differences in C sequestration in woody biomass between AEP and IWP mangroves (Fig. 13.5A). C sequestration in mangrove woody AGB for natural mangroves globally average 4.23 MgC ha^{-1} yr^{-1}, ranging from 0.12 to 19.97 MgC ha^{-1} yr^{-1} (Xiong et al., 2019). Precipitation and temperature seasonality have been shown to positively influence DBH growth and wood production in natural mangrove forests, with edaphic

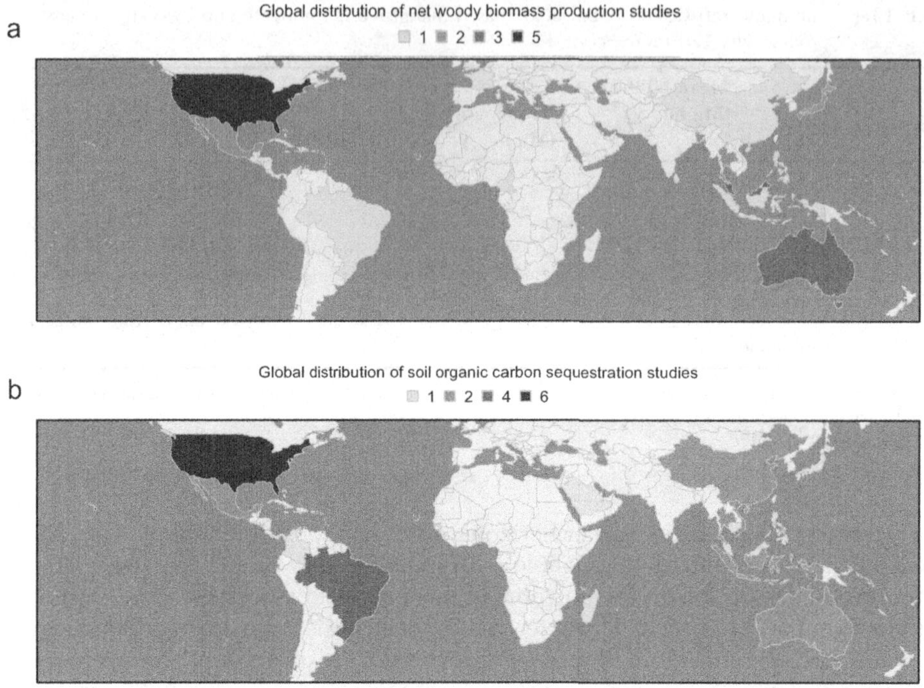

FIGURE 13.4 Number of studies per country on net above-ground woody biomass productivity (A) and soil organic carbon sequestration rates (B) in mangrove soils.

(e.g., nutrient) and biological (e.g., species composition) characteristics playing a secondary role under specific climate regimes (Xiong et al., 2019).

However, the reduced number of highly variable observations of C sequestration in woody biomass prevents robust testing for differences across distinct coastal morphologies (Fig. 13.5C). Moreover, the high variability in wood production across sites globally may be partly explained by differences between climate zones where these mangroves occur. For example, in regions where annual precipitation exceeds evapotranspiration rates wood production would be expectedly higher; however, if this climate balance occurs in a subtropical or warm temperate mangrove zone, lower temperatures or extreme events could hinder biomass productivity. For example, near certain ecosystem distribution range limits, severe winter temperatures and sporadic frosts (Cavanaugh et al., 2019, 2015; Schaeffer-Novelli et al., 1990b) constrain mangrove growth (Chen et al., 2017; Lugo and Zucca, 1977) and biomass allocation (Cavanaugh et al., 2019, 2018; Feher et al., 2017; Saenger and Snedaker, 1993). Future efforts should strive to account for the role of coastal morphology and ecological type (e.g., fringe vs. interior mangroves) to clarify how scale-dependent processes control biomass accumulation in mangrove forests across broad geographical gradients.

The number of studies on organic C sequestration in mangrove soils is equal between AEP ($N = 13$) and IWP ($N = 13$) regions, however, distributed over a disparate number of countries (4 countries in AEP and 12 IWP) (Fig. 13.4B, Table 13.5). There were no differences in

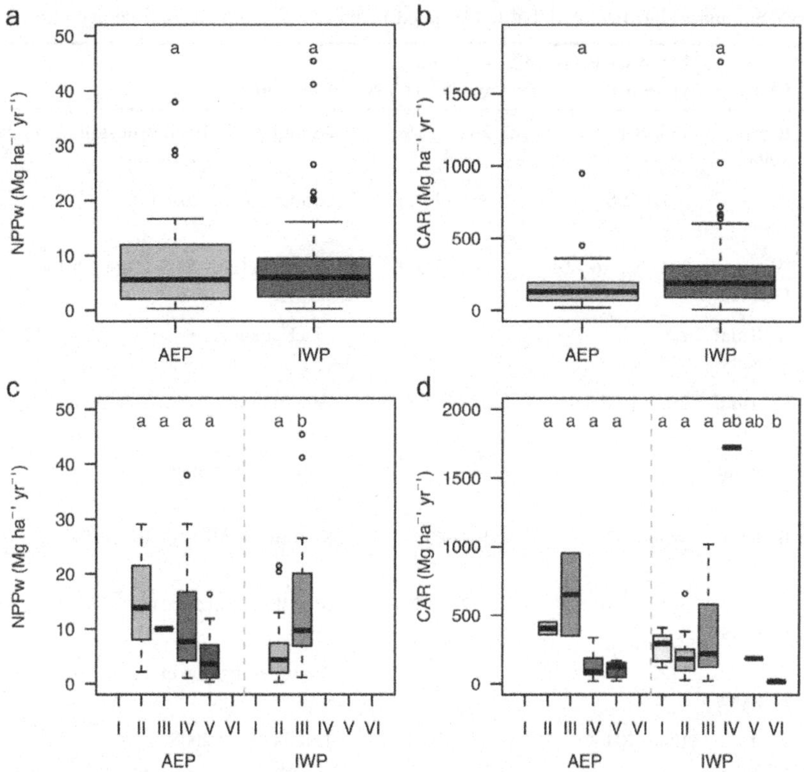

FIGURE 13.5 Net above-ground woody biomass productivity (NPPw) and organic carbon accretion rates (CAR) in mangrove soils across biogeographic regions and coastal environmental settings worldwide. There were no differences in NPPw between Atlantic East Pacific (AEP) and Indo West Pacific (IWP) biogeographic regions (A). NPPw varies widely across coastal environmental settings (C). CAR is also consistent between AEP and IWP mangroves (B) but highly variable across coastal environmental settings (D). Different lowercase letters on top of groups denote statistical difference ($P < .05$). I—Large rivers, II—Small deltas, III—Tidal systems, IV—Lagoons, V—Carbonate, VI—Arheic.

organic C sequestration in mangrove soils between AEP and IWP regions (Fig. 13.5B). In terms of diversity of coastline types, while studies in the IWP region cover the diversity of coastal environmental settings, the AEP lacks data from river-dominated (type I) and arheic systems (type VI) (Fig. 13.5D). Combining AEP and IWP estimates due to the reduced number of observations shows that organic C sequestration in mangrove soils was higher in tide- (type III) and river-dominated (type I) systems, averaging 350.6 and 269.5 gC m^{-2} yr^{-1}, respectively. Lowest organic C accretion was found in carbonate and arheic coastlines at 114.9 and 18.6 gC m^{-2} yr^{-1}, respectively.

Factors that control organic C burial in mangrove soils range from site-specific to regional coastal processes and climate regimes. At a macroscale (e.g., coastal environmental setting, *sensu* Woodroffe et al., 2016), it is reasonable to assume that C burial in mangroves is controlled by similar mechanisms that explain soil C stock distributions (Twilley et al., 2018) since soil C stocks

TABLE 13.5 Summary statistics of global field–based mangrove soil organic carbon accretion rates (CAR).

Country	CES	Mean CAR (gC m^{-2} yr^{-1} ± 1SE)	CAR range (gCm^{-2} yr^{-1}; N)	References
Australia	II. Small deltas	138.78 ± 36.22	26–336; 9	Alongi et al., 1999, Brunskill et al., 2002
Brazil	III. Tidal systems	651 ± 298	353–949; 2	Sanders et al., 2010a
	IV. Lagoons	214.33 ± 26.42	168–337; 6	Sanders et al., 2008, 2010b, 2010c
Cambodia	III. Tidal systems	276		MacKenzie et al., 2020, Chapter 12, this book
China	III. Tidal systems	286.2 ± 98.32	37–1020; 10	Alongi et al., 2005, Yang et al., 2014
Colombia	II. Small deltas	361.5		Ruiz-Fernàndez et al., 2011
Indonesia	II. Small deltas	426 ± 232	194–658; 2	Kusumaningtyas et al. (2019)
	III. Tidal systems	581 ± 65.01	412–717; 4	Brunskill et al., 2004
	IV. Lagoons	1722		Kusumaningtyas et al. (2019)
Japan	III. Tidal systems	126 ± 104	22–230; 2	Tateda et al., 2005
Madagascar	III. Tidal systems	110		Arias-Ortiz et al., 2020
Malaysia	I. Large rivers	309.33 ± 37.83	148–410; 6	Alongi et al., 2004
Mexico	IV. Lagoons	97.39 ± 20.27	53–237; 9	Gonneea et al., 2004, Lynch, 1989
Palau	V. Carbonate	187		MacKenzie et al., 2016
Philippines	II. Small deltas	214.25 ± 57.8	139–384; 4	MacKenzie et al., 2020, Chapter 12, this book
Saudi Arabia	VI. Arheic	18.56 ± 4.1	6.2–38; 7	Cusack et al., 2018
Thailand	II. Small deltas	223.5 ± 21	184–281; 4	Alongi et al., 2001
	III. Tidal systems	350 ± 250	100–600; 2	Tateda et al., 2005
Hawaii, United States	II. Small deltas	450		Soper et al., 2019

TABLE 13.5 Summary statistics of global field–based mangrove soil organic carbon accretion rates (CAR).—cont'd

Country	CES	Mean CAR (gC m^{-2} yr^{-1} ± 1SE)	CAR range (gCm^{-2} yr^{-1}; N)	References
United States	IV. Lagoons	115.75 ± 33.19	20–270; 8	Bianchi et al., 2013, Cahoon and Lynch 1994, unpublished in Chmura et al. (2003)
	V. Carbonate	107.74 ± 17.31	21–168; 10	Callaway et al., 1997, Marchio et al., 2016, Smoak et al., 2013
Vietnam	I. Large rivers	150 ± 30	120–180; 2	Tateda et al., 2005
	III. Tidal systems	237		MacKenzie et al., 2016

Data are organized by country and coastal environmental settings. Mean values are provided along with standard error (±1 SE), range, and number of observations (N).
Modified from Breithaupt et al. (2012).

represent a "snapshot" of the processes that control soil formation. A comprehensive review about abiotic and biotic processes that control organic C burial in mangroves across multiple spatial and temporal scales is provided in Chapter 12, therefore will not be reiterated here.

Lastly, while both biogeographic regions and coastal environmental settings seem to be covered in this data set, there are still huge spatial gaps that remain open. For example, in the AEP, regions such as the entire pacific coast of the Americas, Caribbean islands, and from Venezuela all the way to northern Brazil are devoid of data on organic C sequestration in mangrove soils. The totality of the African continent and the majority of Western and Central Indo-Pacific regions also lack data on mangrove soil organic C sequestration. This limited availability of studies along several representative coastlines within both IWP and AEP regions precludes the development of robust models to predict soil organic C burial for the global mangrove area. It should be noted, however, that in this review we limited our analyses to long-term stable isotope dating techniques but there are many other studies that have used recent accretion methods to quantify contemporary sedimentation and C burial rates, and net surface elevation change in mangrove ecosystems (e.g., Lovelock et al., 2014; Rogers et al., 2006; Swales et al., 2019; Whelan et al., 2009). For a review about pros and cons on short- versus long-term C accretion techniques, we refer the reader to the comprehensive studies by Arias-Ortiz et al. (2018) and Breithaupt et al. (2018).

5. Challenges and opportunities to improve mangrove blue C inventories and mapping

In addition to the spatial data gaps discussed in the sections above, we can draw other conclusions that may be helpful to advance the planning of future blue C research in mangroves. While mangrove AGB and SOC stocks seem to have been reasonably mapped globally, there is still much to do with regards to net above-ground woody biomass productivity, BGB, and

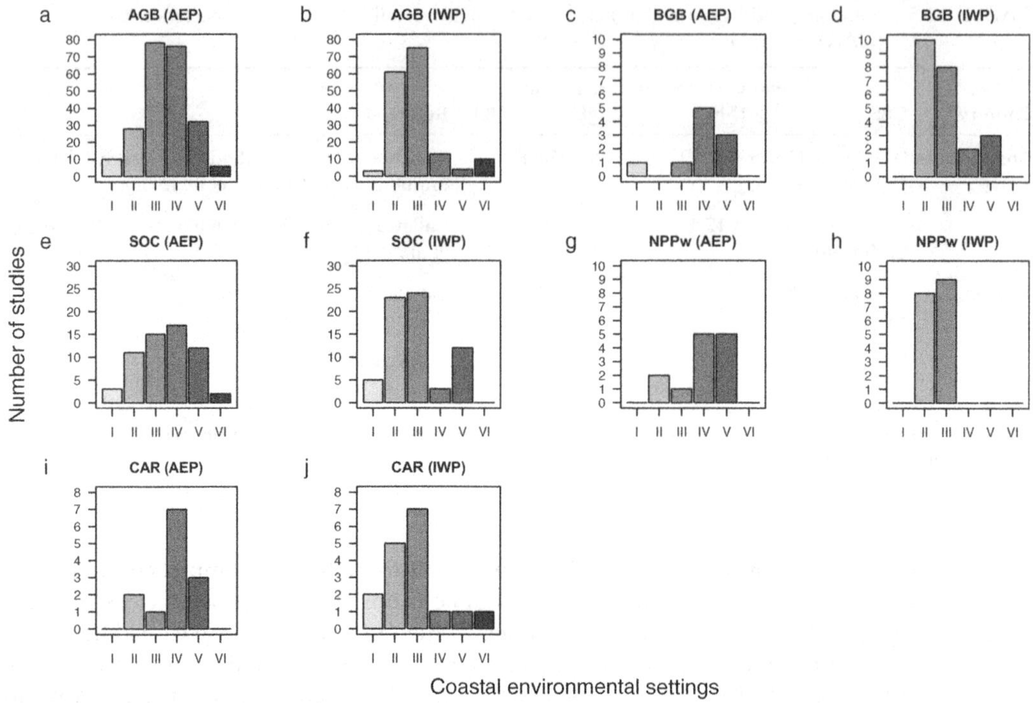

FIGURE 13.6 Number of studies per type of coastal environmental settings across AEP (Atlantic East Pacific) and IWP (Indo West Pacific) biogeographic regions. AGB—above-ground biomass (A, B), BGB—below-ground biomass (C, D), SOC—soil organic carbon density (E, F), NPPw—net woody above-ground biomass productivity (G, H), CAR—carbon accretion rate (I, J). *BGB data modified from Adame et al. (2017), NPPw data modified from Xiong et al. (2019), CAR data modified from Breithaupt et al. (2012).*

CAR in terms of sampling areas representative of both biogeographic regions and of the diversity of coastal environmental settings (Fig. 13.6A–J). The use of the coastal environmental approach is helpful in identifying and addressing apparent differences noted across broad biogeographic regions. The proportion of coastline types across AEP and IWP regions is not equivalent (Dürr et al., 2011), which partly explains the concentration of sampling efforts in certain coastal environmental settings. For example, in the conterminous United States most mangrove forests occur along Florida's peat- (carbonate) and wave-dominated (lagoons) coastlines. Indeed, the neotropics shelter about 57% of the carbonate coastlines worldwide (Dürr et al., 2011), whereas in the IWP region peat-dominated mangroves are constrained for the most of it to Holocene reef tops surrounding oceanic islands (Woodroffe, 1992, 1987). Thus, higher mangrove SOC stocks in the AEP region (Fig. 13.3C) may be explained by most of the sampling being carried out in carbonate and coastal lagoon environments relative to IWP regions (Fig. 13.6E and F). The distribution of coastal environmental settings sampled in C sequestration studies is contrasting (Fig. 13.6I and J), yet no difference is noted between these biogeographic regions due to the high variability of measurements (Fig. 13.5B). In addition, increased sediment load due to erosion caused by land use change and deforestation upstream also contributes to high variability and elevated C burial among distinct coastal morphologies (Galy et al., 2015). Yet, deltaic mangroves sequester between 3 and 5 times more C than peat-dominated and arheic mangroves (Fig. 13.5D). Overall, small deltas (river-dominated but without prograding delta front; see Dürr et al., 2011 for details) and tidal systems are currently the most studied coastal environmental settings globally given the cosmopolitan nature of these coastline types (Fig. 13.6A–J; Tables 13.1–13.5).

Equally important is to determine whether environmental filters such as those imposed by biogeographic or regional (e.g., coastal morphology and regional climate) gradients constrain C storage in mangrove forests worldwide (Twilley et al., 2018). The assessment of adequate predictors of C stocks in mangrove ecosystems is fundamental to improve our capacity to develop not only statistically but also ecologically sound global C estimates. For example, patterns in biomass accumulation and soil properties for distinct coastal environmental settings between IWP and AEP mangroves bring up important questions concerning conceptual and methodological aspects when scaling up ecological attributes and ecosystems services to the global mangrove area. In this respect, the analyses presented here suggest that biogeographical differences between IWP and AEP mangroves may be only apparent. As pointed earlier in this section, apparent disparities between IWP and AEP mangroves traits (e.g., biomass, soil carbon) are likely due to sampling and analytical bias. Some potential confounding factors include geographical sampling bias, different methods used to sample, observer and instrument precision errors, and computation and reporting errors. While new data are needed to safely reject the null hypothesis concerning potential differences between IWP and AEP mangrove functional attributes, the coastal environmental setting approach offers a robust scaling-up framework, particularly considering the fractal property of coastlines with the repetition of coastal processes and resulting landforms across the globe despite discrete spatial boundaries. Moreover, the coastal environmental setting offers a quantitative alternative to binary or categorical data, allowing for the use of global environmental data compilations as predictors of mangrove ecological attributes within a more robust and reproducible modeling framework.

Spatial resolution is another frontier to be expanded in large-scale modeling of blue C research in mangroves. Remote sensing techniques have far improved this issue allowing for modeling of mangrove attributes at higher spatial resolution (Hamilton et al., 2018). The use of finer spatial grid resolution is extremely import particularly for mapping vegetation ground cover, which serves as the base to extrapolate mangrove attributes such as C stocks in biomass and soil, and differences in area coverage may lead to huge discrepancies in global estimates (Friess and Webb, 2014). For example, the Mangrove Forests of the World (MFW) database uses Landsat imagery and considers each 30 × 30 m (approximately, at the equator) pixel with mangroves in it as being entirely covered by mangroves (Giri et al., 2011), while the Continuous Global Mangrove Forest Cover for the 21st Century (CGMFC-21), as opposed to each pixel having a value of "1" for mangrove presence, or a value of "0" for absence, provides each 900 m^2 pixel with a value ranging from 0% to 100% mangrove canopy cover (Hamilton and Casey, 2016). As a result, these data sets have a 40% difference in the global mangrove area (see Hamilton et al., 2018 for details), which is reflected across recent global C stock estimates in biomass and soils (Atwood et al., 2017; Hamilton and Friess, 2018; Hutchison et al., 2014; Jardine and Siikamäki, 2014; Sanderman et al., 2018; Simard et al., 2019). More recently, a new global baseline of mangrove extent has been produced yielding a total area approximately equal to that of Giri et al. (2011), but with significant improvements for several regions worldwide (See Bunting et al., 2018 for details).

Spatial grid resolution used in current global mangrove blue C estimates ranges from as high as 0.0009 km^2 at the equator (Hamilton and Friess, 2018; Jardine and Siikamäki, 2014; Sanderman et al., 2018; Simard et al., 2019) to 100 km^2 (Hutchison et al., 2014), and to 625 km^2 (Rovai et al., 2018). The different degree-cell resolution used across recent global mangrove blue C estimates likely reflects the goals and criteria used in each study (but see Worthington et al., 2020 for an overview of available and forthcoming global datasets to support mangrove research across multiple spatial scales). It appears that the choice about the spatial grid resolution to be used should be based on an attempt to balance the loss of information during the trade-off process of down- and upscaling data with different resolutions (Blackburn and Gaston, 2002). Indeed, recent efforts in macroecology strived to consolidate a database of environmental variables that are thought to be relevant to species' ecology and geographic distribution at a reasonable spatial resolution (0.08333° or approximately 8.3 km at the equator) (Title and Bemmels, 2018). On that note, even the WorldClim database (Hijmans et al., 2005), which has over 3000 citations, has resolutions that range from 1 to 340 km^2.

Moreover, the use of finer spatial grid resolution as a mask to model mangrove functional attributes based on environmental predictors does not necessarily improve the quality of the model's output. This is particularly true for relatively flat coastal floodplains that do not have marked environmental gradients, and finer resolution templates have shown not to always be the best choice to map soil properties across larger spatial scales (Hannam et al., 2012; Harris et al., 2016). For example, recent global estimates produced using very high-resolution mangrove cover area mask (e.g., Jardine and Siikamäki, 2014; Sanderman et al., 2018) do not seem to capture finer-scale variability in SOC along inundation profiles (e.g., patterns between fringe vs. interior sites, see Friess, 2017; Lugo and Snedaker, 1974). Moreover, even though these studies produce high-resolution gridded data, they still use concepts of the coastal environmental setting framework (e.g., coastal process) to explain differences in mangrove traits. This discrepancy between the capacity to predict ecological attributes at high-resolution and the availability of predictors at a higher spatial hierarchical level has led other studies to use a lower-degree cell resolution but still compatible with the scale of coastal processes that control coastal environmental settings and thereby mangrove function (Hutchison et al., 2014; Rovai et al., 2018).

At finer scales, the integration of information on mangrove typology based on local hydrology and topography (for example, fringe vs. interior sites) would potentially allow for more robust local and global estimates; however, the vast majority of studies do not include accurate information on hydroperiod. Accordingly, the spatial resolution of most global compilations on marine and terrestrial environmental variables (see Title and Bemmels, 2018) do not reflect the variability compatible with high-resolution databases such as MFW or CMFGC-21. In order to perform a multiscale spatial analysis, both dependent and independent variables would have to be available at differing resolutions. Moreover, the set of environmental variables would have to hold ecological meaning across different spatial scales, which is unlikely as variables that control, for example, SOC variability in coastal wetlands differ at different scales. While these scale-dependent issues discussed here are perhaps one of the major challenges coastal wetlands scientists will face when upscaling ecological traits across spatial scales, the use of the coastal environmental setting framework offers a promising alternative to reconcile site-specific measurements to global blue C distribution patterns, with soil properties and biomass following close the energetic signature of distinct coastline types.

References

Adame, M.F., Cherian, S., Reef, R., Stewart-Koster, B., 2017. Mangrove root biomass and the uncertainty of below-ground carbon estimations. For. Ecol. Manag. 403, 52–60.

Adame, M.F., Kauffman, J.B., Medina, I., Gamboa, J.N., Torres, O., Caamal, J.P., Reza, M., Herrera-Silveira, J.A., 2013. Carbon stocks of tropical coastal wetlands within the karstic landscape of the Mexican Caribbean. PLoS One 8, e56569. https://doi.org/10.1371/journal.pone.0056569.

Adame, M.F., Neil, D., Wright, S.F., Lovelock, C.E., 2010. Sedimentation within and among mangrove forests along a gradient of geomorphological settings. Estuar. Coast Shelf Sci. 86, 21–30.

Adame, M.F., Reef, R., Santini, N.S., Najera, E., Turschwell, M.P., Hayes, M.A., Masque, P., Lovelock, C.E., 2020. Mangroves in arid regions: ecology, threats, and opportunities. Estuar. Coast Shelf Sci. 106796.

Adame, M.F., Teutli, C., Santini, N.S., Caamal, J.P., Zaldívar-Jiménez, A., Herńndez, R., Herrera-Silveira, J.A., 2014. Root biomass and production of mangroves surrounding a karstic oligotrophic coastal lagoon. Wetlands 34, 479–488.

Alongi, D.M., Trott, L.A., Rachmansyah, Tirendi, F., McKinnon, A.D., Undu, M.C., 2008. Growth and development of mangrove forests overlying smothered coral reefs, Sulawesi and Sumatra, Indonesia. Mar. Ecol. Prog. Ser. 370, 97–109.

Arias-Ortiz, A., Masqué, P., Garcia-Orellana, J., Serrano, O., Mazarrasa, I., Marbá, N., Lovelock, C.E., Lavery, P.S., Duarte, C.M., 2018. Reviews and syntheses: 210Pb-derived sediment and carbon accumulation rates in vegetated coastal ecosystems - setting the record straight. Biogeosciences 15, 6791–6818.

Atwood, T.B., Connolly, R.M., Almahasheer, H., Carnell, P.E., Duarte, C.M., Ewers Lewis, C.J., Irigoien, X., Kelleway, J.J., Lavery, P.S., Macreadie, P.I., Serrano, O., Sanders, C.J., Santos, I., Steven, A.D.L., Lovelock, C.E., 2017. Global patterns in mangrove soil carbon stocks and losses. Nat. Clim. Change 7, 523–528.

Bardou, R., Parker, J.D., Feller, I.C., Cavanaugh, K.C., 2020. Variability in the fundamental versus realized niches of North American mangroves. J. Biogeog. 1–16. https://doi.org/10.1111/jbi.13990, 10.1111/jbi.13990.

Blackburn, T.M., Gaston, K.J., 2002. Scale in macroecology. Global Ecol. Biogeogr. 11, 185–189.

Blanco-Libreros, J., Álvarez-León, R., 2018. HELIO_SP.CO v.1: Hierarchical, Entity-Based and Landscape-Level Information Observatory for Mangrove Species in Colombia, Version 1.

Blanco-Libreros, J.F., Álvarez-León, R., 2019. Mangroves of Colombia revisited in an era of open data, global changes, and socio-political transition: homage to Heliodoro Sánchez-Páez. Rev. Acad. Colomb. Ciencias Exactas, Fis. y Nat. 43, 84–97.

Bouillon, S., Borges, A.V., Castañeda-Moya, E., Diele, K., Dittmar, T., Duke, N.C., Kristensen, E., Lee, S.Y., Marchand, C., Middelburg, J.J., Rivera-Monroy, V.H., Smith, T.J., Twilley, R.R., 2008. Mangrove production and carbon sinks: a revision of global budget estimates. Global Biogeochem. Cycles 22, GB2013.

Boyd, R., Dalrymple, R., Zaitlin, B.A., 1992. Classification of clastic coastal depositional environments. Sediment. Geol. 80, 139–150.

Breithaupt, J.L., Smoak, J.M., Bianchi, T.S., Vaughn, D., Sanders, C.J., Radabaugh, K.R., Osland, M.J., Feher, L.C., Lynch, J.C., Cahoon, D.R., Anderson, G.H., Whelan, K.R.T., Rosenheim, B.E., Moyer, R.P., Chambers, L.G., 2020. Increasing rates of carbon burial in southwest Florida coastal wetlands. J. Geophys. Res. Biogeosci. 125, e2019JG005349.

Breithaupt, J.L., Smoak, J.M., Byrne, R.H., Waters, M.N., Moyer, R.P., Sanders, C.J., 2018. Avoiding timescale bias in assessments of coastal wetland vertical change. Limnol. Oceanogr. 63, S477–S495.

Breithaupt, J.L., Smoak, J.M., Smith III, T.J., Sanders, C.J., Hoare, A., 2012. Organic carbon burial rates in mangrove sediments: strengthening the global budget. Global Biogeochem. Cycles 26, GB3011.

Bunting, P., Rosenqvist, A., Lucas, R.M., Rebelo, L.M., Hilarides, L., Thomas, N., Hardy, A., Itoh, T., Shimada, M., Finlayson, M., 2018. The Global Mangrove Watch - A new 2010 global baseline of mangrove extent. Remote Sens. 10, 1669 https://doi.org/10.3390/rs10101669.

Castañeda-Moya, E., Rivera-Monroy, V.H., Chambers, R.M., Zhao, X., Lamb-Wotton, L., Gorsky, A., Gaiser, E.E., Troxler, T.G., Kominoski, J.S., Hiatt, M., 2020. Hurricanes fertilize mangrove forests in the Gulf of Mexico (Florida Everglades, USA). Proc. Natl. Acad. Sci. U.S.A. 117 (9), 4831–4841, 201908597.

Castañeda-Moya, E., Rivera-Monroy, V.H., Twilley, R.R., 2006. Mangrove zonation in the dry life zone of the Gulf of Fonseca, Honduras. Estuar. Coast 29, 751–764.

Castañeda-Moya, E., Twilley, R.R., Rivera-Monroy, V.H., 2013. Allocation of biomass and net primary productivity of mangrove forests along environmental gradients in the Florida Coastal Everglades, USA. For. Ecol. Manag. 307, 226–241.

Castañeda-Moya, E., Twilley, R.R., Rivera-Monroy, V.H., Marx, B.D., Coronado-Molina, C., Ewe, S.M.L., 2011. Patterns of root dynamics in mangrove forests along environmental gradients in the Florida Coastal Everglades, USA. Ecosystems 14, 1178–1195.

Castañeda-Moya, E., Twilley, R.R., Rivera-Monroy, V.H., Zhang, K., Davis III, S.E., Ross, M., 2010. Sediment and nutrient deposition associated with hurricane Wilma in mangroves of the Florida Coastal Everglades. Estuar. Coast 33, 45–58.

Cavanaugh, K.C., Dangremond, E.M., Doughty, C.L., Park Williams, A., Parker, J.D., Hayes, M.A., Rodriguez, W., Feller, I.C., 2019. Climate-driven regime shifts in a mangrove–salt marsh ecotone over the past 250 years. Proc. Natl. Acad. Sci. U.S.A. 116, 21602–21608.

Cavanaugh, K.C., Kellner, J.R., Forde, A.J., Gruner, D.S., Parker, J.D., Rodriguez, W., Feller, I.C., 2014. Poleward expansion of mangroves is a threshold response to decreased frequency of extreme cold events. Proc. Natl. Acad. Sci. U.S.A. 111, 723–727.

Cavanaugh, K.C., Osland, M.J., Bardou, R., Hinojosa-Arango, G., López-Vivas, J.M., Parker, J.D., Rovai, A.S., 2018. Sensitivity of mangrove range limits to climate variability. Glob. Ecol. Biogeogr. Biogeogr. 27, 925–935.

Cavanaugh, K.C., Parker, J.D., Cook-Patton, S.C., Feller, I.C., Williams, A.P., Kellner, J.R., 2015. Integrating physiological threshold experiments with climate modeling to project mangrove species' range expansion. Global Change Biol. 21, 1928–1938.

CCRCN, 2020. Coastal Carbon Atlas [WWW Document]. Coast. Carbon Res. Coord. Netw. URL. https://ccrcn.shinyapps.io/CoastalCarbonAtlas.

Chapin III, F.S., Vitousek, P.M., Van Cleve, K., 1986. The nature of nutrient limitation in plant communities. Am. Nat. 127, 48–58.

Chapman, S.K., Hayes, M.A., Kelly, B., Adam Langley, J., 2019. Exploring the oxygen sensitivity of wetland soil carbon mineralization. Biol. Lett. 15, 11–15.

Chapman, V.J., 1975. Mangrove biogeography. In: Walsh, G., Snedaker, S., Teas, H. (Eds.), Proceedings of International Symposium on Biology and Management of Mangroves. Institute of Food and Agricultural Sciences, University of Florida, Gainesville, FL, pp. 3–22.

Chen, L., Wang, W., Li, Q.Q., Zhang, Y., Yang, S., Osland, M.J., Huang, J., Peng, C., 2017. Mangrove species' responses to winter air temperature extremes in China. Ecosphere 8, e01865.

Chen, R., Twilley, R.R., 1999a. A simulation model of organic matter and nutrient accumulation in mangrove wetland soils. Biogeochemistry 44, 93–118.

Chen, R., Twilley, R.R., 1999b. Patterns of mangrove forest structure and soil nutrient dynamics along the Shark River Estuary, Florida. Estuaries 22, 955–970.

Cintrón, G., Lugo, A.E., Pool, D.J., Morris, G., 1978. Mangroves of arid environments in Puerto Rico and adjacent islands. Biotropica 10, 110–121.

Cormier, N., Twilley, R.R., Ewel, K.C., Krauss, K.W., 2015. Fine root productivity varies along nitrogen and phosphorus gradients in high-rainfall mangrove forests of Micronesia. Hydrobiologia 750, 69–87.

Cragg, S.M., Friess, D.A., Gillis, L.G., Trevathan-Tackett, S.M., Terrett, O.M., Watts, J.E.M., Distel, D.L., Dupree, P., 2020. Vascular plants are globally significant contributors to marine carbon fluxes and sinks. Ann. Rev. Mar. Sci. 12, 16.1–16.29.

Crase, B., Liedloff, A., Vesk, P.A., Burgman, M.A., Wintle, B.A., 2013. Hydroperiod is the main driver of the spatial pattern of dominance in mangrove communities. Global Ecol. Biogeogr. 22, 806–817.

Donato, D.C., Kauffman, J.B., Murdiyarso, D., Kurnianto, S., Stidham, M., Kanninen, M., 2011. Mangroves among the most carbon-rich forests in the tropics. Nat. Geosci. 4, 293–297.

Drakare, S., Lennon, J.J., Hillebrand, H., 2006. The imprint of the geographical, evolutionary and ecological context on species-area relationships. Ecol. Lett. 9, 215–227.

Duarte, C.M. Al, Cebrián, J., Cebrih, J., 1996. The fate of marine autotrophic production. Limnol. Oceanogr. 41, 1758–1766.

Duke, N.C., Ball, M.C., Ellison, J.C., 1998. Factors influencing biodiversity and distributional gradients in mangroves. Global Ecol. Biogeogr. Lett. 7, 27–47.

Duke, N.C., Kovacs, J.M., Griffiths, A.D., Preece, L., Hill, D.J.E., Van Oosterzee, P., Mackenzie, J., Morning, H.S., Burrows, D., 2017. Large-scale dieback of mangroves in Australia's Gulf of Carpentaria: a severe ecosystem response, coincidental with an unusually extreme weather event. Mar. Freshw. Res. 68, 1816–1829.

Duke, N.C., Lo, E., Sun, M., 2002. Global distribution and genetic discontinuities of mangroves – emerging patterns in the evolution of Rhizophora. Trees 16, 65–79.

Dürr, H.H., Laruelle, G.G., van Kempen, C.M., Slomp, C.P., Meybeck, M., Middelkoop, H., 2011. Worldwide typology of nearshore coastal systems: defining the estuarine filter of river inputs to the oceans. Estuar. Coast 34, 441–458.

Ellison, A.M., 2002. Macroecology of mangroves: large-scale patterns and processes in tropical coastal forests. Trees 16, 181–194.

Ellison, A.M., Farnsworth, E.J., Merkt, R.E., 1999. Origins of mangrove ecosystems and the mangrove biodiversity anomaly. Global Ecol. Biogeogr. 8, 95–115.

Ewel, K.C., Hauff, R.D., Cole, T.G., 2003. Analyzing mangrove forest structure and species distribution on a Pacific island. Phytocoenologia 33, 251–266.

Feher, L.C., Osland, M.J., Griffith, K.T., Grace, J.B., Howard, R.J., Stagg, C.L., Enwright, N.M., Krauss, K.W., Gabler, C.A., Day, R.H., Rogers, K., 2017. Linear and nonlinear effects of temperature and precipitation on ecosystem properties in tidal saline wetlands. Ecosphere 8, e01956.

Feller, I.C., Whigham, D.F., McKee, K.L., Lovelock, C.E., 2003. Nitrogen limitation of growth and nutrient dynamics in a disturbed mangrove forest, Indian River Lagoon, Florida. Oecologia 134, 405–414.

Fosberg, F.R., 1957. Description and occurrence of atoll phosphate rock in Micronesia. Am. J. Sci. 255, 584–592.

Friess, D.A., 2017. J.G. Watson, inundation classes, and their influence on paradigms in mangrove forest ecology. Wetlands 37, 603–613.

Friess, D.A., Webb, E.L., 2014. Variability in mangrove change estimates and implications for the assessment of ecosystem service provision. Global Ecol. Biogeogr. 23, 715–725.

Galy, V., Peucker-Ehrenbrink, B., Eglinton, T., 2015. Global carbon export from the terrestrial biosphere controlled by erosion. Nature 521, 204–207.

Giri, C., Ochieng, E., Tieszen, L.L., Zhu, Z., Singh, A., Loveland, T., Masek, J., Duke, N., 2011. Status and distribution of mangrove forests of the world using earth observation satellite data. Global Ecol. Biogeogr. 20, 154–159.

Hamilton, S.E., Casey, D., 2016. Creation of a high spatio-temporal resolution global database of continuous mangrove forest cover for the 21st century (CGMFC-21). Global Ecol. Biogeogr. 25, 729–738.

Hamilton, S.E., Castellanos-Galindo, G.A., Millones-Mayer, M., Chen, M., 2018. Remote sensing of mangrove forests: current techniques and existing databases. In: Makowski, C., Finkl, C.W. (Eds.), Threats to Mangrove Forests. Springer, Cham, pp. 497–520.

Hamilton, S.E., Friess, D.A., 2018. Global carbon stocks and potential emissions due to mangrove deforestation from 2000 to 2012. Nat. Clim. Change 8, 240–244.

Hannam, J., Cavazzi, S., Fealy, R., Corstanje, R., Mayr, T., 2012. Are fine resolution digital elevation models always the best choice in digital soil mapping? Geoderma 195–196, 111–121.

Harris, J.A., Corstanje, R., Evans, K.L., Hancock, S., Warren, P.H., Grafius, D.R., 2016. The impact of land use/land cover scale on modelling urban ecosystem services. Landsc. Ecol. 31, 1509–1522.

Hayes, M.A., Jesse, A., Tabet, B., Reef, R., Keuskamp, J.A., Lovelock, C.E., 2017. The contrasting effects of nutrient enrichment on growth, biomass allocation and decomposition of plant tissue in coastal wetlands. Plant Soil 416, 193–204.

Hayes, M.A., Jesse, A., Welti, N., Tabet, B., Lockington, D., Lovelock, C.E., 2019. Groundwater enhances aboveground growth in mangroves. J. Ecol. 107, 1120–1128.

Hijmans, R.J., Cameron, S.E., Parra, J.L., Jones, P.G., Jarvis, A., 2005. Very high resolution interpolated climate surfaces for global land areas. Int. J. Climatol. 25, 1965–1978.

Holdridge, L.R., 1967. Life Zone Ecology. Tropical Science Center, San Jose, Costa Rica.

Howard, J., Hoyt, S., Isensee, K., Pidgeon, E., Telszewski, M., 2014. Coastal Blue Carbon: methods for assessing carbon stocks and emissions factors in mangroves, tidal salt marshes, and seagrass meadows. In: Conservation International, Intergovernmental Oceanographic Commission of UNESCO, International Union for Conservation of Nature, Arlington, Virginia, USA.

Howard, J., Sutton-Grier, A., Herr, D.D., Kleypas, J., Landis, E., Mcleod, E., Pidgeon, E., Simpson, S., 2017. Clarifying the role of coastal and marine systems in climate mitigation. Front. Ecol. Environ. 15, 42–50.

Hutchings, P., Saenger, P., 1987. Ecology of Mangroves. University of Queensland Press, St Lucia, Qld.

Hutchison, J., Manica, A., Swetnam, R., Balmford, A., Spalding, M., 2014. Predicting global patterns in mangrove forest biomass. Conserv. Lett. 7, 233–240.

Imbert, D., 2018. Hurricane disturbance and forest dynamics in east Caribbean mangroves. Ecosphere 9, e02231.

Jardine, S.L., Siikamäki, J.V., 2014. A global predictive model of carbon in mangrove soils. Environ. Res. Lett. 9, 104013.

Jennerjahn, T.C., 2020. Relevance and magnitude of 'Blue Carbon' storage in mangrove sediments: Carbon accumulation rates vs. stocks, sources vs. sinks. Estuar. Coast. Shelf Sci. 107027 https://doi.org/10.1016/j.ecss.2020.107027.

Jennerjahn, T.C., Ittekkot, V., 2002. Relevance of mangroves for the production and deposition of organic matter along tropical continental margins. Naturwissenschaften 89, 23–30.

Kauffman, J.B., Adame, M.F., Arifanti, V.B., Schile-Beers, L.M., Bernardino, A.F., Bhomia, R.K., Donato, D.C., Feller, I.C., Ferreira, T.O., Jesus-Garcia, M.C., MacKenzie, R.A., Megonigal, J.P., Murdiyarso, D., Simpson, L., Hernández-Trejo, H., 2020. Total ecosystem carbon stocks of mangroves across broad global environmental and physical gradients. Ecol. Monograp. 90 (2), e01405 https://doi.org/10.1002/ecm.1405.

Kauffman, J.B., Bernardino, A.F., Ferreira, T.O., Bolton, N.W., Gomes, L.E. de O., Nóbrega, G.N., Nobrega, G.N., 2018. Shrimp ponds lead to massive loss of soil carbon and greenhouse gas emissions in northeastern Brazilian mangroves. Ecol. Evol. 8, 1–11.

Komiyama, A., Havanond, S., Srisawatt, W., Mochida, Y., Fujimoto, K., Ohnishi, T., Ishihara, S., Miyagi, T., 2000. Top/root biomass ratio of a secondary mangrove (*Ceriops tagal* (Perr.) C.B. Rob.) forest. For. Ecol. Manag. 139, 127–134.

Komiyama, A., Ogino, K., Aksornkoae, S., Sabhasri, S., 1987. Root biomass of a mangrove forest in southern Thailand. 1. Estimation by the trench method and the zonal structure of root biomass. J. Trop. Ecol. 3, 97–108.

Komiyama, A., Poungparn, S., Kato, S., 2005. Common allometric equations for estimating the tree weight of mangroves. J. Trop. Ecol. 21, 471–477.

Kristensen, E., Bouillon, S., Dittmar, T., Marchand, C., 2008. Organic carbon dynamics in mangrove ecosystems: a review. Aquat. Bot. 89, 201–219.

Kusumaningtyas, M.A., Hutahaean, A.A., Fischer, H.W., Pérez-Mayo, M., Ransby, D., Jennerjahn, T.C., 2019. Variability in the organic carbon stocks, sources, and accumulation rates of Indonesian mangrove ecosystems. Estuar. Coast Shelf Sci. 218, 310–323.

Lo, E.Y.Y., Duke, N.C., Sun, M., 2014. Phylogeographic pattern of Rhizophora (Rhizophoraceae) reveals the importance of both vicariance and long-distance oceanic dispersal to modern mangrove distribution. BMC Evol. Biol. 14, 83.

Lovelock, C.E., Adame, M.F., Bennion, V., Hayes, M., O'Mara, J., Reef, R., Santini, N.S., 2014. Contemporary rates of carbon sequestration through vertical accretion of sediments in mangrove forests and saltmarshes of South East Queensland, Australia. Estuar. Coast 37, 763–771.

Lovelock, C.E., Feller, I.C., Adame, M.F., Reef, R., Penrose, H.M., Wei, L., Ball, M.C., 2011. Intense storms and the delivery of materials that relieve nutrient limitations in mangroves of an arid zone estuary. Funct. Plant Biol. 38, 514–522.

Lovelock, C.E., Feller, I.C., Reef, R., Hickey, S., Ball, M.C., 2017. Mangrove dieback during fluctuating sea levels. Sci. Rep. 7, 1–8.

Lugo, A.E., Snedaker, S.C., 1974. The ecology of mangroves. Annu. Rev. Ecol. Systemat. 5, 39–64.

Lugo, A.E., Zucca, P., 1977. The impact of low temperature stress on mangrove structure and growth. Trop. Ecol. 18, 149–161.

Mandelbrot, B., 1967. How long is the coast of Britain? Statistical self-similarity and fractional dimension. Science 156, 636–638 (80-.).

McGowan, T., 2006. An Assessment of Mangrove Forest Structure in the Las Perlas Islands, Panama. Heriot-Watt University.

McKee, K.L., Cahoon, D.R., Feller, I.C., 2007. Caribbean mangroves adjust to rising sea level through biotic controls on change in soil elevation. Global Ecol. Biogeogr. 16, 545–556.

McLeod, E., Chmura, G.L., Bouillon, S., Salm, R., Björk, M., Duarte, C.M., Lovelock, C.E., Schlesinger, W.H., Silliman, B.R., 2011. A blueprint for blue carbon: toward an improved understanding of the role of vegetated coastal habitats in sequestering CO_2. Front. Ecol. Environ. 9, 552–560.

Mooney, H.A., 1972. The carbon balance of plants. Annu. Rev. Ecol. Systemat. 3, 315–346.

Morris, J.T., Bowden, W.B., 1986. A mechanistic, numerical model of sedimentation, mineralization, and decomposition for marsh sediments. Soil Sci. Soc. Am. J. 50, 96–105.

Osland, M.J., Feher, L.C., Griffith, K.T., Cavanaugh, K.C., Enwright, N.M., Day, R.H., Stagg, C.L., Krauss, K.W., Howard, R.J., Grace, J.B., Rogers, K., 2017. Climatic controls on the global distribution, abundance, and species richness of mangrove forests. Ecol. Monogr. 87, 341–359.

Ouyang, X., Lee, S.Y., Connolly, R.M., 2017. The role of root decomposition in global mangrove and saltmarsh carbon budgets. Earth Sci. Rev. 166, 53–63.

Phang, V.X.H., Chou, L.M., Friess, D.A., 2015. Ecosystem carbon stocks across a tropical intertidal habitat mosaic of mangrove forest, seagrass meadow, mudflat and sandbar. Earth Surf. Process. Landforms. 40, 1387–1400.

Ribeiro, R.D.A., Rovai, A.S., Twilley, R.R., Castañeda-Moya, E., 2019. Spatial variability of mangrove primary productivity in the neotropics. Ecosphere 10 (8), e02841. https://doi.org/10.1002/ecs2.2841.

Rivera-Monroy, V.H., Danielson, T.M., Castañeda-Moya, E., Marx, B.D., Travieso, R., Zhao, X., Gaiser, E.E., Farfan, L.M., 2019. Long-term demography and stem productivity of Everglades mangrove forests (Florida, USA): resistance to hurricane disturbance. For. Ecol. Manag. 440, 79–91.

Robertson, A.I., 1986. Leaf-burying crabs: their influence on energy flow and export from mixed mangrove forests (Rhizophora spp.) in northeastern Australia. J. Exp. Mar. Bio. Ecol. 102, 237–248.

Robertson, A.I., Daniel, P.A., 1989. The influence of crabs on litter processing in high intertidal mangrove forests in tropical Australia. Oecologia 78, 191–198.

Rogers, K., Kelleway, J.J., Saintilan, N., Megonigal, J.P., Adams, J.B., Holmquist, J.R., Lu, M., Schile-Beers, L., Zawadzki, A., Mazumder, D., Woodroffe, C.D., 2019. Wetland carbon storage controlled by millennial-scale variation in relative sea-level rise. Nature 567, 91–95.

Rogers, K., Wilton, K.M., Saintilan, N., 2006. Vegetation change and surface elevation dynamics in estuarine wetlands of southeast Australia. Estuar. Coast Shelf Sci. 66, 559–569.

Rougerie, F., Jehl, C., Trichet, J., 1997. Phosphorus pathways in atolls: interstitial nutrient pool, cyanobacterial accumulation and Carbonate-Fluoro-Apatite (CFA) precipitation. Mar. Geol. 139, 201–217.

Rovai, A.S., Riul, P., Twilley, R.R., Castañeda-Moya, E., Rivera-Monroy, V.H., Williams, A.A., Simard, M., Cifuentes-Jara, M., Lewis, R.R., Crooks, S., Horta, P.A., Schaeffer-Novelli, Y., Cintrón, G., Pozo-Cajas, M., Pagliosa, P.R., 2016. Scaling mangrove aboveground biomass from site-level to continental-scale. Global Ecol. Biogeogr. 25, 286–298.

Rovai, A.S., Twilley, R.R., Castañeda-Moya, E., Riul, P., Cifuentes-Jara, M., Manrow-Villalobos, M., Horta, P.A., Simonassi, J.C., Fonseca, A.L., Pagliosa, P.R., 2018. Global controls on carbon storage in mangrove soils. Nat. Clim. Change 8, 534–538.

Saenger, P., Snedaker, S.C., 1993. Pantropical trends in mangrove above-ground biomass and annual litterfall. Oecologia 96, 293–299.

Saintilan, N., Khan, N.S., Ashe, E., Kelleway, J.J., Rogers, K., Woodroffe, C.D., Horton, B.P., 2020. Thresholds of mangrove survival under rapid sea level rise. Science 386, 1118–1121. https://doi.org/10.1126/science.aba2656.

Saintilan, N., Rogers, K., Mazumber, D., Woodroffe, C., 2013. Allochthonous and autochthonous contributions to carbon accumulation and carbon store in southeastern Australian coastal wetlands. Estuar. Coast Shelf Sci. 128, 84–92.

Sanderman, J., Hengl, T., Fiske, G., Solvik, K., Adame, M.F., Benson, L., Bukoski, J.J., Carnell, P., Cifuentes-Jara, M., Donato, D., Duncan, C., Eid, E.M., zu Ermgassen, P., Ewers, C., Glass, L., Gress, S., Jardine, S.L., Jones, T., Macreadie, P., Nsombo, E.N., Rahman, M.M., Sanders, C., Spalding, M., Landis, E., 2018. A global map of mangrove forest soil carbon at 30 m spatial resolution. Environ. Res. Lett. 13, 055002.

Sasmito, S.D., Kuzyakov, Y., Lubis, A.A., Daniel, M., Hutley, L.B., Bachri, S., Friess, D.A., Martius, C., Borchard, N., 2020. Organic carbon burial and sources in soils of coastal mudflat and mangrove ecosystems. CATENA 187, 104414. https://doi.org/10.1016/j.catena.2019.104414.

Scales, I.R., Friess, D.A., 2019. Patterns of mangrove forest disturbance and biomass removal due to small-scale harvesting in southwestern Madagascar. Wetlands Ecol. Manage. 27, 609–625.

Schaeffer-Novelli, Y., Cintrón-Molero, G., Adaime, R.R., Camargo, T.M., 1990a. Variability of mangrove ecosystems along the Brazilian coast. Estuaries 13, 204–218.

Schaeffer-Novelli, Y., Mesquita, H. de S.L., Cintrón-Molero, G., 1990b. The Cananéia lagoon estuarine system, São Paulo, Brazil. Estuaries 13, 193–203.

Schlichting, C.D., 1986. The evolution of phenotypic plasticity in plants. Annu. Rev. Ecol. Systemat. 17, 667–693.

Simard, M., Fatoyinbo, T., Smetanka, C., Rivera-Monroy, V., Castaneda-Moya, E., Thomas, N., Van der Stocken, T., 2019. Mangrove canopy height globally related to precipitation, temperature and cyclone frequency. Nat. Geosci. 12, 40–45.

Smith, T.J., Boto, K.G., Frusher, S.D., Giddins, R.L., 1991. Keystone species and mangrove forest dynamics: the influence of burrowing by crabs on soil nutrient status and forest productivity. Estuar. Coast Shelf Sci. 33, 419–432.

Song, J., Wan, S., Piao, S., Knapp, A.K., Classen, A.T., Vicca, S., Ciais, P., Hovenden, M.J., Leuzinger, S., Beier, C., Kardol, P., Xia, J., Liu, Q., Ru, J., Zhou, Z., Luo, Y., Guo, D., Adam Langley, J., Zscheischler, J., Dukes, J.S., Tang, J., Chen, J., Hofmockel, K.S., Kueppers, L.M., Rustad, L., Liu, L., Smith, M.D., Templer, P.H., Quinn

Thomas, R., Norby, R.J., Phillips, R.P., Niu, S., Fatichi, S., Wang, Y., Shao, P., Han, H., Wang, D., Lei, L., Wang, J., Li, X., Zhang, Q., Li, X., Su, F., Liu, B., Yang, F., Ma, G., Li, G., Liu, Y., Liu, Y., Yang, Z., Zhang, K., Miao, Y., Hu, M., Yan, C., Zhang, A., Zhong, M., Hui, Y., Li, Y., Zheng, M., 2019. A meta-analysis of 1,119 manipulative experiments on terrestrial carbon-cycling responses to global change. Nat. Ecol. Evol. 3, 1309–1320.

Spalding, M., Kainuma, M., Collins, L., 2010. World Atlas of Mangroves. Earthscan, New York.

Swales, A., Reeve, G., Cahoon, D.R., Lovelock, C.E., 2019. Landscape evolution of a fluvial sediment-rich *Avicennia marina* mangrove forest: insights from seasonal and inter-annual surface-elevation dynamics. Ecosystems 22, 1232–1255.

Syvitski, J.P.M., Vörösmarty, C.J., Kettner, A.J., Green, P., 2005. Impact of humans on the flux of terrestrial sediment to the global coastal ocean. Science 308, 376–380 (80-.).

Thom, B.G., 1982. Mangrove ecology - a geomorphological perspective. In: Clough, B.F. (Ed.), Mangrove Ecosystems in Australia: Structure, Function and Management. Australian National University Press, Canberra, Australia, pp. 3–17.

Title, P.O., Bemmels, J.B., 2018. ENVIREM: an expanded set of bioclimatic and topographic variables increases flexibility and improves performance of ecological niche modeling. Ecography 41, 291–307.

Tomlinson, P.B., 2016. The Botany of Mangroves, second ed. Cambridge University Press.

Twilley, R., Chen, R., Hargis, T., 1992. Carbon sinks in mangroves and their implications to carbon budget of tropical coastal ecosystems. Water Air Soil Pollut. 64, 265–288.

Twilley, R., Lugo, A.E., Patterson-Zucca, C., 1986. Litter production and turnover in basin mangrove forests in southwest Florida. Ecology 67, 670–683.

Twilley, R.R., Castañeda-Moya, E., Rivera-Monroy, V.H., Rovai, A.S., 2017. Productivity and carbon dynamics in mangrove wetlands. In: Rivera-Monroy, V.H., Lee, S.Y., Kristensen, E., Twilley, R.R. (Eds.), Mangrove Ecosystems: A Global Biogeographic Perspective. Springer International Publishing, p. 446.

Twilley, R.R., 1995. Properties of mangrove ecosystems related to the energy signature of coastal environments. In: Hall, C.A.S. (Ed.), Maximum Power: The Ideas and Applications of H. T. Odum. University Press of Colorado, Denver, CO, pp. 43–62.

Twilley, R.R., Rovai, A.S., Riul, P., 2018. Coastal morphology explains global blue carbon distributions. Front. Ecol. Environ. 16, 1–6.

Vitousek, P.M., Farrington, H., 1997. Nutrient limitation and soil development: experimental test of a biogeochemical theory. Biogeochemistry 37, 63–75.

Whelan, K.R.T., Smith, T.J., Anderson, G.H., Ouellette, M.L., 2009. Hurricane Wilma's impact on overall soil elevation and zones within the soil profile in a mangrove forest. Wetlands 29, 16–23.

Woodroffe, C., 1992. Mangrove sediments and geomorphology. In: Robertson, A.I., Alongi, D.M. (Eds.), Coastal and Estuarine Studies-Tropical Mangrove Ecosystems. American Geophysical Union, Washington, D. C., pp. 7–41

Woodroffe, C.D., 1988. Relict mangrove stand on last interglacial terrace, Christmas Island, Indian Ocean. J. Trop. Ecol. 4, 1–17.

Woodroffe, C.D., 1987. Pacific island mangroves: distribution and environmental settings. Pac. Sci 41, 166–185.

Woodroffe, C.D., 2002. Coasts: Form, Process and Evolution, p. 623.

Woodroffe, C.D., Rogers, K., McKee, K.L., Lovelock, C.E., Mendelssohn, I.A., Saintilan, N., 2016. Mangrove sedimentation and response to relative sea-level rise. Ann. Rev. Mar. Sci. 8, 243–266.

Worthington, T.A., Andradi-Brown, D.A., Bhargava, R., Buelow, C., Bunting, P., Duncan, C., Fatoyinbo, L., Friess, D.A., Goldberg, L., Hilarides, L., Lagomasino, D., Landis, E., Longley-Wood, K., Lovelock, C.E., Murray, N.J., Narayan, S., Rosenqvist, A., Sievers, M., Simard, M., Thomas, N., van Eijk, P., Zganjar, C., Spalding, M., 2020. Harnessing big data to support the conservation and rehabilitation of mangrove forests globally. One Earth 2, 429–443. https://doi.org/10.1016/j.oneear.2020.04.018.

Worthington, T.A., zu Ermgassen, P.S.E., Friess, D.A., Krauss, K.W., Lovelock, C.E., Thorley, J., Tingey, R., Woodroffe, C.D., Bunting, P., Cormier, N., Lagomasino, D., Lucas, R., Murray, N.J., Sutherland, W.J., Spalding, M., 2020. A global biophysical typology of mangroves and its relevance for ecosystem structure and deforestation. Sci. Rep. 10, 14652 https://doi.org/10.1038/s41598-020-71194-5.

Würtz, P., Annila, A., 2008. Roots of diversity relations. J. Biophys. 2008, 654672.

Xiong, Y., Cakir, R., Phan, S.M., Ola, A., Krauss, K.W., Lovelock, C.E., 2019. Global patterns of tree stem growth and stand aboveground wood production in mangrove forests. For. Ecol. Manag. 444, 382–392.

Yan, Y.-B., Duke, N.C., Sun, M., 2016. Comparative analysis of the pattern of population genetic diversity in three Indo-West Pacific Rhizophora mangrove species. Front. Plant Sci. 7, 1434.

CHAPTER

14

State of biogeochemical blue carbon in South Asian mangroves

Raghab Ray[1], *Sahadev Sharma*[2], *Tapan Kumar Jana*[3]

[1]Atmosphere and Ocean Research Institute, The University of Tokyo, Kashiwa, Chiba, Japan; [2]Institute of Ocean and Earth Sciences, University of Malaya, Kuala Lumpur, Selangor, Malaysia; [3]Department of Marine Science, The University of Calcutta, Kolkata, West Bengal, India

OUTLINE

1. **Introduction** 336
   - 1.1 *Concept of blue C: global role of mangroves* 336
   - 1.2 *Carbon burial in mangrove sediment* 337
   - 1.3 *Carbon export from sediment to water* 338
   - 1.4 *Carbon emission from sediment to atmosphere* 339
2. **Overview of past and current research in South Asia** 341
   - 2.1 *South Asian distribution of mangrove blue C* 341
   - 2.2 *State of C stock and sources in South Asian countries* 343
     - 2.2.1 Total blue C stock in mangroves 343
     - 2.2.2 Sources of OC in sediment 346
     - 2.2.3 Litterfall 347
3. **Mangrove blue C in the Indian Sundarbans** 348
   - 3.1 *A model mangrove site in South Asia* 348
   - 3.2 *Organic and inorganic C storage* 350
     - 3.2.1 TOC in mangrove plant 350
     - 3.2.2 TOC in mangrove sediment 352
     - 3.2.3 Inorganic C as carbonate in sediment 353
   - 3.3 *Biogeochemical C exchange flux: source or sink?* 353
     - 3.3.1 Forest−atmosphere coupled system 353
     - 3.3.2 Forest−sediment coupled system 354
     - 3.3.3 Sediment−atmosphere coupled system 354
     - 3.3.4 Sediment−water coupled system 354

Dynamic Sedimentary Environments of Mangrove Coasts
https://doi.org/10.1016/B978-0-12-816437-2.00021-5

3.4 *Budget and fate of blue C* 355
3.5 *Sensitivity of the blue C to climate change* 356
4. Research gaps and future directions 357
4.1 *Improving laboratory-based data by applying new methods* 357
4.2 *Research on fine root production* 358
4.3 *Microenvironmental heterogeneity* 358
4.4 *Other gaps and challenges* 359
References 360

1. Introduction

1.1 Concept of blue C: global role of mangroves

Shallow coastal habitats are among the most productive areas on earth that support surrounding coastal communities through providing several provisioning, regulatory, cultural, and supporting ecosystem services (Barbier et al., 2008; Kuwae and Hori, 2018). Coastal habitats like mangroves, saltmarshes, seagrasses, and macroalgae provide numerous benefits such as coastal protection from storm surges, sea-level rise and coastal erosions, regulating water quality and nutrient recycling, supporting fisheries, timber products, and cultural service (Othman, 1994; Primavera, 1998; Rivera-Monroy et al., 2011; Richards and Friess, 2015; Dasgupta et al., 2019). These ecosystems play very important role protecting coastal communities from the impact of cyclone (Mandal and Hosaka, 2020), hurricane (Zhang et al., 2012), storm surges, and tsunamis (Kathiresan and Rajendran, 2005). Recent discussions have further highlighted another important role, i.e., sequestration of atmospheric and oceanic carbon dioxide (CO_2) and storage as organic carbon (OC) within their biomass and sediment for centuries to millennia (that stored carbon is termed as coastal "blue carbon") (Duarte et al., 2005; Alongi, 2014). The term "blue C" was first introduced a decade ago to describe the large quantity of OC present in shallow coastal habitats (Nellemann et al., 2009). Blue C meets all three essential elements of sustainability, economic, environmental, and social, and should be considered as an intrinsic part of blue economy. However, if the ecosystems are degraded or damaged, their C sink capacity will be lost or adversely affected, eventually C stored will be released as CO_2, thus contribute to climate change. For instance, conversion of mangroves to different land-use types such as aquaculture ponds, paddy field, tree removal, pasture, etc., has resulted in blue C stocks 2 to 8 times lower than intact mangrove forests (Boone Kauffman et al., 2017; Sasmito et al., 2019; Sharma, 2020). Due to land cover change between 1996 and 2016, a global reduction of 1.8% of mangrove C stock has been reported (Richards et al., 2020). Therefore, quantifying blue C as an ecosystem service is an important task that provides added value for wetland restoration and serves as an useful management tool when implementing plans for sustained ecosystem health and productivity (Sheehan et al., 2019).

Among the shallow coastal habitats, mangroves have the largest stock of blue C (Fig. 14.1). The mean global C stock of mangrove ecosystems is estimated to be 956 Mg C/ha, which is much higher than those of rainforests, peat swamps, saltmarshes, and seagrass meadows (Alongi, 2014). Except for tropical peatlands (storage of 88.6 Pg C, Page et al., 2011), mangroves are the most productive ecosystems (storage of 15 Pg C in the live biomass and

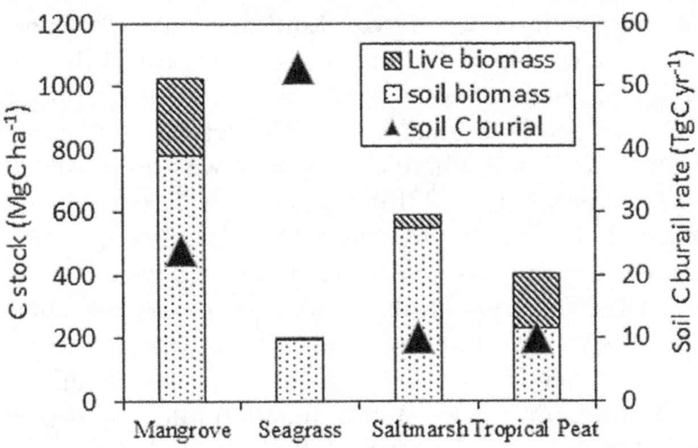

FIGURE 14.1 Based on global estimates, C stock and burial rate in sediment within the shallow coastal habitats and tropical peats swamps are provided. *Data source: mangrove: Donato et al., 2011; Alongi, 2014. seagrass stock: Fourqurean et al., 2012. saltmarsh: Ouyang and Lee, 2014; Duarte et al., 2005. Peat swamp stock: Page et al., 2011.*

soil) and sequester at a faster rate (174 g C/m^2 yr, Donato et al., 2011) than any other coastal habitats, e.g., seagrasses (138 g C/m^2 yr, Fourqurean et al., 2012) highlighting their importance as most proficient C fixers on earth (Donato et al., 2011). For example, world's largest, Sundarbans mangroves at the land ocean boundary of the Gangetic delta and the Bay of Bengal act as blue C sink (Ganguly et al., 2008; Ray et al., 2011). However, according to a recent assessment, such blue C capacity by mangroves as a climate change mitigation option is very much significant for the national scale, but it is still of limited importance globally (Taillardat et al., 2018).

Mangrove sediment is the largest depository of OC (Fig. 14.1). Carbon accumulated in the above-ground and below-ground biomass (as AGB and BGB, respectively) is much lower than in the sediment of the tropical mangroves (0.5–3 m in depth representing ~49%–98% of C storage, Donato et al., 2011) owing to their efficiency in trapping suspended sediments and associated organic matter (OM), high autochthonous and allochthonous inputs, and low decomposition rates of OM under anoxic conditions in the sediment. Resultantly their advantage over terrestrial forests is much higher in C accumulation in sediments/soils (McLeod et al., 2011). In the subsequent paragraphs, magnitude of C fluxes occurring within the sediment and fate of blue C into the coastal water and atmosphere has been discussed.

1.2 Carbon burial in mangrove sediment

In mangrove C cycle, the fraction of mangrove-derived OM, which escapes degradation or export, is buried under bottom few meters depth for long-term basis (Twilley et al., 1992; Lallier-Vergès et al., 1998; Dittmar et al., 2001). The fraction of buried C strongly depends on many environmental conditions, i.e., OM input, forest age, deforestation and degradation rate, sediment accretion and topography, and tidal regime (Alongi, 2014). Water-logged anoxic condition diagenetically favors long-term C storage in mangrove sediment. This C sink capacity by mangrove sediment will increase if input of OC via litterfall or benthic algal sources exceeds the loss via export as DOC and POC or benthic emission as CO_2. With mangrove deforestation (globally ~1% per year, Kauffman et al., 2017), reduction in C storage capacity of sediments, and, consequently, increase of C export to the adjacent coastal water and atmosphere would be intense (Lovelock et al., 2011; Breithaupt et al., 2012; Atwood et al., 2017); however, sometimes C gain was also observed after mangrove deforestation, e.g., in a eutrophic waters in Brazil where algal source was responsible for C accumulation even after mangrove loss (Pérez et al., 2018).

Several studies relied on direct measurement of C burial rates by combining sedimentary C content and soil accumulation rates estimated from radioisotopes, ^{210}Pb and ^{137}Cs, or natural markers like volcanic ashes (Sanders et al., 2010). Burial rates are calculated based on OC presently available for measurement and not the amount originally deposited, meaning OC burial rates are partially dependent on the time scale of interest, and consequently on the dating methods used to measure sediment accumulation rates (Breithaupt et al., 2012; Perez et al., 2018; Sasmito et al., 2020). Contrast to this classical approach, (Marchand, 2017) applied another strategy by determining the C stock of a forest of known age, such as C burial = [(C stock of the stand)−(C stock of the previous stand)]/[(age of the stand)−(age of the previous stand)].

Although mangroves occupy only 0.5% of the global coverage, yet they sequester ~25% of the total C burial rate by the vegetated habitats like saltmarshes, seagrasses, and tropical peat (Duarte et al., 2005). Only very few studies in past three decades have come up with direct estimation of OC burial rate ranging from 100 g C/m^2 yr (Twilley et al., 1992) to 226 g C/m^2 yr (McLeod et al., 2011) depending on study sites. The most up-to-date mean value of local C burial rate is given by Breithaupt et al. (2012) as 163 g C/m^2 yr globally equating to 26 Tg C/yr, which is very close to earlier estimates (24 Tg C/yr, Twilley et al., 1992; 23 Tg C/yr, (Jennerjahn and Ittekott, 2002); 29 Tg C/yr, Alongi, 2009; 27.8 Tg C/yr, Duarte et al., 2005; 18.4 Tg C/yr, Bouillon et al. (2008) (Fig. 14.1).

Unlike OC burial, sequestration of inorganic form of carbon (IC) in mangrove sediment has been less appreciated until recently when Saderne et al. (2019) came up with first ever global estimation IC burial rate as 0.8 Tg C/yr that is quite lower than the burial rate by the seagrass ecosystems (15%−62%). According to author, IC burial rate (as 12% of $CaCO_3$ mass) could represent a source of CO_2 to the atmosphere (as calcification produces CO_2 with a ratio of ~0.6 mol of CO_2 per mol of $CaCO_3$ precipitated, Ware et al., 1992) which may partially offset the role of blue C ecosystems.

1.3 Carbon export from sediment to water

Despite such a potential storage capacity, a significant fraction of mangrove-derived C is lost through tidal export of dissolved and particulate organic C (DOC and POC, respectively) to the estuaries and ocean (Bouillon et al., 2008). Mangrove-derived DOC contributing ~10% could account for terrestrial DOC exported to the ocean globally, despite their small area relative to other habitats (Dittmar et al., 2006). Similarly, mangrove-derived POC could account for 11% of the total input of terrestrial C into the ocean and 15% of the total C accumulating in modern marine sediment (Jennarjahn and Ittekkot, 2002). However, the magnitude of changes in mangrove-derived C fluxes is very uncertain, mainly because of the variability in tidal (spring versus neap) and seasonal (wet versus dry) patterns (Maher et al., 2013; Taillardat et al., 2018), differences in sampling sites and geomorphic settings (Call et al., 2019;

Twilley et al., 2018; Ray et al., 2018), and mangrove stand structure and productivity (Ray et al., 2018).

Our understanding of OC export from the mangroves has developed from a relatively small number of quantitative estimates (<15) for DOC, POC, and total OC export from various geographical settings (Twilley, 1985; Dittmar et al., 2001; Rezende et al., 2007; Maher et al., 2013; Li et al., 2018; Ray et al., 2018; Santos et al., 2018; Taillardat et al., 2018; Call et al., 2019; Ray et al., 2020). Alongi (2014) assumed that world mangroves transport 86 Tg C/yr as dissolved inorganic carbon (DIC) to the adjacent coastal waters, which seemed to be considerably greater than the estimated total export of POC (28 $\pm$ 21 Tg C/yr), DOC (15 $\pm$ 13 Tg C/yr) (Ray et al., 2018), endorsing the relative importance of lateral exchange of DIC and total alkalinity as a long-term atmospheric C sink (Maher, 2018). The area weighted global export of mangrove-derived DOC and POC widely vary from 14 to 1460 g C/m^2 yr and 4–2116 g C/m^2 yr depending on mangrove settings, biogeochemical processes, and other environmental factors (Twilley, 1985; Ray et al., 2018; Li et al., 2018). So far, DOC export from the French Guiana mangroves has been reported maximum (678 mmol/m^{2}d, Ray et al., 2020) follwed by the Indian Sundarbans (161 mmol/m^2 d, Ray et al., 2018), Amazon, Floridan, and Australian mangroves (all together averaging ~135 mmol m^2 d, Fig. 14.2). Mangroves like French Guiana or Indian Sundarbans have very high ranges than the rest of available DOC flux estimates owing to high water discharge from big rivers like Amazon or Ganges-Hooghly, as well as very high DOC concentrations measured within the creeks (300–800 μM) (Ray et al., 2018, 2020). Maximum POC and DIC export are recorded for the mangrove in Fly delta, Papua New Guinea (20 mmol/m^2 d, Robertson and Alongi, 1995), and Moreton Bay, East Australia (262 mmol/m^2 d, Maher et al., 2013), respectively.

Recently mangroves are recognized to export a substantial amount of exchangeable DOC (EDOC, volatile and semivolatile OC) which corresponds to 60% of the global EDOC flux from the ocean to the atmosphere (Sippo et al., 2016). Unlike many other studies, a recent evidence by Ray et al. (2020) even showed marine mud-derived DIC import to the Sinnamary mangroves of French Guiana. However, such estimates of mangrove-derived organic and inorganic C are still limited to a small number of studies and more seasonal data collection and multiple time series studies are required to improve the resolution of C fluxes (Sippo et al., 2016).

1.4 Carbon emission from sediment to atmosphere

Fate of blue C in mangrove sediment will largely depend upon OM decomposition favored by heterotrophic microorganism. Enhanced heterotrophic respiration in combination with autotrophic respiration (live BGB) leads to high CO_2 efflux from the sediment to the atmosphere or export to adjacent tidal creeks through porewater seepage (Bouillon et al., 2008; Leopold et al., 2015; Thi-Hien et al., 2018). CO_2 fluxes at the sediment–air interface can be highly variable in mangrove sediments, depending on many factors such as forest stand age and live or dead stocks, leaf area index, presence of biogenic structures and algal biofilm, anthropogenic impact, seasonal and tidal variability, and sediment properties (redox potential, temperature, moisture, OC loading, grain size) (Lovelock, 2008; Chen et al., 2012; Leopold et al., 2013; Bulmer et al., 2015).

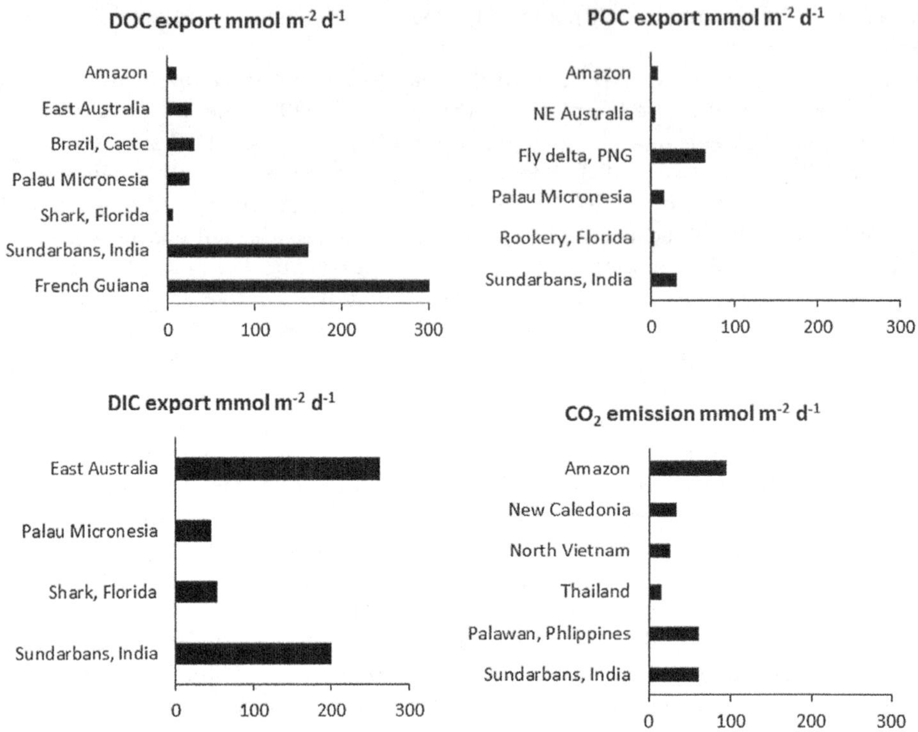

FIGURE 14.2 Examples of carbon export as DOC, POC, and DIC and emission as CO_2 from selected mangroves around the world. From this graph, their estimated global mean ± SD (mmol/m^2 d) are as follows: DOC = 43 ± 58, POC = 21 ± 23, DIC = 140 ± 108, CO_2 = 47 ± 29. *Data for Amazon and Brazil: Dittmar et al., 2001; Dittmar et al., 2006; Metcalfe et al., 2007. French Guiana: Ray et al., 2020. Florida: Romigh et al., 2006; Ho et al., 2017. Micronesia: Call et al., 2019. Sundarbans: Ray et al., 2011, 2018; Chanda et al., 2014. East Australia: Maher et al., 2013. New Caledonia: Leopold et al., 2015. Vietnam: Thi-Hien et al., 2018. Philippines: Castillo et al., 2017.*

Soil CO_2 efflux is generally higher in the studies using chamber-based techniques, where CO_2 is measured in situ over a short period (incubation time ~3–4 min), unlike in the laboratory by incubating sediment cores retrieved from the field (Kristensen et al., 2008). Therefore, proper method applied to determine benthic CO_2 emission as well as capturing larger temporal and spatial variability must be considered while updating or upscaling data inventory. According to Alongi (2014), global rates of CO_2 efflux (mean ± SE, n = 140) are significantly greater from exposed sediment surface (69 ± 8 mmol/m^2 d) than from inundated conditions (49 ± 6 mmol/m^2 d) reflecting gases to have a faster molecular diffusion than in liquid phase. Rates of CO_2 efflux under dark condition are typically higher (61 ± 46 mmol/m^2 d, n = 82) than in light condition when photosynthesis stimulates CO_2 uptake by the microphytobenthos (MPB) (−15 ± 54 mmol/m^2 d, n = 14) (Bouillon et al., 2008). The biofilm constituted by MPB at the surface sediment has been demonstrated to significantly reduce CO_2 efflux (Chen et al., 2019; Leopold et al., 2015), and is a strong driver for efflux variability (Yang et al., 2018). However, rate of CO_2 loss by benthic emission is generally two- to fourfold lower compared to mangrove C export (DOC, POC, and DIC) (Fig. 14.2), but there are exceptions as well, e.g., in the Amazonian system, emission is ~5 times higher than the combined export of DOC and POC, but for the Indian Sundarbans C export is much stronger than benthic emission (Table 14.2).

On spatial-scale basis, significantly lower CO_2 efflux is observed from the non-vegetated mangrove site than the vegetated mangrove sites ($P < 0.05$, Thi Hien et al., 2018) because mangrove development induces an enrichment in OM, thereby activating heterotrophic microbial processes. However, sediment CO_2 fluxes are defined as the sum of autotrophic and heterotrophic respirations, and the latter, according to Lovelock et al. (2015), can account for majority of the sediment respiration in fringing forests. Troxler et al. (2015) measured highest CO_2 efflux from sediment with the presence of pneumatophores in the Everglades. On seasonal basis, higher CO_2 efflux is generally observed during the no-rainfall periods, for example, in Indian climate, lower emission rate from the mangrove sediment was consistently examined during the monsoon period when rainfall increased sediment water saturation leading to reduction in benthic emission (Gnanamoorthy et al., 2019; Chanda et al., 2014).

2. Overview of past and current research in South Asia

2.1 South Asian distribution of mangrove blue C

Mangroves occur in mostly warm regions, the tropics and subtropics spanning between about 35 degrees north latitude and about 40 degrees south latitude with the most northerly distribution in Bermuda (32°20′N) and the most southerly distribution in Corner Inlet, Australia (38°45′S). Several global distribution maps of mangroves have been drawn and can be downloaded from Internet (Spalding et al., 2010; Giri et al., 2011). Globally, mangroves account for <1% of the total area of tropical forests and for <0.4% of the total forested area. Global mangroves coverage of 18.1 million ha (Spalding et al., 2010) has been revised downward to 13.78 million ha (Giri et al., 2014), and then to ~8.35 million ha (Hamilton and Casey, 2016). Such decrease in coverage does not evidence a substantial loss of mangrove or any error by either set of authors as noted by Hamilton and Casey (2016), but this is more due to a differing definition of mangroves used in such analyses. Anyhow, it is more or less common in all mapping assessments that most of mangrove forests are located in a broad region stretching from northeast India through Southeast Asia (~46%) into northern Australia (4.3%), South and Central America (16.4%), West and Central Africa (20%), and sparse distribution in the Middle East (0.83%), and South Asia (6.5%) (Fig. 14.3). Indonesia has the largest coverage of mangroves (20% of global area). According to mangrove coverage, biomass, that is the mass of living biological organisms in a given ecosystem at a given time, also changes proportionately (Hutchinson, 2013). From Fig. 14.3, data chart shows that Africa and South Asia hold maximum AGB at continental scale (231 and 314, respectively, in Mg/ha). Whereas South Asia which is our subject of discussion here, despite accommodating world's largest mangrove tract, i.e., the Sundarbans (in India and Bangladesh), are recognized to have poor data biomass inventory and blue C stock.

Mangrove forests in South Asia occur on the tidal sea edge of Bangladesh (411,487 ha), India (343,065 ha), Pakistan (76,616 ha), and Sri Lanka (21,433 ha) and represent approximately

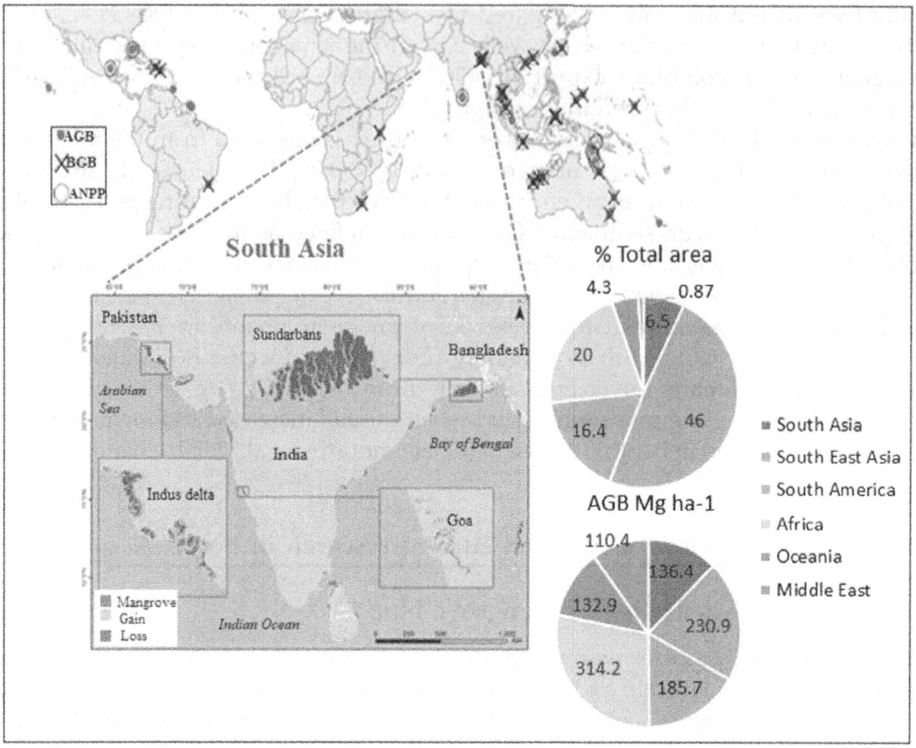

FIGURE 14.3 Global locations of mangroves where C stock data are available. Chart showing South East Asia to score maximum mangrove coverage % and AGB (Spalding et al., 2010). *Based on global coverage from Hutchinson (2013), and South Asian coverage from Giri, C., Long, J., Abbas, S., Murali, R.M., Qamer, F.M., Pengra, B., Thau, D., 2014. Distribution and dynamics of mangrove forests of South Asia. J. Environ. Manag. 148, 101–111.*

6.5% of the global total (Giri et al., 2014). Mangrove species diversity in the region ranges from 8 to 10 species in Pakistan, 22 to 66 in Bangladesh (Hoque and Datta, 2005), 18 to 20 Sri in Lanka, and 30 to 35 in India (Polidoro et al., 2010). The South Asian mangroves are under threat from both natural and anthropogenic stressors with deforestation rate of 7677 ha/yr. However, as a result of raising awareness among local people and public sector helping more mangrove forest protection initiatives and restorations, total mangrove area has increased between 2000 and 2012 with reforestation rate of 6705 ha/yr, resulting into a net loss of 972 ha/yr (Giri et al., 2014).

In India, the majority of mangroves (60%) is found on the east coast compared to only 14% along the west coast due to the huge sediment transported by numerous big rivers (Ganges, Mahanadi, Godavari) forming suitable terrain and providing nutrients in the east coast. The remainder (26%) are found in the Andaman and Nicobar Islands. The current assessment shows that mangrove cover in the country has seen a net increase of 54 km^2 in the mangrove cover of the country as compared to 2017 assessment (Chapter 3, India State of Forest Report, 2019). According to the same report, very dense, moderately dense, and open mangroves constitute 1476 km^2 (29.6%), 1479 km^2 (29.7%), and 2020 km^2 (40.6%), respectively. West

Bengal has the largest area of mangroves (the Sundarbans) (42.5% of total mangrove cover) followed by Gujarat (23.6%) and Andaman and Nicobar (12.4%) (Report by ISFR, 2019).

In Pakistan, approximately 95% of mangroves are distributed in the Indus Delta, in a few patches westward along the Balochistan coast along the Arabian Sea with other major areas as backwater Sandspit, Sonmiani, Kalmat Khor, and Jiwani. The Indus River Delta–Arabian Sea mangroves are a large mangrove ecoregion on the Arabian Sea coast of Sindh Province, Pakistan. The flora of the delta is not diverse as few plants can survive the severe conditions of heat and salinity. 95% of the trees were gray mangroves (*Avicennia marina*), with the remaining 5%, growing on small areas of higher ground, being red mangrove (*Rhizophora mucronata*), Indian mangrove (*Ceriops tagal*), and black mangrove (*Aegiceras corniculatum*). During early 1980s, mangroves were found along the 240 km long coastline in Pakistan with estimated area to be 600,000 acres that was approximately 40% of the entire coast and 10% of the Indus Delta (Wells and Coleman, 1985). During that period they were considered fifth or sixth largest mangrove forests in the world and largest in arid climate (Snedaker, 1984). However, due to extreme tampering of the environment, both upstream and in the delta mangroves and overexploitation by population, these mangroves disappeared quite fast in Indus delta (Saifullah, 1997).

In Bangladesh, mangroves are located in the Sundarbans with areal coverage of about 6300 km^2. Past study revealed that from 1973 to 2010, vegetated land has decreased by 4.01%, indicating the decreasing tree density and canopy closure in Sundarbans (Mondal and Debnath, 2017). Shrimp farming has led to destruction of scattered mangroves in Bangladesh while it is also the most important sector for their economy (Ahmed et al., 2017). Mangroves in Sri Lanka are restricted to estuaries along the coasts with largest mangrove patches found in the Puttalam Lagoon–Dutch Bay Portugal Bay complex, Batticaloa lagoon, and Trincomalee (Giri et al., 2011).

2.2 State of C stock and sources in South Asian countries

2.2.1 Total blue C stock in mangroves

Published data compiled from only few regional sources of South Asian countries suggest that the average total blue C stock (above-ground biomass and sediment including below-ground biomass) in the mangrove forest ranges from 104 to 506.3 Mg C/ha, around 35%–68% of that was stored in the sediment (Table 14.1), in line with global percentages for the tropical mangroves (49%–98%, Donato et al., 2011). However, overall total blue C stock (Mg C/ha) in South Asian mangroves is very poor compared to the SE Asian mangroves (countries like Indonesia: 488, Murdiyarso et al., 2009; Indonesia: 1083 Murdiyarso et al., 2015; Vietnam: 700–820, Vinh et al., 2019; Micronesia: 897.8, Kauffman et al., 2011; Singapore: 497, Phang et al., 2015; Philippines 548, Salmo et al., 2019; Cambodia: 600–796, Sharma et al., 2020. These differences in mangrove C stock would occur due to the amount of peat soil and OC being much higher in the Indo-Pacific mangroves (total OC, TOC ~2%–18%, Donato et al., 2011), above-ground vegetation density (Murdiyarso et al., 2009), tree age (Walcker et al., 2018), wood density (Baker et al., 2004), dominant species (Ksawani et al., 2007), sediment profile, salinity, and nutrients (Adame et al., 2015).

Among the four countries, comprehensive blue C data atleast at few state-wise basis are achieved from India only, while for Bangladesh, Sri Lanka, and Pakistan, these data are

TABLE 14.1 Total blue C stock in AGB, BGB, and sediment of various species-dominated mangrove ecosystems in South Asian countries.

Country	Dominant mangrove sp.	AGB-C Mg C/ha	BGB-C Mg C/ha	Sediment-C Mg C/ha	Total blue C Mg C/ha	Source reference
India						
Sundarbans, West Bengal	*Avicennia marina*	29–50	7–14	68–86	104–150	Ray et al. (2011), This review
Kadalundi, Kerala	*Rhizophora mucronata*	83.3 ± 11	34.9 ± 4.3	63.8 ± 8.6	182	Vinod et al. (2018)
Bhitarkanika, Odisha	*Avicennia alba*	45–70	15–25	61–134	121–229	Bhomia et al. (2016)
Mahanadi, Odisha	*Avicennia marina*	62.5 ± 7.5	27.2 ± 3.7	57.6 ± 11	147.3	Sahu et al. (2016)
Pichavaram, Tamil Nadu	*Rhizophora apiculata*	–	–	95.6 ± 28.4	–	Gnanamoorthy et al. (2019)
Bangladesh						
Sundarbans	*Sonneratia apetala*	45–152	11–62	90–134	146–348	Rahman et al. (2015)
Sri Lanka						
Batticaloa Lagoon	*Excoecaria agallocha*	131.6 ± 1.6	26.9 ± 0.3	347.8 ± 33.8	506.3	Perera and Amarasinghe (2018)
Pakistan						
Sandpit Backwater	*Avicennia marina*	93.4 ± 2.6	39 ± 4.5	51.6 ± 3.9	184	Nazim et al. (2018)

limited to either single study or only limited to very few for each country. From Table 14.1, Sri Lanka is recognized to have highest total blue C stock. This is due to very high aboveground biomass in the Batticaloa lagoon of Sri Lanka (246.4 Mg/ha) which could be ranked relatively high among other published reports on mangrove areas in the Asian region (Perera and Amarasinghe, 2018). It is reasonable to believe that due to very high TOC% in the sediment of Batticaloa lagoon (5.2%–6.1%), being much higher than the Indian, Bangladesh, and Pakistan mangroves sites (generally between 0.5% and 2.5%), total blue C stock in Sri Lankan mangroves sediment is highest among SE Asian countries. Whereas in Pakistan, mangroves of Sandspit Backwater, populated by the 12- to 20-year-old *Avicennia marina,* have estimated live biomass C stock of 132.3 Mg C/ha which is higher than the Indian mangroves (36–118.2 Mg C/ha) and play an important role in overall C sequestration by acting as sink for C (Nazim et al., 2018). Total blue C stock in the Sundarbans of Bangladesh is higher than the Indian part because of higher forest productivity and sediment C content in the former than latter (Rahman et al., 2015; Ray et al., 2011).

Although India has the second-largest mangrove forested area in South Asia after Bangladesh, blue C stock in Indian mangroves is relatively low, mainly because India has

TABLE 14.2 Compilation of sediment core C and N data (stable isotopes and elemental ratio) in South, South East, and Middle East mangrove locations. Mean ± SD (n = number of records).

Mangroves in Asia	$\delta^{13}C‰$	TOC %	TOC:TN	$\delta^{15}N‰$	Data source
South Asia					
Sundarbans, Bangladesh	−16.73 ± 1.90 (13)	1.3 ± 0.18 (25)	19.5 ± 2.4 (19)	5.8 ± 1.1 (5)	1
Sundarbans, India	−23.78 ± 2.36 (4)	1.0 ± 0.28 (4)	17.8 ± 5.4 (4)	3.9 ± 0.5 (4)	1,2,3,4
Pichavaram, India	−27.15 ± 1.77 (10)	1.2 ± 0.15 (35)	14.2 ± 1.9 (35)	3.7 ± 1.1 (10)	5,6
Andhra Pradesh, India	−20.22 ± 0.92 (32)	0.9 ± 0.70 (31)	11.5 ± 2 (18)	5.5 ± 1.0 (25)	7
North of Kochi, India	−25.15 ± 0.24 (15)	1.4 ± 0.32 (15)	15 ± 1.2 (15)	No data found	8
SE Asia					
E. Java, Indonesia	−22.42 ± 0.13 (45)	0.8 ± 0.02 (44)	14.5 ± 8 (44)	2.6 ± 0.15 (45)	9
Bali, Indonesia	−27.58 ± 0.07 (15)	2.3 ± 0.09 (15)	12.3 ± 9 (15)	3.4 ± 0.08 (15)	10
E. Kalimantan, Indonesia	−27.50 ± 0.40 (16)	5.1 ± 2.3 (16)	27.2 ± 5 (15)	No data found	11
North-Central Vietnam	−25.80 ± 0.30 (16)	1.5 ± 0.05 (20)	No data found	No data found	12
West Papua, Indonesia	−28.3 ± 0.5 (4)	2.5–16.5 (15)	22.6 ± 4	2.6 ± 0.5 (4)	13
Middle East					
Red Sea, Saudi Arabia	−25.50 ± 0.4 (144)	0.3 ± 0.02 (312)	No data found	2.8 ± 0.3 (144)	14

[1]Prasad et al. (2017), [2]Sen and Bhadury (2017), [3]Ray et al. (2015), [4] Das et al. (2012), [5] Prasad and Ramanathan (2009), [6]Ranjan (2011), [7]Bouillon (2003), [8]Resmia et al. (2016), [9]Sidik et al. (2016), [10]Sidik et al. (2019), [11]Kusumaningtyas et al. (2019), [12]Pham et al. (2017), [13]Sasmito et al. (2020), [14]Almahasheer et al. (2017).

experienced the greatest mangrove losses (~58,000 ha) between the years 2000–12 (Giri et al., 2014) due to agriculture, aquaculture, and coastal development, and diversion of freshwater away from the estuaries. Bhomia et al. (2016) showed the difference between blue C stock under aquaculture land use (~60 Mg C/ha) and scrub or dense mangroves (~240 Mg C/ha) and planted mangroves (100 Mg C/ha) as indicative of C loses that occurred when mangroves are severely altered. Lower C stock allows Indian mangroves for their rapid turn-over and faster export to the sea, as observed for the Sundarbans where although blue C stock is low, turnover time is much faster (10 years) than the Amazon forest (50 years), indicating the advantage of Indian mangroves over terrestrial forests for rapid C sequestration (Ray et al., 2011; Ray et al., 2018).

However, source of variation in blue C estimates due to different method adopted would create inconsistent trend of stock and overall C budget. For instance, Ray et al. (2011) estimated AGB-C stock by developing mixed species allometric model and by performing tree counting in 10 m × 10 m quadrates from multiple points in the Indian Sundarbans. Whereas for the Bangladesh Sundarbans, Rahman et al. (2015) adopted widely known protocol by Kauffman and Donato (2012) that measures biomass and C density from plot scale study (125 m transect covering six plots having 7 m radius each plot). Sediment core of different depths could alter blue C stock estimations. Also, measuring all component C concentration such as live above- and below-ground C, standing dead above- and below-ground C, dry coarse wood debris

(CWD), downed CWD, MPB, and soil C are necessary to quantify ecosystem C stock. There are many missing C pool data at present in the South Asian regions, such as standing dead biomass, CWD, and MPB. By far the most comprehensive methodical way to estimate total blue C stock in SE Asia was provided in a regional analysis in the east coast of India by Bhomia et al. (2016). For blue C estimate, authors considered wide range of mangrove forest structures (aquaculture, scrub, planted, naturally dense), C in classified downed wood debris (finer to coarser) and long core sediment, having highest pool in dense mangroves.

2.2.2 *Sources of OC in sediment*

To delineate the sources of OC, and subsequently infer the primary producers, three methods are currently used: stable isotopes of C, N (as $\delta^{13}C$ and $\delta^{15}N$), molecular biomarkers, and environmental DNA (eDNA) (Miyajima and Hamaguchi, 2018; Ortega et al., 2020; Bouillon and Boschker, 2006). Among these methods, stable isotopes are widely used in blue C research, and this is particularly convenient for tracing mangrove-derived OC, because mangroves show vascular plant (C3 type) signature with depleted $\delta^{13}C$ (~−28‰) in contrast to other primary producers like marine phytoplankton (~−22‰). Mangrove sediment receives multiple sources of OC, e.g., plant, marine phytoplankton, MPB, seagrass, or other C4 plants, anthropogenic, etc. The $\delta^{13}C$ can be used to evaluate the relative importance of each source using mixing models either with $\delta^{15}N$ or C:N ratios (like Iso-source mixing, Bayesian mixing) and stochastic simulation (e.g., Phillips et al., 2005; Parnell et al., 2010). Their applicability for the mangrove can be found in many recent works (Gontharet et al., 2014; Ray and Shahraki, 2016; Kusumaningtyas et al., 2019; Sasmito et al., 2020). Sometimes these results can be dubious, for instance, unlike $\delta^{13}C$, $\delta^{15}N$ or C:N ratios are relatively less conservative, meaning they are more prone to changes upon sediment diagenetic reactions (N_2 fixation, or nitrification, or denitrification etc.), that could result into biases in source apportionment by appyling such mixing models which were developed basically for food-web studies not for dynamic soils like mangroves. Another thing is, contamination in the MPB pool due to very high abundances of small detritus fragments in mangrove sediment samples could lead to significant errors in the stable isotope values. Therefore, effective isolation and purification of endmember sources are essential for such studies.

Molecular biomarkers such as lignin and lignin-derived phenols present in marine sediment are more conservative tracer than $\delta^{15}N$ or C:N ratios, and provide a better proxy for OC sources, especially the fraction derived from vascular plants. A combination of molecular biomarkers and compound-specific isotope analysis (CSIA) would have greater potential for provenance analysis of sediment OC, because the use of isotopic signatures such as $\delta^{13}C$ allows a wider range of organic molecules to be used as mangrove or seagrass-specific tracers, including common fatty acids, sterols, and hydrocarbons (Miyajima and Hamaguchi, 2018).

Recently, eDNA techniques has gained lot of attentions as an efficient indicator of source evaluation in blue C ecosystem (also known as "Fingerprinting of blue C ecosystems"). Recent assessments (Geraldi et al., 2019; Ortega et al., 2020) reveal that particular use of C and N isotopes of lipids along with eDNA has great potential to identify the sources and quantify the contributions of different primary producers to sedimentary OC in marine ecosystems. While eDNA techniques is applied successfully for the seagrass meadows and macrophytes (Reef et al., 2017; Hamaguchi et al., 2018; Ortega et al., 2020), this technique has yet to be explored for the mangroves and other blue C habitats.

Among these OC proxies, stable isotopes data of C and N and their ratios are greatly used for the mangrove sediments of South Asian regions (Table 14.2). From data compilation, it is quite evident that the $\delta^{13}C‰$ values are more negative in South East and Middle East Asian mangroves than the South Asian regions. Countries like in Indonesia, TOC% in mangrove sediment is recorded to be very high, being largely contributed by plant litter, whereas biogeochemical C cycle in Indian and Bangladesh mangroves is governed by both litter and marine POM input.

In India, contrasting sedimentary sources of OC are quite prominent among mangroves in Pichavaram (C3 type, Prasad and Ramanathan, 2009), Andhra Pradesh ("algal" type, Bouillon, 2003), and the Sundarbans ("mixture" type, Prasad et al., 2017). According to Prasad and Ramanathan (2009), OM deposited within the Pichavaram mangrove zone was nearly 60%–100% of terrigenous supply because of overwhelming contribution of isotopically depleted C3 plant debris. In the same mangrove sediment, authors employed molecular biomarkers and measured low yield of total lignin (suggesting its poor preservation) which was originated from nonwoody angiosperm tissues and inferred sedimentary OM as moderately degraded. Although from $\delta^{13}C‰$ values in Table 14.2, it may seem that distinct sources of OC are present in both Indian and Bangladesh Sundarbans mangrove, but this is very apparent, and rather three end-member mixing model reinforces that the marine POC inputs are dominating over terrestrial/mangrove inputs in both parts of the Sundarbans, especially the downward estuarine streches being directly exposed to marine algal assemblages from the Bay of Bengal (Prasad et al., 2017). Biomarker further analysis reveals that total lignin is generally preserved in Sundarbans mangrove during diagenesis (contrary to Pichavaram) and OM, mainly mediated by the oxidation of propyl side chain, was predominantly derived from the nonwoody angiosperms.

2.2.3 *Litterfall*

Litter produced in the canopy of mangrove forests represents a major source of OM and nutrients for outwelling to adjacent coastal waters (Odum and Heald, 1972; Twilley, 1985). Export of leaf litter and detritus from mangroves is linked to the hydrology and geophysical energy of mangrove forests, increasing in response to the increase in seasonal fluctuation in sea level, rainfall, and volume of tidal water inundating the forest each month (Twilley et al., 1986). Observations also suggested that most of the leaf litter (between 28% and 79%, Robertson, 1986) on forest floor is harvested by the macrobenthos like Sesarmid and Ocipodid crabs (Cannicci et al., 2008), and transported to sediment burrows, thus shielding organic materials from flushing by ebb tides (Lee, 2008) and enhancing retention of organic C within the benthic system (Andreetta et al., 2014). The export of organic C as POC from mangroves generally ranges from 10% to 50% of litterfall (Twilley et al., 1992).

Data compilation from the South Asian mangrove locations (Table 14.3) reveals higher litterfall production (Mg/ha yr) in India and Bangladesh mangroves (6.5–17, and 10.1, respectively) compared to Sri Lanka (3.7–6.2). No published data of litter production are recorded for mangrove in Pakistan. These estimates are well within range of global litter production database (2–16 Mg/ha yr, Twilley et al., 1992). Assuming a constant % value of C in litter (41%, Lee, 2016), the estimated C input to sediment via litterfall would be 2.6 to 6.9, 4.1, and 1.5–2.5 Mg C/ha yr for India, Bangladesh, and Sri Lanka, respectively. Other than mangrove settings, climatological factors, litter production might also change depending

TABLE 14.3 Rate of annual litterfall (Mg/ha) from monotonic to mixed vegetations in Sri Lanka, India, and Bangladesh.

Site	Vegetation	Litterfall Mg/ha yr	References
Sri Lanka	*Rhizophora*	6.23	Amarasinghe and Balasubramaniam (1992)
Sri Lanka	Mixed	5.52	Amarasinghe and Balasubramaniam (1992)
Sri Lanka	*Rhizophora*	4.41	Amarasinghe and Balasubramaniam (1992)
Sri Lanka	*Avicennia*	3.74	Amarasinghe and Balasubramaniam (1992)
India	Mixed	6.72	Ray et al. (2011)
India	*Rhizophora*	11.7	Wafar et al. (1997)
India	*Rhizophora*	11.8	Wafar et al. (1997)
India	*Sonneratia*	17.0	Wafar et al. (1997)
India	*Avicennia*	10.2	Wafar et al. (1997)
India	*Avicennia*	6.53	Dehairs et al. (2000)
India	*Excoecaria*	13.6	Dehairs et al. (2000)
Bangladesh	Mixed	10.1	Kamruzzaman et al. (2019)

on species compositions, as can be seen from Table 14.3. However, it is difficult to discern the reasons of species-wise changes in litter production, and sometimes it is also very complex to separate wet litters of individual species in a diverse forest, that is why mixed species data are often presented.

3. Mangrove blue C in the Indian Sundarbans

3.1 A model mangrove site in South Asia

The Sundarbans, a UNESCO world heritage site and the largest mangrove ecosystem, is situated both in India and Bangladesh at the estuarine phase of the River Ganges–Brahmaputra and the Bay of Bengal. The Sundarbans cover a mangrove forest area of 10,200 km^2 with a region of around 4200 km^2 of reserved forest located in India (Ray et al., 2015), forming an archipelago of 102 islands out of which 54 are impacted by anthropogenic activities and human inhabitation. The biosphere of the Indian Sundarbans (9600 km^2) constitutes 44% of forest covert, and 19% of the estuarine waterways comprising 891 km^2 as river water domain (close to blind rivers, creeks, canals) and 983 km^2 as tidal influenced aquatic area (Ray et al., 2018). Most of the rivers at the Sundarbans biosphere region flow north to south and are influenced by the tides from the Bay of Bengal. Recovery of the Sundarbans mangrove ecosystem from the past disturbances like salinity stress has been reported (Nandi et al., 1983). Salinity structure of the Sundarbans was changed after the commissioning of Farakka Barrage in 1975, which regulates freshwater flow through the Hooghly estuary to the Bay of Bengal. In the Sundarbans,

TABLE 14.4 Environmental settings of the Indian Sundarbans. (refers to Ray and Shahraki, 2016).

Geology	Latitude/longitude	21 degrees 32′-21 degrees 42′ North, 88 degrees 05′-89 degrees East
	Land coverage (km^2)	4200
	Elevation (m)	3–8
	Sediment texture	Silt ~ 80%, sand plus clay ~ 20%
	Sedimentation (mm/yr)	
Hydrology	Estuarine type	Well mixed
	Tidal nature	Semidiurnal, meso–macrotidal
	Tidal height (m)	3–6
	Water current (cm/s)	108–117 in high/low tide
	Salinity	14–26
	River discharge (m^3/s)	3000 ± 1000 (in monsoon)
Climate	Season	Southwest monsoon (July–September) Post-monsoon (October–January) Pre-monsoon (February–May)
	Rainfall (mm/yr)	1800–2000 (1990), 1550–2500 (2009–11)
	Evaporation (cm/yr)	100–120
	Air temperature (°C)	25-32 Pre- and monsoon, 18–22 post-monsoon
Vegetation	Dominant species	*Avicennia alba, Avicennia marina, Avicennia officinalis, Ceriops decandra, Sonneratia apetala,*
	Height (m)	8–10
	Circumference (cm)	10–20
	Wood density (g/cm^3)	0.6–0.8

changes in river flows have reduced sediment inputs, which caused the loss of over 75 km^2 of coastline in the last 37 years (Bhargava et al., 2020). Hydrogeological settings, climate, and vegetation types of the Indian Sundarbans are described in Table 14.4.

Compared to any other South Asian mangroves, biogeochemical and ecological data from the Indian Sundarbans are robust, long-term, and comprehensive. In past two decades, research articles on C dynamics in and around the Indian Sundarbans have reached more than 50 (web of search, January 2020), which is superior in number than any other mangroves of the South Asian regions. The Sundarbans, known to act as CO_2 sink, are also identified in removing gaseous pollutants by soaking up N-gases (NO_x, NH_3) from the atmosphere delineating their role in keeping a low-ozone state (Ganguly et al., 2009), emitting S-gases (H_2S) and increasing non–sea sulfate and increasing albedo (i.e., decrease in the amount of incoming solar radiation) (Ganguly et al., 2018). The Sundarbans maintain the balance of heat and energy budget, e.g., the extent of warming effect by methane (CH_4) and sensible heat flux was found to be partially counterbalanced by the cooling effect due to the processes such as photosynthesis, evapotranspiration, and albedo, effectively curbing greenhouse effects on the extent of local temperature (Ganguly et al., 2008).

Therefore, given the specificities in terms of geographic settings, huge forest coverage, species diversity, dynamic tidal influences, human habitation and exploitation of the mangrove, and the role in greenhouse gas and energy budget, Indian Sundarbans is considered here as model mangrove region for the entire South Asia. Therefore, in the subsequent text, we give sole focus on these mangroves to understand the biogeochemical state and fate of blue C.

3.2 Organic and inorganic C storage

3.2.1 TOC in mangrove plant

Applying mixed-species allometric model equations developed out of 100 trees covering more than 10 species, their height, wood specific gravity, circumference at breast height (cbh), their total weight, Ray et al. (2011) quantified total C stock in the above-ground and live below-ground biomass from various study locations in the Sundarbans (Fig. 14.4). TOC concentrations are measured to be 43.0%–45.1% in roots, 42.4%–43.1% in stems, 42.1%–46.9% in leaves, and 40%–42.8% in litter. Compared to live plant organs, C:N:P in litter and sediment are lower (leaf: 1200:40:1; litter: 940:35:1, sediment: 183:8:1; Ray et al., 2017), owing to N enrichment in the detritus after OM decomposition.

The East and Western part of the Indian Sundarbans hold comparable C stock in AGB (50.7 ± 13.6 Mg DW C/ha versus 29.2 ± 4.4 Mg DW C/ha, respectively, DW: dry weight), being 4 times higher than live BGB (7.9 ± 1.2 Mg DW C/ha at West Sundarbans, versus 14.1 ± 8.2 Mg DW C/ha at East Sundarbans) and the percentage ratio of BGB and AGB to total biomass (24%) is comparable with the average value reported for world mangroves (19%, Alongi, 2009). Above-ground C stock by Ray et al. (2011) is similar to another observation in the Indian Sundarbans by Mitra et al. (2011).

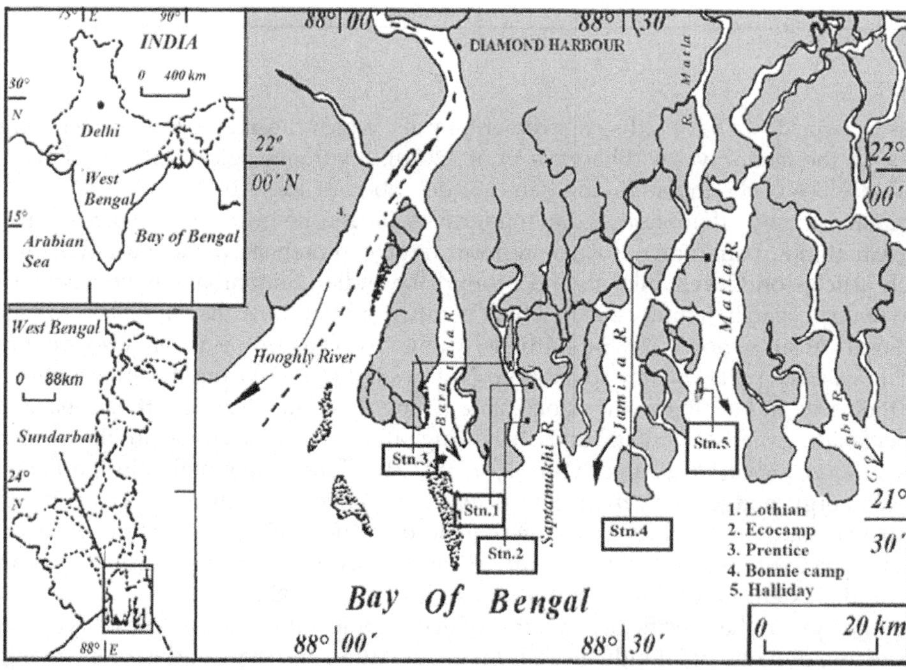

FIGURE 14.4 Map of the Indian Sundarbans showing site locations of biomass study (Ray et al., 2011).

TABLE 14.5 Comparison of C stock in AGB, BGB, and sediment between the Sundarbans and other mangroves.

Mangroves site	AGB-C Mg C/ha	BGB −C Mg C/ha	Sediment-C Mg C/ha	Source
Indian Sundarbans	29–50	7–14	68–86	Ray et al. (2011), this study
Bangladesh Sundarbans	45–152	11–62	90–134	Rahman et al. (2015)
Berau, Indonesia	130	–	485	Kusumaningtyas et al. (2019)
Micronesia	104–169	80–144	150–236	Kauffman et al. (2011)
Mindoro, Philippines	71	27.5	450	Salmo et al. (2019)
Cambodia	62–66	40–42	497–688	Sharma et al. (2020)
West Coast, Malaysia	74.5	9.8	488	Hossain (2013)
Chek Jawa, Singapore	138	52	307	Phang et al. (2015)
Vietnam	90–119	27–45	584–655	Tue et al. (2014), Vinh et al. (2019)
Red Sea, UAE	2–37[a]	1.5–25[a]	102–153[a], 43[b]	[a]Schile et al. (2017), [b]Almahasheer et al. (2017)
Caribbean Coast	196	–	381	Bhomia et al. (2016)
West-Central Africa	102	–	278	Kauffman and Bhomia (2017)
Mexico	421	–	506	Adame et al. (2015)
French Guiana, Latin America	88–218	31.5–64	85–282[c] 4.8–107[d]	[c]Walcker et al. (2018) [d]Marchand (2017)

Indian Sundarbans showed considerably lower blue C stock in the biomass (39.93 ± 14.05 Mg DW C/ha) compared to other tropical mangroves, especially SE Asian regions (Table 14.5). Even Bangladesh part of the Sundarbans has higher mangrove C stock in AGB and BGB (45–152 and 11–62 Mg DW C/ha, respectively, Rahman et al., 2015) than the Indian part. Such difference might arise from various approaches applied to estimate C stock. For example, Rahman et al. (2015) considered generic value plant TOC concentration (as 50%) to derive C stock, but in the Indian Sundarbans, measured TOC concentration (averaging ~ 42%) was used, hence final value becomes 1.2 times lower. Results might also vary from different allometry models used for the Indian and Bangladesh part of the Sundarbans. In the Indian part, allometric equations of AGB and BGB were developed from 100 harvested trees (Ray et al., 2011), but Rahman et al. (2015) applied allometric equations of Chave et al. (2005) for AGB and Komiyama et al. (2005) for BGB. During field measurement, authors in the Indian Sundarbans performed 10 m × 10 m quadrate-based tree counting and measurement of cbh, but for the Bangladesh, authors relied on plot-scale monitoring that is now widely used as blue C protocol (Kauffman and Donato, 2012; Howard, 2014).

Mangroves exhibit wide fluctuations of C sequestration, showing an increase with increase wood density. Application of wood density and chemical constituents of plants (cellulose and lignin) is established for better characterizing C fixation rates by the Indian Sundarbans

(Ray et al., 2012). Authors showed that C sequestration rate increased with density varied between 0.088 and 0.171 μg C/kg (AGB)/s, and *Avicennia marina* showed the maximum value and *Bruguiera gymnorrhiza* the minimum. Applying Fourier Transform Infra-Red spectroscopy (FT-IR) and wet chemical analysis, authors further concluded that hardwood mangrove (like *Bruguiera sp, Ceriops sp*) with high lignin content (11%–46% DW) sequesters more C than softwood cellulose-rich mangrove (20%–62% DW, *Avicennia sp*).

3.2.2 TOC in mangrove sediment

Sedimentary OC is generally a mixture of different endmember sources with significant contribution from the C3 type mangrove plants ($\delta^{13}C = -24$ to $-25‰$, mean $-24 \pm 2.4‰$; $\delta^{15}N = 3.5–4.5‰$, mean $3.9 \pm 0.5‰$; TOC:TN = 15–20, mean 17.8 ± 5.4, Table 14.2). However, compared to other mangroves where highly depleted $\delta^{13}C$ were recorded in the sediment (such as French Guiana mangroves: c.a. $-28‰$, Marchand et al., 2005; Indonesian East Kalimantan: $-27‰$, Kusumaningtyas et al., 2019), Sundarbans sediment is not entirely terrigenous rather marine or benthic algal sources are other substantive contributors to the OC pool.

TOC (%) in the sediment is generally low (~0.5–1.5, mean 1.25 ± 0.2, Table 14.2) compared to global average (2.2, Kristensen et al., 2008) and much lower than the average for the Indo-Pacific mangroves (1.8–18, Donato et al., 2011, West Papua, mean 15% interior mangrove by Sasmito et al., 2020), but similar to few other mangroves (Iranian mangroves: 0.4%–0.8%, Ray and Weigt, 2018; Papua New Guinea: 0.4%–2.2%, Alongi et al., 1993; French Guiana: 0.7%–2.0%, Ray et al., 2018. More than terrigenous input, such low TOC could be the result of marine sedimentation and mixing processes at the sediment water interface, and rates of delivery of TOC as suspended particles by strong tidal action favored by gentle slope of the mangrove settings. Besides that, microbial-induced intense OM remineralization might be another possible reason of such low OC in the sediment (Das et al., 2012). Resultantly, long-term TOC storage within top 30 cm of the humic layer (16.4–22.4 Mg C/ha with a mean of 18.8 ± 2.7 Mg C/ha, Ray et al., 2013) appears lower than the global data (Table 14.5). These estimates require more careful interpretation, such that, in the stock calculation, authors use constant bulk density of 2.5 g/cm^3 that is seemingly very high for silty-muddy sediment like the Sundarbans, and also TOC per humic layer (as long-term accumulation) that is slightly different from the classical C stock estimation formula where instead of humic layer, TOC present in bulk sediment at each depth that extends to 50 cm to 2 m depth is used (stock = TOC% × bulk density × depth). That said, sediment C stock data of the Indian Sundarbans by Ray et al. (2011) needs revisit. Using bulk density of 1.2 g/cm^3 (Authors unpublished data) and OC profile up to 30 cm from many cores by the authors, TOC stock in the bulk sediment of the Indian Sundarbans is revised to 60–75 Mg C/ha. This OC stock is further combined with IC data (next Subsection 3.2.3) to estimate total sedimentary C stock as 68–86 Mg C/ha equivalent to 26.6 to 33.7 Tg C for the whole mangroves. At first approximation, this revised estimate appears convincing for the upper layers of the sediments, but, reevaluation of C stock is still very much recommended accounting long core sediment retrieval i.e., at depth 1 m or more, that would reflect long-term C sink in many other sites.

Occurrence of humic substances (humic acid as HA, and fulvic acid as FA) in the sediment apart from sand, silt, and clay particles (Ray et al., 2011) controls the sediment characteristic and long-term C preservation. FA concentration (630–1300 μg/g) exceeds threefold than the

HA (216–458 μg/g, Ray et al., 2011). However, higher C was measured for HA (54.3%) over fulvic acid (46.2%), while carbohydrate and protein content are comparable in both substances. Effects of new nutrients on the C storage of sediment indicated humification of OM increases with respective chemical properties of the mangrove sediment. Effects of porewater pH and salinity were also observed to be determinant properties of live C stock. For instance, Ray et al. (2011) explained that variation of soil pH could be attributable to oxidation of OM promoted by sulfate reduction, and also by the release of organic acids from mangrove roots. According to Mitra et al. (2011), owing to varying degree of salinity, spatial variations of biomass and C storage capacity of different mangrove species were observed, e.g., hypersalinity of the central part of Indian Sundarbans might explain the stunted growth and relatively low C stock in the AGB. Similar attributes are given for the Sundarbans in Bangladesh where the variations of C stock in different locations depend on the adverse impact of increased salinity, thus conducive for mangrove shunting and biomass productivity (Wahid et al., 2007; Rahman et al., 2015).

3.2.3 *Inorganic C as carbonate in sediment*

Organic C generally represents a minor fraction (2%–8%, though with uncertainty) of the buried C in mangrove sediment, the remainder being carbonate particles and minerals. Recent global assessment of calcium carbonate ($CaCO_3$) occurring as biogenic calcifying organism (calcified algae, foraminifera, snails, corals) in the blue C habitats indicated the role of sedimentary IC in sequestration (Saderne et al., 2019). However, unlike seagrasses, where IC exceeds OC concentration (Mazarrasa et al., 2015), estimation of the former in mangroves is still very scarce.

In the Sundarbans benthic system, a hoard of benthic fauna (macrobenthos, meiobenthos, and microbenthos), both infauna (sessile, semisessile, and burrowing) and epifauna (crawlers and creepers) are the happy residents. Among the intertidal macrobenthic fauna, 26 species of brachyuran crabs belonging to 15 genera and 5 families were recorded from the deltaic Sundarbans estuarine complex (Ray et al., 2018). IC measured from very few sediment cores in the Indian Sundarbans showed stable concentration of ~0.4% (unpublished data, Ray et al.), equivalent to the lower limit of OC concentration, suggesting the potential role of carbonate chemistry on blue C for the environment. Globally, $CaCO_3$ (% dry weight) in the Red Sea mangroves, Everglades, and Australian mangroves are 75, 22, and 1 to 5, respectively (Saderne et al., 2019; Smoak et al., 2013; Sanders et al., 2016). Therefore, rough estimates of IC stock for the Indian Sundarbans would range from around 8 to 11 Mg C/ha (computed from the cores collected by Ray et al., 2011).

3.3 Biogeochemical C exchange flux: source or sink?

3.3.1 *Forest–atmosphere coupled system*

Sundarbans mangrove forest acts as sink for atmospheric CO_2 and is potentially sensitive to increasing atmospheric CO_2. Past research showed very clear daytime uptake and nighttime emission of CO_2 between biosphere and the atmosphere in the mangrove sites (Ganguly et al., 2008; Ray et al., 2011). Applying gradient method by using micrometeorological parameters and air CO_2 (Flux = $V_{C.}\ \Delta\chi$, where V_C or exchange velocity defined as $1/(r_a+r_s)$, aerodynamic (r_a) and surface layer (r_s) resistance, CO_2 concentration difference in ppm (χ) between 10 and 20 m; Barrett, 1998), authors reported significant seasonality in CO_2 exchange over the year with minimum uptake during monsoon. Mean uptake rate of CO_2 (81.8 ± 54 mmol CO_2/m^2 h) exceeds the emission rate (75.3 ± 36 mmol CO_2/m^2 h) suggesting C sink capacity of these mangroves. Extrapolating, net ecosystem productivity (i.e., photosynthesis—respiration) of 6.5 mmol CO_2/m^2 h would amount to annual sequestration of 2.79 Tg C/yr or equivalent to 6.54 Mg C/ha yr (Ray et al., 2011).

3.3.2 *Forest–sediment coupled system*

Litterfall plays pivotal role as C sink pathway for the mangroves. A 2-year comprehensive monitoring by Ray et al. (2013) in the Sundarbans showed monthly variations of litterfall in *Avicennia*-dominated mangrove sites (35 ± 18 g C DW/m^2 month) with highest flux observed during the post-monsoon. These values are comparable with the fringe and shrub mangroves in southwest Florida (13–75 g C DW/m^2 month, Twilley et al., 1986). Annual litterfall in the Indian Sundarbans is estimated to be 2.07 Tg C/yr or 4.85 Mg DW C/ha yr.

3.3.3 *Sediment–atmosphere coupled system*

Mangrove sediment is generally known to act as net emitter of CO_2 to the atmosphere with global average flux between 0.25 and 10 mmol CO_2/m^2 h (Leopold et al., 2015; Vinh et al., 2019; Chen et al., 2014). Comprehensive yearly data of day-time CO_2 emission by closed chamber technique ranged between 0.5 and 10.5 mmol CO_2/m^2 h from the Sundarbans sediment (mean: 2.5 ± 0.5 mmol C/m^2 h, Ray et al., 2013; Chanda et al., 2014). Both authors observed low CO_2 emission during monsoon due to high sediment water saturation leading to anoxic conditions, unfavorable for OM decomposition. According to Chanda et al. (2014), old growth mangroves at the landward sites emitted more CO_2 than the bare mud flats or seashore sites. Authors figured out that temperature at top few centimeters of the sediment mainly controlled CO_2 emission but its strong dependence on organic C and sediment texture could not be ruled out. Overall data extrapolation for the entire mangroves estimated total CO_2 converted C emission of 1.12 Tg C/yr or 2.7 Mg C/ha yr, at the sediment–atmosphere interphase.

3.3.4 *Sediment–water coupled system*

Fate of blue C in mangroves depends on outwelling efficiency of these habitats at the sediment–water interphase. Generally, mangroves are known for exporting dissolved organic and inorganic carbon (DOC, POC, and DIC) to the coastal water (Twilley, 1985; Dittmar et al., 2001; Rezende et al., 2007; Maher et al., 2013; Ray et al., 2018; Call et al., 2019). A recent study by Ray et al. (2018) identified Sundarbans mangroves as a major source of carbon (DOC, POC, and DIC) transporting into the Bay of Bengal, with export rates of 3.03, 0.58, and 3.7 Tg C/yr, respectively, altogether amounting to 7.3 Tg C/yr. The area weighted net export of mangrove-derived DOC, POC, and DIC are 6.7, 1.3, and 8.4 mmol C/m^2 h, respectively. These estimates are generally higher than the range of global estimates (e.g., DOC as 0.08–1.5 mmol/m^2 h, Kristensen et al., 2008) suggesting C loss via outwelling as one of the most vital components of C cycle, that may explain poor C stock in the sediment of the Indian Sundarbans. Almost all of the estuaries in the Indian part of the Sundarbans are macrotidal with regular tidal ranges between 4 and 5 m with very high water current (0.5–1.5 m/s). As a consequence of the estuarine geometry in the Sundarbans a "funneling effect" tends to amplify the tide as it propagates toward the head of the channel resulting in a further amplification of the tidal range thus inducing a faster material transport (Chatterjee et al., 2013).

3.4 Budget and fate of blue C

Attempt has been made to give a descriptive account of the C cycle for the Indian Sundarbans considering Box model approach that originally was first outlined by Bolin and Eriksson (1959) for the ocean. The box model approach has been used for budgeting various biogeochemical processes in the ocean (Berner, 1994) and estuaries (Mukhopadhyay et al., 2006). In the box model, non-proportionalities between fluxes and reservoirs contents are considered (Rodhe and Bjorkstrom, 1979, Fig. 14.5). There are four coupled C reservoirs in the model: (1) C in the form of CO_2 in the atmosphere, (2) C in the form of organic matter in above- and below-ground biomass, (3) C in the form of organic matter in sediments, and (4) C in the form of dissolved and particulate organic and inorganic in estuarine water. Out of the total area of Indian Sundarbans, mangrove surrounding water and forest roughly covers about 18% and 44%, respectively.

C budget of the Indian Sundarbans has been revisited from previous data compilation (Fig. 14.5, Ray et al., 2018, 2011, 2013). From the revised stock estimation, sediment appears to be the maximum storehouse of C, in line with observed trend for the tropical mangrove (Donato et al., 2011). C accumulation rate (CAR) of 0.25 Tg C/yr has been calculated

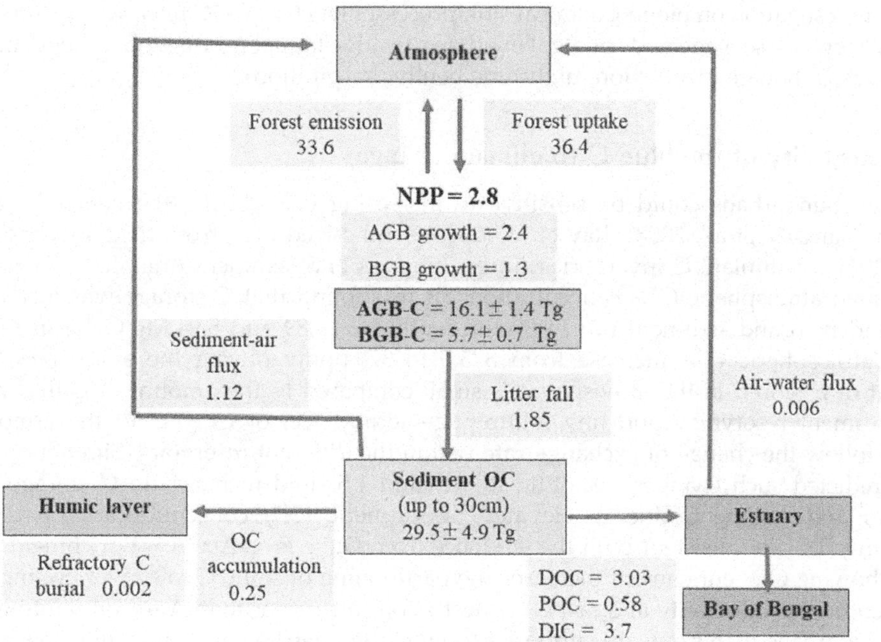

FIGURE 14.5 Box model representation of C cycle in the Indian Sundarbans. Stocks are represented in Tg C and flux in Tg C/yr.

applying mean TOC% (0.8), mass accumulation rate (0.75 g/cm^2 yr, only published sedimentation data applying ^{210}Pb for the Indian Sundarbans, Banerjee et al., 2011). Mean area-weighted CAR in the Sundarbans sediment is recalculated as 60±17 g C/m^2 yr, which is generally in lower ranges of the global value (163±26 g/m^2 yr, Breithaupt et al., 2012). Furthemore using the data by Banerjee et al. (2011), long-term refractory C burial in the humic layer as HA (mean: 202.5 μg C/g) and FA (mean: 496 μg C/g) is estimated to be 0.0008 and 0.001 Tg C/yr, respectively, totaling ~0.002 Tg C/yr, that is comparable to previous indirect estimate (0.005 Tg C/yr, Ray et al., 2011).

Net primary productivity (NPP) calculated from both approaches, i.e., CO_2 exchange flux between forest and air and live biomass growth subtracted from litterfall, does not result much difference (2.8 and 1.9 Tg C/yr, respectively) leading to a mean NPP approximation of 2.4 Tg C/yr (as AGB growth in Fig. 14.5). Maximum C exchange occurs in the form of CO_2 at the forest–atmosphere interphase (33.6–36.4 Tg C/yr) and minimum at the water–atmosphere interphase (0.006 Tg C/yr). According to flux estimates, 60.5% of the OC supplied as litter (1.85 Tg C/yr) is respired and emitted as CO_2 to the atmosphere, while 13.5% is buried within the sediment, and 31.5% is exported as POC to the water. Burial of refractory C (0.002 Tg C/yr) in the humic layer accounts for less than 1% of the OC accumulation in sediment, referring to a "modern" sediment with high labile OC pool. This is generally a well-balanced budget of sedimentary or particulate OC for the Indian Sundarbans with an excess of ~4% as 'unaccounted' OC. However, total dissolved fraction of the exported C (as DOC and DIC) creates a "missing sink" of 6.73 Tg C/yr which is now slightly improved from the previous calculation (7.3 Tg C/yr, Ray et al., 2018), but still requires improvement and further investigation on biogeochemical sink processes (such as DOC released via litter leaching or porewater seepage and/or algal exudates, benthic faunal production, lithogenous, and biogenous carbonate production, nighttime benthic respiration).

3.5 Sensitivity of the blue C to climate change

Indian Sundarbans could be sensitive to increasing CO_2 during the course of climate change (Ganguly et al., 2008). Ray et al. (2013) synthesized data from field surveys during 2009–2011 to simulate C inventories (forest biomass and sediment organic C) in response to elevated atmospheric CO_2 concentrations. It was found that C storage was increased in AGB and BGB and sediment (up to 30 cm depth) by 18.89 and 5.94 Mg C/ha in response to the atmospheric CO_2 increase from 373.5 to 378 ppmv during the study period. The amount of C stored in the atmosphere is small compared to the amount of C in the forest and sediment reservoirs, and any anthropogenic addition of CO_2–C to the atmosphere would follow the change of exchange rate within the different reservoirs. Simulation model well predicted such changes, resulting in 1.1- and 1.57-fold increases in C storage in live biomass and sediment, due to elevated atmospheric CO_2 concentrations from 367 to 580 ppmv. This is consistent with the evidence from Duke Free-Air CO_2 Enrichment experiment showing CO_2 enrichment increases the partitioning of soil OC to slower cycling pools.

Given high productivity and fast C sequestration rate (faster than Amazon rainforest), Indian Sundarbans mangrove are supposed to take safe-guard against the atmospheric pollution and CO_2 enrichment from point sources like thermal power plant in Kolaghat and the heavily populated metropolis of Kolkata, both being present within 150 km from the mangroves. Nevertheless, there is a gap in understanding the actual capacity of the Indian Sundarbans to mitigate all the CO_2 emitted from the proximate thermal power plant. Another study by Ray and Jana (2017) shows that about 4328 km^2 mangrove coverage (i.e., 1.5% increase) is needed to sequester all that CO_2 emitted from the same thermal power plant, and that the rate of C sequestration was 0.64% of the annual fossil fuel emission, and 1.54% of total C emissions from coal-based power plants in India for the year 2010. Although the capacity of storing anthropogenic CO_2 by these mangroves seems poor, authors concluded that further comprehensive research is required in other mangroves of the world in order to better understand their biosequestration potential and come up with a significant database in the "green C" inventories.

4. Research gaps and future directions

4.1 Improving laboratory-based data by applying new methods

In comparison to other global mangroves, rate of C accumulation in sediment from the South Asian locations is very limited. Apart from one study showing estimates of sedimentation rate in the Indian Sundarbans (^{210}Pb dating by Banerjee et al., 2011), there are no published data available from the entire coverage of the South Asia that largely limit our understanding on C burial rate, thereby blue C sequestration potential by these mangroves. This is particularly important for these mangrove locations which are influenced by the big rivers like Ganges, Padma, and Indus that bring huge sediment loads to the inner shore and influence sedimentation rates. In future, sea-level rise would also impact mangroves sediment C burial rate. Two methods conventionally used to determine the sedimentation rate from intact sediment cores are radioactive carbon (^{14}C) dating by accelerator mass spectrometry and $^{210}Pb/^{137}Cs$ dating by γ-spectroscopy. It is known that ^{14}C dating is used for the sediment cores of hundreds to thousands of years old (like marine sediment), but modern sediment like mangroves (less than 100 years), $^{210}Pb/^{137}Cs$ method is more efficient. Therefore, instead of using past literature data, new estimates of sedimentation rate from a range of locations are required by $^{210}Pb/^{137}Cs$ dating, and that will allow us to estimate C burial rate of these mangroves.

Contrary to spatial scale, reports on seasonal variability of OC source compositions in the mangrove sediments are still very rare, although this is an important aspect to examine. This is especially important in the South Asian mangrove regions where intense monsoon period prevails for a longer period of the year, driving loads of terrigenous OM from upstream via river and surface run off as well as mangrove materials to the sea. Another point is, during riverine transport in the monsoon, more OC would be derived from the upstream sources (known as "green" C) and deposited at the deltaic sediment. Therefore, the sediment should be considered as composite of both "blue" and "green" C, but stable isotopes alone are not helpful to separate blue C from green C sources, unless other methods (e.g., eDNA method, Ortega et al., 2020) are developed and applied. This is still a knowledge gap in our blue C studies and warrants more research to undertake on eDNA in future.

4.2 Research on fine root production

In spite of being recognized as important contributor to sediment OC pool, fine root production in mangroves receives very little attention (Fujimoto et al., 1999; Xiong et al., 2016) compared to above-ground and litter production. Because of their slow decomposition rate (Middleton and McKee, 2001), and their possible role in increasing soil volume, thus ability by the coastal forest to keep pace with sea-level rise, studies on fine root production is unarguably a very important subject in blue C. While extensive researches have been conducted on root biomass in terrestrial forests (review by Yuan and Chen, 2010), fine root production and turnover in mangroves remain poorly understood due to the logistical difficulties of working in mangrove and the intensive labor required in fine root processing (Adame et al., 2014). From the limited data, it is even ambiguous to understand the factors controlling mangrove fine root production and turnover rate. For instance, porewater salinity was positively correlated to higher root production by Adame et al. (2014), but their inverse relationship was examined by Ball (2002). Xiong et al. (2016) showed that mangrove sites in China with higher soil nutrients and lower salinity favored fine root production, thus soil C accumulation. According to same author, higher proportion of root C was retained in the soil than above-ground litter C, and thus higher allocation to root than above-ground litter production favored soil C enrichment. Other controlling factors, like hydroperiod, soil temperature, soil water pressure, and seasonal conditions (rainfall), are only partially recognized to play role in root production (Poungparn et al., 2016), but the mechanisms are yet to be fully understood. Therefore, this important yet overlooked component of mangrove C cycle requires more detailed investigations in future, especially in macrotidal regions like the Sundarbans where litter export is high (Ray et al., 2011) that might be compensated by enhanced fine root production.

4.3 Microenvironmental heterogeneity

The term "microenvironmental heterogeneity," used by Aller (2014), should be specially applied to mangrove sediment due to its highly heterogenous conditions impacted by the presence of abundant mud dwelling faunal assemblages and their sediment reworking activities. POC uptake via fresh leaves by the macrobenthos fauna like crabs is a C sink mechanism that has been largely overlooked as a component of mangrove C cycle despite its relevance in blue C science. Large population of such bivalves and crab burrows (also known as bioirrigators or "ecosystem engineers") common in the intertidal flats could drive significant modifications in the sedimentation rate, sediment texture, and surface topography. They could modify porewater flow and solute transport like DOC, DIC, and nutrients (Stieglitz et al., 2000; Santos et al., 2014).

In Kenyan mangroves, Andreetta et al. (2014) observed highest OC concentration and stock in the 20–40 cm interval of the core due to differences in crab population composition other than fine root activities. However, role of such biogenic structures as biogeochemical functioning in mangrove sediments has been documented only in few comprehensive studies in Africa (Andreetta et al., 2014; Penha-Lopes et al., 2009), Latin America (Aschenbroich et al., 2016), and Australia (Kristensen and Alongi, 2006). Surprisingly there have been no attempt till date to assess a linkage between bioturbating animals (crabs) and soil C storage for the largest mangrove-covered Asian regions. Supply of labile OC from either MPB or leaf litter and thereafter OC removal as food consumption by the crabs or storage within the burrowing pockets for future food supply should be the subject matter of high interest within blue C communities. Indian mangrove ecosystems shelter a total of 3985 biological species, 77% of which are dominated by the faunal species (Kathiresan, 2000). Despite such large living resources, there is no published record on C stock in terms of faunal production or C burial rate by faunal activities. Therfore, if future field studies do not consider faunal bioturbation as a critical factor, a likely underestimation of vertical soil-CO_2 fluxes and, ultimately, overestimation of carbon retention rates and net carbon storage would result (Agusto, 2020).

4.4 Other gaps and challenges

A number of publications and reports should be considered on blue C ecosystems that have described 10 key fundamental questions on blue C science like how climate change, restoration, and disturbance would impact blue C mitigation potential (Macreadie et al., 2019). Quantification of emissions and removals of blue C based on mangrove geomorphology and land use land cover types as well as disturbed, degraded, and restored blue C habitats is a major area not just to promote conservation and protection of these ecosystems but to support also sustainable management and climate change mitigation plan (Richards et al., 2020). The uncertainties vis-à-vis limitation in blue C data (storage and accumulation rate) contribute further to a lack of focus on the matter in the South Asian coastal region. A proper blue C analysis is needed in the South Asian countries to achieve improved scientific and technical understanding of blue C stocks, fluxes and monitoring their changes over time.

Therefore, there is a crucial need to improve scientific understanding of the blue C habitats in the region at appropriate scales that would help in future policy development. Achieving improved scientific understanding on C stocks and to monitor the changes would require practical tools and guidance to enable the conduct of proper C analyses. Furthermore, it is necessary to develop an appropriate framework for blue C assessment and monitoring mechanisms in this region. The Clean Development Mechanism under the Kyoto Protocol Methodologies has been developed for monitoring, reporting, and verifying C from mangrove afforestation and reforestation. However, there are still large gaps in present knowledge C storage and sequestration potential of blue C habitats. It has been largely acknowledged by researchers and policy makers that the conservation, restoration, and sustainable use of the coastal habitats would support climate change mitigation efforts, in addition to benefiting the coastal communities from other ecosystem services they provide such as fisheries resource and coastal protection. Countries like Indonesia have therefore begun to address the land-use land cover changes and degradation of blue C habitats through their policies and management plans.

Emphasis and interest on blue C are also emerging in South Asian countries like India and Bangladesh, especially as a means to mitigate impacts of climate change, India has pledged to use coastal habitats as NDC (Nationally determined contributions) in Paris agreement 2015. Additionally, national obligations to the United Nations Framework Convention on Climate Change (UNFCCC) and related mechanism such as the Reducing Emissions from Deforestation and Forest Degradation (REDD+) also indirectly support some of the activities and efforts in mitigating climate change. Specific work undertaken on blue C in terms of the national level inventory is however still limited at mostly at its infant stage in many countries, for example, only Indonesia (Murdiyarso et al., 2015) and Cambodia (Sharma, 2020) have quantified national-level mangrove blue C stocks. There are still various gaps and challenges such as long-term monitoring to understand sequestration potential that would need to be filled. In general, there would firstly be a need to focus on national inventory on emissions and removals of blue C reservoirs to comprehensively include into the national-level mitigation activities and increasing capacity on the technical, policy, and institutional aspects at national level. Key areas of uncertainty and specific actions needed to address blue C sustainability have also started to get attention for addressing at various regional and national levels with an emphasis on research agenda to prepare a greenhouse gas offset protocol for coastal ecosystems restoration and management and an action plan to guide protocol development.

References

Adame, M.F., Teutli, C., Santini, N.S., Caamal, J.P., Zaldívar-Jiménez, A., Hernández, R., Herrera-Silveira, J.A., 2014. Root biomass and production of mangroves surrounding a karstic oligotrophic coastal lagoon. Wetlands 34, 479–488.

Adame, M.F., Santini, N.S., Tovilla, C., Vázquez-Lule, A., Castro, L., Guevara, M., 2015. Carbon stocks and soil sequestration rates of tropical riverine wetlands. Biogeosciences 12, 3805–3818.

Agusto, L.E., et al., 2020. Fiddling with the carbon budget: fiddler crab burrowing activity increases wetland's carbon flux. ESSOAr. https://doi.org/10.1002/essoar.10503504.1.

Ahmed, N., Cheung, W.W., Thompson, S., Glaser, M., 2017. Solutions to blue carbon emissions: shrimp cultivation, mangrove deforestation and climate change in coastal Bangladesh. Mar. Pol. 82, 68–75.

Aller, R.C., 2014. Sedimentary diagenesis, depositional environments, and benthic fluxes. In: Holland, H.D., Turekian, K.K. (Eds.), Treatise on Geochemistry, vol. 8, pp. 293–334.

Almahasheer, H., Oscar Serrano, O., Duarte, C.M., Arias-Ortiz, A., Masque, P., Irigoien, X., 2017. Low carbon sink capacity of Red Sea mangroves. Sci. Rep. 7, 9700. https://doi.org/10.1038/s41598-017-10424-9.

Alongi, D.M., 2009. The Energetics of Mangrove Forest, vol. 24. Springer, p. 11.

Alongi, D.M., 2014. Carbon cycling and storage in mangrove forests. Annu. Rev. Mar. Sci. 6, 195–219.

Alongi, D.M., Christoffersen, P., Tirendi, F., 1993. The influence of forest type on microbialnutrient relationships in tropical mangrove sediments. J. Exp. Mar. Biol. Ecol. 171, 201–223.

Amarasinghe, M.D., Balasubramaniam, S., 1992. Structural properties of two types of mangrove stands on the Northwestern Coast of Sri Lanka. Hydrobiologia 247, 17–27.

Andreetta, A., Fusi, M., Cameldi, I., Cimò, F., Carnicelli, S., Cannicci, S., 2014. Mangrove carbon sink. Do burrowing crabs contribute to sediment carbon storage? Evidence from a Kenyan mangrove system. J. Sea Res. 85, 524–533.

Aschenbroich, A., Michaud, E., Stieglitz, T., Fromard, F., Gardel, A., Tavares, M., Thouzeau, G., 2016. Brachyuran crab, community structure and associated sediment reworking activities in pioneer and young mangroves of French Guiana, South America. Estuar. Coast. Shelf Sci. 182, 60–71.

Atwood, T.B., Connolly, R.M., Almahasheer, H., Carnell, P.E., Duarte, C.M., Lewis, C.J.E., Irigoien, X., Kelleway, J.J., Lavery, P.S., Macreadie, P.I., Serrano, O., Sanders, C.J., Santos, I., Steven, A.D.L., Lovelock, C.E., 2017. Global patterns in mangrove soil carbon stocks and losses. Nat. Clim. Change 7, 523–528.

Baker, T.R., Phillips, O.L., Malhi, Y., et al., 2004. Variation in wood density determines spatial patterns in Amazonian forest biomass. Glob. Change Biol. 10, 545–562.

Ball, M.C., 2002. Interactive effects of salinity and irradiance on growth implications for mangrove forest structure along salinity gradients. Trees 16, 126–139.

Banerjee, K., Senthilkumar, B., Purvaja, R., Ramesh, R., 2011. Sedimentation and trace metal distribution in selected locations of Sundarbans mangroves and Hooghly estuary, Northeast Coast of India. Environ. Geochem. Health 34, 27–42.

Barbier, E.B., Koch, E.W., Silliman, B.R., Hacker, S.D., Wolanski, E., Primavera, J., et al., 2008. Coastal ecosystem–based management with nonlinear ecological functions and values. Science 319, 321–323.

Barrett, K, 1998. Oceanic ammonia emissions in the Europe and their trans boundary fluxes. Atmos. Environ 32, 381–391.

Berner, R.A., 1994. GEOCARB II: a revised model of atmospheric CO_2 over Phanerozoic time. Am. J. Sci. 294, 56–91.

Bhargava, R., Sarkar, D., Friess, D., 2020. Cloud computing-based approach to mapping mangrove erosion and progradation: case studies from the Sundarbans and French Guiana. Estuarine, Coastal and Shelf Science. https://doi.org/10.1016/j.ecss.2020.106798.

Bhomia, R.K., MacKenzie, R.A., Murdiyarso, D., Sasmito, S.D., Purbopuspito, J., 2016. Impacts of land use on Indian mangrove forest carbon stocks: implications for conservation and management. Ecol. Appl. 26, 1396–1408.

Bolin, B., Eriksson, E., 1959. Changes in the carbon dioxide content of the atmosphere and sea due to fossil fuel combustion. In: Rossby Memorial Volume. Rockefeller Institute Press, New York, pp. 130–142.

Bouillon, S., Boschker, H.T.S., 2006. Bacterial carbon sources in coastal sediments: a cross-system analysis based on stable isotope data of biomarkers. Biogeosciences 3, 175–185.

Bouillon, S., et al., 2003. Inorganic and organic carbon biogeochemistry in the Gautami Godavari estuary (Andhra Pradesh, India) during pre-monsoon: the local impact of extensive mangrove forests. Glob. BioGeoChem. Cycle. https://doi.org/10.1029/2002GB002026.

Bouillon, S., et al., 2008. Mangrove production and carbon sinks: a revision of global budget estimates. Glob. Biogeochem. Cycles 22, GB2013. https://doi.org/10.1029/2007GB003052.

Breithaupt, J.L., Smoak, J.M., Smith III, T.J., Sanders, C.J., Hoarel, A., 2012. Organic carbon accumulation rates in mangrove sediments: strengthening the global budget. Glob. Biogeochem. Cycles GB3011.

Bulmer, R.H., Lundquist, C.J., Schwendenmann, L., 2015. Sediment properties and CO_2 efflux from intact and cleared temperate mangrove forests. Biogeosciences 12, 6169–6180.

Call, M., Sanders, C.J., Macklin, P.A., Santos, I.R., Maher, D.T., 2019. Carbon outwelling and emissions from two contrasting mangrove creeks during the monsoon storm season in Palau, Micronesia. Estuar. Coast. Shelf Sci. 218, 340–348.

Cannicci, S., Burrows, D., Fratini, S., Smith III, T.J., Offenberg, J., Dahdouh-Guebas, F., 2008. Faunal impact on vegetation structure and ecosystemfunction inmangrove forests: a review. Aquat. Bot. 89, 186–200.

Castillo, J.A., Apana, A.A., Maraseni, T.N., Salmo III, S.G., 2017. Soil greenhouse gas fluxes in tropical mangrove forests and in land uses on deforested mangrove lands. Catena 159, 60–69.

Chanda, A., Akhand, A., Manna, S., Dutta, S., Das, I., Hazra, S., Rao, K.H., Dadhwal, V.K., 2014. Measuring daytime CO_2 fluxes from the inter-tidal mangrove soils of Indian Sundarbans. Environ. Earth Sci. 72, 417–427.

Chatterjee, M., Shankar, D., Sen, G.K., et al., 2013. Tidal variations in the Sundarbans Estuarine System, India. J. Earth Syst. Sci. 122, 899–933.

Chave, J., Andalo, C., Brown, S., Cairns, M.A., Chambers, J.Q., Eamus, D., Fölster, H., Fromard, F., Higuchi, N., Kira, T., Lescure, J.P., Nelson, B.W., Ogawa, H., Puig, H., Riéra, B., Yamakura, T., 2005. Tree allometry and improved estimation of carbon density and balance in tropical forests. Oecologia 145, 87–99.

Chen, G.C., Tam, N.F.Y., Ye, Y., 2012. Spatial and seasonal variations of atmospheric N_2O and CO_2 fluxes from a subtropical mangrove swamp and their relationships with soil characteristics. Soil Biol. Biochem. 48, 175–181.

Chen, G.C., Ulumuddin, Y.I., Pramudji, S., Chen, S.Y., Chen, B., Ye, Y., et al., 2014. Rich soil carbon and nitrogen but low atmospheric greenhouse gas fluxes from North Sulawesi mangrove swamps in Indonesia. Sci. Total Environ. 487, 91–96.

Chen, S.Y., Chmura, G.L., Wang, Y., Yu, D., Ou, D.Y., Chen, B., Ye, Y., Chen, G.C., 2019. Benthic microalgae offset the sediment carbon dioxide emission in subtropical mangrove in cold seasons. Limnol. Oceanogr. 64, 1297–1308.

Das, S., De, M., Ray, R., Chowdhury, C., Jana, T.K., De, T.K., 2012. Microbial ecosystem in Sunderban mangrove forest sediment, North-East Coast of Bay of Bengal, India. Geomicrobiol. J. 29, 656–666.

Dasgupta, S., Islam, M.S., Huq, M., Khan, Z.H., Hasib, M.R., 2019. Quantifying the protective capacity of mangroves from storm surges in coastal Bangladesh. PLoS One 14, e0214079.

Dehairs, F., Rao, R.G., Chandra Mohan, P., Raman, A.V., Marguillier, S., Hellings, L., 2000. Tracing mangrove carbon in suspended matter and aquatic fauna of the Gautami-Godavari Delta, Bay of Bengal (India). Hydrobiologia 431, 225–241.

Dittmar, T., Lara, R.J., Kattner, G., 2001. River or mangrove? Tracing major organic matter sources in tropical Brazilian coastal waters. Mar. Chem. 73, 253–271.

Dittmar, T., Hertkorn, N., Kattner, G., Lara, R.J., 2006. Mangroves, a major source of dissolved organic carbon to the oceans. Glob. Biogeochem. Cycles 20, GB1012. https://doi.org/10.1029/2005GB002570.

Donato, D.C., Kauffman, J.B., Murdiyarso, D., Kurnianto, S., Stidham, M., Kanninen, M., 2011. Mangroves among the most carbon-rich forests in the tropics. Nat. Geosci. 4, 293–297.

Duarte, C.M., Middelburg, J., Caraco, N., 2005. Major role of marine vegetation on the oceanic carbon cycle. Biogeosciences 1, 1–8.

Fourqurean, J.W., et al., 2012. Seagrass ecosystems as a globally significant carbon stock. Nat. Geosci. 5, 505–509.

Fujimoto, K., Imaya, A., Tabuchi, R., Kuramoto, S., Utsugi, H., Murofushi, T., 1999. Belowground carbon storage of Micronesian mangrove forests. Ecol. Res. 14, 409–413.

Ganguly, D., Dey, M., Mandal, S.K., De, T.K., Jana, T.K., 2008. Energy dynamics and its implication to biosphere-atmosphere exchange of CO_2, H_2O and CH_4 in a tropical mangrove forest canopy. Atmos. Environ. 42, 4172–4184.

Ganguly, D., Dey, M., Sen, S., Jana, T.K., 2009. Biosphere-atmosphere exchange of NOx in the tropical mangrove forest. J. Geophys. Res. (Biogeosci.) 114, G04014.

Ganguly, D., Ray, R., Majumdar, N., Chowdhury, C., Jana, T.K., 2018. Biogenic hydrogen sulphide emissions and non-sea sulfate aerosols over the Indian Sundarban mangrove forest. J. Atmos. Chem. 75, 319–333.

Geraldi, N.R., Ortega, A., Serrano, O., Macreadie, P.I., Lovelock, C.E., Krause-Jensen, D., et al., 2019. Fingerprinting blue carbon: rationale and tools to determine the source of organic carbon in marine depositional environments. Front. Mar. Sci. 6 https://doi.org/10.3389/fmars.2019.00263.

Giri, C., Ochieng, E., Tieszen, L.L., Zhu, Z., Singh, A., Loveland, T., Masek, J., Duke, N., 2011. Status and distribution of mangrove forests of the world using earth observation satellite data. Glob. Ecol. Biogeogr. 20, 154–159.

Giri, C., Long, J., Abbas, S., Murali, R.M., Qamer, F.M., Pengra, B., Thau, D., 2014. Distribution and dynamics of mangrove forests of South Asia. J. Environ. Manag. 148, 101–111.

Gnanamoorthy, P., Selvam, V., Ramasubramanian, R., Chakraborty, S., Pramit, D., Karipot, A., 2019. Soil organic carbon stock in natural and restored mangrove forests in Pichavaram South-East Coast of India. Indian J. Geo. Mar. Sci. 48, 801–808.

Gontharet, S., Mathieu, O., Lévêque, J., Lesourd, S., Caillaud, J., Gardel, A., Sarrazin, M., Proisy, C., 2014. Distribution and sources of bulk organic matter (OM) on a tropical intertidal mud bank in French Guiana from elemental and isotopic proxies. Chem. Geol. 376, 1–10.

Hamaguchi, M., Shimabukuro, H., Hori, M., Yoshida, G., Terada, T., Miyajima, T., 2018. Quantitative real-time PCR and droplet digital PCR duplex assays for detecting *Zostera marina* DNA in coastal sediments. Limnol. Oceanogr. Methods 16, 253.

Hamilton, S.E., Casey, D., 2016. Creation of a high spatio-temporal resolution global database of continuous mangrove forest cover for the 21st century (CGMFC-21). Glob. Ecol. Biogeogr. 25, 729–738.

Ho, D.T., Ferrón, S., Engel, V.C., Anderson, W.T., Swart, P.K., Price, R.M., Barbero, L., 2017. Dissolved carbon biogeochemistry and export in mangrove dominated rivers of the Florida Everglades. Biogeosciences 14, 2543–2559.

Hoque, A.K.F., Datta, D.K., 2005. The mangroves of Bangladesh. Int. J. Ecol. Environ. Sci. 31, 245–253.

Hossain, M., 2013. Carbon pools and fluxes in *Bruguiera parviflora* dominated naturally growing mangrove forest of Peninsular Malaysia. Wetlands Ecol. Manag. https://doi.org/doi:0.1007/s11273-013-9318-2.

Howard, J., et al., 2014. Coastal blue carbon: Methods for assessing carbon stocks and emissions factors in mangroves, tidal salt marshes, and seagrass meadows. Arlington, Virginia, USA: Conservation International, International Union for Conservation of Nature.

Hutchinson, J., et al., 2013. Predicting global patterns in mangrove forest biomass. Conserv. Lett. https://doi.org/10.1111/conl.12060.

India State of Forest Report, 2019. Mangrove Cover. Chapter Three. Forest Survey of India Ministry of Environment, Forest & Climate Change. Government of India.

Jennerjahn, T.C., Ittekott, V., 2002. Relevance of mangroves for the production and deposition of organic matter along tropical continental margins. Naturwissenschaften 89, 23–30.

Kamruzzaman, M., Basak, K., Paul, S.K., Ahmed, S., Osawa, A., 2019. Litterfall production, decomposition and nutrient accumulation in Sundarbans mangrove forests, Bangladesh. Forest Sci. & Technol. 15, 24–32.

Kathiresan, K., 2000. A review of studies on Pichavaram mangroves, Southeast India. Hydrobiologia 430, 185–205.

Kathiresan, K., Rajendran, N., 2005. Coastal mangrove forests mitigated tsunami. Estuar. Coast. Shelf Sci. 65, 601–606.

Kauffman, J.B., Bhomia, R.K., 2017. Ecosystem carbon stocks of mangroves across broad environmental gradients in West-Central Africa: global and regional comparisons. PLoS One 12 (11), e0187749. https://doi.org/10.1371.

Kauffman, J.B., Donato, D., 2012. Protocols for the Measurement, Monitoring and Reporting of Structure, Biomass and Carbon Stocks in Mangrove Forests. Center for International Forestry Research (CIFOR), Bogor, Indonesia.

Kauffman, J.B., Heider, C., Cole, T.G., Dwire, K.A., Donato, D.C., 2011. Ecosystem carbon stocks of micronesian mangrove forests. Wetlands 31, 343–352.

Kauffman, J.B., Arifanti, V.B., Hernández, H., García, M., Norfolk, J., Cifuentes, M., Hadriyanto, T., Murdiyarso, D., 2017. The jumbo carbon footprint of a shrimp: carbon losses from mangrove deforestation. Front. Ecol. Environ. 15, 183–188.

Komiyama, A., Poungparn, S., Kato, S., 2005. Common allometric equations for estimating the tree weight of mangroves. J. Trop. Ecol. 21, 471–477.

Kristensen, E., Alongi, D.M., 2006. Control by fiddler crabs (*Uca vocans*) and plant roots (*Avicennia marina*) on carbon,-iron, and sulfur biogeochemistry in mangrove sediment. Limnol. Oceanogr. 51, 1557–1571.

Kristensen, E., Bouillon, S., Dittmar, T., Marchand, C., 2008. Organic carbon dynamics in mangrove ecosystems: a review. Aquat. Bot. 89, 201–219.

Ksawani, I., Kmarusaman, J., Nurum-Nadhirah, M.I., 2007. Biological diversity assessment of Tok Bali mangrove forest, Klantan, Malaysia. WSEAS Trans. Environ. Dev. 3, 7–44.

Kusumaningtyas, M.A., Hutahaeanc, A.A., Fischerd, H.W., Pérez-Mayod, M., Ransby, D., Jennerjahn, T.C., 2019. Variability in the organic carbon stocks, sources, and accumulation rates of Indonesian mangrove ecosystems. Estuar. Coast. Shelf Sci. 218, 310–323.

Kuwae, T., Hori, M., 2018. Blue Carbon in Shallow Coastal Ecosystm. Springer, ISBN 978-981-13-1294-6.

Lallier-Verges, E., Perrussel, B.P., Disnar, J.-R., Baltzer, F., 1998. The relationship between environmental conditions and the diagenetic evolution of organic matter derived from higher plant in a present mangrove swamp system (Guadeloupe, French West Indies). Org. Geochem. 29, 1663–1686.

Lee, S.Y., 2008. Mangrove macrobenthos: assemblages, services, and linkages. J. Sea Res. 59, 16–29.

Lee, S.Y., 2016. From blue to black: anthropogenic forcing of carbon and nitrogen influx to mangrove-lined estuaries in the South China Sea. Mar. Pollut. Bull. 109, 682–690.

Leopold, A., Marchand, C., Deborde, J., Chaduteau, C., Allenbach, M., 2013. Influence of mangrove zonation on CO_2 fluxes at the sediment-air interface (New Caledonia). Geoderma 202–203, 62–70.

Leopold, A., Marchand, C., Deborde, J., Allenbach, M., 2015. Temporal variability of CO_2 fluxes at the sediment-air interface in mangroves (New Caledonia). Sci. Total Environ. 502, 617–626.

Li, S.-B., Chen, P.-H., Huang, J.-S., Hsueh, M.-L., Hsieh, L.-Y., Lee, C.-L., Lin, H.-J., 2018. Factors regulating carbon sinks in mangrove ecosystems. Glob. Change Biol. 24, 4195–4210.

Lovelock, C.E., 2008. Soil respiration and belowground carbon allocation in mangrove forests. Ecosystems 11, 342–354.

Lovelock, C.E., Ruess, R.W., Feller, I.C., 2011. CO_2 efflux from cleared mangrove peat. PLoS One 6, e21279.

Lovelock, C.E., Simpson, L.T., Duckett, L.J., Feller, I.C., 2015. Carbon budgets for Caribbean mangrove forests of varying structure and with phosphorus enrichment. Forests 6, 3528–3546.

Macreadie, P.I., Anton, A., Raven, J.A., Beaumont, N., Connolly, R.M., Friess, D.A., Kelleway, J.J., Kennedy, H., Kuwae, T., Lavery, P.S., 2019. The future of blue carbon science. Nat. Commun. 10, 1–13.

Maher, D.T., et al., 2018. Beyond burial: lateral exchange is a significant atmospheric carbon sink in mangrove forests. Biol. Lett. 14. https://doi.org/10.1098/rsbl.2018.0200.

Maher, D.T., Santos, I.R., Golsby-Smith, L., Gleeson, J., Eyre, B.D., 2013. Groundwater-derived dissolved inorganic and organic carbon exports from a mangrove tidal creek: the missing mangrove carbon sink? Limnol. Oceanogr. 58, 475–488.

Mandal, M.S.H., Hosaka, T., 2020. Assessing cyclone disturbances (1988–2016) in the Sundarbans mangrove forests using Landsat and Google Earth Engine. Natural Hazards 102, 133–150. https://doi.org/10.1007/s11069-020-03914-z. In this issue.

Marchand, C., 2017. Soil carbon stocks and burial rates along a mangrove forest chronosequence (French Guiana). Forest Ecol. Manag. 384, 92–99.

Marchand, C., Disnar, J.-R., Lallier-Verge`s, E., Lottier, N., 2005. Early diagenesis of carbohydrates and lignin in mangrove sediments subject to variable redox conditions (French Guiana). Geochimica et Cosmochimica Acta 69, 131–142.

Mazarrasa, I., et al., 2015. Seagrass meadows as a globally significant carbonate reservoir. Biogeosciences 12, 4993–5003.

McLeod, E., Chmura, G.L., Bouillon, S., Salm, R., Björk, M., Duarte, C.M., Lovelock, C.,E., Schlesinger, W.H., Silliman, B.R., 2011. A blueprint for blue carbon: toward an improved understanding of the role of vegetated coastal habitats in sequestering CO_2. Front. Ecol. Environ. 9, 552–560.

Metcalfe, D.B., Meir, P., Aragão, L.E.O.C., Malhi, Y., et al., 2007. Factors controlling spatio-temporal variation in carbon dioxide efflux from surface litter, roots, and soil organic matter at four rain forest sites in the Eastern Amazon. J. Geophys. Res. 112, G04001. https://doi.org/10.1029/2007JG00044.

Middleton, B.A., McKee, K.L., 2001. Degradation of mangrove tissues and implications for peat formation in Belizean island forests. J. Ecol. 89, 818–828.

Mitra, A., Sengupta, K., Banerjee, K., 2011. Standing biomass and carbon storage of above-ground structures in dominant mangrove trees in the Sundarbans. Forest Ecol. Manag. 261, 1325–1335.

Miyajima, T., Hamaguchi, M., 2018. Carbon sequestration in sediment as an ecosystem function of seagrass meadows. In: Kuwae, T., Hori, M. (Eds.), Chapter Three, Blue Carbon in Shallow Coastal Ecosystm. Springer, ISBN 978-981-13-1294-6, pp. 33–63.

Mondal, S.H., Debnath, P., 2017. Spatial and temporal changes of Sundarbans reserve forest in Bangladesh. Environ. & Nat. Resour. J. 15, 51–61.

Mukhopadhyay, S.K., Biswas, H., De, T.K., Jana, T.K., 2006. Fluxes of nutrients from the tropical river Hooghly at the land ocean boundary of Sundarbans, NE coast of Bay of Bengal, India. J. Mar. Syst. 62, 9–21.

Murdiyarso, D., Donato, D., Kauffman, J.B., Kurnianto, S., Stidham, M., Kanninen, M., 2009. Carbon Storage in Mangrove and Peatland Ecosystems: A Preliminary Account from Plots in Indonesia. CIFOR, 35 pp.

Murdiyarso, D., Purbopuspito, J., Kauffman, J.B., Warren, M.W., Sasmito, S.D., Donato, D.C., Manuri, S., Krisnawati, H., Taberima, S., Kurnianto, S., 2015. The potential of Indonesian mangrove forests for global climate change mitigation. Nat. Clim. Change 5, 1089.

Nandi, A.C., Bagchi, M.M., Majumder, S.K., 1983. Ecological changes in the Hooghly estuary due to water release from Farakka Barrage. Mahasagar Bull. Natl. Inst. Oceanogra. 16, 209–220.

Nazim, K., Ahmed, M., Khan, M.U., Duranni, A.T.H., Khokhar, A., 2018. Carbon sequestration potential of sandspit backwater mangrove and sediment, Pakistan. Fuuast J. Biol. 8, 209–215.

Nellemann, C., Corcoran, E., Duarte, C.M., Valdés, L., De Young, C., Fonseca, L., Grimsditch, G., 2009. Blue Carbon: A Rapid Response Assessment. United Nations Environment Programme, GRID-Arendal.

Odum, W.E., Heald, E.J., 1972. Trophic analysis of an estuarine mangrove community. Bull. Mar. Sci. 22, 671–738.

Ortega, A., Geraldi, N.R., Duarte, C.M., 2020. Environmental DNA identifies marine macrophyte contributions to Blue Carbon sediments. Limnol. Oceanogr 9999. https://doi.org/10.1002/lno.11579.

Othman, M.A., 1994. Value of mangroves in coastal protection. Hydrobiologia 285, 277–282.

Ouyang, X., Lee, S.Y., 2014. Updated estimates of carbon accumulation rates in coastal marsh sediments. Biogeosciences 11, 5057–5071, 201.

Page, S.E., Rieley, J.O., Banks, C.J., 2011. Global and regional importance of the tropical peatland carbon pool. Glob. Change Biol. https://doi.org/10.1111/j.1365-2486.2010.02279.x.

Parnell, A.C., Inger, R., Bearhop, S., Jackson, A.L., 2010. Source partitioning using stable isotopes: coping with too much variation. PLoS One 5, 9672.

Penha-Lopes, G., Bartolini, F., Limbu, S., Cannicci, S., Kristensen, E., Paula, J., 2009. Are fiddler crabs potentially useful ecosystem engineers in mangrove wastewater wetlands? Mar. Pollut. Bull. 58, 1694–1703.

Perera, K.A.R.S., Amarasinghe, D., 2018. Ecosystem carbon stock of mangroves at the Batticaloa Lagoon, Sri Lanka. OUSL J. 13, 81–100.

Pérez, A., Machado, W., Gutiérrez, D., Borges, A.C., Patchineelam, S.R., Sanders, C.J., 2018. Carbon accumulation and storage capacity in mangrove sediments three decades after deforestation within a eutrophic bay. Mar. Pollut. Bull. 126, 275–280.

Pham, V.-H., Luu, V.-D., Nguyen, T.-T., Koji Omoric, K., 2017. Will restored mangrove forests enhance sediment organic carbon and ecosystem carbon storage? Regional Stud. Mar. Sci. 14, 43–52.

Phang, V.X.H., Chou, L.M., Friess, D.A., 2015. Ecosystem carbon stocks across a tropical intertidal habitat mosaic of mangrove forest, seagrass meadow, mudflat and sandbar. Earth Surf. Process. Landforms 40, 1387–1400.

Phillips, D.L., Newsome, S.D., Gregg, J.W., 2005. Combining sources in stable isotope mixing models: alternative methods. Oecologia 144, 520–527.

Polidoro, B.A., Carpenter, K.E., Collins, L., Duke, N.C., Ellison, A.M., Ellison, J.C., Farnsworth, E.J., Fernando, E.S., Kathiresan, K., Koedam, N.E., 2010. The loss of species: mangrove extinction risk and geographic areas of global concern. PLoS One 5, e10095.

Poungparn, S., Charoenphonphakdi, T., Sangtiean, T., Patanaponpaiboon, P., 2016. Fine root production in three zones of secondary mangrove forest in Eastern Thailand. Trees 30, 467–474.

Prasad, M.B.K., Ramanathan, A.L., 2009. Organic matter characterization in a tropical estuarine-mangrove ecosystem of India: preliminary assessment by using stable isotopes and lignin phenols. Estuar. Coast. Shelf Sci. 84, 617–624.

Prasad, M.B.K., Kumar, A., Ramanathan, A.L., Datta, D.K., 2017. Sources and dynamics of sedimentary organic matter in Sundarban mangrove estuary from Indo-Gangetic delta. Ecol. Processes 6, 8.

Primavera, J., 1998. Mangroves as nurseries: shrimp populations in mangrove and non-mangrove habitats. Estuar. Coast. Shelf Sci. 46, 457–464.

Rahman, M.M., Khan, M.N.I., Hoque, A.F., Ahmed, I., 2015. Carbon stock in the Sundarbans mangrove forest: spatial variations in vegetation types and salinity zones. Wetlands Ecol. Manag. 23, 269–283.

Ray, R., Jana, T.K., 2017. Carbon sequestration by mangrove forest: one approach for managing carbon dioxide emission from coal-based power plant. Atmos. Environ. 171, 140–154.

Ray, R., Shahraki, M., 2016. Multiple sources driving the organic matter dynamics in two contrasting tropical mangroves. Sci. Total. Environ. 571, 218–227.

Ray, R., Thouzeau, G., Walcker, R., Vantrepotte, V., Morvan, S., Devesa, J., le Goff, M., Michuad, E., 2020. Mangrove-derived organic and inorganic carbon export to the Atlantic Ocean? Time series study in the Guianese estuarine system, Latin America. J. Geophy. Res.: BioGeoSci. 125. https://doi.org/10.1029/2020JG005739.

Ray, R., Weigt, M., 2018. Seasonal and habitat-wise variations of organic carbon in the arid environment of the Persian Gulf, Iran. Cont. Shelf Res. 165, 60–70.

Ray, R., Ganguly, D., Chowdhury, C., Dey, M., Das, S., Dutta, M.K., Mandal, S.K., Majumder, N., De, T.K., Mukhopadhyay, S.K., Jana, T.K., 2011. Carbon sequestration and annual increase of carbon stock in a mangrove forest. Atmos. Environ. 45, 5016–5024.

Ray, R., Baum, A., Gleixner, G., Rixen, T., Jana, T.K., 2018a. Exportation of dissolved (inorganic and organic) and particulate carbon from mangroves and their implications to the carbon budget in the Indian Sundarbans. Sci. Total Environ. 621, 535–547.

Ray, R., Chowdhury, C., Majumdar, N., Dutta, M.K., Mukhopadhyay, S.K., Jana, T.K., 2013. Improved model calculation of atmospheric CO_2 increment in affecting carbon stock of tropical mangrove forest. Tellus B: Chem. & Phys. Meteorol. 65, 18981.

Ray, R., Michaud, E., Aller, R.C., Vantrepotte, V., Gleixner, G., Walcker, R., Devesa, J., Le Goff, M., Morvan, S., Thouzeau, G., 2018b. Sources and distribution of carbon (DOC, POC, DIC) in a mangrove dominated estuary (French Guiana, South America). Biogeochemistry 138, 297–321.

Ray, R., Rixen, T., Baum, A., Malik, A., Gleixner, G., Jana, T.K., 2015. Distribution, sources and biogeochemistry of organic matter in a mangrove dominated estuarine system (Indian Sundarbans) during the pre-monsoon. Estuar. Coast. Shelf Sci. 167, 404–413.

Ray, R., Majumder, N., Chowdhury, C., Das, S., Jana, T.K., 2017. Phosphorus budget of the Sundarban mangrove ecosystem: box model approach. Estuar. Coast. 41, 1036–1049.

Ranjan, R.K., et al., 2011. Elemental and stable isotope records of organic matter input and its fate in the Pichavaram mangrove–estuarine sediments (Tamil Nadu, India). Marine Chem. 126, 163–172.

Ray, R., et al., 2012. Wood chemistry and density: An analog for response to the change of carbon sequestration in mangroves. Carbohydrate Polymers 90, 102–108.

Reef, R., Atwood, T.B., Samper-Villarreal, J., Adame, M.F., Sampayo, E.M., Lovelook, C.E., 2017. Using eDNA to determine the source of organic carbon in seagrass meadows. Limnol. Oceanogr. 62, 1254–1265.

Resmia, P., Manjua, M.N., Gireeshkumar, T.R., Ratheesh Kumar, C.S., Chandramohanakumara, N., 2016. Source characterisation of sedimentary organic matter in mangrove ecosystems of Northern Kerala, India: inferences from bulk characterisation and hydrocarbon biomarkers. Regional Stud. Mar. Sci. 7, 43–54.

Rezende, C.E., Lacerda, L.D., Ovalle, A.R.C., Silva, L.F.F., 2007. Dial organic carbon fluctuations in a mangrove tidal creek in Sepetiba bay, Southeast Brazil. Braz. J. Biol. 67, 673–680.

Richards, D.R., Benjamin, S.T., Lahiru, W., et al., 2020. Quantifying net loss of global mangrove carbon stocks from 20 years of land cover change. Nat. Comm. 11. https://doi.org/10.1038/s41467-020-18118-z.

Richards, D.R., Friess, D.A., 2015. A rapid indicator of cultural ecosystem service usage at a fine spatial scale: content analysis of social media photographs. Ecol. Indic. 53, 187–195.

Rivera-Monroy, V.H., Twilley, R.R., Davis III, S.E., Childers, D.L., Simard, M., Chambers, R., Jaffe, R., Boyer, J.N., Rudnick, D.T., Zhang, K., 2011. The role of the Everglades Mangrove Ecotone Region (EMER) in regulating nutrient cycling and wetland productivity in South Florida. Crit. Rev. Environ. Sci. & Technol. 41, 633–669.

Robertson, A.I., 1986. Leaf-burying crabs: their influence on energy flow and export from mixed mangrove forests (Rhizophora spp.) in Northeastern Australia. J. Exp. Mar. Biol. Ecol. 102, 237–248.

Robertson, A.I., Alongi, D.M., 1995. Role of riverine mangrove forest in organic carbon export to the tropical coastal ocean: A preliminary mass balance for the Fly Delta (Papua New Guinea). Geo-Marine Lett. https://doi.org/10.1007/BF01204454.

Rodhe, H., Bjorkstrom, A., 1979. Some consequence of nonproportionality between fluxes and reservoir contents in natural systems. Tellus 31, 269–278.

Romigh, M.A., David, S.E., Rivera-Monroy, V.H., Twilley, R.R., 2006. Flux of organic carbon in a riverine mangrove wetland in the Florida coastal Everglades. Hydrobiologia 569, 505–516.

Saderne, V., Geraldi, N.R., Macreadie, P.I., 2019. Role of carbonate burial in Blue Carbon budgets. Nat. Commun. 10, 1106. https://doi.org/10.1038/s41467-019-08842-6.

Sahu, S.C., Kumar, M., Ravindranath, N.H., 2016. Carbon stocks in natural and planted mangrove forests of Mahanadi mangrove wetland, East Coast of India. Curr. Sci. 110, 12.

Saifullah, S., 1997. Management of the Indus delta mangroves. In: Coastal Zone Management Imperative for Maritime Developing Nations. Springer, pp. 333–346.

Salmo III, S.G., Malapit, V., Garcia, M.C.A., Pagkalinawan, H.M., 2019. Establishing rates of carbon sequestration in mangroves from an earthquake uplift event. Biol. Lett. 15, 20180799.

Sanders, C.J., Smoak, J.M., Naidu, A.S., Sanders, L.M., Patchineelam, S.R., 2010. Organic carbon burial in a mangrove forest, margin and intertidal mud flat. Estuar. Coast. Shelf Sci. 90, 168–172.

Sanders, C.J., Maher, D.T., Tait, D.R., Williams, D., Holloway, C., Sippo, J.Z., Santos, I.R., 2016. Are global mangrove carbon stocks driven by rainfall? J. Geophys. Res. Biogeosci. 121, 2600–2609.

Santos, I.R., Bryan, K.R., Pilditch, C.A., Tait, D.R., 2014. Influence of porewater exchange on nutrient dynamics in two New Zealand estuarine intertidal flats. Mar. Chem. 167, 57–70.

Santos, I.R., Maher, D.M., Larkin, R., Webb, J.R., Sanders, C.J., 2018. Carbon outwelling and outgassing vs. burial in an estuarine tidal creek surrounded by mangrove and saltmarsh wetlands. Limnol. Oceanogr. 9999, 1–18.

Sasmito, S.D., Taillardat, P., Clendenning, J.N., Cameron, C., Friess, D.A., Murdiyarso, D., Hutley, L.B., 2019. Effect of land-use and land-cover change on mangrove blue carbon: a systematic review. Glob. Change Biol. 25, 4291–4302.

Sasmito, S.D., Kuzyakov, Y., Lubisg, A.A., Murdiyarso, D., Hutley, L.B., Bachrii, S., Friess, D.A., Martiusk, C., Borchard, N., 2020. Organic carbon burial and sources in soils of coastal mudflat and mangrove ecosystems. Catena 104414.

Schile, L.M., Kauffman, J.B., Crooks, S., Fourqurean, J.W., Glavan, J., Megonigal, J.P., 2017. Limits on carbon sequestration in arid blue carbon ecosystems. Ecol. Appl. 27, 859–887.

Sen, A., Bhadury, P., 2017. Intertidal foraminifera and stable isotope geochemistry from world's largest mangrove, Sundarbans: assessing a multiproxy approach for studying changes in sea-level. Estuar. Coast. Shelf Sci. 192, 128–136.

Sharma, S., et al., 2020. The impacts of degradation, deforestation and restoration on mangrove ecosystem carbon stocks across Cambodia. Sci. Tot. Environ. https://doi.org/10.1016/j.scitotenv.2019.135416.

Sheehan, L., Sherwood, E., Moyer, R.P., Radabaugh, K.R., Simposon, S., 2019. Blue carbon: an additional driver for restoring and preserving ecological services of coastal wetlands in Tampa Bay (Florida, USA). Wetlands 39, 1317–1328.

Sidik, F., Neil, D., Lovelock, C.E., 2016. Effect of high sedimentation rates on surface sediment dynamics and mangrove growth in the Porong River, Indonesia. Mar. Pollut. Bull. 107, 355–363.

Sidik, F., Adame, M.F., Lovelock, C.E., 2019. Carbon sequestration and fluxes of restored mangroves in abandoned aquaculture ponds. J. Indian Ocean Region 15, 177–192.

Sippo, J.Z., Maher, D.T., Tait, D.R., Holloway, C., Santos, I.R., 2016. Are mangroves drivers or buffers of coastal acidification? Insights from alkalinity and dissolved inorganic carbon export estimates across a latitudinal transect. Glob. Biogeochem. Cy. 30 https://doi.org/10.1002/2015GB005324.

Smoak, J., Breithaupt III, J., Smith, T., Sanders, C., 2013. Sediment accretion and organic carbon burial relative to sea-level rise and storm events in two mangrove forests in Everglades National Park. Catena 104, 58–66.

Snedaker, S., 1984. Mangroves: a summary of knowledge with emphasis on Pakistan. In: Haq, B.U., Milliman, J.D. (Eds.), Marine Geology and Oceanography of Arabian Sea and Coastal Pakistan. Van Nostrand Reinhold Co., NY.

Spalding, M., Kainuma, M., Collins, L., 2010. World Atlas of Mangroves. Earthscan.

Stieglitz, T., Ridd, P., Muller, P., 2000. Passive irrigation and functional morphology of crustacean burrows in a tropical mangrove swamp. Hydrobiologia 421, 69–76.

Taillardat, P., Ziegler, A.D., Friess, D.A., Widory, D., Truong Van, V., David, F., Thành- Nho, N., Marchand, C., 2018. Carbon dynamics and inconstant porewater input in a mangrove tidal creek over contrasting seasons and tidal amplitudes. Geochim. Cosmochim. Ac. 237, 32–48.

Thi-Hien, H., Marchand, C., Aimé, J., Kim Cuc, N.T., 2018. Seasonal variability of CO_2 emissions from sediments in planted mangroves (Northern Viet Nam). Estuar. Coast. Shelf Sci. 213, 28–39.

Troxler, T.G., Barr, J.G., Fuentes, J.D., Engel, V., Anderson, G., Sanchez, C., Lagomasino, D., Price, R., Davis, S.E., 2015. Component-specific dynamics of riverine mangrove CO_2 efflux in the Florida coastal Everglades. Agric. Forest Meteorol. 213, 273–282.

Tue, N.T., Dung, L.V., Nhuan, M.T., Omori, K., 2014. Carbon storage of a tropical mangrove forest in Mui Ca Mau National Park, Vietnam. Catena 121.

Twilley, R.R., 1985. The exchange of organic carbon in basin mangrove forests in a Southwest Florida estuary. Estuar. Coast. Shelf Sci. 20, 543–557.

Twilley, R.R., Lugo, A.E., Patterson-Zucca, C., 1986. Production, standing crop, and decomposition of litter in basin mangrove forests in Southwest Florida. Ecology 67, 670–683.

Twilley, R.R., Chen, R.H., Hargis, T., 1992. Carbon sinks in mangroves and their implications to carbon budget to tropical coastal ecosystems. Water Air Soil Pollut. 64, 265–288.

Twilley, R.R., Rovai, A.S., Riul, P., 2018. Coastal morphology explains global blue carbon distributions. Front. Ecol. & Environ. 16, 503–508.

Vinh, T.V., Allenbach, M., Joanne, A., Marchand, C., 2019. Seasonal variability of CO_2 fluxes at different interfaces and vertical CO_2 concentration profiles within a *Rhizophora* mangrove forest (Can Gio, VietNam). Atmos. Environ. 201, 301–309.

Vinod, K., Koya, A.A., KunhiKoya, V.A., Silpa, P.G., Asokan, P.A., Zacharia, P.U., Joshi, K.K., 2018. Biomass and carbon stocks in mangrove stands of Kadalundi estuarine wetland, South-Eest Coast of India. Indian J. Fish. 65, 89–99.

Wafar, S., Untawale, A.G., Wafar, M., 1997. Litter fall and energy flux in a mangrove ecosystem. Estuar. Coast. Shelf Sci. 44, 111–124.

Wahid, S.M., Mukand, S.B., Bhuiyan, A.R., 2007. Hydrologic monitoring and analysis in the Sundarbans mangrove ecosystem, Bangladesh. J. Hydrol. 332, 381–395.

Walcker, R., Gandois, L., Proisy, C., Corenblit, D., Mougin, É., Ray, R., Fromard, F., 2018. Control of 'blue carbon' storage by mangrove ageing: evidence from a 66-year-chronosequence in French Guiana. Glob. Change Biol. 24, 2325–2338.

Ware, J.R., Smith, S.V., Reaka-kudla, M.L., 1992. Coral reefs: sources or sinks of atmospheric CO2? Coral Reefs 11, 127–130.

Wells, J.T., Coleman, J.M., 1985. Deltaic morphology and sedimentology, with special reference to the Indus River delta. In: Louisiana State Univ Baton Rouge Coastal Studies Inst.

Xiong, Y., Liu, X., Guan, W., Liao, B., Chen, Y., Li, M., Zhong, C., 2016. Fine root functional group based estimates of fine root production and turnover rate in natural mangrove forests. Plant Soil 413, 83–95.

Yang, P., Zhang, Y., Lai, D.Y.F., Tan, L., Jin, B., Tong, C., 2018. Fluxes of carbon dioxide and methane across the water–atmosphere interface of aquaculture shrimp ponds in two subtropical estuaries: the effect of temperature, substrate, salinity and nitrate. Sci. Total Environ. 635, 1025–1035.

Yuan, Z.Y., Chen, H.Y.H., 2010. Fine root biomass, production, turnover rates, and nutrient contents in boreal forest ecosystems in relation to species, climate, fertility, and stand age: literature review and meta-analyses. Crit. Rev. Plant Sci. 29, 204–221.

Zhang, K., Liu, H., Li, Y., Xu, H., Shen, J., Rhome, J., Smith III, T.J., 2012. The role of mangroves in attenuating storm surges. Estuar. Coast. Shelf Sci. 102–103, 11–23.

CHAPTER

15

Quantity and quality of organic matter in mangrove sediments

Martin Zimmer[1,2,3], *Véronique Helfer*[1]

[1]ZMT - Leibniz Centre for Tropical Marine Research, Bremen, Germany; [2]University of Bremen, Faculty of Biology/Chemistry, Bremen, Germany; [3]IUCN-SSC Mangrove Specialist Group

OUTLINE

1. **Organic matter in mangrove sediments—setting the scene** 369
   - *1.1 Mangrove ecosystems: sources or sinks of organic matter?* 371
   - *1.2 Dynamics of organic matter turnover* 373
2. **Quantity versus quality of SOM** 374
3. **Origin of SOM** 376
   - *3.1 Stable isotope signatures* 376
   - *3.2 Biomarkers and untargeted environmental metabolomics* 378
   - *3.3 Environmental DNA metabarcoding and DNA metagenomics* 381
4. **Mangroves worldwide—alike but distinct** 382
   - *4.1 Implications for mangrove management?* 387
5. **Conclusion** 388

References 388

1. Organic matter in mangrove sediments—setting the scene

Mangrove **sediments** are derived from a plethora of different terrestrial, limnetic, and marine sources. Upon deposition of remnants of organisms and physical, chemical, and biotic processes acting on this material, sediments become enriched in organic matter (OM). The difference between a mangrove sediment and a terrestrial forest **soil** is that "soil" is generated

Dynamic Sedimentary Environments of Mangrove Coasts
https://doi.org/10.1016/B978-0-12-816437-2.00015-X

on site through both weathering of a bed rock and the deposition and **decomposition** of OM, whereas "sediment" is derived from a weathered mother rock that, upon erosion and transport, is deposited off its original site. A vegetated "sediment" in the upper intertidal or **supratidal zone** may at some point be considered a "soil" upon OM deposition on stabilized sediment layers (through physical deposition and biotic formation)—however, the mineral component will still be allochthonous, rendering this substratum a sediment *sensu strictu*; the latter would also apply to sediment found in seagrass beds or near coral reefs.

In the era of climate change and "blue carbon" attention has mostly been focusing on the role of mangrove ecosystems as carbon (C) stores. However, OM is composed of many more elements than only C (H, O, N, P, S, etc.), and in the context of climate-active gases (mostly carbon dioxide, dinitrogen oxide, and methane), nitrogenous compounds may prove as important as carbonic compounds. Many of these elements present in OM are **nutrients** that are taken up by organisms either incorporated in organic molecules (e.g., upon decay of macromolecules: amino acids, simple sugars) or as inorganic compounds upon mineralization (e.g., ammonium, nitrate, phosphate). The role of OM as store of nitrogen (N) (as compound of a potential climate-active gas) or as source of elemental nutrients other than C is still understudied. Hereafter, we will refer to sediment organic matter (SOM); when focusing on organic C exclusively, we will explicitly refer to sediment organic carbon (SOC).

This chapter discusses the role of mangroves and their sediments as sources and sinks of OM with an emphasis on the dynamics of OM turnover that depends on OM quality that is, in turn determined by its origin. Along this line, we introduce novel approaches to better understand the underlying processes of transport and degradation.

Decay—*the degradation of detritus through physicochemical (e.g., leaching, light-driven and atmospheric degradation) and microbial processing.*

Decomposition— *the entire process of detrital breakdown includes* ***decay****, promotion of microbial activity and leaching (e.g., through fragmentation) by* ***detritivores****, and direct contributions of detritivores, such as feeding and digestion, and* ***mineralization****.*

Mineralization—*the final steps of* ***decomposition*** *result in the formation of inorganic compounds (e.g., nitrate and nitrite, or the climate-active gases carbon dioxide (through respiration) and dinitrogen oxide) from OM and, thus, the release of soluble* ***nutrients*** *into forms that are available to plants. The processes of nitrification (aerobic transformation, under oxic conditions, of ammonium/ammonia into nitrite and nitrate) and denitrification (anaerobic transformation, under anoxic conditions, of nitrate into nitrite, nitric oxide, nitrous oxide, and finally molecular nitrogen) are the result of concerted actions of specific microbial communities.*

Nutrients—*according to the classical definition, a nutrient is a chemical element required by an organism for its metabolism and growth. Along this line, C, H, N, O, P, S (macronutrients), as well as, e.g., Ca, K, Mg, Fe (micronutrients) are nutrients, whereas organic compounds—as parts of OM, such as carbohydrates, proteins, amino acids, lipids (fats), or vitamins—or inorganic compounds, such as ammonia, nitrate, or phosphate, are not. OM relates to nutrients as being a source of organic precursors of (elemental) nutrients through being transformed into inorganic (mineral) compounds upon decomposition, decay, and mineralization. Within this chapter, we use the term nutrients sensu lato, including organic compounds and inorganic compounds, such as ammonium, nitrate, and phosphate.*

Sediment*—mineral material derived from weathering and erosion of a mother rock that is then transported through the action of wind, water, ice, or by gravity, and deposited at a different place. Upon deposition and stabilization, sediments can become enriched in organic material of various allochthonous or autochthonous origins. As this process is typical for soils, the appearance of the stable and OM–rich forest floor of a mangrove forest growing high in the intertidal may appear as a soil, but the allochthonous origin of the mineral portion determines this still as a sediment.*

Soil*—a product of the simultaneous weathering of the bed rock (from below) and biotic activity, resulting in incorporation of OM (from above), together referred to as pedogenesis. The distinction from sediments is that the mineral compartment of a soil is autochthonous in origin and remains in place.*

Supratidal*—the zone immediately above the high intertidal is not prone to regular tidal submersion; however, it is still influenced by the marine environment, e.g., through spray of seawater or extreme tidal and storm events.*

1.1 Mangrove ecosystems: sources or sinks of organic matter?

Over decades, it has been controversially discussed whether mangroves are sources or sinks of OM, and the plethora of studies performed provided contrasting insights and evidence, largely depending on the geographical region and the geomorphological type of mangrove. Most studies agree that exchanges of OM among coastal ecosystems are common; some provide evidence for enhanced C storage in connected versus isolated ecosystems (e.g., Saavedra-Hortua et al., 2020; Huxham et al., 2018; Cragg et al., 2020), but our understanding of the contribution of OM of different origins is still in its infancy, partly because techniques for unambiguously disentangling different sources of OM and following decay processes and chemical degradation in detail have been developed and started being applied to studies on dynamics of OM in mangroves only recently.

Hence, OM deposited on and in the mangrove sediment may be autochthonous, stemming from mangrove plants, algae thereon, animal inhabitants (detritus and its products of decay and decomposition) or from dead and/or alive microbes (e.g., archaea, bacteria, fungi, protists), or allochthonous, being deposited in mangroves as **spatial subsidies**. The latter, in turn, can be of marine origin, having been washed in by incoming tides, of riverine origin, reaching the mangrove sediment upon inundation or through porewater flow, or of terrestrial origin, being washed into the mangrove from the hinterland by surface runoff, riverine transport, or groundwater discharge (Fig. 15.1). On the other side, OM originating from mangrove ecosystems, be it particulate (POM) or dissolved (DOM), may also be washed out the mangrove by outgoing tides or porewater exchange (Figs. 15.1 and 15.2). Above all, migrating animals that take up food in one habitat and defecate in an adjacent habitat significantly contribute to the distribution of allochthonous sources of OM (spatial subsidies) among ecosystems. While naturally both import and export will take place simultaneously, which of these processes prevail, rendering mangroves either a sink or a source of OM, depends on local and regional settings (including topography, hydrology, and biotic communities in place) or extreme events.

Once the OM is incorporated in the sediment, it becomes part of a substratum that is much more stable than tidal or riverine waters or surface runoff, and export of OM, rendering

FIGURE 15.1 Schematic sketch of possible trajectories of (vegetal) OM fluxes and exchange among mangroves and adjacent ecosystems. Individual components of sketch obtained from http://ian.umces.edu. *OM*, organic matter.

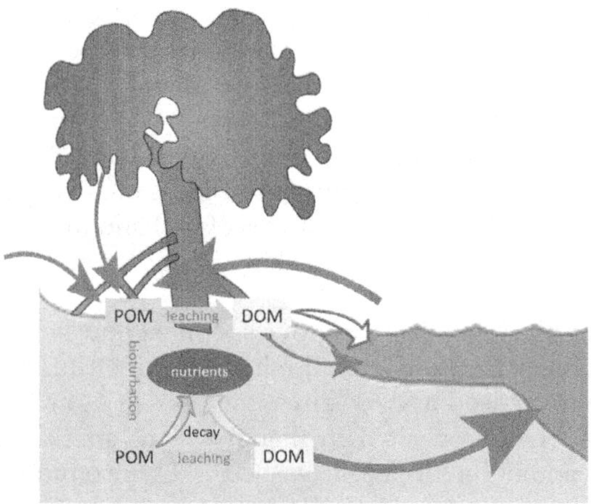

FIGURE 15.2 Schematic sketch of potential inflows of POM into and outflows of DOM out of mangroves, and dynamics of POM and DOM in mangroves sediments. Individual components of sketch obtained from http://ian.umces.edu. *DOM*, dissolved organic matter; *POM*, particulate organic matter.

mangroves a source, becomes less important (unless sediments are destabilized and removed upon extreme currents, waves or storm events, or after mangrove degradation or felling).

One peculiarity of most mangrove sediments is the saline and anoxic conditions that hamper and slow down microbial decay processes, thus rendering OM stable against mineralization. However, dense networks of roots that connect the below-ground realm and the air, such as pneumatophores of mangroves of the genera *Avicennia* or *Sonneratia*, as well as bioturbation through the action of burrowing crabs may counteract this anoxic stabilization through oxygenating the sediment and boosting aerobic decay processes (Sakho et al., 2015). Therefore, the OM storage capacity of mangrove ecosystems depends strongly on the biotic communities in place.

***Spatial Subsidies** (cross-boundary subsidies)—OM, be it alive (organisms) or dead (detritus), that is passively or actively transported into a habitat or an ecosystem from an allochthonous source and affects autochthonous populations, communities, and consumer-resource and food web dynamics. The distinction between the active movement of motile organisms and the passive transport of their mobile propagules (seeds, seedlings, larvae, or other dispersal stages) or (dissolved or suspended particulate) OM through the action of waves, tides, or currents links this process to the concept of connectivity, where the movement among habitats is not only a function of an organism's attributes but also depends on land- or seascape characteristics. Accordingly, connectedness is the degree to which a sea- or landscape impedes or facilitates movement among habitats, emphasizing both a structural and a behavioral element to this concept.*

1.2 Dynamics of organic matter turnover

OM can be categorized according to various characteristics. From an ecological point of view, the most relevant distinction is between readily degraded labile compounds (that are also used as sources of nutrients by many organisms) and recalcitrant compounds that remain essentially unmodified and stable for extended time (c.f., Schmidt et al., 2011). The differential recalcitrance of different classes of organic compounds, however, has rarely been taken into account in studies of marine or coastal OM dynamics (Cragg et al., 2020). OM turnover is driven by abiotic and biotic processes (for summary: Zimmer, 2019). Particulate organic matter (POM) is prone to leaching of water-soluble compounds (e.g., phenolic compounds, amino acids, simple sugars) into the porewater, through which they become part of SOM s.l. as dissolved organic matter (DOM: Fig. 15.2). The remaining detritus exhibits increasingly low quality as food and nutrient source over time (c.f. Treplin and Zimmer, 2012). In a recent review, Cragg et al. (2020) stress the difference in chemical composition of detritus from vascular plants (terrestrial) versus macroalgae (marine), the former being characterized by lignocellulose, a highly recalcitrant complex of cellulose, hemicelluloses, and lignins. In terms of transport and exchange of OM among neighboring ecosystems, the distinction between DOM and POM is essential. DOM is dissolved in the water masses that run off the mangroves during outgoing ebb tide, and will, thus, be exported from mangroves into coastal waters. The release of small soluble molecules from the lignin moiety after decay of cellulose and hemicelluloses, feeding the pool of DOM, is very slow, particularly in anoxic sediments (Cragg et al., 2020). POM, by contrast, is partly retained by above-ground vegetation structures or inside crab burrows, rendering mangroves a sink.

Fragmentation of POM, e.g., through the feeding activity of detritivorous animals or through wave action, increases the particle surface and facilitates further leaching. As a purely physical process, fragmentation per se does not result in any chemical modification of the OM. On the sediment surface, POM is subjected to UV-/light-driven and atmospheric degradation, e.g., autoxidation. Along with the degradation of chemical compounds upon microbial activity, these processes are referred to as **decay** (Zimmer, 2019), resulting in size reduction of detrital chemical compounds and, at the same time, relative accumulation of larger and more recalcitrant compounds (e.g., lignins). Feeding, digestion and defecation by **detritivores**, and the related changes in OM chemistry, add another level of OM dynamics. Together with microbial activity inevitably connected with detritivory, this results in **decomposition** (Zimmer, 2019). Another pathway of transformation of OM of vegetal origin is

herbivory. Through defecation, herbivorous animals (such as insects, snails, crabs) also contribute to the pool of OM (Zimmer and Topp, 2002, in terrestrial forests), but this direct link between living plant biomass and SOM, without the pathway through plant detritus, is essentially unstudied in mangroves. Bats and kangaroos, frugivores and herbivores, respectively, that feed in terrestrial ecosystems and seek shelter in mangroves drop fecal pellets at their roosting sites, and thus, contribute to nutrient dynamics inside the mangrove (Reef et al., 2014; Lizard Island off the North-East Australian coast). Waterbirds, mostly piscivorous, nesting and roosting in mangroves, also markedly contribute to the nutrient status of mangroves (McFadden et al., 2016; in Pacific Central America). Such direct or indirect effects of **carnivory** are even less studied.

As the majority of decay processes that result in chemical modification of OM compounds depend on aerobic organisms and the presence of oxygen, these processes are markedly slowed down under saline and anoxic conditions of mangrove sediments. Hence, rather than being successively degraded and eventually mineralized, a large proportion of the OM that gets incorporated (see above) in the sediment remains there essentially unmodified over decades and centuries, leading to OM-accumulation in the sediment. The next section (Section 2) will deal with this particular feature of mangrove ecosystems.

Beyond these environmental conditions that mediate the fate, turnover rates and dynamics of OM (both particulate and dissolved), both the processes that drive these dynamics and the resulting products depend on OM structure and quality, i.e., its origin (Schmidt et al., 2011, for terrestrial soils). Thus, in addition to the above particular conditions in the sediment, the recalcitrance of the OM determines the stability, and thus, size of the SOM stock. These aspects will be discussed in Section 3.

Carnivory—organisms *that consume animals, either partly (partial carnivory) or entirely, are carnivores. Some (opportunistic) carnivores also occasionally consume vegetal food sources; then they are considered omnivorous.*

Detritivory—*animals that consume dead material of vegetal or animal origin are detritivores. Alike* ***herbivores****, the food of detritivores is often of low quality, but deterrent or toxic compounds are usually leached off the detritus before consumption. Many detritivores also occasionally consume animal food sources to compensate for the low quality of vegetal detritus; then they are considered omnivorous.*

Herbivory—*animals that consume fresh living plant or algal tissues are herbivores. Herbivores are often nitrogen-limited and face food sources of low quality and high contents of recalcitrant, deterrent, or toxic compounds.*

2. Quantity versus quality of SOM

Several methods have been used to estimate the quantity of SOM in mangrove (and other) ecosystems. Here we review the different approaches and point out their differences.

Loss on Ignition (LOI) is the most widespread and simplest approach to total OM content in any environmental sample. The dry weight of the material is compared to what is left after combustion (usually at 550°C for 3–4 h), the difference being considered the amount of OM (that gets lost through ignition). Surprisingly often, this value is converted into organic carbon (C_{org}) content by multiplying with the van Bemmelen factor of 0.58 (or by dividing by 1.724); similarly, C_{org} contents have repeatedly been converted mathematically into OM contents. However, the global use of such a factor has been heavily discussed and proven too high in several empirical studies (c.f. Pribyl, 2010, and references herein), as it does not take into account the specific chemical composition of OM of different origins (c.f. Ouyang and Lee, 2019).

First insight into OM quality, beyond pure quantity, can be achieved by stepwise combustion, i.e., the sequential temperature increase in subsequent steps of ignition. Based on compound-specific thermostability, LOI at low temperatures is a proxy for the content of labile OM, whereas recalcitrant organic compounds are combusted at high temperatures only (e.g., Cheng et al., 2013; Reyerson et al., 2016). Applied to mangrove sediments, this approach provides insight into the (potential) stability of different OM fractions (V. Patire, V. Helfer, M. Zimmer, V. Hatje, unpubl.) and potentially the age of these fractions (Cheng et al., 2013), when used in combination with ^{14}C dating.

Element-specific quantification of OM (obtained with an element analyzer) is often used for rough insight into its chemical composition, such as H, O, C, and N contents. Most commonly, C_{org} and total N contents are used to calculate a C:N ratio as proxy for OM quality (as food or nutrient source; the lower the ratio, the higher the quality is usually considered). Parallel determination of total (C_{tot}) or inorganic (C_{in}) carbon contents will yield additional insight into the (carbonic vs. siliciclastic) origin of the sediment. Even though nitrogenous compounds generally seem to be less prone to leaching than carbonic compounds (e.g., Pirainen et al., 2007), the pure quantity of C or N in any sample (even as C:N ratio) does, however, not allow for deducing the quality, chemical composition, or recalcitrance of the OM. Thus, a combination of different methods is highly recommended.

Water-soluble **aromatic compounds** (e.g., phenolics) are leached off detritus more readily than less water-soluble **aliphatic compounds** (e.g., alkanes and terpenes). The latter accumulate in POM, whereas the former contribute to DOM, both potentially becoming SOM upon entering the sediments. Upon microbial decay, stable aromatic compounds are formed through polymerization (leading to "**humic substances**"), and the aromatic:aliphatic ratio rises (c.f. Stewart et al., 2011). Hence, monitoring the ratio of aromatic:aliphatic compounds (A280/A460, through spectrophotometry) over time (i.e., over depth in sediment cores) can serve as a measure of chemical changes of OM (aromatic *vs.* aliphatic carbonic compounds) during decay (Steinberg, 2003; Treplin and Zimmer, 2012).

Humic substances are the major organic component in most soils and sediments and can be extracted from SOM, but we are far from understanding their chemical structure or formation in any detail. Through a series of differential extraction steps, the two major components of humic substances, **humic acids** (HAs) and **fulvic acids** (FAs), can be separated and distinguished (Steinberg, 2003; Lamar et al., 2014), allowing for gaining insight into different compositions of SOM through determining the HA:FA ratio (Steinberg, 2003; Stewart et al., 2011).

Aliphatic versus aromatic compounds—*among* ***organic compounds****, hydrocarbons (being composed of only carbon and hydrogen) are either* ***aromatic*** *or* ***aliphatic.*** ***Aromatics*** *are always cyclic and characterized by a specific arrangement of double bonds between neighboring C atoms. Thus, aromatics are always unsaturated. Many aromatic compounds contain other elements (e.g., oxygen) bound to a carbon atom. In some aromatic compounds, one or several C atoms are replaced by oxygen, nitrogen, or sulfur (heterocyclic compounds). A typical example of aromatics are phenolic compounds s.l.* ***Aliphatics*** *are saturated or unsaturated chains (noncyclic) or cyclic compounds, but their double bonds (if any, i.e., when they are unsaturated) do not follow the rules for aromatics. Typical examples of aliphatics are (cyclo-)alkanes, alkenes, fatty acids, and terpenes.*

Fulvic acids (FAs)—*when differentially extracting components of* ***humic substances****, FAs remain in solution after removal of HAs upon acidification, because they are soluble at all pH conditions. They differ from* ***HAs*** *mostly in their lower degree of polymerization (i.e., their smaller molecular size), their higher oxygen content, and higher acidity (i.e., larger number of proton-donating functional groups, such as carboxyl groups).* ***FAs*** *are characterized by lower contents of lignin-derived phenols, and thus, overall lower aromaticity. They form particularly strong bonds with* Al^{3+} *and* Cu^{2+} *ions.*

Humic acids (HAs)—*major component of* ***humic substances*** *that can be precipitated through acidification. They differ from* ***FAs*** *mostly in their higher degree of polymerization (i.e., their* larger *molecular size) and their lower acidity (i.e.,* lower *number of proton-donating functional groups, such as carboxyl groups).* ***HAs*** *are characterized by higher contents of nitrogen, and higher contents of lignin-derived phenols, and thus overall higher aromaticity. They* bind cations, *such as* Mg^{2+}, Ca^{2+}, Fe^{2+} *or* Fe^{3+}*, into stable complexes.*

Humic substances—*structurally heterogeneous components of the soil/sediment organic matter, being compounds of biological origin and varying from small- to medium-sized, that autoassemble in supramolecular associations upon abiotic and biotic reactions. Thus, we consider humic substance mixtures of a multitude of organic molecules, many of which with aromatic nuclei and phenolic* or *carboxylic functional groups that contribute to their characteristics and reactivity.*

Organic compounds—*generally, chemical compounds that contain C are considered organic (with few exceptions, such as carbonates and corresponding derivates (CO,* CO_2, $CO_3^{\ddagger}$*) or cyanides (*CN^-*)). Because carbon can form chains with other C atoms and can covalently bind to four other atoms, the variety of organic compounds is essentially unlimited.*

3. Origin of SOM

Several techniques, embracing the quality of SOM (see Section 2), have been used to infer the origin of OM in mangrove sediments, stable isotopes thus far being the most commonly used one. Recently, new developments in molecular and chemical techniques are opening new avenues for tracing back the origin of SOM with higher resolution and more accuracy. Pros and cons of each approaches, summarized in Table 15.1, will be discussed in more detail hereafter.

3.1 Stable isotope signatures

Uncountable studies have analyzed potential origins of SOM through comparing the stable isotope signature (either only $\delta^{13}C$, or $\delta^{13}C$ and $\delta^{15}N$) of SOM with those of potential sources (e.g., phytoplankton, phytobenthos, macroalgae, and seagrass, mangrove, saltmarsh, or

TABLE 15.1 Strengths and weaknesses of techniques for determining the origin of SOM as discussed herein.

	Stable isotope analysis	(Pyrolysis-)GC/MS-omics	eDNA-omics
Status	+established	+/−partly established	+/−partly established
Cost	−moderately expensive	+inexpensive	−expensive
Sample number	−low throughput	+high throughput	+high throughput
Sample preparation	+simple	+simple	−demanding
Taxonomic resolution	−low	+taxon specific	+taxon specific
Qualitative versus quantitative	+semiquantitative	−qualitative (difficult to quantify)	+semiquantitative
Data analysis	− difficult to interpret	−demanding (bioinformatics)	−demanding (bioinformatics)

GC, gas chromatography; *MS*, mass spectrometry; *SOM*, sediment organic matter.

terrestrial vegetation). Using a variety of mixing models, some estimates of the relative contributions of these different sources can be provided, and such approaches have also been used to describe changes in these source-specific contributions over time from deep sediment cores. It is essential for an accurate outcome of these models that reference material (potential sources), stemming from the exact same region and having been sampled at the exact same time—which is impossible for ancient sources in deep cores—is always coanalyzed. However, broad-scale differentiation between the contribution of terrestrial versus marine sources to coastal SOM (e.g., Prasad et al., 2017, and references therein) is a valuable contribution to our understanding of OM exchange and dynamics in changing coastal environments. Further, the contribution of animal tissue to the diets of herbivorous or detritivorous consumers, such as crabs, can clearly be demonstrated through stable isotope analyses (e.g., Kristensen et al., 2010; Hübner et al., 2015), deepening our understanding of the chemical composition, origin, and dynamics of SOM.

On the other hand, both microbial decay and digestion of detritus by detritivores change the isotopic signature of OM, and limitations to predicting these changes accurately may render unambiguous estimates of contributions of mangrove sediments as sinks and sources of OM in a network of coastal ecosystems based on stable isotope signatures difficult, if not impossible (e.g., Bouillon et al., 2008). Through a large-scale and long-term comparison of stable isotope signatures of plants from terrestrial, aquatic, and intertidal habitats, Cloern et al. (2002) determined the variability in these signatures within species and among microhabitats, as well as changes over annual cycles of plant growth and senescence, as obscuring source-specific isotopic signatures. Along the same line, the large dissimilarity between the signatures of primary producers and the OM derived from their decay and sequestration into sediments also confounded the application of C- and N-isotopes for identifying the origin of OM. The same applies to the construction of food webs based on stable isotope signatures, where trophic conversion or enrichment factors have to be included in the mixing models to take into account that heavy isotopes get metabolically enriched, thus allowing for the detection of trophic relationships by shifts in the isotopic signature. While, however, feeding experiments under controlled laboratory conditions can provide accurate estimates of these enrichment factors (although this has rarely been done, most studies refer to general, and thus rough, estimates of trophic enrichment), this is essentially impossible for decay and decomposition processes that transform detritus into SOM. Hence, the resolution of these mixing models is constrained by the accuracy of determining (1) the signatures of potential sources and (2) the enrichment of heavy isotopes upon decay (and decomposition) processes. Further, it is essentially impossible to distinguish OM derived from mangroves and that derived from terrestrial trees through their isotopic signature. Hence, the contribution of mangroves to coastal SOM cannot be dissociated from that of terrestrial sources.

3.2 Biomarkers and untargeted environmental metabolomics

Recent developments in organic chemistry revealed the promising avenue of using **biomarker** compounds. The identification of taxon-specific biomarkers from reference species can also yield insight into the origin of SOM (e.g., Ladd et al., 2019), and the similarity of different SOM and potential sources (coanalyzed as references) can be depicted. The advantages over the use of stable isotope signatures are that (1) references need not necessarily stem from the same region (although, taking into account the possibility of environmental effects on the chemical composition of source organisms, renders this still highly advisable), (2) many biomarkers are taxon specific, and thus, yield a high taxonomic resolution of source contribution.

Decay and decomposition of OM result in changes in its chemical composition. By comparing different sources of OM, and mostly aiming for finding differences in their chemical composition, the biomarker approach explicitly targets those compounds that are recalcitrant and remain (essentially) unchanged over time (upon decay). While the overall chemical composition of mangrove leaves is changed upon digestion by detritivorous mangrove crabs, Bakkar et al. (2017) could still clearly assign crab feces to a distinct food source through (dis-)similarity analysis of their chemical profiles. The resolution of the approach was even high enough to distinguish between the fecal pellets of male and female crabs (Bakkar et al., 2017). This study used the high-throughput approach of pyrolysis-GC/MS (gas chromatography coupled with mass spectrometry; e.g., Tolu et al., 2015) for comparing the source-specific peak patterns of chromatograms ("fingerprints"), without necessarily trying to identify the chemical compounds that underlie these peaks (metabolomic fingerprints: Du and Zeisel, 2013; or untargeted environmental metabolomics: Viant and Sommer, 2013). This approach has been used for analyzing the chemical composition of OM (Buurman and Roscoe, 2011) and inferring changes in climatic conditions and vegetation composition over time (Vancampenhout et al., 2009; Schellekens et al., 2011), potentially indicating different routes of decay or decomposition (Buurman et al., 2009; Vancampenhout et al., 2009; Schellekens and Buurman, 2011) that can be correlated with ecosystem processes and environmental conditions in time and space. In a study on mangrove sediments in North-Eastern Brazil, the pyrolysis approach yielded biomarkers that unambiguously demonstrate the terrigenous origin of most SOM, along with autochthonous input from local mangroves. In contrast to the general approach (see above), some of the biomarkers used therein seemed to decay quickly in the mangrove sediment (Kumar et al., 2019). Along the same line, pathways of leaf senescence and litter decay can be monitored for changes in the organic compound composition, providing insight into drivers of OM turnover through depicting species- versus region-specific patterns of change (V. Helfer, C. Hassenrück, S. Wöhle, M. Zimmer, unpubl.).

In an investigation of a mangrove—seagrass—coral reef ecotone, more than 200 (unidentified) organic compounds have been found to be mangrove-specific, 20 of them unambiguously being assigned to one species, *Rhizophora mangle* (Bakkar, 2018). Similarly, several compounds specific to macroalgae could be detected. Based on those specific compounds, the distribution of mangrove- or macroalgae-derived SOM from the mangrove-dominated coastline through the barrier reef could be mapped (Fig. 15.3).

This approach potentially also allows for estimating the stability of SOM and its resistance against microbial or photochemical decay. Amended by lab experiments under controlled conditions, and based on existing databases that will yield IDs of compounds, this approach will inform about the potential retention of OM of different origins in sediments, and will thus contribute to our understanding of SOM dynamics and the potential of long-term Blue Carbon storage in mangrove (and other) sediments.

A more targeted approach making use of the same high-resolution and high-throughput technique, following the same logic of focusing on stable and recalcitrant compounds, is the specific analysis of *n*-alkanes, aliphatic hydrocarbons of high relevance in the chemical taxonomy of plants. Long-chained *n*-alkanes are mostly produced by terrestrial (higher) plants, medium-length alkane chains by aquatic macrophytes, and short chains are typical for algae and phytoplankton. Hence, the relative contribution of different sources to the total SOM can be inferred from the composition of the *n*-alkane mixture found in sediments (e.g., Charriau et al., 2009; He et al., 2016; Yu et al., 2016). Applying this approach to coastal vegetated areas, however, has only recently been started (e.g., Carreira et al., 2015; Guimarães et al., 2019), and we are still lacking basal background knowledge on *n*-alkanes in mangroves. Furthermore, in some studies, variation among habitat type and plant communities is confounded, and it remains unclear whether the observed *n*-alkane patterns are due to environmental conditions or to the taxonomic or genetic background of the investigated plant communities.

Similarly high taxonomic resolutions can be achieved by focusing on other specific biomarkers, such as lipids, sterols, or saccharides. Saccharides dominated soils of arable fields, whereas n-alkanols were dominant in soils under coniferous and deciduous vegetation in a study in the United States (Rushdi et al., 2016). Among polysaccharides of marine angiosperms, arabinogalactans are typical for mangroves, while galactans occur in seagrasses (Aquino et al., 2005). Lipids are particularly suitable for distinguishing floral, phytoplanktonic, and bacterial sources of OM (e.g., Rushdi et al., 2006), and typical mangrove compounds, such as taraxerol (e.g., Koch et al., 2011), and known degradation products thereof (here: dihydrolacunosic acid) can provide evidence for decay processes (e.g., Jaffé et al., 2006). Besides the potential of identifying sources of OM, the high taxon-specificity of biomarkers also allows for detecting changes in environmental conditions or land-use (reflected by changes in the vegetation) over time, when analyzing deep sediment cores (e.g., Wang et al., 2018).

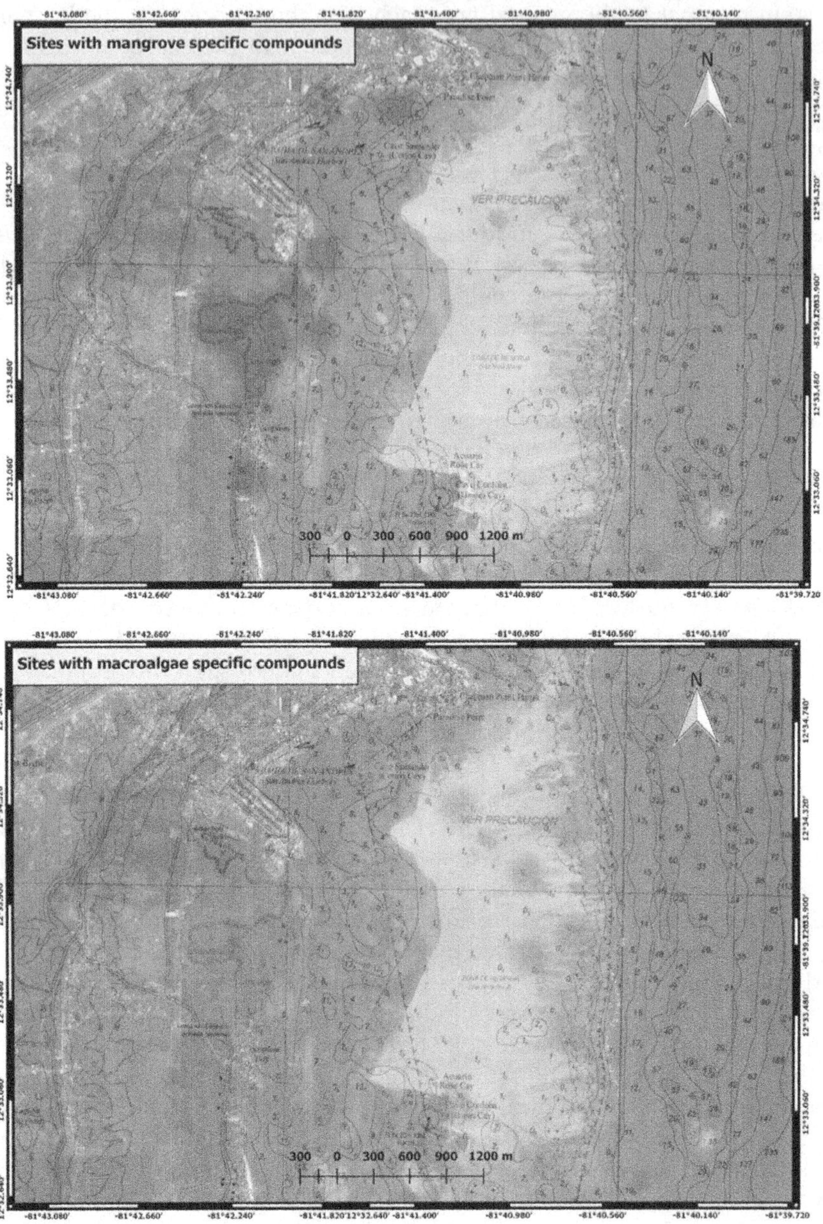

FIGURE 15.3 Distribution of sediment samples containing mangrove- (top) versus macroalgae-derived (bottom) OM in the upper sediment layer across a mangrove–seagrass–reef ecotone (Bakkar, 2018). The position of colored circles indicates the location of sampling spots where specific biomarkers were detected; the area of the circle depicts the OM content of the sediment. *OM*, organic matter.

Biomarker—*any kind of organic compound or its derivatives that identify the presence of an organism, or a group of organisms, in an ancient or recent environmental sample (e.g., sediments).*

3.3 Environmental DNA metabarcoding and DNA metagenomics

The highest taxonomic resolution and accuracy (Table 15.1) can probably be derived from sequence data obtained via **environmental DNA** (eDNA) metabarcoding (that relies on PCR amplification of a target gene) or eDNA metagenomic sequencing (direct shotgun sequencing of the total eDNA pool; see, e.g., Cowart et al., 2018). Initially used to characterize biodiversity patterns of microbial, faunal, or floral communities in space and time (e.g., Bohmann et al., 2014; Thomsen and Willerslev, 2015; Ruppert et al., 2019), eDNA sequencing has very recently been proposed to trace the source of OM in marine sediments (Reef et al., 2017; Helfer and Zimmer, 2018).

Using *rbcL* as DNA metabarcoding marker, Reef et al. (2017) detected 150 different floral contributors to the SOM in seagrass beds in central East Australia, but the vast majority of almost 90% of the eDNA in the upper sediments was derived from seagrass. These findings suggest higher autochthonous contribution to SOM than did stable isotope mixing models, a result attributed to the ambiguity of the stable isotope approach due to a lack of distinctiveness of the isotopic signatures of the potential sources (see above). In addition, the primer used in their study, originally designed for land plants, did not amplify a broad range of macroalgae, and thus, their contribution to the total OM could not be assessed accurately using this specific barcode.

While eDNA metabarcoding has been shown to reliably provide exact information on which species is represented in SOM (presence vs. absence), this approach yields at best semiquantitative information on the contribution of different sources to the total SOM (Table 15.1), mostly due to the many biases linked to the PCR-amplification step (Kelly et al., 2019; Piñol et al., 2019) and other steps of the analysis workflow (McLaren et al. 2020). Hence, the reliability of interpretations based on a semiquantitative approach is still debated (Lamb et al. 2019; Pawlowski et al., 2018; Ruppert et al., 2019). Further, semiquantitative statements on the relative abundance of DNA of different sources can only be deduced—if at all—when using the same barcode (i.e., the same primer pair) for all potential sources—this can be challenging, when a wide range of potential sources (e.g., macroalgae, seagrass, and mangroves) has to be considered, as exemplified by Reef et al. (2017). This challenge can be circumvented by referring only to the presence (*vs.* absence) of particular species in the eDNA pool but not their (relative) abundance.

eDNA metagenomic sequencing bypasses the caveats of the PCR-amplification step and could thereby provide a priori more reliable quantitative estimates of the contribution of various sources to SOM. Nonetheless, metagenomic data consist mainly of sequences related to prokaryotes (e.g., ca 65% in Stat et al., 2017). Therefore, the approach is limited, when the aim is to quantify the relative contribution of a restricted set of eukaryotic organisms, namely plants, to SOM. Nonetheless, a recently published metagenomic pipeline (Ji et al. 2020), mapping shotgun sequencing reads to barcode reference databases or mitogenomes, appears as a promising framework for deriving reliable quantitative estimates of the contribution of different plant sources to SOM.

Alike environmental metabolomics—but different from the stable isotope approach that can take chemical change upon microbial decay or digestive processes explicitly into account (e.g., Meyers, 1994), if and when (trophic) enrichment factors are known—metagenomics of eDNA relies on the stability of the target molecules over time. While this approach has been proven highly valuable in identifying organismic remains in environmental samples thousands of years old (ancient DNA) from permafrost soils (Pedersen et al., 2015), high temperatures in tropical regions might constrain the use of eDNA for analyzing SOM at tropical coasts to more recent deposits. However, the anoxic and often highly saline conditions in mangrove sediments appear promising with regard to DNA preservation probably over decades if not centuries (Helfer and Zimmer, 2018, and references therein).

A considerable advantage of the eDNA approach is that the analysis of sequence data is based on publicly available databases (such as ENA, SILVA) and does not require the collection of reference material from all potential sources of OM from the very exact region. Provided that the DNA sequence of potential source taxa is already archived in those databases, the origin of the OM will be traceable.

One limiting factor in the use of eDNA sequencing is the relatively high costs. Metabarcoding has recently become less expensive, but this approach has some limitations regarding its potential for a quantitative assessment of the origin of SOM (see above). Metagenomics, on the other hand, which circumvents the issues related to PCR bias and would allow for more reliable estimates of the relative contribution of different sources to the SOM (see above), is still relatively expensive. Further, the bioinformatics workflows for the analysis of the resulting data are way more complex and challenging than for metabarcoding. The avenue of using eDNA-sequencing for analyzing OM dynamics in mangrove sediments, particularly when combined with environmental metabolomics, is promising and wide open, but the road is not yet paved and many obstacles have still to be overcome.

Environmental DNA (eDNA)—*DNA that can be recovered from environmental samples (e.g., soil, sediments, water, permafrost, feces), without prior isolation of specific organisms; as such, eDNA is composed of a mixture of genomic DNA from various organisms, and composed of cellular DNA (from living cells) and degraded extracellular DNA (from dead cells) released upon cell structure destruction* (Taberlet et al. 2012).

From Helfer and Zimmer (2018).

4. Mangroves worldwide—alike but distinct

One of the most cited references for **organic carbon stocks** (erroneously coined **storage** therein) in mangroves is Donato et al. (2011). It is mostly based on their data that people claim that mangroves contain a greater amount of C_{org} in their sediment than most other ecosystems. This paper is built on cores of varying depths from 25 sites covering a large portion of the Indo-West Pacific. What, however, can we actually deduce from this data set for mangroves in general and worldwide? Both earlier and later studies showed huge variability of carbon (and OM) stocks at different spatial scales (for an overview, see: Kusumaningtyas et al., 2019), suggesting that mangrove SOM stocks differ with environmental and geographical factors (Ouyang and Lee, 2019). When including (living) biomass, both above- and below-ground, in global estimates of C_{org} or OM stored in mangrove stands, often neither species-specific nor regional peculiarities are taken into account, adding remarkably to the inherent uncertainties. Further, not clearly distinguishing between C_{org} and OM in sediments causes additional confusion (see above). Considering this, Ouyang and Lee (2019) recently reestimated global C_{org} stocks in mangrove sediments (upper 1 m) and potentially corrected the previously published values on average by about −20%.

Within and among biogeographical regions, mangroves are clearly distinct from each other, both in terms of species composition and with respect to structure, density and biomass, as well as in regard to sediment characteristics that might affect communities responsible for the decay and decomposition, and therefore the fate, of OM (e.g., grain size, inundation patterns, salinity). An extreme example are the arid mangrove patches of Baja California in Northern Mexico (Fig. 15.4). There, stunted trees of *R. mangle, Laguncularia racemosa,* and *Avicennia germinans* (along a gradient from sea to land) of mostly less than 2 m height grow in direct vicinity to desert vegetation. Similarly, many mangroves of Senegal, composed of *R. mangle, R. racemosa,* and *A. germinans,* gradually transform into the typical step vegetation of arid Northwestern Africa (Fig. 15.5). There, the upper 5–10 cm of the riverine mangrove sediments along the Saloum River contain about 4% of C_{org} and 12% of OM, mostly of intermediate recalcitrance (unpublished data from the DiSeMiNation project, coordinated by the authors). The same three species as in Mexico occur in Colombia (Fig. 15.6) and Northern Brazil (Fig. 15.7). There, however, trees grow 25 m and taller, and SOM contents range from 11% (pure stands of *A. germinans*) to 14% (*R. mangle*) with mostly high recalcitrance in Brazil, and reach up to 95% with remarkably lower recalcitrance in Colombia. Accordingly, SOC contents are around 3% in Brazil and range from 20% to 55% in Colombia (DiSeMiNation, unpubl.).

FIGURE 15.4 Mixed stands of *Rhizophora mangle, Laguncularia racemosa,* and *Avicennia germinans* in direct vicinity to desert vegetation in the arid zone of Baja California, Mexico. *Credit: The authors.*

FIGURE 15.5 Dense riverine stands of stunted *Rhizophora mangle* on the edge to the African savannah of Senegal. *Credit: The authors.*

FIGURE 15.6 Dense inland stand of *Rhizophora mangle* on San Andrés Island (Caribbean Colombia). *Credit: The authors.*

Similarly, large variations, albeit on a much larger geographical scale, characterize the Indo-West Pacific region. In South Africa, extended old-grown stands of *Avicennia marina* and *Bruguiera gymnorhiza* (Fig. 15.8) harbor between 5 and almost 40% SOM, with relatively even distribution of labile and recalcitrant compounds, and about 8% SOC (with large within-region variation from 2% to 18% but no species-specific differences) in their sediments.

FIGURE 15.7 One of the authors getting prepared for taking deep sediment cores in an old-grown stand of *Rhizophora mangle* in North Brazil. *Credit: The authors.*

FIGURE 15.8 Old-grown estuarine stand of *Avicennia marina* in South Africa. *Credit: The authors.*

Estuarine mangroves composed of almost pure stands of *B. gymnorhiza* and partly mixed *Rhizophora stylosa* and *R. samoensis* stands in Fiji (Fig. 15.9) contain 10%–30% of mostly recalcitrant SOM and ca 5% (2%–10%) SOC (DiSeMiNation, unpubl.).

The OM content of the upper sediment layer represents the recent OM-storage of the last years (depending on the local sedimentation rate), and, thus, can serve as a proxy for annual sequestration rates. In a global comparison, Kusumaningtyas et al. (2019) found variations by

FIGURE 15.9 Dense stands of *Rhizophora stylosa*, mixed with some *Bruguiera gymnorhiza*, on the southern coast of the main island of Fiji, Viti Levu. *Credit: The authors.*

a factor of almost 20 in OC accumulation rates among regions (e.g., Mexico and Indonesia), and such huge differences have been reported repeatedly for both **sequestration** and **stocks** over the last decades. Hence, generalizations and global extrapolations have to be made with great caution. Even within the same lagoon, OC-accumulation rates differed by factor 3 to 4 (Kusumaningtyas et al., 2019).

In a global approach, Twilley et al. 2018 stress the importance of coastal geomorphology for explaining variations in SOC in mangroves, distinguishing between terrigenous coastal settings of estuaries, deltas, and lagoons, rich in sediments from the hinterland, and coastal platforms of carbonate settings, characterized by in situ sediment deposition. Mangroves of small deltas and tidal systems predominate worldwide but are relatively poor in SOC. Their often low SIC (sediment C_{in}) contents (i.e., negligible differences between C_{org} and C_{tot} contents) indicate siliciclastic sediments, mostly of terrigenous origin. Sediments of carbonate systems (high SIC content) are mostly of biogenic marine origin, such as corals, calcifying algae, or molluscs. Despite differences in accumulation rates of C_{in} (carbonate sediments), depending on environmental settings that drive SOC:SIC ratios of mangrove sediments, such as proximity to reefs or the prevailing bed rock, SOC accumulation rates remain the best predictor of overall sediment accretion rates (Breithaupt et al., 2017). Based on another global comparison, Balke and Friess (2016) conclude that the average tidal amplitude, in turn, is a prime predictor of SOM, with high SOM stocks prevailing where the tidal amplitude is low. Along this line, the export of mangrove material into coastal waters was mostly driven by tidal transport and riverine runoff, the latter being more important during the rainy season, in Gazi Bay of Kenya (Signa et al., 2017). Within the Indo-West Pacific, Donato et al. (2011) had shown before, on the other hand, that estuarine mangroves contain remarkably higher amounts of SOC (but see above: Twilley et al., 2018) than oceanic mangroves (despite the fact that both above- and below-ground biomass of mangroves is on average higher by at least factor 2 in the latter).

Along the same line, interior mangroves sequester almost twice as much OC as fringe mangroves (Sasmito et al., 2020; West Papua). In comparison to the inland mangrove on San Andrés Island (Fig. 15.6), a basin mangrove of *A. germinans* and a fringe mangrove of *R. mangle* (c.f. Fig. 15.3) on the same coast of the island and only a few km apart contained only about half the amount of OC in their sediments (DiSeMiNation, unpubl.). In some cases, it might be difficult to disentangle confounding effects of site characteristics, such as geomorphology, from species effects, with, e.g., *R. mangle* producing larger amounts of leaf litter that contributes to OM-sequestration (Quadros et al., 2019) and retaining more POM among their roots (Gillis et al., 2016) than *A. germinans*. Beyond these quantitative differences, we also expect species-specific differences in OM quality that would be reflected in its recalcitrance and stability.

The above examples mostly refer to the uppermost sediment layer, representing only the last few years of sedimentation and OM sequestration. The greatest amounts of OM are probably stored in deeper sediment layers (c.f., e.g., Donato et al., 2011), representing the accumulation of SOM over centuries and millennia. However, we do often not know whether the OM stored underneath current mangrove stands actually originated from mangroves or whether these sites had been covered by other vegetation types—e.g., saltmarshes—back then. The above methods of environmental metabolomics and metagenomics, applied to deep sediment cores, could help answer these open questions and improve our understanding of how the quality of SOM drives its long-term stability and accumulation.

Carbon sequestration *(carbon storage)—the process of* C_{org} *uptake into, and* ***storage*** *in, the sediment has to be distinguished from the state of the* ***carbon stock*** *(or store).*

Carbon stock *(carbon store)—the amount of* C_{org} *stored in the sediment (and/or above- and/or below-ground biomass) as a result of* ***carbon sequestration****.*

4.1 Implications for mangrove management?

Regardless of the controversial discussion about the exact values, huge amounts of climate-active gases stored in mangrove sediments (in form of SOM) are being released into the atmosphere (e.g., Hamilton and Friess, 2018) upon clear-cutting of mangroves. Whereas the annual global area loss of mangroves has lately been reduced from 1% to 3% during the late 20th century to a mere 0.2%–0.7% over the last decade (Feller et al., 2017; Friess et al., 2020), the estimated amounts of annually released greenhouse gases are still remarkable. In deeper sediment layers that remain protected from disturbance and aeration (e.g., Elwin et al., 2019), SOM stocks may continue being stable, but this stability will partly depend on the recalcitrance (see above) of the OM stored in the sediment.

Even though regrowing or replanted mangroves restore SOM stocks over time, the OC stored in the surface sediments of newly established mangroves will be comparable to that of natural mangrove forests only after 20–25 years (Elwin et al., 2019, reporting on mangrove-rehabilitated aquaculture ponds). Interestingly, this time interval roughly coincides with the duration of 25 years of the forestry cycles—from clear-cutting through regrowing to the next clear-cutting—in the Malaysian Matang Mangrove Forest Reserve (Ariffin & Nik Mohd Shah, 2013). Restoring mangroves, however, does not replenish the entire previous stocks, and new SOM stocks need to build up, and—just alike mangrove stands have to develop over probably 30–40 years for becoming comparable to mature forests (Sillanpää et al., 2017; West Papua, Indonesia)—a freshly planted mangrove stand is not a replacement of an old-grown mangrove forest when it comes to SOM stocks, neither in terms of quantity nor with respect to quality and stability. Hence, while biomass recovery in previously degraded mangroves takes about 40 years, SOM reestablishment may occur only at the decadal or even century time scale (Sasmito et al., 2019). Hence, protecting and conserving existing mangroves has a much greater impact on climate change mitigation, through storing climate-active gases as SOM, than restoring degraded or lost mangrove stands.

5. Conclusion

While origin and fluxes of OM have been assessed mainly using stable isotopes, that prove of little efficiency for determining the origin of OM from distinct terrestrial plants (including mangroves, saltmarshes, and hinterland vegetation), we hold that the recent developments in environmental metabolomics and metagenomics will help achieve a better picture of the exchange of OM not only between solely mangrove ecosystems and other coastal and/or marine systems but also between mangroves and terrestrial ecosystems in the hinterland. The resulting information on the origin of the OM will prove essential for a deeper insight into long-term dynamics of OM stores in the sediment of both old-grown and rehabilitated mangrove forests and should inform sustainable and adaptive management strategies of networks of coastal ecosystems.

References

Ariffin, R., Shah, N.M., 2013. A Working Plan for the Matang Mangrove Forest Reserve, Perak: Sixth Revision of the First 10-year Period (2010–2019) of the Third Rotation. State Forestry Department of Perak, 229 pp.

Aquino, R.S., Landeira-Fernandez, A.M., Valente, A.P., Andrade, L.R., Mourao, P.A.S., 2005. Occurrence of sulphated galactans in marine angiosperms: evolutionary implications. Glycobiology 15, 11–20.

Bakkar, T., 2018. The Mangrove Connection – Organic Matter Exchange in Tropical Coastal Ecosystems. MSc thesis University of Bremen, "Marine Biodiversity and Conservation EMBC+". 75 pp.

Bakkar, T., Helfer, V., Himmelsbach, R., Zimmer, M., 2017. Chemical changes in detrital matter upon digestive processes in a sesarmid crab feeding on mangrove leaf litter. Hydrobiologia 803, 307–315.

Balke, T., Friess, D.A., 2016. Geomorphic knowledge for mangrove restoration: a pan-tropical categorization. Earth Surf Process Landf 41 (2), 231–239.

Bohmann, K., Evans, A., Gilbert, M.T.P., Carvalho, G.R., Creer, S., Knapp, M., Yu, D.W., de Bruyn, M., 2014. Environmental DNA for wildlife biology and biodiversity monitoring. Trends Ecol. Evol. 29, 358–367.

Bouillon, S., Connolly, R.M., Lee, S.Y., 2008. Organic matter exchange and cycling in mangrove ecosystems: recent insights from stable isotope studies. J. Sea Res. 59, 44–58.

Breithaupt, J.L., Smoak, J.M., Rivera-Monroy, V.H., Castañeda-Moya, E., Moyer, R.P., Simard, M., Sanders, C.J., 2017. Partitioning the relative contributions of organic matter and mineral sediment to accretion rates in carbonate platform mangrove soils. Mar. Geol. 390, 170–180.

Buurman, P., Nierop, K.G.J., Kaal, J., Senesi, N., 2009. Analytical pyrolysis and thermally assisted hydrolysis and methylation of EUROSOIL humic acid samples – a key to their source. Geoderma 150, 10–22.

Buurman, P., Roscoe, R., 2011. Different chemical composition of free light, occluded light and extractable SOM fractions in soils of Cerrado and tilled and untilled fields, Minas Gerais, Brazil: a pyrolysis–GC/MS study. Eur. J. Soil Sci. 62, 253–266.

Carreira, R., de Albergaria-Barbosa, A.C.R., Arguelho, M.L.P.M., Garcia, C., 2015. Evidence of sewage input to inner shelf sediments in the NE coast of Brazil obtained by molecular marker distribution. Mar. Pollut. Bull. 90, 312–316.

Charriau, A., Bodineau, L., Ouddane, B., Fischer, J.-C., 2009. Polycyclic aromatic hydrocarbons and *n*-alkanes in sediments of the upper Scheldt River basin: contamination levels and source appointment. J. Environ. Monit. 11, 1086–1093.

Cheng, P., Zhou, W., Wang, H., Lu, X., Du, H., 2013. ^{14}C dating of soil organic carbon (SOC) in loess-paleosol using sequential pyrolysis and accelerator mass spectrometry (AMS). Radiocarbon 55, 563–570.

Cloern, J.E., Canuel, E.A., Harris, D., 2002. Stable carbon and nitrogen isotope composition of aquatic and terrestrial plants of the San Francisco Bay estuarine system. Limnol. Oceanogr. 47, 713–729.

Cowart, D.A., Murphy, K.R., Cheng, C.C., 2018. Metagenomic sequencing of environmental DNA reveals marine faunal assemblages from the West Antarctic Peninsula. Mar. Genom. 37, 148–160.

Cragg, S.M., Friess, D.A., Gillis, L.G., Trevathan-Tackett, S.M., Terrett, O.M., Watts, E.M., Distel, D.L., Dupree, P., 2020. Vascular plants are globally significant contributors to marine carbon fluxes and sinks. Annu. Rev. Mar. Sci. 2020, 1–29.

Donato, D.C., Kauffman, J.B., Murdiyarso, D., Kurnianto, S., Stidham, M., Kanninen, M., 2011. Mangroves among the most carbon-rich forests in the tropics. Nat. Geosci. 4, 293–297.

Du, X., Zeisel, S.H., 2013. Spectral deconvolution for gas chromatography mass spectrometry-based metabolomics: current status and future perspectives. Comput. Struct. Biotechnol. J. 4, e201201013.

Elwin, A., Bukoski, J.J., Jintana, V., Robinson, E.J.Z., Clark, J.M., 2019. Preservation and recovery of mangrove ecosystem carbon stocks in abandoned shrimp ponds. Sci. Rep. 9, 18275.

Feller, I.C., Friess, D.A., Krauss, K.W., Lewis III, R.R., 2017. The state of the world's mangroves in the 21st century under climate change. Hydrobiologia 803, 1–12.

Friess, D.A., Yando, E.S., Abuchahla, G.M.O., Adams, J.B., Cannicci, S., Canty, S.W.J., Cavanaugh, K.C., Connolly, R.M., Cormier, N., Dahdouh-Guebas, F., Diele, K., Feller, I.C., Fratini, S., Jennerjahn, T.C., Lee, S.Y., Ogurcak, D.E., Ouyang, X., Rogers, K., Rowntree, J.K., Sharma, S., Sloey, T.M., Wee, A.K.S., 2020. Mangroves give cause for conservation optimism, for now. Curr. Biol. 30, 153–154.

Gillis, L.G., Zimmer, M., Bouma, T.J., 2016. Mangrove leaf transportation: do mimic *Avicennia* and *Rhizophora* roots retain or donate leaves? Mar. Ecol. Prog. Ser. 551, 107–115.

Guimarãesa, L.M., de Abreu, C., Limaa, T., De Françab, E.J., de Arrudab, G.N., de Souzac, R.B.B., de Albergaria-Barbosa, A.C.R., 2019. Impactos da mudança de vegetação local no aporte de matéria organica para um estuário tropical reservado (Estuário do Rio Itapicuru). Quim. Nova 42, 611–618.

Hamilton, S.E., Friess, D.A., 2018. Global carbon stocks and potential emissions due to mangrove deforestation from 2000 to 2012. Nat. Clim. Change 8, 240–244.

He, J., Zhang, S., Zhang, X., Qian, Y., He, H., Wu, H., 2016. Composition and distribution characteristics and geochemical significance of n-alkanes in core sediments in the northern part of the South Yellow Sea. J. Chem. https://doi.org/10.1155/2016/4741939.

Helfer, V., Zimmer, M., 2018. High-throughput techniques as support for knowledge-based spatial conservation prioritization in mangrove ecosystems. In: Makowski, C., Finkl, C.W. (Eds.), Threats to Mangrove Forests: Hazards, Vulnerability and Management. Springer, pp. 539–554.

Hübner, L., Pennings, S.C., Zimmer, M., 2015. Sex- and habitat-specific movement of an omnivorous semi-terrestrial crab controls habitat connectivity and subsidies: a multi-parameter approach. Oecologia 178, 999–1015.

Huxham, M., Whitlock, D., Githaiga, M., Dencer-Brown, A., 2018. Carbon in the coastal seascape: how interactions between mangrove forests, seagrass meadows and tidal marshes influence carbon storage. Curr. For. Rep. https://doi.org/10.1007/s40725-018-0077-4.

Jaffé, R., Rushdi, A.I., Medeiros, P.M., Simoneit, B.R.T., 2006. Natural product biomarkers as indicators of sources and transport of sedimentary organic matter in a subtropical river. Chemosphere 64, 1870–1884.

Ji, Y., Huotari, T., Roslin, T., 2020. SPIKEPIPE: a metagenomic pipeline for the accurate quantification of eukaryotic species occurrences and intraspecific abundance change using DNA barcodes or mitogenomes. Mol. Ecol. Res. 20, 256–267.

Kelly, R.P., Shelton, A.O., Gallego, R., 2019. Understanding PCR processes to draw meaningful conclusions from environmental DNA studies. Sci. Rep. 9, 12133.

Koch, B.P., Souza, P.W.M., Behling, H., Cohen, M.C.L., Kattner, G., 2011. Triterpenols in mangrove sediments as a proxy for organic matter derived from the red mangrove (*Rhizophora mangle*). Org. Geochem. 42, 62–73.

Kristensen, D., Kristensen, E., Mangion, P., 2010. Food partitioning of leaf-eating mangrove crabs (Sesarminae): experimental and stable isotope (13C and 15N) evidence. Estuar. Coast Shelf Sci. 87, 583–590.

Kumar, M., Boski, T., Lima-Filho, F.P., Bezerra, F.H.R., González-Vila, F.J., Bhuiyan, K.A., González-Pérez, J.A., 2019. Biomarkers as indicators of sedimentary organic matter sources and early diagenetic transformation of pentacyclic triterpenoids in a tropical mangrove ecosystem. Estuar. Coast Shelf Sci. 229, 106403.

Kusumaningtyasa, M.A., Hutahaeanc, A.A., Fischer, H., Pérez-Mayod, M., Ransbyd, D., Jennerjahn, T.C., 2019. Variability in the organic carbon stocks, sources, and accumulation rates of Indonesian mangrove ecosystems. Estuar. Coast Shelf Sci. 218, 310–323.

Ladd, M.P., Giannone, R.J., Abraham, P.E., Wullschleger, S.D., Hettich, R.L., 2019. Evaluation of an untargeted nano-liquid chromatography-mass spectrometry approach to expand coverage of low molecular weight dissolved organic matter in Arctic soil. Sci. Rep. 9, 5810. https://doi.org/10.1038/s41598-019-42118-9.

Lamar, R.T., Olk, D.C., Mayhew, L., Bloom, P.R., 2014. A new standardized method for quantification of humic and fulvic acids in humic ores and commercial products. J. AOAC Int. 97, 721–730.

Lamb, P.D., Hunter, E., Pinnegar, J.K., Creer, S., Davies, R.G., Taylor, M.I., 2019. How quantitative is metabarcoding: a meta-analytical approach. Mol. Ecol. 28, 420–430.

McFadden, T.N., Kauffman, B., Bhomia, R.K., 2016. Effects of nesting waterbirds on nutrient levels in mangroves, Gulf of Fonseca, Honduras. Estuar. Coast 24, 217–229.

McLaren, M.R., Willis, A.D., Callahan, B.J., 2020. Consistent and correctable bias in metagenomic sequencing experiments. bioRxiv 559831.

Meyers, P.A., 1994. Preservation of elemental and isotopic source identification of sedimentary organic matter. Chem. Geol. 114, 289–302.

Ouyang, X., Lee, S.Y., 2019. Improved estimates on global carbon stock and carbon pools in tidal wetlands. Nat. Commun. https://doi.org/10.1038/s41467-019-14120-2.

Pawlowski, J., Kelly-Quinn, M., Altermatt, F., 2018. The future of biotic indices in the ecogenomic era: integrating (e) DNA metabarcoding in biological assessment of aquatic ecosystems. Sci. Total Environ. 637–638, 1295–1310.

Pedersen, M.W., Overballe-Petersen, S., Ermini, L., Sarkissian, C.D., Haile, J., Hellstrom, M., Spens, J., Thomsen, P.F., Bohmann, K., Cappellin, E., Schnell, I.B., Wale, N.A., Carø, C., Campos, P.F., Schmidt, A.M., Gilbert, M.T., Hansen, A.J., Orlando, L., Willerslev, E., 2015. Ancient and modern environmental DNA. Phil. Trans. Roy. Soc. Lond. B Biol. Sci. 370, 20130383.

Piñol, J., Senar, M.A., Symondson, W.O.C., 2019. The choice of universal primers and the characteristics of the species mixture determine when DNA metabarcoding can be quantitative. Mol. Ecol. 28, 407–419.

Pirainen, S., Finér, L., Mannerkosi, H., Starr, M., 2007. Carbon, nitrogen and phosphorous leaching after site preparation at a boreal forest clear-cut area. For. Ecol. Manag. 243, 10–18.

Prasad, M.B.K., Kumar, A., Ramanathan, A.L., Kumar, D., 2017. Sources and dynamics of sedimentary organic matter in Sundarban mangrove estuary from Indo-Gangetic delta. Ecol. Processes 6, 8. https://doi.org/10.1186/s13717-017-0076-6.

Pribyl, D.W., 2010. A critical review of the conventional SOC to SOM conversion factor. Geoderma 156, 75–83.

Quadros, A.F., Nordhaus, I., Reuter, H., Zimmer, M., 2019. Modelling of mangrove annual leaf litterfall with emphasis on the role of vegetation structure. Estuar. Coast Shelf Sci. 218, 292–299.

Reef, R., Atwood, T.B., Samper-Villarreal, J., Adame, M.F., Sampayo, E.M., Lovelock, C.E., 2017. Using eDNA to determine the source of organic carbon in seagrass meadows. Limnol. Oceanogr. 62, 1254–1265.

Reef, R., Feller, I.C., Lovelock, C.E., 2014. Mammalian herbivores in Australia transport nutrients from terrestrial to marine ecosystems via mangroves. J. Trop. Ecol. 30, 179–188.

Reyerson, P.E., Alexandre, A., Harutyunyan, A., Corbineau, R., Martinez De La Torre, H.A., Badeck, F., Cattivelli, L., Santos, G.M., 2016. Unambiguous evidence of old soil carbon in grass biosilica particles. Biogeosciences 13, 1269–1286.

Ruppert, K.M., Kline, R.J., Rahman, M.S., 2019. Past, present, and future perspectives of environmental DNA (eDNA) metabarcoding: a systematic review in methods, monitoring, and applications of global eDNA. Global Ecol. Conserv. 17, e00547.

Rushdi, A.I., DouAbul, A.A., Mohammed, S.S., Simoneit, B.R.T., 2006. Compositions and sources of extractable organic matter in Mesopotamian marshland surface sediments of Iraq: II. Polar compounds. Environ. Geol. 50, 1171–1181.

Rushdi, A.I., Oros, D.R., Al-Mutlaq, K.F., He, D., Medeiros, P.M., Simoneit, B.R.T., 2016. Lipid, sterol and saccharide sources and dynamics in surface soils during an annual cycle in a temperate region. Appl. Geochem. 66, 1–13.

Saavedra-Hortua, D.A., Friess, D.A., Zimmer, M., Gillis, L.G., 2020. Sources of particulate organic matter across mangrove forests and adjacent ecosystems in different geomorphic settings. Wetlands 40, 1047–1059 (in press).

Sakho, I., Mesnage, V., Copard, Y., Deloffre, J., Faye, G., 2015. A cross-section analysis of sedimentary organic matter in a mangrove ecosystem under dry climate conditions: the Somone estuary, Senegal. J. Afr. Earth Sci. 101, 220–231.

Sasmito, S.D., Kuzyakov, Y., Lubis, A.A., Murdiyarso, D., Hutley, L.B., Bachri, S., Friess, D.A., Martius, C., Borchard, N., 2020. Organic carbon burial and sources in soils of coastal mudflat and mangrove ecosystems. Catena 187, 104414.

Sasmito, S.D., Taillardat, P., Clendenning, J.N., Cameron, C., Friess, D.A., Murdiyarso, D., Hutley, L.B., 2019. Effect of land-use and land-cover change on mangrove blue carbon: a systematic review. Global Change Biol. https://doi.org/10.1111/gcb.14774.

Schellekens, J., Buurman, P., 2011. n-Alkane distributions as palaeoclimatic proxies in ombrotrophic peat: the role of decomposition and dominant vegetation. Geoderma 164, 112–121.

Schellekens, J., Buurman, P., Fraga, I., Martínez-Cortizas, A., 2011. Holocene vegetation and hydrologic changes inferred from molecular vegetation markers in peat, Penido Vello (Galicia Spain). Palaeogeogr. Palaeoclimatol. Palaeoecol. 299, 56–69.

Schmidt, M.W.I., Torn, M.S., Abiven, S., Guggenberger, G., Janssens, I.A., Kleber, M., Kögel-Knabner, I., Lehmann, J., Manning, D.A.C., Nannipieri, P., Rasse, D.P., Weiner, S., Trumbore, 2011. Persistence of soil organic matter as an ecosystem property. Nature 478, 49–56.

Signa, G., Mazzola, A., Kairo, J., Vizzini, S., 2017. Small-scale variability in geomorphological settings influences mangrove-derived organic matter export in a tropical bay. Biogeosciences 14, 617–629.

Sillanpää, M., Vantellingen, J., Friess, D.A., 2017. Vegetation regeneration in a sustainably harvested mangrove forest in West Papua, Indonesia. For. Ecol. Manag. 390, 137–146.

Stat, M., Huggett, M.J., Bernasconi, R., 2017. Ecosystem biomonitoring with eDNA: metabarcoding across the tree of life in a tropical marine environment. Sci. Rep. 7, 12240.

Steinberg, C.E.W., 2003. Ecology of Humic Substances in Freshwaters. Springer, Berlin.

Stewart, C.E., Neff, J.C., Amatangelo, K.L., Vitousek, P.M., 2011. Vegetation effects on soil organic matter chemistry of aggregate fractions in a Hawaiian forest. Ecosystems 14, 382–397.

Taberlet, P., Coissac, E., Hajibabaei, M., Rieberg, L.H., 2012. Environmental DNA. Mol. Ecol. 21, 1789–1793.

Thomsen, P.F., Willerslev, E., 2015. Environmental DNA – an emerging tool in conservation for monitoring past and present biodiversity. Biol. Conserv. 183, 4–18.

Tolu, J., Gerber, L., Boily, J.-F., Bindler, R., 2015. High-throughput characterization of sediment organic matter by pyrolysis–gas chromatography/mass spectrometry and multivariate curve resolution: a promising analytical tool in (paleo)limnology. Anal. Chim. Acta 880, 93–102.

Treplin, M., Zimmer, M., 2012. Drowned or dry: a cross-habitat comparison of detrital breakdown processes. Ecosystems 15, 477–491.

Twilley, R.R., Rovai, A.S., Riul, P., 2018. Coastal morphology explains global blue carbon distributions. Front. Ecol. Environ. 16 (9), 503–508.

Vancampenhout, K., Wouters, K., De Vos, B., Buurman, P., Swennen, R., Deckers, J., 2009. Differences in chemical composition of soil organic matter in natural ecosystems from different climatic regions – a pyrolysis–GC/MS study. Soil Biol. Biochem. 41, 568–579.

Viant, M.R., Sommer, U., 2013. Mass spectrometry based environmental metabolomics: a primer and review. Metabolomics 9, S144–S158.

Wang, S., Liu, G., Yuan, Z., Da, C., 2018. n-Alkanes in sediments from the Yellow River estuary, China: occurrence, sources and historical sedimentary record. Ecotoxicol. Environ. Saf. 150, 199–206.

Yu, Y., Li, Y., Guo, Z., Zou, H., 2016. Distribution and sources of n-alkanes in surface sediments of Taihu Lake, China. Arch. Environ. Protect. 42, 49–55.

Zimmer, M., 2019. Detritus. In: Jorgensen, S.E., Fath, B.D. (Eds.), Encyclopedia of Ecology, second ed., vol. 3. Elsevier, pp. 292–301.

Zimmer, M., Topp, W., 2002. The role of coprophagy in nutrient release from the feces of phytophagous insects. Soil Biol. Biochem. 34, 1093–1099.

CHAPTER

16

Relevance of allochthonous input from an agriculture-dominated hinterland for "Blue Carbon" storage in mangrove sediments in Java, Indonesia

Tim C. Jennerjahn[1,2]

[1]Leibniz Centre for Tropical Marine Research, Bremen, Germany; [2]Faculty of Geoscience, University of Bremen, Bremen, Germany

OUTLINE

1. Introduction 394

2. Material and methods 395

2.1 *Study area* 395

2.2 *Sampling* 398

2.3 *Analytical methods* 398

3. Results 399

3.1 *Organic carbon and total nitrogen content and stable isotope composition of plants and soils* 399

3.2 *Organic carbon and total nitrogen content and stable isotope composition of surface sediments* 400

3.3 *Organic carbon and total nitrogen content and stable carbon and nitrogen isotope composition of mangrove sediments* 400

3.4 *Grain-size composition of mangrove sediments* 400

4. Discussion 400

4.1 *Sources of organic matter* 400

4.2 *Autochthonous versus allochthonous contributions to mangrove sedimentary organic matter related to hydrodynamics and hinterland input* 405

4.3 *Temporal variations of organic matter composition* 406

4.4 *Relevance of autochthonous versus allochthonous OM sources for*

https://doi.org/10.1016/B978-0-12-816437-2.00017-3

carbon budgets in times of global environmental change 408

5. Conclusions 410

References 410

1. Introduction

Climate change and its outcomes are expected to alter environmental conditions on planet Earth to an extent that they will impair the well-being of humans, animals, and plants in many places in the near future, as stated in the reports of the Intergovernmental Panel on Climate Change (IPCC) (IPCC, 2014). The man-made emissions of the greenhouse gas carbon dioxide are a major driver in this context. The history of IPCC reports and political actions in the past decades have shown that recent and planned emission reductions are not yet sufficient to meet the goal of "keeping a global temperature rise this century well below 2°C above preindustrial levels and to pursue efforts to limit the temperature increase even further to 1.5°C," as outlined in the Paris Agreement (United Nations, 2015). In this context it is of utmost importance to identify, to quantify, and to preserve existing natural carbon sinks.

In the context of the "Blue Carbon" concept (Nellemann et al., 2009), carbon storage in the so-called vegetated coastal habitats (mangrove forests, seagrass meadows, saltmarshes) gained major attention in the past decade because of their capability of sequestering larger amounts of carbon per unit area than many other ecosystems (e.g., Alongi, 2014). Mangrove forests have been identified as a major carbon sink, and it has been shown that the majority of the carbon is stored belowground in the sediments (Donato et al., 2011), contributing up to 15% of the total carbon accumulation in marine sediments (Jennerjahn and Ittekkot, 2002). Mangrove forests have a much higher carbon accumulation in sediments/soils than terrestrial forests, which is mainly due to high autochthonous and allochthonous inputs and relatively low decomposition rates of organic matter (OM), because of the mostly anoxic conditions in the sediment (Donato et al., 2011; Kristensen, 2000). The global average rate of organic carbon accumulation in mangrove ecosystems is on the order of 22 ± 6 Tg C/yr (Jennerjahn et al., 2017).

The global mangrove carbon stock estimate of 4–20 Pg (Donato et al., 2011) underscores the large uncertainties in the magnitude of this sink. It is well known that factors like climate, coastal morphology, and the type of environmental setting largely determine the distribution of mangrove species, transport and deposition of sediment and the efflux, accumulation, storage, and composition of OM (Thom, 1982; Woodroffe, 1992; Saintilan et al., 2013; Rovai et al., 2018; Twilley et al., 2018). Regional variability can be high from large to small spatial scale. Carbon accumulation rate estimates for the two major biogeographic regions are on the order of 265 ± 161 g C/m^2/yr in the IWP (Indo-West Pacific) and 194 ± 155 g C/m^2/yr in the AEP (Atlantic East Pacific) regions, respectively, but are also fraught with large uncertainties

(Breithaupt et al., 2012; Jennerjahn et al., 2017). Even less is known on the small-scale variability of carbon accumulation within mangrove ecosystems. Moreover, coastal development, land use and land cover change, regulations of hydrology, and climate change as well as the interaction of those factors can all alter the carbon accumulation in mangrove ecosystems (e.g., Chowdhury et al., 2017; Huxham et al., 2017; Jennerjahn et al., 2017). A recent global analysis has shown that the direct conversion of mangroves into other land uses decreased mangrove soil carbon stocks by more than 50% (Sasmito et al., 2019).

In order to evaluate the effectiveness of the "Blue Carbon" sink, it is not only important to obtain a better understanding of its magnitude but also to identify the geographically varying OM sources, because autochthonous versus allochthonous contributions can become relevant in the context of carbon emission offsets and related trade schemes (Saintilan et al., 2013; Watanabe and Kuwae, 2015). While autochthonous OM is produced from newly fixed CO_2 in the mangrove ecosystem, allochthonous OM is relocated from other terrestrial or marine reservoirs and is at least partly generated from CO_2 fixed some time ago. In this context, the quantitatively most important coastal environmental settings are small deltas and tidal systems, which are characterized by a generally high allochthonous input (Twilley et al., 2018). This, in turn, has consequences for the amount and composition of the OM being stored in the respective system.

In order to illustrate the relevance of these processes and the large spatial variability even within one "system", this chapter presents a data set from the mangrove-fringed Segara Anakan Lagoon (SAL) in Java, Indonesia. It highlights the differences in the sources and in the amount and composition of carbon stored in mangrove sediments in different compartments of this coastal lagoon that is fed by a medium-size river, which drains an agriculture-dominated hinterland, is in continuous exchange with the Indian Ocean, and suffers also from overexploitation of resources, mainly logging and fisheries.

2. Material and methods

2.1 Study area

The elongated, mangrove-fringed SAL stretches over 25 km along the south coast of the densely populated island of Java, Indonesia (Fig. 16.1). It has two inlets that connect it to the Indian Ocean at its western and eastern ends. It is separated from the ocean by the rocky mountainous Nusakambangan Island. The climate of the region is dominated by the monsoons with heavy precipitation during the wet NW monsoon (austral summer, November to March) and much less precipitation during the SE monsoon, which brings drier and cooler air masses during austral winter with the driest months being July to September. The SAL receives high freshwater input mainly from the Citanduy River and a few other rivers in its western and central parts, in particular during the wet season, while freshwater input is low in the eastern part, which is dominated by tidal exchange with the Indian Ocean. Formerly, an annual average discharge of the Citanduy of 227 m^3/s (dry season 171 m^3/s, wet season 283 m^3/s) was reported (Ludwig, 1985). More recent measurements resulted in an average 78 m^3/s during the dry season and individual daily measurements during the wet season ranged between 50 and 350 m^3/s (Holtermann et al., 2009). The SAL has a typical

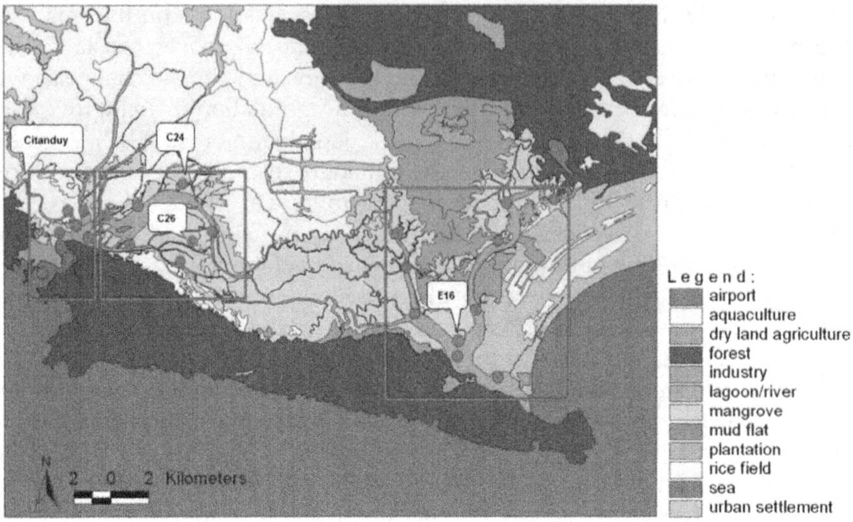

FIGURE 16.1 Map of the Segara Anakan Lagoon including land use/cover. The *red (gray in print version) rectangles and dots* delineate the western, central, and eastern (W, C, E) working areas and the respective sampling sites. C24, C26, and E16 are the locations of the mangrove sediment cores. *Map modified from Jennerjahn, T.C., Nasir, B., Pohlenga, I., 2009. Spatio-temporal variation of dissolved inorganic nutrients in the Segara Anakan lagoon, Java, Indonesia. Reg. Environ. Change 9, 259–274.*

tidal range of 1.4 m during spring tide and 0.4 m during neap tide, respectively (Holtermann et al., 2009).

The lagoon is fringed by the largest remaining single mangrove forest on the south coast of Java (Fig. 16.2), which had an extent of 21,185 ha in 1980 and of 9238 ha in 2006 (Ardli and Wolff, 2009). Vegetation in the mangrove forest area consists of 21 mangrove tree species and 5 understory genera with a higher density and diversity in the eastern than in the central and western parts of the lagoon (Hinrichs et al., 2009). Ten years later the number of species was reduced to 15 and the dominance shifted (Nordhaus et al., 2019, Table 16.1). Besides the mangrove forest, the waterbody of the lagoon has shrunk drastically and became shallower due to heavy sedimentation from rivers, mainly the Citanduy River, in the past 150 years (Yuwono et al., 2007; Lukas, 2014, 2017). The catchment area of the lagoon is about 450 km^2 in size and comprises a range of land use and land cover types, including irrigated rice agriculture, rainfed agriculture with a diversity of crops, settlements with house gardens, forests, and plantations (Lukas, 2017). The city of Cilacap is located at the eastern end of the lagoon. It has about 234,000 inhabitants (BPS Cilacap, 2018) and harbors the largest oil refinery of Indonesia in its port.

As a result of collaborative interdisciplinary research in the frame of the SPICE program (Science for the Protection of Indonesian Coastal Ecosystems; www.leibniz-zmt.de/en/research/research-projects/spice-i-and-ii.html, www.leibniz-zmt.de/en/research/research-projects/spice.html), the SAL can be subdivided into three geographical areas according to differences in freshwater input, hydrodynamics, flora and fauna, and sediment redox conditions (Table 16.1; Yuwono et al., 2007; Hinrichs et al., 2009; Holtermann et al., 2009; Jennerjahn et al., 2009; Nordhaus et al., 2009; Geist et al., 2012).

FIGURE 16.2 Characteristics of the Segara Anakan Lagoon and its mangrove forest. (A) Citanduy River plume enters the western lagoon; (B) typical *Rhizophora* trees in the central lagoon; (C) medium stature mangroves in the central lagoon; (D) Nypa palms mainly in the central lagoon; (E) low-gradient tidal channel in the eastern lagoon; (F) logged area in the eastern lagoon, (G, H) sandy, small stature mangroves in the eastern lagoon.

TABLE 16.1 Major environmental setting features of the western, central, and eastern parts of the Segara Anakan Lagoon.

	Western	Central	Eastern
Freshwater input	High river input, mainly by Citanduy, avg. dry season 110 m^3/s, avg. wet season 270 m^3/s	Strongly influenced by river input, but less than in SAL W	No river input, significant surface runoff input during wet season
Hydrodynamics	River dominance throughout year, strong tidal currents in Plawangan channel	Tidal dominance with strong river influence from SAL W	Tidal dominance
Water residence time	0–3 days dry season, 0–1 days wet season	3–12 days dry season, 1–6 days wet season	2–10 days dry season, 1–7 days wet season
Salinity	1–33	5–27	12–34
Sediment texture	Silty clay	Silty clay	Silty sand
Sediment redox conditions	Oxic to hypoxic	Hypoxic to anoxic	Oxic
Vegetation	*Nypa fruticans, Aegiceras corniculatum, Sonneratia sp.*		*Aegiceras corniculatum, Rhizophora apiculata, Nypa fruticans*
Benthic fauna	Low species number, gastropods dominant	Medium species number, gastropods dominant	High species number, polychaetes dominant

2.2 Sampling

In order to characterize the present day spatial distribution of biogeochemical properties, surface sediments from the SAL were collected by using a small Ekman grab in January 2006. An aliquot of the upper cm was collected into precombusted glass vials (450°C, 5 h). Sediment cores of up to 60 cm length were collected in three mangrove-covered areas of the SAL at low tide with a handheld corer equipped with plexiglass tubes in January 2006, in order to characterize possible changes of biogeochemical properties over time. Four replicate cores were taken in each area in 30 m distance from each other along a 90 m transect perpendicular to the coastline. Cores were sampled in 1 cm intervals in the upper 10 cm, in 2 cm intervals between 10 and 20 cm core depth, and in 5 cm intervals in the rest of the core. Samples were collected into precombusted glass vials. Plant and soil samples from the mangrove areas and the hinterland were also collected in precombusted glass vials. Samples were stored at 4°C after sampling until further processing. All samples were dried at 40°C in a drying oven and ground and homogenized to a fine powder prior to analysis. Wet sediment core samples for grain-size analysis were collected in plastic bags and stored at 4°C until analysis.

2.3 Analytical methods

The grain-size distribution of sediment was determined by a wet-sieving analysis (fractions > 2000, 600, 212, 62 μm), followed by an Atterberg analysis (McCave and Syvitski,

2007) for the other fractions (20, 6.3, 2 μm). Concentrations of total carbon, total nitrogen (N), and organic carbon (C_{org}) were determined by high temperature combustion in a Carlo Erba NA 2100 element analyzer. The stable carbon and nitrogen isotope composition ($\delta^{13}C_{org}$, $\delta^{15}N$) of sediments was determined in a Thermo Finnigan Delta Plus mass spectrometer after high temperature combustion in a Flash EA 1112 element analyzer. Prior to the determination of C_{org} and $\delta^{13}C_{org}$, carbonate was removed by acidification with 1 N HCl and samples were dried at 40°C before measurement. The analytical errors are as follows: C_{org} <0.1%, N <0.01%, $\delta^{13}C_{org}$, and $\delta^{15}N$ < 0.2‰.

3. Results

3.1 Organic carbon and total nitrogen content and stable isotope composition of plants and soils

The C_{org} and N concentrations of yellow leaves of six dominant "true" mangrove species span a range of 37.6%–48.0% and 0.4%–1.2%, respectively, and result in a highly variable C/N ratio. Similarly, the $\delta^{13}C_{org}$ and $\delta^{15}N$ vary largely from −30.4‰ to −27.6‰ and −1.0‰ to 5.1‰, respectively (Table 16.2). The values for the understory species *Acanthus ilicifolius* and *Derris trifoliata* are in the same ranges. Yellow senescent leaves were chosen,

TABLE 16.2 Biogeochemical properties of plants and soils from the mangrove forest and the hinterland of the Segara Anakan Lagoon.

Plants and soils	C (%)	N (%)	C/N	$\delta^{13}C_{org}$ (‰)	$\delta^{15}N$ (‰)	Data source
Avicennia alba yellow	44.0	1.2	35.7	−27.6	5.1	1
Aegiceras corniculatum yellow	48.0	0.8	57.7	−30.4	4.4	1
Ceriops decandra yellow	45.3	0.7	68.6	−30.4	4.5	1
Rhizophora apiculata yellow	44.9	0.8	54.7	−28.6	3.4	1
Sonneratia caseolaris yellow	37.6	1.1	36.7	−29.9	5.0	1
Bruguiera gymnorrhiza yellow	41.6	0.4	97.5	−28.6	3.8	2
Xylocarpus granatum yellow	41.8	1.1	37.5	−29.3	−1.0	2
Acanthus ilicifolius yellow	39.2	0.8	53.1	−27.2	4.4	2
Derris trifoliata yellow	46.7	1.1	44.7	−26.9	−0.5	2
Rice plant young field	37.2	3.2	11.5	−29.6	6.4	2
Rice plant mature field	37.0	1.3	28.5	−29.5	10.4	2
Rice soil wet young field	1.6	0.1	10.8	−27.9	4.2	2
Rice soil dry young field	1.4	0.1	11.5	−25.6	5.8	2
Rice soil dry mature field	3.2	0.2	12.8	−24.9	6.1	2

Data sources: 1 – Nordhaus, I., Salewski, T., Jennerjahn, T.C., 2011. Food preferences of mangrove crabs related to leaf nitrogen compounds in the Segara Anakan Lagoon, Java, Indonesia. J. Sea Res. 65, 414–426; 2 - this study.

because they represent the possible sources for the C_{org} deposited in sediments, rather than green leaves which do not fall from the trees and have a very different biochemical composition (Nordhaus et al., 2017). Rice plants display major differences in N content and $\delta^{15}N$ between freshly sewn and mature fields. All values are higher in soils of mature fields than in those of young dry and wet fields.

3.2 Organic carbon and total nitrogen content and stable isotope composition of surface sediments

In general, surface sediments display a large variability in biogeochemical properties and grain-size distribution. However, when grouped into the three areas (W, C, E), there are substantial differences. While the C_{org} and N content as well as the contribution of coarser grained material increases from West to East, the $\delta^{13}C_{org}$ decreases and the $\delta^{15}N$ does not display any regional trend (Table 16.3).

3.3 Organic carbon and total nitrogen content and stable carbon and nitrogen isotope composition of mangrove sediments

The cores display variability of biogeochemical properties downcore, as well as between sites. Replicates from the three sites displayed quite large variability, but no statistically significant differences between individual cores (Pohlenga, 2007, Fig. 16.3). The C_{org} content is similarly high at sites E16 and C24, except for a mid-depth maximum at E16 because of a higher contribution of root tissue, and it is higher than at site C26. While the C/N ratio similarly ranges roughly between 10 and 15 at sites C24 and C26, it is higher and displays more variability at E16. The $\delta^{13}C_{org}$ displays little downcore variation at E16, where it is lower than at C26, but higher than in the upper part of C24. In the latter, $\delta^{13}C_{org}$ increases strongly below 30 cm core depth. $\delta^{15}N$ is lowest at E16 and displays a slight downcore decrease, and it is higher at sites C24 and C26 (Fig. 16.3).

3.4 Grain-size composition of mangrove sediments

Sediments at sites C24 and C26 have quite similar grain-size compositions with a much higher portion of the fraction <2 μM (ca. 50%–70%) than sediments at site E16 (ca. 20%). The silt fraction (2–63 μM) contributions display little differences between sites, whereas the portion of the fraction >63 μM is almost zero at sites C24 and C26, but amounts to approximately 50% at site E16 (Fig. 16.4).

4. Discussion

4.1 Sources of organic matter

The large differences in freshwater input and hydrodynamics between the western and central parts of the lagoon on the one hand and the eastern part of the lagoon on the other hand are an important determinant of the supply of substances from the hinterland

TABLE 16.3 Biogeochemical properties and grain-size distribution of surface sediments from the Segara Anakan Lagoon with area averages and standard deviations.

Stations	C_{org} (%)	N (%)	C/N	$\delta^{13}C_{org}$ (‰)	$\delta^{15}N$ (‰)	<2 μm (%)	2–63 μm (%)	>63 μm (%)
SA_01_06_W_06	1.35	0.12	11.2	−25.5	3.9	20.9	17.9	61.2
SA_01_06_W_07	0.26	0.02	11.4	−25.3	4.5	25.5	24.5	50.1
SA_01_06_W_08	0.89	0.12	7.7	−26.1	4.1	36.0	22.0	42.1
SA_01_06_W_09	n.m.	n.m.	n.m.	n.m.	n.m.	17.6	22.6	59.8
SA_01_06_W_10	1.50	0.09	16.7	−24.7	4.2	29.4	32.4	38.2
SA_01_06_W_11	0.19	0.02	8.4	−24.5	4.1	41.2	38.4	20.4
SA_01_06_W_17	0.40	0.03	14.8	−25.9	4.9	59.1	37.9	2.9
SA_01_06_C_18	0.83	0.09	9.4	−25.1	4.1	53.2	45.8	1.0
SA_01_06_C_19	1.24	0.12	10.1	−26.7	4.2	2.6	6.9	90.6
SA_01_06_C_20	0.71	0.11	6.6	−25.2	9.3	58.3	40.2	1.5
SA_01_06_C_21	1.01	0.09	11.9	−27.3	4.0	58.2	39.5	2.3
SA_01_06_C_22	1.22	0.10	11.7	−26.7	3.8	20.3	22.6	57.1
SA_01_06_E_01	0.96	0.04	22.1	−27.1	3.6	0.8	5.0	94.2
SA_01_06_E_02	2.35	0.19	12.5	−26.4	3.7	5.8	37.0	57.1
SA_01_06_E_03	2.43	0.19	13.0	−26.9	3.5	15.7	82.0	2.3
SA_01_06_E_04	1.24	0.08	14.9	−26.1	3.7	15.2	75.2	9.6
SA_01_06_E_05	1.48	0.08	17.8	−26.7	3.3	n.m.	n.m.	n.m.
SA_01_06_E_12	1.27	0.09	14.9	−26.8	3.7	9.1	70.7	20.2
SA_01_06_E_13	0.45	0.04	12.5	−26.6	4.9	4.8	32.4	62.8
SA_01_06_E_14	0.78	0.05	15.7	−27.1	3.0	n.m.	n.m.	n.m.
SA_01_06_E_15	0.16	0.02	9.9	−25.9	3.0	4.5	25.2	70.3
Area W average	0.77	0.07	11.7	−25.3	4.3	32.8	28.0	39.2
Area C average	1.00	0.10	9.9	−26.2	5.1	38.5	31.0	30.5
Area E average	1.24	0.09	14.8	−26.6	3.6	7.0	40.9	39.6
Area W SD	0.57	0.05	3.5	0.6	0.4	14.2	8.2	21.2
Area C SD	0.23	0.02	2.1	1.0	2.3	25.6	16.0	41.3
Area E SD	0.77	0.06	3.5	0.4	0.6	6.0	31.7	35.9

N.m., not measured.

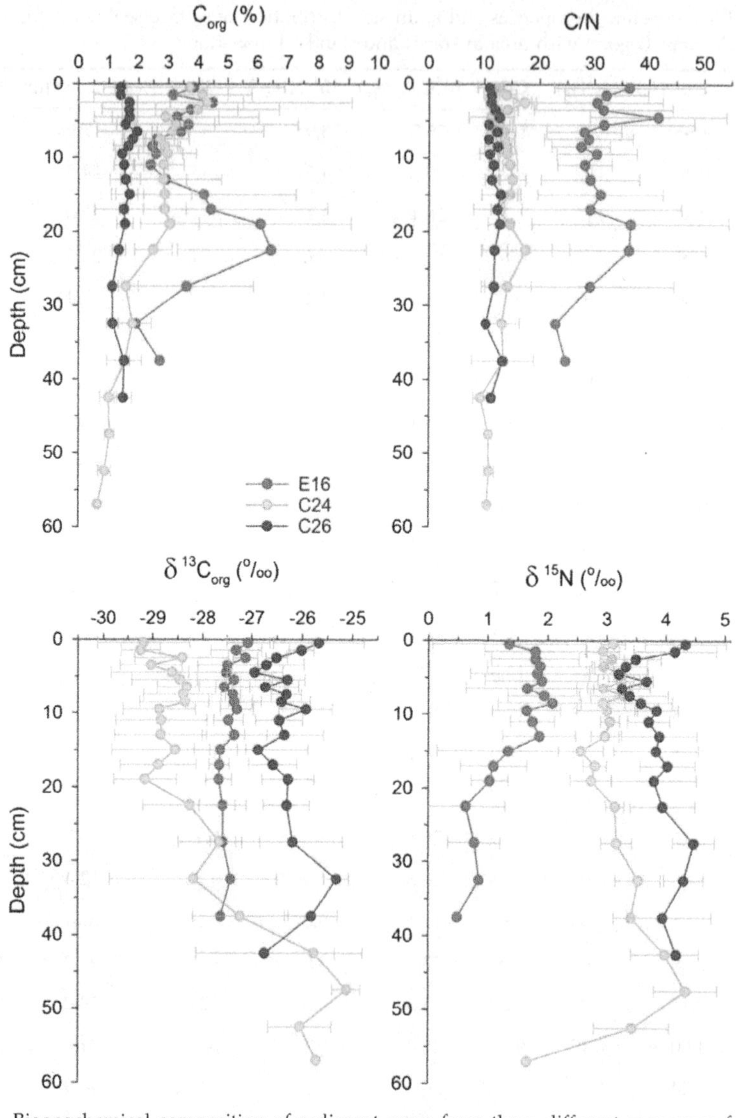

FIGURE 16.3 Biogeochemical composition of sediment cores from three different mangrove forest sites in the Segara Anakan Lagoon. Displayed are averages of four cores per site, the *gray bars* denote the standard deviation.

(Holtermann et al., 2009; Jennerjahn et al., 2009). While dry upland agriculture and forest are the dominant land uses/cover in the eastern hinterland and no rivers drain into the eastern part of the lagoon, the hinterland of the western and central parts of the SAL is dominated by rice cultivation under irrigation, and it is connected to the lagoon through the Citanduy River and a few smaller rivers providing the opportunity to transport substances derived from the hinterland into the lagoon (White et al., 1989; Ardli and Wolff, 2009). Surface sediments in the

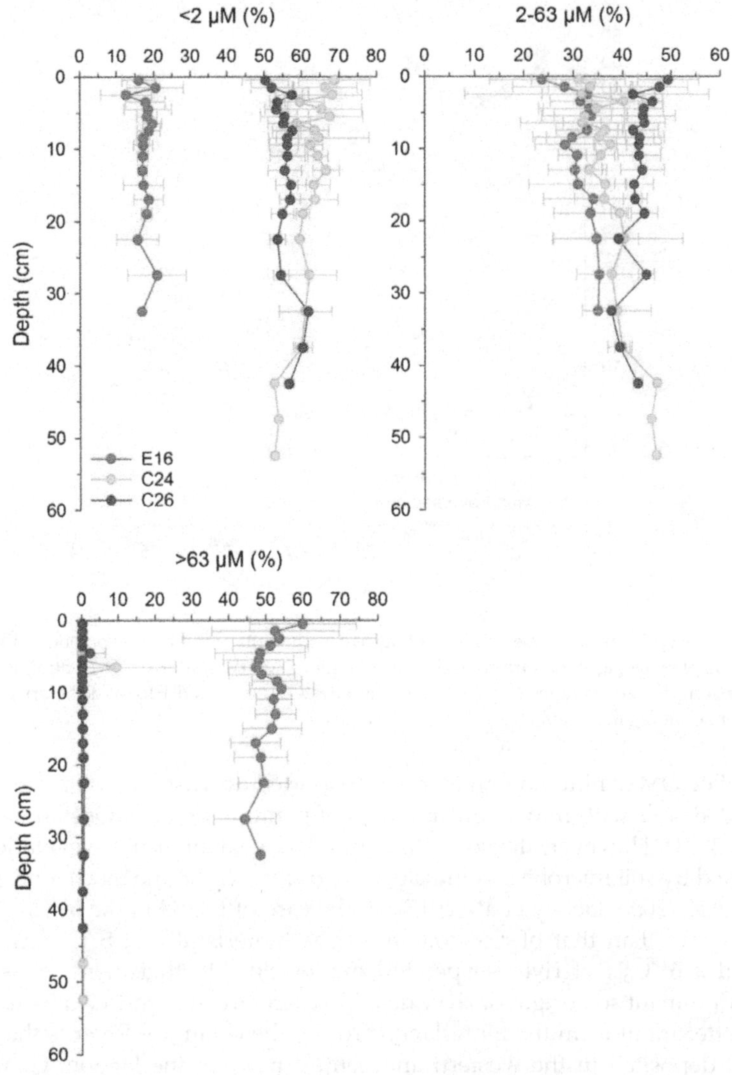

FIGURE 16.4 Grain-size composition of sediment cores from three different mangrove forest sites in the Segara Anakan Lagoon. Displayed are averages of four cores per site, the *gray bars* denote the standard deviation.

western and central parts of the SAL display fairly similar biogeochemical characteristics as the rice soils, which have an average $\delta^{13}C_{org}$ of $-26.1‰ \pm 1.6‰$, an average $\delta^{15}N$ of $5.3‰ \pm 1.0‰$, and a C/N ratio of 11–13. In contrast, the senescent leaves of mangrove trees and shrubs bordering the lagoon and covering the lagoon islands display a wide range of $\delta^{13}C_{org}$ values, but which are much lower than those of the rice soils, and they have much higher C/N ratios and similar $\delta^{15}N$ values. Like mangroves, rice is a C3 plant and therefore rice plants from the SAL hinterland have a $\delta^{13}C_{org}$ close to the average of mangrove and shrub leaves of $-28.8‰ \pm 1.3‰$.

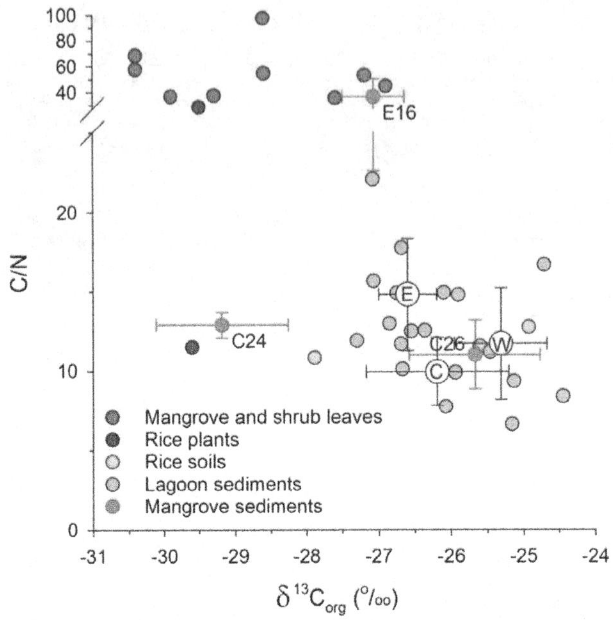

FIGURE 16.5 Biogeochemical properties of mangrove sedimentary OM and potential OM sources. The numbered circles display averages for surface sediments per subregion (W = West, C = Central, E = East). The *pink* (gray in print version) denote averages for whole cores at sites C24, C26, and E16, and the error bars denote one standard deviation. OM, organic matter.

The particulate OM of hinterland material transported downstream by rivers often consists mainly of eroded soils with minor contributions of plant tissue (e.g., Onstad et al., 2000; Jennerjahn et al., 2004). However, the soil OM is to a large extent also of vegetation origin, but usually modified by soil microbes, as indicated by differences in the biochemical composition (e.g., Onstad et al., 2000; Jacoby et al., 2017). In the case of the SAL, the $\delta^{13}C_{org}$ of rice plants that is much lower than that of rice soils from the hinterland and SAL surface sediments (Fig. 16.5) and a $\delta^{13}C_{org}$ of river suspended matter close to that of the soils (Moll, 2011) indicate a predominant soil origin of riverine OM. Therefore, it is conceivable that suspended particulate matter input from the hinterland through the Citanduy River is the predominant source of OM deposited in the western and central parts of the lagoon. Compared to the western and central parts of SAL, surface sediments from the eastern SAL have a $\delta^{13}C_{org}$ that is slightly lower, but it is slightly higher than in mangrove sediments at site E16. Similarly, the C/N ratio is considerably higher than in the central and western SAL, but much lower than in the mangrove sediments at E16. Moreover, the portion of the silt (2–63 μM) and sand (>63 μM) fractions is much higher in the eastern than in the central and western SAL. Taken together this indicates that input from the agriculture-dominated hinterland is less and export from mangroves into the lagoon is more relevant than in the central and western parts of the SAL. In general, the differences in the biogeochemical composition between all lagoon surface sediments and averages of mangrove cores at sites C24 and E16 show that OM in the latter is largely of autochthonous origin.

4.2 Autochthonous versus allochthonous contributions to mangrove sedimentary organic matter related to hydrodynamics and hinterland input

Despite some variability, the average biogeochemical composition of sediment cores from site C26 indicates the same source as for the surrounding lagoon parts (Fig. 16.4). Site C26 is located on an island in the lagoon. The SAL started to shrink since the middle of the 19th century. Increasing human modifications of the landscape, mainly compulsory coffee cultivation, large-scale extraction of timber, and railway construction, promoted erosion and increased sediment deposition in the lagoon in the second-half of the 19th century and the beginning 20th century (Lukas, 2017). The increase in rainfed agriculture on peasants' private lands, building of infrastructure (roads, trails, settlements), land tenure conflicts over state forests and plantations, state forest management practices, and river channel and floodplain modifications all contributed to an accelerated erosion in the hinterland and sediment deposition in the lagoon in the second-half of the 20th century (Lukas, 2017). Under these conditions the majority of the islands in the SAL formed since the 1970s. Site C26 is located on such an island and its biogeochemical composition indicates that the sedimentary OM there is mainly of allochthonous origin from the hinterland, despite the island's cover with mangrove vegetation (Hinrichs et al., 2009).

Differences in hydrodynamics and morphology play an important role in the dispersal and deposition of particulate matter, as indicated by the grain-size composition of sediments. Although being largely variable, on average the fraction >63 μM is >30% in all SAL surface sediments and the fraction <2 μM is particularly small in the eastern SAL (7%, Table 16.3), where freshwater input from the hinterland is low and the tides dominate material transport (Holtermann et al., 2009). There, mangrove sediments have a high portion of the >63 μM fraction of ca. 50% and sediments display erosional features. The site is located directly off the channel connecting the lagoon to the Indian Ocean where tidal currents are strong (Holtermann et al., 2009). It is conceivable that sedimentation rates there are probably low and little allochthonous material is being deposited. This is corroborated by the fairly high portion of root tissue in the sediments, in particular between 15 and 30 cm depth in cores at site E16 (Fig. 16.3; Pohlenga, 2007).

The hydrodynamics and the sediment dispersal and deposition in the western and central SAL are largely dominated by the river input in combination with the tides and the shallow morphology of the lagoon. While the vast majority of the suspended sediments delivered by the Citanduy are exported to the Indian Ocean through the Plawangan channel in the West, the incoming tide pushes part of the river plume toward the East (White et al., 1989; Holtermann et al., 2009). Bottom friction promotes the slowing down of currents in the shallow central lagoon and allows for the deposition of fine-grained material. The increased sediment deposition ultimately led and leads to the formation of islands and a southward migration of mangroves from the northern border of the central SAL. Therefore, sediments at C24 and C26 consist almost exclusively of the fine fractions <2 and 2–63 μM (Fig. 16.4). This is supported by a high sedimentation rate of 36 ± 22 mm/yr that was measured at site C24 in a recent study (Kusumaningtyas et al., 2019).

Ultimately, sedimentary carbon burial in aquatic systems strongly depends on the so-called oxygen exposure time, which includes factors such as primary productivity, the flux of organic carbon through the water column, sedimentation rate, organic carbon

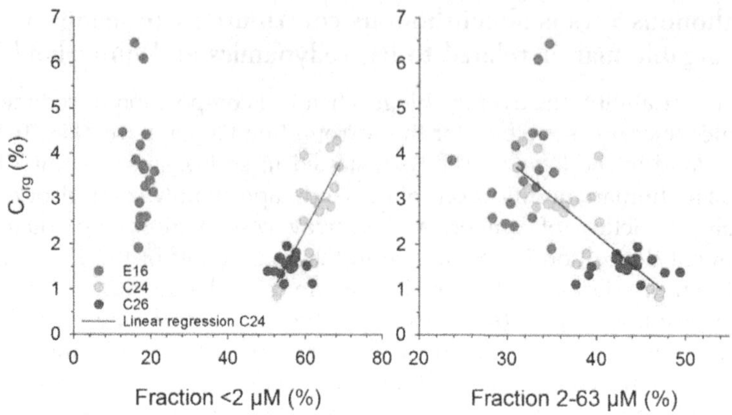

FIGURE 16.6 Organic carbon concentration versus portion of clay- (<2 μm, left panel) and silt-sized (2–63 μm, right panel) fractions of Segara Anakan mangrove sediments. Statistically significant linear relationships exist for C24 sediments (left: r = 0.78, $P < .001$; right: r = 0.79, $P < .001$).

degradation rate, and bottom-water oxygen concentration (Hartnett et al., 1998). Another important factor in the burial of carbon in sediments as well as in soils is the strong relationship to mineral surface area (Mayer, 1994a). On the long term, organic carbon burial in marine sediments is related to a monolayer coating of mineral surface area. Statistically significant relationships between grain size and organic carbon concentration have been observed in numerous sediments around the globe, i.e., organic carbon concentrations increase with increasing portions of clay- (<2 μm) and fine and medium silt-sized (2–6.3 μm, 6.3–20 μm) sediment fractions (e.g., Mayer, 1994b; Wiesner et al., 1990; Jennerjahn and Ittekkot, 1997).

Exceptions from these trends are observed in areas with very high riverine sediment inputs, organic pollution, and low bottom water oxygen (Mayer, 1994b). In the case of mangrove sediments from the SAL such a trend cannot be observed, except for core C24 (Fig. 16.6). The positive correlation of C_{org} with the clay-size fraction and its negative correlation with the silt-size fraction largely correspond to the change in sedimentation from lagoon to mangrove at mid-depth of the core. The more dynamic lagoon environment and the predominant hinterland input with a relatively lower carbon concentration resulted in coarser deposits with a lower C_{org} content in the lower part of the core. When mangroves established at the location of C24, a slowing down of tidal currents through increasing bottom friction and root growth allowed for deposition of more clay-sized material, providing more mineral surface area for organic carbon monolayer coating. Moreover, the supply of mangrove carbon through litter fall and root biomass contributed to the higher C_{org} content in the upper part of the core.

4.3 Temporal variations of organic matter composition

The rather low downcore variation of grain size indicates that the supply and the dispersal and deposition mechanisms did not change largely during the period of time that is represented by the cores, which is on the order of several decades in the central SAL, and of decades to centuries in the eastern SAL, according to a recent study (Kusumaningtyas et al., 2019). The observed downcore biogeochemical changes therefore appear to reflect temporal variations in sources and degradation of the OM. At site E16 with its low sedimentation rate and low external supply of OM, the below-ground root biomass is higher than at the other two sites. This is reflected in the high C/N ratio of around 30, which is much higher than the C/N ratio of marine particulate OM (approximately 5–10; Redfield, 1934; Jennerjahn and Ittekkot, 1997) and of rice soils from the SAL hinterland (Table 16.2), as well of soils from other parts of the world (ca. 10–30; Tian et al., 2010; Kirkby et al., 2011; Cools et al., 2014). The fairly low $\delta^{15}N$ of sedimentary OM at site E16 also points toward a high contribution of relatively fresh autochthonous material.

The $\delta^{15}N$ provides information on the nitrogen source and freshness of the OM. Fractionation during uptake can lead to an enrichment of ^{14}N. During OM production this leads to a lower $\delta^{15}N$, whereas $\delta^{15}N$ increases during OM decomposition because of the preferential removal of the lighter isotope (e.g., Wada and Hattori, 1990). The fixation of atmospheric nitrogen, which has a $\delta^{15}N$ of 0‰, by nitrogen-fixing bacteria can also lead to a low $\delta^{15}N$ of the produced OM. Variations of $\delta^{15}N$ are therefore caused by a number of processes and their interactions (e.g., Liu and Kaplan, 1989; Wada and Hattori, 1990; Schäfer and Ittekkot, 1993; Holmes et al., 1998; Ostrom et al., 1998; Carpenter et al., 1999; Holguin et al., 2001; Voss et al., 2001; Stewart et al., 2002). The $\delta^{15}N$ of nitrate varies around 6‰ in the ocean (e.g., Liu and Kaplan, 1989) and is somewhat higher in temperate estuaries (6–12; e.g., Middelburg and Nieuwenhuize, 2001; Dähnke et al., 2008). Plankton in the temperate Childs River estuary (Massachusetts, USA), which receives agriculture effluents and wastewater, has an average $\delta^{15}N$ of 5.9‰ (McClelland et al., 1997). $\delta^{15}N$ of suspended particulate matter from the Brantas River estuary in East Java ranges between 5‰ and 10‰ (Jennerjahn et al., 2004). Nitrogen fixation by heterotrophic bacteria is a key process and a major source of nitrogen in mangrove forests (e.g., Holguin et al., 2001 and references therein), as, for example, demonstrated in the nitrogen budget for the Missionary Bay mangrove forest in Australia (Alongi et al., 1992).

In the SAL, the $\delta^{15}N$ of mangrove leaves of different species and in various stages of aging varies roughly between 4‰ and 5‰ (Table 16.2; Nordhaus et al., 2011). Tissue of below-ground mangrove roots at site C24 in the SAL has a $\delta^{15}N$ of 1.3‰–3.4‰ (Pohlenga, 2007). In mangrove forests under the influence of external nutrient supply, sediment $\delta^{15}N$ roughly varies between 3‰ and 9‰, for example, mangrove sediment in the Brantas River estuary has a $\delta^{15}N$ of 3.7‰ (Jennerjahn et al., 2004), mangrove sediments in Moreton Bay, Australia, display a range of 3‰–5‰ (Sanders et al., 2016), and in Brazil $\delta^{15}N$ in sediments of an impacted mangrove forest varies between 4‰ and 9‰ (Sanders et al., 2014). The low $\delta^{15}N$ of mangrove sediments at site E16 in the SAL in combination with the high C/N ratio therefore indicates a mixture of relatively fresh and mainly autochthonous sedimentary OM. It consists to a large extent of root tissue and has additions of primary produced lagoon OM fueled by abundant dissolved inorganic nutrients, which mainly originate from the agriculture-dominated hinterland (Jennerjahn et al., 2009).

Large differences in the amount and composition of OM at C24 below 20 cm core depth indicate substantial changes in OM sources over time. Station C24 is located at the northern edge of the central lagoon, the most dynamic area of the lagoon in terms of sediment infilling over the past decades (Lukas, 2017). While the upper part of the sediment core contains fairly high amounts of C_{org} and $\delta^{13}C_{org}$ indicates a mainly mangrove source of the deposited OM, C_{org} decreases strongly below 20 cm core depth and $\delta^{13}C_{org}$ increases strongly, with values being in the range of hinterland soils and lagoon surface sediments of the western and central SAL, as well as of C26 sediments on the recently developed lagoon island. It indicates that site C24, at the time when the OM was deposited, was still lagoon and not yet covered by mangroves. This is in line with satellite image–based coastline reconstructions that suggest that C24 became a mangrove site in the 1990s (Lukas, 2014) and the fact that sedimentation rates in the central part of the lagoon are in the range of 1–3 cm/yr in the past ca. 50 years (Hapsari et al., 2020).

4.4 Relevance of autochthonous versus allochthonous OM sources for carbon budgets in times of global environmental change

The observed variations in the amount and composition of sedimentary OM in the mangrove-fringed SAL demonstrate the large variability on small spatial scales, i.e., in one mangrove ecosystem, indicating that a good spatial coverage of samples and solid understanding of the underlying processes are required to come up with robust and representative carbon budgets from mangrove ecosystems. When putting mangrove carbon budgets in relation to other reservoirs in space and time and the recent carbon dioxide emissions, the autochthonous versus allochthonous source of the OM becomes relevant. While only the autochthonous OM results from CO_2 fixed in the mangrove forest, allochthonous OM results from the fixation of CO_2 in other areas and possibly during other times.

In times of global change, carbon fixation and storage in sediments may also change because of changing inputs from the hinterland as well as of productivity changes and inputs from coastal waters. It has been demonstrated that erosion rates are very sensitive to land use change. The conversion of natural landscapes to cultivated land can increase river sediment loads by orders of magnitude (e.g., Walling, 1999). The erosion rate of agricultural soils is at least one order of magnitude larger than that of natural soils (Montgomery, 2007). It can also change the amount and composition of riverine OM and nutrients (e.g., Filoso et al., 2003; Jennerjahn et al., 2004; Lu et al., 2012; Jennerjahn et al., 2013). For example, human activities in the watershed of the Taiwanese Lanyang River have strongly altered the sediment load of the river. Road construction in the 1950s and 1960s and expanding agriculture in the 1970s and 1980s increased the sediment load of the river from 2.1×10^6 to 358×10^6 t/yr, 65% of which is related to the human activities (Kao and Milliman, 2008).

On the island of Java deforestation and conversion to agriculture land increased since the 1930s and peaked in the 1990s (Nibbering and de Graaff, 1998; Lavigne and Gunnell, 2006). Consequently, soil loss rates in the Serayu and Merawu catchments, which are located not far east of the Segara Anakan catchment, increased from <2 mm/yr in the 1990s to 4–14 mm/yr in 2002 (Lavigne and Gunnell, 2006). Similarly, land use change in the Segara Anakan catchment increased strongly after the 1920s and the resulting increased river sediment loads enhanced lagoon infilling (Lukas, 2017). Such drastic land use changes often include the erosion of deeper soil layers including very old OM. In the Taiwanese Lanyang River ^{14}C age-dating of suspended sediments revealed an average carbon age of 10,000–16,000 years or a fraction of modern carbon around 25% or even less (Kao and Liu, 1996), indicating that the massive erosion caused by road construction and agriculture expansion contributed high amounts of fossil carbon to the river load.

The fraction of allochthonous carbon in mangrove sediments can vary strongly within systems on a spatial scale and with depth, and it can also vary among environmental settings. For example, in studies on Western Australian mangroves it was inferred that carbon accumulation was dominated by allochthonous input at surface and by root growth and decomposition in deeper layers. Given the continuous sediment accretion in those systems, it appears that autochthonous processes dominate carbon accumulation over longer time scales, i.e., centuries, there (Saintilan et al., 2013). In general, however, the overall long-term proportion of allochthonous and autochthonous carbon accumulation will strongly depend on the rate of allochthonous input and the resulting sedimentation rate. OM content and composition varied with depth in 2 m long sediment cores from mangrove forests in different stages of development along the coast of French Guiana. While there was generally a significant input of allochthonous refractory detrital carbon from the Amazon River plume, downcore variations were to a large extent controlled by the maturity of the forest and, hence, the diagenesis of the autochthonous OM (Marchand et al., 2003). Two sediment cores taken from a mangrove forest in the Ba Lat Estuary of the Vietnamese Red River also displayed variations of OM content and composition with depth. While lower parts of the cores were dominated by allochthonous marine algal OM, the upper parts had larger proportions of autochthonous mangrove OM. With moderate sedimentation rates of 0.8 and 1.2 cm/yr the cores covered the history of development of the mangrove ecosystem in the past 60–100 years, during which proportions of autochthonous and allochthonous OM inputs changed significantly (Tue et al., 2011).

The fraction of allochthonous carbon also varies strongly in mangrove sediments of the SAL, but can be very high, in particular in the western and central parts of the lagoon where river input from the agriculture-dominated hinterland is high. This large fraction of allochthonous carbon is not the result of CO_2 fixation in the mangrove ecosystem, but results from CO_2 fixation in the hinterland and is probably to some extent not recently fixed, it may date back decades or even centuries. With respect to balancing recent emissions, which are on the order of $36{,}153 \times 10^6$ t CO_2 yr^{-1} in 2017 (Friedlingstein et al., 2019), only the autochthonous production by mangrove plants themselves and microphytobenthos in the mangrove ecosystem is direct CO_2 fixation. Quantitatively, most relevant with respect to allochthonous sediment and carbon accumulation are the mangroves in river- and tide-dominated environmental settings (Woodroffe et al., 2016; Twilley et al., 2018), which receive high inputs from the hinterland on the one hand and from the ocean on the other hand.

The global total carbon storage in mangrove sediments of $22 \pm 6 \times 10^6$ t C/yr accounts for only 0.2% of the annual emissions (Jennerjahn et al., 2017) and is even less when the allochthonous contribution is not considered. Storing allochthonous and possibly old carbon only means relocation of carbon from one sink to another, an aspect that needs to be considered for carbon budgeting, in particular in the case when calculations are made to offset annual CO_2 emissions, for example, for carbon emission trade schemes. Another factor that contributes to increased carbon storage in mangroves is the eutrophication of coastal waters. An increase of nutrients, primary production, settling OM, heterotrophic activity, and hypoxia has been observed in many coastal regions around the globe (e.g., Rabalais et al., 2009).

Through tidal exchange, nutrients and OM can be imported into mangroves. There, the enhanced nutrient load stimulates primary production of the mangrove plants themselves, as well as that of benthic algae. At least part of this autochthonous produced carbon can be stored in sediments, as well as part of the imported OM. For example, such an enhanced carbon accumulation in mangrove sediments has been observed in southeastern Brazil. There, the carbon, nitrogen, and phosphorus accumulation rates were two-, four-, and eight-fold higher in a mangrove forest affected by anthropogenic nutrient input than in a pristine mangrove forest (Sanders et al., 2014). As such, coastal eutrophication can contribute to enhanced storage of autochthonous and allochthonous carbon in mangrove sediments.

5. Conclusions

There is large variability in the carbon storage in mangrove sediments and therefore in global mangrove carbon budgets owing to the following factors: (i) spatial variability within one mangrove ecosystem (local), between mangrove ecosystems (regional), and between biogeographic regions (AEP vs. IWP, global); (ii) temporal variability because of changes in hydrodynamics, environmental settings, mangrove forest development, climate, and land use/cover change in the hinterland amount and composition; and (iii) differences in the autochthonous versus allochthonous origin of deposited carbon. While not affecting total carbon budgets as such, the latter may become relevant when calculations are made to offset annual CO_2 emissions, e.g., in carbon emission trade. However, identifying and quantifying the autochthonous versus allochthonous contributions of sedimentary carbon is seldomly considered and remains a challenge.

While coastal eutrophication owing to land-based human activities may increase carbon storage in mangrove sediments, direct conversion of mangroves to other uses can decrease it. The example from the SAL in Java, Indonesia, shows that increased river inputs resulting from deforestation and conversion to agriculture in the hinterland may increase carbon and sediment accumulation rates in such a coastal lagoon and its bordering mangrove forests. Such a dominance of hinterland input may strongly increase the allochthonous portion of deposited carbon.

Despite the described large variability, mangrove ecosystems store large amounts of carbon per unit area in their sediments because of high productivity, high external inputs, and the mostly anoxic sediments which support preservation of OM (Alongi, 2009). These unique features, including the capability of storing also CO_2 that was fixed somewhere else, make mangrove ecosystems such an important natural carbon sink.

References

Alongi, D.M., 2009. The Energetics of Mangrove Forests. Springer, Heidelberg, Germany.

Alongi, D.M., 2014. Carbon cycling and storage in mangrove forests. Ann. Rev. Mar. Sci. 6, 195–219.

Alongi, D.M., Boto, K.G., Robertson, A.I., 1992. Nitrogen and phosphorus cycles. In: Robertson, A.I., Alongi, D.M. (Eds.), Tropical Mangrove Ecosystems, vol. 41. AGU, Washington, Coastal and Estuarine Studies, pp. 251–292.

Ardli, E.R., Wolff, M., 2009. Land use and land cover change affecting habitat distribution in the Segara Anakan lagoon, Java, Indonesia. Reg. Environ. Change 9, 235–243.

BPS Cilacap, 2018. Tabel Jumlah Penduduk, Kepadatan per KM persegi, dan Persentasenya menurut Kecamatan di Kabupaten Cilacap, 2010–2016, Badan Pusat Statistik Cilacap. https://cilacapkab.bps.go.id. (Accessed 2 October 2018).

Breithaupt, J.L., Smoak, J.M., Smith III, T.J., Sanders, C.J., Hoare, A., 2012. Organic carbon burial rates in mangrove sediments: strengthening the global budget. Global Biogeochem. Cycles 26, GB3011.

Carpenter, E.J., Montoya, J.P., Burns, J., Mulholland, M.R., Subramanian, A., Capone, D.G., 1999. Extensive bloom of a N_2-fixing diatom/cyanobacterial association in the tropical Atlantic Ocean. Mar. Ecol. Prog. Ser. 185, 273–283.

Chowdhury, R.R., Uchida, E., Chen, L., Osorio, V., Yoder, L., 2017. Chapter 9: anthropogenic drivers of mangrove loss: geographic patterns and implications for livelihoods. In: Rivera-Monroy, V., Lee, S.Y., Kristensen, E., Twilley, R.R. (Eds.), Mangrove Ecosystems: A Global Biogeographic Perspective – Structure, Function and Services. Springer Publishing Company, New York, USA, pp. 275–300.

Cools, N., Vesterdal, L., De Vos, B., Vanguelova, E., Hansen, K., 2014. Tree species is the major factor explaining C:N ratios in European forest soils. For. Ecol. Manag. 311, 3–16.

Dähnke, K., Bahlmann, E., Emeis, K., 2008. A nitrate sink in estuaries? An assessment by means of stable nitrate isotopes in the Elbe estuary. Limnol. Oceanogr. 53, 1504–1511.

Donato, D.C., Kauffman, J.B., Murdiyarso, D., Kurnianto, S., Stidham, M., Kanninen, M., 2011. Mangroves among the most carbon-rich forests in the tropics. Nat. Geosci. 4, 293–297.

Filoso, S., Martinelli, L.A., Williams, M.R., Lara, L.B., Krusche, A., Ballester, M.V., Victoria, R., De Camargo, P.B., 2003. Land use and nitrogen export in the Piracicaba River basin, Southeast Brazil. Biogeochemistry 65, 275–294.

Friedlingstein, P., Jones, M.W., O'Sullivan, M., Andrew, R.M., Hauck, J., Peters, G.P., Peters, W., Pongratz, J., Sitch, S., Le Quéré, C., Bakker, D.C.E., Canadell, J.G., Ciais, P., Jackson, R.B., Anthoni, P., Barbero, L., Bastos, A., Bastrikov, V., Becker, M., Bopp, L., Buitenhuis, E., Chandra, N., Chevallier, F., Chini, L.P., Currie, K.I., Feely, R.A., Gehlen, M., Gilfillan, D., Gkritzalis, T., Goll, D.S., Gruber, N., Gutekunst, S., Harris, I., Haverd, V., Houghton, R.A., Hurtt, G., Ilyina, T., Jain, A.K., Joetzjer, E., Kaplan, J.O., Kato, E., Klein Goldewijk, K., Korsbakken, J.I., Landschützer, P., Lauvset, S.K., Lefèvre, N., Lenton, A., Lienert, S., Lombardozzi, D., Marland, G., McGuire, P.C., Melton, J.R., Metzl, N., Munro, D.R., Nabel, J.E.M.S., Nakaoka, S.-I., Neill, C., Omar, A.M., Ono, T., Peregon, A., Pierrot, D., Poulter, B., Rehder, G., Resplandy, L., Robertson, E., Rödenbeck, C., Séférian, R., Schwinger, J., Smith, N., Tans, P.P., Tian, H., Tilbrook, B., Tubiello, F.N., van der Werf, G.R., Wiltshire, A.J., Zaehle, S., 2019. Global Carbon Budget 2019. Earth Syst. Sci. Data 11, 1783–1838.

Geist, S.J., Nordhaus, I., Hinrichs, S., 2012. Occurrence of species-rich crab fauna in a human-impacted mangrove forest questions the application of community analysis as an environmental assessment tool. Estuar. Coast Shelf Sci. 96, 69–80.

Hapsari, K.A., Jennerjahn, T.C., Lukas, M.C., Karius, V., Behling, H., 2020. Intertwined effects of climate and land use change on environmental dynamics and carbon accumulation in a mangrove-fringed coastal lagoon in Java, Indonesia. Global Change Biol. 26, 1414–1431. https://doi.org/10.1111/gcb.14926.

Hartnett, H.M., Keil, R.G., Hedges, J.I., Devol, A.H., 1998. Influence of oxygen exposure time on organic carbon preservation in continental margin sediments. Nature 391, 572–574.

Hinrichs, S., Nordhaus, I., Geist, S.J., 2009. Status, diversity and distribution patterns of mangrove vegetation in the Segara Anakan lagoon, Java, Indonesia. Reg. Environ. Change 9, 275–289.

Holguin, G., Vazquez, P., Bashan, Y., 2001. The role of sediment microorganisms in the productivity, conservation, and rehabilitation of mangrove ecosystems: an overview. Biol. Fertil. Soils 33, 265–278.

Holtermann, P., Burchard, H., Jennerjahn, T.C., 2009. Hydrodynamics of the Segara Anakan lagoon. Reg. Environ. Change 9, 259–274.

Holmes, M.E., Müller, P.J., Schneider, R.R., Segl, M., Wefer, G., 1998. Spatial variations in euphotic zone nitrate utilization based on $\delta^{15}N$ in surface sediments. Geo Mar. Lett. 18, 58–65.

Huxham, M., Dencer-Brown, A., Diele, K., Kathiresan, K., Nagelkerken, I., Wanjiru, C., 2017. Chapter 8: mangroves and people: local ecosystem services in a changing climate. In: Rivera-Monroy, V., Lee, S.Y., Kristensen, E., Twilley, R.R. (Eds.), Mangrove Ecosystems: A Global Biogeographic Perspective – Structure, Function and Services. Springer Publishing Company, New York, USA, pp. 245–274.

IPCC, 2014. Climate change 2014: synthesis report. In: Core Writing Team, Pachauri, R.K., Meyer, L.A. (Eds.), Contribution of Working Groups I, II and III to the Fifth Assessment Report of the Intergovernmental Panel on Climate Change. IPCC, Geneva, Switzerland, p. 151.

Jacoby, R., Peukert, M., Succurro, A., Koprovova, A., Kopriva, S., 2017. The role of soil microorganisms in plant mineral nutrition – current knowledge and future directions. Front. Plant Sci. 8, 1617.

Jennerjahn, T.C., Ittekkot, V., 1997. Organic matter in sediments in the mangrove areas and adjacent continental margins of Brazil: I. Amino acids and hexosamines. Oceanol. Acta 20, 359–369.

Jennerjahn, T.C., Ittekkot, V., 2002. Relevance of mangroves for the production and deposition of organic matter along tropical continental margins. Naturwissenschaften 89, 23–30.

Jennerjahn, T.C., Ittekkot, V., Klöpper, S., Adi, S., Nugroho, S.P., Sudiana, N., Yusmal, A., Prihartanto, Gaye-Haake, B., 2004. Biogeochemistry of a tropical river affected by human activities in its catchment: Brantas River estuary and coastal waters of Madura Strait, Java, Indonesia. Estuar. Coast Shelf Sci. 60, 503–514.

Jennerjahn, T.C., Nasir, B., Pohlenga, I., 2009. Spatio-temporal variation of dissolved inorganic nutrients in the Segara Anakan lagoon, Java, Indonesia. Reg. Environ. Change 9, 259–274.

Jennerjahn, T.C., Jänen, I., Propp, C., Adi, S., Nugroho, S.P., 2013. Environmental impact of mud volcano inputs on the anthropogenically altered Porong River and Madura Strait coastal waters, Java, Indonesia. Estuar. Coast Shelf Sci. 130, 152–160.

Jennerjahn, T.C., Gilman, E., Krauss, K.W., Lacerda, L.D., Nordhaus, I., Wolanski, E., 2017. Chapter 7: mangrove ecosystems under climate change. In: Rivera-Monroy, V., Lee, S.Y., Kristensen, E., Twilley, R.R. (Eds.), Mangrove Ecosystems: A Global Biogeographic Perspective – Structure, Function and Services. Springer Publishing Company, New York, USA, pp. 211–244.

Kao, S.-J., Liu, K.-K., 1996. Particulate organic carbon export from a subtropical mountainous river (Lanyang Hsi) in Taiwan. Limnol. Oceanogr. 41, 1749–1757.

Kao, S.-J., Milliman, J., 2008. Water and sediment discharge from small mountainous rivers, Taiwan: The roles of lithology, episodic events, and human activities. J. Geology 116, 431–448.

Kirkby, C.A., Kirkegaard, J.A., Richardson, A.E., Wade, L.J., Blanchard, C., Batten, G., 2011. Stable soil organic matter: a comparison of C:N:P:S ratios in Australian and other world soils. Geoderma 163, 197–208.

Kristensen, E., 2000. Organic matter diagenesis at the oxic/anoxic interface in coastal marine sediments, with emphasis on the role of burrowing animals. Hydrobiologia 426, 1–24.

Kusumaningtyas, M.A., Hutahaean, A.A., Fischer, H.W., Pérez-Mayo, M., Pittauer, D., Jennerjahn, T.C., 2019. Variability in the organic carbon stocks, sources, and accumulation rates of Indonesian mangrove ecosystems. Estuar. Coast Shelf Sci. 218, 310–323.

Lavigne, F., Gunnell, Y., 2006. Land cover change and abrupt environmental impacts on Javan volcanoes, Indonesia: a long-term perspective on recent events. Reg. Environ. Change 6, 86–100.

Liu, K.-K., Kaplan, I.R., 1989. The eastern tropical Pacific as a source of ^{15}N-enriched nitrate in seawater off southern California. Limnol. Oceanogr. 34, 820–830.

Lu, X.X., Li, S., He, M., Zhou, Y., Li, L., Ziegler, A.D., 2012. Organic carbon fluxes from the upper Yangtze basin: an example of the Longchuanjiang River, China. Hydrol. Process. 26, 1604–1616.

Ludwig, H.F., 1985. Segara Anakan Environmental Monitoring and Optimal Use Planning Project. Institute of Hydraulic Engineering, Agency for Research and Development and Ministry of Public Works, Bandung, Indonesia, p. 285. Final Report.

Lukas, M.C., 2014. Cartographic reconstruction of historical environmental change. Cartogr. Perspect. 78, 5–24.

Lukas, M.C., 2017. Widening the scope: linking coastal sedimentation with watershed dynamics in Java, Indonesia. Reg. Environ. Change 17, 901–914.

Marchand, C., Lallier-Vergès, E., Baltzer, F., 2003. The composition of sedimentary organic matter in relation to the dynamic features of a mangrove-fringed coast in French Guiana. Estuar. Coast Shelf Sci. 56, 119–130.

Mayer, L.M., 1994a. Relationships between mineral surfaces and organic carbon concentrations in soils and sediments. Chem. Geol. 114, 347–363.

Mayer, L.M., 1994b. Surface area control of organic carbon accumulation in continental shelf sediments. Geochem. Cosmochim. Acta 58, 1271–1284.

McCave, I.N., Syvitski, J.P.M., 2007. Principles and methods of geological particle size analysis. In: Syvitski, J.P.M. (Ed.), Principles, Methods and Application of Particle Size Analysis. Cambridge University Press, Cambridge, UK, pp. 3–21.

McClelland, J.W., Valiela, I., Michener, R.H., 1997. Nitrogen-stable isotope signatures in estuarine food webs: a record of increasing urbanization in coastal watersheds. Limnol. Oceanogr. 42, 930–937.

Middelburg, J.J., Nieuwenhuize, J., 2001. Nitrogen isotope tracing of dissolved inorganic nitrogen behaviour in tidal estuaries. Estuar. Coast Shelf Sci. 53, 385–391.

Moll, R., 2011. Impact of Mangroves and an Agriculture-Dominated Hinterland on the Carbon and Nutrient Biogeochemistry in the Segara Anakan Lagoon, Java, Indonesia (Ph.D. thesis). University of Bremen, p. 164. https://elib.suub.uni-bremen.de/edocs/00101939-1.pdf.

Montgomery, D.R., 2007. Soil erosion and agricultural sustainability. Proc. Natl. Acad. Sci. U.S.A. 104, 13268–13272.

Nellemann, C., Corcoran, E., Duarte, C.M., Valdés, L., De Young, C., Fonseca, L., Grimsditch, G. (Eds.), 2009. Blue Carbon. A Rapid Response Assessment. United Nations Environment Programme, GRID-Arendal. www.grida.no.

Nibbering, J.W., de Graaff, J., 1998. Simulating the past: reconstructing historical land use and modeling hydrological trends in a watershed area in Java. Environ. Hist. 4, 251–278.

Nordhaus, I., Hadipudjana, F.A., Janssen, R., Pamungkas, J., 2009. Spatio-temporal variation of macrobenthic communities in the mangrove-fringed Segara Anakan Lagoon, Indonesia, affected by anthropogenic activities. Reg. Environ. Change 9, 291–313.

Nordhaus, I., Salewski, T., Jennerjahn, T.C., 2011. Food preferences of mangrove crabs related to leaf nitrogen compounds in the Segara Anakan Lagoon, Java, Indonesia. J. Sea Res. 65, 414–426.

Nordhaus, I., Salewski, T., Jennerjahn, T.C., 2017. Interspecific variations in mangrove leaf litter decomposition are related to labile nitrogenous compounds. Estuar. Coast Shelf Sci. 192, 137–148.

Nordhaus, I., Toben, M., Fauziyah, A., 2019. Impact of deforestation on mangrove tree diversity, biomass and community dynamics in the Segara Anakan lagoon, Java, Indonesia: a ten-year perspective. Estuar. Coast Shelf Sci. 227, 106300.

Onstad, G.D., Canfield, D.E., Quay, P.D., Hedges, J.I., 2000. Sources of particulate organic matter in rivers from the continental USA: lignin phenol and stable carbon isotope compositions. Geochem. Cosmochim. Acta 64, 3539–3546.

Ostrom, N.E., Knoke, K.E., Hedin, L.O., Robertson, G.P., Smucker, A.J.M., 1998. Temporal trends in nitrogen isotope values of nitrate leaching from an agricultural soil. Chem. Geol. 146, 219–227.

Pohlenga, I., 2007. Nutrient Cycles in Mangrove Sediments of the Segara Anakan Lagoon (Java). Unpublished diploma thesis, University of Bremen, p. 63 (in German).

Rabalais, N.N., Turner, R.E., Diaz, R.J., Justic, D., 2009. Global change and eutrophication of coastal waters. ICES J. Mar. Sci. 66, 1528–1537.

Redfield, A.C., 1934. On the proportions of organic derivatives in sea water and their relation to the composition of plankton. In: James Johnstone Memorial Volume. Liverpool University Press, pp. 176–192.

Rovai, A.S., Twilley, R.R., Castañeda-Moya, E., Riul, P., Cifuentes-Jara, M., Manrow-Villalobos, M., Horta, P.A., Simonassi, J.C., Fonseca, A.L., Pagliosa, P.R., 2018. Global controls on carbon storage in mangrove soils. Nat. Clim. Change 8, 534–538.

Saintilan, N., Rogers, K., Mazumder, D., Woodroffe, C., 2013. Allochthonous and autochthonous contributions to carbon accumulation and carbon store in southeastern Australian coastal wetlands. Estuar. Coast Shelf Sci. 128, 84–92.

Sanders, C.J., Eyre, B.D., Santos, I.R., Machado, W., Luiz-Silva, W., Smoak, J.M., Breithaupt, J.L., Ketterer, M.E., Sanders, L., Marotta, H., Silva-Filho, E., 2014. Elevated rates of organic carbon, nitrogen, and phosphorus accumulation in a highly impacted mangrove wetland. Geophys. Res. Lett. 41, 2475–2480.

Sanders, C.J., Santos, I.R., Maher, D.T., Breithaupt, J.L., Smoak, J.M., Ketterer, M., Call, M., Sanders, L., Eyre, B.D., 2016. Examining $^{239+240}Pu$, ^{210}Pb and historical events to determine carbon, nitrogen and phosphorus burial in mangrove sediments of Moreton Bay, Australia. J. Environ. Radioact. 151, 623–629.

Sasmito, S.D., Taillardat, P., Clendenning, J.N., Cameron, C., Friess, D.A., Murdiyarso, D., Hutley, L.B., 2019. Effect of land-use and land-cover change on mangrove blue carbon: a systematic review. Global Change Biol. 25, 4291–4302.

Schäfer, P., Ittekkot, V., 1993. Seasonal variability of ^{15}N in settling particles in the Arabian Sea and its paleogeochemical significance. Naturwissenschaften 80, 511–513.

Stewart, G.R., Aidar, M.P.M., Joly, C.A., Schmidt, S., 2002. Impact of point source pollution on nitrogen isotope signatures ($\delta^{15}N$) of vegetation in SE Brazil. Oecologia 131, 468–472.

Thom, B.G., 1982. Mangrove ecology: a geomorphological perspective. In: Clough, B.F. (Ed.), Mangrove Ecosystems in Australia: Structure, Function and Management. Australia National University Press, Canberra, pp. 3–17.

Tian, H., Chen, G., Zhang, C., Melillo, J.M., Hall, C.A.S., 2010. Pattern and variation of C:N:P ratios in China's soils: a synthesis of observational data. Biogeochemistry 98, 139–151.

Tue, N.T., Hamaoka, H., Sogabe, A., Quy, T.D., Nhuan, M.T., Omori, K., 2011. The application of $\delta^{13}C$ and C/N ratios as indicators of organic carbon sources and paleoenvironmental change of the mangrove ecosystem from Ba Lat Estuary, Red River, Vietnam. Environ. Earth Sci. 64, 1475–1486.

Twilley, R.R., Rovai, A.S., Riul, P., 2018. Coastal morphology explains global blue carbon distributions. Front. Ecol. Environ. https://doi.org/10.1002/fee.1937.

United Nations, 2015. The Paris Agreement. In: https://unfccc.int/process-and-meetings/the-paris-agreement/the-paris-agreement. (Accessed 22 November 2019).

Voss, M., Dippner, J.W., Montoya, J.P., 2001. Nitrogen isotope patterns in the oxygen-deficient waters of the eastern tropical North Pacific ocean. Deep-Sea Res. I 48, 1905–1921.

Wada, E., Hattori, A., 1990. Nitrogen in the Sea: Forms, Abundances, and Rate Processes. CRC Press, Boca Raton, p. 224.

Walling, D.E., 1999. Linking land use, erosion and sediment yields in river basins. Hydrobiologia 410, 223–240.

Watanabe, K., Kuwae, T., 2015. How organic carbon derived from multiple sources contributes to carbon sequestration processes in a shallow coastal system? Global Change Biol. 21, 2612–2623.

White, A.T., Martosubroto, P., Sadorra, M.S.M., 1989. The coastal environment profile of Segara Anakan – Cilacap, south Java, Indonesia. In: ICLARM, Association of Southeast Asian Nations, United States Coastal Resources Management Project, Technical Publication Series 4.

Wiesner, M.G., Haake, B., Wirth, H., 1990. Organic facies of surface sediments in the North Sea. Org. Geochem. 15, 419–432.

Woodroffe, C.D., 1992. Mangrove sediments and geomorphology. In: Robertson, A.I., Alongi, D.M. (Eds.), Tropical Mangrove Ecosystems. American Geophysical Union, Washington, DC, pp. 7–41.

Woodroffe, C.D., Rogers, K., McKee, K.L., Lovelock, C.E., Mendelssohn, I., Saintilan, N., 2016. Mangrove Sedimentation and Response to Relative Sea-Level Rise. Ann. Rev. Mar. Sci. 8, 243–266.

Yuwono, E., Jennerjahn, T.C., Nordhaus, I., Riyanto, E.A., Sastranegara, M.H., Pribadi, R., 2007. Ecological status of Segara Anakan , Indonesia: a mangrove-fringed lagoon affected by human activities. Asian J. Water Environ. Pollut. 4, 61–70.

CHAPTER

17

Potential carbon loss in sediment through methane production during early development stage of mangrove regeneration in restored mangroves

Yaya I. Ulumuddin[1,2], *Sara Beavis*[2], *Michael Roderick*[3], *Stephen Eggins*[3], *Irawan Sugoro*[4], *Sukristijono Sukardjo*[1]

[1]Research Center for Oceanography, Indonesian Institute of Sciences., Jakarta, Indonesia; [2]Fenner School of Environment and Society, Australian National University, Acton, ACT, Australia; [3]Research School of Earth Sciences, Australian National University, Acton, ACT, Australia; [4]National Nuclear Energy Agency of Indonesia, Jakarta, Indonesia

OUTLINE

1. **Introduction** 416
2. **Study area and sampling sites** 418
3. **Sample collection and analytical methods** 420
   - *3.1 Porewater dissolved CH_4* 420
   - *3.2 Soil surface CH_4 fluxes* 421
   - *3.3 Physicochemistry of porewater* 421
4. **Results** 422
   - *4.1 Stand basal area, porewater dissolved CH_4, and CH_4 fluxes* 422
   - *4.2 Physicochemical parameters* 423
   - *4.3 Relationships between porewater CH_4 and stand basal area* 424
5. **Discussion** 427
   - *5.1 Effects of seasons and mangrove regrowth on dissolved CH_4* 427
   - *5.2 Potential mangrove contribution to CH_4 emissions* 431
   - *5.3 Indication of CH_4 releases through mangrove trees* 434
6. **Conclusion** 436

Appendix 1: soil surface CH_4 fluxes 436

https://doi.org/10.1016/B978-0-12-816437-2.00020-3

Appendix 2: CH_4 fluxes from tree stem and soil surface in the literature 438

Acknowledgments 439

References 439

1. Introduction

Regeneration of degraded mangrove forests is thought to be an option to recover ecosystem carbon stocks, possibly contributing to atmospheric carbon abatement and thus alleviating climate change. Mangrove regrowth, natural or planted, could increase biomass carbon (Cameron et al., 2019b; Sasmito et al., 2019). Biomass carbon increases rapidly during early regrowth and slowly in the latter stages (Sasmito et al., 2019), which has also been conceptualized by Odum (1969) and Alongi (2011). However, there is no clear pattern in soil carbon recovery, and the estimates of the mangrove net carbon balance based on available literature remain uncertain (Sasmito et al., 2019), leading them to suggest the need for long-term monitoring of carbon stocks and fluxes following mangrove loss and regeneration. This is a key point for maximizing carbon gain and minimizing loss in mangrove ecosystems (Macreadie et al., 2017). In this context, greenhouse gas (GHG) GHG assessment is required in carbon balance estimations.

In the context of GHG assessment, besides carbon dioxide (CO_2), methane (CH_4) is also of importance. An estimate of 3% of fixed carbon in freshwater wetlands is released back into the atmosphere as CH_4 (Aselmann and Crutzen, 1989; Whiting and Chanton, 1993). Furthermore, CH_4 emissions are of global importance given the positive feedback effect this gas has on climate change. CH_4, together with other GHGs such as CO_2, causes an increase of Earth's surface temperature, and this temperature increase could, in turn, accelerate methanogenesis, generating a further increase in CH_4 emissions (Gedney et al., 2004). Additionally, CH_4, on a mass basis, is 25 times more potent in heating the Earth's surface than CO_2 (Forster et al., 2007). Its atmospheric concentration is steadily increasing and is projected to be 2200 ppb by 2030 (IPCC, 2013). The contribution of wetland CH_4 emissions is about 23% of global CH_4 sources (Reeburgh, 2013) and represents around 60% of emissions from all natural sources (Kirschke et al., 2013).

Wetland contributions to CH_4 emissions may be increasing because of warmer temperatures (Gedney et al., 2004) as well as another climate change phenomenon (e.g., sea-level rise) (Ardon et al., 2018; Lu et al., 2018; Helton et al., 2019). Climate change modifies three main factors that control CH_4 emissions from wetlands: (1) soil temperature (controlling bacterial activity rates) (Inamori et al., 2007; Datta et al., 2013; Lofton et al., 2014); (2) water inundation (changing oxic–anoxic soil layers) (Zhang and Ding, 2011; Marín-Muñiz et al., 2015); and (3) soil carbon availability (providing favorable substrates for methanogens) (Yu et al., 2013; Konnerup et al., 2014).

Like other wetlands, mangrove ecosystems naturally produce and emit CH_4 because they accumulate soil organic matter and exist in primarily anaerobic and hydrodynamically complex environments (Bouillon et al., 2008; Weston et al., 2014). However, the relative contribution of mangrove ecosystems to atmospheric CH_4 concentrations is still debatable (Chen et al., 2016; Cabezas et al., 2018). Studies measuring CH_4 emissions from mangrove ecosystems indicate a large range, from zero (Giani et al., 1996; Alongi et al., 2004) to around 80 mg/m^2/h (Chen et al., 2010).

To date, studies on CH_4 in mangrove ecosystems have focused primarily on soil surface CH_4 emissions (Giani et al., 1996; Chauhan et al., 2008; Krithika et al., 2008; Chen et al., 2016; Cabezas et al., 2018). Other studies have sought to explain the variability of CH_4 emissions from mangrove ecosystems. For example, CH_4 produced in the soil can be transported to the atmosphere through the vascular tissue (i.e., stems) of herbaceous and woody plants (Van Der Nat et al., 1998; Bazhin, 2004; Pangala et al., 2013, 2017; Maier et al., 2018). A recent study of Amazon wetlands reported that CH_4 emissions from woody stems accounted for around 40% of total emissions and far exceeded soil surface emissions (~15% of the total) (Pangala et al., 2017). The remaining emissions (~45%) were from open water surfaces and the stems of herbaceous plants. The results from this recent report provide evidence that previous estimates of mangrove contributions to ecosystem CH_4 emissions based only on soil surface CH_4 fluxes are likely to be underestimated.

This chapter presents our study that investigated dissolved CH_4 in mangrove porewater as a source of mangrove CH_4 emissions. This may be more representative of the contribution of mangrove ecosystems to global CH_4 emissions for two reasons. First, as aforementioned, mangrove tree-mediated CH_4 emissions may be mainly responsible for ecosystem CH_4 emissions. Indeed, recent literature reported this pathway of CH_4 emissions (Jeffrey et al., 2019). Second, although sulfate-reducing bacteria (SRB) may outcompete methanogens because of an extensive sulfate supply from seawater, methylotrophic methanogens—which use a different substrate (methylated compounds, rather than acetate or CO_2/H_2)—can keep producing CH_4 under conditions of high sulfate concentrations (Chuang et al., 2016; Jing et al., 2016). In that case, CH_4 production will be reflected in the porewater dissolved concentrations of CH_4. In addition, empirical results have indicated a strong relationship between ecosystem CH_4 fluxes and porewater CH_4 (Bartlett et al., 1987; Wassmann et al., 2000; Yoshikawa et al., 2014).

In this study, we assessed dissolved CH_4 heterogeneity at sites with different stages of mangrove regeneration and changes of dissolved CH_4 throughout different seasons. Regarding this, Cameron et al. (2019a) recently reported that CH_4 emissions, in the rehabilitation sites at Tanakeke Island, Indonesia, increased following regrowth of mangrove forest in abandoned aquaculture ponds. Due to both studies being undertaken in the same area, our study can be considered to be complementary to Cameron et al. (2019a) and can be used to improve understandings of how the state of mangrove regrowth controls dissolved porewater CH_4. We also measured several physicochemical parameters (pH, redox potential, temperature, and salinity) that could help to explain heterogeneity in porewater CH_4 concentrations along with mangrove regrowth. The results provided seasonal data on porewater dissolved CH_4 in the natural regeneration of mangroves and identified possible factors controlling its variability. Conceptualization of the major finding is also provided to highlight the importance of including CH_4 emissions in a GHG accounting budget for mangrove rehabilitation projects.

2. Study area and sampling sites

Research was conducted at Dusun Lantangpeo, Tanakeke Island, South Sulawesi, Indonesia (5° 27′ S and 119° 17′ E) (Fig. 17.1). In this area, Blue Forest Foundation conducted mangrove rehabilitation in 2010 (Brown et al., 2014), where Cameron et al. (2019a) undertook their study on CH_4 emissions. Next to the rehabilitation site, natural regeneration of mangrove trees also occurs in the disused aquaculture ponds, where the sampling sites for the current study were located.

Tanakeke is a coralline island located around 10 km from the mainland of Sulawesi and falling within the Takalar (administrative) District, Province of South Sulawesi. It is about 40 km southwest of Makassar, the capital city of the province. The island has a tropical climate, with two distinct seasons. It experiences a wet season from November to April, with occasional dry season rainfalls also occurring (Fig. 17.2). The annual average precipitation is approximately 3000 mm, with around 80% of precipitation typically falling in the wet season. Air temperature is reasonably stable with respect to small diurnal or seasonal fluctuations (Fig. 17.2). The annual mean minimum and maximum air temperatures are around 25 and 32°C, respectively, with an overall annual mean air temperature close to 28°C.

In Tanakeke, mangrove vegetation grows in sandy sediments, adjacent to seagrass beds and coral reefs, which are a dominant feature of the island. The mangrove vegetation is inundated by tides on a daily basis. There is approximately one low and one high tide per day, with a maximum range of 1 m difference between low and high tide. At the highest high tide, there is around 0.5 m depth of seawater covering the soil surface. The mangrove

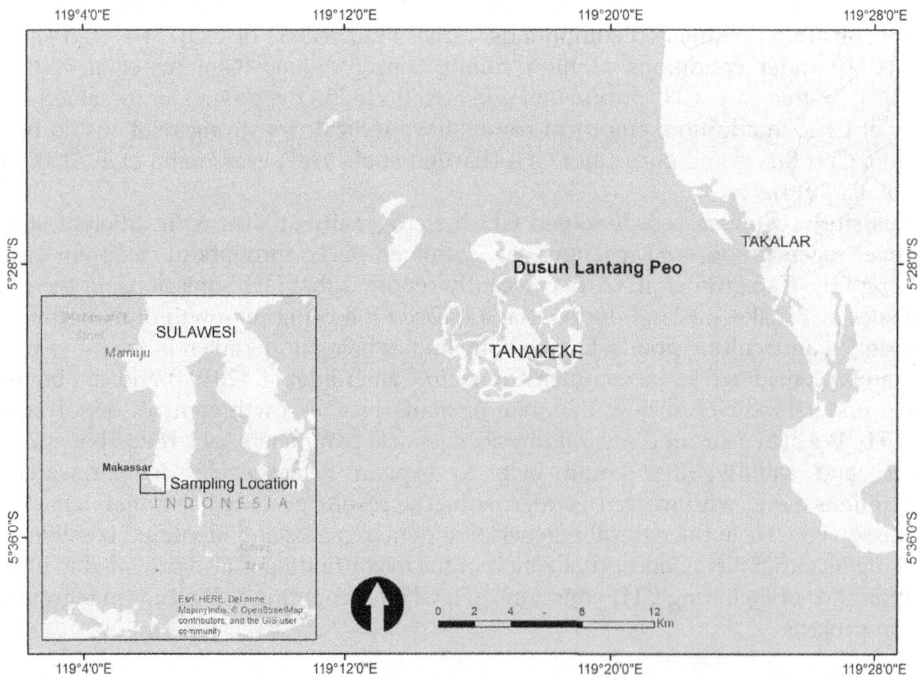

FIGURE 17.1 Map showing location of the field site at Lantangpeo, Tanakeke Island.

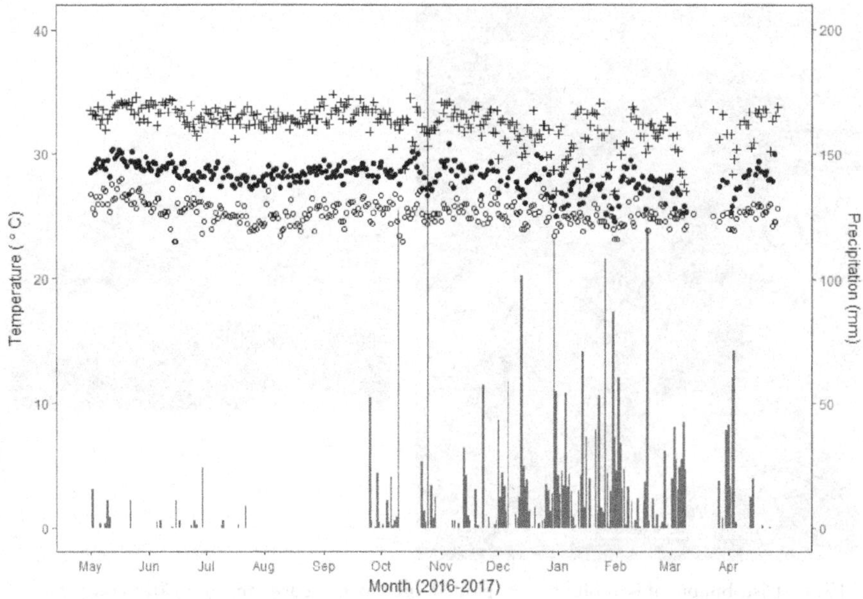

FIGURE 17.2 Climate data of the observation period (May 2016 to April 2017) showing daily precipitation (bars, right axis) and daily air temperature (left axis). *Open circles, closed circles,* and *plus symbols* indicate minimum, mean, and maximum temperature, respectively (Meteorological, Climatological, and Geophysical Agency—Stasiun Meteorologi Maritim Paotere, Makassar).

vegetation is characterized as overwash mangroves dominated by *Rhizophora stylosa* (Setiawan and Mursidin, 2018). A further 10 species occur on the island, including *Avicennia alba, Bruguiera gymnorrhiza, Excoecaria agallocha, Gymnanthera paludosa, Heritiera littoralis, Lumnitzera racemosa, Pemphis acidula, Rhizophora apiculata, Rhizophora mucronata,* and *Sonneratia alba.*

We used a contrasting approach to investigate the changes of dissolved CH_4 in the porewater of mangrove sediment due to natural regeneration. The sampling locations were selected and grouped into three sites, with contrasting canopy cover. Sites 1 to 3 represented new regeneration sites, while Sites 4 to 6 represented old regeneration sites. These two groupings had distinct canopy cover identifiable from Google Earth imagery. The third group comprised three additional sites (Sites 7 to 9), as natural mangroves representing control sites. We also selected a further two control sites in unvegetated areas (Sites A and B)—mangrove creeks (Fig. 17.3). All sites can be seen in Google Earth imagery (Fig. 17.3, left) with Sites 1 to 6 all located in abandoned fishponds. In addition, all the sites were in a similar position in relation to mean sea level (MSL), ranging from −11.8 to 6.9 cm relative to MSL. In the study area, the maximum tidal range was 112 cm and the average flooding period was 13 h per day. This implies that long inundation periods occurred, and thus soil at the sites contained a large volume of porewater (Cameron et al., 2019a).

To confirm the sequence of sites based on their successional stages, we undertook a forest structure assessment using measurements following an international protocol for assessing

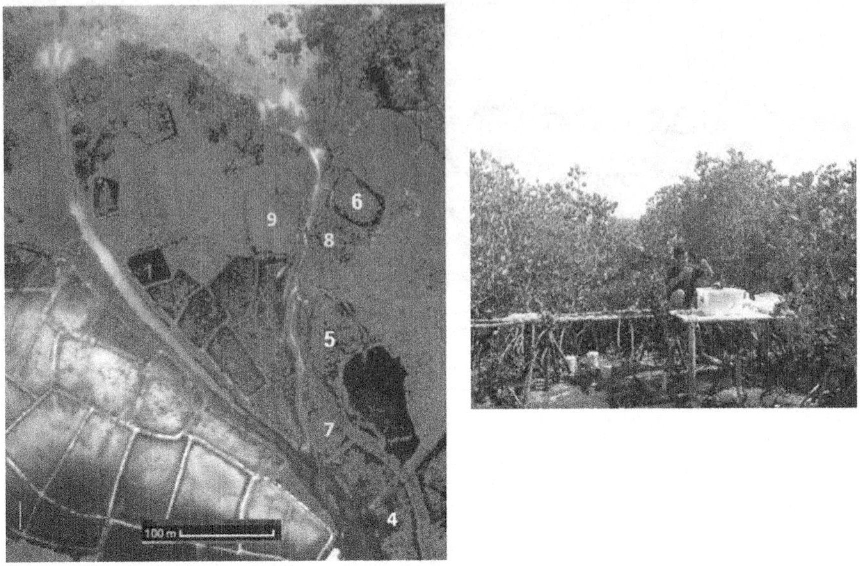

FIGURE 17.3 Distribution of sampling sites plotted on Google Earth imagery, depicting the landscape of aquaculture ponds and mangroves with two creeks splitting the landscape. White and brown (dark gray in print version) color spotted by green color (light gray in print version) (lower left) depicts recently abandoned ponds; patches of green (light gray in print version) color surrounded by a brown (black in print version) line depict long abandoned ponds; green (light gray in print version) color blocks show undisturbed mangrove vegetation (left). Photograph at Site 4 shows the elevated work bench (right).

carbon stocks and forest structure in mangroves (Kauffman and Donato, 2012). We selected stand basal area (BA) as a measure of forest structure reflecting successional stages, rather than canopy cover or stand density. This is because BA increases logarithmically with successional stages (Lohbeck et al., 2012; Cheng et al., 2013; Gosper et al., 2013; Pszwaro et al., 2016), while tree density increases in the early successional stages, reaches a peak, and then decreases through natural self-thinning (Schulze et al., 2005).

For CH_4 monitoring, at each site, we deployed one multilevel porewater sampler and constructed an elevated work bench to avoid disturbing the mangrove soil while collecting samples (Fig. 17.3, right). At the nine sites, seasonal variations were observed by repeated sampling undertaken every month from May 2016 to March 2017, during spring tides (Fig. 17.2). However, heavy rain during January 2017 caused a cancellation of the fieldwork in that month.

3. Sample collection and analytical methods

3.1 Porewater dissolved CH_4

The concentration of porewater dissolved CH_4 was determined by using gas chromatography (GC). At the selected sites, porewater samples were pumped out from mangrove sediments by using a recently developed porewater sampler (Ulumuddin, 2019), which is a modification of Martin et al. (2003). With this sampler, we collected porewater from five different depths of 10, 20, 30, 40, and 50 cm. Gas extractions from the porewater samples were undertaken following Magen et al. (2014). The extracted gas was transferred into aluminum gas-tight bags (Shanghai Sunrise Instrument Co., Ltd., Shanghai, $15 \times 15\ cm^2$, 30 mL) while in the field. These gas bags were transported (by boat and aircraft) from the field site to the Greenhouse Gases Laboratory, Research Institute for Agricultural Environment, Indonesian Agency for Agricultural Research and Development, Pati, Central Java. The gas samples were analyzed by using Shimadzu type 14A GC equipped with a flame ionization detector and a packed column of Porapak Q. The temperatures of the GC were set at 25°C (injector), 50°C (column), and 300°C (detector). This instrument has a limit of detection of 0.12 ppmv for CH_4. According to GC measurements, CH_4 concentrations in the porewater samples were calculated as explained by Magen et al. (2014).

3.2 Soil surface CH_4 fluxes

Methane fluxes ($Flux_{CH4}$) from soil surfaces in $mg/m^2/h$ were estimated by using the gas transfer equation (Sarmiento and Gruber, 2006; Wanninkhof et al., 2009; Wania et al., 2010). The reasons for, and assumptions of, choosing this model are explained in Appendix 1. For comparison, CH_4 fluxes were also measured by using the static chamber method. However, unfortunately, the results indicated that the CH_4 fluxes were much too small to be realistic. We then used estimations of CH_4 fluxes from the soil surface in mangrove sites in Tanakeke, as reported by Cameron et al. (2019a), including two rehabilitation sites and mature mangrove forests, which accounted for CH_4 fluxes of 0.14 ± 0.0; 0.09 ± 0.0; and $0.45 \pm 0.09\ mg/m^2/h$, respectively.

3.3 Physicochemistry of porewater

Physicochemical parameters of porewater were also observed to help understand porewater CH_4 variability across sites and seasons. These included physical parameters (e.g., pH, temperature, and redox potential) and ion species that are closely related to methanogenesis (i.e., Fe^{2+}, Fe^{3+}, SO_4^{2-}, and S^{2-}). All physicochemical parameters were determined by measuring porewater samples stored in HDPE bottles. The redox potential (ORP), temperature, and pH of porewater samples were measured using an Mi180 pH/ORP/Conductivity/TDS/NaCl/Temperature Laboratory Bench Meter by Milwaukee Instruments, Inc., United States. This instrument has ±0.01 pH, ±0.4°C, and 0.2 mV accuracy, and it can measure within the ranges of −2.00 to 16.00 pH, −20 to 120°C, and −699.9 to 699.9 mV. The ORP value (Eh) was normalized with respect to a standard hydrogen electrode by adding +200 mV (Jardim, 2014; Bourgeois et al., 2019). Ion content including Fe^{2+}, Fe^{3+}, SO_4^{2-}, and S^{2-} was measured using a DR 2700 Hach Portable Spectrophotometer. This instrument can detect concentrations within the ranges of 0.02–3.00 (±0.01) mg/L[1] of Fe^{2+}, 2–70 (±10) mg/L of SO_4^{2-}, and 5–800 (±16) μg/L of S^{2-}.

4. Results

4.1 Stand basal area, porewater dissolved CH_4, and CH_4 fluxes

Analysis of BA, as a measure of forest structure, confirmed the successional characteristics of sites, with two exceptions. First, during the fieldwork program, local inhabitants collected firewood at Site 7 and this site was subsequently excluded from the analysis. Second, the BA at Site 6 indicated that this was a much older regeneration site. Accordingly, Fig. 17.4 displays the porewater dissolved CH_4 concentrations for each month, at five depths across the (eight) sites based on the sequential BA. Unvegetated sites are not displayed in this figure, because it did not cover the same period of measurements as the other sites (i.e., only December 2016).

Porewater CH_4 concentrations of ~400 samples (5 depths × 8 sites × 10 months) ranged from 0.04 to 59.87 μmol/L, with an average of 7.3 μmol/L (Fig. 17.5). Compared with methane concentration in equilibrium with the atmosphere (i.e., 1.75×10^{-3} μmol/L), CH_4 saturation levels ranged from 20 to 30,000 times the mean atmospheric values. Therefore,

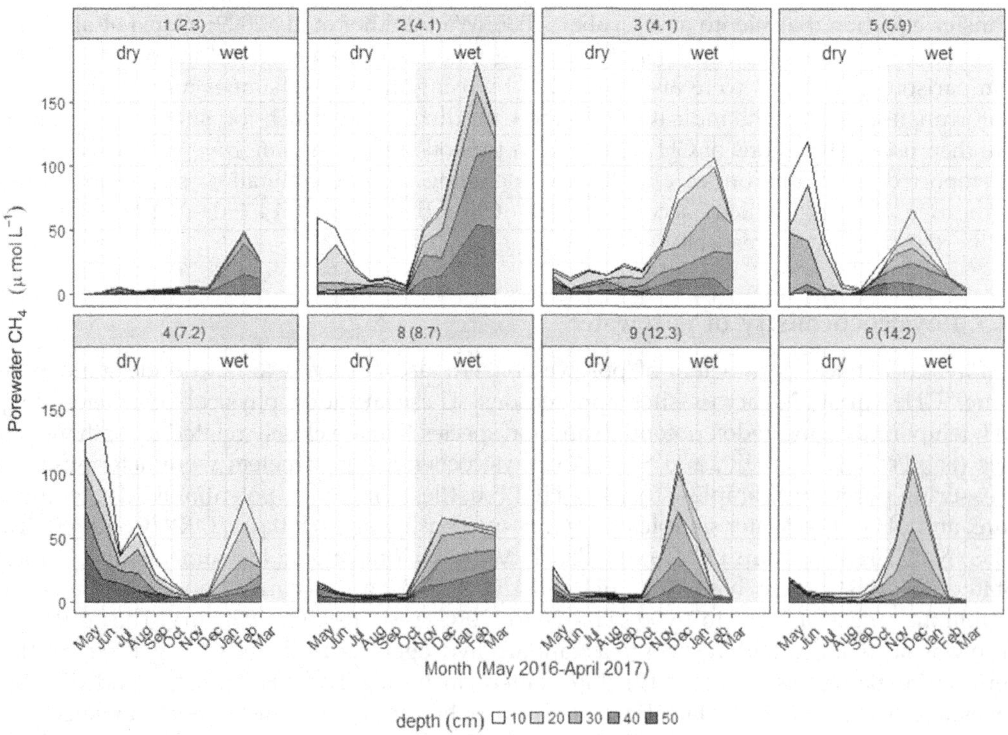

FIGURE 17.4 Seasonal variability of porewater CH_4 concentrations at the eight sites over a year (dry season = May–October and wet season = November–April [shaded]). Stacked area chart (gray gradations) indicates the porewater CH_4 concentration at different depths from 10 to 50 cm. The top of each panel shows the site number (see Fig. 17.3) with the BA (m^2/ha) in brackets. The panels are ordered by their BA, ranging from 2.3 to 14.2 m^2/ha. Unvegetated sites are not displayed, while the control sites are panels 8 (8.7) and 9 (12.3).

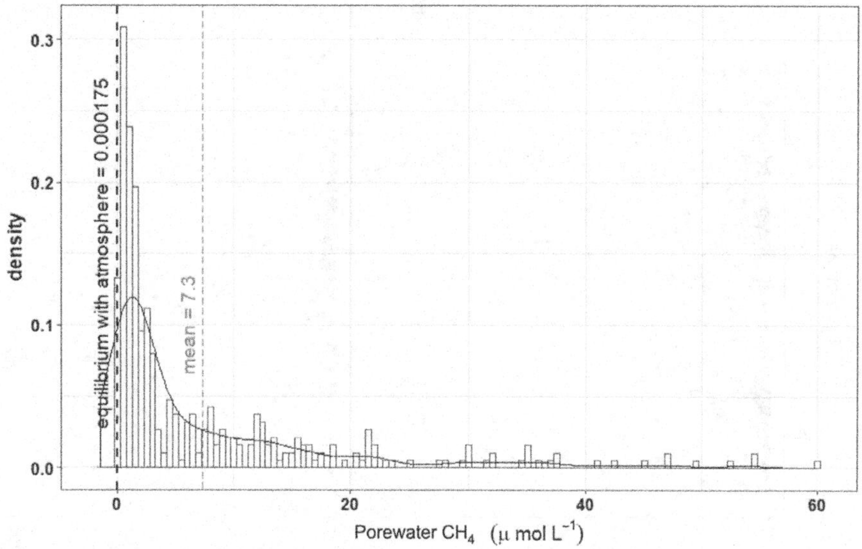

FIGURE 17.5 Data distribution of porewater CH_4 concentrations, collected from ~400 porewater samples.

all porewater samples were supersaturated with CH_4 with respect to the atmosphere. In addition, according to CH_4 concentrations of porewater at 10 cm depth, CH_4 fluxes from the soil surface in the study area ranged from 0.02 to 17 mg/m/h, with an average of 2.1 mg/m/h.

Porewater CH_4 concentrations varied considerably among sites and seasons. Generally, the sites with low (Sites 1, 2, and 3) and high (control sites 8/9 and Site 6) BA exhibited higher porewater CH_4 concentrations in the wet season (November to April) than those in the dry season (May to October). However, sites with intermediate BA (Sites 5 and 4, see Fig. 17.4) showed the highest porewater CH_4 concentrations in the dry season (May to October). Seasonal differences up to two orders of magnitude are evident in CH_4 concentrations. For example, at Site 3 (BA $= 4.1\ m^2/ha$), the range of CH_4 concentrations was 0.6–11.3 and 0.05–35.9 μmol/L in the dry and wet season, respectively. As noted above, Sites 4 and 5 showed an opposite trend, with the dissolved CH_4 concentrations being greater in the dry season. In addition to this, those two sites also had extremely high seasonal variability in porewater CH_4 concentrations.

4.2 Physicochemical parameters

Overall, porewater temperatures ranged from 27 to 35°C (Fig. 17.6A), while pH ranged from 6.2 to 7.7 (Fig. 17.6B). Within this range of soil pH and temperature, there was no indication that the pH or temperature significantly reduced CH_4 production. Moreover, at the same pH or temperature values, porewater CH_4 was observed over the full range. Therefore, there was no reason to expect that these two parameters were salient factors that influenced CH_4 production and emission at the study sites.

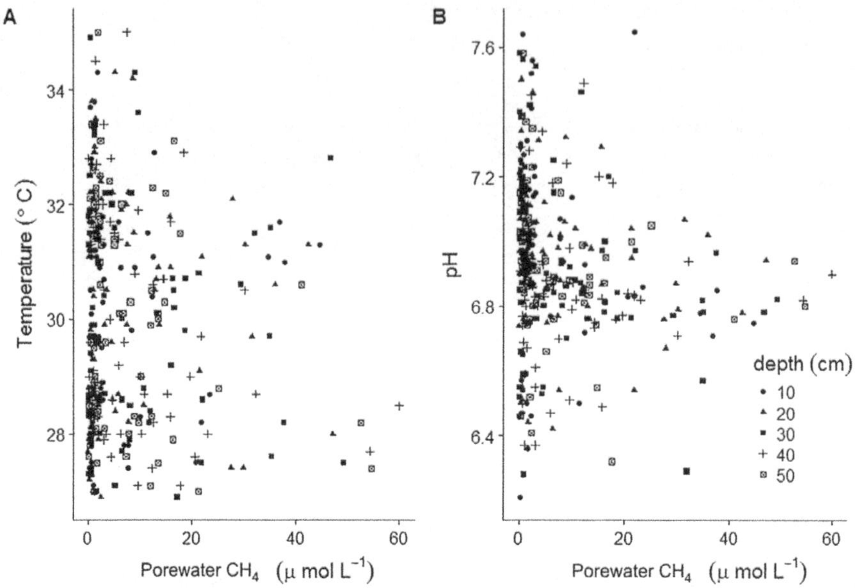

FIGURE 17.6 Relationship between porewater CH_4 concentrations and (A) temperature ($r = -0.09$, $P = .08$) and (B) pH ($r = -0.17$, $P = .0009$) at the eight sites. pH and temperature were measured in the field at the time of collection, and approximate the in situ conditions. Legend denotes the depth.

ORP values were generally at the optimum range for methanogenesis. Only 13 samples exhibited ORP >100 mV and they were associated with the nearly zero porewater CH_4 (Fig. 17.7). Most samples were found in the top soil layer which was easily oxidized. Some were taken from a depth where O_2 might be sourced from roots or animal burrows. However, the vast majority of data on porewater ORP were in the range of −200 and 0 mV (Fig. 17.7). In this range, high CH_4 concentrations were consistently detected, although some low concentrations were also observed. This finding indicated that ambient ORP was perfect for methanogenesis.

Among measured ions, the data for Fe^{2+} and Fe total are not presented because the concentrations were low (<1 ppm) and most were below the instrumental detection limits. Porewater SO_4^{2-} ranged between 3.1 and 59.3 mmol/L, while S^{2-} ranged between 0.3 and 1622 μmol/L (Fig. 17.8). In relation to dissolved CH_4 concentration, there was no clear pattern in SO_4^{-2} ($r = 0.25$, $P = 3.4 \times 10^{-4}$). While S^{2-} showed a positive relationship with CH_4, there was considerable noise in low concentrations ($r = 0.67$, $P < .001$). Therefore, the relationships between CH_4 and SO_4^{2-}/S^{2-} indicated that CH_4 production was minimally affected by the abundance of SO_4^{2-}.

4.3 Relationships between porewater CH_4 and stand basal area

The relationship between dissolved porewater CH_4 and BA showed a plotted "hump" shape, with a rise of CH_4 concentrations from sites with low mangrove cover—Site 1

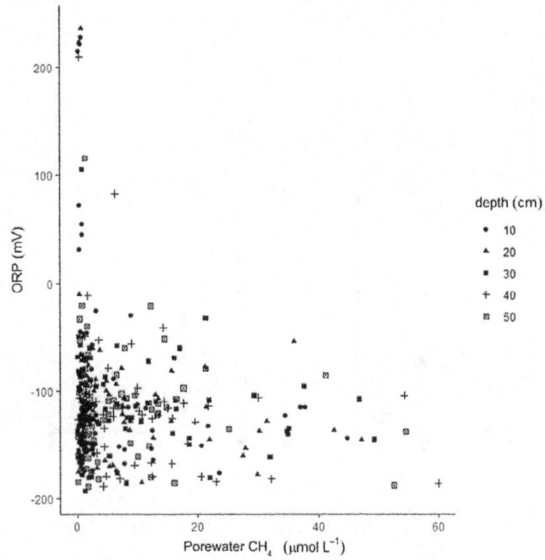

FIGURE 17.7 Relationship between porewater CH_4 concentrations and ORP ($r = -0.15$, $P = .004$). ORP was measured in the field at the time of data collection and approximates the in situ conditions. Legend denotes the depth.

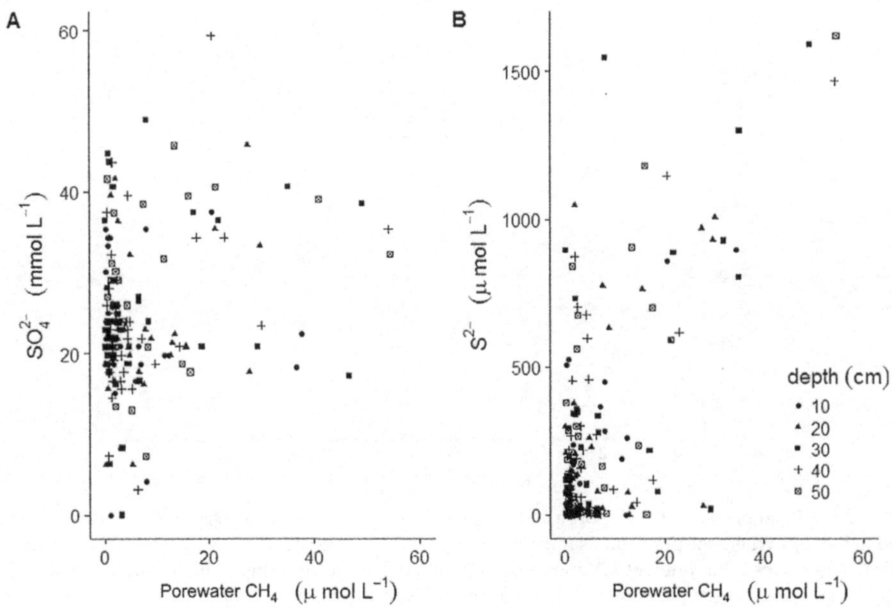

FIGURE 17.8 Relationship between porewater CH_4 concentrations and (A) SO_4^{2-} and (B) S^{2-} at the eight sites. SO_4^{2-} and S^{2-} were measured in the field at the time of data collection, and approximate the in situ conditions. Legend denotes the depth.

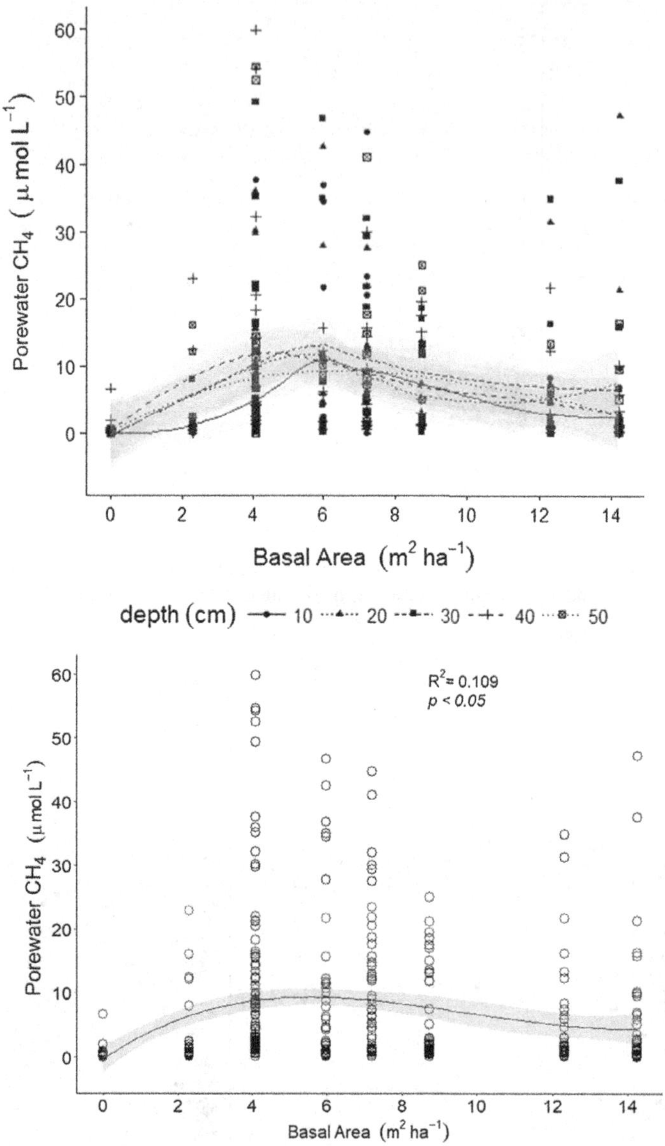

FIGURE 17.9 Relationship between porewater CH_4 concentrations and mangrove stand basal area (BA). BA was measured in the sites where the porewater samplers were located at the center of the site. (Top) CH_4 concentrations at five depths, across sites with different BA and general pattern of the relationship, with locally estimated scatterplot smoothing (span = 0.8). Legend denotes the depth. (Bottom) Relationship model for the whole data set, $y = 4.02x - 0.5x^2 + 0.02x^3 - 0.4$.

(2.1 m^2/ha)—to a peak at intermediate BA (Sites 2 and 3), and then started to decrease at higher BA, which included the control sites and Site 6 (Fig. 17.9). Although there was much scatter, we fitted a smooth curve to the data (Fig. 17.9), and the porewater CH_4 at all depths showed a peak at an intermediate BA of around 6 m^2/ha. By using a linear model, the porewater and BA relationship was fitted to a polynomial model, $y = 4.02x - 0.5x^2 + 0.02x^3 - 0.4$, $R^2 = 0.109$, $P < .05$.

5. Discussion

In the literature, there are only a handful of studies on porewater dissolved CH_4 from mangrove ecosystems (Giani et al., 1996; Lee et al., 2008; Strangmann et al., 2008; Dutta et al., 2015; Schile et al., 2017). In addition, when researchers did examine CH_4 in porewater, they measured it without investigating variation with depth, season, or stand age/status. Our study provides data on the porewater CH_4 measured on a monthly basis at different depths (up to 50 cm) over a full year. Accordingly, the results help to understand the effects of mangrove regrowth on porewater CH_4 concentrations, which may in turn be responsible for atmospheric CH_4 emissions.

5.1 Effects of seasons and mangrove regrowth on dissolved CH_4

In general, the range of porewater CH_4 concentration (0.04–59.9 μmol/L^{-1}) in this study is comparable to previous works in mangrove ecosystems (Table 17.1). These values are similar to those in oceanic mangrove islands in Belize (Lee et al., 2008) and pristine mangroves at Balandra, Mexico (Strangmann et al., 2008). The CH_4 concentrations are lower than those reported from the coastal Sabkha ecosystem in the United Arab Emirates (UAE) (Schile et al., 2017), but are much higher than in a study of Balandra mangroves (Giani et al., 1996). Interestingly, porewater CH_4 in the current study could be up to one order of magnitude higher in concentration than that from an earlier study involving pristine mangroves in Sundarbans, India (Biswas et al., 2007; Dutta et al., 2013, 2017). However, it is lower than the values in most nonmangrove wetlands, such as peatland and brackish marsh (Gross et al., 1993; Pangala et al., 2013).

Although many studies have found porewater physicochemical factors to be responsible for controlling CH_4 turnover (Inamori et al., 2007; Poffenbarger et al., 2011; Zhang and Ding, 2011; Dutta et al., 2013; Marín-Muñiz et al., 2015), it may be less important here. The temperature and ORP found here (Figs. 17.6 and 17.7) are within the optimum range for methanogen bacteria to convert organic matter to CH_4 (Wang et al., 1993; Le Mer and Roger, 2001). Furthermore, the ranges of pH (Fig. 17.6B) are favorable for methane production and reach a maximum at a range of 6.8–6.9, which agrees with Wang et al. (1993). In addition, abundance of SO_4^{2-} is unlikely to affect CH_4 production. A weak correlation between SO_4^{2-} and porewater CH_4 is an indication of no competition for the organic matter between methanogens and SRB (Fig. 17.8). Thus, in general, these physicochemical factors indicate favorable conditions for CH_4 production.

The only measured parameter that corresponds well with porewater CH_4 concentrations is stand BA (Fig. 17.9). We assumed that soil organic carbon (SOC) in the study sites is mainly from autochthonous sources, because the sites are in a coralline island and no riverine input from the mainland of Sulawesi. Therefore, SOC availability may be the most likely

TABLE 17.1 Ranges of porewater CH_4 (μmol/L) in this study compared with worldwide data from mangrove and other wetland ecosystems.

No.	References	Locations	Ecosystem	Porewater CH_4 μmol/L
1	Current study	EMR[a] site, Tanakeke, Indonesia	Mangrove	0.04–59.9
2	Schile et al. (2017)	Coastal sabkha ecosystem, UAE		125–875
3	Lee et al. (2008)	Oceanic mangrove islands, Belize		0–80
4	Strangmann et al. (2008)	Pristine mangrove soils, Balandra, Mexico		0–77.7
5	Dutta et al. (2017)	Mangrove, Sundarban, India		2.0–6.0
6	Dutta et al. (2015)	Mangrove, Sundarban, India		2.8–4.0
7	Giani et al. (1996)	Balandra and Ensenada La Paz mangrove, Mexico		0–2.1
8	Terazawa et al. (2007)	Forest floodplain, northern Japan	Forest floodplain	5.6–28.4
9	Pangala et al. (2013)	Sebangau River catchment, Sumatra, Indonesia	Peatland	113.0–1539.0
10	Pangala et al. (2015)	Forested peatland, Flitwick, Bedfordshire, United Kingdom		0–450.0
11	Gross et al. (1993)	Delaware brackish marsh, United States	Brackish marsh	0–700.0
12	Tong et al. (2015b)	Oligohaline marsh, Min River estuary, southeastern China		17.1–94.5
13	Biswas et al. (2007)	Sundarban mangrove water, India	Estuarine	0.8–1.5
14	Chuang et al. (2016)	Mangrove-dominated coastal lagoon, Mexico	Lagoon	20–3500
15	Ding et al. (2004)	Sanjiang Mire Wetland Experimental Station, northeast China	Freshwater marsh	8–232

[a]*EMR, Ecological Mangrove Rehabilitation, is a technique which prioritizes the restoration of hydrological regimes to facilitate multispecies recruitment of mangrove seedlings enabling establishment at natural tidal inundation levels (Lewis, 2005; Lewis and Brown 2014; Brown et al., 2014).*

explanation for a regulatory factor of porewater CH_4 concentration. This can also be inferred from findings by Cameron et al. (2019b) that SOC in the newly abandoned ponds is lower than the sites with mangrove regrowth post-EMR in Tanakeke, where we also did our study. Sasmito et al. (2019) further provide evidences that soil carbon content tends to increase following the successional or regeneration age ($R^2 = 0.441$, $P < .05$), while BA has a linear positive correlation with the regeneration age ($R^2 = 0.667$, $P < .001$). Also, Sasmito et al. (2019) found that belowground carbon increases until it reaches a plateau at some stage.

The abundances of carbon sources are a conducive environment supporting methanogen bacteria to produce CH_4. This was reflected in the fact that the full range of CH_4 concentrations was observed at the same value of some physicochemical factors—temperature, pH,

ORP, and SO_4^{2-} (Figs. 17.6–17.8). This meant that high porewater CH_4 concentration could be detected when substrates for methanogen bacteria were available. Conversely, it could be low if the sites had a lack of organic carbon, even if the physicochemical environments at the sites were optimal for methanogenesis. In addition, this idea was supported by Tong et al. (2015a) when studying CH_4 production variability at a small scale, who reported that the best predictor of CH_4 concentration was the availability of organic carbon. Furthermore, Marinho et al. (2012) reported a similar idea, where they found that methane concentration in porewater was a function of the amount of carbon, mainly associated with the occurrence of vegetation. In the sampling sites of this current study, mangrove vegetation was likely to be the major source of soil organic matter because the study area was located on a small island, separated from the mainland by about 10 km, which precludes allochthonous carbon from riverine sources. In addition, the base rock was coralline rubble and sand, both of which lack organic matter (Cameron et al., 2019b).

Therefore, SOC supply from vegetation is highly dependent on photosynthesis, and the allocation of photosynthates or SOC in our sites is more autochthonous than allochthonous. Forests actively fix CO_2 by photosynthesis, whereby some amount of the fixed carbon is emitted back to the atmosphere by autotrophic respiration, and the rest is allocated to biomass and root exudates. Litterfall and root biomass are thought to be the main sources of soil organic matter (Lallier-Vergès et al., 2008; Marchand et al., 2008; Marchand, 2017; Ha et al., 2018). Accumulation in the soil occurs because of slow decomposition rates, particularly in waterlogged soils. Meanwhile, root exudates exhibit a rapid turnover and are generally thought to account for a small proportion of soil organic matter (Allen et al., 2010).

Regarding methane production, a further explanation of the soil organic matter might be related to its type or source. Some studies show a straightforward relationship between total soil organic matter and methane flux (Xiang et al., 2015; Yu et al., 2013; Konnerup et al., 2014) and several studies have specifically examined the response of CH_4 production to different carbon sources (Lu et al., 2000, 2018; Lin et al., 2015; Girkin et al., 2018). Those studies found that labile organic matter, which mainly comes from root exudates, is preferable for methanogen bacteria. In addition, the exudates can induce decomposition of large and complex organic compounds (van Nugteren et al., 2009) through activating microbial metabolisms that can decompose insoluble and recalcitrant soil organic matter (Kuzyakov, 2010; Mason-Jones and Kuzyakov, 2017). This eventually leads to the production of labile organic matter from decomposition of recalcitrant organic matter.

The body of knowledge outlined above provides a basis for interpreting the pattern of porewater CH_4 reported here, where seasonal variability is largely dependent on the stage of mangrove regeneration (see Fig. 17.4). In Fig. 17.4, sites are arranged based on the BA of the mangrove stand, which is a representation of successional stages. The figure indicates that sites in the early (Sites 1 to 3) and late (control sites 6 and 8 and Site 9) successional/regeneration stages had similar patterns of porewater CH_4. Those sites all exhibited low concentrations during the dry season, yet high concentrations during the wet season. However, the intermediate stage sites (Sites 4 and 5) showed a reverse pattern (Fig. 17.4). Thus, we speculate that there is a shift in the relative abundance of labile to the total soil organic matter, which would then cause differences in the seasonal pattern of dissolved CH_4 among sites. Rocha et al. (2015) suggested that root exudates—a form of labile organic matter—are produced seasonally, and production is higher during the growing season. Similarly, as a major source of soil carbon in mangrove ecosystems, fine root production is also closely related to seasons (Poungparn et al., 2016; Xiong et al., 2017). While root exudates are readily digested by methanogen bacteria as soon as they are released to the soil (King et al., 2002; Dorodnikov et al., 2011; Tokida et al., 2011), fine roots need several months of active decomposition before the residues can form labile organic matter (Robertson and Alongi, 2016; Liu and Xiong, 2017).

The slow decomposition rate of fine roots might affect a shift in the abundance and sources of labile organic matter for methane production. At the intermediate stage sites (4 and 5), methanogens might use labile organic matter from fine root decompositions as the main substrates (see Fig. 17.4). However, slow decomposition could cause a lag in methane production during the peak production periods of fine roots. Fine root necromass, which is abundantly produced during the wet season, decomposes more rapidly in the dry than the wet season (Poungparn et al., 2016). Hence, in the dry season, methane production might increase, even though root exudate releases are in low concentrations. Meanwhile, in the control sites/mature forests (Sites 6, 8, and 9), much slower decomposition could cause methanogens to produce CH_4 mostly using root exudates. Therefore, at these sites, porewater CH_4 concentrations in the wet season are greater than in the dry season. The slow decomposition causes a gradual carbon accumulation in soils over periods of time; hence, the more developed mangrove stands also have higher soil carbon reserves (Marchand, 2017; Sasmito et al., 2019). Zhang et al. (2013) found a much slower decomposition rate in late successional vegetation, which is consistent with an increase in soil carbon. Therefore, in the control sites/mature forests, root exudates might be more abundant than labile organic matter from decomposition, especially in the wet season. In contrast, having similar seasonal patterns in porewater CH_4 with the control sites/mature forests, the early successional stages (Sites 1, 2, and 3) might rely on labile organic matter from both root exudates and decomposition. During production periods of fine roots and exudates, the decomposition rate in these sites would be faster than in the intermediate stages (Sites 4 and 5), as Zhang et al. (2013) indicated. Thus, together with root exudates, labile organic matter from decomposition will amplify methane production during the wet season.

As fine roots and root exudates are integral components of forest production, net primary productivity (NPP) could also be a proxy to explain porewater CH_4 variability across sites in the current study. From freshwater wetland studies, a review paper concluded that a positive correlation between NPP and CH_4 emission is evident (Aselmann and Crutzen, 1989). Interestingly, in our study, porewater CH_4 concentrations across sites seemed to follow the successional stages (Figs. 17.9 and 17.10). The general pattern of forest NPP across successional stages has long been known (Day et al., 1996; Chen et al., 2002). In brief, NPP increases rapidly during the young ages, peaks in the middle ages, and declines slowly as vegetation stand matures (Chen et al., 2002; Wang et al., 2011; He et al., 2012). Likewise, this very general pattern is followed in mangrove ecosystems (Okimoto et al., 2008). Specifically, Aksornkoae (1993) concluded that NPP of *R. apiculata* stands starts to decline at the age of 5–6 years after they exhibit an exponential increase in the initial growth phase. Meanwhile, Ong et al. (1985) noted that the decline starts at 10 years. These conclusions support the inference here that mangrove productivity at the study sites would lead to changes in dissolved CH_4 concentrations in the porewater.

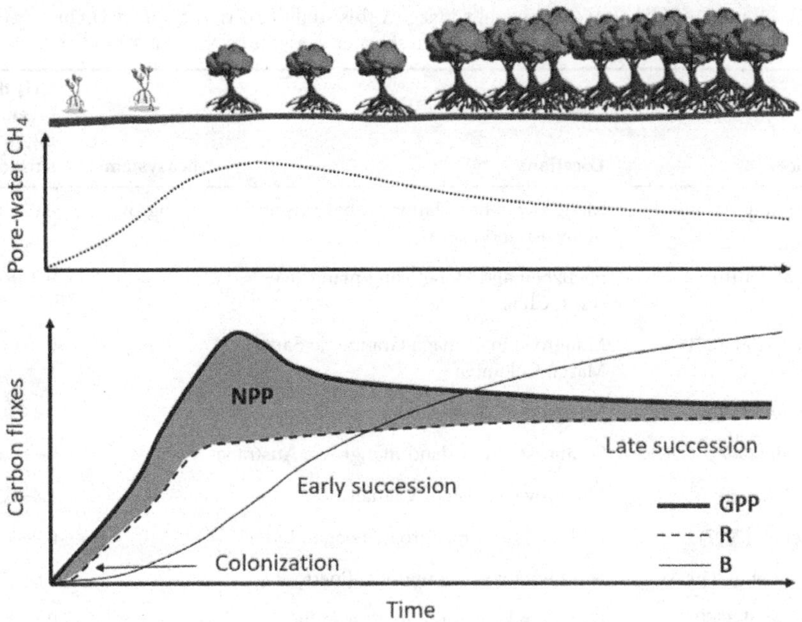

FIGURE 17.10 (A) Conceptual model of porewater dissolved CH_4 in mangrove rehabilitations. Vertical axis represents porewater dissolved CH_4 and horizontal axis represents time (current study). (B) Forest successional development. *B*, biomass; *GPP*, gross primary productivity; *NPP*, net primary productivity; *R*, respiration (Odum, 1969; Alongi, 2011). *From Alongi, D.M., 2011. Carbon payments for mangrove conservation: ecosystem constraints and uncertainties of sequestration potential. Environ. Sci. Policy 14, 462–470, Modified from Odum, E.P., 1969. The strategy of ecosystem development. Science 164, 262–270.*

Porewater CH_4 variability in this study could be conceptualized as a simple relationship between the porewater CH_4 and successional stages, which is in accordance with the successional pattern of NPP (see Fig. 17.10). In freshwater wetlands, based on the relationships between CH_4 emissions and NPP, 3% of fixed carbon (NPP) is released back into the atmosphere as CH_4 (Aselmann and Crutzen, 1989; Whiting and Chanton, 1993). Therefore, assuming that this fraction applies to mangrove wetlands, our findings might reconcile missing carbon in the carbon budget calculations in mangrove forests (Bouillon et al., 2008; Alongi, 2009; Twilley et al., 2017).

5.2 Potential mangrove contribution to CH_4 emissions

Based on values calculated using the gas transfer model, soil surface CH_4 fluxes in the current study could be higher than in most mangrove ecosystems that have been reported, as well as comparable with those recorded at freshwater wetlands. Mangroves in the current study area covered low and high emitters (0.02–17 mg m^2/h), although the majority of CH_4 fluxes were less than 10 $mg/m^2/h$, with an average and median value of 2.1 and 0.6 $mg/m^2/h$. This is in agreement with most CH_4 studies in mangrove ecosystems (Table 17.2, No. 6 to 24), particularly Cameron et al. (2019a) who measured CH_4 flux at the

TABLE 17.2 Ranges of CH_4 fluxes from soil surface in this study (gas transfer model) compared with worldwide data from mangrove and freshwater wetland ecosystems (static chamber method).

No.	References	Locations	Ecosystem	CH_4 flux ranges [average mg/m^2 (ground surface) h^{-1}
1	Current study	Mangrove rehabilitation, Tanakeke, South Sulawesi, Indonesia	Mangrove	0.02–17 [2.1]
2	Chen et al. (2010)	Shenzhen and Hong Kong mangroves, south China		0.2–82.7 []
3	Konnerup et al. (2014)	Mangrove in Ciénaga Grande de Santa Marta, Colombia		0–31.6 []
4	Wang et al. (2016)	Mangrove, Fujian, China		0.4–30.8 []
5	Allen et al. (2007)	Southeast Queensland mangrove, Australia		0–17.4 []
6	Arai et al. (2016)	Mangrove, Ca Mau, Vietnam		4.4–7.3 []
7	Biswas et al. (2007)	Lothian Island mangrove, Bengal, India		0.9–3.5 []
8	Sotomayor et al. (1994)	Southwest coast mangrove, Puerto Rico		0.01–3.4 []
9	Chauhan et al. (2008)	Bhitarkanika mangrove, East India		0.1–3.2 []
10	Purvaja et al. (2004)	Pichavaram mangrove, South India		0.03–2.7 []
11	Chauhan et al. (2015)	Bhitarkanika mangrove, India		0.1–2.3 []
12	Allen et al. (2011)	Subtropical mangrove, Southeast Queensland, Australia		0.05–1.6 []
13	Krithika et al. (2008)	Muthupet mangrove, South India		0.8–1.6 []
14	Lyimo et al. (2002)	Mtoni mangrove, Tanzania		0.02–1.3 []
15	Chauhan et al. (2008)	Pichavaram mangrove, South India		0.6–0.9 []
16	Dutta et al. (2013)	Sundarban mangrove biosphere, India		0.2–0.4 []
17	Kreuzwieser et al. (2003)	Queensland mangrove, Australia		0.02–0.4 []
18	Cameron et al. (2019a)	Mangrove rehabilitation, Tanakeke, South Sulawesi, Indonesia		0.09–0.45 []
19	Lu et al. (1999)	Dongzhai Harbor Mangrove, China		0–0.2 []
20	Chang and Yang (2003)	Kang-nan wetland mangrove, northern Taiwan		0.04–0.1 []
21	Cabezas et al. (2018)	Southwest Florida, US		0.01–0.1 []
22	Chen et al. (2010)	Guangdong, China		0–0.1 []
23	Lekphet et al. (2005)	Ranong, Thailand		0–0.01 []
24	Chen et al. (2014)	North Sulawesi, Indonesia		0–0.01 []

TABLE 17.2 Ranges of CH_4 fluxes from soil surface in this study (gas transfer model) compared with worldwide data from mangrove and freshwater wetland ecosystems (static chamber method).—cont'd

No.	References	Locations	Ecosystem	CH_4 flux ranges [average mg/m^2 (ground surface) h^{-1}
25	Alongi et al. (2005)	Fujian, China		0–0.01 []
26	Aselmann and Crutzen (1989), a review	Bogs	Freshwater wetlands	0.04–2.1 [0.6]
		Lakes		0.7–3.7 [1.8]
		Fens		1.2–9 [3.3]
		Swamps		2.4–4.6 [3.5]
		Floodplain		2.4–8.3 [4.2]
		Marsh		5.7–16.6 [10.5]
		Rice paddies		7.5–18.3 [12.3]

same study location. Interestingly, the average CH_4 flux is similar to that observed in freshwater wetlands, which have been relatively well studied and hence included in the global carbon budget (Table 17.2, No. 25). As a comparison to the average CH_4 flux in the current study (2.1 ± 3.6 mg/m^2/h), the averages from lakes, fens, swamps, and floodplains account for 1.8, 3.3, 3.5, and 4.2 mg/m^2/h, respectively.

In the literature presented in Table 17.2, four studies (No. 2–5) reported relatively high CH_4 fluxes from mangrove environments, accounting for greater than 10 mg/m^2/h. All the studies explicitly agreed with the explanation that high CH_4 fluxes were due to high discharges of sewage to the study sites, except Wang et al. (2016). However, the study site of Wang et al. was located at the estuary of Jiulong River, China, and had been experiencing sewage discharges from agriculture, industry, and settlements (Cao et al., 2005; Wu et al., 2017). The mechanism behind the discharge of sewage is nutrient enrichment in mangrove waters, which may lead to eutrophication, triggering algal blooms (Pacheco et al., 2014). In general, eutrophication causes increased methane emissions for two main reasons. First, increasing algal biomass provides an abundance of carbon sources for methanogen bacteria, particularly labile organic carbon (West et al., 2012; DelSontro et al., 2018). Second, with organic matter accumulation because of eutrophication, dissolved oxygen decreases to decompose the organic matter, thereby generating anoxic environments that are conducive to methane production (Howarth et al., 2011; Jenny et al., 2016).

Finally, comparing CH_4 fluxes from the soil surface in the current study with the worldwide mangrove data indicated that mangrove ecosystems are generally low methane emitters; however, anthropogenic activities may amplify CH_4 emissions in these ecosystems through eutrophication. Moreover, the magnitude of CH_4 fluxes in the eutrophic mangroves could be as high as an anthropogenic wetland (i.e., rice paddies) or even greater. For example, in a global data synthesis, CH_4 fluxes from the soil surface of rice paddies ranged from 7.5 to

18.3, with an average of 12.3 $mg/m^2/h$ (Aselmann and Crutzen, 1989), while mangroves could range from 0.2 to 82.7 $mg/m^2/h$ (Chen et al., 2010)—see Table 17.2.

5.3 Indication of CH_4 releases through mangrove trees

Methane dissolved in the porewater can be removed from sediments by either consumption or transport. In the current study area, porewater geochemistry indicated that CH_4 consumption might be apparent only at the top 10 cm depth. This could be the reason why CH_4 flux measurements and estimations gave too small a result, which agreed with Cameron et al. (2019a) who measured CH_4 fluxes in the same location. Unfortunately, we were not measuring porewater CH_4 in the top 10 cm depth. Thus, methane transports could be an alternative explanation for porewater CH_4 removals as there are emerging insights in tidal pumping transporting porewater CH_4 horizontally (e.g., Call et al., 2015) and CH_4 releases to the atmosphere through lenticels of mangrove stems (e.g., Jeffrey et al., 2019).

Although there are no direct measurements on tidal pumping of porewater CH_4 and CH_4 releases through mangrove stems or pneumatophores, tidal cycles, porewater geochemistry, and porewater CH_4 concentrations (Figs. 17.4–17.8) could help to understand the CH_4 egress from soil compartments. Anoxic environments were mostly detected in all porewater samples, obviously not favorable for methane oxidation at the depth <10 cm (Fig. 17.7). In addition, anaerobic methanotrophy coupled with sulfate reduction was unlikely because we did not observe a strong relationship between CH_4 and SO_4^{2-} (Fig. 17.8). In the mechanism of anaerobic methanotrophy, SRB uses H_2 as an electron donor to reduce sulfate, and this creates a more favorable environment for CH_4 oxidation (Hanson and Hanson, 1996; Chowdhury and Dick, 2013). Then, if CH_4 oxidation exists, SO_4^{2-} and CH_4 would disappear simultaneously, which was not detected in our experiments. Although another anaerobic methanotrophy pathway could be occurring, which is driven by nitrate, but we did not explore this parameter. Horizontal (tidal pumping) CH_4 transports would be unlikely in our study sites because (1) all the sites were located in a similar position at low-lying elevations, susceptible to flooding and inundated for around 13 h per day. If this is the case, porewater could not be drained during low tide—see Susilo et al. (2005) and Xia and Li (2012) for the relationship between porewater, tidal cycles, and tidal pumping; and (2) ORP was low and stable under anoxic conditions, even during low tide (Fig. 17.7), thereby indicating that porewater seepage might be very limited. Therefore, it is conceivable that CH_4 transport through lenticels on the surface of mangrove stems could primarily contribute to changes in the depth profile of porewater CH_4.

The importance of woody plants as a pathway for CH_4 transport in vegetated wetlands has very recently attracted much attention in CH_4 flux studies. The idea of CH_4 transport through the woody plant was first proposed by Zeikus and Ward (1974), who found that new borer holes of a cotton wood tree stem, *Populus deltoides*, released high-pressure flammable gas. However, there was a lack of interest in this idea until the 1990s, when Pulliam (1992) observed CH_4 fluxes from knee roots of bald cypress (*Taxodium distichum*), and Rusch and Rennenberg (1998) experimented with black alder seedlings (*Alnus glutinosa*). In fact,

TABLE 17.3 CH_4 fluxes from tree stem and soil surface in the literature, presented in the same unit (milligram CH_4 per meter square ground surface per hour).

No.	References	Locations	CH$_4$ flux mg/m^2 (ground surface) h^{-1}	
			Stem	**Soil**
Flooded				
1	Pangala et al. (2017)	Negro River, floodplain forest	3.9 ± 1.3	0.6 ± 0.5
		Amazon River, floodplain forest	6.9 ± 1.8	0.4 ± 1.5
		Solimões River, floodplain forest	15.6 ± 3.9	0.7 ± 0.9
		Madeira River, floodplain forest	6.9 ± 2.7	2.1 ± 2.4
		Tapajos River, floodplain forest	23.9 ± 6.3	3.8 ± 4.7
Nonflooded				
2	Pangala et al. (2013)	Sebangau River catchment, tropical peatland	0.2 ± 0.02	0.03 ± 0.01
3	Terazawa et al. (2007)	Northern Japan, floodplain forest	0.003	−0.004
4	[a] Pitz et al. (2018)	Upland forest	0.01	−0.06 ± 0.01
		Transitional	0.03	0.01 ± 0.03
		Wetland forest	0.08	0.2 ± 0.1
5	[a] Terazawa et al. (2015)	Floodplain forest	0.01–0.18	
6	[a] Gauci et al. (2010)	Freshwater wetlands	0.004–0.01	0.3–0.8
7	[a] Maier et al. (2018)	Upland forest	0.01	−0.07

[a]In the original paper, fluxes were reported per unit stem surface area. This was converted to per unit ground surface area using an approximate relation (see Appendix 2, Table A2.1).

while the fieldwork for our study was being conducted, there were several additional reports of methane fluxes through the woody stems of trees (Pangala et al., 2015, 2017; Terazawa et al., 2015; Maier et al., 2018; Pitz et al., 2018).

Table 17.3 presents a summary and comparison of CH_4 fluxes through the sediment surface and woody stems. It is no surprise that CH_4 emissions are higher in flooded than nonflooded forests. This is because continuous soil flooding is known to cause anaerobic conditions and hence an increase in CH_4 production, as is well-documented in studies of CH_4 and water management in rice paddies (Wassmann et al., 2000; Xu et al., 2015; Islam et al., 2018). Interestingly, in flooded forests, methane fluxes through woody stems appear to be much larger (by a factor of 5–20) than soil fluxes. In contrast, unflooded forests exhibit lower CH_4 fluxes through tree stems than the soil surface (Table 17.3, No. 6). In addition, some observations have indicated that there is CH_4 absorption by soils, but CH_4 emission by tree stems (Table 17.3, No. 4, 5, and 7). Nevertheless, the previous studies identified in this table highlight the possibility of a CH_4 flux pathway through tree stems, which can be higher than soil surface CH_4 flux in flooded soils.

6. Conclusion

In this study, we tracked for the first time the annual cycle of porewater CH_4 concentrations at mangrove regeneration sites. Such data are essential to refine GHG emissions from restored mangroves given the increased interest in blue carbon ecosystem restoration/rehabilitation as a mechanism of GHG abatement. The porewater at all eight sites was supersaturated with CH_4, with a wide range of CH_4 fluxes from the soil surface, ranging between 0.02 and 17 mg m^2/h. The variability of the porewater CH_4 showed a seasonal pattern that correlated with the successional stages (Figs. 17.9 and 17.10). The implication is that the concentration of CH_4 in porewater is likely to be largely controlled by the availability of carbon substrates, which varies with NPP and hence the stage of mangrove regrowth. Future research should consider (1) CH_4 fluxes through woody stems in combination with porewater CH_4 to confirm CH_4 pathways in a soil–plant–atmosphere system and mangrove contributions to CH_4 emissions, and (2) root exudates and NPP in relation to porewater CH_4. With that knowledge, we can better characterize the complete carbon budget on mangrove ecosystems.

Appendix 1: soil surface CH_4 fluxes

Due to limited data of CH_4 depth profile at depth 0–10 cm, we could not estimate soil surface CH_4 fluxes using Fick's first law, which is widely implemented in the calculation of molecular diffusion of chemical species either in fluid or solid media. Consequently, we used a gas transfer equation (Sarmiento and Gruber, 2006; Wanninkhof et al., 2009; Wania et al., 2010). Fick's law assumes that soil surface CH_4 flux occurs due to molecular diffusions within a few centimeters at the top layer of soils (Van Der Nat and Middelburg, 2000) or at soil–air or soil–water interfaces (Man et al., 2004; Dutta et al., 2015) (see Fig. 17.1). In our case, we assumed that dissolved CH_4 in the top layer of soil (at 10 cm) was well mixed and the soils were saturated with water, where the water level remains at the soil surface (see the section on study location and sampling sites). The molecular diffusion through the soil compartment within a 10 cm depth was assumed to be zero because of no gradient concentration (Fig. A1.1, dashed line). Thus, the diffusion primarily occurs as a result of gradient CH_4 concentration in the boundary layer (Fig. A1.1).
If one used Fick's first law for CH_4 diffusion across the boundary layer in Fig. A1.1, difficulties determining the boundary conditions would be encountered because gas diffusivities change by at least 10^4 times (Wania et al., 2010). Thus, we used a model flux in Eq. (A.1), as suggested by Wania et al. (2010), which is basically a modification of Fick's equation. Further explanations of this can be found in Sarmiento and Gruber (2006) and Wanninkhof et al. (2009).

$$Flux\ CH4 = k([CH4]PW - \beta[CH4]a) \tag{A.1}$$

where k is a transfer coefficient and $[CH_4]_a$ is the concentration in the near-surface air. $[CH_4]_{pw}$ in mol/m^3 is the concentration of porewater dissolved CH_4 in the top layer of soils or at 10 cm depth. The use of Wania et al.'s (2010) model was due to our limited data of CH_4 depth profile in the boundary layers.

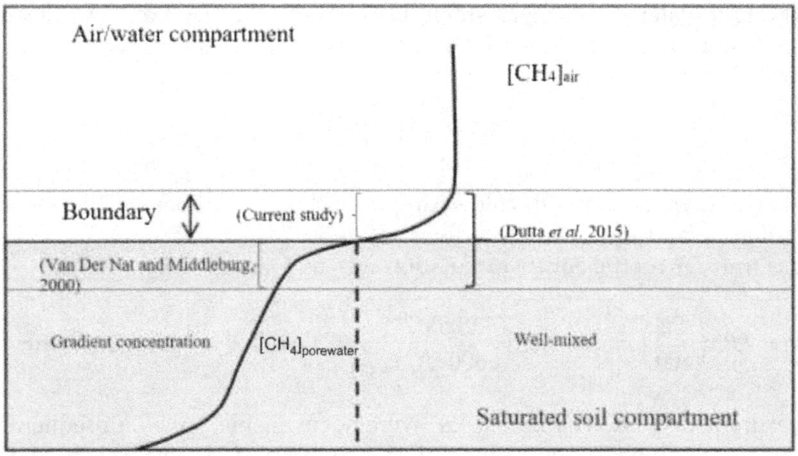

FIGURE A1.1 Illustration of the conceptual model of gas transfer through an interface, and assumptions underlying the mathematical model used in the current study compared with previous studies.

Eq. (A.1) contains a gas transfer coefficient (k, in cm/h), and the generic equation to estimate k is as follows:

$$k = k_{600}\left(\frac{Sc}{600}\right)^{n} \tag{A.2}$$

where S_c is a Schmidt number that has been presented in Eq. (A.3), and n is a Schmidt exponent.

$$S_c(CH_4) = 2101.2 - 131.54\,T + 4.4931\,T^2 - 0.08676\,T^3 + 0.000707\,T^4 \tag{A.3}$$

In the current study, the exponent of −0.5 was suitable for mangrove areas because of high turbulence, instead of −2/3, which is commonly used for open ocean waters at wind speeds below 3 m/s (Borges et al., 2004). The gas transfer coefficient at a Schmidt number of 600 (k_{600}, cm/h) was used to normalize the gas transfer coefficient k. Many equations of k_{600} have been proposed, yet mostly for gas flux calculation in lakes or oceans (Wanninkhof et al., 2009). Thus, we evaluated two equations that could be appropriate for the conditions of our study site. First, k_{600} is expressed as a function of wind speed at 10 m height (U_{10}, m/s) (Wania et al., 2010):

$$k_{600} = 2.07 + 0.215\,U_{10}^{1.7} \tag{A.4}$$

Second, k_{600} is a function of speed of the background water current (v, m/s):

$$k_{600} = 2.03 + 0.43\,v \tag{A.5}$$

Eq. (A.5) is an empirical model for CH_4 flux calculations from surface water in mangrove creeks (Rosentreter et al., 2017). Finally, we assumed that U_{10} and v were 0 because the porewater–atmosphere interface in mangrove soils is less influenced by wind, and the speed of the background water current was small. Hence, both Eqs. (A.4 and A.5) were identical. We then calculated the gas transfer coefficient (k) using the following equation.

$$k = 2.03\left(\frac{Sc}{600}\right)^{-0.5} \tag{A.6}$$

To apply Eq. (A.1), we began with calculating the Schmidt number (Sc) by using Eq. (A.3). The value of temperature to calculate Sc was set to 30°C.
Thus, the gas transfer coefficient (k) calculation was as follows:

$$k = 2.03\left(\frac{Sc}{600}\right)^{-0.5} = 2.03\left(\frac{428.6}{600}\right)^{-0.5} = 2.4\ \text{cm h}^{-1} = 2.4 \times 10^{-2}\ \text{mh}^{-1}$$

Because measurements of $[CH_4]_a$ in Eq. (A.1) were not available, it was subsequently assumed to be equal to the global atmospheric CH_4 concentration and set to 1.8 ppmv or 7.2×10^{-5} mol/m at 30°C. Finally, by using these data, we calculated soil surface CH_4 flux as follows:

$$Flux_{CH_4} = k([CH_4]_{PW} - \beta[CH_4]_a)$$

$$= (2.4 \times 10^{-2}\ \text{m h}^{-1})[(3.7 \times 10^{-2}\ \text{mol m}^{-3}) - (0.02424)(7.2 \times 10^{-5}\ \text{mol m}^{-3})]$$

$$= (9 \times 10^{-4}\ \text{mol m}^{-2}\text{h}^{-1})(16\ \text{g mol}^{-1})(10^3\text{mg g}^{-1}) = 14\ \text{mg m}^{-2}\text{h}^{-1}$$

Appendix 2: CH_4 fluxes from tree stem and soil surface in the literature

TABLE A2.1 CH_4 fluxes from tree stem and soil surface in the literature.

			CH_4 fluxes		
			Stem	**Stem[a]**	**Soil**
No.	**References**	**Locations**	**mg/m² (stem surface) h⁻¹**	**mg/m² (ground surface) h⁻¹**	
1	Pitz et al. (2018)	Upland forest	0.07 ± 0.01	0.01	−0.06 ± 0.01
		Transitional	0.2 ± 0.06	0.03	0.01 ± 0.03
		Wetland forest	0.6 ± 0.2	0.08	0.2 ± 0.1
2	Terazawa et al. (2015)	Floodplain forest	0.1−1.3	0.01−0.18	
3	Gauci et al. (2010)	Freshwater wetlands	0.03−0.1	0.004−0.01	0.3−0.8
4	Maier et al. (2018)	Upland forest	0.1	0.01	−0.07

Note the different area basis of expression.
[a]*Rough estimates according to the ratio between an average of stem surface area of mature trees in 13 plots (565 m²) and the ground surface area of the plot (50 × 80 m²) in a CH_4 study of the Amazon floodplain (Pangala et al., 2017).*

Acknowledgments

This manuscript is part of Yaya Ulumuddin's PhD thesis which was sponsored by Australian Awards Scholarship 2015–19. Our sincere gratitude goes to Dr. Ghuangcheng Chen who provided the aluminum gas-tight bags. Also, special thanks go to Blue Carbon Foundation staff for field work assistance, including Yusran, Akhzan, and their team.

References

Aksornkoae, S., 1993. Ecology and Management of Mangroves. IUCN, Bangkok, Thailand.

Allen, D.E., Dalal, R.C., Rennenberg, H., Meyer, R.L., Reeves, S., Schmidt, S., 2007. Spatial and temporal variation of nitrous oxide and methane flux between subtropical mangrove sediments and the atmosphere. Soil Biol. Biochem. 39 (2), 622–631.

Allen, D.E., Dalal, R.C., Rennenberg, H., Schmidt, S., 2011. Seasonal variation in nitrous oxide and methane emissions from subtropical estuary and coastal mangrove sediments, Australia. Plant Biology 13 (1), 126–133. https://doi.org/10.1111/j.1438-8677.2010.00331.x.

Allen, D.E., Pringle, M.J., Page, K.L., Dalal, R.C., 2010. A review of sampling designs for the measurement of soil organic carbon in Australian grazing lands. Rangel. J. 32 (2), 227–246.

Alongi, D.M., 2009. The Energetics of Mangrove Forests. Springer Netherlands, Dodrecht.

Alongi, D.M., 2011. Carbon payments for mangrove conservation: ecosystem constraints and uncertainties of sequestration potential. Environ. Sci. Pol. 14 (4), 462–470.

Alongi, D.M., Sasekumar, A., Chong, V.C., Pfitzner, J., Trott, L.A., Tirendi, F., Dixon, P., Brunskill, G.J., 2004. Sediment accumulation and organic material flux in a managed mangrove ecosystem: estimates of land-ocean-atmosphere exchange in peninsular Malaysia. Mar. Geol. 208 (2–4), 383–402.

Alongi, D.M., Pfitzner, J., Trott, L.A., Tirendi, F., Dixon, P., Klumpp, D.W., 2005. Rapid sediment accumulation and microbial mineralization in forests of the mangrove *Kandelia candel* in the Jiulongjiang Estuary, China. Estuar. Coast Shelf Sci. 63 (4), 605–618.

Arai, H., Yoshioka, R., Hanazawa, S., Minh, V.Q., Tuan, V.Q., Tinh, T.K., Phu, T.Q., Jha, C.S., Rodda, S.R., Dadhwal, V.K., Mano, M., Inubushi, K., 2016. Function of the methanogenic community in mangrove soils as influenced by the chemical properties of the hydrosphere. Soil Sci. Plant Nutr. 62 (2), 150–163.

Ardon, M., Helton, A.M., Bernhardt, E.S., 2018. Salinity effects on greenhouse gas emissions from wetland soils are contingent upon hydrologic setting: a microcosm experiment. Biogeochemistry 140, 217–232.

Aselmann, I., Crutzen, P.J., 1989. Global distribution of natural freshwater wetlands and rice paddies, their net primary productivity, seasonality and possible methane emissions. J. Atmos. Chem. 8 (4), 307–358.

Bartlett, K.B., Bartlett, D.S., Harriss, R.C., Sebacher, D.I., 1987. Methane emissions along a salt marsh salinity gradient. Biogeochemistry 4 (3), 183–202.

Bazhin, N.M., 2004. Influence of plants on the methane emission from sediments. Chemosphere 54 (2), 209–215. https://doi.org/10.1016/S0045-6535(03)00571-X.

Biswas, H., Mukhopadhyay, S.K., Sen, S., Jana, T.K., 2007. Spatial and temporal patterns of methane dynamics in the tropical mangrove dominated estuary, NE coast of Bay of Bengal, India. J. Mar. Syst. 68 (1–2), 55–64.

Borges, A.V., Delille, B., Schiettecatte, L., Gazeau, F., Abril, G., Frankignoulle, M., Gazeau, F., 2004. Gas transfer velocities of CO_2 in three European estuaries. Limnol. Oceanogr. 49 (5), 1630–1641.

Bourgeois, C., Alfaro, A.C., Dencer-Brown, A., Duprey, J.L., Desnues, A., Marchand, C., 2019. Stocks and soil-plant transfer of macro-nutrients and trace metals in temperate New Zealand estuarine mangroves. Plant Soil 436, 565–586.

Bouillon, S., Borges, A.V., Castañeda-Moya, E., Diele, K., Dittmar, T., Duke, N.C., Kristensen, E., Lee, S.Y., Marchand, C., Middelburg, J.J., Rivera-Monroy, V.H., Smith, T.J., Twilley, R.R., 2008. Mangrove production and carbon sinks: a revision of global budget estimates. Global Biogeochem. Cycles 22 (2), 1–12.

Brown, B., Fadillah, R., Nurdin, Y., Soulsby, I., Ahmad, R., 2014. Community based ecological mangrove rehabilitation (CBEMR) in Indonesia. Sapiens 7 (2), 1–13.

Cabezas, A., Mitsch, W.J., MacDonnell, C., Zhang, L., Bydałek, F., Lasso, A., 2018. Methane emissions from mangrove soils in hydrologically disturbed and reference mangrove tidal creeks in southwest Florida. Ecol. Eng. 114, 57–65.

Call, M., Maher, D.T., Santos, I.R., Ruiz-Halpern, S., Mangion, P., Sanders, C.J., Erler, D.V., Oakes, J.M., Rosentreter, J., Murray, R., Eyre, B.D., 2015. Spatial and temporal variability of carbon dioxide and methane fluxes over semi-diurnal and spring-neap-spring timescales in a mangrove creek. Geochem. Cosmochim. Acta 150, 211–225.

Cameron, C., Hutley, L.B., Friess, D.A., Munksgaard, N.C., 2019a. Hydroperiod, soil moisture and bioturbation are critical drivers of greenhouse gas fluxes and vary as a function of landuse change in mangroves of Sulawesi, Indonesia. Sci. Total Environ. 654, 365–377.

Cameron, C., Hutley, L.B., Friess, D.A., Brown, B., 2019b. Community structure dynamics and carbon stock change of rehabilitated mangrove forests in Sulawesi, Indonesia. Ecol. Appl. 29 (1), 1–20.

Cao, W., Hong, H., Yue, S., 2005. Modelling agricultural nitrogen contributions to the Jiulong River estuary and coastal water. Global Planet. Change 47, 111–121.

Chang, T.C., Yang, S.S., 2003. Methane emission from wetlands in Taiwan. Atmos. Environ. 37 (32), 4551–4558.

Chauhan, R., Datta, A., Ramanathan, A.L., Adhya, T.K., 2015. Factors influencing spatio-temporal variation of methane and nitrous oxide emission from a tropical mangrove of eastern coast of India. Atmos. Environ. 107, 95–106.

Chauhan, R., Ramanathan, A.L., Adhya, T.K., 2008. Assessment of methane and nitrous oxide flux from mangroves along Eastern coast of India. Geofluids 8 (4), 321–332.

Chen, G.C., Chen, B., Yu, D., Tam, N.F.Y., Ye, Y., Chen, S., 2016. Soil greenhouse gas emissions reduce the contribution of mangrove plants to the atmospheric cooling effect. Environ. Res. Lett. 11, 1–10.

Chen, G.C., Ulumuddin, Y.I., Pramudji, S., Chen, S.Y., Chen, B., Ye, Y., Ou, D.Y., Ma, Z.Y., Huang, H., Wang, J.K., 2014. Rich soil carbon and nitrogen but low atmospheric greenhouse gas fluxes from North Sulawesi mangrove swamps in Indonesia. Sci. Total Environ. 487 (1), 91–96.

Chen, G.C., Tam, N.F.Y., Ye, Y., 2010. Summer fluxes of atmospheric greenhouse gases N_2O, CH_4 and CO_2 from mangrove soil in South China. Sci. Total Environ. 408 (13), 2761–2767.

Chen, W., Chen, J.M., Price, D.T., Cihlar, J., 2002. Effects of stand age on net primary productivity of boreal black spruce forests in Ontario, Canada. Can. J. For. Res. 32 (5), 833–842.

Cheng, C.H., Hung, C.Y., Chen, C.P., Pei, C.W., 2013. Biomass carbon accumulation in aging Japanese cedar plantations in Xitou, central Taiwan. Bot. Stud. 54 (1), 1–9.

Chowdhury, T.R., Dick, R.P., 2013. Ecology of aerobic methanotrophs in controlling methane fluxes from wetlands. Appl. Soil Ecol. 65, 8–22.

Chuang, P.C., Young, M.B., Dale, A.W., Miller, L.G., Herrera-Silveira, J.A., Paytan, A., 2016. Methane and sulfate dynamics in sediments from mangrove-dominated tropical coastal lagoons, Yucatan, Mexico. Biogeosciences 13 (10), 2981–3001.

Datta, A., Yeluripati, J.B., Nayak, D.R., Mahata, K.R., Santra, S.C., Adhya, T.K., 2013. Seasonal variation of methane flux from coastal saline rice field with the application of different organic manures. Atmos. Environ. 66, 114–122.

Day Jr., J.W., Coronado-Molina, C., Vera-Herrera, F.R., Twilley, R.R., Rivera-Monroy, V.H., Alvarez-Guillen, H., Day, R., Conner, W., 1996. A 7 year record of above-ground net primary production in a southeastern Mexican mangrove forest. Aquat. Bot. 55, 39–60.

DelSontro, T., Beaulieu, J.J., Downing, J.A., 2018. Greenhouse gas emissions from lakes and impoundments: upscaling in the face of global change. Limnol. Oceanogr. Lett. 3, 64–75.

Ding, W., Cai, Z., Tsuruta, H., 2004. Methane concentration and emission as affected by methane transport capacity of plants in freshwater marsh. Water Air Soil Pollut. 158 (1), 99–111.

Dorodnikov, M., Knorr, K.H., Kuzyakov, Y., Wilmking, M., 2011. Plant-mediated CH_4 transport and contribution of photosynthates to methanogenesis at a boreal mire: a ^{14}C pulse-labeling study. Biogeosciences 8 (8), 2365–2375.

Dutta, M.K., Bianchi, T.S., Mukhopadhyay, S.K., 2017. Mangrove methane biogeochemistry in the Indian Sundarbans: a proposed budget. Front. Mar. Sci. 4, 1–15.

Dutta, M.K., Chowdhury, C., Jana, T.K., Mukhopadhyay, S.K., 2013. Dynamics and exchange fluxes of methane in the estuarine mangrove environment of the Sundarbans, NE coast of India. Atmos. Environ. 77, 631–639.

Dutta, M.K., Mukherjee, R., Jana, T.K., Mukhopadhyay, S.K., 2015. Biogeochemical dynamics of exogenous methane in an estuary associated to a mangrove biosphere; the Sundarbans, NE coast of India. Mar. Chem. 170, 1–10.

Forster, P., Ramaswamy, V., Artaxo, P., Berntsen, T., Betts, R., Fahey, D.W., Haywood, J., Lean, J., Lowe, D.C., Myhre, G., Nganga, J., Prinn, R., Raga, G., Schulz, M., Dorland, R. Van, 2007. Changes in atmospheric constituents and in radiative forcing. In: Solomon, S., Qin, D., Manning, M., Chen, Z., Marquis, M., Averyt, K.B., Tignor, M., Miller, H.L. (Eds.), Climate Change 2007: The Physical Science Basis. Contribution of Working Group I to the Fourth Assessment Report of the Intergovernmental Panel on Climate Change, vol. 30, pp. 129–234, 22.

Gauci, V., Gowing, D.J.G., Hornibrook, E.R.C., Davis, J.M., Dise, N.B., 2010. Woody stem methane emission in mature wetland alder trees. Atmos. Environ. 44 (17), 2157–2160.

Gedney, N., Cox, P.M., Huntingford, C., 2004. Climate feedback from wetland methane emissions. Geophys. Res. Lett. 31 (20), 1–4.

Giani, L., Bashan, Y., Holguin, G., Strangmann, A., 1996. Characteristics and methanogenesis of the Balandra lagoon mangrove soils, Baja California Sur, Mexico. Geoderma 72 (1–2), 149–160.

Girkin, N.T., Turner, B.L., Ostle, N., Craigon, J., Sjögersten, S., 2018. Root exudate analogues accelerate CO_2 and CH_4 production in tropical peat. Soil Biol. Biochem. 117, 48–55.

Gosper, C.R., Prober, S.M., Yates, C.J., 2013. Multi-century changes in vegetation structure and fuel availability in fire-sensitive eucalypt woodlands. For. Ecol. Manag. 310, 102–109.

Gross, M.F., Hardisky, M.A., Wolf, P.L., Klemas, V., 1993. Relationships among Typha Biomass, Pore Water Methane, and Reflectance in a Delaware (U.S.A.) Brackish Marsh. J. Coast Res. 9 (2), 339–355.

Ha, T.H., Marchand, C., Aimé, J., Dang, H.N., Phan, N.H., Nguyen, X.T., Nguyen, T.K.C., 2018. Belowground carbon sequestration in a mature planted mangrove (Northern Viet Nam). For. Ecol. Manag. 407, 191–199.

Hanson, R.S., Hanson, T.E., 1996. Methanotrophic bacteria. Microbiol. Mol. Biol. Rev. 60 (2), 439–471.

He, L., Chen, J.M., Pan, Y., Birdsey, R., Kattge, J., 2012. Relationships between net primary productivity and forest stand age in U.S. forests. Global Biogeochem. Cycles 26 (3), 1–19.

Helton, A.M., Ardon, M., Bernhardt, E.S., 2019. Hydrologic context Alters greenhouse gas feedbacks of coastal wetland salinization. Ecosystems 22, 1108–1125.

Howarth, R., Chan, F., Conley, D.J., Garnier, J., Doney, S.C., Marino, R., Billen, G., 2011. Coupled biogeochemical cycles: Eutrophication and hypoxia in temperate estuaries and coastal marine ecosystems. Front. Ecol. Environ. 9 (1), 18–26.

Inamori, R., Gui, P., Dass, P., Matsumura, M., Xu, K.Q., Kondo, T., Ebie, Y., Inamori, Y., 2007. Investigating CH_4 and N_2O emissions from eco-engineering wastewater treatment processes using constructed wetland microcosms. Process Biochem. 42 (3), 363–373.

IPCC, 2013. In: Stocker, T.F., Qin, D., Plattner, G.K., Tignor, M.M.B., Allen, S.K., Boschung, J., Nauels, A., Xia, Y., Bex, V., Midgley, P.M. (Eds.), Climate Change 2013: The Physical Science Basis. Cambridge University Press, New York.

Islam, S.F.U., van Groenigen, J.W., Jensen, L.S., Sander, B.O., de Neergaard, A., 2018. The effective mitigation of greenhouse gas emissions from rice paddies without compromising yield by early-season drainage. Sci. Total Environ. 612, 1329–1339.

Jardim, W.F., 2014. Medição e interpretação de valores do potential redox (*Eh*) em matrizes ambientais. Quim. Nova 37 (7), 1233–1235.

Jeffrey, L.C., Reithmaier, G., Sippo, J.Z., Johnston, S.G., Tait, D.R., Maher, D.T., Harada, Y., 2019. Are methane emissions from mangrove stems a cryptic carbon loss pathway? Insights from a catastrophic forest mortality. New Phytol. 1–9.

Jenny, J.P., Normandeau, A., Francus, P., Taranu, Z.E., Gregory-Eaves, I., Lapointe, F., Jautzy, J., Ojala, A.E.K., Dorioz, J.M., Schimmelmann, A., Zolitschka, B., 2016. Urban point sources of nutrients were the leading cause for the historical spread of hypoxia across European lakes. Proc. Natl. Acad. Sci. U.S.A. 113 (45), 12655–12660.

Jing, H., Cheung, S., Zhou, Z., Wu, C., Nagarajan, S., Liu, H., 2016. Spatial variations of the methanogenic communities in the sediments of tropical mangroves. PLoS One 11 (9), 1–19.

Kauffman, J., Donato, D., 2012. Protocols for the Measurement, Monitoring and Reporting of Structure, Biomass and Carbon Stocks in Mangrove Forests. Working Paper 86. CIFOR, Bogor. Indonesia.

King, J.Y., Reeburgh, W.S., Thieler, K.K., Kling, G.W., Loya, W.M., Johnson, L.C., Nadelhoffer, K.J., 2002. Pulse-labelling studies of carbon cycling in Arctic tundra ecosystems: the contribution of photosynthates to methane emission. Global Biogeochem. Cycles 16 (4), 48-1–48-8.

Kirschke, S., Bousquet, P., Ciais, P., Saunois, M., Canadell, J.G., Dlugokencky, E.J., Bergamaschi, P., Bergmann, D., Blake, D.R., Bruhwiler, L., Cameron-Smith, P., Castaldi, S., Chevallier, F., Feng, L., Fraser, A., Heimann, M., Hodson, E.L., Houweling, S., Josse, B., Fraser, P.J., Krummel, P.B., Lamarque, J.F., Langenfelds, R.L., Le Quéré, C., Naik, V., O'doherty, S., Palmer, P.I., Pison, I., Plummer, D., Poulter, B., Prinn, R.G., Rigby, M., Ringeval, B., Santini, M., Schmidt, M., Shindell, D.T., Simpson, I.J., Spahni, R., Steele, L.P., Strode, S.A., Sudo, K., Szopa, S., Van Der Werf, G.R., Voulgarakis, A., Van Weele, M., Weiss, R.F., Williams, J.E., Zeng, G., 2013. Three decades of global methane sources and sinks. Nat. Geosci. 6 (10), 813–823.

Konnerup, D., Betancourt-Portela, J.M., Villamil, C., Parra, J.P., 2014. Nitrous oxide and methane emissions from the restored mangrove ecosystem of the Ciénaga Grande de Santa Marta, Colombia. Estuar. Coast Shelf Sci. 140, 43–51.

Kreuzwieser, J., Buchholz, J., Rennenberg, H., 2003. Emissions of methane and nitrous oxide by Australian mangrove ecosystem. Plant Biol. 5, 9.

Krithika, K., Purvaja, R., Ramesh, R., 2008. Fluxes of methane and nitrous oxide from an Indian mangrove. Curr. Sci. 94 (2), 218–224.

Kuzyakov, Y., 2010. Priming effects: interactions between living and dead organic matter. Soil Biol. Biochem. 42 (9), 1363–1371.

Lallier-Vergès, E., Marchand, C., Disnar, J.R., Lottier, N., 2008. Origin and diagenesis of lignin and carbohydrates in mangrove sediments of Guadeloupe (French West Indies): Evidence for a two-step evolution of organic deposits. Chem. Geol. 255 (3–4), 388–398.

Le Mer, J., Roger, P., 2001. Production, oxidation, emission and consumption of methane by soils: a review. Eur. J. Soil Biol. 37 (1), 25–50.

Lee, R.Y., Porubsky, W.P., Feller, I.C., McKee, K.L., Joye, S.B., 2008. Porewater biogeochemistry and soil metabolism in dwarf red mangrove habitats (Twin Cays, Belize). Biogeochemistry 87 (2), 181–198.

Lekphet, S., Nitisoravut, S., Adsavakulchai, S., 2005. Estimating methane emissions from mangrove area in Ranong Province, Thailand. Songklanakarin J. Sci. Technol. 27 (1), 153–163.

Lewis III, R.R., 2005. Ecological engineering for successful management and restoration of mangrove forests. Ecological Engineering 24 (4), 403–418. https://doi.org/10.1016/j.ecoleng.2004.10.003.

Lewis III, R.R., Brown, B., 2014. Ecological mangrove rehabilitation – a field manual for practitioners. Mangrove Action Project, Canadian International Development Agency and OXFAM.

Lin, Y., Liu, D., Ding, W., Kang, H., Freeman, C., Yuan, J., Xiang, J., 2015. Substrate sources regulate spatial variation of metabolically active methanogens from two contrasting freshwater wetlands. Appl. Microbiol. Biotechnol. 99 (24), 10779–10791.

Liu, X., Xiong, Y., 2017. Relative contributions of leaf litter and fine roots to soil organic matter accumulation in mangrove forests. Plant Soil 421, 493–503.

Lofton, D.D., Whalen, S.C., Hershey, A.E., 2014. Effect of temperature on methane dynamics and evaluation of methane oxidation kinetics in shallow Arctic Alaskan lakes. Hydrobiologia 721 (1), 209–222.

Lohbeck, M., Poorter, L., Paz, H., Pla, L., van Breugel, M., Martínez-Ramos, M., Bongers, F., 2012. Functional diversity changes during tropical forest succession. Perspect. Plant Ecol. Evol. Systemat. 14 (2), 89–96.

Lu, C.Y., Wong, Y.S., Tam, N.F.Y., Ye, Y., Lin, P., 1999. Methane flux and production from sediments of a mangrove wetland on Hainan Island, China. Mangroves Salt Marshes 3 (1), 41–49.

Lu, X., Zhou, Y., Zhuang, Q., Prigent, C., Liu, Y., Teuling, A., 2018. Increasing methane emissions from natural land ecosystems due to sea-level rise. J. Geophys. Res.: Biogeosci. 123 (5), 1756–1768.

Lu, Y., Wassmann, R., Neue, H.U., Huang, C., Bueno, C.S., 2000. Methanogenic responses to exogenous substrates in anaerobic rice soils. Soil Biol. Biochem. 32 (11–12), 1683–1690.

Lyimo, T.J., Pol, A., Op den Camp, H.J.M., 2002. Sulfate reduction and methanogenesis in sediments of Mtoni mangrove forest. Ambio 31 (7), 614–616.

Macreadie, P.I., Nielsen, D.A., Kelleway, J.J., Atwood, T.B., Seymour, J.R., Petrou, K., Connolly, R.M., Thomson, A.C.G., Trevathan-Tackett, S.M., Ralph, P.J., 2017. Can we manage coastal ecosystems to sequester more blue carbon? Front. Ecol. Environ. 15 (4), 206–213.

Man, K.W., Zheng, J., Leung, A.P.K., Lam, P.K.S., Lam, M.H.W., Yen, Y.F., 2004. Distribution and behavior of trace metals in the sediment and porewater of a tropical coastal wetland. Sci. Total Environ. 327 (1–3), 295–314.

Magen, C., Lapham, L.L., Pohlman, J.W., Marshall, K., Bosman, S., Casso, M., Chanton, J.P., 2014. A simple headspace equilibration method for measuring dissolved methane. Limnol Oceanogr. Methods 12, 637–650.

Maier, M., Machacova, K., Lang, F., Svobodova, K., Urban, O., 2018. Combining soil and tree-stem flux measurements and soil gas profiles to understand CH_4 pathways in *Fagus sylvatica* forests. J. Plant Nutr. Soil Sci. 181 (1), 31–35.

Marchand, C., 2017. Soil carbon stocks and burial rates along a mangrove forest chronosequence (French Guiana). For. Ecol. Manag. 384, 92–99.

Marchand, C., Lallier-Vergès, E., Disnar, J.R., Kéravis, D., 2008. Organic carbon sources and transformations in mangrove sediments: a Rock-Eval pyrolysis approach. Org. Geochem. 39 (4), 408–421.

Marinho, C., Campos, E., Guimarães, J., Esteves, F., 2012. Effect of sediment composition on methane concentration and production in the transition zone of a mangrove (Sepetiba Bay, Rio de Janeiro, Brazil). Braz. J. Biol. 72 (3), 429–436.

Marín-Muñiz, J.L., Hernández, M.E., Moreno-Casasola, P., 2015. Greenhouse gas emissions from coastal freshwater wetlands in Veracruz Mexico: effect of plant community and seasonal dynamics. Atmos. Environ. 107 (26), 107–117.

Martin, J.B., Hartl, K.M., Corbett, D.R., Swarzenski, P.W., Cable, J.E., 2003. A multilevel pore-water sampler for permeable sediments. Journal of Sedimentary Research 73 (1), 128–132.

Mason-Jones, K., Kuzyakov, Y., 2017. 'Non-metabolizable' glucose analogue shines new light on priming mechanisms: triggering of microbial metabolism. Soil Biol. Biochem. 107, 68–76.

Odum, E.P., 1969. The strategy of ecosystem development. Science 164 (3877), 262–270.

Okimoto, Y., Nose, A., Ikeda, K., Agarie, S., Oshima, K., Tateda, Y., Ishii, T., Nhan, D.D., 2008. An estimation of CO_2 fixation capacity in mangrove forest using two methods of CO_2 gas exchange and growth curve analysis. Wetl. Ecol. Manag. 16 (2), 155–171.

Ong, J.E., Gong, W.K., Wong, C.H., 1985. Seven years of productivity studies in a Malaysian managed mangrove forest then what? In: Beardsley, K.N., Davie, J.D.S., Woodroffe, C.D. (Eds.), *Coasts and Tidal Wetlands of the Australian Monsoon Region—NARU Mangrove Monographs* 1. Australian National University Press, Canberra, pp. 213–223.

Pacheco, F.S., Roland, F., Downing, J.A., 2014. Eutrophication reverses whole-lake carbon budgets. Inland Waters 4 (1), 41–48.

Pangala, S.R., Moore, S., Hornibrook, E.R.C., Gauci, V., 2013. Trees are major conduits for methane egress from tropical forested wetlands. New Phytol. 197 (2), 524–531.

Pangala, S.R., Hornibrook, E.R.C., Gowing, D.J., Gauci, V., 2015. The contribution of trees to ecosystem methane emissions in a temperate forested wetland. Global Change Biol. 21 (7), 2642–2654.

Pangala, S.R., Enrich-Prast, A., Basso, L.S., Peixoto, R.B., Bastviken, D., Hornibrook, E.R.C., Gatti, L.V., Marotta, H., Calazans, L.S.B., Sakuragui, C.M., Bastos, W.R., Malm, O., Gloor, E., Miller, J.B., Gauci, V., 2017. Large emissions from floodplain trees close the Amazon methane budget. Nature 552 (7684), 230–234.

Pitz, S.L., Megonigal, J.P., Chang, C.H., Szlavecz, K., 2018. Methane fluxes from tree stems and soils along a habitat gradient. Biogeochemistry 137 (3), 307–320.

Poffenbarger, H.J., Needelman, B.A., Megonigal, J.P., 2011. Salinity influence on methane emissions from tidal marshes. Wetlands 31, 831–842.

Poungparn, S., Charoenphonphakdi, T., Sangtiean, T., Patanaponpaiboon, P., 2016. Fine root production in three zones of secondary mangrove forest in eastern Thailand. Trees Struct. Funct. 30 (2), 467–474.

Pszwaro, J.L., D'Amato, A.W., Burk, T.E., Russell, M.B., Palik, B.J., Strong, T.F., 2016. Analysis of stand basal area development of thinned and unthinned *Acer rubrum* forests in the upper Great Lakes region, USA. Can. J. For. Res. 46 (5), 645–655.

Pulliam, W.M., 1992. Methane emissions from cypress knees in a southeastern floodplain swamp. Oecologia 91 (1), 126–128.

Purvaja, R., Ramesh, R., Frenzel, P., 2004. Plant-mediated methane emission from an Indian mangrove. Global Change Biol. 10 (11), 1825–1834.

Reeburgh, W.S., 2013. Global methane biogeochemistry. In: Holland, H.D., Turekian, K.K. (Eds.), Treatise on Geochemistry, second ed. Elsevier, Oxford, pp. 71–94.

Robertson, A.I., Alongi, D.M., 2016. Massive turnover rates of fine root detrital carbon in tropical Australian mangroves. Oecologia 180 (3), 841–851.

Rocha, A.C.S., Almeida, C.M.R., Basto, M.C.P., Vasconcelos, M.T.S.D., 2015. Influence of season and salinity on the exudation of aliphatic low molecular weight organic acids (ALMWOAs) by *Phragmites australis* and *Halimione portulacoides* roots. J. Sea Res. 95, 180–187.

Rosentreter, J.A., Maher, D.T., Ho, D.T., Call, M., Barr, J.G., Eyre, B.D., 2017. Spatial and temporal variability of CO_2 and CH_4 gas transfer velocities and quantification of the CH_4 microbubble flux in mangrove dominated estuaries. Limnol. Oceanogr. 62 (2), 561–578.

Rusch, H., Rennenberg, H., 1998. Black alder (*Alnus glutinosa* (L.) Gaertn.) trees mediate methane and nitrous oxide emission from the soil to the atmosphere. Plant Soil 201 (1), 1–7.

Sarmiento, J.L., Gruber, N., 2006. Ocean Biogeochemical Dynamics. Princeton University Press, Princeton, Woodstock.

Sasmito, S.D., Taillardat, P., Clendenning, J.N., Cameron, C., Friess, D.A., Murdiyarso, D., Hutley, L.B., 2019. Effect of land-use and land-cover change on mangrove blue carbon: A systematic review. Global Change Biol. 00, 1–12.

Schile, L.M., Kauffman, J.B., Crooks, S., Fourqurean, J.W., Glavan, J., Megonigal, J.P., 2017. Limits on carbon sequestration in arid blue carbon ecosystems. Ecol. Appl. 27 (3), 859–874.

Schulze, E.D., Wirth, C., Mollicone, D., Ziegler, W., 2005. Succession after stand replacing disturbances by fire, wind throw, and insects in the dark Taiga of Central Siberia. Oecologia 146 (1), 77–88.

Setiawan, H., Mursidin, 2018. Ecological characteristic and health of mangrove forest at Tanakeke Island, South Sulawesi. Jurnal Penelit. Kehutan. Wallacea 7 (1), 47–58.

Sotomayor, D., Corredor, J.E., Morell, J.M., 1994. Methane flux from mangrove sediments along the southwestern coast of Puerto Rico. Estuaries 17 (1), 140.

Strangmann, A., Bashan, Y., Giani, L., 2008. Methane in pristine and impaired mangrove soils and its possible effect on establishment of mangrove seedlings. Biol. Fertil. Soils 44 (3), 511–519.

Susilo, A., Ridd, P.V., Thomas, S., 2005. Comparison between tidally driven groundwater flow and flushing of animal burrows in tropical mangrove swamps. Wetl. Ecol. Manag. 13 (4), 377–388.

Terazawa, K., Ishizuka, S., Sakata, T., Yamada, K., Takahashi, M., 2007. Methane emissions from stems of *Fraximum mandshurica* var. *japonica* trees in a floodplain forest. Soil Biol. Biochem. 39, 2689–2692.

Terazawa, K., Yamada, K., Ohno, Y., Sakata, T., Ishizuka, S., 2015. Spatial and temporal variability in methane emissions from tree stems of *Fraxinus mandshurica* in a cool-temperate floodplain forest. Biogeochemistry 123 (3), 349–362.

Tokida, T., Adachi, M., Cheng, W., Nakajima, Y., Fumoto, T., Matsushima, M., Nakamura, H., Okada, M., Sameshima, R., Hasegawa, T., 2011. Methane and soil CO_2 production from current-season photosynthates in a rice paddy exposed to elevated CO_2 concentration and soil temperature. Global Change Biol. 17 (11), 3327–3337.

Tong, C., Huang, J., Jia, Y., 2015a. Small-scale spatial variability of soil methane production potential and porewater characteristics in an estuarine *Phragmites australis* marsh. J. Coast Res. 314 (4), 994–1004.

Tong, C., She, C.X., Yang, P., Jin, Y.F., Huang, J.F., 2015b. Weak correlation between methane production and abundance of methanogens across three Brackish Marsh Zones in the Min River Estuary, China. Estuar. Coast 38 (6), 1872–1884.

Twilley, R.R., Castañeda-Moya, E., Rivera-Monroy, V.H., Rovai, A., 2017. Productivity and carbon dynamics in mangrove wetlands. In: Rivera-Monroy, V.H., Yip Lee, S., Kristensen, E., Twilley, R.R. (Eds.), Mangrove Ecosystems: A Global Biogeographic Perspective. Springer International Publishing, Cham, Switzerland, pp. 113–162.

Ulumuddin, Y.I., 2019. Pore-water Methane Dynamics as an Indicator of Ecosystem Functioning Following Ecological Mangrove Rehabilitation (EMR) in South Sulawesi, Indonesia. Fenner School of Environment and Society, Australian National University, Canberra.

Van Der Nat, F., Middelburg, J.J., Meteren, D. Van, Wielemakers, A., 1998. Diel methane emission patterns from *Scirpus lacustris* and *Phragmites australis*. Biogeochemistry 41 (1), 1–22.

Van Der Nat, F., Middelburg, J.J., 2000. Methane emission from tidal freshwater marshes. Biogeochemistry 49 (2), 103–121.

van Nugteren, P., Moodley, L., Brummer, G.J., Heip, C.H.R., Herman, P.M.J., Middelburg, J.J., 2009. Seafloor ecosystem functioning: the importance of organic matter priming. Mar. Biol. 156 (11), 2277–2287.

Wang, H., Liao, G., D'Souza, M., Yu, X., Yang, J., Yang, X., Zheng, T., 2016. Temporal and spatial variations of greenhouse gas fluxes from a tidal mangrove wetland in Southeast China. Environ. Sci. Pollut. Control Ser. 23 (2), 1873–1885.

Wang, S., Zhou, L., Chen, J., Ju, W., Feng, X., Wu, W., 2011. Relationships between net primary productivity and stand age for several forest types and their influence on China's carbon balance. J. Environ. Manag. 92 (6), 1651–1662.

Wang, Z.P., DeLaune, R.D., Patrick, W.H., Masscheleyn, P.H., 1993. Soil redox and pH effects on methane production in a flooded rice soil. Soil Sci. Soc. Am. J. 57 (2), 382.

Wania, R., Ross, I., Prentice, I.C., 2010. Implementation and evaluation of a new methane model within a dynamic global vegetation model: LPJ-WHyMe v1.3.1. Geosci. Model Dev. 3 (2), 565–584.

Wanninkhof, R., Asher, W.E., Ho, D.T., Sweeney, C., McGillis, W.R., 2009. Advances in quantifying air-sea gas exchange and environmental forcing. Ann. Rev. Mar. Sci. 1 (1), 213–244.

Wassmann, R., Buendia, L.V., Lantin, R.S., Bueno, C.S., Lubigan, L.A., Umali, A., Nocon, N.N., Javellana, A.M., Neue, H.U., 2000. Mechanisms of crop management impact on methane emissions from rice fields in Los Banos, Philippines. Nutr. Cycl. Agroecosyst. 58, 107–119.

West, W.E., Coloso, J.J., Jones, S.E., 2012. Effects of algal and terrestrial carbon on methane production rates and methanogen community structure in a temperate lake sediment. Freshw. Biol. 57 (5), 949–955.

Weston, N.B., Neubauer, S.C., Velinsky, D.J., Vile, M.A., 2014. Net ecosystem carbon exchange and the greenhouse gas balance of tidal marshes along an estuarine salinity gradient. Biogeochemistry 120 (1–3), 163–189.

Whiting, G.J., Chanton, J.P., 1993. Primary production control of methane emission from wetlands. Nature 367 (6463), 794–795.

Wu, Z., Ren, D., Zhou, H., Gao, H., Li, J., 2017. Sulfate reduction and formation of iron sulfide minerals in nearshore sediments from Qi'ao Island, Pearl River Estuary, Southern China. Quat. Int. 452, 137–147.

Xia, Y.Q., Li, H.L., 2012. A combined field and modeling study of groundwater flow in a tidal marsh. Hydrol. Earth Syst. Sci. 16 (3), 741–759.

Xiang, J., Liu, D., Ding, W., Yuan, J., Lin, Y., 2015. Invasion chronosequence of *Spartina alterniflora* on methane emission and organic carbon sequestration in a coastal salt marsh. Atmos. Environ. 112, 72–80.

Xiong, Y., Liu, X., Guan, W., Liao, B., Chen, Y., Li, M., Zhong, C., 2017. Fine root functional group based estimates of fine root production and turnover rate in natural mangrove forests. Plant Soil 413 (1–2), 83–95.

Xu, Y., Ge, J., Tian, S., Li, S., Nguy-Robertson, A.L., Zhan, M., Cao, C., 2015. Effects of water-saving irrigation practices and drought resistant rice variety on greenhouse gas emissions from a no-till paddy in the central lowlands of China. Sci. Total Environ. 505, 1043–1052.

Yoshikawa, C., Hayashi, E., Yamada, K., Yoshida, O., Toyoda, S., Yoshida, N., 2014. Methane sources and sinks in the subtropical South Pacific along 17°S as traced by stable isotope ratios. Chem. Geol. 382, 24–31.

Yu, B., Stott, P., Yu, H., Li, X., 2013. Methane emissions and production potentials of forest swamp wetlands in the eastern great Xing'an Mountains, Northeast China. Environ. Manag. 52 (5), 1149–1160.

Zeikus, J.G., Ward, J.C., 1974. Methane formation in living trees: a microbial origin. Source Sci. New Series 184 (4142), 1181–1183.

Zhang, K., Cheng, X., Dang, H., Ye, C., Zhang, Y., Zhang, Q., 2013. Linking litter production, quality and decomposition to vegetation succession following agricultural abandonment. Soil Biol. Biochem. 57, 803–813.

Zhang, Y., Ding, W., 2011. Diel methane emissions in stands of *Spartina alterniflora* and *Suaeda salsa* from a coastal salt marsh. Aquat. Bot. 95 (4), 262–267.

CHAPTER

18

Blue carbon storage comparing mangroves with saltmarsh and seagrass habitats at a warm temperate continental limit

Sinegugu P. Banda[1], *Janine B. Adams*[1], *Anusha Rajkaran*[2], *Jaime Leigh Johnson*[2], *Jacqueline L. Raw*[1]

[1]**Nelson Mandela University, Port Elizabeth, Eastern Cape, South Africa;** [2]**University of the Western Cape, Bellville, Cape Town, South Africa**

OUTLINE

1. Introduction 448

2. Materials and methods 450

2.1 Study site 450

2.2 Soil collection 451

2.3 Soil analysis 452

2.3.1 Dry bulk density and organic matter 452

2.3.2 Organic and inorganic carbon analysis 452

2.3.3 Total soil carbon 453

2.4 Biomass collection 453

2.5 Biomass analysis 453

2.5.1 Organic and inorganic carbon 454

2.6 Data analysis 455

2.7 Limitations to the study 455

3. Results 455

3.1 Soil carbon 455

3.2 Above- and belowground biomass 458

4. Discussion 460

5. Conclusion 466

References 466

https://doi.org/10.1016/B978-0-12-816437-2.00008-2

1. Introduction

Previously, coastal habitats such as mangroves, saltmarshes, and seagrasses have been recognized for their high productivity and biodiversity; however, interest has recently been shifted toward the ecosystem services that these habitats provide and their responses to changing climate and sea level rise (Duarte et al., 2013; Rogers et al., 2014; Kelleway et al., 2017). An important service includes climate regulation through the sequestration and storage of carbon and has been termed "blue carbon habitats" or "blue forests" (Mcleod et al., 2011). Blue carbon can be defined as the carbon stored in the soil, living and nonliving above- and belowground biomass of mangrove, saltmarsh, and seagrass habitats (Nellemann et al., 2009). Although these habitats cover less than 2% of the area of the oceans worldwide, they capture up to 70% of carbon (~770 Gt), making these areas more intensive carbon stores than other habitats (Nellemann et al., 2009; Macreadie et al., 2014).

Despite their importance as carbon stores, large areas of these blue forests continue to be degraded due to human-induced pressures which have reduced their ability to sequester atmospheric carbon dioxide for climate change mitigation (Donato et al., 2011; Pidgeon et al., 2011; Fourqurean et al., 2012). Carbon offset mechanisms were introduced in 1997 to decrease global carbon dioxide emissions through the conservation of coastal vegetated habitats, and this increased the assessment of carbon storage in these systems. Thus, substantial research effort has been put into blue carbon quantification with attempts to improve carbon sequestration and storage estimates over local, regional, and global scales. While most of these studies have been largely focused around North America, Europe, and Australia (Duarte, 2017), Africa has also made some significant strides. Carbon assessments and trading projects have become increasingly popular in the Western Indian Ocean (WIO) region (Bosire et al., 2016). The Mikoko Pamoja Project in Gazi Bay, Kenya, and The Blue Forests Initiative in Madagascar are two eminent projects in the WIO region. Their establishment was to enhance mangrove ecosystem services (mainly climate regulation via carbon sequestration) through working with local communities and to implement carbon financing schemes while conserving mangroves (Wylie et al., 2016).

The African region also boasts some sound published scientific research on carbon storage potential in mangroves habitats, while earlier studies concentrated mainly on biomass and productivity (Steinke et al., 1995; Kirui et al., 2006; Kairo et al., 2008, 2009; Bosire et al., 2012). Studies on mangrove carbon stock assessments emerged from Madagascar (Jones et al., 2014; Benson et al., 2017), Mozambique (Sitoe et al., 2014; Shapiro et al., 2015; Stringer et al., 2015), Kenya (Gress et al., 2016), and Tanzania (Njana et al., 2015). Yet, blue carbon research for seagrass habitats is still lacking despite the region containing vast meadows that may contribute greatly to global carbon stores (Duarte et al., 2011). To date, only one quantitative assessment of soil organic and inorganic carbon has been done for Zanzibar, Tanzania mainland, and Mozambique (Gullström et al., 2018) along with a study on the effects of shading and grazing on carbon sequestration on seagrass meadows in Zanzibar (Dahl et al., 2016). For saltmarsh habitats, the situation is even more dire, with no record of published data on carbon stocks and accumulation rates for Africa, India, and South America (Chastain et al., 2018). The majority of the saltmarsh studies are concentrated in Eastern and Northern United States (Drake et al., 2015; Tripathee and Schafër, 2015; Thorhaugh et al., 2019), Europe (Beaumont et al., 2014; Sousa et al., 2017), and Australia (Macreadie et al., 2014, 2017; Kelleway et al., 2016).

In South Africa, blue carbon habitats occur in sheltered estuarine environments where at present, very little is known about their blue carbon dynamics and only three studies have been done in the country thus far. The first study evaluated soil carbon storage across a mangrove–saltmarsh ecotone at Nahoon Estuary, which is the range limit for mangroves along the east coast of Africa. Soil carbon stocks were similar between the mangrove, ecotone, and saltmarsh habitats, despite a mangrove range expansion at this estuary over the past years (Raw et al., 2019). At the Swartkops Estuary (a nationally important estuary in the country), Els (2019) reported that soil organic carbon concentrations ranged from 212 to 247 Mg C/ha for saltmarsh and seagrass habitats. This was much higher than stated values internationally, possibly due to high levels of pollution at this estuary (Els, 2019). The first comprehensive assessment of carbon storage for mangroves in a warm temperate estuary was done by (Johnson et al., 2020) who found that the average carbon storage for mangrove soil ranged from 179.91 to 262 Mg C/ha and for all the carbon pools being 239 Mg C/ha, which was within the range for mangroves in the southern hemisphere.

At present, there has been no study comparison between mangrove, saltmarsh, and seagrass blue carbon dynamics. The paucity of these data prompted the present study, which is a component of a larger study on Climate Change and Blue Carbon Habitats (Adams et al., 2019). This chapter shows our study that aims to assess soil characteristics associated with blue carbon storage between habitat types at warm temperate estuary in the Eastern Cape of South Africa and to compare these finding to global data sets. The main objectives of the study are (1) to determine the differences in soil carbon storage between habitat types within the estuary (mangroves, saltmarsh, and seagrass) as well as establish the relationship between soil organic matter and organic carbon; (2) to compare the carbon stocks of above- and belowground biomass between the three habitats; and (3) to establish allometric equations for just the saltmarsh aboveground biomass. For the mangrove component, data for the soil and aboveground biomass carbon storage were taken from (Johnson et al., 2020) with the relationship between organic matter and organic carbon also reported in that study.

Even though a large portion of the carbon in blue carbon habitats is stored in the soil and biomass carbon is considered a small component, it is still a significant carbon pool that requires accurate quantification. Aboveground biomass is usually measured using metrics that can be easily determined, related to total plant biomass (i.e., canopy volume and stem height) and used to form species-specific allometric equations (Kauffman and Donato, 2012; Howard et al., 2014; Radabaugh et al., 2017). Species-specific allometric equations required for biomass carbon estimates are generally limited in the literature except for well-studied species (e.g., *Spartina alterniflora*) because creating these equations is laborious and destructive. Thus, when an allometric equation has been determined, it can be used in future studies to quantify biomass quickly with less destruction (Radabaugh et al., 2017). This study will also contribute to the lack of species-specific carbon data and allometric equations globally and will be the first to provide an allometric equation for the saltmarsh species *Salicornia tegetaria* (S. Steffen, Mucina, and G. Kadereit; Piirainen and G. Kadereit) in South Africa. The scope of this study did not include developing an allometric equation for mangroves (details in Johnson, 2019) and an equation for seagrasses, which was done in Els (2019).

This chapter will also include discussions on some of the potential factors which could influence carbon storage in the estuary because blue carbon storage has also been linked to a whole suite of global, regional, and local factors including topography/elevation (Alongi, 2018), soil depth, deposition, plant productivity, forest age, and soil type (Kelleway et al., 2016).

2. Materials and methods

2.1 Study site

Nxaxo Estuary is a permanently open estuary located at Wavecrest near Butterworth in the Eastern Cape of South Africa (32°35′S; 28°31′E) (Fig. 18.1) within the warm temperate bioclimatic region (Harrison, 2004). The Eastern Cape is also the southern distributional limit for mangroves along the east coast of South Africa (Steinke et al., 1995; Adams et al., 2004). Nxaxo Estuary was selected for study as it has the largest area of saltmarsh (10.9 ha) co-occurring with mangroves of a similar area (9.5 ha) (saltmarsh:mangrove ratio < 1) (Adams et al., 2016) and contains patches of seagrass.

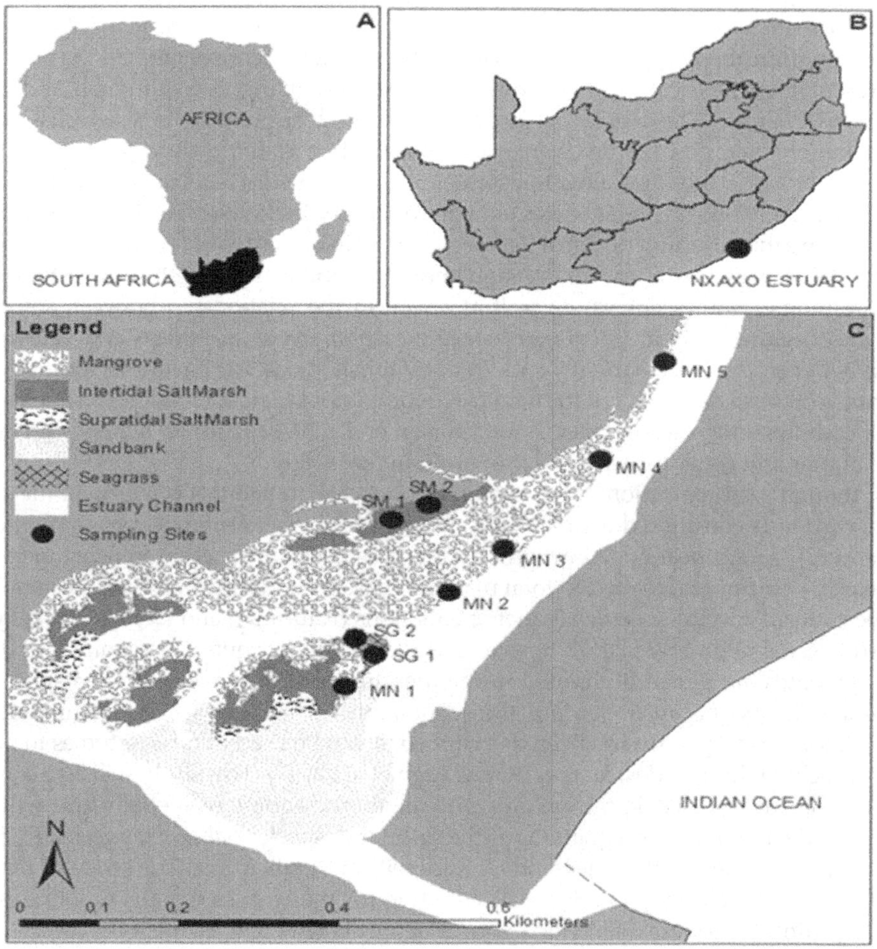

FIGURE 18.1 A map of the Nxaxo Estuary in the Eastern Cape province of South Africa and the location of the mangrove (MN), saltmarsh (SM), and seagrass (SG) sampling sites within the estuary. *Adapted from Johnson, J., 2019. A Quantitative Assessment of Blue Carbon for Mangroves in a Warm- Temperate Estuary of South Africa (M.Sc. thesis). Nelson Mandela University, Port Elizabeth.*

This estuary has three mangroves species, namely *Bruguiera gymnorrhiza* (L.) Lamk, *Rhizophora mucronata* (L.), and *Avicennia marina* (Forssk.), the latter which is dominant and found fringing the channels from the mouth to the upper reaches (Hoppe-Speer and Adams, 2015). There are four saltmarsh species including *S. tegetaria* and *Salicornia natalensis* (Bunge ex. Ung.-Sternb.) in the lower intertidal area and *Bassia diffusa* (Thunb.) Kuntze and *Sporobolus virginicus* (L.) Kunth in the supratidal area (Hoppe-Speer, 2013). There are also two species of seagrass, *Halophila ovalis* (R. Brown) J. D Hooker and the endangered lower intertidal seagrass species *Zostera capensis* Setchell (also known as the Cape dwarf-eelgrass). *Z. capensis* co-occurs with *A. marina* in most estuaries along the east coast of South Africa due to their salinity preferences, and at Nxaxo Estuary specifically, *Z. capensis* may be involved in the facilitation of mangrove recruitment as it is sometimes found growing with germinating propagules (Adams, 2016). The *A. marina* trees within the estuary are stunted due to previous harvesting, heavy browsing, and trampling by cattle during drought periods (Hoppe-Speer and Adams, 2015; Mbense, 2017). We defined the mangrove population at this estuary as seedlings (<50 cm), juveniles (101–150 cm), and adults (>151 cm). Some adult trees along the main channel are taller (3–4 m in height), whereas the stunted adults occur more landward (Hoppe-Speer and Adams, 2015).

The sampling design for the mangroves was outlined in Johnson (2019). For the mangroves, existing *A. marina* sites along the main channel were used for this study (MN 1–5). Two sites were strategically selected within the *Salicornia*-dominated saltmarsh area and the *Z. capensis* patches. The saltmarsh plots (SM 1–2) were located behind the mangrove area while the seagrass plots (SG 1–2) were in front of the mangroves on an island close to the main channel of the estuary (Fig. 18.1).

2.2 Soil collection

Two saltmarsh and seagrass sites were randomly chosen within the estuary, and two 25-m^2 quadrats were laid in each plot. Within each quadrat, six soil cores of approximately 1 m depth were extracted randomly using PVC tube (11 cm diameter) at different areas within the quadrat. The core was divided into eight depth intervals (0–5, 5–10, 10–20, 20–30, 30–40, 40–50, 50–70, 70–100 cm) (Fig. 18.2A). A subsample was taken at the middle of each depth interval by inserting a syringe into predrilled holes in the PVC tube (Fig. 18.2B) (adapted from Howard et al., 2014). Individual soil increments were then packed into small, prelabeled plastic bags and transferred into a cooler box for storage until further analysis in the laboratory.

Each mangrove site was divided into six (5-m^2) plots which were defined by their location in the landscape—the lower being along the bank of the estuary and the upper zone in the landward edge of the mangrove forest. One soil core was extracted in the center of each plot using a Russian Peat Corer (Fig. 18.2C). Additionally, three cores were extracted in the area between the lower and upper zones. Each core was divided into four sections at depths intervals of 0–15, 15–30, 30–50, and 50–100 cm. Additional soil samples were also taken in the middle of each site at 25, 50, 75 cm, and 1 m depth, for redox measurements and soil analysis (Johnson et al., 2020). Samples were also stored in plastic bags in cooler boxes and processed in the laboratory.

FIGURE 18.2 (A) An example of an extracted soil core using a PVC pipe with predrilled holes. (B) Extracting a subsample from the core using a syringe. (C) A soil core extracted from a mangrove plot using a Russian Peat Core. *Credit: S. Mbense and J. Raw, 2017.*

2.3 Soil analysis

2.3.1 *Dry bulk density and organic matter*

Each soil subsample for each depth increment was dispensed into a glass Petri dish and oven-dried at 60°C until a constant weight was reached. Dry bulk density (DBD) (g/cm^3) was then determined by dividing the mass of the dried soil (g) by the original volume sampled (cm^3), because the sample was taken with a syringe and volume was measured directly (where 1 cc = 1 cm^3) (Howard et al., 2014). The dried soil was then further divided into two subsamples; one subsample was used to determine organic matter using the percent mass loss on ignition (LOI), which provides a quantification of percentage organic matter (OM%). The LOI procedure has been shown to be a cheap and effective predicator of soil organic carbon (Craft et al., 1991; Macreadie et al., 2013). To determine OM%, the ashing method was used and involved heating dried subsamples in a muffle furnace for 8 h at 550°C (Briggs, 1977). OM% was then calculated using the dry mass and mass after ashing.

2.3.2 *Organic and inorganic carbon analysis*

The second subsample was homogenized into a fine powder using a mortar and pestle and was used to determine percentage organic carbon (C_{org}%) and nitrogen (%) through elemental analysis. As elemental analysis reports total carbon (inorganic and organic), samples required correction for inorganic carbon using acid. This included testing a subset of soil samples by treating them with dilute hydrochloric acid (HCL). The acidification process included adding dilute 1 M hydrochloric acid to a known weight of the homogenized subsample until the acid covered the sample. The samples were rinsed three times with distilled water and left to dry overnight at 60°C. The amount of calcium carbonate (In_{org}%) in each sample was calculated according to Howard et al. (2014); however, inorganic carbon in saltmarsh and mangroves was undetectable and thus not reported.

Samples were prepared for elemental analysis following standard protocol. For soil with organic matter between 0.5% and 1%, an amount of 25–30 mg was weighed out, and for soil with organic matter between 2% and 11%, an amount of 9–10 mg of the sample was weighed using a fine scale (UniBloc, Shimadzu, AUWW220D). The samples were then packed into pressed tin capsules (8 × 5 mm) (Elemental Microanalysis, UK). Three replicate samples of the subsample were analyzed for total carbon with a carbon hydrogen and nitrogen (CHN) analyzer at the Wetland Biochemistry laboratory at the Louisiana State University in the United States.

2.3.3 *Total soil carbon*

Carbon density (g/cm^3) was calculated using C_{org}% and DBD for each depth interval from the surface to 1 m depth. The carbon content (Mg C/ha) was determined as the sum of carbon per section over the total sampling depth (~1 m). The total amount of carbon per ecosystem was determined by multiplying the average carbon content (Mg C/ha) by the area of the ecosystem (ha) to determine total carbon for the soil pool for each habitat (Mg C) (Howard et al., 2014).

2.4 Biomass collection

For the vegetative carbon component in both saltmarsh and seagrass plots, the same 25-m^2 plots as aforementioned were divided into 1-m^2 subplots to capture greater variation and increase accuracy in the quantification of each metric (Howard et al., 2014). Within each 1-m^2 plot, three replicates of living aboveground and belowground were collected by inserting the same PVC tube (11 cm diameter) into the surface of the soil (Fig. 18.3A and B). Care was taken to ensure that the core was inserted deep enough to harvest and capture both the aboveground and belowground components. For saltmarsh vegetation, any remaining soil was removed, and plant material was put in labeled plastic bags. For the seagrass, plant material in the core was transferred into a mesh sieve, washed until free of soil, packed into prelabeled plastic bags, and stored in cooler boxes. The processing of biomass samples began within 2 days of collection as seagrasses have a high decay rate.

For the mangroves, the total carbon for the aboveground biomass included living trees, leaf litter, and pneumatophores. The sampling methods were outlined in Johnson et al., (2020) and Fig. 18.3C.

2.5 Biomass analysis

In the laboratory, saltmarsh plants were separated into above- and belowground material. The height of each aboveground *S. tegetaria* branch was measured to the nearest 0.1 cm. The

FIGURE 18.3 Above- and belowground biomass collection within the smaller quadrats for (A) seagrass *Zostera capensis*, (B) saltmarsh *Salicornia tegetaria*, and (C) the mangrove *Avicennia marina* at Nxaxo Estuary. *Credit: S. Mbense and J. Johnson, 2017.*

method of plant measurement usually depends on the growth pattern of the species. For succulent shrubs with a diverse growth pattern, measurements of individual stems within a shrub are required to properly determine a stem height and biomass relationship (Radabaugh et al., 2017). Stems that were slightly dead or senescing were also included in the stem height measurements. Following height measurements, the plant material was oven-drying to a constant weight for approximately 72 h at 60°C to determine biomass (as dry weight).

The seagrass was separated into shoots and rhizomes; the shoots were measured for height and both components were dried using the same protocol as the saltmarsh to determine biomass as dry weight. The above- and belowground carbon of saltmarsh and seagrass was estimated by multiplying the biomass per given area (g/cm^2) by the organic carbon determined from CHN analysis. For mangrove aboveground biomass, the formula from Chave et al. (2014) and methods from Howard et al. (2014), as previously mentioned, have been outlined in detail in (Johnson et al., 2020).

2.5.1 *Organic and inorganic carbon*

A subsample of dried above- and belowground vegetation was ground into a fine powder using a mortar and pestle. The champagne test was performed on a subset of seagrass and saltmarsh samples, and further inorganic carbon removal was deemed unnecessary. The same methods as described above were used for analyzing total carbon (see above).

2.6 Data analysis

Analysis of variance (ANOVA) was used to test for significant differences in soil parameters (organic matter, bulk density, organic and inorganic carbon, and carbon density) with changes in depth (depth section as a fixed factor) using the MASS package in RStudio, R version 3.5.1 (2018, The R Foundation for Statistical Computing).

Equations were created to determine the relationship between soil organic matter (LOI%) and organic carbon concentration (C_{org}%) by plotting the values against each other. Natural logarithmic transformations were performed for *Z. capensis* soil values to improve the fit of the linear relationship between organic matter and organic carbon. The values were then further transformed from negative to positive values using the modulus. A similar linear equation was done for saltmarsh to determine the relationship between stem height and stem biomass (as dry weight).

2.7 Limitations to the study

Our study suggested that there are numerous complicated, local, and regional processes underlying carbon accumulation and standing stocks in blue habitats. Also, these carbon stock assessments may have their limitations especially in terms of the methodologies used to scale up the carbon data (i.e., taking soil samples from a small area using a corer and then scaling up to landscape level), which may lead to over- or underestimation of carbon stocks (Schile et al., 2017). We also acknowledge that as this is a baseline study, there may have been some errors during the field sampling, processing, and in calculating the carbon. For example, separation of the above- and belowground biomass of seagrasses is sometimes ambiguous (Howard et al., 2014) as it is based on how well one can distinguish the "green" stem-like parts from "brown" root structures, which is sometimes difficult in smaller species such as *Z. capensis*. We also acknowledge that the organic carbon stored in the soil of blue habitats is not a true reflection of the present vegetation but is rather a result of habitat changes over time, which will be influenced by sea level rise and soilation rates in the future (Mudd et al., 2009).

3. Results

3.1 Soil carbon

Saltmarsh soil DBD ranged from 0.43 to 0.76 g/cm^3, with a mean of 0.62 ± 0.13 g/cm^3, and DBD was correlated and increased significantly with depth ($P < .05$, $r = 0.63$) (Fig. 18.4A, Table 18.2). In soil underlying the seagrass beds, the bulk density ranged from 0.68 to 0.98 g/cm^3 with a mean of 0.89 ± 0.1 g/cm^3. In mangrove soils, DBD ranged from 0.79 to 0.89 g/cm^3 with a mean of 0.81 ± 0.04 g/cm^3 (Fig. 18.4A). In both seagrass and mangrove soils, DBD showed no significant trend or correlation with depth (Table 18.2). Organic matter for saltmarsh soil decreased from 4% at the surface and to 0.8% at depth with a mean of 1.7% ± 0.34% ($P < .05$) (Fig. 18.4B; Table 18.1). Seagrass soils OM% ranged from 1.58% to 3.06% (Fig. 18.4B) and decreased in amount with increasing depth (Table 18.1). Mangrove soil organic matter was lower than both saltmarsh and seagrass with a mean of 0.63% ± 0.15% and showed no significant change with depth (Fig. 18.4B, Table 18.2).

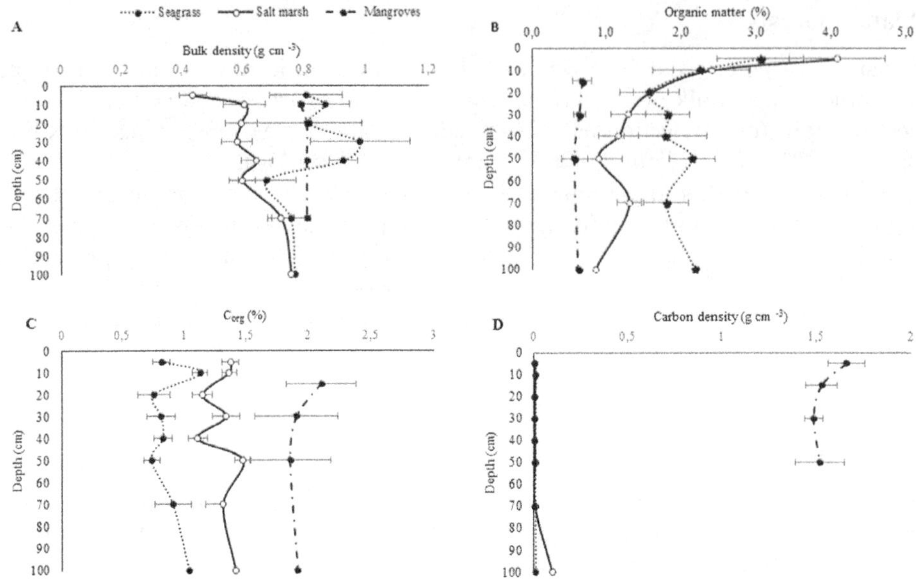

FIGURE 18.4 (A) Dry bulk density (g/cm^3), (B) organic matter (OM%), (C) organic carbon (C_{org}%), and (D) carbon density (g/cm^3) depth profiles for soil cores extracted in saltmarsh, seagrasses, and mangroves at Nxaxo Estuary (mean ± standard error, n = 9 mangroves, n = 12 saltmarsh and seagrass).

TABLE 18.1 Inorganic carbon (%), nitrogen (%), and the ratio between organic carbon and nitrogen of saltmarsh, seagrass, and mangrove soil cores at different depths at Nxaxo Estuary (mean - +standard error).

Depth	Inorganic carbon (%)			Nitrogen (%)			Organic carbon:nitrogen ratio		
	Saltmarsh	Seagrass	Mangroves	Saltmarsh	Seagrass	Mangroves	Saltmarsh	Seagrass	Mangroves
0–5	Undetectable	0.22 ± 0.06		0.08 ± 0.04	0.07 ± 0.01		16.54	11.7	
5–10	Undetectable	0.28 ± 0.1		0.09 ± 0.05	0.09 ± 0.01		15.08	12.02	
10–20	Undetectable	0.16 ± 0.05	Undetectable	0.1 ± 0.06	0.06 ± 0.01	0.17 ± 0.02	11.24	11.9	12.4
20–30	Undetectable	0.17 ± 0.06		0.09 ± 0.05	0.07 ± 0.01		14.61	12.2	
30–40	Undetectable	0.11 ± 0.02	Undetectable	0.07 ± 0.04	0.07 ± 0.01	0.21 ± 0.01	14.97	12.6	9.1
40–50	Undetectable	0.17 ± 0.08		0.09 ± 0.06	0.08 ± 0.01		16.34	11.6	
50–70	Undetectable	0.08 ± 0.05	Undetectable	0.08 ± 0.05	0.06 ± 0.01	0.16 ± 0.07	15.37	15.3	11.6
70–100	Undetectable	0.13 ± 0.06	Undetectable	0.15 ± 0.09	0.08 ± 0.02	0.31 ± 0.02	9.29	12.9	6.19

TABLE 18.2 Results of the Analysis of Variance for saltmarsh, seagrass, and mangrove soil core variables (dry bulk density, organic content, organic and inorganic carbon content, nitrogen, and carbon density), with depth as a fixed factor.

		df	SS	F-ratio	P	r
Sediment core variables						
Dry bulk density (g/cm^3)	Saltmarsh	7	52.95	23.83	<0.05	0.63
	Seagrass	7	40.7	15.93	0.13	0.46
	Mangrove	3	15.32	55.41	0.1	0.36
Organic matter (OM%)	Saltmarsh	7	4.62	16.6	<0.05	0.62
	Seagrass	7	4.99	16.2	<0.05	0.14
	Mangrove	3	8.61	1.47	0.22	0.23
Carbon content (C_{org}%)	Saltmarsh	7	12.6	1	0.49	0.02
	Seagrass	7	31.19	0.99	0.51	0.27
	Mangrove	3	56.23	8.35	0.87	0.36
Inorganic carbon content (In_{org}%)	Saltmarsh	7	N.D	N.D	N.D	N.D
	Seagrass	7	32.29	18.47	0.07	0.61
	Mangrove	3	N.D	N.D	N.D	N.D
Nitrogen (%)	Saltmarsh	7	33.41	21.16	0.62	0.25
	Seagrass	7	19.2	5.98	0.08	0.64
	Mangrove	3	54.89	20.67	0.82	0.34
Carbon density (g/cm^3)	Saltmarsh	7	20.9	17.59	0.49	0.44
	Seagrass	7	55.2	9.91	0.51	0.48
	Mangrove	3	45.98	11.27	0.28	0.38

Soil carbon density was higher in mangroves (1.55 ± 0.09 g/cm^3) and much lower in both saltmarsh and seagrass (~0.01 g/cm^3). In all three habitats, soil carbon density also showed no significant trend with depth. Soil organic carbon showed no significant change with depth in all three habitats (Fig. 18.4C, Table 18.2). Organic carbon ranged from 0.85% to 1.96% with means of $0.86\% \pm 0.07\%$, $1.4\% \pm 0.29\%$, and $1.96\% \pm 0.31\%$ for seagrass, saltmarsh, and mangroves, respectively. In seagrass soils, inorganic carbon ranged from 0.08% to 0. 2% and showed a nonsignificant decrease with depth (Tables 18.1 and 18.2). Soil nitrogen for all three habitats ranged from 0.06% to 0.31% and showed no significant changes with depth. The ratio between carbon and nitrogen ranged from 6.91 to 16.5 and there was no consistent change in C:N with depth in all three habitats.

There was a strong relationship between OM and C_{org} in saltmarsh soil ($R^2 = 0.88$) and approximately 78% of the organic matter was organic carbon (Fig. 18.5A). In both mangrove and seagrass, the relationship between C_{org} and OM was very weak (Fig. 18.5B and C).

Total soil organic carbon (to 1 m depth) was shown to be significantly higher in mangroves (228.05 ± 27.99 Mg C/ha) compared with saltmarsh (2.61 ± 0.19 Mg C/ha) and seagrass (1.67 ± 0.81 Mg C/ha) (Table 18.3) ($F = 180.92$, $df = 2$, $P < .0001$). The total carbon for the entire soil carbon pool was also significantly greater in mangroves (2166.48 ± 265.91 Mg C) compared with saltmarsh (28.5 ± 2.11 Mg C) and seagrass (0.06 ± 0.01 Mg C) ($F = 78.12$, $df = 2$, $P < .0001$).

3.2 Above- and belowground biomass

Total aboveground biomass carbon was significantly higher in mangroves (40.1% ± 0.81%) compared with saltmarsh (25.3% ± 0.62%) and seagrass (22.49% ± 0.75%) (Fig. 18.6A) ($F = 21$, $P < .001$). Nitrogen composition was similar in all three habitat types (~1%) and there were no noticeable differences in nitrogen composition between above- and belowground biomass.

S. tegetaria had a fairly good degree of fit ($R^2 = 0.707$) between stem height and biomass even though they are herbaceous shrubs that generally have a high degree of branching. Stem height and biomass ranged from 11 to 60 cm and 1 to 11 g (Fig. 18.7).

For saltmarsh, above- and belowground biomass ranged from 715 to 1157.62 g/m^2 and 604 to 832 g/m^2, respectively. Also saltmarsh belowground biomass (714 ± 16.7 g/m^2) was significantly higher than seagrass (314.6 ± 76.65 g/m^2) ($F = 143.51$, $df = 1$, $P < .0001$). For

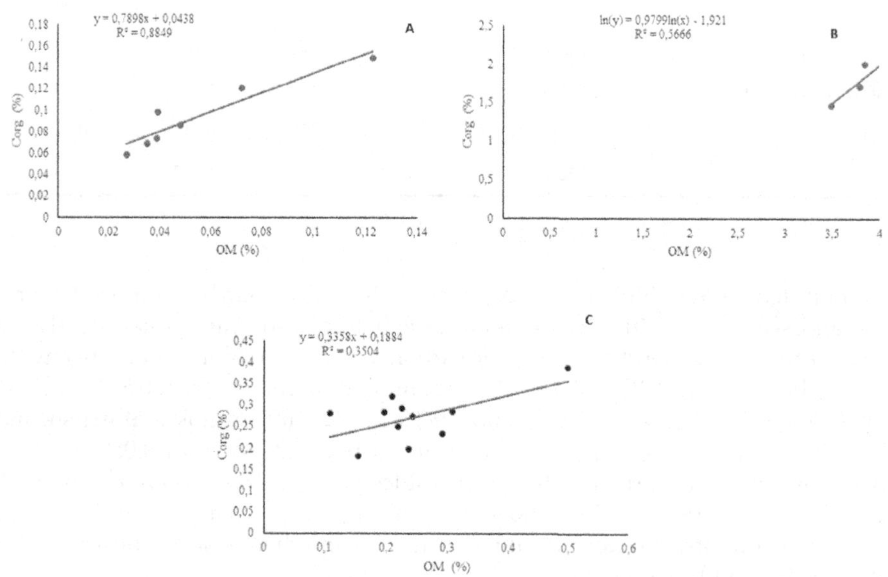

FIGURE 18.5 The relationship between soil carbon concentration (C_{org}%) and organic content (OM%) in (A) saltmarsh, (B) seagrass, and (C) mangroves at Nxaxo Estuary.

TABLE 18.3 Total soil (1 m core) depth, aboveground (ABG) and belowground (BGB) biomass carbon stocks for mangroves, saltmarshes, and seagrasses at Nxaxo Estuary (mean ± standard error).

	Area (ha)	Soil carbon (Mg C/ha)	Total soil carbon pool (Mg C)	Biomass (g/m)		Total biomass carbon (Mg C/ha)	Total biomass carbon pool (Mg C)
				ABG	BGB		
Saltmarsh	10.9	2.61 ± 0.19	28.55 ± 2.1	872.4 ± 39.81	714 ± 16.7	3.96 ± 0.78	39.66 ± 1.05
Seagrass	0.04	1.67 ± 0.01	0.06 ± 0.01	354.5 ± 86.47	314.6 ± 76.65	$1.26 \pm 8.1 \times 10^{-5}$	$0.1 \pm 3.2 \times 10^{-6}$
Mangroves	9.5	228.05 ± 27.99	2166.48 ± 265.91	*6526.3 ± 117[a]	ND	71.45 ± 2.47	678.45 ± 24.15

[a]*Mangrove aboveground biomass included living trees, leaf litter, and pneumatophores.*

both saltmarsh and seagrasses, there were not significant differences between the quadrats. Mangroves had a higher aboveground biomass (6526.3 ± 117 g/m^2) compared with saltmarsh (872.4 ± 39.81 g/m^2) and seagrasses (354.5 ± 86.47 g/m^2) ($F = 4.41$, $df = 2$, $P < .0001$).

Total biomass carbon storage estimates were influenced by biomass estimates and carbon content. Mangrove biomass carbon (71.45 ± 2.47 Mg/ha) was higher than saltmarsh

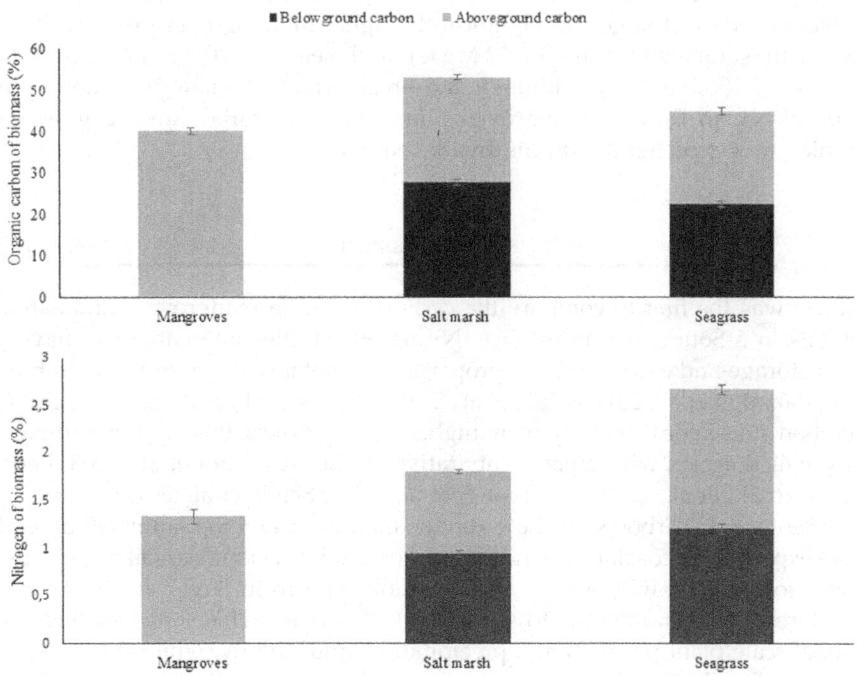

FIGURE 18.6 Percentage organic carbon and nitrogen for saltmarsh, seagrass, and mangrove plant biomass at Nxaxo Estuary (mean ± standard error).

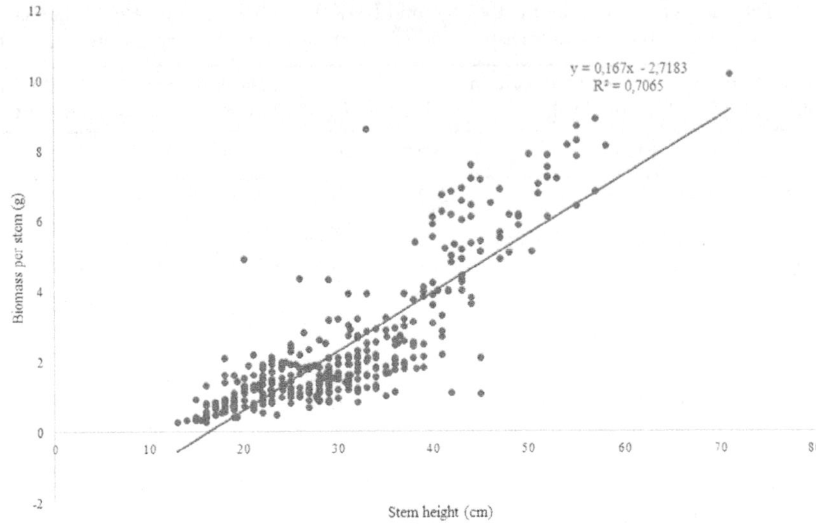

FIGURE 18.7 The relationship between biomass (g) (as dry weight) versus height (cm) for saltmarsh *Salicornia tegetaria* at Nxaxo Estuary.

(2.5 ± 0.05 Mg/ha) and seagrass ($1.26 \pm 8.1 \times 10^{-5}$ Mg/ha) ($F = 59.99$, $df = 2$, $P < .0001$). As a result, overall carbon storage was significantly higher in mangroves (678.75 ± 17.37 Mg C) compared with saltmarsh (27.63 ± 0.62 Mg C) and seagrass ($0.1 \pm 3.2 \times 10^{-6}$) ($F = 38.2$, $df = 2$, $P < .0001$) (Table 18.3). Although the areal extent of mangroves and saltmarshes was similar at Nxaxo Estuary, mangroves stored the most carbon and seagrasses had the lowest (which was expected due to the smallest area).

4. Discussion

This study was the first to compare the carbon stored in mangrove, saltmarsh, and seagrass habitats in a South African estuary. Numerous studies internationally have reported that carbon storage and associated soil properties are notably different between blue carbon habitats (Radabaugh et al., 2017; Schile et al., 2017; Cusack et al., 2018; Lewis et al., 2018). Soil organic carbon stocks for this study were higher in mangroves, followed by saltmarshes and seagrasses, which agrees with other comparative studies (Chmura et al., 2003; Pendleton et al., 2012; Siikamäki et al., 2012; Radabaugh et al., 2017; Schile et al., 2017).

Large differences in carbon stock between the mangrove and the saltmarsh habitats in our study were expected. In coastal wetlands in arid or semiarid climates, soil organic matter has been shown to decrease with a shift in vegetation type from larger and more productive mangrove forests and graminoids to succulent saltmarshes. This shift has been associated with regional scale plant productivity, precipitation, and salinity, while at local scales (estuary and transect), soil organic matter was influenced by topography and biota (Osland et al., 2018). In Central Texas, *Salicornia* spp. has too been linked to lower soil carbon as a result of low rainfall, which caused hypersalinity and increased plant physiological stress (Yando et al., 2016). *Sarcocornia–Sporobolus* sites in Australia similarly had lower carbon stocks compared with other saltmarshes (Saintilan et al., 2013). *Sarcocornia* inhabits the upper intertidal zone, where inundation occurs for shorter periods and less frequently than in the lower intertidal (where mangroves are situated) making vertical accretion of soil limited.

Total soil carbon stocks in our study were like global values reported for mangroves and lower than values reported for saltmarshes and seagrasses. Comparisons of values organic carbon and total carbon stocks from other studies can be found in Table 18.4. Global scale variability in carbon stocks has been linked to factors such as rainfall, temperature, latitude, tidal range, and soil formation (Chmura et al., 2003; Ewers Lewis et al., 2019). Low rainfall and evapotranspiration in the semiarid countries potentially reduce carbon storage in blue carbon habitats (Adame et al., 2013; Ezcurra et al., 2016; Benson et al., 2017; Schile et al., 2017), and this is attributable to differences in decomposition rates between dry and waterlogged soil (Sanders et al., 2016; Radabaugh et al., 2017).

Low carbon stocks in our study could be linked to low rainfall, as Nxaxo Estuary is in the warm temperate region and is close to the southern latitudinal limit for mangroves along the east coast of Africa (Steinke et al., 1995; Adams et al., 2004). This region only receives an annual average rainfall of ~900 mm/yr, which is considerably drier than subtropical and tropical sites. Mangrove and saltmarsh soil carbon stocks differ with latitude and latitudinal gradients (Twilley et al., 1992; Hayes et al., 2017). Subtropical and tropical mangroves tend to have higher productivity compared with warm temperate mangroves and saltmarshes. Mangroves at their southern distributional limit are generally slow growing and their ecological functioning is like that of very young forests (Alongi, 2009; Alongi et al., 2016; Hayes et al., 2017). This has been reported in some studies where there were no differences between mangrove and saltmarsh carbon storage (Kelleway et al., 2016; Hayes et al., 2017; Lewis et al., 2018).

Soil grain size has been highlighted as a strong predictor of carbon storage in coastal habitats as it influences accumulation of organic particles. Numerous studies have directly correlated soil grain size and carbon storage blue carbon habitats (i.e., Dahl et al., 2016; Serrano et al., 2012; van Ardenne et al., 2018). Soil carbon concentration has been connected to sand and silt content along with soil bulk density. Sandy or coarse-grained soils are highly permeable with more aeration, which increases remineralization of carbon due to the environment becoming more reduced (Howarth and Hobbie, 1982; van Ardenne et al., 2018). Fine-grained soil accumulates and preserves more organic matter due to their high surface area. Therefore, soil with fine particles and high organic matter reduces oxygen (which is consumed by detritivores) in the soil, decreasing the permeability and decomposition of organic matter (Dahl et al., 2016 and references therein). While in our study, grain size and other environmental factors were not specifically examined, however, (Hoppe-Speer and Adams, 2015) found that mangrove soil at Nxaxo Estuary was composed mainly of sand (55%) and saltmarsh soil was made up of sand (42%) (Julie, 2018). Therefore, we speculate that sandy soils combined with low rainfall in both mangroves and saltmarshes could be partly responsible for the low carbon stocks at Nxaxo Estuary.

Seagrass carbon storage was also low, and this was expected as *Z. capensis* is a smaller, short lived species which is highly unlikely to accumulate large carbon stores (Fourqurean et al., 2012). Our findings were like Schile et al. (2017) who concluded that low carbon storage

TABLE 18.4 Summary data from literature for organic carbon (C_{org}%) and total carbon stocks (Mg/ha) for different species of mangrove, saltmarsh, and seagrass soil.

	Mangroves			Saltmarsh			Seagrass			
Region/ location	**Species**	**C_{org} (%)**	**Total carbon stock (Mg/ ha)**	**Species**	**C_{org} (%)**	**Total carbon stock (Mg/ha)**	**Species**	**C_{org} (%)**	**Total carbon stock (Mg/ha)**	**Reference**
Global	Multiple	2.2	761.4	Multiple	Data not available	162; 593	Multiple species	2.5	<100	Kristensen et al. (2008); Duarte et al. (2013); Alongi (2018)
United Arab Emirates	*Avicennia marina*	2.2	156.3	Data not available	Data not available	Data not available	Multiple	0.6	49.1	Schile et al. (2017)
Arabian Gulf	*A. marina* and *Rhizophora mucronata*	0.36 −0.87	76	Data not available	Data not available	Data not available	Data not available	Data not available	Data not available	(Cusack et al., 2018)
United States	Multiple	5.09	133.6	Multiple	16.3; 4.7	29.8; 66	Data not available	Data not available	Data not available	Radabaugh et al. (2017); Drake et al. (2015); Macreadie et al. (2013)
Madagascar	Multiple	3.4; 3.5	457; 113 −797							Jones et al. (2014); Benson et al. (2017)
Tanzania, Mozambique							Multiple	0.20 to 1.44	85.6	(Gullström et al., 2018)
Mozambique	Multiple	1.8	484							Stringer et al. (2015)
Australia	*A. marina*	3.9	65.6; 242	Multiple species	2.3; 53	239; 164; 87.1	*Posidonia australis; Zostera muelleri* Multiple	2.24; 1.33	108:24.3	Kelleway et al. (2016); Rozaimi et al. (2016); Hayes et al. (2017); Lewis et al. (2018)
South Africa	*A. marina*	1.96	151.67	*Salicornia tegetaria*	1.3	2.61	*Zostera capensis*	0.86	5.1	This study

was because of smaller sized seagrasses. Studies have also shown a clear difference in soil carbon storage between meadows with larger seagrass species (e.g., *Posidonia* spp. and *Amphibolis* spp.) and those with smaller ones (e.g., *Zostera* spp.) (Dahl et al., 2016). *Posidonia oceanica* meadows, for example, have been reported in several studies for having the largest carbon sequestration and storage capacity (Duarte et al., 2005; Lavery et al., 2013; Röhr et al., 2016). This is because larger seagrass species develop higher belowground biomass compared with smaller sized seagrasses and therefore contribute to higher soilary organic carbon (Duarte and Chiscano, 1999; Gullström et al., 2018).

We found that the *Z. capensis* beds at Nxaxo Estuary occurred in large patches. According to Ricart et al. (2015, 2017), patchy seagrass meadows tend to store less carbon per unit area compared with continuous meadows as a result of less retention of autochthonous material (i.e., leaf detritus). Additionally, in patchy meadows, the age and development of the patch potentially influences carbon storage. Older and more stable patches store larger amounts of carbon than younger and more dynamic patches (Duarte et al., 2013; Greiner et al., 2013; Marbà et al., 2014; Ricart et al., 2015). For this study, there is uncertainty regarding the age and stability of the seagrass that was sampled due to limited availability of past areal data. Also, in South African estuaries, *Z. capensis* beds are highly dynamic and it is often difficult to map their habitat changes over time as area cover alternates in response to freshwater inflow and marine soil input (Adams, 2016). Ricart et al. (2015) also suggested that newly established seagrass patches store carbon mostly in the aboveground biomass, and as the patches got older, carbon storage in the soil increases. Therefore, our findings showed that organic carbon was higher in the aboveground biomass (~20%) compared with the soil (<1%) of the seagrass, which may indicate that the seagrass patches at Nxaxo Estuary are relatively young.

In seagrasses, carbon storage has been related to structural complexity. Those seagrass meadows with a low structural complexity (such as those with one pioneer species that is often small and occurs in patches) are prone to increased erosion and subsequent loss of particulate organic carbon. In this study, our seagrass meadow is of low structural complexity as it has only had one species (*Z. capensis*). This species has short and narrow leaves at shallow, exposed sites as a morphological adaptation to drying out from exposure and long and broad leaves in sites that are calm with deeper depths (Talbot and Bate, 1987; Adams, 2016). Therefore, meadows with low structural complexity such as those at Nxaxo Estuary are associated with low carbon storage. This is aligned with findings by both Rozaimi et al. (2013) and Samper-Villarreal et al. (2016) who reported low carbon storage in meadows with one small pioneer species and linked low structural complexity to increased flow rates, less deposition, and increased carbon loss from erosion (Fonseca and Fisher, 1986; Bos et al., 2007; Samper-Villarreal et al., 2016). Despite the growing body of literature that quantifies carbon storage in seagrasses, many of these studies could not adequately explain the factors that influence carbon accumulation and storage. A recent study on blue carbon storage *Zostera marina* meadows in the United Kingdom showed that even differences in landscapes and habitat conditions may not necessarily influence carbon storage and concluded that factors that determine carbon storage are still poorly understood (Green et al., 2018). It has also been suggested that when global seagrass carbon estimates are made, there should be a clear distinction between estimates for patchy versus continuous meadows and smaller versus larger species for more accurate comparisons.

The relationship between C_{org}% and OM% was strong for saltmarsh soil and like the relationship ($R^2 = 0.99$) reported by Craft et al. (1991) and Howard et al. (2014). We also found that 78% of the organic content was carbon, which was more than double the 46% reported by Howard et al. (2014), and this indicated the importance of region-specific data because using the conversion factor of 0.46 would have further reduced the carbon stock value. For seagrass, this relationship was weak and required natural logarithmic transformation to increase the strength, but it was still less than the global value ($R^2 = 0.56$ vs. $R^2 = 0.96$) (Fourqurean et al., 2012; Howard et al., 2014). In mangroves, the relationship was also weak compared to the global equation ($R^2 = 0.59$ vs. $R^2 = 0.35$) (Kauffman et al., 2011; Howard et al., 2014).

Although much of the carbon in blue forests is stored in the soil, the vegetative component is still a small but significant component that requires accurate quantification (Radabaugh et al., 2017). This study was the first attempt to develop an allometric equation for saltmarsh carbon estimates in South Africa's blue carbon habitats. This contributes to the global availability of allometric equations and carbon data for saltmarsh to aid with quicker and less destructive quantification of aboveground vegetative stocks in future. We found a good relationship between stem height and biomass ($R^2 = 0.706$). However, plant height and biomass may vary seasonally because of changes in growth rates, flowering, and senescence (Reidenbaugh, 1983; Gonzalez Trilla et al., 2013; Radabaugh et al., 2017), which may influence such allometric equations (Radabaugh et al., 2017). The presence (or absence) of an inflorescence at the time of collection has been shown to influence plant height and biomass, thus decreasing the predictive ability of the allometric equation (Gonzalez Trilla et al., 2013; Radabaugh et al., 2017). In this study, there is no deviation from the expected biomass value as a result of flowering because *S. tegetaria* has terminal inflorescences that are generally small (8 mm in length and 2 mm in diameter). Other factors that contribute to the strength of the allometric equation include plant elevation and other abiotic factors (Morris and Haskin, 1990; Radabaugh et al., 2017). For example, *S. alterniflora* is taller in tidal creeks and shorter in more landward areas due to differences in soil parameters such as nutrient availability, salinity, and moisture content (Reidenbaugh, 1983; Clewell, 1997; Wieski and Pennings, 2014). As *S. tegetaria* is usually found in the low- to mid-intertidal zone in most estuaries in South Africa (Steffen et al., 2009), it is generally the same size and the allometric equation we developed would still be accurate to use. Although all factors are important, the variations are usually minor considerations, making the relationship between plant height and biomass still quite robust (Gonzalez Trilla et al., 2013; Radabaugh et al., 2017).

There have been very few published data sets existing for carbon content for saltmarshes (Radabaugh et al., 2017). As a result, the most used value for comparison is 45% from Howard et al. (2014). However, this value is not the most accurate as it is based on a study done in China for terrestrial systems (Fang et al., 1996). We found that saltmarsh organic carbon content for this study was much lower (25.34%) than the value from Howard et al. (2014). Elemental composition of *S. tegetaria* was also less than saltmarsh species in Tampa Bay, where 18 saltmarsh species were found to be composed of 41.1% ± 5.5% carbon. In south east Australia, all saltmarsh species had an average carbon content of 38.3% ± 2.5%, and *Salicornia quinqueflora*, which is a higher intertidal species and that is shrubbier than the succulent low-growing *S. tegetaria*, had a value of 34.4% ± 0.8% (Owers et al., 2018). Saltmarsh aboveground biomass estimates in this study were like studies done in Australia, as both were classified within the temperate saltmarsh group (Adam, 1990; Kelleway et al., 2016).

Our results showed that aboveground biomass carbon stock value was 26.1 ± 2.47 Mg C/ha and this was within the range (22–88 Mg C/ha) reported for Australia (Owers et al., 2018).

For seagrasses, the average global standing stock has been estimated at 460 g DW/m^2; however, this value was derived using very few studies from Africa (Duarte and Chiscano, 1999; Githaiga et al., 2016), which emphasized the need for more comprehensive assessments of seagrass biomass and blue carbon storage in Africa. We found that *Z. capensis* aboveground biomass averaged 354.5 ± 86.47 g/m^2, which was smaller than the global estimate. The biomass in our study was, however, larger than what has been reported for *Z. capensis* in other estuaries in South Africa. Grindley (1976) reported aboveground biomass to be 206 g/m^2 at Knysna Estuary, 217 g/m^2 at Langebaan Lagoon (Christie, 1981), and 55–105 g/m^2 at Kromme Estuary (Hanekom and Baird, 1988). According to Githaiga et al. (2016), seagrass aboveground biomass for South Africa averaged at 413.3 g/m^2, which was larger than the biomass at Nxaxo. Githaiga et al. (2016) also reported that belowground biomass is generally higher (474.6 g/m^2) than aboveground biomass (174.4 g/m^2); however, in this study, seagrass above- and belowground biomass was similar. Biomass was generally less in smaller species such as *Z. capensis* compared with larger species due to a slow turnover of their belowground material (Duarte and Chiscano, 1999; Githaiga et al., 2016). Seagrass biomass carbon in our study was similar to the global average (1.76 Mg C/ha) reported by Fourqurean et al. (2012) but the aboveground biomass carbon in mangroves was much smaller (6.08 ± 1.83 Mg C/ha) compared with other regions.

Mangrove aboveground biomass can be rather variable; the general trend though is that larger mangrove forests in the tropical regions have an aboveground biomass of ~500 Mg C/ha, while smaller dwarfed mangroves have a biomass of ~8 Mg C/ha (Kauffman and Cole, 2010; Kauffman et al., 2014; Howard et al., 2014; Alongi et al., 2016). Data for mangrove aboveground biomass were quite scarce for the temperate region compared with the other two habitats, and this is mainly because the aboveground biomass is difficult to quantify and requires extensive resources (Owers et al., 2018). The available estimates can be quite variable and largely dependent on mangrove forest structure, species composition, and sampling design (Stringer et al., 2015). For example, Gazi Bay in Kenya (a well-studied mangrove area), earlier estimates reported 125 and 226 Mg C/ha (Slim et al., 1996; Kirui et al., 2006), but later only 67 Mg C/ha was reported (Cohen et al., 2013; Stringer et al., 2015), which highlighted that these estimates can be quite variable.

The aboveground biomass carbon stock in our study was slightly less than *A. marina* in the United Arab Emirates (77.6 Mg C/ha) (Schile et al., 2017), and we suspected that this was due to the stunted nature of the trees at Nxaxo Estuary. Stunting has been linked to occurrence at higher elevations (that lack tidal influence), limited nutrients (such as nitrogen and phosphorus) (Feller et al., 2003; Naidoo, 2016), hypersalinity, soil redox potential, and waterlogging (Naidoo, 2016). The stunting at Nxaxo Estuary was further exacerbated by cattle browsing, which further limited growth and caused changes in mangrove morphology because of top-bottom browsing (Hoppe-Speer and Adams, 2015). Although at present, the cattle have been excluded from the mangrove area and the forest has shown signs of recovery (Mbense, 2017), we speculated that the cattle browsing in the past may have contributed to stunting and low aboveground carbon stocks; however, this was not been explicitly tested in this study. In agreement with our findings, (Benson et al., 2017) reported that harvesting pressure by local communities led to major alterations in mangrove forest structure and a significant reduction in the aboveground biomass carbon storage in Madagascar. Harvesting pressure altered dense, closed canopies with tall trees to sparse, short trees with open canopies. The former had significantly more aboveground carbon (127.95 Mg C/ha) and the latter had much less (4.87 Mg C/ha). This verified that degradation can indeed alter mangrove forest structure and ultimately carbon storage.

The overall biomass carbon stock in our study was within range of what was reported for *A. marina* in Australia (495–1047 Mg C) by Owers et al. (2018). We expected similar values to Australia, because mangrove productivity (based on litter fall) is less at higher latitudes with low temperatures (<30°C) (Bouillon et al., 2008; Naidoo, 2016) and resource allocation toward leaf production is limited by the short growth season (Morrisey et al., 2010; Naidoo, 2016). However, because our mangroves are stunted and occur at their distributional limit, resources could be shifted toward the belowground components, which potentially have a greater proportion of carbon belowground (Naidoo, 2009, 2016). However, belowground biomass (i.e., roots) was not measured in this study because it is difficult and destructive to sample.

5. Conclusion

Although blue carbon studies are becoming increasingly topical, only a few studies have done a comprehensive assessment of carbon storage in all three habitats. We found that values of carbon storage for mangroves, saltmarshes, and seagrasses are significantly lower than global means from other studies. However, these findings will still contribute toward the global understanding of blue carbon storage in warm temperate habitats and mangroves at their distributional limits. Granting that, at present, South African policies have not directly addressed the value of carbon storage in coastal wetlands, there is future opportunities for the protection of these habitats through participation in carbon trading, offset schemes, and payments for ecosystems services. As outlined in our findings, carbon storage at higher latitudes was low because of less productivity, thus making these habitats more vulnerable to threats such as global warming and sea-level rise. Sea-level rise and climate change might result in habitat switching and encroachment by mangroves on saltmarsh habitats (Saintilan et al., 2014). Future studies should expand to other estuaries in South Africa for comparative analysis and to create a more robust blue carbon inventory.

References

Adam, P. (Ed.), 1990. Salt Marsh Ecology. Cambridge University Press, Cambridge, UK.

Adame, M.F., Kauffman, J.B., Medina, I., Gamboa, J.N., Torres, O., Caamal, J., et al., 2013. Carbon stocks of tropical coastal wetlands within the karstic landscape of the Mexican Caribbean. PLoS One 8, e56569.

Adams, J.B., Colloty, B.M., Bate, G.C., 2004. The distribution and state of mangroves along the coast of Transkei, Eastern Cape Province, South Africa. Wetl. Ecol. Manag. 12, 531–541.

Adams, J.B., 2016. Distribution and status of *Zostera capensis* in South Africa. South Afr. J. Bot. 107, 63–73.

Adams, J.B., Veldkornet, D., Tabot, P., 2016. Distribution of macrophyte species and habitat in South African estuaries. South Afr. J. Bot. 107, 5–11.

Adams, J.B., Raw, J.L., Mbense, S.P., Bornman, T.G., Rajkaran, A., Van, Niekerk., 2019. Climate change and South Africa's blue carbon ecosystems. Water Research Commission Report No. K5/2769. Water Research Commission. South Africa, Pretoria.

Alongi, D.M., 2018. Introduction. In: Blue Carbon. Springer Briefs in Climate Studies. Springer, Cham. https://doi.org/10.1007/978-3-319-91698-9_1.

Alongi, D.M., 2009. The Energetics of Mangrove Forests. Springer, Amsterdam.

Alongi, D.M., Murdiyarso, D., Fourqurean, Kauffman, J.B., Hutahaean, A., Crooks, S., et al., 2016. Indonesia's blue carbon: a globally significant and vulnerable sink for seagrass and mangrove carbon. Wetl. Ecol. Manag. 24, 3–13.

Beaumont, N.J., Jones, L., Garbutt, A., Hansom, J.D., Toberman, M., 2014. The value of carbon sequestration and storage in coastal habitats. Estuar. Coast Shelf Sci. 32–40.

Benson, L., Glass, L., Jones, T.G., Ravaoarinorotsihoarana, L., Rakotomahazo, C., 2017. Mangrove carbon stocks and ecosystem cover dynamic in Southwest Madagascar and the implications for local management. Forests 8 (190), 1–21.

Blue carbon. In: Nellemann, C., Corcoran, E., Duarte, C.M., Valdés, L., DeYoung, C., Fonseca, L., et al. (Eds.), 2009. A Rapid Response Assessment. GRID-Arendal: United Nations Environment Programme. ISBN: 978-82-7701-060-1.

Bos, A.R., Bouma, T.J., De Kort, G.L.J., Van Katwijk, M.M., 2007. Ecosystem engineering by annual intertidal seagrass beds: soil accretion and modification. Estuar. Coast Shelf Sci. 74, 344–348.

Bosire, J.O., Bandeeira, S., Rafael, J., 2012. Coastal Climate Mitigation and Adaptation through REDD+ Carbon Programs in Mangroves in Mozambique: Plot in the Zambezi Delta. Determination of Carbon Stocks through Localized Allometric Equations Component. WWF, p. 27.

Bosire, J.O., Mangora, M.M., Banderia, S., Rajkaran, A., Ratsimbazafy, R., Appadoo, C., Kairo, J.G. (Eds.), 2016. Mangroves of the Western Indian Ocean: Status and Management. WIOMSA, Zanzibar Town, p. 8.

Bouillon, S., Borges, A.V., Castaneda-Moya, E., Diele, K., Dittmar, T., Duke, N.C., 2008. Mangrove production and carbon sinks: a revision of global budget estimates. Global Biogeochem. Cycles 22, GB2013.

Briggs, D., 1977. Sources and Methods in Geography. Butterworths, London, p. 123.

Chastain, S.G., Kohfeld, K., Pellat, M.G., 2018. Carbon Stocks and accumulation rates in salt marshes of the Pacific Coast of Canada. Biogeosci. Discuss. https://doi.org/10.5194/bg-2018-166.

Chave, J., Réjou-Méchain, M., Búrquez, A., Chidumayo, E., Colgan, M.S., Delitti, W.B., et al., 2014. Improved allometric models to estimate the above ground biomass of tropical trees. Global Change Biol. 20 (10), 3177–3190.

Chmura, G.L., Anisfeld, S.C., Cahoon, D.R., Lynch, C.J., 2003. Global carbon sequestration in tidal, saline wetland soils. Global Biogeochem. Cycles 17, 1111.

Christie, N.D., 1981. Primary production in the Langebaan lagoon. In: Day, J.H. (Ed.), Estuarine Ecology with Particular Reference to Southern Africa, Cape Town, Balkema, pp. 101–115.

Clewell, A.F., 1997. Vegetation. In: Coultas, C.L., Hsieh, Y.-P. (Eds.), Ecology and Management of Tidal Marshes: A Model from the Gulf of Mexico. St. Lucie Press, Delray Beach, pp. 77–110.

Cohen, R., Kaino, J., Okello, J.A., Bosire, J.O., Kairo, J.G., Huxham, M., Mencuccini, M., 2013. Propagating uncertainty to estimates of above-ground biomass for Kenyan mangroves: a scaling procedure from tree to landscape level. For. Ecol. Manag. 310, 968–982.

Craft, C.B., Seneca, E.D., Broome, S.W., 1991. Loss on ignition and Kjeldahl digestion for estimating organic carbon and total nitrogen in estuarine marsh soils: calibration with dry combustion. Estuaries 14, 175–179.

Cusack, M., Saderne, V., Arias – Ortiz, A., Masque, P., Krishnakumar, P.K., Rabaoui, L., Qurban, M.A., Qasem, A.M., Prihartato, P., Loughland, R.A., Elyas, A.,A., Duarte, C.M., 2018. Organic carbon sequestration and storage in vegetated coastal habitats along the western coast of the Arabian Gulf. Environ. Res. Lett. 13.

Dahl, M., Deyanova, D., Lyimo, L.D., Näslund, J., Samuelsson, G.S., Mtolera, M.S.P., Björk, M., Gullström, M., 2016. Effects of shading and simulated grazing on carbon sequestration in a tropical seagrass meadow. J. Ecol. 104, 654–664.

Donato, D.C., Kauffman, J.B., Murdiyarso, D., Kurnianto, S., Stidaham, M., Kanninen, M., 2011. Mangroves among the most carbon-rich forests in the tropics. Nat. Geosci. 4, 293–296.

Drake, K., Halifax, H., Adamowicz, S.C., Craft, C., 2015. Carbon sequestration in tidal salt marshes of Northeast United States. Environ. Manag. 56, 998–1008.

Duarte, C.M., Middelburg, J.J., Caraco, N., 2005. Major role of marine vegetation on the oceanic carbon cycle. Biogeosciences 2, 1–8.

Duarte, C.M., 2017. Reviews and syntheses: hidden forests, the role of vegetated coastal habitats in the ocean carbon budget. Biogeosciences 14, 301–310.

Duarte, C.M., Chiscano, C., 1999. Seagrass biomass and production: a reassessment. Aquat. Bot. 66, 225–239.

Duarte, C.M., Kennedy, H., Marba, N., Hendriks, I., 2011. Assessing the capacity of seagrass meadows for carbon burial: current limitations and future strategies. Ocean Coast Manag. 83, 32–38.

Duarte, C.M., Losada, I.J., Hendriks, I.E., Mazarrasa, I., Marbà, N., 2013. The role of coastal plant communities for climate change mitigation and adaptation. Nat. Clim. Change 3, 961–968.

Els, J., 2019. Carbon and Nutrient Storage of the Swartkops Estuary Salt Marsh and Seagrass Habitats (M.Sc. thesis). Nelson Mandela University, Port Elizabeth, South Africa.

Ewers Lewis, C.J., Young, M.A., Ierodiaconou, D., Baldock, J.A., Hawke, B., Sanderman, P.E., Carnell1, P.E., Macreadie, P.I., 2019. Drivers and modelling of blue carbon stock variability. Biogeosci. Discuss. https://doi.org/10.5194/bg-2019-294.

Ezcurra, P., Ezcurra, E., Garcillán, P.P., Costa, M.T., Aburto-Oropeza, O., 2016. Coastal landforms and accumulation of mangrove peat increase carbon sequestration and storage. Proc. Natl. Acad. Sci. U.S.A. 113, 4404–4409.

Fang, J., Liu, G., Xu, S., 1996. Carbon pools in terrestrial ecosystems in China. In: Wang, G.C., Wen, Y.P. (Eds.), Emissions and Their Relevant Processes of Greenhouse Gases in China. China Environment Science Press, Beijing, pp. 109–128.

Feller, I.C., Whigham, D.F., Mckee, K.L., Lovelock, C.E., 2003. Nitrogen limitation of growth and nutrient dynamics in a disturbed mangrove forest, Indian River Lagoon, Florida. Oecologia 134, 405–414.

Fonseca, M., Fisher, J.S., 1986. A comparison of canopy friction and soil movement between four species of seagrass with reference to their ecology and restoration. Mar. Ecol. Prog. Ser. 29, 15–22.

Fourqurean, J.W., Duarte, C.M., Kennedy, H., Marbà, N., Holmer, M., Mateo, M.A., et al., 2012. Seagrass ecosystems as a globally significant carbon stock. Nat. Geosci. 5, 505–509.

Githaiga, M.N., Gilpin, L., Kairo, J.G., Huxham, M., 2016. Biomass and productivity of seagrasses in Africa. Bot. Mar. 59 (2–3), 173–186.

Gonzalez Trilla, G.G., Marta Borro, M., Soledad Morandeira, N., Schivo, F., Kandus, P., Marcovecchio, J., 2013. Allometric scaling of dry weight and leaf area for *Spartina densiflora* and *Spartina alterniflora* in two southwest Atlantic saltmarshes. J. Coast Res. 29, 1373–1381.

Green, A., Chadwick, M.A., Jones, P.J.S., 2018. Variability of UK seagrass soil carbon: implications for blue carbon estimates and marine conservation management. PLoS One 13 (9), e0204431. https://doi.org/10.1371/journal.pone.0204431.

Greiner, J.T., McGlathery, K.J., Gunnell, J., McKee, B.A., 2013. Seagrass restoration enhances "blue carbon" sequestration in coastal waters. PLoS One 8, e72469. https://doi.org/10.1371/journal.pone.0072469.

Gress, S.K., Huxham, M., Kairo, J.G., Mugi, L.M., Briers, R.A., 2016. Evaluating, predicting and mapping below ground carbon stores in Kenyan mangroves. Global Change Biol. 23, 224–234.

Grindley, J.R., 1976. Report on Ecology of Knysna Estuary and Proposed Braamekraal Marina (Internal report). School of Environmental Studies, University of Cape, Town, South Africa, p. 133.

Gullström, M., Lyimo, L.D., Dahl, M., et al., 2018. Blue Carbon Storage in Tropical Seagrass Meadows Relates to Carbonate Stock Dynamics, Plant–Sediment Processes, and Landscape Context: Insights from the Western Indian Ocean. Ecosystems 21, 551–566.

Hanekom, N., Baird, D., 1988. Distribution and variations in seasonal biomass of eelgrass *Zostera capensis* in the Kromme Estuary, St Francis Bay, South Africa. South African Journal of Marine Science 7:1, 51–59. https://doi.org/10.2989/025776188784379099.

Harrison, T.D., 2004. Physico-chemical characteristics of South African estuaries in relation to the zoogeography of the region. Estuar. Coast Shelf Sci. 61, 73–87.

Hayes, M.A., Jesse, A., Hawke, B., Baldock, J., Tabet, B., Lockington, D., Lovelock, C.E., 2017. Dynamics of sediment carbon stocks across intertidal wetland habitats of Moreton Bay, Australia. Global Change Biol. 23, 4222–4234.

Hoppe-Speer, S., 2013. Response of Mangroves in South Africa to Anthropogenic and Natural Impacts. Nelson Mandela Metropolitan University, South Africa, p. 174.

Hoppe-Speer, S.C., Adams, J.B., 2015. Cattle browsing impacts on stunted *Avicennia marina* mangrove trees. Aquat. Bot. 121, 9–15.

Howard, J., Hoyt, S., Isensee, K., Telszewski, M., Pidgeon, E., 2014. Coastal blue carbon: methods for assessing carbon stocks and emissions factors in mangroves, tidal salt marshes, and seagrass meadows. In: Conservation International, Intergovernmental Oceanographic Commission of UNESCO, International Union for Conservation of Nature. Arlington, Virginia, USA.

Howarth, R.W., Hobbie, J.E., 1982. The Regulation of decomposition and heterotrophic microbial activity in salt marsh soils: a review. In: Kennedy, V.S. (Ed.), Estuarine Comparisons. Academic Press, pp. 183–207. https://doi.org/10.1016/B978-0-12-404070-0.50017-X.

Johnson, J., 2019. A Quantitative Assessment of Blue Carbon for Mangroves in a Warm- Temperate Estuary of South Africa (M.Sc. thesis). Nelson Mandela University, Port Elizabeth.

Johnson, J.L., Raw, J.L., Adams, J.B., 2020. First report on carbon storage in a warm-temperate mangrove forest in South Africa. Estuarine. Coastal and Shelf Science 235, 106556.

Jones, T.G., Ratsimba, H.R., Ravaoarinorotsihoarana, L., Cripps, G., Bey, A., 2014. Ecological variability and carbon stock estimates of mangrove ecosystems in Northwestern Madagascar. Forests 5, 177–205.

Julie, C.L., 2018. Carbon storage across a salt marsh – mangrove ecotone (A comparative study at the Nxaxo and Nahoon estuaries). BSc Honours Research Project. Nelson Mandela University, Port Elizabeth, South Africa.

Kairo, J.G., Bosire, J., Langat, J., Kirui, B., Koedam, N., 2009. Allometry and biomass distribution in replanted mangrove plantations at Gazi Bay, Kenya. Aquat. Conserv. 19, S63–S69.

Kairo, J.G., Joseph, K.S.L., Dahdouh-Guebas, F., Bosire, J., Karachi, M., 2008. Structural development and productivity of replanted mangrove plantations in Kenya. For. Ecol. Manag. 255, 2670–2677.

Kauffman, J.B., Cole, T.G., 2010. Micronesian mangrove forest structure and tree responses to a severe typhoon. Wetlands 30 (6), 1077–1084.

Kauffman, J.B., Donato, D.C., 2012. Protocols for the measurement, monitoring and reporting of structure, biomass and carbon stocks in mangrove forests. In: Working Paper 86. Center for International Forestry Research, Bogor.

Kauffman, J.B., Heider, C., Cole, T.G., Dwire, K.A., Donato, D.C., 2011. Ecosystem carbon stocks of Micronesian mangrove forests. Wetlands 343–352.

Kauffman, J.B., Heider, C., Norfolk, J., Payton, F., 2014. Carbon stocks of intact mangroves and carbon emissions arising from their conversion in the Dominican Republic. Ecol. Appl. 24, 518–527.

Kelleway, J.J., Saintilan, N., Macreadie, P.I., Baldock, J.A., Ralph, P.J., 2017b. Soil and carbon deposition vary among vegetation assemblages in a coastal salt marsh. Biogeosciences 14, 3763–3779.

Kelleway, J.J., Saintilan, N., Macreadie, P.I., Ralph, P.J., 2016. Soilary factors are key predictors of carbon storage in SE Australian saltmarshes. Ecosystems 19 (5), 865–880.

Kirui, B., Kairo, J.G., Karachi, M., 2006. Allometric equations for estimating above ground biomass of *Rhizophora mucronata* mangroves at Gazi Bay Kenya. WIO J. Mar. Sci. 55 (1), 27–34.

Kristensen, E., Bouillon, S., Dittmar, T., Marchand, C., 2008. Organic carbon dynamics in mangrove ecosystems: a review. Aquat. Bot. 89, 201–219.

Lavery, P.S., Mateo, M.-A.:., Serrano, O., Rozaimi, M., 2013. Variability in the carbon storage of seagrass habitats and its implications for global estimates of blue carbon ecosystem service. PLoS One 8 (9), e73748. http://10.1371/journal.pone.0073748.

Lewis, C.J.E., Carnell, P.E., Sanderman, J., Baldock, J.A., Macreadie, P.I., 2018. Variability and vulnerability of coastal 'blue carbon' stocks: a case study from Southeast Australia. Ecosystems 21, 263–279.

Macreadie, P.I., Baird, M.E., Trevathan-Tackett, S.M., Larkum, A.W.D., Ralph, P.J., 2014. Quantifying and modelling the carbon sequestration capacity of seagrass meadows–a critical assessment. Mar. Pollut. Bull. 83 (2), 430–439.

Macreadie, P.I., Hughes, A.R., Kimbro, D.L., 2013. Loss of 'blue carbon' from coastal salt marshes following habitat disturbance. PLoS One 8 (7), e69244. https://doi.org/10.1371/journal.pone.0069244.

Macreadie, P.I., Ollivier, Q.R., Kelleway, J.J., Serrano, O., Carnell, P.E., Lewis, C.E., et al., 2017. Carbon sequestration by Australian tidal marshes. Sci. Rep. 7, 44071pp.

Marbà, N., Díaz-Almela, E., Duarte, C.M., 2014. Mediterranean seagrass (*Posidonia oceanica*) loss between 1842 and 2009. Biol. Conserv. 176, 183–190.

Mbense, S., 2017. The Growth and Recovery of Mangroves at Three South African Study Sites (M.Sc. thesis). Nelson Mandela Metropolitan University, Port Elizabeth.

Mcleod, E., Chmura, G.L., Bouillon, S., Salm, R., Björk, M., Duarte, C.M., et al., 2011. A blueprint for blue carbon: toward an improved understanding of the role of vegetated coastal habitats in sequestering CO_2. Front. Ecol. Environ. 9 (10), 552–560.

Morris, J.T., Haskin, B., 1990. A 5-yr record of aerial primary production and stand characteristics of Spartina alterniflora. Ecology 71 (6), 2209–2217.

Morrisey, D.J., Swales, A., Dittmann, S., Morrison, M.A., Lovelock, C.E., Beard, C.M., 2010. The ecology and management of temperate mangroves. In: Gibson, R.N., Atkinson, R.J.A., Gordon, J.D.M. (Eds.), Oceanography and Marine Biology: an Annual Review, vol. 48. Crc Press-Taylor and Francis Group, Boca Raton, pp. 43–160.

Mudd, S.M., Howell, S.M., Morris, J.T., 2009. Impact of dynamic feedbacks between soilation, sea-level rise, and biomass production on near-surface marsh stratigraphy and carbon accumulation. Estuar. Coast Shelf Sci. 82 (3), 377–389.

Naidoo, G., 2009. Differential effects of nitrogen and phosphorus enrichment on growth of dwarf *Avicennia marina* mangroves. Aquat. Bot. 90, 184–190.

Naidoo, G., 2016. The mangroves of South Africa: an ecophysiological review. South Afr. J. Bot. 107, 101–113.

Njana, M.A., Eid, T., Zahabu, E., Malimbwi, R., 2015. Procedures for quantification of below ground biomass of three mangrove tree species. Wetl. Ecol. Manag. 23, 749–764.

Osland, M.J., Gabler, C.A., Grace, J.B., Day, R.H., McCoy, M.L., McLeod, J.L., Andrew, S., From, A.S., Enwright, N.M., Feher, L.C., Stagg, C.L., Hartley, S.B., 2018. Climate and plant controls on soil organic matter in coastal wetlands. Global Change Biol. 24, 5361–5379. https://doi.org/10.1111/gcb.14376.

Owers, C.J., Rogers, K., Woodroffe, C.D., 2018. Spatial variation of above-ground carbon storage in temperate coastal wetlands. Estuar. Coast Shelf Sci. 210, 55–67.

Pendleton, L., Donato, D.C., Murray, B.C., Crooks, S., Jenkins, W.A., Sifleet, S., Craft, C., Fourqurean, J.W., Kauffman, J.B., Marbà, N., Megonigal, P., Pidgeon, E., Herr, D., Gordon, D., Baldera, A., 2012. Estimating global "blue carbon" emissions from conversion and degradation of vegetated coastal ecosystems. PLoS ONE 7, e43542.

Pidgeon, E., Herr, D., Fonseca, L., 2011. Minimizing Carbon Emissions and Maximizing Carbon Sequestration and Storage by Seagrasses, Tidal Marshes, and Mangroves. Conservation International, Washington, DC.

Radabaugh, K.R., Powell, C.E., Bociu, I., Clark, B.C., Moyer, R.P., 2017. Plant size metrics and organic carbon content of Florida salt marsh vegetation. Wetl. Ecol. Manag. 25, 443–455.

Raw, J.L., Julie, C.L., Adams, J.B., 2019. A comparison of soil carbon pools across a mangrove-salt marsh ecotone at the southern African warm-temperate range limit. South Afr. J. Bot. 127, 301–307.

Reidenbaugh, T.G., 1983. Productivity of cordgrass, *Spartina alterniflora*, estimated from live standing crops, mortality, and leaf shedding in a Virginia salt-marsh. Estuaries 6, 57–65.

Ricart, A.M., Pérez, M., Romero, J., 2017. Landscape configuration modulates carbon storage in seagrass soil. Estuar. Coast Shelf Sci. 69–76.

Ricart, A.M., York, P.H., Rasheed, M.A., Pérez, M., Romero, J., Bryant, C.V., Macreadie, P.I., 2015. Variability of soilary organic carbon in patchy seagrass landscapes. Mar. Pollut. Bull. 100, 476–482.

Rogers, K., Saintilan, N., Woodroffe, C.D., 2014. Surface elevation change and vegetation distribution dynamics in a subtropical coastal wetland: implications for coastal wetland response to climate change. Estuar. Coast Shelf Sci. 149, 46–56.

Rozaimi, M., Lavery, P.S., Serrano, O., Kyrwood, D., 2016. Long-term carbon storage and its recent loss in an estuarine *Posidonia australis* meadow (Albany, Western Australia). Estuar. Coast Shelf Sci. 171, 58–65.

Rozaimi, M., Serrano Gras, O., Lavery, P.S., 2013. Comparison of carbon stores by two morphologically different seagrasses. J. Roy. Soc. West Aust. 96, 81–83.

Röhr, M.E., Boström, C., Canal – Vergés, P., Holmer, M., 2016. Blue carbon stocks in Baltic Sea eelgrass (*Zostera marina*) meadows. Biogeosciences 13, 6139–6153.

Saintilan, N., Rogers, K., Mazumder, D., Woodroffe, C., 2013. Allochthonous and autochthonous contributions to carbon accumulation and carbon store in southeastern Australian coastal wetlands. Estuar. Coast Shelf Sci. 128, 84–92.

Saintilan, N., Wilson, N.C., Rogers, K., Rajkaran, A., Krauss, K.W., 2014. Mangrove expansion and salt marsh decline at mangrove poleward limits. Global Change Biol. 20, 147–157.

Samper-Villarreal, J., Lovelock, C.E., Saunders, M.I., Roelfsema, C., Mumby, P.J., 2016. Organic carbon in seagrass soils is influenced by seagrass canopy complexity, turbidity, wave height, and water depth. Limnol. Oceanogr. https://doi.org/10.1002/lno.10262.

Sanders, C.J., Maher, D.T., Tait, D.R., Williams, D., Holloway, C., Sippo, J.Z., Santos, I.R., 2016. Are global mangrove carbon stocks driven by rainfall? J. Geophys. Res. Biogeosci. 121, 2600–2609.

Serrano, O., Mateo, M.A., Renom, P., Brugues, J.R., 2012. Characterization of soils beneath a *Poisidonia oceanica* meadow. Geoderma 185–189, 26–36.

Schile, L.M., Kauffman, J.B., Crooks, S., Fourqurean, J.W., Glavan, J., Megonigal, J.P., 2017. Limits on carbon sequestration in arid blue carbon habitats. Ecol. Appl. 27, 859–874.

Shapiro, A., Trettin, C., Küchly, H., Alavinapanah, S., Bandeira, S., 2015. The mangroves of the Zambezi Delta from 1994 to 2013: increase in extent observed via satellite. Rem. Sens. 7 (12), 16505–16518.

Siikamäki, J., Sanchirico, J.N., Jardine, S., McLaughlin, D., Morris, D.F., 2012. Blue Carbon: Global Options for Reducing Emissions from the Degradation and Development of Coastal Ecosystems. Resources for the Future, Washington, DC, p. 20036.

Sitoe, A.A., Mandlate, L.J.C., Guedes, B.S., 2014. Biomass and carbon stocks of Sofala Bay mangrove forests. Forests 5, 1967–1981.

Slim, F.J., Gwada, P.M., Kodjo, M., Hemminga, M.A., 1996. Biomass and litterfall of *Ceriops tagal* and *Rhizophora mucronata* in the mangrove forest of Gazi Bay, Kenya. Mar. Freshw. Res. 47, 999–1007.

Sousa, A.I., Santos, D.B., Ferreira da Silva, E., Sousa, L.P., Cleary, D.F.R., Soares, A.M.V.M., Lillebø, A.I., 2017. Blue carbon and nutrient stocks of salt marshes at a temperate coastal lagoon (Ria de Aveiro, Portugal). Sci. Rep. 7.

Steffen, S., Mucina, L., Kadereit, G., 2009. Three new species of Sarcocornia (Chenopodiaceae) from South Africa. Kew Bulletin 64 (3), 447–459.

Steinke, T.D., Ward, C.J., Rajh, A., 1995. Forest structure and biomass of mangroves in the Mgeni estuary, South Africa. Hydrobiologia 295, 159–166.

Stringer, C.E., Trettin, C.C., Zarnoch, S.J., Tang, W., 2015. Carbon stocks of mangroves within the Zambezi river delta, Mozambique. For. Ecol. Manag. 354, 139–148.

Talbot, M.M.B., Bate, G.C., 1987. The distribution and biomass of the seagrass *Zostera capensis* in a warm-temperate estuary. Bot. Mar. 30, 91–99.

Thorhaugh, A.L., Poulos, H.M., López-Portillo, J., Barr, J., Lara-Domínguez, A.L., Ku, T.C., Berlyn, G.P., 2019. Gulf of Mexico estuarine blue carbon stock, extent and flux: mangroves, marshes, and seagrasses: a North American hotspot. Sci. Total Environ. 653, 1253–1261.

Tripathee, R., Schäfer, K.V.R., 2015. Above-and below ground biomass allocation in four dominant salt marsh species of the eastern United States. Wetlands 35 (1), 21–30.

Twilley, R.R., Chen, R.H., Hargis, T., 1992. Carbon sinks in mangroves and their implications to carbon budget of tropical coastal ecosystems. Water Air Soil Pollut. 64, 265–288.

van Ardenne, L.B., Jolicouer, S., Bérubé, D., Burdick, D., Chmura, G.L., 2018. The importance of geomorphic context for estimating the carbon stock of salt marshes. Goederma 330, 264–275.

Wieski, K., Pennings, S.C., 2014. Climate drivers of Spartina alterniflora saltmarsh production in Georgia, USA. Ecosystems 17, 473–484.

Wylie, L., Sutton-Grier, A.E., Moore, A., 2016. Keys to successful blue carbon projects: lessons learned from global case studies. Mar. Pol. 65, 76–84.

Yando, E.S., Osland, M.J., Willis, J.M., Day, R.H., Krauss, K.W., Hester, M.W., 2016. Salt marsh-mangrove ecotones: using structural gradients to investigate the effects of woody plant encroachment on plant–soil interactions and ecosystem carbon pools. Journal of Ecology 104, 1020–1031.

CHAPTER

19

Mangrove carbon sequestration and sediment deposition changes under cordgrass invasion

Luzhen Chen[1], *Yining Chen*[2], *Yihui Zhang*[1], *Hongyu Feng*[1]

[1]Key Laboratory of the Ministry of Education for Coastal and Wetland Ecosystems, College of the Environment and Ecology, Xiamen University, Xiamen, Fujian, China; [2]Second Institute of Oceanography, Ministry of Natural Resources, Hangzhou, Zhejiang, China

OUTLINE

1. Introduction 474

2. Cordgrass invasion status in Chinese mangroves 475
- *2.1 Mangroves of China* 475
- *2.2 Cordgrass and its biological characteristics* 477
- *2.3 Invasive status along Chinese mangroves* 478

3. Biological factors affecting the sediment carbon flux 480
- *3.1 Sediment accumulations* 480
- *3.2 Biomass allocation and productivity* 485
- *3.3 Carbon source changes* 487
- *3.4 Soil CO_2 and methane flux* 487

4. Vegetation–sediment dynamic interactions and the consequent sedimentary processes 488
- *4.1 Hydrodynamic modifications by vegetation: a comparison between mangrove and cordgrass* 489
- *4.2 Sediment settling and trapping by vegetation: a comparison between mangroves and cordgrass* 493
- *4.3 Sediment stability associated with carbon release from the sediments* 495

5. Conceptual model and evaluation of regional carbon storage 496
- *5.1 An evaluation of regional carbon storage changes after cordgrass invasion in China* 496

Dynamic Sedimentary Environments of Mangrove Coasts
https://doi.org/10.1016/B978-0-12-816437-2.00019-7

5.2 *A conceptual model of carbon dynamics in mangroves and saltmarsh ecotones* 498

6. Conclusions and perspectives 500

Acknowledgments 501

References 501

1. Introduction

Mangroves and saltmarshes are economically important ecosystems because they protect shorelines from storms and other coastal hazards, exchange nutrients between land and sea, and provide organic matter and nursery grounds for fish and animals. Mangroves are distributed in tropical to subtropical regions, while saltmarshes colonize tropical to arctic climates where mangroves are absent (Kelleway et al., 2017). However, these two vegetation types can form ecotonal communities in the subtropical intertidal zones where their latitudinal distributions overlap, such as the coasts of China, Africa, Australia, and the United States (see Kelleway et al., 2017).

Mangroves and tidal saltmarshes are also known for their considerable capacity to store and sequestrate carbon (C). In such ecosystems, below-ground C pools are usually the largest pool composed of living and decomposing roots, rhizomes, and leaf litter. In mangrove ecosystems, the below-ground soil C pool comprises 50% to over 90% of the total ecosystem C stocks (Donato et al., 2011). In tidal saltmarshes, the proportional contribution of sediment C, which reaches more than 90% (Johnson et al., 2016), is even higher than that in mangroves. Globally, on average, mangroves store more C than tidal saltmarshes due to their different biomass and architecture (Taillardat et al., 2018).

In the context of climate change, the distribution of mangroves has shown poleward expansion and encroachment into saltmarshes on some coasts in Australia and the United States, which has enhanced the C storage of the saltmarshes (Kelleway et al., 2016; Yando et al., 2016). Furthermore, the soil volume and surface elevation have increased due to the accumulation of mineral sediment and organic matter through the biophysical processes of these wetland plants, thereby leading to ecogeomorphological feedbacks (Friess et al., 2012; Saintilan et al., 2013; Rogers et al., 2014).

Biological and biophysical processes were changed in the mangrove–saltmarsh ecotones of China after the introduction of *Spartina alterniflora* (cordgrass) (Chen, 2019). Since its introduction to China from the East Coast of the United States in 1979, cordgrass has been extensively invading mangrove wetlands on the southern coasts of China by transforming large areas of mudflats into tidal saltmarshes in the 1990s (An et al., 2007; Zuo et al., 2012). It is of wide concern that the advance of cordgrass replacing mangroves will cause serious environmental problems, such as changing C/N dynamics or decreasing the biodiversity of coastal regions (Zhang et al., 2012; Feng et al., 2014, 2017), and change ecosystem services as mentioned in other mangrove–saltmarsh ecotones (Kelleway et al., 2017).

According to the framework of Friess et al. (2012), the ecological interactions between species-specific life-history traits of mangroves–saltmarsh ecotones in China have also been well known (Zhang et al., 2012; Liu et al., 2017a), while biophysical interactions and biogeomorphological feedback processes have received less attention (Chen et al., 2014). The C stock changes in mangroves invaded by cordgrass have piqued interest (Wang et al., 2019), while the surface sediment dynamics at the regional and landscape levels regarding wetland stability and their function as C pools have not been clear. The focus of this chapter is the C dynamics and associated sedimentary processes within mangrove wetlands under cordgrass invasion in China. We review the biochemical and ecogeomorphological feedbacks related to surface sediment C and discuss the biological factors affecting sediment C flux, vegetation–sediment interaction, and the consequent sedimentary processes. A conceptual model of C dynamics and the evaluation of regional C storage in typical mangrove–*Spartina* ecotones are also presented here based on the current evidence.

2. Cordgrass invasion status in Chinese mangroves

2.1 Mangroves of China

Mangroves dominate the coasts of tropical and subtropical regions of the world, covering only 0.1% of the continental surface of the Earth (Spalding et al., 2010). Mangroves in China occur in the Indo West Pacific subregions with a high diversity of species. A total of 27 true mangrove species (including nature hybrids) and 10 mangrove associates are naturally found in China (Table 19.1), accounting for one-third of the mangrove species in the world. Almost all the mangroves are located on the subtropical coasts of China, which are also the northern limits for many mangrove species. In China, mangroves covered >50,000 ha in the 1950s, but the area has been reduced to approximately 25,000 ha currently (Chen et al., 2009; Liao and Zhang, 2014). Similar to the challenges of mangroves globally (Thomas et al., 2017; Friess et al., 2019), the mangrove areas in China shrank or fragmented during the last four decades, and this situation has resulted in great risk for their long-term survival (Chen et al., 2009). The main sources of mangrove deforestation have been linked to the growing economy since the 1950s (He et al., 2014). Along the southeast China coasts where mangroves are distributed, agriculture, aquaculture, and urbanization were the three major pressures on the mangroves from 1950 to the 1990s.

Since 1995, the Biodiversity Conservation Action Plan of China has been calling for increased mangrove conservation areas, and the majority of the natural mangroves have been protected as nature reserves (Chen et al., 2009). Initiated after the 2000s, several programs began reforestation to compensate for the loss of mangrove areas. There are still some ongoing national projects for mangrove restoration in China. Substantial efforts have been made for the protection and restoration of mangroves by governmental and private sectors (Liu et al., 2016a).

Although these actions helped to protect and increase mangrove areas, forest degradation remains a major threat to mangroves in China (Peng et al., 2016). Urbanization with the stresses of eutrophication, noise and insect pests from a large population, and the proximity of human settlements threatens mangroves (Chowdhury et al., 2017). Some mangrove reserves are surrounded by skyscrapers, and others have become tourist sightseeing areas. Eutrophication is especially severe near urban areas and affects the health of ecosystems. Plant diseases and insect pests also reduce the reproduction of mangroves.

TABLE 19.1 Mangrove species and mangrove associate species in China.

No.	Families with mangroves	Species	Mangrove species/Mangrove associate
1	Acanthaceae	*Acanthus ebracteatus*	Mangrove species
2		*Acanthus ilicifolius*	
3	ex Verbenaceae, or Avicenniaceae	*Avicennia marina*	
4	Arecaceae	*Nypa fruticans*	
5	Combretaceae	*Lumnitzera littorea*	
6		*Lumnitzera racemosa*	
7	Euphorbiaceae	*Excoecaria agallocha*	
8	Lythraceae (or Sonneratiaceae)	*Pemphis acidula*	
9		*Sonneratia alba*	
10		*Sonneratia caseolaris*	
11		*Sonneratia ovata*	
12		*Sonneratia* × *gulngai*	
13		*Sonneratia* × *hainanensis*	
14	Malvaceae	*Heritiera littoralis*	
15	Meliaceae	*Xylocarpus granatum*	
16	Myrsinaceae	*Aegiceras corniculatum*	
17	Pteridaceae	*Acrostichum aureum*	
18		*Acrostichum speciosum*	
19	Rhizophoraceae	*Bruguiera gymnorrhiza*	
20		*Bruguiera sexangula*	
21		*Bruguiera* × *rhynchopetala*	
22		*Ceriops tagal*	
23		*Kandelia obovata*	
24		*Rhizophora apiculata*	
25		*Rhizophora stylosa*	
26		*Rhizophora* × *lamarckii*	
27	Rubiaceae	*Scyphiphora hydrophyllacea*	
28	Apocynaceae	*Cerbera manghas*	Mangrove associate
29	Bignoniaceae	*Dolichandrone spathacea*	

TABLE 19.1 Mangrove species and mangrove associate species in China.—cont'd

No.	Families with mangroves	Species	Mangrove species/Mangrove associate
30	Compositae	*Pluchea indica*	
31	Hernandiaceae	*Hernandia nymphaeifolia*	
32	Lecythidaceae	*Barringtonia racemosa*	
33	Leguminosae	*Pongamia pinnata*	
34	Malvaceae	*Hibiscus tiliaceus*	
35		*Thespesia populnea*	
36	Verbenaceae	*Clerodendrum inerme*	
37		*Premna serratifolia*	

Furthermore, biological invasion is common in mangroves in China as another threat to native species and local ecosystems (Chen et al., 2009). *S. alterniflora* (cordgrass), a C_4 grass, was first introduced to China for coastal protection purposes but became a severe threat to mangroves as one of the consequences. The mudflats suitable for mangrove reproduction and waterfowl feeding were occupied by this invasive cordgrass (Fig. 19.1).

2.2 Cordgrass and its biological characteristics

Cordgrass is a long-lived clonal plant. It is an intertidal native species in low-energy environments along the Atlantic and Gulf Coasts of North America and was introduced from the United States (native populations occur in North Carolina, Georgia, and Florida) (An et al.,

FIGURE 19.1 Mangrove–*Spartina* ecotone in Zhangjiang Estuary in China.

2007). Since its introduction to China in 1979, cordgrass has widely spread by natural dispersal and deliberate planting from a midlatitude site in Luoyuan, Fujian, in 1980 (Xu and Zhuo, 1985). To date, it has vigorously grown in China and has spread over most of the Chinese coasts, southward to Hainan Island (19.7 °N) and northward well into Liaoning (39.9 °N), a range of more than 20 degrees of latitude (Liu et al., 2016b). This invasion is the largest and most recent of many substantial invasions of cordgrass around the world (Strong and Ayres, 2013).

Within its native range (~27–45 °N) in North America, cordgrass varies considerably in phenotype across latitude, with smaller plants in high latitudes (Turner, 1976; Kirwan et al., 2009). Different populations across the geographic range vary in stem height and diameter, leaf size, and other traits, and these variations correlate with wide variation in abiotic conditions such as annual temperature, annual growing degree days, total precipitation, tidal range, soil organic content, and water content (Turner, 1976; Kirwan et al., 2009; McCall and Pennings, 2012; Crosby et al., 2017).

In its introduced range (19–40 °N) in China, the phenotypic plasticity of cordgrass in vegetative traits, such as stature, foliage, and biomass, is displayed in this species as in the native range, facilitating the exploitation of similarly variable habitats on the two continents (Strong and Ayres, 2013; Liu et al., 2016b). A geographical survey spanning the entire latitudinal gradients along the east coast of China showed that the plant height in cordgrass varied from 1 to 3 m, the shoot density varied from 100 to 300 shoot m^{-2}, and the above-ground biomass varied from 1000 to 4000 g m^{-2}. Both the plant height and the above-ground biomass displayed a parabolic relationship with latitude, reaching their maximums at intermediate latitudes (~30 °N) (Liu et al., 2016b, 2017a). Moreover, interbreeding among the multiple sources generated a "hybrid swarm" that spread throughout the coast of China. The genetic admixture appeared to have broken the negative correlation between plant height and shoot regeneration, which was genetically based in the native range, and has facilitated the evolution of supercompetitive genotypes in the invasive range (Qiao et al., 2019).

The invasion of cordgrass has caused considerable effects on the invaded regions, such as the transformation of mudflats by trapping sediment, increasing primary productivity, changing the diet of benthic macrofauna, altering C and nitrogen cycling, and threatening native plants in China (Li et al., 2009; Feng et al., 2018). Moreover, the introduced range of cordgrass in China extends to lower latitudes (to ~20 °N) than those in the native range, including extensive areas where mangroves are common (Chen et al., 2009; Zhang et al., 2012), raising the possibility for more complicated latitudinal patterns that have been documented in the native range.

2.3 Invasive status along Chinese mangroves

On the southern coast of China, cordgrass has aggressively invaded stands of native mangroves and formed a mangrove–*Spartina* ecotone during the last two decades (Chen et al., 2009). Cordgrass was introduced to China in 1979 with the goal of protecting the coast from erosion (Xu and Zhuo, 1985). By planting and natural dispersal, cordgrass has spread to cover vast expanses of intertidal mudflat in China (An et al., 2007).

Along the latitudinal gradient, the mangrove–*Spartina* ecotone spans northward to Zhejiang (28.3 °N) and southward to Hainan Island (19.7 °N). The plant height and aboveground biomass of cordgrass decreased with decreasing latitude, whereas the growth, abundance, and biodiversity of mangroves increased with decreasing latitude. High temperatures can reduce cordgrass productivity because the thermal optimum for photosynthesis in cordgrass is approximately 30 °C (Giurgevich and Dunn, 1979; Dunn et al., 1987; Wieski and Pennings, 2014), and the competitive interactions between native mangroves and cordgrass might also resist cordgrass invasion.

A field survey in the Zhangjiang Estuary (117.5 °E, 23.9 °N) at four tidal sites (oligohaline upstream, mesohaline, polyhaline, and euhaline downstream) showed that mangroves and cordgrass broadly cooccurred along the salinity gradient (Zhang et al., 2012). From the oligohaline site upstream of the mesohaline and polyhaline sites in the middle of the estuary, cordgrass occurred on the margins of mangrove forests, within mangrove woodlands with more open canopies, and in disturbed areas. Both mangroves and cordgrass were most productive at the mesohaline site in the Zhangjiang Estuary, perhaps because of the combination of relatively low stress and higher nutrients provided by the turbidity maximum in brackish waters (Loomis and Craft, 2010; Wieski et al., 2010). The exotic cordgrass had a wider distribution range along the estuarine salinity gradient than that of the native mangroves. Cordgrass occurred in the saline coastal areas outside the river mouth, but mangroves could not survive in this habitat. The downstream distribution of estuarine vegetation is limited by tolerance to physiological stressors, with salinity being the major effect (Tomlinson, 1986; Adam, 1990). Cordgrass is known to be quite tolerant of salt stress, both in its original habitat in North America (Hester et al., 2001; Vasquez et al., 2006) and in the introduced habitat in China (An et al., 2007; Li et al., 2009).

Across the intertidal elevation gradient, the field survey at the mesohaline site in the Zhangjiang Estuary showed that cordgrass has invaded mudflats and formed dense stands that extend up to the lower intertidal border of mangroves (Zhu et al., 2019). Cordgrass also occurs in gaps in mangrove stands but cannot invade intact mangrove stands (Zhang et al., 2012; Li et al., 2014). Throughout the mangrove–*Spartina* ecotone in China, cordgrass has not invaded mangrove areas with a closed canopy but has been established in the mangrove zone where the canopy was opened due to human disturbance. The biotic resistance of mangrove forests to plant invasions has generally been thought to be driven by competition (Janzen, 1985; Lugo, 1986; Biswas et al., 2007; Krauss et al., 2008; Fourqurean et al., 2010). As elsewhere, mangrove stands with a closed canopy in southern China often exclude understory plants, including cordgrass and other marsh grasses (Chen et al., 2009). This hypothesis was also supported by a study showing that cordgrass transplants were not established in the understory of mangrove stands with closed canopies (Zhang et al., 2012). Additionally, although cordgrass seeds germinate in the mangrove understory, the survival of seedlings is low (Li et al., 2014), and the herbivory effect of a native rodent (*Rattus losea*) can enhance the resistance of mangrove forests to cordgrass invasion in southern China (Zhang et al., 2018).

Compared to the adult mangrove forest, cordgrass suppressed the growth of the native mangrove species seedlings in transplantation experiments along latitudinal, estuarine salinity, and intertidal elevation gradients (Zhang et al., 2012; Peng et al., 2018). In stressful habitats, cordgrass facilitates the survival of juvenile mangroves in hypersaline habitats (Zhang et al., 2012) or at low intertidal levels (Peng et al., 2018) by protecting them from waves and improving soil oxygenation (McKee et al., 2007a; Friess et al., 2012; Peterson and Bell, 2012; Guo et al., 2013), thereby improving early establishment. However, cordgrass competes with juvenile mangroves elsewhere in more hydrodynamically benign habitats. By suppressing mangrove regeneration, cordgrass poses a geographically widespread threat to these important Chinese plants, to mariculture practiced around them and to the other species that the mangrove ecosystems support.

3. Biological factors affecting the sediment carbon flux

With remarkable CO_2 assimilating capacity in the living biomass and high burial rates, mangroves and tidal saltmarshes are regarded as the most efficient and low-cost ways to remove CO_2 in the atmosphere and mitigate climate change (Nellemann et al., 2009). Dissimilar to terrestrial forests, coastal wetlands store large amounts of C belowground (Mcleod et al., 2011). At the same time, the buried organic matter is slowly decomposed because of the anoxia and saline stress by tidal inundation. These processes make mangroves an effective ecosystem for storing organic C. Thus, the C stored in these ecosystems can be classified into two sources: autochthonous C and allochthonous C (Howard et al., 2014).

The autochthonous C refers to carbon produced and deposited in the same location. Plants assimilate CO_2 through photosynthesis and convert it to organic C, which can be used or stored in leaves, stems, roots, or rhizomes and is allocated to the plant biomass. The dead biomass decomposes at the sediment surface or within the sediment. Waterlogged and anoxic states in the sediment result in slow decomposition rates and low soil C turnover rates, which lead to high autochthonous C burial in these ecosystems (Kennedy, 1995).

The allochthonous C is the carbon produced in one location and deposited in another. Allochthonous C is a unique compound that differs among terrestrial ecosystems. Owing to the complex aerial root systems in mangroves or the dense plant community structure of saltmarsh, these two ecosystems can efficiently trap particles associated with organic C, which are transported by tidal currents or river runoff from adjacent terrestrial or marine ecosystems (Middelburg et al., 1997; Bouillon et al., 2003). This part of C can be deposited as surface sediments and stored in the belowground for long periods of time.

However, the ratios of autochthonous C and allochthonous C are different between these two coastal ecosystems. Even within mangroves or tidal saltmarsh ecosystems, the actual C burial rates are highly site specific. Moreover, C emissions from surface sediment over time also influence C stocks in different ecosystems (Webb et al., 2013; Lovelock et al., 2015; Molly and Törbjorn, 2019). Broad biological and biogeographic drivers (e.g., the factors affecting plant biomass allocation), the surface structure (e.g., the complex root systems of mangroves and the community density of tidal saltmarshes), and the bottom-up effects from plant invasion all affect sediment trapping and result in different sediment C stocks (Kirwan and Megonigal, 2013). In addition, plant community changes after the cordgrass invades mangroves were also proven to change CO_2 and methane flux from the sediment surface (Chen et al., 2014; Tong et al., 2018). Thus, understanding how mangrove C sequestration and sediment deposition changes under cordgrass invasion is important when accounting for the ecosystem C stocks along shorelines.

3.1 Sediment accumulations

A surface elevation table—marker horizon (SET-MH) system consists of a benchmark rod and a portable measuring arm (Cahoon et al., 1995, 2002), which can provide a direct measure

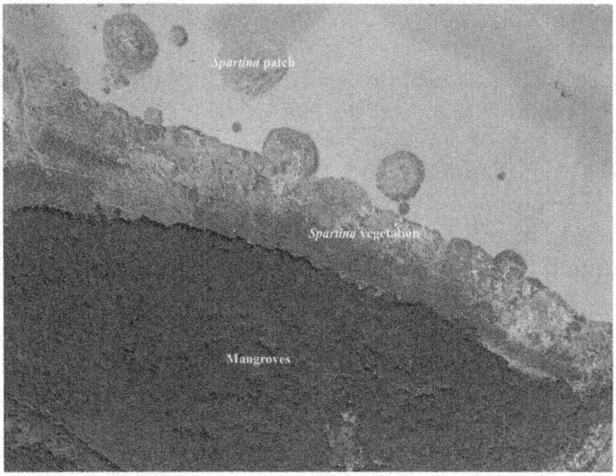

FIGURE 19.2 Typical mangrove–*Spartina* ecotone mixed in southern China. The invasive cordgrass occupies tidal mudflats on the sea edge of mangroves. The circles are the cordgrass clones newly spread to mudflats, which will mix and form continuous vegetation within 2 or 3 years.

of surface-elevation change (SEC) relative to the base of a benchmark rod, alongside the change in vertical accretion (VA) relative to the benchmark feldspar layer. Shallow subsidence, as shallow subsurface processes on elevation change, can be calculated by comparing accretion and elevation change (accretion minus elevation change; Cahoon et al., 1995). The SET-MH system provides high-resolution measurements of wetland sediment-elevation changes with an accuracy of 1.3 mm (Cahoon et al., 2002) and allows repetitive measurements at temporal scales (e.g., seasonally or yearly) to determine an average C gain/loss (Howard et al., 2014). The system has been extensively used in mangroves and tidal saltmarshes to measure SEC, VA, and shallow subsidence/expansion, as shown in Fig. 19.2 (Cahoon et al., 1995; Krauss et al., 2011; Lovelock et al., 2015).

Organic matter, mineral matter, and water or gas-filled pore space are the three major components contributing to sediment-elevation change (Kirwan and Megonigal, 2013), within which organic matter accounts for between 1% and 80% of the dry mass of tidal wetland soil and further forms peat soils (Neubauer, 2008). Therefore, the elevation gain/loss within the surface and subsurface layers may reflect the organic C processes across the whole environment.

Vegetative structures, such as trunks and the complex aerial root systems of mangroves, have been shown to alter water flows and capture sediment in water as they move through the system (Furukawa et al., 1997; Mazda et al., 2005; Mazda and Wolanski, 2009). Through these physical processes, mangrove ecosystems have the capacity to influence elevation dynamics and the sediment C stocks (Krauss et al., 2014). In the mangroves of China, still roots, pneumatophores, knee roots, and plank roots are the four major types of aerial roots, similar to the types mentioned in Krauss et al. (2014). A previous study showed that the VA of different aerial root-type forests ranged from 8.3 to 11.0 mm yr^{-1} in Micronesia, within which the pneumatophores contributed greatly to the elevation gain (Krauss et al., 2003). Furthermore, the aerial root density is positively related to VA in some mangrove settings in New

Zealand and Micronesia (Young and Harvey, 1996; Krauss et al., 2003). Thus, the C gain with VA can also be regarded as an additional C gain along with the changes in soil volume (Lovelock et al., 2014). Moreover, biological processes, such as plant litter and woody debris deposition, root accumulation, and algal mat development on the soil surface, also contribute to sediment C gain (Krauss et al., 2014). McKee (2011) found that the accretion rates of leaf litter in Florida mangroves ranged from 1.1 to 3.4 mm yr^{-1} where tidal flushing was low.

Using the SET-MH technique, scientists found a high variation in VA changes in different mangrove hydrogeomorphic zones, where fringe forests range from +1.6 to +8.6 mm yr^{-1} in mangroves in the United States and Australia (Krauss et al., 2014). For those mangroves in China invaded by cordgrass, the forest fringe was the most dynamic zone for both biological and physical processes (Fig. 19.2). The measurements of SEC and VA within the mangrove–cordgrass fringe may detect the C dynamic during the invading process.

Cordgrass is good at trapping sediments and accumulating particulate C with its dense above-ground biomass (Möller et al., 2014). Its roles as an ecological engineer are very significant. Some current evidence has shown geomorphological changes in coastal ecosystems after being invaded by cordgrass (Callaway and Josselyn, 1992; Crooks, 2002). Simultaneously, the success of cordgrass invasion in China threatens the capacity of C sequestration in mangrove wetlands (Wang et al., 2019). In addition, there is still a lack of evidence on quantifying the "new carbon" gained in fringe mangrove forests driven by cordgrass invasion.

SET results from tidal saltmarsh indicated a range of 1.4~58 mm yr^{-1} of VA rates and 0–40 mm yr^{-1} of SEC on the global average (Table 19.2). However, the data from mangroves and tidal saltmarsh ecotones were very limited, showing the variation among sites and vegetation composition (Table 19.2).

Cordgrass invasion created land acting as an ecosystem engineer, which is found especially on the coasts in northern China (Zhang et al., 2004). As a typical tidal saltmarsh plant, cordgrass attenuates wave forces and protects the coasts from erosion during storms and high tides (Strong and Ayres, 2013). Our in *situ* observation in the transect from mangroves to sea edge cordgrass marshes in the Zhangjiang Estuary also suggested the significant role of cordgrass in increasing the surface elevation (Chen et al., unpublished data). In this location, mangroves were located from landward to the middle tidal zones, and the open mudflat was invaded by cordgrass since the early 2000s and formed a mangrove–*Spartina* ecotone (Zhang et al., 2012). Transects of SET-MH were set from the fringe forests to the cordgrass patches at their early stage of invasion. The patches expanded and formed closed vegetation (Fig. 19.2) and aided in the trapping and retention of detrital particles and mineral sediment. From our observation, the "new" deposition accumulated by VA contributes to 80% of the elevation gain (Fig. 19.3; Chen et al., unpublished data). The mudflat plots were invaded by cordgrass soon after 2 years of monitoring. Furthermore, the increased elevation on the forest fringe owing to cordgrass expansion changed the hydrological condition by blocking tidal water flow into the forests (Chen et al., unpublished data).

In the former case on the mangrove–*Spartina* transect in the Zhangjiang Estuary, cordgrass vegetation maintained higher SEC rates than those of mangroves, even though they both had high VA rates. The dense roots of cordgrass and its fast clone growth helped to hold the surface elevation. By fitting the current observation data on the marsh equilibrium model for cordgrass (Morris et al., 2002), we predicted the future patterns of the mangrove–*Sprartina*

TABLE 19.2 Surface-elevation change (SEC) and vertical accretion (VA) in the mangrove–tidal saltmarsh ecotones. Vegetation type (Veg.) and species are presented. The distribution in the intertidal zones was categorized as seaward, landward, and middle zone.

Location	Veg./Zones	Dominant species	SEC (mm yr^{-1})	VA (mm yr^{-1})	Source
Moreton Bay, Brisbane, AUS	M/Seaward	*Avicennia marina, Rhizophora stylosa*	1.4~5.9	8.5~9.6	Lovelock et al. (2011)
	TS/Landward	*Juncus kraussii, Sarcocornia,* and *Suaeda* genus	−0.3~1.8	0.8~1.5	
Minnamurra River, Sydney, AUS	M/Seaward	*Avicennia marina, Aegiceras corniculatum*	0.62	–	Oliver et al. (2012)
	TS/Landward	*Sporobolus virginicus, Sarcocornia quinqueflora*	0.29	1.4	
Kooragang Island, AUS	M/Seaward	*Avicennia marina*	2.85	2.23	Rogers et al. (2013a)
	Mix/Middle	*Aegiceras corniculatum, Sporobolus virginicus, Sarcocornia quinqueflora*	2.81	0.66	
	TS/Landward	*Sporobolus virginicus, Sarcocornia quinqueflora*	1.72	1.39	
Sydney, AUS	M/Seaward	Seaward/no mention	0.8	0.82	Rogers et al. (2013b)
	Mix/Middle	Middle/no mention	0.41	2.75	
	TS/Landward	Landward/no mention	1.875	3.125	
Tweed River, New South Wales and	M/Seaward	Seaward/*Aegiceras corniculatum, Avicennia marina, Bruguiera gymnorrhiza, Excoecaria agallocha, Rhizophora stylosa*	1.37~1.43	0.74~0.82	Rogers et al. (2014)
Queensland, AUS	TS/Landward	Landward/*Sporobolus virginicus, Juncus kraussii, Casuarina glauca*	0.08~0.26	0.08~0.25	
Belize Barrier Reef, Tampa Bay, USA	M/Seaward	Seaward/*Rhizophora mangle, Laguncularia racemosa, Avicennia germinans*	2.10~6.60	3.60~9.09	Krauss et al. (2017)
	Mix/Middle	Landward/*Rhizophora mangle, Laguncularia racemosa, Avicennia germinans, Spartina alterniflora*	3.90~10.89	4.05~13.85	
Port Fourchon, Louisiana, USA	M/Seaward	Landward/*Avicennia americana, Avicennia germinans*	4.1	11.4	McKee & Vervaeke (2017)
	Mix/Middle	Middle/*Avicennia americana, Avicennia germinans, Spartina alterniflora*	6.1	10.4	
	TS/Landward	Seaward/*Spartina alterniflora*	4.6	10.4	
Average	**Mangroves**		**1.87~3.64**	**4.15~5.56**	
	Mixture		**4.00~7.11**	**9.13~10.41**	
	Tidal saltmarsh		**3.43~4.77**	**6.46~7.24**	

M, mangroves; *Mix*, mixture of mangroves and tidal saltmarsh; *TS*, tidal saltmarsh.

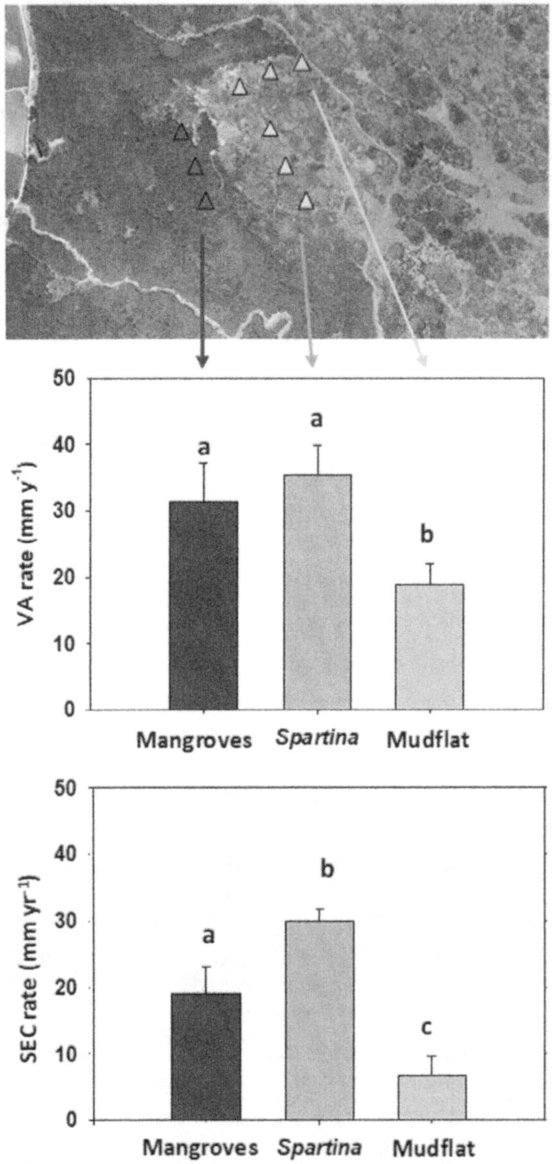

FIGURE 19.3 Surface-elevation change (SEC) rates and vertical accretion (VA) rates in a typical mangrove–*Spartina* transect in the Zhangjiang Estuary in China (Chen et al., unpublished data).

ecotone (Fig. 19.4). The positive effects of cordgrass cloning growth and sediment deposition would greatly accelerate elevation increases, which may create land in the near future owing to the geomorphic effects of cordgrass as an environmental engineer.

Multiple SET observation networks have been established in the mangrove–*Spartina* transects and in native tidal saltmarsh–*Spartina* transects along the east coast of China to discover

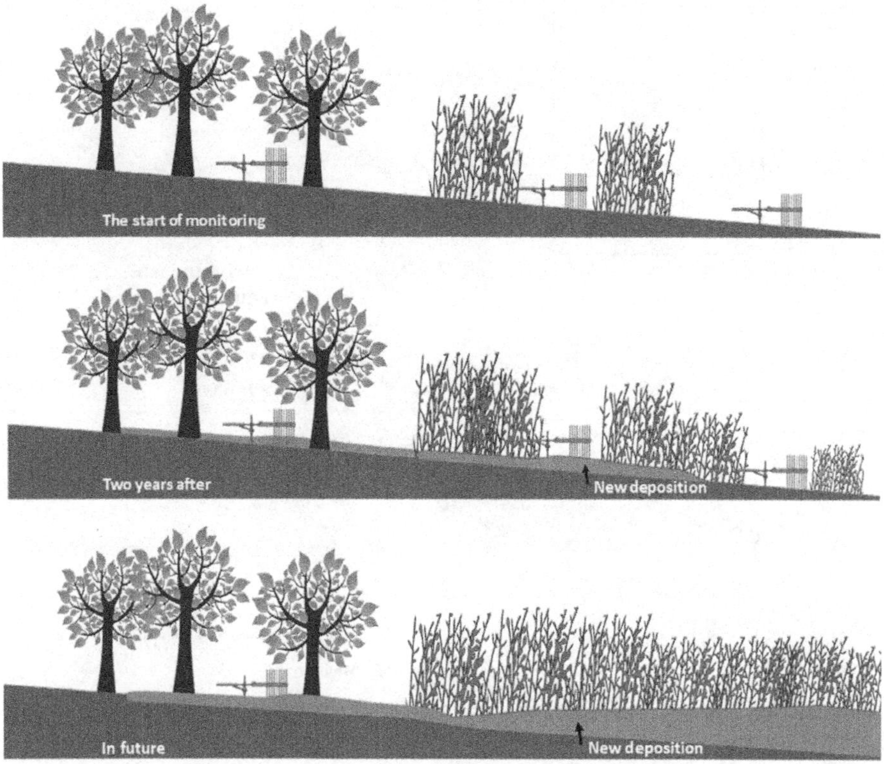

FIGURE 19.4 Diagrams of surface-elevation change in a typical mangrove–*Spartina* transect in the Zhangjiang Estuary in China: current status and future prediction by the marsh equilibrium model for cordgrass.

their contribution to geomorphological changes (Fig. 19.5). Thus, the coupling of ecological models of the marsh–forest margin with geomorphic models can be a future direction in spatial landscape-scale studies (Kirwan and Megonigal, 2013).

3.2 Biomass allocation and productivity

Root production and decomposition are two significant biological contributors driving subsurface processes in mangroves belowground (Krauss et al., 2014). SETs in biogenic settings in Florida and the Caribbean mangroves suggested a direct contribution of root accumulation to SEC (Cahoon et al., 2003; McKee et al., 2007b; McKee, 2011). Due to their refractory nature and anaerobic soil environment, root decomposition in mangroves was slow, which remarkably increased the accumulation of below-ground C (Saintilan et al., 2013), leading to a high contribution of below-ground C in mangrove ecosystems (Donato et al., 2011).

As a type of colonizing tidal saltmarsh, cordgrass vegetation has high root productivity ranging between 1512 and 4509 g m^{-2} yr^{-1} (Hill and Roberts, 2012). With lower aboveground biomass, the invasion of cordgrass in mangroves may have less effects on increasing

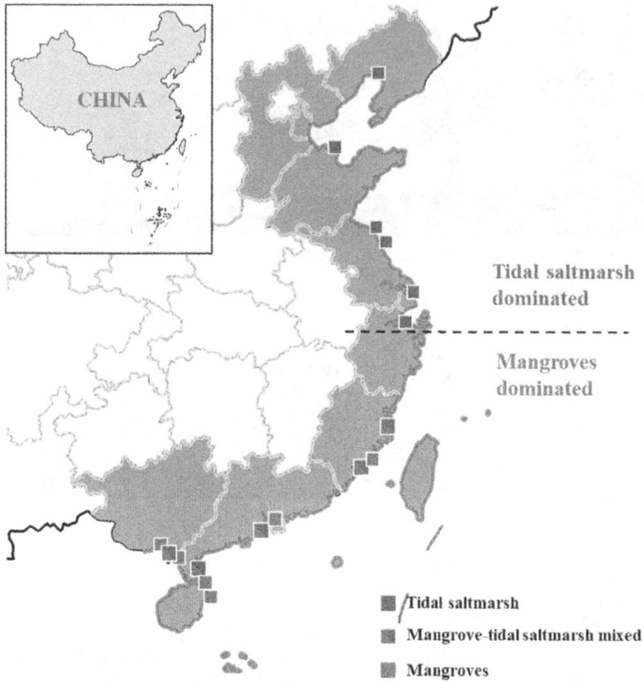

FIGURE 19.5 SET sites in coastal wetlands along the shorelines of China.

above-ground C, whereas it still contributes substantially to below-ground C. Johnson et al. (2016) found a higher proportional contribution of sediment C in tidal saltmarsh than that in mangroves. A preliminary study with a "marsh organ" setting using young vegetation suggested that the invasive cordgrass had much higher biomass gain belowground than that of mangrove seedlings (Peng et al., 2018). The performances of the native mangrove *Kandelia obovata* and the cordgrass across elevation were also assessed (Peng et al., 2018). Both species displayed parabolic patterns of biomass across elevation, with poor performance at low elevations due to flooding and poor performance at high elevations due to drought (Mendelssohn and Morris, 2002; Morris et al., 2002; Krauss et al., 2008; Fagherazzi et al., 2017). Cordgrass performed better at slightly lower elevations, consistent with its distribution as a low-marsh species in its native range (Mendelssohn and Morris, 2002; Morris et al., 2002). *K. obovata* performed better at slightly higher elevations, consistent with its distribution as a high-intertidal mangrove in China (Chen et al., 2009).

Under field conditions, approximately 65%–95% of below-ground C sequestration occurred at low and middle elevations, whereas above-ground biomass became a more significant C pool in the high-elevation marshes (Elsey-Quirk et al., 2011). Nevertheless, this pattern may change with the plantation time because the below-ground C contribution could also be driven by the age of the vegetation (Lunstrum and Chen, 2014). Direct evidence combining in situ measurements for below-ground biological processes and SET observations is needed to predict the below-ground C pool changes under cordgrass invasion.

3.3 Carbon source changes

Vegetation contributes to a large amount of organic matter in the sediments of coastal wetlands. Changes in the dominant species within a plant community may greatly alter material cycling processes. The invasion of cordgrass dramatically influenced the net ecosystem production, soil nutrients, and quality that altered the process of organic matter formation and turnover in coastal wetlands of China (Liao et al., 2007; Yang et al., 2016). As C_4 plant invasion, the contribution of cordgrass to the local C_3 community organic C stocks could be detected with the ^{13}C stable isotope technique (Cheng et al., 2006; Yu et al., 2015; Yang et al., 2016).

The ^{13}C isotope technique is based on differences in stable C isotope ratios between C_3 and C_4 plants. Due to different photosynthetic pathways, C_3 and C_4 plants make different relative contributions to the net primary productivity in the surface soil, and the $\delta^{13}C$ values of soil organic carbon (SOC) vary (Gebrehiwet et al., 2008). The $\delta^{13}C$ values of C_3 plants are between −35 and −20‰, whereas the $\delta^{13}C$ values of C_4 plants range between −19 and −9‰ (Fry, 2014). This difference can be detected and distinguished from the contribution of C sources. Using this method, studies on cordgrass invasion in tidal saltmarshes successfully distinguished the C sources between cordgrass and native marshes (Cheng et al., 2006; Wang et al., 2018). When using an invasion chronosequence, this technique could also explain the history of C storage changes (Wang et al., 2018). Invasive cordgrass affected the C storage capacity in typical C_3 mangrove mudflats by contributing 42.6–62.2 of organic C in the Zhangjiang Estuary in China (Feng et al., 2017). With this technique, the effects on the organic C gain of this C_4 invader can be quantified in habitats with native C_3 marshes or mangroves (Wang et al., 2018).

When we focused on the profiles of soil $\delta^{13}C$ value distribution in the invaded mudflats, we found that the deeper sediment layers had lower contributions from the invasive cordgrass (Cheng et al., 2006; Feng et al., 2017). Without the input by C_4 signals in roots and rhizomes, the deep sediment was weakly influenced by cordgrass, which showed less negative $\delta^{13}C$ values at greater depth. Even inside the native communities (tidal saltmarshes and mangroves), the $\delta^{13}C$ values of sediment in the surface layer could be less negative, which allowed the qualification of the allochthonous C gain in the systems with significant input of the C_4 signal from nearby cordgrass vegetation. This new C source was showed to be passing into a higher trophic level in mangroves (e.g., macrobenthos or waterfowl) owing to bottom-up effects caused by vegetation changes after cordgrass invasion (Feng et al., 2014).

3.4 Soil CO_2 and methane flux

In coastal wetlands, multiple biological factors are often emphasized as important controls on sediment C flux, especially on greenhouse gas (GHG) emissions. Studies investigating the soil GHG emissions from cordgrass-invaded tidal saltmarsh showed higher methane efflux from the invasive cordgrass marsh than that in the native saltmarshes (e.g., Tong et al., 2012; Yin et al., 2015). With high biomass production and poor litter quality, cordgrass slowed decomposition rates and stimulated high methane emissions compared to those of native tidal marsh plants (Liao et al., 2007; Tong et al., 2012). In mangrove–*Spartina* ecotones, it has been proven that invasive cordgrass remarkably promoted methane emissions with various values across different study sites (Chen et al., 2015; Yuan et al., 2015; Gao et al., 2018). Methane emissions from sediments after cordgrass invasion reached 2261.23 $\mu g\ m^{-2}\ h^{-1}$, which was approximately 275 times the value in the mudflat and 5–57 times higher than those in mangroves due to the high discrepancy in sediment functional microbial abundance (Gao et al., 2018).

In contrast, cordgrass invasion could decrease sediment CO_2 emissions. Although sediment CO_2 emissions in mangroves might stem from below-ground fine-root respiration and heterotrophic microbial respiration sources, or even from the transformation of methane, the combinations of heterotrophic microbial respiration and anaerobic methane oxidation were the main factors regulating CO_2 reduction (Gao et al., 2018). As this case study shows, the net C flux of stimulating methane emissions and reducing CO_2 efflux by cordgrass invasion is likely to increase the global warming potential (GWP) of mangroves, even though they assimilate CO_2 through photosynthesis. For the integrated function of C flux changes in mangrove ecosystems driven by cordgrass invasion, we still need to assess C for the whole mangrove ecosystem and look for methods for future Blue C management.

4. Vegetation–sediment dynamic interactions and the consequent sedimentary processes

Natural coastal wetlands are widely recognized for their ability to trap sediments (Perillo et al., 2009; Barbier et al., 2011; Möller et al., 2014), as well as for their defense potential against the actions of waves and tidal flows (Temmerman et al., 2013; Horstman et al., 2015; Carus et al., 2016). Intertidal habitats are widely found on low-energy coasts, often occurring in the high intertidal zone, and are characterized by the presence of halophytic (salt-tolerant) plants on a muddy substrate (Allen and Pye, 1992). Mangroves are dominated by halophytic trees and shrubs, thriving on dynamic coastlines controlled by physical forces and geomorphic evolutionary processes (Mitsch and Gosselink, 2007; Rovai et al., 2018).

Vegetation occurring along intertidal flats can stabilize, trap, and bind sediments. Plants have been observed to directly increase the erosion threshold of bed sediments (Chen et al., 2012) and indirectly trap sediments by providing additional drag force (Leonard and Luther, 1995; Temmerman et al., 2005; Chen et al., 2016), which mediates flow patterns and consequently enhances local sediment deposition (Neumeier and Amos, 2006; Horstman et al., 2015). The enhancement of sediment deposition by coastal wetland vegetation has received more extensive attention in recent decades, in part due to the predicted impacts of accelerated sea-level rise and the C storage ability in this type of particular environment (van der Wal and Pye, 2004; Cahoon et al., 2006; Nellemann et al., 2009; Arkema et al., 2013; Woodroffe et al., 2016; Rovai et al., 2018).

Saltmarsh grasses (particularly cordgrass) exhibit a considerable ability in flow reduction and turbulent energy dissipation by the presence of above-ground biomass (Leonard and Luther, 1995; Christiansen et al., 2000; Nepf, 2012; Chen et al., 2016). Mangrove trees have also been shown to alter flows due to the presence of canopies, trunks, and aerial roots (Wolanski et al., 1992; Furukawa et al., 1997; Mazda et al., 2005; Mazda and Wolanski, 2009; Chen et al., 2016; Mullarney et al., 2017). As a consequence, both saltmarshes and mangroves are able to trap sediments to accelerate deposition or decrease surface sediment erosion. In this section, a comparison of the vegetation–hydrodynamic interactions between saltmarsh plants and mangrove trees will be discussed. On this basis, the subsequent sediment-trapping processes by woody and herbaceous plants will be compared. C storage is highly associated with sediment dynamics and geomorphological processes under vegetation impacts. C storage changes associated with sedimentary processes in saltmarshes and mangroves will be addressed in the last part.

4.1 Hydrodynamic modifications by vegetation: a comparison between mangrove and cordgrass

Most of the saltmarshes and mangrove swamps occur on mudflats under the forces of flows and waves. The presence of vegetation is able to considerably dampen flows and waves by providing extra drag forces compared to those of unvegetated beds (Mazda et al., 2005; Nepf, 2012; Chen et al., 2016; Mullarney et al., 2017).

The classical fluid dynamic theory describes the vertical velocity profile above a bare mudflat bed using the Law of the Wall, which means the water flow above the bed follows a logarithmic shape (Soulsby, 1997, Fig. 19.6). An early flume study insightfully looked into the unique pattern for this vertical velocity profile within saltmarsh vegetation canopies (Shi et al., 1995). In a more recent review, Nepf (2012) summarized the two types of vegetation canopies in relation to vegetation–hydrodynamic interactions: emergent canopies and submerged canopies. Detailed laboratory and field observations reveal a modified vertical flow structure into a J-shape by emergent canopies, alongside an S-shape by submerged canopies (Shi et al., 1995; Neumeier and Ciavola, 2004; Leonard and Croft, 2006; Nepf, 2012, Fig. 19.6). Compared to saltmarsh canopies, the flow structure altered by mangrove trees has been less understood. Owing to the unique biophysiological characteristics of mangroves, such as their canopy structure, networks of trunks, and their aerial roots, the flow reduction, and energy dissipation caused by mangroves are very complicated (Massel et al., 1999; Mazda and Wolanski, 2009). *In situ* observation has been conducted to measure the vertical velocity profile within a mangrove swamp in southeast China, which revealed a C-shaped flow structure as affected by mangrove trees, remarkably different from that of herbaceous canopies (Chang et al., 2019, Fig. 19.6).

Numerous studies have shown that saltmarsh grass canopies (e.g., cordgrass canopy) can significantly reduce the mean flow velocity (Shi et al., 1995; Fonseca, 1996; Bouma et al., 2005; Neumeier, 2007). Cordgrass canopies were measured to dampen 40%–90% of the mean flow speed compared to that of mudflat (Table 19.3). The efficiency of flow reduction depends on a number of parameters, including species, density, height, and vegetation width. In addition to fine-scale measurements, observations at larger scales have also revealed a decrease in the flow velocity of up to one order of magnitude between a saltmarsh and mudflat (Bouma et al., 2005). Field observations also found direct evidence of wave attenuation by saltmarsh grasses, and the attenuation effectiveness depended on the species, density, and vegetation width (Knutson et al., 1982; Möller, 2006; Möller et al., 1999, 2014). Among the saltmarsh species, *S. alterniflora*, being an effective species to resist waves, appears to be a good choice to protect coasts from wave actions (Table 19.4).

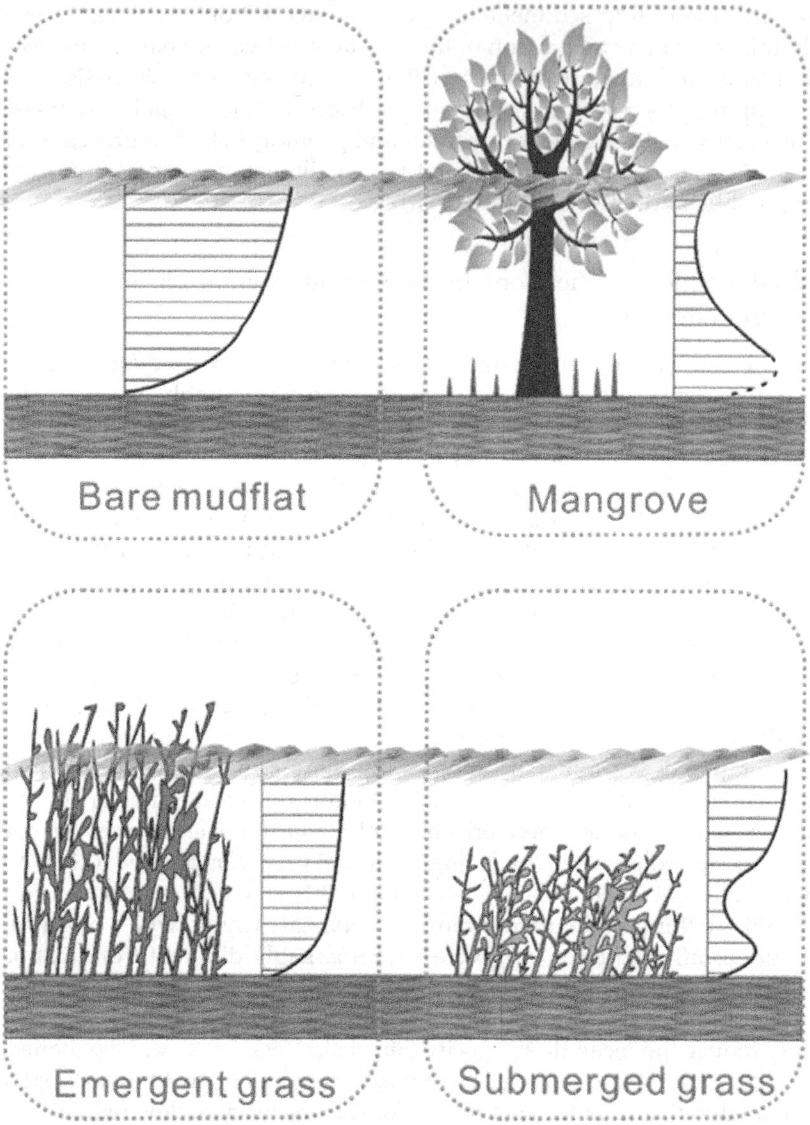

FIGURE 19.6 Modification of the vertical velocity profile by the presence of vegetation species. Top-left panel: the logarithmic velocity profile over a bare mudflat; top-right panel: the velocity profile modified by mangrove trees; bottom-left panel: the velocity profile modified by emergent grass; and bottom-right panel: the velocity profile modified by submerged grass.

The available field measurements of flow variation due to the presence of mangrove trees show a reduction in the flow speed of approximately 50%–70% only over a vegetation width of 10–15 m, together with a gradual change in the flow direction (Mazda, 2003; Kobashi and Mazda, 2005; Chen et al., 2016). More attention has been paid to the wave attenuation by

TABLE 19.3 The influence of *Spartina* grasses on flow speed.

Source	Species	Shoot density (standings m^{-2})	Vegetation width (m)	Flow reduction rate
Leonard and Luther (1995)	*Spartina alterniflora*	180	10–20	44%
Widdows and Brinsley (2002)	*Spartina anglica*	400	–	75%
		2000	–	90%
Neumeier and Ciavola (2004)	*Spartina anglica*	2500	20	75%
Chen et al. (2016)	*Spartina alterniflora*	580	10	40%

TABLE 19.4 Wave attenuation over saltmarsh edges.

Source	Vegetation width (m)	Species	Wave attenuation rate
Knutson et al. (1982)	10	*Spartina alterniflora*	65%
	20	*Spartina alterniflora*	87%
	30	*Spartina alterniflora*	94%
Yang (1998)	10	*Scirpus mariqueter*	57%
Möller et al. (2001)	180	*Spartina anglica, Salicornia europaea*	61%
Möller and Spencer (2002)	163	*Suaeda salsa, Salicornia europaea*	65%
	10	*Suaeda salsa, Salicornia europaea*	25%
Möller (2006)	10	*Spartina anglica, Salicornia europaea*	15%–20%
	10	*Spartina anglica, Salicornia europaea*	11%–17%
	10	*Spartina anglica, Salicornia europaea*	10%–12%
Ysebaert et al. (2011)	>50	*Spartina alterniflora, Scirpus mariqueter*	80%
Yang et al. (2012)	51	*Spartina alterniflora*	79%

mangroves regarding their coastal defense function. In general, mature mangrove forests exert radical protection against wave actions, with a reduction between 20% and 95% (Table 19.5). In particular, mangrove forests are able to mitigate the damage of storms and other coastal hazards (Barbier, 2016; McIvor, 2012; Doughty et al., 2017). Based on bulk drag coefficient calculations, Pinsky et al. (2013) estimated a wave reduction of 87% by a 200 m-wide mangrove habitat under storm conditions.

TABLE 19.5 Field observations of wave attenuation within mangroves.

Source	Species	Vegetation width (m)	Wave attenuation
Brinkman et al. (1997)	*Rhizophora stylosa, Bruguiera gymnorrhiza*	150	45%~80%
Mazda et al. (1997)	*Kandelia candel*	100	20%
Mazda et al. (1997)	*Kandelia candel*	1500	95%
Mazda et al. (1997)	*Sonneratia caseolaris* (2 months)	100	0%
Massel et al. (1999)	*Sonneratia caseolaris*	100	50%
Quartel et al. (2007)	*Kandelia candel,* *Sonneratia caseolaris*	20	55%

From an energy perspective, plants can affect the low-frequency energy (mean flow) and high-frequency turbulent energy of the flow. Previous studies on the structure and intensity of turbulence have found that the turbulent energy inside the plant canopy of saltmarshes (such as the canopy of cordgrass) becomes consistent in the vertical direction, whereas the water turbulence above the canopy is not affected by the presence of plants (Leonard and Luther, 1995; Neumeier and Amos, 2006). Several studies have proposed that plants decrease the relatively large turbulence length on the bare flats by changing the turbulence length (associated with the plant characteristics), which dissipates the turbulent energy (Nepf, 2012).

Fourier-transformed spectral information in the frequency domain can conceal information about the temporal variability of hydrodynamic energy; thus, the energy spectrum has been widely adopted by oceanic scientists to measure the characteristics of hydrodynamics (Leonard and Luther, 1995; Liu, 2000; Chang et al., 2019). The turbulent reduction by mangroves is different from that by cordgrass canopies at a scale of energy spectrum (Chang et al., 2019) because mangrove trees are better at dissipating low-frequency energy, while cordgrass shows a better ability to suppress high-frequency energy (Fig. 19.7). A direct comparison between the mangrove trees and cordgrass in terms of turbulent energy dissipation has been conducted in an ecotone of these two habitat types (Chen et al., 2016). This field observation indicated the importance of the relative water level to the vegetation canopy height, which determines the relative ability of the turbulent energy damping between those habitat types. When the water level was below the mangrove canopies, cordgrass was found to be more efficient than mangrove trees in dissipating energy, but mangrove trees could be more efficient once the water level reached their canopies (Chen et al., 2016). Another study (Doughty et al., 2017) in Florida (US) also compared wave attenuation by saltmarshes and mangroves through numerical modeling over a large scale. This simulation revealed a similar pattern: mangroves were found to be better than saltmarshes in dissipating wave energy (Doughty et al., 2017).

Overall, both cordgrass canopies and mangrove trees considerably affect tidal currents and waves as well as high-frequency turbulence dissipation. The settling of suspended sediments within the vegetation canopy is related to the dissipation of turbulence (Shi et al., 1995; Nepf, 1999). Therefore, intertidal vegetation, such as saltmarsh grasses and mangrove trees, has

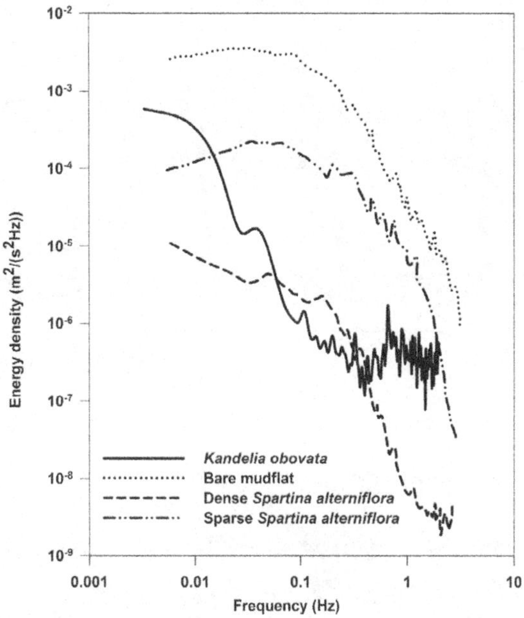

FIGURE 19.7 Comparison of the energy spectrum between bare mudflats and the seabeds covered by a variety of vegetation species. *Adapted from Chang, Y., Chen, Y., Li, Y., 2019. Flow modification associated with mangrove trees in a macro-tidal flat, Southern China. Acta Oceanol. Sin. 38, 1–10.*

been recognized for its ability to reduce hydrodynamics and consequently enhance local sediment deposition (Neumeier and Amos, 2006; Horstman et al., 2015; Chen et al., 2018).

4.2 Sediment settling and trapping by vegetation: a comparison between mangroves and cordgrass

A number of studies have indicated the presence of accelerated sedimentation rates, up to one order of magnitude, by intertidal vegetation (Allen, 2000). In regions with limited sediment supply, such as Europe and Australia, the sedimentation rates of mangroves and saltmarshes normally reach a few millimeters per year (Pethick, 1981; Stevenson et al., 1986; Cahoon et al., 1996; Roman et al., 1997; Rogers et al., 2006; Lovelock et al., 2015). However, once the sediment supply is abundant, for example, in some tidal flats in China, the sedimentation rates within saltmarshes and mangroves are estimated to be several centimeters per year (Yang and Chen, 1994; Zhang et al., 1996; Li and Yang, 2007; Chen et al., 2018). A more fundamental question associated with sedimentation is how the sedimentation process affects the competition between plants. Saintilan and Rogers (2006) might be the first study looking into the relationship between mangrove encroachment into saltmarshes and the sedimentation rate changes. Their study nicely described the importance of SEC in determining the competition between mangroves and saltmarshes. A similar case can be discussed using the example in the Zhangjiang Estuary in China, an ecotone of native mangrove species and

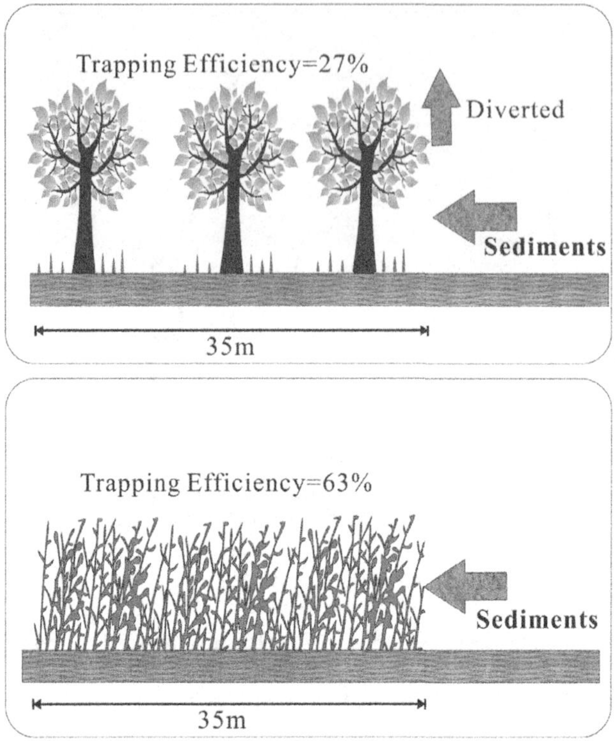

FIGURE 19.8 A direction comparison in sediment-trapping abilities between saltmarsh and mangrove boundaries in the Zhangjiang Estuary, China. *Modified from Chen, Y., Li, Y., Thompson, C., Wang X., Cai, T., Chang, Y., 2018. Differential sediment trapping abilities of mangrove and saltmarsh vegetation in a subtropical estuary. Geomorphology 318, 270–282.*

exotic cordgrass. In the maximum turbidity zone of this estuary, ecologists have found no invasion of cordgrass into local mangroves based on long-term observations (Zhang et al., 2012). A further investigation of the sediment dynamic processes in this area revealed a higher sediment-trapping capacity in the front edge of the saltmarsh than that in its neighboring mangrove front (Chen et al., 2018, Fig. 19.8). However, once the seedlings of each species are transplanted into a habitat dominated by other species, the sediment-trapping ability of the seedlings was decreased and resulted in the death of the seedlings caused by burial. The cause of this phenomenon might be associated with different hydrodynamic alterations and subsequent sediment-trapping processes by mangroves and cordgrass in this region under the condition of high sediment supply, such as in the Zhangjiang Estuary. However, in the regions with less sediment supply, modeling results indicated that conversion from saltmarshes to mangroves in Florida might increase erosion prevention by mangroves with aerial roots (Doughty et al., 2017).

A complete understanding of the mechanisms mediating sediment accumulation by tidal flat vegetation, particularly the possible impact of vegetation on suspended particles, is still elusive (Graham and Manning, 2007). The trapping of suspended particles can be caused both by changes to hydraulic forces and by the direct presence of vegetation leaves and stems (Chen et al., 2018). The settling velocities of suspended particles in a flocculent form within saltmarshes and mangrove swamps have been suggested to be a magnitude of 10^{-1}–10^{0} mm s^{-1} (French and Spencer, 1993; Wolanski, 1995; Voulgaris and Meyers, 2004; Graham and Manning, 2007; Chen et al., 2018). A direct comparison between neighboring mangroves and cordgrass saltmarshes showed an insignificant difference in the settling velocities for those two habitats (Chen et al., 2018). However, tidal asymmetry in particle settling was found to be the key factor in understanding the difference between those two ecosystems: the tidal asymmetry was more pronouncedly altered by cordgrass canopies than by mangrove trees, and this situation resulted in a higher sedimentation tendency in cordgrass saltmarshes (Chen et al., 2018).

On the other hand, direct trapping by plant bodies is also worth investigating, particularly in the field. Traditional laboratory experiments on sediment trapping by vegetation used to exclude the effect of precipitation. Most of the saltmarsh grasses and mangrove trees become inundated and emerge alternatively. Once the plants are inundated, the interactions between the plants and water flows cause sediment settling as well as direct sediment trapping by vegetation (such as leaves and aerial roots). When the plants emerge for several hours during a tidal cycle, possible interactions between the water and plants include rinsing off by rain drops. A field measurement implies the potential contribution to sediment deposition caused by a combined process of direct trapping by plants during inundation and the subsequent rinse of rain drops during emerged periods (Chen et al., 2018). Along the subtropical coasts of China, dominated by wet and dry season cycles, precipitation helps to provide an extra 0.1 cm a^{-1} annual deposition rate to local mangroves and 0.5 cm a^{-1} to cordgrass saltmarshes (Chen et al., 2018). It appears that precipitation creates a pathway for particles to be deposited into the mudflat, although its details have not yet been fully understood.

4.3 Sediment stability associated with carbon release from the sediments

As a component of particles in the seabed, C storage is determined by both sediment deposition and sediment erosion processes. Sediment deposition has a positive effect on C storage, while erosion processes cause the release of C from the seabed. The sediment accumulation by mangroves and saltmarshes and the subsequent accelerated C storage have been well studied, as summarized in Section 3.1 of this chapter. In contrast, the erosion reduced by intertidal plants in association with net C fluxes has been less understood, particularly with a background of saltmarsh net loss under relative sea-level rises.

The erosional processes, which shape tidal flat geomorphology, are associated with the balance between erosive forces, e.g., flow-induced shear stress and the stability of sediments. The erosion threshold (e.g., critical shear stress) is defined as a value below which little or no erosion occurs, whereas once exceeded, significant erosion occurs (Teisson et al., 1993). The erosion threshold, together with the erosion rate, constitutes the criteria for sediment erodibility or sediment stability (Amos et al., 1992), although it is somewhat arbitrary to provide the definition of "stability."

Research undertaken to date has demonstrated that biota play an important role in determining the erosion potential or stability of intertidal sediments (Amos et al., 1998; Friend et al., 2003; Chen et al., 2012). Friend et al. (2003) reported that the erosion threshold of the saltmarsh surface between individual cordgrass plants could exceed 9.1 Pa, which was only 0.1~1 Pa in natural abiotic sediments. More importantly, the root systems of intertidal plants have been revealed to contribute additional stability to bed sediments by more than one order of magnitude (Chen et al., 2012, 2019).

The mechanisms controlling erosion fall into three groups, namely, subaerial erosion, scour, and mass failure (Rutherfurd et al., 1999). Subaerial erosion and scour form direct erosion on the saltmarsh surface, but other geomorphological processes within tidal flats also cause a net loss of deposited sediments by mass failure. Cliffs and tidal creek systems have been extensively observed within a number of erosive or depositional saltmarshes (van Eerdt, 1985; Knighton et al., 1992; Gao and Collins, 1997; Wang et al., 1999; Allen, 2000; Chen et al., 2011; Zhao et al., 2017; Gong et al., 2018). The bank failure of cliffs and the expansion of tidal creek systems both cause erosion, even within depositional environments. As a result, the buried C in soil pools has the possibility to be released into the air or washed away by water. Therefore, the estimate of C storage within mangroves or saltmarshes should take this process into account.

Unlike surface sediment erosion or deposition rates, which have been monitored by standard methods, bank erosion and subsequent mass failure are less quantified for C storage ability estimates. However, the release of stored C by bank failure can be achieved even in low-energy environments by long-term wetting–drying cycles, which is relevant to relative sea-level rise (Chen et al., 2011). Sea-level rise will be an increasing threat to coastal wetlands under climate change. Global mean sea level rose at a mean rate of 1.7 (1.5–1.9) mm yr^{-1} between 1900 and 2010 and at a rate of 3.2 (2.8–3.6) mm yr^{-1} from 1993 to 2010 (IPCC Fifth Assessment Report). Accelerating sea-level rise (Church and White, 2006) will have a pronounced effect on those wetland areas that do not have sufficient sediment supply to vertically accrete at the same rate (Allen, 2000) and may lead to considerable coastal squeeze and C storage decrease. Half of the world's coastal wetlands could be lost with a sea-level rise of just 1 m (Nicholls, 2004). The consequent net loss of coastal wetlands is made up of surface erosion or inundation, together with bank processes such as mass failure. Both of these factors are highly related to the estimate of sediment C storage amounts within mangroves or saltmarshes.

5. Conceptual model and evaluation of regional carbon storage

5.1 An evaluation of regional carbon storage changes after cordgrass invasion in China

Both tidal saltmarshes and mangrove habitats have been seen as C-rich ecosystems, with a large potential for blue C storage (Pendleton et al., 2012; Alongi, 2014). The introduction of cordgrass into mudflats in China has been reported to increase the C storage of the bare mudflat through changes in net primary production, litter decomposition, and soil C stock (Liao et al., 2007; Liu et al., 2017b). After the invasion of cordgrass, an increase of 1.7~5.3 times in topsoil C content was revealed compared to that of native tidal saltmarshes and bare mudflats (Yang et al., 2013). More importantly, a considerable increase in long-term sedimentation rates within cordgrass saltmarshes has been widely recorded, with an acceleration of at least 1–2 cm yr^{-1} (Wang et al., 2005; Chen et al., 2018). As a consequence, the soil C storage ability is enhanced, together with better preservation of underground C. A debate on the relative SOC stock between the exotic cordgrass and native saltmarsh species indicates that the exotic species is more efficient in increasing topsoil C storage, but the native species increases the C storage at a deeper layer (Liu et al., 2013, 2017b). This problem might be explained by accelerated sedimentation rates after the introduction of cordgrass. High sedimentation can result in C accumulation on the top sediment layer within a short term, but due to the restriction of short life history in China, this exotic species is unable to affect the deeper layer, which can be remarkably modified by native species in the long term.

Remote sensing image analyses provide an approach to measure the area of cordgrass tidal saltmarshes along the coasts of China and the geographical pattern (Table 19.6). Previous studies based on remote sensing images have revealed that the total area of cordgrass tidal saltmarshes in China increased from 34,451 ha (Zuo et al., 2012) in 2007 to 54,580 ha in 2015 (Liu et al., 2018). The C density of cordgrass marshes in China is still elusive due to a lack of data. In this study, combining the below-ground C density from Liu et al. (2017b) and the above-ground C density from Liu et al. (2013), a C density of 80.41 MgC ha^{-1} was adopted for the following estimates. The rapid expansion of cordgrass marshes over Chinese tidal flats contributed an additional blue C storage of 1.62 Tg C between 2007 and 2015 (Table 19.5). It is also noted that Jiangsu, Shanghai, and Zhejiang are the main habitats for cordgrass in China (Table 19.6). Shanghai and Zhejiang also showed the most significant expansion of cordgrass saltmarshes during this period (Table 19.6). On the one hand, this

TABLE 19.6 Changes in *Spartina* tidal saltmarsh areas and the subsequent soil carbon storage. The data in 2007 are from Zuo et al. (2012), while the data in 2015 are from Liu et al. (2018). The carbon density of *Spartina saltmarsh* = 80.41 Mg C ha^{-1} is estimated using the data from Liu et al. (2013) and Liu et al. (2017b).

Province	*Spartina* area 2007 (ha)	Carbon storage 2007 (Mg C)	*Spartina* area 2015 (ha)	Carbon storage 2015 (Mg C)
Liaoning	57	4583	0	0
Tianjin	163	13,106	426	34,255
Hebei	474	38,114	26	2091
Shandong	686	55,161	2484	199,738
Jiangsu	18,711	1,504,552	18,363	1,476,569
Shanghai	4741	381,224	10,109	812,865
Zhejiang	4812	386,933	14,282	1,148,416
Fujian	4166	334,988	7267	584,339
Guangdong	541	43,904	780	62,720
Guangxi	95	7639	843	67,786
Total	34,451	2.77 Tg C	54,580	4.39 Tg C

TABLE 19.7 Mangrove area along Chinese coasts (data are mainly from Chen et al. (2017), but the area data for Zhejiang Province are from a most recent field survey, unpublished) together with carbon density data (Liu et al., 2014) and the estimated carbon storage. The carbon density is 270.39 MgC ha^{-1} on average (Liu et al., 2014).

Province	Mangrove area (ha)	Carbon density (Mg C ha^{-1})	Carbon storage (Mg C)
Zhejiang	113 (updated)	376.33	42,525.29
Fujian	675	348.00	234,900
Taiwan	410	348.00	142,680
Guangdong	8136	332.05	2,701,558.8
Hong Kong	544	332.05	180,635.2
Macao	13	332.05	4316.65
Guangxi	6849	313.20	2,145,106.8
Hainan	3667	418.06	1,533,026.02
Total	20,407	–	6,984,748.76 **(6.98 Tg C)**

expansion might have caused the degradation of native ecosystems, but on the other hand, it greatly increased the blue C storage.

However, once cordgrass marshes invade into mangroves (see the full details from the ecotone map of Fig. 19.5), a decrease in C storage is expected due to a higher C density (~350 Mg C ha^{-1} on average) of mangrove swamps. Most recently, the overall area of the mangroves in China was 20,407 ha (Table 19.7). Most (>70%) of the mangrove swamps are located within Guangdong and Guangxi Provinces (Table 19.7), where the tidal ecosystems are also threatened by cordgrass. The current estimate of the C storage by mangroves indicates a total of 6.98 Tg C, higher than that of cordgrass saltmarshes (4.39 Tg C). Therefore, mangroves have become an important sink for C along Chinese coasts. If those mangroves are replaced by cordgrass saltmarshes, a net loss of 1.64 Tg C can be estimated, although accelerated sedimentation by cordgrass saltmarshes might compensate in the long term if the sediment supply is assumed to be unlimited.

In summary, after the invasion of cordgrass, the C storage of Chinese mudflats has been increased via higher primary production, litter decomposition, and accelerated sedimentation processes compared to that of bare mudflats and native saltmarshes. The rapid expansion of cordgrass marshes has significantly contributed to C storage over the past several decades. However, the invasion of this exotic species into the local mangroves is expected to decrease the C storage ability of the original ecosystems by a magnitude of more than 1 Tg C.

5.2 A conceptual model of carbon dynamics in mangroves and saltmarsh ecotones

Mangroves and tidal saltmarshes as blue C ecosystems have the potential to contribute to climate change mitigation. Based on the former knowledge of C flux with biological and

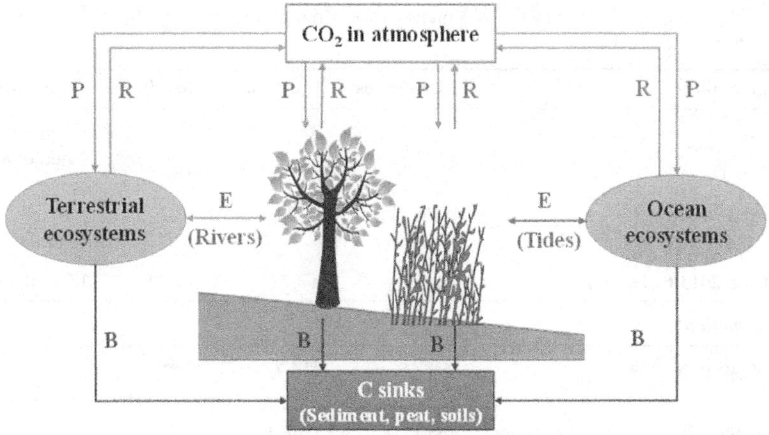

FIGURE 19.9 A conceptual model for C cycling in the coastal wetlands including mangrove ecosystems invaded by cordgrass at the seaward edge. The model demonstrates C exchange with the atmosphere by photosynthesis and respiration and exchanges across boundaries with terrestrial and oceanic ecosystems. *P* represents photosynthesis or gross productivity; *R* represents respiration; *E* represents exchange; and *B* represents burial. *Modified from Twilley, R.R., Castañeda-Moya, E., Rivera-Monroy, V.H., Rovai, A., 2017. Productivity and carbon dynamics in mangrove wetlands. In: Rivera-Monroy, V.H., Lee, S.Y., Kristensen, E., Twilley, R.R. (Eds.), Mangrove Ecosystems: A Global Biogeographic Perspective. Springer, Cham, pp. 113–162.*

geomorphic processes on the surface and subsurface sediments, we develop a simple conceptual model for C dynamics in mangrove ecosystems under cordgrass invasion (Fig. 19.9). In this model, the SET technique combined with the direct measurements of sediment cores provides the qualification of C gain and loss on the surface sediment and C dynamics in the vertical direction. Flux methods are also used to estimate the GHG flux between the soil and atmosphere and estimate the CO_2 exchange between vegetation and the atmosphere in the vertical direction. These exchanges across boundaries with terrestrial and oceanic ecosystems can be qualified by a stable C isotope tracing approach and hydrological models in the lateral direction.

The stock–difference method, gain–loss method, and flux method are the three major approaches recommended to determine these changes in coastal ecosystem C stocks (Howard et al., 2014) and have been used previously to understand mangrove C loss in Indonesia. The former two approaches estimate the difference in C stocks at two points of time, except the gain–loss method combines the emission factors for specific activities, such as restoration or deforestation. We can also use the flux methods as a direct method in C assessment, whereas the assumptions about the emission variations over time are essential if we need to extrapolate these measurements at an annual scale/long-term scale (Poffenbarger et al., 2011). However, the exchanges that occur in the atmosphere–sediment boundary and in the sediment–water boundary should be included in the flux methods owing to the effect of hydrological processes.

Here, for our case in a mangrove–*Spartina* ecotone in the Zhangjiang Estuary, we estimated the C stock changes on the atmosphere–sediment boundary in the vertical direction according to the dataset from this area with the stock–difference method following Howard et al. (2014). The temporal scale was set to cover 2002–2014 according to the historical dataset

TABLE 19.8 The C stock change in the sediment of the Zhangjiang Estuary affected by cordgrass invasion during 2013–14.

Parameters/Vegetation	Mangroves	Cordgrass	Mudflat	Data sources
SEC rates (mm yr^{-1})	19	32	7	Chen, unpublished data; Chen et al. (2018)
VA rates (mm yr^{-1})	31	35	18	
C density of surface sediment (g cm^{-3}, 0–10 cm depth)	0.158	0.129	0.07	
Area changed from 2013 to 14 (ha)	+0.56	+4.72	−5.28	Li et al. (2017)
Stock–difference method				
Net C stock changes in sediment (MgC ha^{-1})	300.2	412.8	−49	
The C stock changes with vegetation changes (MgC ha^{-1})	168.1	1948.4	−258.7	
The C stock change in the sediment of Zhangjiang Estuary in 2014 (Mg C)				**2375.3**

C, carbon; *SEC*, surface-elevation change; *VA*, vertical accretion

provided by Li et al. (2017), which indicated that the first cordgrass invasion was detected in 2002. The expansion of both mangroves and cordgrass covered the area of the mudflats. Combining SET monitoring from 2015 to 17 and in situ sediment dynamic observations in 2014 (Chen et al., 2018), we obtained the SEC rates of different vegetation (Table 19.8). In 2014, a total of 2375.3 Mg C was gained in the sediment in the Zhangjiang Estuary in China.

6. Conclusions and perspectives

In coastal wetland ecosystems, there is a broad range of biogeographic drivers and biotic and abiotic factors controlling the C dynamics. In our case, the invasion of cordgrass has caused remarkable changes in mangrove C sequestration and sediment deposition. The interaction between biological processes and sediment dynamics processes drives the C gain in the surface sediment. The dense biomass from fast-growing cordgrass substantially contributes to the below-ground C pool and accelerates the sedimentation rates through hydrodynamic condition modification and direct trapping of suspended particles. The surface elevation was significantly increased in cordgrass vegetation compared to that in native mangrove forests. With the expansion of cordgrass to mudflats, the total sediment C gain in a typical mangrove–*Spartina* ecotone has increased. This situation can be regarded as an advantage for sediment C deposition. Using the current conceptual model for C cycling in the mangrove–*Spartina* ecotones and the stock–difference method, we are able to estimate the net C deposition rates of these ecosystems. By multiplying the rates by the invasive area, we can upscale the total C deposition contributed by the invasion of cordgrass.

However, there are still some disadvantages of cordgrass marshes sequestering C in sediment. The high methane efflux from surface sediment induced by microbiome activity in the cordgrass marshes, which was approximately 275 times that in mudflats and 5–57 times that in mangroves, has greatly increased GWP. Owing to the uncertainty of C exchange on the boundaries between sediment water and sediment atmosphere, there is an urgent need for robust and applicable models for C flux evaluation in such ecotones. Multidisciplinary collaborations with integrated approaches to biology, hydrology, sediment dynamics, and geomorphology are needed to resolve this major problem.

Acknowledgments

We acknowledge partial financial from the National Natural Science Foundation of China (NSFC) (31770579, 41776096 and 41476071), the National Key Research and Development Program of China (2017YFC0506103), and the Fundamental Research Funds for the Central Universities of China (20720180118).

References

Adam, P., 1990. Saltmarsh Ecology. Cambridge University Press, London, UK.

Allen, J.R.L., Pye, K., 1992. Saltmarshes: Morphodynamics, Conservation and Engineering Significance. Cambridge University Press, London, UK.

Allen, J.R.L., 2000. Morphodynamics of holocene salt marshes: a review sketch from the Atlantic and Southern North Sea coasts of Europe. Quat. Sci. Rev. 19, 1155–1231.

Alongi, D.M., 2014. Carbon cycling and storage in mangrove forests. Annu. Rev. Mar. Sci. 6, 195–219.

Amos, C.L., Brylinsky, M., Sutherland, T.F., O'Brien, D., Lee, S., Cramp, A., 1998. The stability of a mud flat in the Humber estuary, South Yorkshire, U.K. In: Black, K.S., Paterson, D.M., Cramp, A. (Eds.), Sedimentary Processes in the Intertidal Zone, Special Publications. Geological Society, London, pp. 25–43.

Amos, C.L., Daborn, G.R., Christian, H.A., Atkinson, A., Robertson, A., 1992. In situ erosion measurements on fine-grained sediments from the Bay of Fundy. Mar. Geol. 108, 175–196.

An, S., Gu, B., Zhou, C., Wang, Z., Deng, Z., Zhi, B., Li, H., Chen, L., Yu, D., Liu, Y., 2007. *Spartina* invasion in China: implications for invasive species management and future research. Weed Res. 47, 183–191.

Arkema, K.K., Guannel, G., Verutes, G., Wood, S.A., Guerry, A., Ruckelshaus, M., Kareiva, P., Lacayo, M., Silver, J.M., 2013. Coastal habitats shield people and property from sea-level rise and storms. Nat. Clim. Change 3, 913–918.

Barbier, E.B., Hacker, S.D., Kennedy, C., Koch, E.W., Stier, A.C., Silliman, B.R., 2011. The value of estuarine and coastal ecosystem services. Ecol. Monogr. 81, 169–193.

Barbier, E.B., 2016. The protective service of mangrove ecosystems: a review of valuation methods. Mar. Pollut. Bull. 109, 676–681.

Biswas, S., Choudhury, J., Nishat, A., Rahman, M., 2007. Do invasive plants threaten the Sundarbans mangrove forest of Bangladesh? For. Ecol. Manag. 245, 1–9.

Bouillon, S., Frankignoulle, M., Dehairs, F., Velimirov, B., Eiler, A., Abril, G., Etcheber, H., Borges, A.V., 2003. Inorganic and organic carbon biogeochemistry in the Gautami Godavari estuary (Andhra Pradesh, India) during pre-monsoon: the local impact of extensive mangrove forests. Global Biogeochem. Cycles 17, 1114.

Bouma, T.J., De Vries, M.B.D., Low, E., Kusters, L., Herman, P.M.J., Tánczos, I.C., Temmerman, S., Hesselink, A., Meire, P., van Regenmortel, S.V., 2005. Flow hydrodynamics on a mudflat and in salt marsh vegetation: identifying general relationships for habitat characterisations. Hydrobiologia 540, 259–274.

Brinkman, R.M., Massel, S.R., Ridd, P.V., Furukawa, K., 1997. Surface wave attenuation in mangrove forests. In: Pacific Coasts and Ports '97: Proceedings of the 13th Australasian Coastal and Ocean Engineering Conference and the 6th Australasian Port and Harbour Conference, vol. 2, pp. 941–946.

Cahoon, D.R., Hensel, P., Rybczyk, J., McKee, K.L., Proffitt, C.E., Perez, B.C., 2003. Mass tree mortality leads to mangrove peat collapse at Bay Islands, Honduras after Hurricane Mitch. J. Ecol. 91, 1093–1105.

Cahoon, D.R., Hensel, P.F., Spencer, T., Reed, D.J., McKee, K.L., Saintilan, N., 2006. Coastal wetland vulnerability to relative sea-level rise: wetland elevation trends and process controls. In: Verhoeven, J.T.A., Beltman, B., Bobbink, R., Whigham, D.F. (Eds.), Wetlands and Natural Resource Management, Ecological Studies -Analysis and Synthesis. Springer, Berlin, Heidelberg, pp. 271–292.

Cahoon, D.R., Lynch, J.C., Perez, B.C., Segura, B., Holland, R.D., Stelly, C., Stephenson, G., Hensel, P., 2002. High-precision measurements of wetland sediment elevation: II. The rod surface elevation table. J. Sediment. Res. 72, 734–739.

Cahoon, D.R., Lynch, J.C., Powell, A.N., 1996. Marsh vertical accretion in a Southern California Estuary, U.S.A. Estuar. Coast Shelf Sci. 43, 19–32.

Cahoon, D.R., Reed, D.J., Day, J.W., 1995. Estimating shallow subsidence in microtidal salt marshes of the Southeastern United States: Kaye and Barghoorn revisited. Mar. Geol. 128, 1–9.

Callaway, J.C., Josselyn, M.N., 1992. The introduction and spread of smooth cordgrass (*Spartina alterniflora*) in south San Francisco Bay. Estuaries 15, 218–226.

Carus, J., Paul, M., Schroder, B., 2016. Vegetation as self-adaptive coastal protection: reduction of current velocity and morphologic plasticity of a brackish marsh pioneer. Ecol. Evol. 6, 1579–1589.

Chang, Y., Chen, Y., Li, Y., 2019. Flow modification associated with mangrove trees in a macro-tidal flat, Southern China. Acta Oceanol. Sin. 38, 1–10.

Chen, G., Ulumuddin, Y.I., Pramudji, S., Chen, S., Chen, B., Ye, Y., Ou, D., Ma, Z., Huang, H., Wang, J., 2014. Rich soil carbon and nitrogen but low atmospheric greenhouse gas fluxes from North Sulawesi mangrove swamps in Indonesia. Sci. Total Environ. 487, 91–96.

Chen, L., 2019. Invasive plants in coastal wetlands: patterns and mechanisms. In: An, S., Verhoeven, J.T.A. (Eds.), Wetlands: Ecosystem Services, Restoration and Wise Use. Springer Cham, New York, pp. 97–128.

Chen, L., Wang, W., Zhang, Y., Lin, G., 2009. Recent progresses in mangrove conservation, restoration and research in China. J. Plant Ecol. 2, 45–54.

Chen, Y., Chen, G., Ye, Y., 2015. Coastal vegetation invasion increases greenhouse gas emission from wetland soils but also increases soil carbon accumulation. Sci. Total Environ. 526, 19–28.

Chen, Y., Collins, M.B., Thompson, C.E.L., 2011. Mechanism of creek enlargement in a low-energy degrading saltmarsh, Southern England. Earth Surf. Process. Landforms 36, 767–778.

Chen, Y., Li, Y., Cai, T., Thompson, C., Li, Y., 2016. A comparison of biohydrodynamic interaction within mangrove and saltmarsh boundaries. Earth Surf. Process. Landforms 41, 1967–1979.

Chen, Y., Thompson, C.E.L., Collins, M.B., 2012. Saltmarsh creek bank stability: biostabilisation and consolidation with depth. Continent. Shelf Res. 35, 64–74.

Chen, Y., Li, Y., Thompson, C., Wang, X., Cai, T., Chang, Y., 2018. Differential sediment trapping abilities of mangrove and saltmarsh vegetation in a subtropical estuary. Geomorphology 318, 270–282.

Chen, Y., Thompson, C.E.L., Collins, M.B., 2019. Controls on creek margin stability by the root systems of saltmarsh vegetation, Beaulieu Estuary, Southern England. Anthropocene Coasts 2, 21–38.

Chen, B., Xiao, X., Li, X., Pan, L., Doughty, R., Ma, J., Dong, J., Qin, Y., Zhao, B., Wu, Z., Sun, R., Lan, G., Xie, G., Clinton, N., Giri, C., 2017. A mangrove forest map of China in 2015: Analysis of time series Landsat 7/8 and Sentinel-1A imagery in Google Earth Engine cloud computing Platform. ISPRS Journal of Photogrammetry and Remote Sensing, 131, 104–120.

Cheng, X., Luo, Y., Chen, J., Lin, G., Chen, J., Li, B., 2006. Short-term C_4 plant *Spartina alternifora* invasions change the soil carbon in C_3 plant-dominated tidal wetlands on a growing estuarine Island. Soil Biol. Biochem. 38, 3380–3386.

Christiansen, T., Wiberg, P.L., Milligan, T.G., 2000. Flow and sediment transport on a tidal salt marsh surface. Estuar. Coast Shelf Sci. 50, 315–331.

Chowdhury, R.R., Uchida, E., Chen, L., Osorio, V., Yoder, L., 2017. Anthropogeic drivers of mangrove loss: geographic patterns and implications for livelihoods. In: Rivera-Monroy, V.H., Lee, S.Y., Kristensen, E., Twilley, R.R. (Eds.), Mangrove Ecosystems: A Global Biogeogrphic Perspective. Springer, pp. 275–300.

Church, J.A., White, N.J., 2006. A 20th century acceleration in global sea level rise. Geophys. Res. Lett. 33, L01602.

Crooks, J.A., 2002. Predators of the invasive mussel, *Musculista senhousia* (Mollusca: Mytilidae). Pac. Sci. 56, 49–56.

Crosby, S.C., Angermeyer, A., Adler, J.M., Bertness, M.D., Deegan, L.A., Sibinga, N., Leslie, H.M., 2017. *Spartina alterniflora* biomass allocation and temperature: implications for salt marsh persistence with sea-level rise. Estuar. Coast 40, 213–223.

Donato, D.C., Kauffman, J.B., Murdiyarso, D., Kurnianto, S., Stidham, M., Kanninen, M., 2011. Mangroves among the most carbon-rich forests in the tropics. Nat. Geosci. 4, 293–297.

Doughty, C.L., Cavanaugh, K.C., Hall, C.R., Feller, I.C., Chapman, S.K., 2017. Impacts of mangrove encroachment and mosquito impoundment management on coastal protection services. Hydrobiologia 803, 105–120.

Dunn, R., Thomas, S.M., Keyss, A.J., Long, S.P., 1987. A comparison of the growth of the C_4 grass *Spartina anglica* with the C_3 grass *Lolium perenne* at different temperatures. J. Exp. Bot. 38, 433–441.

Elsey-Quirk, T., Seliskar, D.M., Sommerfield, C.K., Gallagher, J.L., 2011. Salt marsh carbon pool distribution in a mid-Atlantic coastal lagoon, USA: sea level rise implications. Wetlands 31, 87–99.

Fagherazzi, S., Bryan, K.R., Nardin, W., 2017. Buried alive or washed away: The challenging life of mangroves in the Mekong Delta. Oceanography 30, 48–59.

Feng, J., Guo, J., Huang, Q., Jiang, J., Huang, G., Yang, Z., Lin, G., 2014. Changes in the community structure and diet of benthic macrofauna in invasive *Spartina alterniflora* wetlands following restoration with native mangroves. Wetlands 34, 673–683.

Feng, J., Huang, Q., Chen, H., Guo, J., Lin, G., 2018. Restoration of native mangrove wetlands can reverse diet shifts of benthic macrofauna caused by invasive cordgrass. J. Appl. Ecol. 55, 905–916.

Feng, J., Zhou, J., Wang, L., Cui, X., Ning, C., Wu, H., Zhu, X., Lin, G., 2017. Effects of short-term invasion of *Spartina alterniflora* and the subsequent restoration of native mangroves on the soil organic carbon, nitrogen and phosphorus stock. Chemosphere 184, 774–783.

Fonseca, M.S., 1996. The role of seagrass in nearshore sedimentary processes: a review. In: Nordstrom, K.F., Roman, C.T. (Eds.), Estuarine Shores: Evolution, Environments and Human Alterations. John Wiley and Sons, Chichester, pp. 261–286.

Fourqurean, J.W., Smith, T.J., Possley, J., Collins, T.M., Lee, D., Namoff, S., 2010. Are mangroves in the tropical Atlantic ripe for invasion? Exotic mangrove trees in the forests of South Florida. Biol. Invasions 12, 2509–2522.

French, J.R., Spencer, T., 1993. Dynamics of sedimentation in a tide-dominated back barrier salt marsh, Norfolk, UK. Mar. Geol. 110, 315–331.

Friend, P.L., Ciavola, P., Cappucci, S., Santos, R., 2003. Bio-dependent bed parameters as a proxy tool for sediment stability in mixed habitat intertidal areas. Continent. Shelf Res. 23, 1899–1917.

Friess, D.A., Krauss, K.W., Horstman, E.M., Balke, T., Bouma, T.J., Galli, D., Webb, E.L., 2012. Are all intertidal wetlands naturally created equal? Bottlenecks, thresholds and knowledge gaps to mangrove and saltmarsh ecosystems. Biol. Rev. 87, 346–366.

Friess, D.A., Rogers, K., Lovelock, C.E., Krauss, K.W., Hamilton, S.E., Lee, S.Y., Lucas, R., Primavera, J., Rajkaran, A., Shi, S., 2019. The state of the world's mangrove forests: past, present and future. Annu. Rev. Environ. Resour. 44, 89–115.

Fry, J.E., 2014. A plant trait-based approach to evaluate the ability of native C_3 and C_4 garsses to restore functionality to a remnant bluegrass savanna-woodland in Kentucky, USA. Theses Diss. Biol. 20, 1–145.

Furukawa, K., Wolanski, E., Mueller, H., 1997. Currents and sediment transport in mangrove forests. Estuar. Coast Shelf Sci. 44, 301–310.

Gao, G., Peng, F., Zhi, J., Qin, Y., Zhang, X., Ghoto, K., Zhu, X., Zheng, H., 2018. Exotic *Spartina alterniflora* invasion increases CH_4 while reduces CO_2 emissions from mangrove wetland soils in southeastern China. Sci. Rep. 8, 9243–9253.

Gao, S., Collins, M.B., 1997. Formation of salt-marsh cliffs in an accretional environment, Christchurch Harbour, Southern England. In: Wang, P.X., Bergran, W. (Eds.), Proceedings of the 30th International Geological Congress. VSP Press, Amsterdam, pp. 95–100.

Gebrehiwet, T., Koretsky, C.M., Krishnamurthy, R.V., 2008. Influence of *Spartina* and *Juncus* on saltmarsh sediments. III. Organic geochemistry. Chem. Geol. 255, 114–119.

Giurgevich, J., Dunn, E., 1979. Seasonal patterns of CO_2 and water vapor exchange of the tall and short height forms of *Spartina alterniflora* Loisel in a Georgia salt marsh. Oecologia 43, 139–156.

Gong, Z., Zhao, K., Zhang, C., Dai, W., Coco, G., Zhou, Z., 2018. The role of bank collapse on tidal creek ontogeny: a novel process-based model for bank retreat. Geomorphology 311, 13–26.

Graham, G.W., Manning, A.J., 2007. Floc size and settling velocity within a *Spartina anglica* canopy. Continent. Shelf Res. 27, 1060–1079.

Guo, H., Zhang, Y., Lan, Z., Pennings, S.C., 2013. Biotic interactions mediate the expansion of black mangrove (*Avicennia germinans*) into salt marshes under climate change. Global Change Biol. 19, 2765–2774.

He, D., Lu, Y., Li, N., 2014. Spatial and temporal changes in coastal wetlands in Guangxi Beibu Gulf in recent 20 years. Wetland Sci. Manag. 10, 37–41.

Hester, M.W., Mendelssohn, I.A., McKee, K.L., 2001. Species and population variation to salinity stress in *Panicum hemitomon, Spartina* patens, and *Spartina alterniflora*: morphological and physiological constraints. Environ. Exp. Bot. 46, 277–297.

Hill, T.D., Roberts, B.J., 2012. Effects of seasonality and environmental gradients on *Spartina alterniflora* allometry and primary production. Ecol. Evol. 7, 9676–9688.

Horstman, E.M., Dohmen-Janssen, C.M., Bouma, T.J., Hulscher, S.J.M.H., 2015. Tidal-scale flow routing and sedimentation in mangrove forests: combining field data and numerical modelling. Geomorphology 228, 244–262.

Howard, J., Hoyt, S., Isensee, K., Telszewski, M., Pidgeon, E., 2014. Coastal blue carbon: methods for assessing carbon stocks and emissions factors in mangroves, tidal salt marshes, and seagrasses. J. Am. Hist. 14, 4–7.

Janzen, D.H., 1985. Mangroves: where's the understory? J. Trop. Ecol. 1, 89–92.

Johnson, B.J., Lovelock, C.E., Dorothée, H., 2016. Climate regulation: salt marshes and blue carbon. In: Finlayson, C.M., Everard, M., Irvine, K., Mcinnes, R.J., Middleton, B.A., van Dam, A.A., Davidson, N.C. (Eds.), The Wetland Book. Springer, Dordrecht, pp. 1–12.

Kelleway, J.J., Santilan, N., Macreadie, P.I., Skilbeck, C.G., Zawadzki, A., Ralph, P.J., 2016. Seventy years of continuous encroachment substantially increases 'blue carbon' capacity as mangroves replace intertidal salt marshes. Global Change Biol. 22, 1097–1109.

Kelleway, J.J., Carvanaugh, K., Rogers, K., Feller, I.C., Ens, E., Doughty, C., Saintilan, N., 2017. Review of the ecosystem service implications of mangrove encroachment into salt marshes. Global Change Biol. 23, 3967–3983.

Kennedy, D., 1995. Risk analysis, animal health and trade. Aust. Vet. J. 72, 37–38.

Kirwan, M.L., Megonigal, J.P., 2013. Tidal wetland stability in the face of human impacts and sea-level rise. Nature 504, 53–60.

Kirwan, M.L., Guntenspergen, G.R., Morris, J.T., 2009. Latitudinal trends in *Spartina alterniflora* productivity and the response of coastal marshes to global change. Global Change Biol. 15, 1982–1989.

Knighton, A.D., Woodroffe, C.D., Mills, K., 1992. The evolution of tidal creek networks, Mary River, Northern Australia. Earth Surf. Process. Landforms 17, 167–190.

Knutson, P.L., Brochu, R.A., Seelig, W.N., Inskeep, M., 1982. Wave damping in *Spartina alterniflora* marshes. Wetlands 2, 87–104.

Kobashi, D., Mazda, Y., 2005. Tidal flow in riverine-type mangroves. Wetl. Ecol. Manag. 13, 615–619.

Krauss, K.W., Lovelock, C.E., McKee, K.L., Lopez-Hoffman, L., Ewe, S.M.L., Sousa, W.P., 2008. Environmental drivers in mangrove establishment and early development: A review. Aquat. Bot. 89, 105–127.

Krauss, K.W., Allen, J.A., Cahoon, D.R., 2003. Differential rates of vertical accretion and elevation change among aerial root types in Micronesian mangrove forests. Estuar. Coast Shelf Sci. 56, 251–259.

Krauss, K.W., Cormier, N., Osland, M.J., Kirwan, M.L., Stagg, C.L., Nestlerode, J.A., Russell, M.J., From, A.S., Spivak, A.C., Dantin, D.D., 2017. Created mangrove wetlands store belowground carbon and surface elevation change enables them to adjust to sea-level rise. Sci. Rep. 7, 1030–1041.

Krauss, K.W., From, A.S., Doyle, T.W., Doyle, T.J., Barry, M.J., 2011. Sea-level rise and landscape change influence mangrove encroachment onto marsh in the ten thousand Islands region of Florida, USA. J. Coast Conserv. 15, 629–638.

Krauss, K.W., Mckee, K.L., Lovelock, C.E., Cahoon, D.R., Saintilan, N., Reef, R., Chen, L., 2014. How mangrove forests adjust to rising sea level. New Phytol. 202, 19–34.

Leonard, L.A., Luther, M.E., 1995. Flow hydrodynamics in tidal marsh canopies. Limnol. Oceanogr. 40, 1474–1484.

Leonard, L.A., Croft, A.L., 2006. The effect of standing biomass on flow velocity and turbulence in *Spartina alterniflora* canopies. Estuar. Coast Shelf Sci. 69, 325–336.

Li, B., Liao, C., Zhang, X., Chen, H., Wang, Q., Chen, Z., Gan, X., Wu, J., Zhao, B., Ma, Z., Cheng, X., Jiang, L., Chen, J., 2009. *Spartina alterniflora* invasions in the Yangtze River estuary, China: an overview of current status and ecosystem effects. Ecol. Eng. 35, 511–520.

Li, H., Yang, S., 2007. A review of influences of saltmarsh vegetation on physical processes in intertidal wetlands. Adv. Earth Sci. 22, 583–591 (in Chinese with English abstract).

Li, Y., Chen, Y., Li, Y., 2017. Remote sensing analysis of the changes in the ecotone of mangrove forests and *Spartina alterniflora* saltmarshes. Mar. Sci. Bull. 36, 348–360 (in Chinese with English abstract).

Li, Z., Wang, W., Zhang, Y., 2014. Recruitment and herbivory affect spread of invasive *Spartina alterniflora* in China. Ecology 95, 1972–1980.

Liao, B., Zhang, Q., 2014. Area, distribution and species composition of mangroves in China. Wetland Sci. 12, 435–440.

Liao, C., Luo, Y., Jiang, L., Zhou, X., Wu, X., Fang, C., Chen, J., Li, B., 2007. Invasion of *Spartina alterniflora* enhanced ecosystem carbon and nitrogen stocks in the Yangtze Estuary, China. Ecosystems 10, 1351–1361.

Liu, P.C., 2000. Is the wind wave frequency spectrum outdated? Ocean Eng. 27, 577–588.

Liu, H., Ren, H., Hui, D., Wang, W., Liao, B., Cao, Q., 2014. Carbon stocks and potential carbon storage in the mangrove forests of China. J. Environ. Manag. 133, 86–93.

Liu, J., Han, R., Su, H., Wu, Y., Zhang, L., Richardson, C.J., Wang, G., 2017b. Effects of exotic, *Spartina alterniflora*, on vertical soil organic carbon distribution and storage amount in coastal salt marshes in Jiangsu, China. Ecol. Eng. 106, 132–139.

Liu, M., Mao, D., Wang, Z., Li, L., Man, W., Jia, M., Ren, C., Zhang, Y., 2018. Rapid invasion of *Spartina alterniflora* in the coastal zone of mainland China: new observations from Landsat OLI images. Rem. Sens. 10, 1–19.

Liu, W., Maung-Douglass, K., Strong, D.R., Pennings, S.C., Zhang, Y., 2016b. Geographical variationin vegetative growth and sexual reproduction of the invasive *Spartina alterniflora* in China. J. Ecol. 104, 173–181.

Liu, W., Strong, D.R., Pennings, S.C., Zhang, Y., 2017a. Provenance-by-environment interaction of reproductive traits in the invasion of *Spartina alterniflora* in China. Ecology 98, 1591–1599.

Liu, Y., Li, X., Yan, Z., Chen, X., He, Y., Guo, W., Sun, P., 2013. Biomass and carbon storage of *Phragmites australis* and *Spartina alterniflora* in Jiuduan Shoal Wetland of Yangtze Estuary, East China. Chin. J. Appl. Ecol. 24, 2129–2134 (in Chinese with English abstract).

Liu, Z., Cui, B., He, Q., 2016a. Shifting paradigms in coastal restoration: Six decades' lessons from China. Sci. Total Environ. 566–567, 205–214.

Loomis, M., Craft, C., 2010. Carbon sequestration and nutrient (nitrogen, phosphorus) accumulation in river-dominated tidal marshes, Georgia, USA. Soil Sci. Soc. Am. J. 74, 1028–1037.

Lovelock, C.E., Adame, M.F., Bennion, V., Hayes, M., O'Mara, J., Reef, R., Santini, N.S., 2014. Contemporary rates of carbon sequestration through vertical accretion of sediments in mangrove forests and saltmarshes of South East Queensland, Australia. Estuar. Coast 37, 763–771.

Lovelock, C.E., Bennion, V., Grinham, A., Cahoon, D.R., 2011. The role of surface and subsurface processes in keeping pace with sea level rise in intertidal wetlands of Moreton bay, Queensland, Australia. Ecosystems 14, 745–757.

Lovelock, C.E., Cahoon, D.R., Friess, D.A., Guntenspergen, G.R., Krauss, K.W., Reef, R., Rogers, K., Saunders, M.L., Sidik, F., Swales, A., Saintilan, N., Thuyen, L.X., Triet, T., 2015. The vulnerability of Indo-Pacific mangrove forests to sea-level rise. Nature 526, 559–563.

Lugo, A.E., 1986. Mangrove understory: an expensive luxury? J. Trop. Ecol. 2, 287–288.

Lunstrum, A., Chen, L., 2014. Soil carbon stocks and accumulation in young mangrove forests. Soil Biol. Biochem. 75, 223–232.

Massel, S.R., Furukawa, K., Brinkman, R.M., 1999. Surface wave propagation in mangrove forests. Fluid Dynam. Res. 24, 219–249.

Mazda, Y., 2003. The Role of Physical Processes in Mangrove Environments. Eur. J. Biochem. 52, 9–16.

Mazda, Y., Kobashi, D., Okada, S., 2005. Tidal-Scale Hydrodynamics within Mangrove Swamps. Wetl. Ecol. Manag. 13, 647–655.

Mazda, Y., Wolanski, E., 2009. Hydrodynamics and modeling of water flow in mangrove areas. In: Gerardo, M.E., Wolanski, E., Cahoon, D.R., et al. (Eds.), Coastal Wetlands: An Integrated Ecosystem Approach. Elsevier, Amsterdam, pp. 231–261.

Mazda, Y., Wolanski, E., King, B., Sase, A., Ohtsuka, D., Magi, M., 1997. Drag force due to vegetation in mangrove swamps. Mangroves Salt Marshes 1, 193–199.

McCall, B.D., Pennings, S.C., 2012. Geographical variation in salt marsh structure and function. Oecologia 170, 777–787.

McIvor, C.C., 2012. Community ecology of mangrove rivulus at two west Florida locations. Integr. Comp. Biol. 52, 294.

McKee, K.L., 2011. Biophysical controls on accretion and elevation change in Caribbean mangrove ecosystems. Estuar. Coast Shelf Sci. 91, 475–483.

McKee, K.L., Rooth, J.E., Feller, I.C., 2007a. Mangrove recruitment after forest disturbance is facilitated by herbaceous species in the Caribbean. Ecol. Appl. 17, 1678–1693.

McKee, K.L., Cahoon, D.R., Feller, I.C., 2007b. Caribbean mangroves adjust to rising sea level through biotic controls on change in soil elevation. Global Ecol. Biogeogr. 16, 545–556.

McKee, K.L., Vervaeke, W.C., 2017. Will fluctuations in salt marsh–mangrove dominance alter vulnerability of a subtropical wetland to sea-level rise? Global Change Biol. 24, 1224–1238.

Mcleod, E., Chmura, G.L., Bouillon, S., Salm, R., Björk, M., Duarte, C.M., Lovelock, C.E., Schlesinger, W.H., Silliman, B.R., 2011. A blueprint for blue carbon: toward an improved understanding of the role of vegetated coastal habitats in sequestering CO_2. Front. Ecol. Environ. 9, 552–560.

Mendelssohn, I.A., Morris, J.T., 2002. Eco-physiological controls on the productivity of *Spartina alterniflora* Loisel. In: Weinstein, M.P., Kreeger, D.A. (Eds.), Concepts and Controversies in Tidal Marsh Ecology. Kluwer Academic Publishers, New York, pp. 59–80.

Middelburg, J.J., Nieuwenhuize, J., Lubberts, R.K., van de Plassche, O., 1997. Organic carbon isotope systematics of coastal marshes. Estuar. Coast Shelf Sci. 45, 681–687.

Mitsch, W.J., Gosselink, J.G., 2007. Wetlands. John Wiley and Sons, Chichester, p. 619.

Möller, I., Kudella, M., Rupprecht, F., Spencer, T., Paul, M., van Wesenbeeck, B.K., Wolters, G., Jensen, K., Bouma, T.J., Miranda-Lange, M., Schimmels, S., 2014. Wave attenuation over coastal salt marshes under storm surge conditions. Nat. Geosci. 7, 727–731.

Möller, I., Spencer, T., 2002. Wave dissipation over macro-tidal saltmarshes: Effects of marsh edge typology and vegetation change. J. Coast Res. 36, 506–521.

Möller, I., 2006. Quantifying saltmarsh vegetation and its effect on wave height dissipation: results from a UK East coast saltmarsh. Estuar. Coast Shelf Sci. 69, 337–351.

Möller, I., Spencer, T., French, J.R., Leggett, D.J., Dixon, M., 1999. Wave transformation over salt marshes: A field and numerical modelling study from north Norfolk, England. Estuar. Coast Shelf Sci. 49, 411–426.

Möller, I., Spencer, T., French, J.R., Leggett, D.J., Dixon, M., 2001. The sea-defence value of salt marshes: field evidence from north Norfolk. J. Chatered Inst. Water Environ. J. 15, 109–116.

Molly, E.K., Törbjorn, E.T., 2019. Measuring rates of present-day relative sea-level rise in low-elevation coastal zones: a critical evaluation. Ocean Sci. 15, 61–73.

Morris, J.T., Sundareshwar, P.V., Nietch, C.T., Kjerfve, B., Cahoon, D.R., 2002. Responses of coastal wetlands to rising sea level. Ecology 83, 2869–2877.

Mullarney, J.C., Henderson, S.M., Reyns, J.A.H., Norris, B.K., Bryan, K.R., 2017. Spatially varying drag within a wave-exposed mangrove forest and on the adjacent tidal flat. Continent. Shelf Res. 147, 102–113.

Nellemann, C., Corcoran, E., Duarte, C.M., Valdés, L., De Young, C., Fonseca, L., Grimsditch, G., 2009. Blue Carbon. A Rapid Response Assessment. Birkeland Trykkeri AS, Norway.

Nepf, H.M., 1999. Drag, turbulence, and diffusion in flow through emergent vegetation. Water Resour. Res. 35, 479–489.

Nepf, H.M., 2012. Flow and transport in regions with aquatic vegetation. Annu. Rev. Fluid Mech. 44, 123–142.

Neubauer, S.C., 2008. Contributions of mineral and organic components to tidal freshwater marsh accretion. Estuar. Coast Shelf Sci. 78, 78–88.

Neumeier, U., 2007. Velocity and turbulence variations at the edge of saltmarshes. Continent. Shelf Res. 27, 1046–1059.

Neumeier, U., Amos, C.L., 2006. Turbulence reduction by the canopy of coastal *Spartina* salt-marshes. J. Coast Res. SI39, 433–439.

Neumeier, U., Ciavola, P., 2004. Flow resistance and associated sedimentary processes in a *Spartina* maritime saltmarsh. J. Coast Res. 20, 435–447.

Nicholls, R.J., 2004. Coastal Flooding and wetland loss in the 21st century: changes under the SRES climate and socioeconomic scenarios. Global Environ. Change 14, 69–86.

Oliver, T.S.N., Rogers, K., Chafer, C.J., Woodroffe, C.D., 2012. Measuring, mapping and modelling: an integrated approach to the management of mangrove and saltmarsh in the Minnamurra River estuary, southeast Australia. Wetl. Ecol. Manag. 20, 353–371.

Pendleton, L., Donato, D.C., Murray, B.C., Crooks, S., Jenkins, W.A., Sifleet, S., Craft, C., Fourqurean, J.W., Kauffman, J.B., Marbà, N., Megonigal, P., Pidgeon, E., Herr, D., Gordon, D., Baldera, A., 2012. Estimating global' blue carbon' emissions from conversion and degradation of vegetated coastal ecosystems. PloS One 7, e43542.

Peng, D., Chen, L., Pennings, S.C., Zhang, Y., 2018. Using a marsh organ to predict future plant communities in a Chinese estuary invaded by an exotic grass and mangrove. Limnol. Oceanogr. 63, 2595–2605.

Peng, Y., Zheng, M., Zheng, Z., Wu, G., Chen, Y., Xu, H., 2016. Virtual increase or latent loss? A reassessment of mangrove populations and their conservation in Guangdong, southern China. Mar. Pollut. Bull. 109, 691–699.

Perillo, G.E.M., Wolanski, E., Cahoon, D.R., Brinson, M.M., 2009. Coastal Wetlands: An Integrated Ecosystem Approach. Elsevier, Amsterdam.

Peterson, J.M., Bell, S.S., 2012. Tidal events and salt-marsh structure influence black mangrove (*Avicennia germinans*) recruitment across an ecotone. Ecology 93, 1648–1658.

Pethick, J.S., 1981. Long term accretion rates on tidal salt marshes. J. Sediment. Petrol. 51, 571–577.

Pinsky, M.L., Guannel, G., Arkema, K.K., 2013. Quantifying wave attenuation to inform coastal habitat conservation. Ecosphere 4, 95–111.

Poffenbarger, H.J., Needelman, B.A., Megonigal, J.P., 2011. Salinity Influence on Methane Emissions from Tidal Marshes. Wetlands 31, 831–842.

Qiao, H., Liu, W., Zhang, Y., Zhang, Y., Li, Q., 2019. Genetic admixture accelerates invasion via provisioning rapid adaptive evolution. Mol. Ecol. 28, 4012–4027.

Quartel, S., Kroon, A., Augustinus, P.G.E.F., van Santen, P., Tri, N.H., 2007. Wave attenuation in coastal mangroves in the Red River Delta, Vietnam. J. Asian Earth Sci. 29, 576–584.

Rogers, K., Saintilan, N., Copeland, C., 2013b. Modelling wetland surface elevation dynamics and its application to forecasting the effects of sea-level rise on estuarine wetlands. Ecol. Model. 244, 148–157.

Rogers, K., Saintilan, N., Howe, A.J., Rodríguez, J.F., 2013a. Sedimentation, elevation and marsh evolution in a southeastern Australian estuary during changing climatic conditions. Estuar. Coast Shelf Sci. 133, 172–181.

Rogers, K., Saintilan, N., Woodroffe, C.D., 2014. Surface elevation change and vegetation distribution dynamics in a subtropical coastal wetland: Implications for coastal wetland response to climate change. Estuar. Coast Shelf Sci. 149, 46–56.

Rogers, K., Wilton, K.M., Saintilan, N., 2006. Vegetation change and surface elevation dynamics in estuarine wetlands of southeast Australia. Estuar. Coast Shelf Sci. 66, 559–569.

Roman, C.T., Peck, J.A., Allen, J.R., King, J.W., Appleby, P.G., 1997. Accretion of a New England (U.S.A.) salt marsh in response to inlet migration, storms, and sea-level rise. Estuar. Coast Shelf Sci. 45, 717–727.

Rovai, A.S., Twilley, R.R., Castaneda-Moya, E., Piul, R., Cifuentes-Jara, M., Manrow-Villalobos, M., Horta, P.A., Simonassi, J.C., Fonseca, A.L., Pagliosa, P.R., 2018. Global controls on carbon storage in mangrove soils. Nat. Clim. Change 8, 534–538.

Rutherfurd, I., Abernethy, B., Prosser, I., 1999. Stream Erosion. In: Siwan, L., Phil, P. (Eds.), Riparian Land Management Technical Guidelines, Volume Two: On-Ground Management Tools and Techniques. LWRRDC, Canberra, pp. 61–78.

Saintilan, N., Rogers, K., 2006. Coastal Wetland Elevation Trends in Southeastern Australia. In: Gilman, E. (Ed.), Proceedings of the Symposium on Mangrove Responses to Relative Sea-Level Rise and Other Climate Change Effects. Western Pacific Regional Fishery Management Council, Honolulu, USA, pp. 42–54.

Saintilan, N., Rogers, K., Mazumder, D., Woodroffe, C., 2013. Allochthonous and autochthonous contributions to carbon accumulation and carbon store in southeastern Australian coastal wetlands. Estuar. Coast Shelf Sci. 128, 84–92.

Shi, Z., Pethick, J.S., Pye, K., 1995. Flow structure in and above the various heights of a saltmarsh canopy: a laboratory flume study. J. Coast Res. 11, 1204–1209.

Soulsby, R., 1997. Dynamics of Marine Sands: A Manual for Practical Applications. Thomas Telford, London.

Spalding, M., Kainuma, M., Collins, L., 2010. World Atlas of Mangroves. Earthscan, London.

Stevenson, J.C., Ward, L.G., Kearney, M.S., 1986. Vertical accretion in marshes with varying rates of sea level rise. In: Wolfe, D.A. (Ed.), Estuarine Variability. Academic Press, San Diego, pp. 241–259.

Strong, D.R., Ayres, D.R., 2013. Ecological and evolutionary misadventures of *Spartina*. Annu. Rev. Ecol. Evol. Syst. 44, 389–410.

Taillardat, P., Friess, D.A., Lupascu, M., 2018. Mangrove blue carbon strategies for climate change mitigation are most effective at the national scale. Biol. Lett. 14, 20180251.

Teisson, C., Ockenden, M., Lehir, P., Kranenburg, C., Hamm, L., 1993. Cohesive sediment transport processes. Coast. Eng. 21, 129–162.

Temmerman, S., Bouma, T.J., Govers, G., Wang, Z., De Vries, M.B., Herman, P.M.J., 2005. Impact of vegetation on flow routing and sedimentation patterns: Three-dimensional modeling for a tidal marsh. J. Geophys. Res. 110, F04019.

Temmerman, S., Meire, P., Bouma, T.J., Herman, P.M.J., Ysebaert, T., De Vriend, H.J., 2013. Ecosystem-based coastal defence in the face of global change. Nature 504, 79–83.

Thomas, N., Lucas, R.M., Bunting, P., Hardy, A., Rosenqvist, A., Simard, M., 2017. Distribution and drivers of global mangrove forest change, 1996–2010. PloS One 12, e0179302.

Tomlinson, P., 1986. The Botany of Mangroves. Cambridge University Press, Cambridge, UK.

Tong, C., Morris, J.T., Huang, J., Xu, H., Wan, S., 2018. Changes in pore-water chemistry and methane emission following the invasion of Spartina alterniflora into an oliogohaline marsh. Limnol. Oceanogr. 63, 384–396.

Tong, C., Wang, W., Huang, J., Gauci, V., Zhang, L., Zeng, C., 2012. Invasive alien plants increase CH_4 emissions from a subtropical tidal estuarine wetland. Biogeochemistry 111, 677–693.

Turner, R., 1976. Geographic variations in salt marsh macrophyte production: a review. Contrib. Mar. Sci. 20, 47–68.

Twilley, R.R., Castañeda-Moya, E., Rivera-Monroy, V.H., Rovai, A., 2017. Productivity and carbon dynamics in mangrove wetlands. In: Rivera-Monroy, V.H., Lee, S.Y., Kristensen, E., Twilley, R.R. (Eds.), Mangrove Ecosystems: A Global Biogeographic Perspective. Springer, Cham, pp. 113–162.

van der Wal, D., Pye, K., 2004. Patterns, rates and possible causes of saltmarsh erosion in the Greater Thames area (UK). Geomorphology 61, 373–391.

van Eerdt, M.M., 1985. Salt marsh cliff stability in the Oosterschelde. Earth Surf. Process. Landforms 10, 95–106.

Vasquez, E.A., Glenn, E.P., Guntenspergen, G.R., Brown, J.J., Nelson, S.G., 2006. Salt tolerance and osmotic adjustment of *Spartina alterniflora* (Poaceae) and the invasive M haplotype of *Phragmites australis* (Poaceae) along a salinity gradient. Am. J. Bot. 93, 1784–1790.

Voulgaris, G., Meyers, S.T., 2004. Temporal variability of hydrodynamics, sediment concentration and sediment settling velocity in a tidal creek. Continent. Shelf Res. 24, 1659–1683.

Wang, A., Gao, S., Jia, J., Pan, S., 2005. Contemporary sedimentation rates on saltmarshes at Wanggang, Jiangsu, China. J. Geogr. Sci. 15, 199–209.

Wang, M., Wang, Q., Sha, C., Chen, J., 2018. *Spartina alterniflora* invasion affects soil carbon in a C_3 plant-dominated tidal marsh. Sci. Rep. 8, 628–639.

Wang, W., Sardans, J., Wang, C., Zeng, C., Tong, C., Chen, G., Huang, J., Pan, H., Peguero, G., Vallicrosa, H., Peñuelas, J., 2019. The response of stock of C, N, and P to plant invasion in the coastal wetlands of China. Global Change Biol. 25, 733–743.

Wang, Y., Zhang, R., Gao, S., 1999. Velocity variations in salt marsh creeks, Jiangsu, China. J. Coast Res. 15, 471–477.

Webb, E.L., Friess, D.A., Krauss, K.W., Cahoon, D.R., Guntenspergen, G.R., Phelps, J., 2013. A global standard for monitoring coastal wetland vulnerability to accelerated sea-level rise. Nat. Clim. Change 3, 458–465.

Widdows, J., Brinsley, M., 2002. Impact of biotic and abiotic processes on sediment dynamics and the consequences to the structure and functioning of the intertidal zone. J. Sea Res. 48, 143–156.

Wieski, K., Guo, H., Craft, C.B., Pennings, S.C., 2010. Ecosystem functions of tidal fresh, brackish, and salt marshes on the Georgia coast. Estuar. Coast 33, 161–169.

Wieski, K., Pennings, S.C., 2014. Climate drivers of *Spartina alterniflora* saltmarsh production in Georgia, USA. Ecosystems 17, 473–484.

Wolanski, E., 1995. Transport of sediment in mangrove swamps. Hydrobiologia 295, 31–42.

Wolanski, E., Mazda, Y., Ridd, P., 1992. Mangrove hydrodynamics. In: Robertson, A.I., Alongi, D. (Eds.), Tropical Mangrove Ecosystems. American Geophysical Union, Washington DC, pp. 43–62.

Woodroffe, C.D., Rogers, K., McKee, K.L., Lovelock, C.E., Mendelssohn, I.A., Saintilan, N., 2016. Mangrove sedimentation and response to relative sea-level rise. Annu. Rev. Mar. Sci. 8, 243–266.

Xu, G., Zhuo, R., 1985. Preliminary studies of introduced *Spartina alterniflora* Loisel in China. J. Nanjing Univ. 40, 212–225 (in Chinese with English abstract).

Yando, E.S., Osland, M.J., Willis, J.M., Day, R.H., Krauss, K.W., Hester, M.W., 2016. Salt marsh-mangrove ecotones: using structural gradients to investigate the effects of woody plant encroachment on plant–soil interactions and ecosystem carbon pools. J. Ecol. 104, 1020–1031.

Yang, S., 1998. The role of *Scirpus* marsh in attenuation of hydrodynamics and retention of fine sediment in the Yangtze Estuary. Estuar. Coast Shelf Sci. 47, 227–233.

Yang, S., Chen, J., 1994. The role of vegetation in the mud coast processes. Oceanol. Limnol. Sinica 25, 631–635 (in Chinese with English abstract).

Yang, S., Shi, B., Bouma, T.J., Ysebaert, T., Luo, X., 2012. Wave attenuation at a salt marsh margin: A case study of an exposed coast on the Yangtze Estuary. Estuar. Coast 35, 169–182.

Yang, W., Yan, Y., Jiang, F., Leng, X., Cheng, X., An, S., 2016. Response of the soil microbial community composition and biomass to a short-term *Spartina alterniflora* invasion in a coastal wetland of eastern China. Plant Soil 408, 443–456.

Yang, W., Zhao, H., Chen, X., Yin, S., Cheng, X., An, S., 2013. Consequences of short-term C_4 plant *Spartina alterniflora* invasions for soil organic carbon dynamics in a coastal wetland of eastern China. Ecol. Eng. 61, 50–57.

Yin, S., An, S., Deng, Q., Zhang, J., Ji, H., Cheng, X., 2015. *Spartina alternifora* invasions impact CH_4 and N_2O fuxes from a salt marsh in eastern China. Ecol. Eng. 81, 192–199.

Young, B.M., Harvey, E.L., 1996. A spatial analysis of the relationship between mangrove (*Avicennia marinavar* var *australasica*) physiognomy and sediment accretion in the Hauraki Plains, New Zealand. Estuar. Coast Shelf Sci. 42 (2), 231–246.

Ysebaert, T., Yang, S.L., Zhang, L.Q., He, Q., Bouma, T.J., Herman, P.M.J., 2011. Wave attenuation by two contrasting ecosystem engineering salt marsh macrophytes in the intertidal pioneer zone. Wetlands 31, 1043–1054.

Yu, X., Yang, J., Liu, L., Tian, Y., Yu, Z., 2015. Effects of *Spartina alternifora* invasion on biogenic elements in a subtropical coastal mangrove wetland. Environ. Sci. Pollut. Res. 22, 3107–3115.

Yuan, J., Ding, W., Liu, D., Kang, H., Freeman, C., Xiang, J., Lin, Y., 2015. Exotic *Spartina alterniflora* invasion alters ecosystem-atmosphere exchange of CH_4 and N_2O and carbon sequestration in a coastal salt marsh in China. Global Change Biol. 21, 1567–1580.

Zhang, Q., Wen, X., Song, J., Liu, S., 1996. The measurement and study on sedimentation rates in mangrove tidal flats. Trop. Oceanol. 15, 57–62 (in Chinese with English abstract).

Zhang, R., Shen, Y., Lu, L., Yan, S., Wang, Y., Li, J., 2004. Formation of *Spartina alterniflora* salt marshes on the coast of Jiangsu Province, China. Ecol. Eng. 23, 95–105.

Zhang, Y., Huang, G., Wang, W., Chen, L., Lin, G., 2012. Interactions between mangroves and exotic *Spartina* in an anthropogenically disturbed estuary in southern China. Ecology 93, 588–597.

Zhang, Y., Meng, H., Wang, Y., He, Q., 2018. Herbivory enhances the resistance of mangrove forest to cordgrass invasion. Ecology 99, 1382–1390.

Zhao, Y., Yu, Q., Wang, D., Wang, Y., Wang, Y., Gao, S., 2017. Rapid formation of marsh-edge cliffs, Jiangsu coast, China. Mar. Geol. 385, 260–273.

Zhu, X., Meng, L., Zhang, Y., Weng, Q., Morris, J., 2019. Tidal and meteorological influences on the growth of invasive *Spartina alterniflora*: Evidence from UAV remote sensing. Rem. Sens. 11, 1208.

Zuo, P., Zhao, S., Liu, C., 2012. Distribution of *Spartina* spp. along China's coast. Ecol. Eng. 40, 160–166.

PART IV

Mangrove management and restoration

CHAPTER

20

A framework for the quantitative assessment of mangrove resilience

Wei Jian Ong, Joanna C. Ellison

School of Geography, Planning and Spatial Sciences, University of Tasmania, Launceston, Tasmania, Australia

OUTLINE

1. Introduction 513
   1.1 Definitions of resilience 514
   1.2 Adaptive capacity 515
   1.3 Absorptive capacity 516
   1.4 Transformative capacity 516
   1.5 Resilience assessment 516
   1.6 Objective 517

2. Methods 517
   2.1 Study areas 517
   2.2 Resilience assessment methodology 517

3. Results and discussion 521
   3.1 Stressor components 522
   3.2 Ecosystem robustness components 526
   3.3 Management components 527
   3.4 Improvement of resilience 529

4. Conclusions 530

References 532

1. Introduction

Resilience in coastal areas has been recognized as a core requirement to sustainable coastal management in the last decade (Rumson et al., 2019), and coastal ecosystems are providers of valuable services and resources (Lu et al., 2018). Mangroves are important for coastal protection, from waves and during storms, particularly from the destructive winds and storm surge associated with typhoons or cyclones (Blankespoor et al., 2016). Mangroves have high rates of

https://doi.org/10.1016/B978-0-12-816437-2.00016-1

tree and plant growth, and waterlogged soils slow decomposition, resulting in long-term carbon storage (Murdiyarso et al., 2015). With an estimated global mangrove area of 137,760 square kilometers (Bunting et al., 2018), carbon sink values are appreciable. However, despite these values that mangroves provide, the worldwide mangrove area has been coarsely estimated to have reduced by more than 50% relative to what it was just over a century ago, owing to direct human impacts (Spalding et al., 2010; Giri et al., 2011). Human stressors on mangroves include forest logging, coastal land use change, coastal engineering, conversion to aquaculture, pollutant inputs, and mining (He and Silliman, 2019).

Climate change is adding more pressures on mangroves. Global greenhouse gas emissions have set a record high this century, leading to adverse effects on the environment (Doyle, 2019; World Meteorological Organisation, 2019). Rising sea levels due to steric effects and melting of glaciers and ice caps have caused increased flooding, coastal erosion, and saltwater intrusion (Nicholls and Cazenave, 2010). Future projections are of increased sea level rise rates (DeConto and Pollard, 2016; Jevrejaya et al., 2019; World Meteorological Organisation, 2019), which will have impacts on mangroves (Ellison, 2018). Owing to sea level rise and net substrate accretion not keeping up, mangroves of the Southeast Asia/West Pacific region have risk of loss by the end of this century (Lovelock et al., 2015).

Mangrove forest management and rehabilitation is imperative, with potential climate change threats to their sustainability increasingly identified (Lewis et al., 2016). Mangroves have suffered major losses in the last century and future impacts will continue without directed management to increase mangrove system resilience.

1.1 Definitions of resilience

Resilience has evolved into different definitions in the last few decades. Traditionally, resilience is the capacity to tolerate disturbances while maintaining structure, function, and identity (Holling, 1973; Cinner and Barnes, 2019). Resilience has evolved into different domains of applications (Davidson et al., 2016). Ecological resilience is the capacity of natural systems to absorb disturbance without changing function, structure, and identity (Holling, 1973), avoiding change that alters an ecosystem from a state of processes and structures to a different state (Angler and Allen, 2016). Engineering resilience focuses on return to the structural and functional attributes before the system was disturbed (Angeler and Allen, 2016), which may have relevance to mangroves being recognized as ecosystem engineers (Cheong et al., 2013). Two categories of indicators are considered to contribute to mangrove resilience: ecological (Day et al., 2018) and socioeconomic (McLeod and Salm, 2006). Ecological indicators are factors involved in ecological integrity, further classified into two categories of abiotic and biotic factors (Day et al., 2018). Socio-economic indicators include social, economic, and political aspects of mangrove sustainability (Randhir, 2016), operating at different spatial scales over timeframes (Fig. 20.1).

The resilience of ecosystem-based strategies is controlled by biotic, abiotic, social, economic, and political systems as part of a multiscale ecosystem framework that works locally, nationally, and globally (Randhir, 2016). Resilience is recognized as ecological integrity (Day et al., 2018), as indicated by suitable abiotic and ecosystem attributes and support services, some of which, however, are not practical to apply. Although ecosystem resilience is a widely recognized concept, it remains a non-quantifiable system. Resilience is a broad concept that

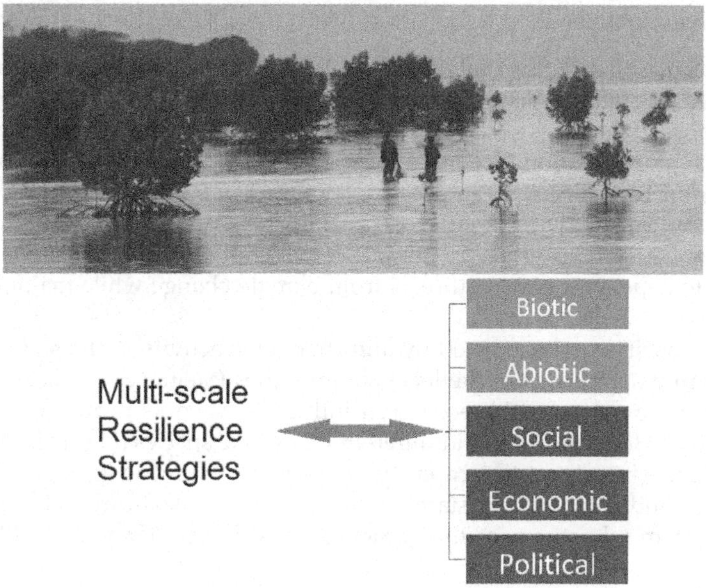

FIGURE 20.1 Multiscale resilience strategy framework for coastal ecosystems that operates over time. *Adapted from Randhir, T.O., 2016. Globalization impacts on local commons multiscale strategies for socioeconomic and ecological resilience. Int. J. Commons 10 (1), 387–404.*

encompasses adaptive capacity (Mumby et al., 2014), absorptive capacity, and transformative capacity (Béné, 2013; White et al., 2018). All these three capacities of resilience progress increasing impacts, costs, and types of responses and identify measures that could be used to monitor resilience (Béné, 2013).

1.2 Adaptive capacity

Adaptive capacity is defined as the conditions by which a system can deal with, accommodate, adapt to, and recover, involving thresholds and coping ranges (Smit and Wandel, 2006). As such, it is applied to ecosystems (Angler et al., 2019) and is also a dimension of vulnerability (Ellison, 2015). Adaptive capacity for mangroves can be improved by improved management, stakeholder involvement, protection status, and provision of migration areas for adaptation to rising sea level (Ellison, 2015).

Adaptive capacity of mangroves can be enhanced by components of multiscale resilience strategies (Fig. 20.1) of Randhir (2016), including biotic and abiotic components as well as human management. Net vertical accretion of mangrove substrates provides adaptive capacity (Ellison and Strickland, 2015; Lovelock et al., 2015). If the net accretion is lower than the relative sea level rise (RSLR), mangroves could adapt to rising sea level through migrating inland (Ellison, 2014). Coastal adaptive capacity can also be measured using social indicators, governance, management plans, participatory planning, and interviews with local communities (Whitney et al., 2017).

1.3 Absorptive capacity

Absorptive capacity is defined as the ability to take protective action against shocks and stress and maintain stability (Jennings and Manlutac, 2016; White et al., 2018). Capacity for ecosystems to absorb threats is a lead approach to detecting ecological resilience and also considers scale of any degradation (Spears et al., 2015). Absorptive capacity has become a dominant dimension of resilience used in practice (White et al., 2018). In mangroves, absorptive capacity has been linked with abilities to tolerate heavy metals (Maiti and Chowdhury, 2013; Chai et al., 2019; Qiu et al., 2019), but for overall resilience is rather the ability for the ecosystem to cope with current threats from climate change while maintaining its structure and function.

Absorptive capacity can be affected by anthropogenic activity such as implementing policies and garnering support from the local communities (Steinbach et al., 2017). Absorptive capacity can also be enhanced by ecological influences such as nearby coastal ecosystems (Henry et al., 2018; Hogarth, 2015) and different mangrove species (He et al., 2018). Indicators of resilience identifying absorptive capacity are of levels of nutrients/eutrophication levels, shoreline change and regeneration, status of fauna/fisheries, soil carbon levels, regeneration potential, ecosystem robustness, or the capacity to resist stress (Day et al., 2018).

1.4 Transformative capacity

Transformative capacity is defined as fundamental restructuring of systems (including cultural, moral, economic, political, technological, and environmental conditions) no longer tenable by governance innovation, such as managed retreat from the coast (White et al., 2018). Transformative capacity can be traced to management science and human organizations and is related to access to resources for capacity development (Broto et al., 2019).

In terms of mangrove resilience, transformative capacity is required when the stressors on the mangrove system become too great and ecosystem robustness becomes threatened and can be applied through management. These management plans involve policies and anthropogenic actions in response to severe stresses and shocks (White et al., 2018), such as the enormous mangrove rehabilitation response after the 2004 Asian tsunami.

1.5 Resilience assessment

Quantification of ecological resilience, such as developed for coral reefs, includes indicators and methods (Mumby et al., 2014). Resilience assessment has been measured using both anthropogenic and ecological factors from adaptive, absorptive, and transformative capacities of resilience (Espana et al., 2015; Lam et al., 2015; O'Leary et al., 2017; Duncan et al., 2018). Resilience indicators for mangroves include vegetation conditions, level surfaces with surplus sediment supply, stable or increasing areas, and no human impacts (Ellison et al., 2017). Resilience has been measured using methods such as Open Source Based Web-GIS assessment of mangrove spatial change (Jayakumar, 2019) and surveying of local communities and measurement of biotic parameters (Moore, 2018). Ecological parameters include submergence vulnerability as shown by accretion rates (Day et al., 2018), plant biodiversity, management plans, and stakeholder involvement (Jonsson, 2017), as well as areal maintenance, landward migration, and/or shoreline retreat (Duncan et al., 2018). Effective management is also critical to managing resilience in mangroves (McLeod and Salm, 2006). More quantitative assessment of resilience in mangroves would allow effective and efficient threat management (Duncan et al., 2018) and is important for coastal communities around the world that rely heavily on mangroves for resources and protection (Jonsson, 2017).

1.6 Objective

This study adapts the multiscale resilience strategy framework of Randhir (2016), ecological integrity factors of Day et al. (2018), in the context of the resilience framework of White et al. (2018), to develop measures of mangrove resilience for mangroves. The intention is to encompass the ecological and anthropogenic components that contribute to mangrove resilience. The objective is to allow targeted implementation of mangrove management plans to build resilience. With climate change, mangroves are more threatened than ever before, and with a usable low-cost resilience assessment plan, efficient management plans can be implemented.

2. Methods

2.1 Study areas

Mangrove resilience assessment was conducted at four locations in Southeast Asia/West Pacific, selected for these reasons:

1. The largest areas and highest diversity of mangroves exist in the region of Southeast Asia/West Pacific.
2. The region has sea level trends that are variable from the global sea level trends.
3. Although all sites are protected areas, little research is available on the resilience of the sites.
4. Sites are recognized to be of international importance.
5. All sites are surrounded with agricultural and fishing activities.
6. The sites provide values for ecotourism and ecological research.
7. Each mangrove area is the largest or a larger area in the country.
8. Resilience of mangroves is influenced by the governance of a nearby community.

2.2 Resilience assessment methodology

The presented methodology considered resilience as centered between dimensions of ecosystem robustness, strength of stressors, and effectiveness of management actions (Fig. 20.2), adapted from the multiscale resilience strategies of Randhir (2016) (Fig. 20.1) and vulnerability assessment (Ellison, 2015). Dimensions are broad concepts, to which several components contribute (Fig. 20.2), and each component has methods of measurement. Some components encompass absorptive capacity through measures of ecosystem robustness and adaptive capacity through management capacity (Spears et al., 2015; Day

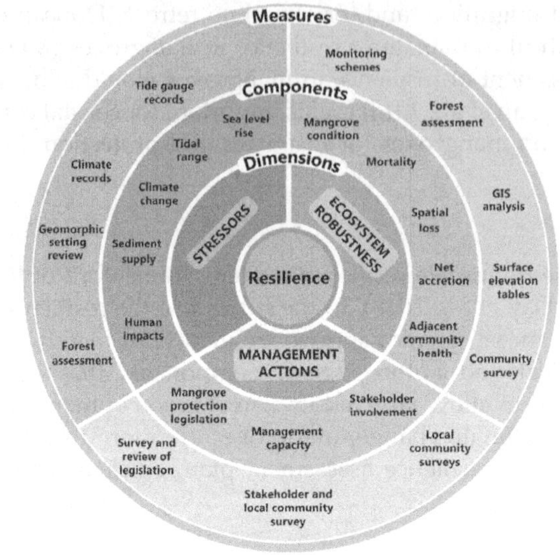

FIGURE 20.2 Dimensions, components, and measurements of mangrove resilience. *Drawn by Michael Helman.*

et al., 2018). Reviewed literature has used the term "indicator" of resilience as reviewed above, and we refine that to a component and its measurement (Fig. 20.2). Each component of resilience can be measured using existing techniques (Fig. 20.2), and results ranked (Table 20.1), which allows comparison of qualitative and quantitative information (Gornitz, 1991; Ellison, 2015).

Ways to allocate ranks using Table 20.1 are outlined below following the order of components in Column 1. The greatest stressor to mangroves from climate change is from rising sea level (Ellison, 2018; He and Silliman, 2019), and RSLR data are available from the National Oceanic and Atmospheric Administration (NOAA, 2019). Tide gauge stations nearest to the study sites were used, and paleo-reconstructions were reviewed where tide gauge records were not long. Tidal range is a stressor with RSLR, with the necessary landward relocation being greater for microtidal sites and less for macrotidal sites. For example, assuming zero accretion, a 0.5 m sea level rise in a 1 m tidal range requires a 100% mangrove habitat relocation, but only 25% habitat relocation in a 4 m tidal range. Tidal range is also important for vertical adjustment and accommodation space, where microtidal mangroves are also disadvantaged.

Of atmospheric climate changes, rainfall trends are the main potential stressor to mangroves to reduce resilience. Reduced rainfall and humidity are expected to cause reduction in mangrove diversity, photosynthesis, productivity, and growth rates along with substrate subsidence (Waycott et al., 2011). In arid areas, high evaporation and soil salinity in upper mangrove zones can lose mangroves to salt pans as found on leeward coast of Viti Levu, Fiji. During drought periods, mangrove mortality in such arid areas can occur (Duke et al., 2017; Lovelock et al., 2017). Mangroves, however, have a high tolerance to heat stress relative to other plants, and the IPCC-projected temperature increases are below those known to cause detrimental effects (Gilman et al., 2008; Ellison and Cannicci, 2016). Rainfall trends

TABLE 20.1 Criteria for ranking mangrove resilience across the three dimensions of mangrove resilience as shown in Fig. 20.2: stressors, ecosystem robustness, and management actions. Each dimension (shown in italics) has subcategories of components, which can be measured to allow ranking.

Resilience rank	1 (Low)	2	3	4	5 (High)
Stressors					
Sea level rise	Site rapidly subsiding	Site slowly subsiding	Site stable	Site slightly uplifting	Site uplifting
Tidal range (m)	<1	1–1.5	1.5–2	2–3	>3
Climate change	Significantly drier	Moderately drier	Somewhat drier	Rainfall unchanged	Becomes wetter
Sediment supply	Low	Fairly low	Medium	Fairly high	High
Human impacts	Severe impact	High impact	Rather high impact	Moderate impact	Slight impact
Ecosystem robustness					
Mangrove condition	Very poor	Poor	Moderate	Fairly good	Good
Mortality	>30%	20%–30%	10%–20%	4%–10%	<4%
Spatial loss	>30%	20%–30%	10%–20%	4%–10%	<4%
Net accretion	>1 mm less than relative sea level rise (RSLR)	<1 mm less than RSLR	Equal to RSLR	<1 mm greater than RSLR	>1 mm greater than RSLR
Adjacent community health	Very poor	Poor	Moderate	Fairly good	Good
Management actions					
Mangrove protection legislation	None	Poor	Moderate	Fairly good	Good
Management capacity	None	Poor	Moderate	Fairly good	Good
Stakeholder involvement	None	Poor	Moderate	Fairly good	Good

and projections were obtained from meteorological data and research available on rainfall patterns at sites and then ranked using criteria in Table 20.1.

Sediment supply is critical to mangrove resilience to sea level rise (Di Nitto et al., 2014) and varies with geomorphic settings of mangrove sites (Ellison, 2019). Table 20.2 categorizes these settings and identifies potential stressors in each that may contribute to reduction in resilience and allocation of rank following Table 20.1. For sites, these guide review of information on processes and supply rates (Chaudhuri et al., 2019) and can include information on natural disasters, land use changes, and man-made infrastructures affecting sediment supply to mangroves.

TABLE 20.2 Mangrove geomorphic settings, attributes, and potential stressors to sediment supply resilience.

Attributes	Tide-dominated	River-dominated	Wave-dominated	Low island
Setting	Estuarine with elongated islands	Deltaic distributaries	Barrier islands, bars, and lagoons	Marine-dominated
Dominant processes	Tidal currents, allochthonous sedimentation	River discharge, allochthonous sedimentation	Wave energy, autochthonous sedimentation inside lagoons	Autochthonous input from roots and vegetation
Dominant process	Tidal currents	Freshwater discharge	Wave energy	Sea level
Mangrove locations	Tidal creeks and islands	Seaward edge and distributaries	Inside lagoons	Fringing or basin
Stressor to resilience	Increased tidal action, reduced sediment supply	Reduced discharge and sediment supply	Increased wave action, reduced sediment supply	Reduced mangrove productivity

Adapted from Ellison J.C., 2015. Vulnerability assessment of mangroves to climate change and sea level rise impacts. Wetl. Ecol. Manag. 23, 115–137, 2019).

Human impacts result stressors of direct impacts, such as reclamation, cut trees, broken canopy, garbage, and pollution, and evaluation of levels of such stressors is an established mangrove monitoring method (English et al., 1997, 140–141; Ellison et al., 2012, 12–16). Site reports and publications were reviewed to determine levels of human impacts that cause reduction of mangrove resilience.

For ecosystem robustness components (Fig. 20.2), mangrove condition is a key indicator, which can be directly measured using established monitoring methods such as direct measurement of mangrove community structure, tree height, tree diameter, and biomass per area (English et al., 1997, 160–184; Ellison et al., 2012, 17–20). Permanent plots can show evidence of detailed change related to climate change impacts, such as preferential decline and mortality of one species and succession by another. Permanent plots are a well-described standard technique, so results can be compared with other sites and published results. Mangrove condition, mortality, and spatial loss of mangroves (Sippo et al., 2018) are indicative of impacts such as natural hazards, deforestation, and urban expansion (Gandhi and Jones, 2019). Mangrove condition and evidence of mortality from sites were reviewed from reports and publications available and where possible trends over time evaluated.

Spatial loss of mangroves includes retreat from the seaward edge owing to stress with rising sea level or direct impacts on mangroves from the landward edge. Measurement of spatial changes in mangroves uses comparison over time of historical and recent aerial photography and satellite imagery. The analysis involves rectification and comparison using GIS (Geographic Information Systems) software and methods such as the US Geological Survey's Digital Shoreline Analysis System (Theiler and Danforth, 1994; Ellison and Zouh, 2012). Review of spatial change analyses can analyze results to rank resilience using guidance in Table 20.1.

While sediment supply reduction is a potential stressor (Table 20.2), actual net accretion rates are an indication of ecosystem robustness (Table 20.1, Fig. 20.2). This is best measured by use of surface elevation tables (SETs) with a sediment surface marker horizon, and a time series of monitoring shows net accretion over time (Rybczyk and Callaway, 2009; Webb et al., 2013; Lovelock et al., 2015). Resilience is ranked by comparison of these results with the rate of RSLR as shown by local tide gauge data (NOAA, 2019, Table 20.1).

Adjacent community health as a component contributes to mangroves through co-benefits of food chain, carbon and nutrient linkages (Huxham et al., 2018), and sediment supplies such as from seagrass and coral reefs (Guannel et al., 2016). For different mangrove settings, this component could include coral reefs, seagrass, or river catchments separately, to allow flexibility in the assessment method. Monitoring results from reefs or seagrass using established methods (English et al., 1997) can allow assessment of rank, but owing to variability and range of methods, the ranking criteria given in Table 20.1 are qualitative. Review of management plans for these natural areas can contribute to ranking assessment.

The contribution of management actions to maintain mangrove resilience can be ranked from components of management capacity, stakeholder involvement, and existence and effectiveness of mangrove protection legislation (Fig. 20.2). Local assessments (Mcleod et al., 2015) include tools such as Local Early Action Planning and Management Planning, which gains information on existence and effectiveness of management, legislation, and stakeholder contributions from communities adjacent to mangrove areas (Sohl et al., 2019). Coastal protection legislation can also be evaluated by reviewing current and future coastal management plans and protection policies. Data reviewed were at national and regional levels and those directly related to the mangrove sites.

The resilience assessment method was designed to identify and evaluate the ecological and socioeconomic components that affect the resilience of mangroves. Results were ranked with 1 as the lowest resilience score and 5 as the highest resilience score, with values shown in Table 20.1. An overall resilience score of each study site was calculated using a formula adapted from Ellison (2015):

$$\text{Overall resilience rank} = \frac{\text{Total component rank scores}}{\text{Number of components completed}}$$

3. Results and discussion

Sites selected for the study using the selection criteria are located in Fig. 20.3, and site characteristics are provided in Table 20.3.

Maps showing mangrove distributions of sites relative to urban development and offshore ecosystems are shown in Fig. 20.4. Each can be described as fitting the selection criteria, with these eight criteria in common. Current and projected sea level rise is variable across the Asian region (Ward et al., 2016), with the Southeast Asia/West Pacific region showing sea level trends that vary from global sea level trends (Strassburg et al., 2015; Annamalai et al., 2014; Devlin et al., 2017). All sites are adjacent to agricultural activities (Victor et al., 2006; Razak et al., 2018; Thiagarajah et al., 2015; Baderan et al., 2018) and are recognized to be of international importance (Scott, 1993; Beh et al., 2014; Friess et al., 2016a). Results from resilience ranking for each site are shown in Table 20.4.

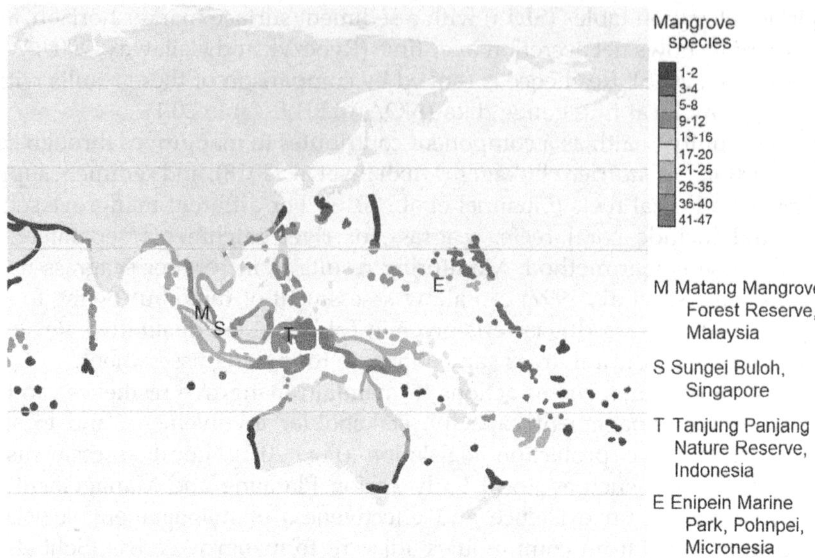

FIGURE 20.3 Map showing mangrove species diversity in the Asia/Pacific region and locations of study areas. The top left corner of each study site's letter is on the site's location. *Adapted from Spalding, M., Kainuma, M., Collins, L., 2010. World Atlas of Mangroves. Earthscan, London and Washington D.C.*

3.1 Stressor components

The Asian sites' tide gauge records were long enough to indicate a trend, showing a high resilience rank result for relative sea level change as the sites were either stable or uplifting (NOAA, 2019) (Table 20.4). Pohnpei's tide gauge record showed sea level rise with interference from ENSO fluctuations (Annamalai et al., 2014), and long-term records from stratigraphy indicate island subsidence (Bloom, 1970; Fujimoto et al., 2015). Tidal range resilience ranking was lowest for Pohnpei, with a microtidal range causing mangrove zones to occur in low magnitude elevation brackets, hence will be more stressed by sea level rise, unless offset by net accretion. All sites are microtidal so would expect greater relocation requirements with sea level rise, except Matang, which has a 4 m tidal range.

Rainfall has remained unchanged for the Singapore and Malaysian sites through recent years (Meteorological Service Singapore, 2019; Tan, 2018). The Indonesia site has rainfall variation with ENSO-related variability, but no overall change (Lee, 2015). Recent modeling results for climate projections for Pohnpei (TNC, 2017) show average rainfall is projected to increase (~3% increase by 2030; ~6% increase by 2050; ~12% increase by 2090). There will still be wet and dry years and decades due to ENSO-related variability, but most models show that the long-term average is expected to be wetter.

Sediment supply can be inferred from geomorphic settings of mangroves (Table 20.2), and all sites (Table 20.3) are mostly influenced by river discharge. Mangrove sites in Asia are predominantly minerogenic (Balke and Friess, 2016), and major rivers have reduced sediment discharge to the coast with dams, which cause sediment deficit to coastal mangroves (Ward

TABLE 20.3 Environmental and socioeconomic settings of the study sites in Singapore, Malaysia, Indonesia, and Micronesia.

Characteristics	Sungei Buloh Wetland Reserve, Singapore	Matang Mangrove Forest, Reserve, Malaysia	Tanjung Panjang Nature Reserve, Indonesia	Enipein Marine Park, Pohnpei, Micronesia
Latitude	1.49°N	4.15°N	0.477°N	6.51°N
Longitude	103.72°E	100.45°E	121.45°E	158.13°E
Mangrove area (km^2)	1.3	432.92	75	7.55
Tidal range (m)	2.0	4.0	2.0	1.0
Precipitation patterns	Equatorial	Equatorial	Equatorial	Equatorial
Annual rainfall (mm)	2340	2408	2460	3429
Air temperatures (°C)	22–31	19–33	22–33	21–27
Average humidity	84%	73%	73%	80%
Socioeconomic aspects				
Population	5.804 million	1.767 million	19.57 million	36,196
Population density (/km^2)	8292	1684	103.8 km^2	108
Human Development Index 2017	0.932	0.802	0.694	0.627
Major economic activities	Agriculture, tourism	Tourism, agriculture, education, aquaculture	Aquaculture, ecotourism	Tourism, fishing
Site management	National Parks Board, Singapore	Forestry Department of Perak, Malaysia	North Sulawesi Regional Government	Pohnpei Government

Using information from: Scott, D.A., 1993. A directory of Wetlands in Oceania. International Waterfowl and Wetlands Research Bureau, Slimbridge, U.K. and Asian Wetland Bureau, Kuala Lumpur, Malaysia; Beh, B.C., Sim, C.K., Syahreza, S., Jafri, M.Z.M., 2014. Discrimination of mangrove species in Matang Mangrove Forest Reserve, Perak using in-situ measurement of hyperspectral leaf reflectance. AIP Conference Proceedings 1657, 110004, https://doi.org/10.1063/1.4915223; Friess, D.A., Richards, D.R., Phang, V.X.H., 2016a. Mangrove forests store high densities of carbon across the tropical urban landscape of Singapore. Urban Ecosyst. 19 (2), 795–810; UNDP, 2018. Global Human Development Indicators. United Nations Development Program, Human Development Reports. Available online: http://hdr.undp.org/en/countries (Accessed 18.11.19); FSM Statistics, 2019. FSM Division of Statistics. www.fsmstatistics.fm. (accessed 9.11.19); Corbin, J., 2013. Restoring Mangrove Forests in Indonesia's Tanjung Panjang Area through the Use of Market-Based Incentives: Lessons Learned from international case studies. Master's Thesis, Dalhouise University, Halifax, Nova Scotia; Rompas P.T.D, Tanunaumang, H., Sangari, F.J., 2016. Study on marine current with approach of a numerical model for marine current power plant (PLTAL) in the Bangka Strait North Sulawesi. International Seminar on Application for Technology of Information and Communication (ISemantic), Semarang, 104–110; Brown, B., Fadillah, R., Nurdine, Y., Soulsby, I. and Ahmad, R., 2014. Case Study: Community Ecological Mangrove Rehabilitation (CBEMR) in Indonesia (2014). Available from: https://journals.openedition.org/sapiens/1589 (accessed 25.3.2020); Baderan, D.W., Utina, R., Lapolo, N., 2018. Vegetation structure, species diversity, and mangrove zonation patterns in the tanjung panjang nature reserve area, Gorontalo, Indonesia. Int. J. Appl. Biol. 2 (2), 1–12.

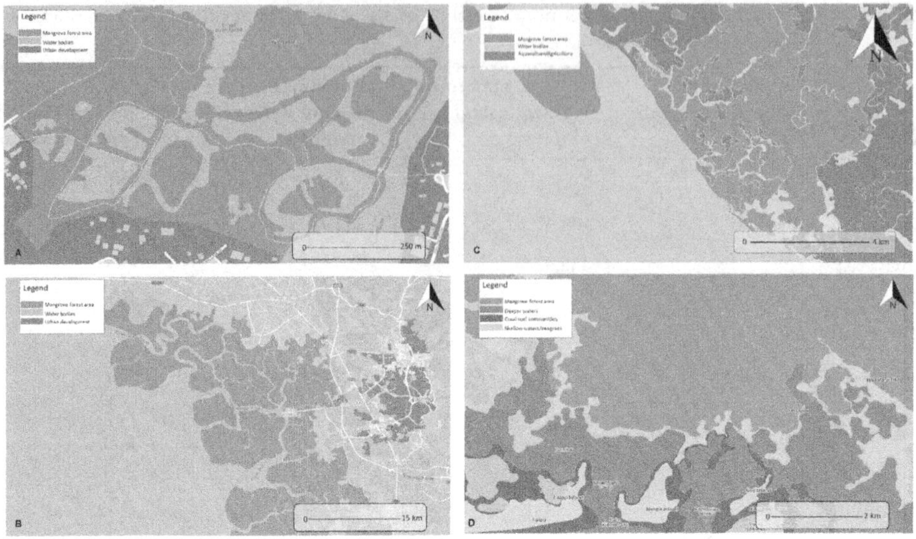

FIGURE 20.4 Maps showing locations of mangroves for each study site, relative to urban development and roads, and offshore ecosystems of coral reefs. (A) Sungei Buloh Wetland Reserve, Singapore. (B) Matang Mangrove Forest, Reserve, Malaysia. (C) Tanjung Panjang Nature Reserve, Indonesia. (D) Enipein Marine Park, Pohnpei, Micronesia. *Adapted from Google Maps.*

et al., 2016). River damming has occurred at the Singapore site (Friess et al., 2012). Hence, sediment supply for the three Asian sites was determined to be low (Table 20.4), with catchments adjacent to the Indonesian site being relatively small. For Micronesia's Enipein, catchment clearance has caused soil erosion resulting in turbid fluvial outflow into the mangrove areas (Victor et al., 2006), and catchment clearance for sakau planting has continued this effect until recently (Sohl et al., 2019). While turbid discharge has a poor effect on the close-by seagrass and coral communities, affecting scores in the adjacent ecosystem robustness rank results (Table 20.4), the minerogenic sediment supply is beneficial to mangrove accretion, hence offsetting the higher RSLR rank.

Human impacts for Enipein were evaluated by local community members in adjacent communities of Wone and Seinwar by Sohl et al. (2019). Degradation by dredging and solid waste and reduction in fish and crabs, along with reduction in river water quality, are described from Seinwar to the northwest of Enipein. Wone community members to the northeast agreed with degradation of river discharge by pig pen contamination and soil erosion and reduction in fish and crabs, however, considering the condition of the mangrove forest to be very healthy. Human impacts have in the past occurred at Sungei Buloh (Friess et al., 2012), and at least three studies on mangrove structure are available although some time ago. The site has become well-managed and studied, hence given a high resilience rank for low human impacts (Table 20.4). The Matang forest reserve is used for forest operations (Otero et al., 2018) but is well-monitored and managed (Tanaka and Choo, 2000; Otero et al., 2019), hence given a good rank for resilience owing to low human impacts (Table 20.4). The Tanjung Panjang Nature Reserve in Indonesia has heavy human impacts on mangroves (Baderan et al., 2018).

TABLE 20.4 Results of mangrove resilience assessment ranking for study sites, with overall resilience score determined in the lowest row (in bold).

Resilience components	Sungei Buloh Wetland Reserve, Singapore	Matang Mangrove Forest Reserve, Malaysia	Tanjung Panjang Nature Reserve, Indonesia	Enipein Marine Park, Pohnpei, Micronesia	Sources
Sea level rise	Stable (3)	Slightly uplifting (4)	Subsiding (2)	Subsiding (2)	Bloom (1970); Siow (2012); NOAA (2019).
Tidal range (m)	2.0 (4)	4.0 (5)	2.0 (4)	1.5 (2)	Chan et al. (2006); Victor et al. (2006); Blaber (2008); Rompas et al., (2016).
Climate change	Rainfall unchanged (4)	Rainfall unchanged (4)	Rainfall unchanged (4)	Rainfall increasing (5)	Meteorological Service Singapore (2019); Tan (2018), TNC (2017); Lee (2015).
Sediment supply	Fairly low (2)	Fairly low (2)	Fairly low (2)	Fairly high (4)	Victor et al. (2006); Friess et al. (2012); Ward et al. (2016).
Human impacts	Slight impact (5)	Slight impact (5)	Severe impact (1)	Moderate impact (4)	Devoe and Cole (1998); Cole et al. (1999); Anson (1993); Tanaka and Choo (2000); Victor et al. (2006); Friess et al. (2012); Friess (2017), Hashim et al. (2017): Sohl et al. (2019); Otero et al. (2019); Lapolo et al. (2018); Brown et al. (2014); Rompas et al. (2016); Baderan et al. (2018).
Mangrove condition	Good (5)	Good (5)	Very poor (1)	Fairly good (4)	
Mortality	<4% (5)	<4% (5)	>30% (1)	<4% (5)	
Spatial loss	c. 6% (4)	<4% (5)	c. 25% (2)	<4% (5)	
Net accretion	>1 mm less than RSLR (1)	>1 mm less than RSLR (1)	>1 mm less than RSLR (1)	<1 mm less than RSLR (2)	Day et al., 2018; Lovelock et al. (2015); NOAA 2019.
Adjacent community health (coral reef)	Moderate (3)	Moderate (3)	Low (2)	Moderate (3)	Turak and Devantier (2005); George et al. (2008); Goldberg et al. (2008); Praveena et al. (2012); Ng et al. (2016); Crawford et al. (2004).
Adjacent community health (seagrass)	High (4)	Moderate (3)	Moderate (3)	Moderate (3)	Victor et al. (2006); Abdullah and Anscelly (2015); McKenzie et al. (2016); Fortes et al. (2018); Waggey (2017).
Management capacity	Good (5)	Good (5)	Poor (2)	Moderate (3)	National Parks Board Singapore (2010); Goessens et al. (2014); Jusoff and Taha (2009); Sohl et al. (2019), Crawford et al. (2004); Corbin (2013).
Mangrove protection legislation	Good (5)	Good (5)	Poor (2)	Moderate (3)	
Stakeholder involvement	Good (5)	Moderate (3)	Poor (2)	Moderate (3)	Raynor and Kosta (2003); MEWR&MND (2016); Albotoush and Shau-Hwai, 2019; Sohl et al. (2019); Crawford et al. (2004); Corbin (2013).
Overall resilience score	**3.9**	**3.9**	**2.0**	**3.5**	

3.2 Ecosystem robustness components

Mangrove condition, mortality rates, and reduction in mangrove area were ranked well at both Singapore and Malaysian sites, with reversal of some past trends (Friess et al., 2012) to become well-maintained mangrove reserves that are now recognized internationally (Friess, 2017; Hashim et al., 2017). Poor management has caused the Indonesia site mangroves to degrade in health rapidly (Corbin, 2013). The Sulawesi government had leased the mangrove areas away to aquaculture farmers, establishing ponds (Lapolo et al., 2018), causing eutrophication, and leading to the high mortality rate and poor mangrove condition (Brown et al., 2014).

Thorough and quantitative assessment of the mangroves of Pohnpei was carried out by the US Forest Service, commencing in June 1990 and completed in May 1992 (Devoe, 1992, 1994), which became the basis for the Pohnpei Mangrove Management Plan (Metz, 1996). Late last century, a mangrove area of 5525 ha was recorded on Pohnpei (Metz, 1996; Devoe and Cole, 1998; Cole et al., 1999). The mangrove area increased in 1976–2006 (Donnegan et al., 2006) to 5658 ha. On the south coast of Pohnpei, the Enipein Marine Park contains among the best examples of the mangrove forests of Pohnpei, featuring extensive areas of tall, mature closed forest especially of the landward mangrove forest association (Scott, 1993). The area has protected status under the Pohnpei Mangrove Management Plan (Metz, 1996).

While Enipein and Matang results show little loss of mangrove spatial area (Table 20.4), the Singapore site ranks lower some reduction in mangrove area. Past reduction of mangrove area has occurred at the Singapore site, but more recently it has been protected and highly valued (Friess et al., 2012). Some coastal loss has occurred more recently, perhaps from boat wake erosion as has occurred further down the coast (Friess et al., 2012). Mangrove spatial area at the Indonesia site has been rapidly decreasing through the years due to converting its area into aquaculture developments (Lapolo et al., 2018).

A range of methods are available to measure mangrove sedimentation rates (Nolte et al., 2013), of which SETs best show net accretion. SET research in Asia shows a mean elevation deficit of 6 mm/year across available study sites including Singapore and Indonesia (Ward et al., 2016). Mangrove accretion rates for Pohnpei are indicated by Lovelock et al. (2015), showing the current rates of sea level rise to exceed the soil surface elevation gain. The net accretion rates relative to RSLR component give the lowest rank for resilience for each of the mangrove sites (Table 20.4).

At the Singapore and Malaysian sites, coral reef community resilience was ranked moderate (Table 20.4) as coral reef communities in Singapore have been in danger due to the busy maritime activities happening in Singapore waters and coral reef rehabilitation efforts have been ongoing since the early 2000s (Ng et al., 2016). Condition of coral reef communities near the Malaysian site has been ranked medium by Praveena et al. (2012). Seagrass community resilience near the Singapore site was ranked high as the communities nearby are diverse and there has been monitoring from 2006 to 2014 (McKenzie et al., 2016). Seagrass condition near the Malaysian site was ranked moderate as the recent coastal land reclamation and development in the area, which lasted from 2003 to 2009, has had an adverse effect on the density of the seagrass community (Abdullah and Anscelly, 2015). Seagrass community resilience rank was also influenced by the lack of knowledge in conservation methods and management policies of seagrass community in Southeast Asia (Fortes et al., 2018).

The Indonesian site's adjacent ecosystem resilience gave poor results (Table 20.4). The seagrass community has experienced eutrophication (Waggey, 2017), and coral mining has been evident around the area (Crawford et al., 2004). At the Indonesian site, both seagrass (Waggey, 2017) and coral communities (Crawford et al., 2004) are poorly managed. Development of aquaculture ponds (Waggey, 2017), ecotourism activities (Crawford et al., 2004), and poor management (Corbin, 2013) has led to the degradation of these communities' health.

Pohnpei is surrounded by barrier reefs, separated from the mangrove margin by shallow lagoons, with extensive cover of seagrass. Coral reef monitoring started in 2005 (George et al., 2008), at which time reef conditions throughout Micronesia were good to excellent (Turgeon et al., 2003), and a follow-up report (Goldburg et al., 2008) found coral reefs to be in good health with few impacts. A baseline survey of reef diversity and condition (Turak and DeVantier, 2005) found coral growth in most shallower sites to be vigorous, though with some disturbances including sediment runoff which reduced water clarity. For example, water visibility was found to be poor southeast of the Enipein mangroves (site 28). Golbuu et al. (2008) assessed reefs at Enipein from the edge of the mangroves to the barrier reef, finding thick layers of terrigenous mud with low diversity coral close to the mangroves where a river was discharging, and declining mud deposits offshore along with increase in coral species richness. Soil loss from the land and resulting nutrient increases, however, have impact on coral reef sites studied at Enipein (Golbuu et al., 2008). In Pohnpei, the Conservation Society of Pohnpei and the Pohnpei State Fisheries have the authority to conduct coral reef research and monitoring (Houk et al., 2015). Pohnpei seagrasses were described by McDermid and Edward (1999) and extensively surveyed in 2005 by McKenzie and Rasheed (2006), finding extensive areas offshore of mangroves in the Pohnpei lagoons, greatest on the eastern shore. Seagrass at Enipein was stated by local community members to be in degraded condition, perceived to be declining because of turbidity (Sohl et al., 2019). For all sites, the adjacent community health component (Table 20.1) was subdivided into coral reef and seagrass measurement results, as they varied at each site (Table 20.4). This shows a site flexibility that the resilience assessment method can incorporate into the overall rank calculation.

The resilience assessment paradigm described in this research could be adapted to directly assess adjacent coastal systems such as coral reefs. Existing resilience assessment of coral reefs has considered ecological factors such as biodiversity, coral growth rates, and nutrient supply (Gibbs and West, 2019; Lam et al., 2017). Socioeconomic criteria as used in this study could be also considered owing to its contribution to management and adaptive capacity.

3.3 Management components

Community management capacity and mangrove protection legislation for both Singapore and Malaysian sites were ranked highest (Table 20.4). The management team for the Singapore site has a well-planned management policy involving the community around the reserve and remediation efforts if needed in the future (National Parks Board Singapore, 2010). Good stakeholder involvement is shown by coastal protection legislation of the Singapore government being effective with 70% of Singapore's coastline protected (Ministry of the Environment and Water Resources and Ministry of National Development [MEWR&MND], 2016). Coastal protection policies in Singapore are currently in effect with higher spending for coastal protection being projected by the MEWR&MND (2016). The Malaysia government has policies that involve community management around the area through restrictions of logging of the reserve (Goessens et al., 2014) and strict legislations that have been in effect since 1978 (Jusoff and Taha, 2009). For Malaysia's coastal protection legislation, only Integrated Coastal Zone Management (ICZM) was implemented where it has been proven effective with stakeholders near the study site (Albotoush and Shau-Hwai, 2019). Resilience rankings also considered the current coastal problems that have risen the past 5 years such as coastal erosion problems in Malaysia described by Friess et al. (2016b). Although important management policies were implemented for Indonesia (Friess et al., 2016b), villagers around the Indonesia site pay no attention to the regulations, and the government instead promoted the clearance of mangrove area for aquaculture ponds (Corbin, 2013). Stakeholders were not involved and regulations for the mangrove preservation was not made clear to the public (Crawford et al., 2004).

Both the Singapore and Malaysian sites are well known to follow international standards in management with frequent review of their management plans (Jusoff and Taha, 2009; MEWR&MND, 2016). Matang Mangrove Forest Reserve is even crowned to be the best managed mangrove reserve in the world (Jusoff and Taha, 2009), with support from local stakeholders for conservation of the forest and to be involved in collaborative management (Martínez-Espinosa et al., 2020). Management agencies of Sungei Buloh Wetland Reserve work closely with the government of Singapore and plan to introduce further coastal protection measures to increase the resilience of the reserve (MEWR&MND, 2016), with strong support from stakeholders for mangrove conservation (Velde et al., 2019). However, Malaysia's coastal protection legislation is lagging. With a large coastal area extent, Malaysian aquaculture and fisheries have been breaching legislations that have either gone unpunished or received lenient punishment (Friess et al., 2016b). Similar ineffective management is also evident near the Indonesian site (Corbin, 2013). Further improvements could be made to Malaysia and Indonesia's coastal protection plans by imposing stricter punishments and improved security. This way, the coastal protection plans can be effective to improve ecosystem resilience.

Pohnpei has a traditional leadership system for management of land and waters, augmented by the Pohnpei Watershed Forest Reserve and Mangrove Protection Act of 1987, designating the Enipein mangroves and a protected area (Raynor and Kostka, 2003). Nongovernmental organizations such as the Conservation Society of Pohnpei (founded 1998) are also very active in mangrove monitoring and management, following capacity building as part of UNEP's South Pacific Biodiversity Conservation Program (Baines et al., 2002). Community surveys near Enipein (Sohl et al., 2019) found that few people living near the mangroves are dependent on mangrove resources for a livelihood; however, management efforts to protect mangroves are poor. Traditional resource management practices are not being implemented (Sohl et al., 2019), resulting in moderate management resilience ranking allocated in Table 20.4. In future, this is likely to improve, as part of the Micronesian Mangrove Adaptation Initiative, the Pacific Islands Climate Change Cooperative (PICCC) is providing support to the Micronesian Conservation Trust to develop and island-wide mangrove adaptation and management plan, which will incorporate findings from comprehensive mangrove vulnerability assessments and extensive stakeholder consultations (PICCC, 2019). Further capacity building for Pohnpei management would improve future mangrove resilience.

3.4 Improvement of resilience

Fig. 20.5 compares rank results from the three resilience dimensions of stressors, ecosystem robustness, and management actions (Table 20.4), for each of the four study sites. Best resilience on this triangle would be shown by a large equilateral triangle encompassing the outer edge, all at rank 5. Lower resilience rank results reduce the triangle area, and departures from an equilateral triangle show differences between dimensions for a site, which identify priorities for resilience building.

Results show the sites to have similar overall resilience ranks for the Singapore, Malaysian sites, and Pohnpei (Table 20.4), though Micronesia scores lower in stressor and management components and higher in ecosystem robustness (Fig. 20.5). Indonesia scored far lower rank results with direct human impacts and poor management action effectiveness. Management component differences are likely owing to variation in socioeconomic capacity (Table 20.3), with the human development index far lower for Micronesia and Indonesia relative to Singapore and Malaysia. Low human development countries have fewer resources to respond to natural hazards and develop contingency plans, and government has relatively little room to invest in environmental protection (UNDP, 2019). Management capacity provides the lowest scores for Pohnpei and Indonesia, the Federated States of Micronesia showing a gross domestic product per capita of $3.058 and Indonesia $3,893, relative to $64,581 in Singapore and $11,239 in Malaysia (World Bank, 2020). These differences indicate a greater need in Micronesia and Indonesia to improve mangrove resilience in conjunction with local managers, which has commenced (PICCC, 2019; Mangroves for the Future, 2020), but needs further ongoing support.

The component of net mangrove substrate accretion rates compared with RSLR overall gave the lowest rank for resilience for each of the mangrove sites (Table 20.4). Substrate accretion is critical with increased rates of sea level rise projected for this century (DeConto

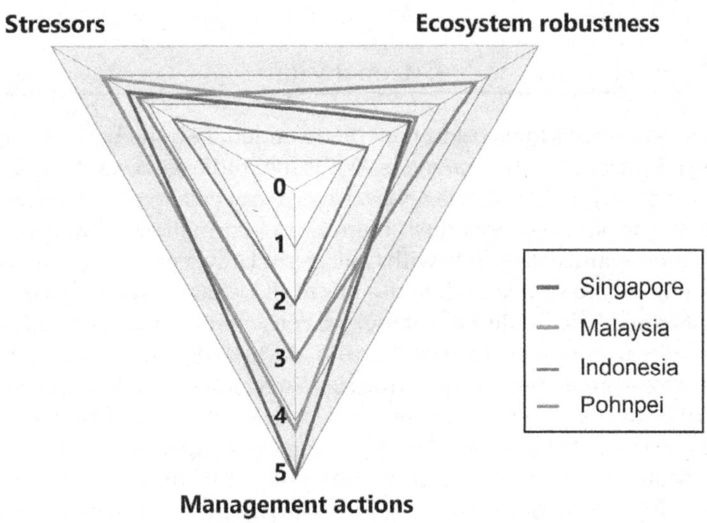

FIGURE 20.5 Resilience triangle showing stressor, ecosystem robustness, and management capacity dimension results for each of the four study sites. *Drawn by Michael Helman.*

and Pollard, 2016; Jevrejaya et al., 2019). Accretion could be boosted through assessment of the sediment budget, identifying sources of sediment supply and potential causes of sediment loss (Bird, 2008). This would have most focus upon minerogenic sources, while autogenic mangrove organic matter accumulation can be boosted by enhancement of mangrove health and productivity. This can be managed by reducing nonclimate impacts on mangroves and replanting degraded areas with mangrove seedlings (Ellison and Strickland, 2015). Fig. 20.6 summarizes ways that mangrove substrate accretion can be enhanced, and further investment in SET research (Lovelock et al., 2015; Ward et al., 2016) would allow improved understanding of site-by-site processes.

This mangrove resilience study has found ecosystem robustness to encompass components that align with absorptive capacity, a dominant dimension of resilience used in practice (White et al., 2018). Positive results for each of the ecosystem robustness and management capacity dimensions (Table 20.4; Fig. 20.5) indicate the capacity of the system to resist stressors in both ecological and engineering resilience concepts. Most ecosystem robustness components are biotic (Fig. 20.2), so fitting with Randhir's model (Fig. 20.1). Components of stressors (Fig. 20.2) have effect on mangroves at different scales and timeframes and are mostly abiotic, so fitting with Randhir's framework model (Fig. 20.1). The magnitude and frequency of stressors (Brunsden and Thornes, 1979) control the level of impact, and several (Table 20.2) operating in the same timescale will reduce resilience. This can be mitigated by the mangrove resilience dimension of management actions (Fig. 20.2) at local, regional, and national levels.

Management capacity supported and maintained by national incomes and/or international aid can allow higher level management capabilities of transformative capacity, such as large-scale mangrove replanting, or planned allocation of migration areas with rising sea levels. While replanting efforts after the Asian tsunami such as by the Mangroves for the Future project show transformative capacity, enhancement of resilience of mangroves can hopefully enable such major efforts to be needed only infrequently.

4. Conclusions

This study has developed a low-cost framework to undertake a site-based mangrove resilience assessment, which is easy to use by managers. While the studies reviewed (Table 20.4) may each be of high technology and high cost, this review process can extend off site if site-based information is not available or omit components if there is no information. Mangrove managers with limited time resources could collaborate with Colleges or Universities to make a student research project of the review, where students could use this resilience assessment process to provide useful results. Results could help guide stakeholder led conservation initiatives, allowing consensus incorporation into local contexts (Velde et al., 2019; Martínez-Espinosa et al., 2020).

The resilience assessment identifies priorities in management and research to guide resource allocation. Priorities for management are any components that rank lower, such as this study found low accretion rates at all sites and lower management capacity in Indonesia and Pohnpei. These results indicate needs for prioritized investment in improvement of these lower components of resilience. Research priorities for sites are shown by the dates of the sources used in the results, some of which are dated (Table 20.3). These indicate a gap in research that needs an updated study, particularly where combined with a low resilience score.

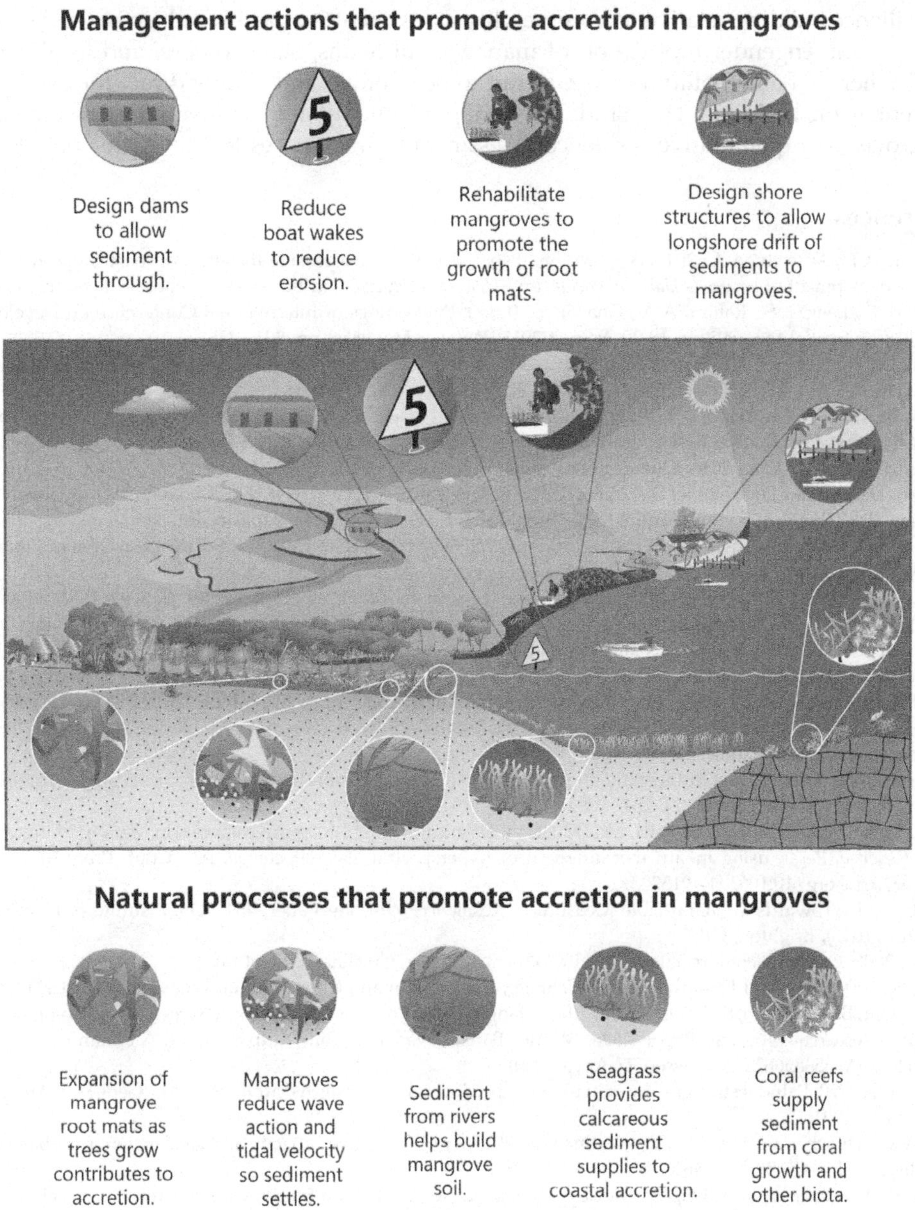

FIGURE 20.6 Management methods to increase net accretion in mangroves. *Drawn by Michael Helman.*

Resilience building relative to vulnerability assessment is a positive objective and outcome better able to engender motivation of management teams, stakeholders, and local communities, whereas vulnerability is more a trait to be avoided and so less direct for use in management team building. This study has adapted quantitative vulnerability assessment of mangroves to a quantitative resilience assessment of mangroves to enable this objective.

References

Abdullah, A.L., Anscelly, A.A., 2015. Application of spatial analysis in assessing the effect of coastal land reclamation and development of seagrass *Halophila ovalis* and *Halophila becarrii* distributions in Penang Waters. In: Samat, N., Mohd, S., Leng, K.S., Rahim, A.A., Ghazali, S. (Eds.), Proceedings of International Conference on Development and Socio Spatial Inequalities, 19-20 August 2015. Penang, Malaysia, pp. 231–236.

Annamalai, H., Hafner, J., Kumar, A., Wang, H., 2014. A Framework for dynamical seasonal prediction of precipitation over the Pacific Islands. J. Clim. 27 (9), 3272–3297.

Albotoush, R., Shau-Hwai, A.T., 2019. Evaluating integrated coastal zone management efforts in Penang Malaysia. Ocean Coast Manag. 181, 104899. https://doi.org/10.1016/j.ocecoaman.2019.104899.

Angeler, D.G., Allen, C.R., 2016. Quantifying resilience. J. Appl. Ecol. 53 (3), 617–624.

Angeler, D.G., Fried-Petersen, H.B., Allen, C.R., Garmestani, A., Twidwell, D., Chuang, W., Donovan, V.M., Eason, T., Roberts, C.P., Sundstrom, S.M., Wonkka, C.L., 2019. Adaptive capacity in ecosystems. In: Bohan, D., Dumbrell, A. (Eds.), Advances in Ecological Research 60, Resilience in Complex Sociological Systems. Academic Press, London, pp. 1–24.

Anson, H.K., 1993. Country report on mangrove ecosystem of Pohnpei, federates States of Micronesia. In: Nakamura, T. (Ed.), Integrated Research on Mangrove Ecosystems in Pacific Islands Region (II). Japan International Association for Mangroves and International Society for mangrove ecosystems, Okinawa, pp. 87–108.

Baderan, D.W., Utina, R., Lapolo, N., 2018. Vegetation structure, species diversity, and mangrove zonation patterns in the tanjung panjang nature reserve area, Gorontalo, Indonesia. Int. J. Appl. Biol. 2 (2), 1–12.

Baines, G., Hunnam, P., Rivers, M.-J., Watson, B., 2002. South Pacific Biodiversity Conservation Programme Terminal Evaluation. United Nations Development Program, New York.

Balke, T., Friess, D.A., 2016. Geomorphic knowledge for mangrove restoration: a pan-tropical categorization. Earth Surf. Process. Landforms 41 (2), 231–239.

Beh, B.C., Sim, C.K., Syahreza, S., Jafri, M.Z.M., 2014. Discrimination of mangrove species in Matang Mangrove Forest Reserve, Perak using in-situ measurement of hyperspectral leaf reflectance. AIP Conf. Proc. 1657, 110004. https://doi.org/10.1063/1.4915223.

Béné, C., 2013. Towards a Quantifiable Measure of Resilience. Working Paper, 434. IDS. Institute of Development Studies (IDS), Brighton, UK.

Bird, E., 2008. Coastal Geomorphology: An Introduction. Wiley, Chichester, West Sussex.

Blaber, S.J., 2008. Tropical Estuarine Fishes: Ecology, Exploitation and Conservation. Wiley, New Jersey, USA.

Blankespoor, B., Dasgupta, S., Lange, G.M., 2016. Mangroves as Protection from Storm Surges in a Changing Climate. Policy Research Working Paper 7596. World Bank. Available from: https://www.ncbi.nlm.nih.gov/pmc/articles/PMC5385662/. (Accessed 27 April 2020).

Bloom, A.L., 1970. Paludal stratigraphy of Truk, Ponape, and Kusaie, eastern Caroline islands. Geol. Soc. Am. Bull. 81 (7), 1895–1904.

Broto, V.C., Trencher, G., Iwaszuk, E., Westman, L., 2019. Transformative capacity and local action for urban sustainability. Ambio 48 (5), 449–462.

Brown, B., Fadillah, R., Nurdine, Y., Soulsby, I., Ahmad, R., 2014. Case Study: Community Ecological Mangrove Rehabilitation (CBEMR) in Indonesia (2014). Available from: https://journals.openedition.org/sapiens/1589. (Accessed 25 March 2020).

Brunsden, D., Thornes, J.B., 1979. Landscape sensitivity and change. Trans. Inst. Br. Geogr. 4 (4), 463–484.

Bunting, P., Rosenqvist, A., Lucas, R.M., Rebelo, L.M., Hilarides, L., Thomas, N., Hardy, A., Itoh, T., Shimada, M., Finlayson, C.M., 2018. The global mangrove watch—a new 2010 global baseline of mangrove extent. Rem. Sens. 10, 1669. https://doi.org/10.3390/rs10101669.

Chai, M., Li, R., Tam, N.F.Y., Zan, Q., 2019. Effects of mangrove plant species on accumulation of heavy metals in sediment in a heavily polluted mangrove swamp in Pearl River Estuary, China. Environ. Geochem. Health 41 (1), 175–189.

Chan, E.S., Tkalich, P., Gin, K.Y.H., Obbard, J.P., 2006. The physical oceanography of Singapore coastal waters and its implications for oil spills. In: Wolanski, E. (Ed.), The Environment in Asia Pacific Harbours. Springer, Dordrecht, pp. 393–412.

Chaudhuri, P., Chaudhuri, S., Ghosh, R., 2019. The role of mangroves in coastal and estuarine sedimentary accretion in Southeast Asia. In: Sedimentary Processes - Examples from Asia, Turkey and Nigeria. IntechOpen. Available from: https://www.intechopen.com/online-first/the-role-of-mangroves-in-coastal-and-estuarine-sedimentary-accretion-in-southeast-asia (accessed 9/12/19).

Cheong, S.M., Silliman, B., Wong, P., van Wesenbeeck, B., Kim, C.K., Guannel, G., 2013. Coastal adaptation with ecological engineering. Nat. Clim. Change 3 (9), 787–791.

Cinner, J.E., Barnes, M.L., 2019. Social dimensions of resilience in social-ecological systems. One Earth 1 (1), 51–56.

Cole, T.G., Ewel, K.C., Devoe, N.N., 1999. Structure of mangrove trees and forests in Micronesia. For. Ecol. Manag. 117 (1–3), 95–109.

Corbin, J., 2013. Restoring Mangrove Forests in Indonesia's Tanjung Panjang Area through the Use of Market-Based Incentives: Lessons Learned from International Case Studies. Master's Thesis. Dalhouise University, Halifax, Nova Scotia.

Crawford, B.R., Siahainenia, A., Rotinsulu, C., Sukmara, A., 2004. Compliance and enforcement of community-based coastal resource management regulations in North Sulawesi, Indonesia. Coast. Manag. 32, 39–50.

Day, R.H., Allen, S.T., Brenner, J., Goodin, K., Faber-Langendoen, D., Ames, K.W., 2018. Ecological resilience indicators for mangrove ecosystems. In: Goodin, K.L., Faber-Langendoen, D., Brenner, J., Allen, S.T., Day, R.H., Congdon, V.M., Shepard, C., Cummings, K.E., Stagg, C.L., Gabler, C.A., Osland, M., Dunton, K.H., Ruzicka, R.R., Semon-Lunz, K., Reed, D., Love, M. (Eds.), Ecological Resilience Indicators for Five Northern Gulf of Mexico Ecosystems. NatureServe, Arlington, Virginia, USA, pp. 91–150.

Davidson, J.L., Jacobson, C., Lyth, A., Dedkorkut-Howes, A., Baldwin, C.L., Ellison, J.C., Holbrook, N.J., Howes, M.J., Serrao-Neumann, S., Singh-Peterson, L., Smith, T.F., 2016. Interrogating resilience: toward a typology to improve its operationalization. Ecol. Soc. 21 (2), 27. https://doi.org/10.5751/ES-08450-210227>.

DeConto, R.M., Pollard, D., 2016. Contribution of Antarctica to past and future sea level rise. Nature 531, 591–597.

Devlin, A.T., Jay, D.A., Zaron, E.D., Talke, S.A., Pan, J., Lin, H., 2017. Tidal variability related to sea level variability in the Pacific Ocean. J. Geophys. Res.: Oceans 122, 8445–8463.

Devoe, N.N., 1992. Mangrove assessments on Yap, Kosrae, and Pohnpei, Federated States of Micronesia. In: Coastal Resources and Systems of the Pacific Basin: Investigation and Steps Toward Protective Management. UNEP Regional Seas Reports and Studies No. 147, pp. 1–19.

Devoe, N.N., 1994. Mangrove exploitation and conservation in the Federated States of Micronesia. J. Micronesian Stud. 2, 67–82.

Devoe, N.N., Cole, T.G., 1998. Growth and yield in mangrove forests of the Federated States OF Micronesia. For. Ecol. Manag. 103 (1), 33–48.

Di Nitto, D., Neukermans, G., Koedam, N., Defever, H., Pattyn, F., Kairo, J.G., Dahdouh-Guebas, F., 2014. Mangroves facing climate change: landward migration potential in response to projected scenarios of sea level rise. Biogeosciences 11, 857–871.

Donnegan, J.A., Butler, S.L., Kuegler, O., Hiserote, B.A., 2006. Federated States of Micronesia's Forest Resources. Resources Bulletin NW-RB-262. U.S. Department of Agriculture, Forest Service, Pacific Northwest Research Station, Portland, OR.

Doyle, A., 2019. Nationally Determined Contributions (NDC) – Global Outlook Report. United Nations Development Programme, Annual Report, New York.

Duke, N.C., Kovacs, J.M., Griffiths, A.D., Preece, L., Hill, D.J., Van Oosterzee, P., Mackenzie, J., Morning, H.S., Burrows, D., 2017. Large-scale dieback of mangroves in Australia's Gulf of Carpentaria: a severe ecosystem response, coincidental with an unusually extreme weather event. Mar. Freshw. Res. 68 (10), 1816–1829.

Duncan, C., Owen, H.J.F., Thompson, J.R., Koldewey, H.J., Primavera, J.H., Pettorelli, N., 2018. Satellite remote sensing to monitor mangrove forest resilience and resistance to sea level rise. Methods Ecol. Evol. 9 (8), 1837–1852.

Ellison, J.C., 2018. Effects of Climate Change on Mangroves Relevant to the Pacific Islands. Centre for Environment, Fisheries and Aquaculture Science, pp. 99–111. Available from: https://www.gov.uk/government/publications/commonwealth-marine-economies-cme-programme-pacific-marine-climate-change-report-card-and-scientific-reviews (accessed 9.12.19).

Ellison, J.C., 2015. Vulnerability assessment of mangroves to climate change and sea level rise impacts. Wetl. Ecol. Manag. 23, 115–137.

Ellison, J.C., 2014. Climate change adaptation: management options for mangrove areas. In: Faridah-Hanum, I., Latiff, A., Hakeem, K.R., Ozturk, M. (Eds.), Mangrove Ecosystems of Asia: Status, Challenges and Management Options. Springer, New York, pp. 391–414.

Ellison, J.C., 2019. Biogeomorphology of mangroves. In: Cahoon, D., Perillo, G.M.E. (Eds.), Coastal Wetlands: an Ecosystem Integrated Approach. Elsevier Science, Amsterdam, pp. 687–715.

Ellison, J.C., Jungblut, V., Anderson, P., Slaven, C., 2012. Manual for Mangrove Monitoring in the Pacific Islands Region. Secretariat of the Pacific Regional Environment Program, Apia, Samoa.

Ellison, J.C., Cannicci, S., 2016. Impacts and effects of ocean warming on mangrove species and ecosystems. In: Laffoley, D., Baxter, J.M. (Eds.), Explaining Ocean Warming: Causes, Scale, Effects and Consequences. International Union for Conservation of Nature and Natural Resources (IUCN), Gland, Switzerland, pp. 135–146.

Ellison, J.C., Mosley, A., Helman, M., 2017. Assessing atoll shoreline condition to guide community management. Ecol. Indicat. 75, 321–330.

Ellison, J.C., Strickland, P., 2015. Establishing relative sea level trends where a coast lacks a long term tide gauge. Mitig. Adapt. Strategies Glob. Change 20, 1211–1227.

Ellison, J.C., Zouh, I., 2012. Vulnerability to climate change of mangroves: assessment from Cameroon, central Africa. Biology 1, 617–638. https://doi.org/10.3390/biology1030617.

English, S., Wilkinson, C., Baker, V., 1997. Survey Manual for Tropical Marine Resources, second ed. Australian Institute of Marine Science, Townsville.

Espana, N.B., Licuanan, W.Y., Alino, P.M., 2015. Resilience to thermal stress of coral communities in Talim Bay, lian, Batangas. Int. J. Philippine Sci. Technol. 8 (2), 37–40.

Fortes, M.D., Ooi, J.L.S., Tan, Y.M., Prathep, A., Bujang, J.S., Yaakub, S.M., 2018. Seagrass in Southeast Asia: a review of status and knowledge gaps, and a road map for conservation. Bot. Mar. 61 (3), 269–288.

Friess, D.A., Phelps, J., Leong, R.C., Lee, W.K., Wee, A.K.S., Sivasothi, N., Oh, R.R.Y., Webb, E.L., 2012. Mandai mangrove Singapore: lessons for the conservation of Southeast Asia's mangroves. Raffles Bull. Zool. 25, 55–65.

Friess, D.A., Richards, D.R., Phang, V.X.H., 2016a. Mangrove forests store high densities of carbon across the tropical urban landscape of Singapore. Urban Ecosyst. 19 (2), 795–810.

Friess, D.A., Thompson, B.S., Brown, B., Amir, A.A., Cameron, C., Koldeway, H.J., Sasmito, S.D., Sidik, F., 2016b. Policy challenges and approaches for the conservation of mangrove forests in Southeast Asia. Conserv. Biol. 30 (5), 933–945.

Friess, D.A., 2017. Mangrove rehabilitation along urban coastlines: a Singapore case study. Reg. Stud. Marine Sci. 16, 279–289.

FSM Statistics, 2019. FSM Division of Statistics. www.fsmstatistics.fm (accessed 9.11.19).

Fujimoto, K., Mochida, Y., Kikuchi, T., Tabuchi, R., Hirata, Y., Lihpai, S., 2015. The relationships among community type, peat layer thickness, belowground carbon storage and habitat age of mangrove forests in Pohnpei Island, Micronesia. Open J. For. 5 (01), 48.

Gandhi, S., Jones, T.G., 2019. Identifying mangrove deforestation hotspots in South Asia, Southeast Asia and Asia-Pacific. Rem. Sens. 11 (6), 728. https://doi.org/10.3390/rs11060728.

George, A., Luckymis, M., Palik, S., Adams, K., Joseph, E., Mathias, D., Malakai, S., Nakayama, M.R., Graham, C., Rikim, K., Marcus, A., 2008. The State of Coral Reef Ecosystems of the Federated States of Micronesia. National Oceanographic and Atmospheric Administration. Available from: https://www.sprep.org/att/IRC/eCOPIES/Countries/FSM/53.pdf (accessed 11.12.19).

Gibbs, D.A., West, J.M., 2019. Resilience assessment of Puerto Rico's coral reefs to inform reef management. PloS One 14 (11), e0224360. https://doi.org/10.1371/journal.pone.0224360.

Gilman, E., Ellison, J.C., Duke, N.C., Field, C., Fortuna, S., 2008. Threats to mangroves from climate change and adaptation options: a review. Aquat. Bot. 89, 237–250.

Giri, C., Ochieng, E., Tieszen, L.L., Zhu, Z., Singh, A., Loveland, T., Masek, J., Duke, N., 2011. Status and distribution of mangrove forests of the world using earth observation satellite data. Global Ecol. Biogeogr. 20, 154–159.

Golbuu, Y., Fabricius, K., Victor, S., Richmond, R.H., 2008. Gradients in coral reef communities exposed to muddy river discharge in Pohnpei, Micronesia. Estuarine. Coast. Shelf Sci. 76 (1), 14–20.

Goldberg, J., Adams, K., Albert, J., Asher, J., Brown, P., Brown, V., Burdick, D., Carroll, B., Craig, P., Fenner, D., Fillmed, C., 2008. Status of coral reef resources in Micronesia and American Samoa. Stat. Coral Reefs World: 2008 1, 199–212.

Goessens, A., Satyanarayana, B., Van der Stocken, T., Quispe Zuniga, M., Mohd-Lokman, H., Sulong, I., 2014. Is Matang Mangrove Forest in Malaysia sustainably rejuvenating after more than a century of conservation and harvesting management? PloS One 9 (8), e105069. https://doi.org/10.1371/journal.pone.0105069.

Gornitz, V., 1991. Global coastal hazards from future sea level rise. Palaeogeogr. Palaeoclimatol. Palaeoecol. 89, 379–398.

Guannel, G., Arkema, K., Ruggiero, P., Verutes, G., 2016. The power of three: coral reefs, seagrasses and mangroves protect coastal regions and increase their resilience. PloS One 11 (7), e0158094.

Hashim, N.I., Rahaman, A.A., Hashim, N.A.M., 2017. Awareness on conservation of mangrove among the community of Kuala Sepetang Taiping. Perak. e-Academia J. 6 (2), 250–263.

He, Z., Peng, Y., Guan, D., Hu, Z., Chen, Y., Lee, S.Y., 2018. Appearance can be deceptive: shrubby native mangrove species contributes more to soil carbon sequestration than fast growing exotic species. Plant Soil 432, 425–436.

He, Q., Silliman, B.R., 2019. Climate change, human impacts, and coastal ecosystems in the Anthropocene. Curr. Biol. 29 (19), R1021–R1035.

Henry, A., Webber, D., Webber, M., 2018. Rapid assessment methods developed for the mangrove forests of the Great Morass, St. Thomas, Eastern Jamaica. In: Dorney, J., Savage, R., Tiner, R.W., Adamus, P. (Eds.), Wetland and Stream Rapid Assessments, Development, Validation, and Application. Academic Press, London, pp. 529–538.

Hogarth, P.J., 2015. The Biology of Mangroves and Seagrasses, third ed. Oxford University Press, Oxford.

Holling, C.S., 1973. Resilience and stability of ecological systems. Ann. Rev. Ecol. Syst. 4, 1–23.

Houk, P., Camacho, R., Johnson, S., McLean, M., Maxin, S., Anson, J., Joseph, E., Nedlic, O., Luckymis, M., Adams, K., Hess, D., 2015. The Micronesia Challenge: assessing the relative contribution of stressors on coral reefs to facilitate science-to-management feedback. PloS One 10 (6), e0130823.

Huxham, M., Whitlock, D., Githaiga, M., Dencer-Brown, A., 2018. Carbon in the coastal seascape: how interactions between mangrove forests, seagrass meadows and tidal marshes influence carbon storage. Curr. Forest. Rep. 4 (2), 101–110.

Jayakumar, K., 2019. Managing mangrove forests using open source-based WebGIS. In: Krishnamurthy, R.R., Jonathan, M.P., Srinivasalu, S., Glaeser, B. (Eds.), Coastal Management: Global Challenges and Innovations. Academic Press, London, pp. 301–321.

Jennings, S., Manlutac, J.I., 2016. Oxfam's Strategy for Resilience in Asia, Guide. Oxfam International, Oxford, United Kingdom.

Jevrejeva, S., Frederikse, T., Kopp, R.E., Le Cozannet, G., Jackson, L.P., van de Wal, R.S.W., 2019. Probabilistic sea level projections at the coast by 2100. Surv. Geophys. 40 (6), 1673–1696.

Jonsson, M., 2017. Social-ecological Resilience of Mangroves and Coastal Households in Batticaloa District, Sri Lanka. Master's Thesis. Norwegian University of Life Sciences, Ås, Norway.

Jusoff, K., Taha, D.H.D.H., 2009. Managing sustainable mangrove forests in peninsular Malaysia. J. Sustain. Dev. 1 (1), 88–96.

Lam, N.S.N., Qiang, Y., Arenas, H., Brito, P., Liu, K., 2015. Mapping and assessing coastal resilience in the Caribbean region. Cartogra. Geograp. Inform. Serv. 42 (4), 315–322.

Lam, V.Y.Y., Doropoulos, C., Mumby, P.J., 2017. The influence of resilience-based management on coral reef monitoring: a systematic review. PloS One 12 (2), e0172064. https://doi.org/10.1371/journal.pone.0172064.

Lapolo, N., Utina, R., Baderanm, D.W.K., 2018. Diversity and density of crabs in degraded mangrove area at tanjung panjang nature reserve in Gorontalo, Indonesia. Biodiversitas 19 (3), 1154–1159.

Lee, H.S., 2015. General rainfall patterns in Indonesia and the potential impacts of local seas on rainfall intensity. Water 7 (4), 1751–1768.

Lewis, R.R., Milbrandt, E.C., Brown, B., Krauss, K.W., Rovai, A.S., Beever III, J.W., Flynn, L.L., 2016. Stress in mangrove forests: early detection and preemptive rehabilitation are essential for future successful worldwide mangrove forest management. Mar. Pollut. Bull. 109 (2), 764–771.

Lovelock, C.E., Cahoon, D.R., Friess, D.A., Guntenspergen, G.R., Krauss, K.W., Reef, R., Rogers, K., Saunders, M.L., Sidik, F., Swales, A., Saintilan, N., Thuyen, L.X., Triet, T., 2015. The vulnerability of Indo-Pacific mangrove forests to sea level rise. Nature 526, 559–563.

Lu, Y., Yuan, J., Lu, X., Su, C., Zhang, Y., Wang, C., Cao, X., Li, Q., Su, J., Ittekkot, V., Garbutt, R.A., Bush, S., Fletcher, S., Wagey, T., Kachur, A., Sweijd, N., 2018. Major threats of pollution and climate change to global coastal ecosystems and enhanced management for sustainability. Environ. Pollut. 239, 670–680.

Lovelock, C.E., Feller, I.C., Reef, R., Hickey, S., Ball, M.C., 2017. Mangrove dieback during fluctuating sea levels. Sci. Rep. 7 (1), 1680.

Maiti, S.K., Chowdhury, A., 2013. Effects of anthropogenic pollution on mangrove biodiversity: a review. J. Environ. Protect. 4, 1428–1434.

Mangroves for the Future, 2020. Indonesia. https://www.mangrovesforthefuture.org/countries/members/indonesia/ (accessed 17.04.20).

Martínez-Espinosa, C., Wolfs, P., Velde, K.V., Satyanarayana, B., Dahdouh-Guebas, F., Hugé, J., 2020. Call for a collaborative management at Matang Mangrove Forest Reserve, Malaysia: an assessment from local stakeholders' view point. For. Ecol. Manag. 458, 117741.

McDermid, K.J., Edward, A., 1999. Seagrass community composition and biomass at Nahpali island, Pohnpei. Micronesica-Agana 31, 255–262.

McKenzie, L.J., Rasheed, M.J., 2006. Seagrasses: Pohnpei Island and and Atoll Marine Assessment: Technical Report of Survey Conducted 26 October–3 November 2005. Seagrass-Watch HQ, DPIF, Cairns.

McKenzie, L.J., Yaakub, S.M., Tan, R., Seymour, J., Yoshida, R.L., 2016. Seagrass habitats of Singapore: environmental drivers and key processes. Raffles Bull. Zool. 34, 60–77.

McLeod, E., Margles weis, S.W., Wongbusarakum, S., Gombos, M., Daze, A., Otzelberger, A., Hammill, A., Agostini, V, Cot, D.U., Wiggins, M., 2015. Community-based climate vulnerability and adaptation tools: A review of tools and their applications. Coast. Manage. Coast. Manag. 43, 439–458.

McLeod, E., Salm, R.V., 2006. Managing Mangroves for Resilience to Climate Change. International Union for the Conservation of Nature and Natural Resources, Gland, Switzerland.

Meteorological Service Singapore, 2019. Past Climate Trends, Singapore. http://www.weather.gov.sg/climate-past-climate-trends/ (accessed 16.10.19).

MEWR&MND- Ministry of the Environment and Water Resources and Ministry of National Development, 2016. Singapore's Climate Action Plan: A Climate-Resilient Singapore, for a Sustainable Future, Singapore. Available from: https://sustainabledevelopment.un.org/content/documents/1549Climate_Action_Plan_Publi cation_Part_2.pdf (accessed 19.10.19).

Metz, W.D., 1996. The Pohnpei Mangrove Management Plan. Division of Resource Management, Office of Agriculture and Forestry, Pohnpei, Federated States of Micronesia.

Moore, J.D., 2018. Assessing Resilience of Community Mangrove Management of Tanakeke Island, Indonesia. Master's Thesis. National University of Singapore, Singapore.

Mumby, P.J., Chollet, I., Bozec, Y., Wolff, N.H., 2014. Ecological resilience, robustness and vulnerability: how do these concepts benefit ecosystem management. Curr. Opin. Environ. Sustain. 7, 22–27.

Murdiyarso, D., Purbopuspito, J., Kauffman, J.B., Warren, M.W., Sasmito, S.D., Donato, D.C., Manuri, S., Krisnawati, H., Taberima, S., Kurnianto, S., 2015. The potential of Indonesian mangrove forests for global climate change mitigation. Nat. Clim. Change 5, 1089–1092.

National Parks Board Singapore, 2010. The Sungei Buloh Wetland Reserve Master Plan, Singapore. Availabe online. https://www.asla.org/2010awards/147.html (accessed 19.9.19).

Ng, C.S.L., Toh, T.C., Chou, L.M., 2016. Coral restoration in Singapore's sediment challenged sea. Reg. Stud. Marine Sci. 8 (3), 422–429.

Nicholls, R.J., Cazenave, A., 2010. Sea level rise and its impact on coastal zones. Science 328 (5985), 1517–1520.

NOAA, 2019. Sea Level Trends. Center for Operational Oceanographic Products and Services, National Oceanic and Atmospheric Administration, United States. Available online. https://tidesandcurrents.noaa.gov/sltrends/. >(accessed 19.9.19).

Nolte, S., Koppenaal, E.C., Esselink, P., Dijkema, K.S., Schuerch, M., De Groot, A.V., Baker, J.P., Temmerman, S., 2013. Measuring sedimentation in tidal marshes: a review on methods and their applicability in biogeomorphological studies. J. Coast Conserv. 17 (3), 301–325.

O'Leary, J.K., Micheli, F., Airoldi, L., Boch, C., Leo, G.D., Elahi, R., Ferretti, F., Graham, N.A.J., Litvin, S.Y., Low, N.H., Lummis, S., Nickols, K.J., Wong, J., 2017. The resilience of marine ecosystems to climatic disturbances. Bioscience 67 (3), 208–220.

Otero, V., Van De Kerchove, R., Satyanarayana, B., Martínez-Espinosa, C., Fisol, M.A.B., Ibrahim, M.R.B., Sulong, I., Mohd-Lokman, H., Lucas, R., Dahdouh-Guebas, F., 2018. Managing mangrove forests from the sky: forest inventory using field data and Unmanned Aerial Vehicle (UAV) imagery in the Matang Mangrove Forest Reserve, peninsular Malaysia. For. Ecol. Manag. 411, 35–45.

Otero, V., Van De Kerchove, R., Satyanarayana, B., Mohd-Lokman, H., Lucas, R., Dahdouh-Guebas, F., 2019. An analysis of the early regeneration of mangrove forests using Landsat time series in the Matang mangrove forest reserve. Peninsular Malaysia. Rem. Sens. 11 (7), 774.

Praveena, S.M., Siraj, S.S., Aris, A.Z., 2012. Coral reef studies and threats in Malaysia: a mini review. Rev. Environ. Sci. Biotechnol. 11 (1), 27–39.

PICCC- Pacific Islands Climate Change Cooperative, 2019. Micronesian Mangrove Adapation Initiative. http://piccc.net/project/micronesian-mangrove-adaptation-initiative/ (accessed 1.12.19).

Qiu, Y.W., Qiu, H.L., Zhang, G., Li, J., 2019. Bioaccumulation and cycling of polybrominated diphenyl ethers (PBDEs) and dechlorane plus (DP) in three natural mangrove ecosystems of South China. Sci. Total Environ. 651, 1788–1795.

Randhir, T.O., 2016. Globalization impacts on local commons multiscale strategies for socioeconomic and ecological resilience. Int. J. Commons 10 (1), 387–404.

Raynor, B., Kostka, M., 2003. Back to the future: using traditional knowledge to strengthen biodiversity conservation in Pohnpei, Federated States of Micronesia. Ethnobot. Res. Appl. 1, 55–63.

Razak, N.A., Afandi, S.H.M., Shuib, A., Ghani, A.N.A., 2018. Visitors travelling time cost for ecotourism at Matang mangrove forest reserve. Int. J. Business Soci. 19 (S1), 117–127.

Rompas, P.T.D., Tanunaumang, H., Sangari, F.J., 2016. Study on marine current with approach of a numerical model for marine current power plant (PLTAL) in the Bangka Strait North Sulawesi. In: International Seminar on Application for Technology of Information and Communication (ISemantic), Semarang, pp. 104–110.

Rumson, A.G., Garcia, P.V., Hallett, S.H., 2019. The role of data within coastal resilience assessments: an East Anglia, UK, case study. Ocean Coast Manag. 105004. https://doi.org/10.1016/j.ocecoaman.2019.105004.

Rybczyk, R.J., Callaway, C., 2009. Surface elevation models. In: Wolanski, E., Cahoon, D., Perillo, G.M.E. (Eds.), Coastal Wetlands: An Ecosystem Integrated Approach. Elsevier Science, Amsterdam, pp. 835–854.

Scott, D.A., 1993. A directory of Wetlands in Oceania. In: International Waterfowl and Wetlands Research Bureau. Slimbridge, U.K. and Asian Wetland Bureau, Kuala Lumpur, Malaysia.

Siow, H., 2012. Case study: relative sea level trends of Singapore. In: Ellison, J.C. (Ed.), Climate Change Vulnerability Assessment and Adaptation Planning for Mangrove Systems. WWF, Washington DC, pp. 54–55. https://www.worldwildlife.org/publications/climate-change-vulnerability-assessment-and-adaptation-planning-for-mangrove-systems.

Sippo, J.Z., Lovelock, C.E., Santos, I.R., Sanders, C.J., Damien, T.M., 2018. Mangrove mortality in a changing climate: an overview. Estuarine. Coast. Shelf Sci. 215, 241–249.

Smit, B., Wandel, J., 2006. Adaptation, adaptive capacity and vulnerability. Global Environ. Change 16 (3), 282–292.

Sohl, F., Gombos, M., Celestine, F., Jonathan, A., Obispo, B., Ray-Ladore, K., 2019. Pohnpei Community Vulnerability Assessments April 2018 – February 2019. Conservation Society of Pohnpei, Kolonia.

Spalding, M., Kainuma, M., Collins, L., 2010. World Atlas of Mangroves. Earthscan, London and Washington D.C.

Spears, B.M., Ives, S.C., Angeler, D.G., Allen, C.R., Birk, S., Carvalho, L., Cavers, S., Daunt, F., Morton, R.D., Pock, M.J.O., Rhodes, G., Thackeray, S.J., 2015. Effective management of ecological resilience – are we there yet? J. Appl. Ecol. 52 (5), 1311–1315.

Steinbach, D., Kaur, N., Manuel, C., Saigal, S., Agrawal, A., Panjiyar, A., Barnwal, A., 2017. Building resilience to climate change – MGNREGS and cyclones in Andhra Pradesh. In: Issue Paper, IIED, London, United Kingdom.

Strassburg, M.W., Hamlington, B.D., Leben, R.R., Manurung, P., Gaol, J.L., Nababan, B., Vigundelli, S., Kim, K.Y., 2015. Sea level trends in Southeast Asian Sea. Clim. Past 11 (5), 743–750.

Tan, C.K., 2018. Rainfall patterns analysis over Ampangan Muda, Kedah from 2007–2016. J. Phys. Conf. 995 (1), 012121.

Tanaka, K., Choo, P.S., 2000. Influences of nutrient outwelling from the mangrove swamp on the distribution of phytoplankton in the Matang Mangrove Estuary, Malaysia. J. Oceanogr. 56 (1), 69–78.

Thiagarajah, J., Wong, S.K.M., Richards, D.R., Friess, D.A., 2015. Historical and contemporary cultural ecosystem service values in the rapidly urbanizing city state of Singapore. Ambio 55 (7), 666–677.

Thieler, E.R., Danforth, W.W., 1994. Historical shoreline mapping (II): application of the Digital shoreline Mapping and analysis systems (DSMS/DSAS) to shoreline change mapping in Puerto Rico. J. Coast Res. 10, 600–620.

TNC- The Nature Conservancy, 2017. Climate Projections and Impacts for Pohnpei. Avaiable online. http://reefresilience.org/wp-content/uploads/ClimateSummary_Pohnpei.pdf (accessed 1.12.19).

Turak, E., DeVantier, L., 2005. Reef-building Corals and Coral Communities of Pohnpei, Federated States of Micronesia: Rapid Ecological Assessment of Biodiversity and Status. Conservation Society of Pohnpei, Kolonia.

Turgeon, D.D., Asch, R.G., Causey, B., Dodge, R.E., Jaap, W., Banks, K., Delaney, J., Keller, B., Spieler, R.E., Matos, C.A., Garcia, J.R., 2003. The State of Coral Reef Ecosystems of the United States and Pacific Freely Associated States. U.S. Department of Commerce, National Oceanic and Atmospheric Administration, National Ocean Service.

UNDP, 2018. Global Human Development Indicators. United Nations Development Program, Human Development Reports. Available online. http://hdr.undp.org/en/countries (Accessed 18.11.19).

UNDP, 2019. Human Development Report 2019: Beyond Income, beyond Averages, beyond Today - Inequalities in Human Development in the 21st Century. UN, New York. https://doi.org/10.18356/838f78fd-en.

Velde, K.V., Huge, J., Friess, D.A., Koedam, N., Dahdouh-Guebas, F., 2019. Stakeholder discourses on urban mangrove conservation and management. Ocean Coast Manag. 178, 104810.

Victor, S., Neth, L., Golbuu, Y., Wolanski, E., Richmond, R.H., 2006. Sedimentation in mangroves and coral reefs in a wet tropical island, Pohnpei, Micronesia. Estuar. Coast Shelf Sci. 66, 409–416.

Waggey, B., 2017. Morphometric analysis of congeneric seagrasses (*cymodocea rotundata* and *cymodocea serrulate*) in the coastal areas of Bunaken National Park, North Sulawesi Indonesia. AACL Bioflux 10 (6). Available online. http://www.bioflux.com.ro/aacl. (Accessed 25 March 2020).

Ward, R.D., Friess, D.A., Day, R.H., MacKenzie, R.A., 2016. Impacts of climate change on mangrove ecosystems: a region by region overview. Ecosys. Health Sustain. 2 (4), e01211.

Waycott, M., McKenzie, L.J., Mellors, J.E., Ellison, J.C., Sheaves, M.T., Collier, C., Schwarz, A.M., Webb, A., Johnson, J.E., Payri, C.E., 2011. Vulnerability of mangroves, seagrasses and intertidal flats in the tropical Pacific to climate change. In: Bell, J.D., Johnson, J.E., Hobday, A.J. (Eds.), Vulnerability of Fisheries and Aquaculture in the Pacific to Climate Change. Secretariat of the Pacific Community, Noumea, pp. 297–368.

Webb, E.L., Friess, D.A., Krauss, K.W., Cahoon, D.R., Guntenspergen, G.R., Phelps, J., 2013. A global standard for monitoring coastal wetland vulnerability to accelerated sea-level rise. Nat. Clim. Change 3, 458–465.

White, I., Glavovic, B., Adams-Hutcheson, G., 2018. Operationalising Resilience: A Heuristic Framework for Analysis. National Science Challenges, New Zealand. Available online. http://resiliencechallenge.nz/governance-programme/operationalising-resilience/ (accessed 20.11.19).

Whitney, C.K., Bennett, N.J., Ban, N.C., Allison, E.H., Armitage, D., Blythe, J.L., Burt, J.M., Cheung, W., Finkbeiner, E.M., Kaplan-Hallam, M., Perry, I., Turner, N.J., Yumaglova, L., 2017. Adaptive capacity: from assessment to action in coastal social-ecological systems. Ecol. Soc. 22 (2), 22.

World Bank, 2020. GDP Per Capita (Current US$). Available online. https://data.worldbank.org/ (Accessed 9.4.20).

World Meteorological Organization, 2019. The Global Climate in 2015-2019. World Meteorological Organization, Geneva. Available online. https://public.wmo.int/en (Accessed 1.12.19).

CHAPTER

21

Assessment of typhoon impacts and post-typhoon recovery in Philippine mangroves: lessons and challenges for adaptive management

Severino G. Salmo III

Institute of Biology, College of Science, University of the Philippines Diliman, Quezon City, Philippines

OUTLINE

1. Introduction 540

2. Materials and methods 541

2.1 Site description 541

2.2 Typhoon history 542

2.3 Field assessment 543

3. Results and discussion 544

3.1 Vegetation damage 544

3.2 Sediment damage 546

3.3 Fauna 547

3.4 Prospects for recovery 551

3.5 Recovery indicators 553

3.5.1 Vegetation 554

3.5.2 Sediment 554

3.5.3 Faunal assemblage and activities 555

4. Summary: vulnerability, research/management implications, and recommendations 556

Acknowledgments 558

References 559

Dynamic Sedimentary Environments of Mangrove Coasts
https://doi.org/10.1016/B978-0-12-816437-2.00022-7

1. Introduction

Mangroves provide several important ecological and socioeconomic services such as fisheries production, biodiversity conservation, natural disaster protection, and climate change adaptation/mitigation (Romañach et al., 2018). Intact mangroves have high primary productivity that results in elevated carbon (C) stores and that provides food for resident mangrove-specialist invertebrate fauna (e.g., crabs, shrimp). These are then fed upon by transient species (e.g., fishes, birds) that migrate into mangroves from adjacent ecosystems.

The loss of mangroves results in the reduction of the delivery of ecosystem services (e.g., biodiversity and C sequestration) and makes coastal communities vulnerable to natural disasters (e.g., storm surge, sea level rise [SLR]; Lovelock et al., 2015). The capacity of mangroves to render ecosystem services depends on their spatial extent, ecosystem health, and stage of forest development (e.g., young vs. mature). International programs have been implemented that either incorporate or help promote mangrove conservation, restoration, and management (e.g., IBCI (International Blue Carbon Initiative), IUCN (International Union for the Conservation of Nature), UNFCCC (United Nations Framework Convention on Climate Change), GMA (Global Mangrove Alliance); Friess et al., 2016). In the Philippines, massive restoration programs have been implemented since the 1990s in an attempt to recover mangrove forest cover and the ecosystem services they provide (Salmo et al., 2007).

Typhoons are natural disturbances that frequently occur especially in the tropics. Severe typhoons bring strong winds, intense rainfall, and large storm surges causing notable damages to the environment, property, and human lives, particularly in coastal areas (Cinco et al., 2016). These larger typhoons are known to cause significant reductions in mangrove forest cover as well as an associated loss of organic matter (OM) in sediments (Salmo et al., 2014). However, the extent and scale of damage within a forest stand are heterogeneous and vary with wind speed and localized site gradients such as geographical position relative to shoreline and tree height (Everham and Brokaw, 1996; Busby et al., 2008).

Mangroves can effectively function as a coastal defense against the impacts of typhoons, storm surges, and SLR. Their vast networks of aerial and underground roots and high structural complexity slow down both water and wind movements by 75% (but only when mangroves have a wide extent of at least 1 km width; McIvor et al., 2012). However, mangroves are vulnerable to typhoons because of their position in the coastal zone (Hogarth, 2007), although vulnerability is also genus/species-specific (Long et al., 2016). The frequency and severity of typhoon shapes the population structure and forest dynamics of mangroves that could define their resilience, survival, and regeneration patterns over longer periods (Ellison and Farnsworth, 1996; Cahoon et al., 2003). With prevailing climate change, several simulations suggest that the frequency of typhoons would decrease, but their severity may increase (Kubota and Chan, 2009).

Catastrophic typhoons are likely to contribute to reductions in habitat quality as well as a decrease in the supply of food for faunal organisms dependent on mangroves. For example, an alteration of mollusk colonization pattern but more particularly a shift in dominant species (e.g., reverting from pre-typhoon mangrove-associated species to post-typhoon pioneer species in mature plantation) may commence after drastic changes in vegetation and sediment conditions. Smith et al. (1994) stressed, however, that the effects of such severe disturbances on the recruitment and colonization dynamics of mangrove-resident fauna had not been adequately studied.

Most studies on the effects of intense storms on mangroves are from the Caribbean region (although see Smith et al., 2009). There are relatively very few reports of typhoon effects on mangroves from South East Asia where typhoons are prevalent (but see recent studies, e.g., Buitre et al., 2019; Salmo and Gianan, 2019; Zhang et al., 2019). Most reports on effects of typhoons on mangroves are mainly focused on the description of "damages" particularly on vegetation, but rarely discuss the effects of typhoons to fauna and faunal activities as well as post-typhoon recoveries. In this chapter, the effects of typhoons on vegetation, sediment, and fauna (e.g., in terms of species composition and faunal activities, and including post-typhoon recovery) were examined and compiled from 15 years of research done in selected mangrove sites in the Philippines. Insights for assessment methods, research needs, and policy needs are also provided to improve the management of Philippine mangroves.

2. Materials and methods

2.1 Site description

The Philippines is a tropical archipelagic country known as one of the most disaster-prone country in the world (Gaillard et al., 2007). An average of around 20 typhoons occur each year (Velasco and Cabanilla, 2003), with intense events (>150 km h^{-1}) having 16–32 years oscillation (Kubota and Chan, 2009). In the last 30 years, the number of typhoons has increased. From 1945 to 2013, the Philippines recorded an average of six landfalls per year (Takagi and Esteban, 2016). Such increased frequency appears to be associated with increases in sea surface temperatures (Webster et al., 2005). The occurrence of high-intensity typhoons in the Philippines threatens the growth, development, and regeneration of both planted and natural mangroves.

The study was gleaned from 15 years of researches done in 10 selected mangrove sites in the Philippines (Fig. 21.1A). These sites were mainly from the Luzon and Visayan areas representing three (of six) marine biogeographic regions. These sites are in Bani, Pangasinan; Masinloc, Zambales; Subic, Zambales; Calapan, Oriental Mindoro; Busuanga, Palawan (in Luzon; north-northwest Philippines); Kalibo, Aklan; Bantayan Island, Cebu; Palompon, Leyte; and Salcedo, Eastern Samar (in Visayas; central-eastern Philippines). Observations were generated from typhoon-disturbed mangroves composed of natural and planted mangrove stands. The natural stands are diverse that are mostly dominated by *Avicennia* and *Sonneratia* spp. while the planted stands are monospecific from the *Rhizophora* genus.

Based on the modified Climate Classification of the Philippines (Lantican, 2001), the sites in Bani, Pangasinan; Subic and Masinloc, Zambales; and Busuanga, Palawan, belong to Type I (with two pronounced seasons—dry from November to April and wet throughout the rest of the year). The sites in Calapan, Oriental Mindoro and Kalibo, Aklan are categorized as Type III (no pronounced seasonality but is relatively dry during November to April and wet from May to October), while Bantayan Island, Cebu and Palompon, Leyte are classified as Type IV (with rainfall that is evenly distributed throughout the year). The sites in Salcedo, Eastern Samar, are categorized as Type II (no dry season with pronounced maximum rainfall from November to January).

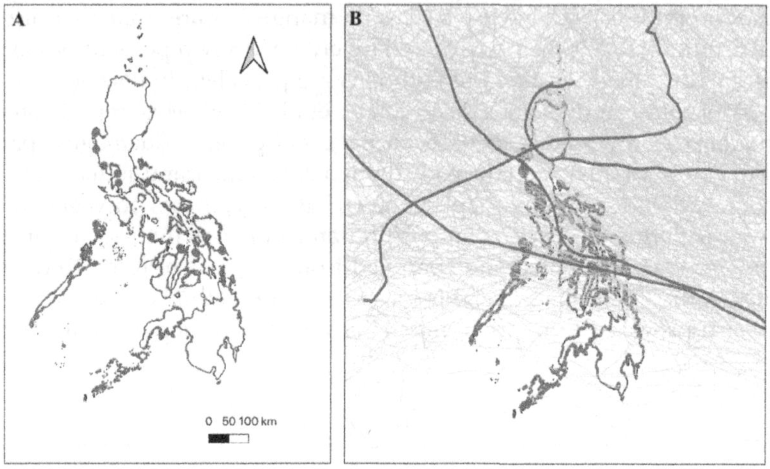

FIGURE 21.1 Distribution of sampling sites with overlaid mangrove extent (A) and incidents of strong typhoons (red [black in print version] lines) relative to the cumulative occurrences (yellow [light gray in print version] lines in B) in the Philippines from 2000 to 2020. *Adapted/modified from NCDC IBTrACS, 2019. US NOAA National Climatic Data Center – International Best Track Archive for Climate Stewardship (IBTrACS) Project (Version 3). https://catalog.data.gov/dataset/ncdc-international-best-track-archive-for-climate-stewardship-ibtracs-project-version-3. (Accessed 1 September 2020).*

2.2 Typhoon history

Typhoons Fengshen, Chan-hom, Koppu, and Haiyan were some of the destructive typhoons that hit the country in ca. 10 years, having maximum sustained winds of 150, 120, 185, and 230 kph and Saffir–Simpson Scale Typhoon category of 3, 2, 4, and 5, respectively (Fig. 21.1B).

Typhoon Fengshen (known locally as Typhoon Frank) was a Category 3 typhoon and the sixth typhoon that hit the Philippines in June 2008. It had strong winds of >200 km h^{-1} with heavy rains that lasted for at least 3 days in the Philippines. Kalibo, and the entire Panay Island, was severely damaged from flooding and mudslides. Typhoon Chan-hom (known locally as Typhoon Emong) was the fifth typhoon that entered the Philippines in the 2009 season. It was formed near South East Vietnam in the West Philippine Sea and crossed the Philippine Area of Responsibility on May 3, 2009. It had strong winds of 85–150 km h^{-1} and was categorized as Typhoon 2 under the Saffir–Simpson Scale. The eye of the typhoon first crossed land at Cape Bolinao on May 7, 2009 bringing strong winds, heavy rains (at least 200 mm in 24 h), and flooding. Sites in Bangrin Mangrove Protected Area in Bani, Pangasinan, were also assessed of the effect of repeated occurrence of typhoons. Bani was damaged by Typhoon Chan-hom in 2009 and Typhoon Koppu in 2015. Two sampling sites in Bani were established in the natural stands of *Avicennia marina* and planted stands of *Rhizophora stylosa*.

The strongest typhoon in modern meteorological history, Super Typhoon Haiyan, impacted the coastline of the Philippines on November 8, 2013. It was categorized as Typhoon Category 5 with maximum strong wind speed of 315 kph and was 16% stronger than the second strongest typhoon, Typhoon Zeb in 1998 (Takagi and Esteban, 2016). The typhoon caused flooding, landslides, and storm surge as high as 4 m. It affected 44 provinces and 9 regions and left more than 6000 casualties and about Php 18.35 billion worth of damage (NDRRMC, 2013). Busuanga, Palawan, was the sixth and final landfall of Super Typhoon Haiyan. The mangrove forest in Busuanga, Palawan, is part of a declared mangrove swamp forest reserve (Presidential Proclamation No. 2152 in 1981) and is considered as one of the most intact mangroves in the Philippines.

2.3 Field assessment

Like other natural disasters, typhoons are unexpected events with heterogeneous effects, and therefore studies that assess their impacts do not comply with optimal ecological experimental designs (Green, 1979; Michener, 1997). The sampling approach adapted in this study was mainly based on the extent and path of typhoons as well as the distribution and extent of typhoon-impacted mangroves. Assessments were done in representative permanent monitoring plots (of 5m radii). All mangrove individuals (composed of trees, saplings, and seedlings) within the plots were tagged and were periodically measured. The site in Bani, Pangasinan had a pre- versus post-typhoon comparisons. The rest of the sites were only assessed after the typhoon (of unequal intervals and frequencies). The findings reported here provided information on extent of damages and also possible recovery trajectory patterns as long as 7-year post-typhoon. A combination of parametric and nonparametric tests was used. General changes in vegetation, sediment, and faunal parameters were analyzed over time per site.

For vegetation assessment, in each plot, tree heights and diameters were measured using calibrated pole and tapes, respectively. Canopy diameters were estimated for each tree. The number of seedlings and saplings was also counted. Canopy cover and gaps (in percentage) were estimated from each plot. Fresh and senescent leaves were collected from the top of the canopy for the analyses of total nitrogen (TN) and total phosphorus (TP) contents.

Porewater variables such as temperature, salinity, redox, and conductivity were measured from the center of the plot using portable water quality monitoring instrument. Sediment samples (3–12 samples per site) were collected and subjected for the analyses of OM (using loss on ignition method), TN (using digestion method), and available phosphorus (AP; using Bray 2 method) contents.

Faunal assemblages (composed of mollusk and nekton) were assessed per site (in Bani, Masinloc, and Kalibo). For mollusk, samples were collected from the arboreal and epifaunal assemblages. Arboreal assemblages were samples attached in mangrove vegetative parts (e.g., stem, root, etc.) collected from 5m radii vegetation plots. Epifaunal assemblages were samples found in sediment surface and collected from three 1×1 m quadrats within the vegetation plot. The arboreal and epifaunal assemblages were composited and reported per site. Specimens were identified at species level and were assessed of abundance and biomass. Nekton specimens (e.g., fishes, crabs, shrimps, squids) were collected using trap nets ($n = 2$ per site in Bani, Masinloc, and Kalibo) that were deployed for 4 days corresponding to a day before, during, and after the spring tide. Individuals were identified at species level and were measured of standard sizes and weights. Species composition and dominant species were assessed per site. In Bantayan Island (Cebu), faunal activities in terms of burrowing and leaf tethering were measured.

3. Results and discussion

3.1 Vegetation damage

The occurrences of typhoons resulted in severe vegetation damages. The vegetation damage was easily the most visible and most reported impacts of typhoons (Salmo et al., 2014), although the damages are not homogenous over all mangrove stands. Similar to published reports, the damage from a typhoon within a forest stand or even at a landscape level is heterogeneous and may be attributed to gradients of wind speed over the path of the typhoon and localized site gradients such as position relative to shoreline, tidal elevation, soil type, and tree height (see for example Everham and Brokaw, 1996; Busby et al., 2008).

High mangrove tree mortalities are commonly reported as impacts. The dead trees resulted to the formation of canopy gaps. However, the occurrences and sizes of canopy gaps have no clear patterns (see for example typhoon-damaged mangroves in Bani, Pangasinan; Carlos, 2014; Fig. 21.2). Canopy gap sizes vary from $<10\ m^2$ to $>100\ m^2$. Within the canopy gaps, trees are completely defoliated, broken stems/trunks, uprooted, or standing dead. Some species have more damages than the other species. The types of stands (as either intact or planted) also manifest different damages. Similar to other tropical forest types, mangrove plantations are perceived to be more vulnerable to typhoons compared to natural stands (Everham and Brokaw, 1996) because of the lower structural complexity and lower

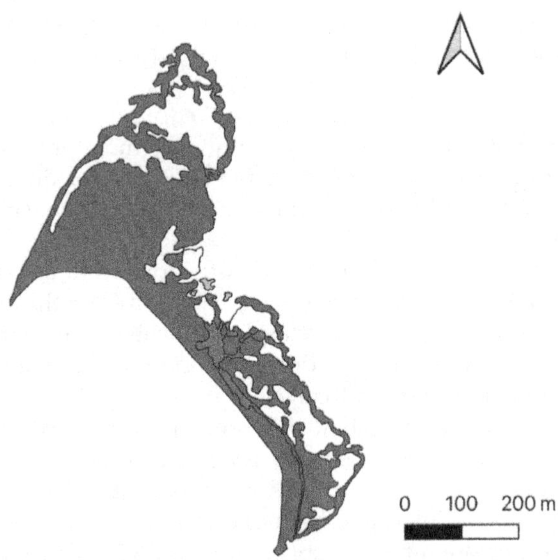

FIGURE 21.2 Canopy gap distribution (of different sizes) in Bani, Pangasinan. Legend: white—gaps; orange—mangrove patch; green—mangroves that survived the typhoon; blue—waterways.

wind firmness (*sensu* Gardiner and Quine, 2000) as defined by high height to diameter at breast height (DBH) ratio (Ancelin et al., 2004). The typhoon-damaged 18-year-old stands were relatively tall with height to DBH ratios of 106.40 and 116.78, respectively. However, catastrophic typhoons such as that of Super Typhoon Haiyan caused significant damage and removal of mangroves within its path regardless of stand types, species, or location.

Consistent with published reports, trees with taller heights and larger diameters had higher mortality than smaller trees (Roth, 1992; Smith et al., 1994; Baldwin et al., 2001). The trees that died had stem diameter >8 cm and height >7 m. The reductions in tree density and basal area in Lingayen Gulf after the typhoon are comparable to losses described in other published reports (Table 21.1). Tree mortalities continued even after 2–5 years post-typhoon, also similar with other published reports (Smith et al., 1994; Sherman et al., 2001). The extent of vegetation damage was also related to the state of sediment damage, primarily temperature, redox, and OM, but may shift over time.

Leaf nutrient contents (primarily nitrogen and phosphorus) were also negatively affected (Garcia, 2017). Nutrients and nutrient recovery are rarely reported but could play an important role in either the hastening or delay of mangrove recovery. Mangrove environments are usually located in nutrient-poor locations (Feller et al., 2017). Mangroves adapt through nutrient conservation mechanism where nutrients are resorbed and remobilized in senescing leaves before the leaves falls to the ground (Killingbeck, 1996). But, when typhoons result to severe defoliation, the nutrient conservation mechanism of the mangroves will be altered (Garcia, 2017). In some cases (in Bani and Kalibo), there were sudden increases in leaf N

TABLE 21.1 Recorded density and basal area reductions (and the mean tree diameters and heights) in some typhoon-damaged mangroves in the Philippines.

Location	Stand/Age (yr)	Density reduction (%)	BA reduction (%)	Tree diameter (cm)	Tree height (m)
Lingayen Gulf	Plantation; 11	61	70	8	7
Lingayen Gulf	Plantation; 18	70	80	11	12
Lingayen Gulf	Natural	60	80	12	12
Kalibo	Plantation; 12	60	65	8	7
Kalibo	Plantation; 17	90	90	12	12
Busuanga	Natural	75	80	15	15
Calapan	Natural	60	60	12	14
Cebu	Plantation; 15	60	70	8	8
Cebu	Natural	60	65	9	8
Leyte	Natural	60	65	10	12
Leyte	Natural	50	70	12	14
Samar	Natural	50	60	8	10
Samar	Planted; 10	80	80	8	8

and P (as a result of reduced salinity and enhanced waterlogged conditions), although the utilization of it is limited due to complete or partial closure of stomatal aperture (Kozlowski and Pallardy, 1997).

3.2 Sediment damage

The damages (and/or changes) in mangrove sediments brought by typhoons are less reported than that of vegetation. But, the damages in vegetation are also related to the damages in the sediments. Similar to vegetation, the damages in the sediments do not have clear patterns and may also be related to the gradients of wind speed and intensity of the typhoon (Salmo et al., 2014).

However, the position of the site relative to the coastal fringes may have a more direct relationship with the effects of the typhoon. For example, sites close to fringes are more likely to suffer erosion especially in typhoons with accompanying storm surges (i.e., Super Typhoon Haiyan). Sites inshore are more likely to experience sediment dumping and hence may increase its surface elevation (Fig. 21.3). In some places (e.g., Florida, USA), typhoon-damaged mangroves experienced high sediment dumping (Breithaupt et al., 2019). In the Philippines, some sites experienced similar phenomenon. However, when typhoons are accompanied by a strong storm surge (i.e., 3–7m in Super Typhoon Haiyan), mangroves showed some sediment losses (ca. 0.5m or even > 1m) than gains. Alternatively, some sites in Leyte Gulf (in Visayas; southern Philippines) had overwash sediments (Soria et al., 2018) where sandy particles were dumped and altered the usually silty-clay mangrove sediments in coastal fringes.

The occurrence of the typhoon also caused significant changes in soil characteristics in the typhoon-affected sites. Changes to soils after intense storms have also been reported from other typhoon-disturbed sites (Smith et al., 1994; Ellison, 1998; Rivera-Monroy et al., 2004). In this study, among the immediate and easily discernible post-typhoon changes are the reductions in OM (40%–53%), more anoxic soils (−10 mV), and increase in nutrients (TN by 30%–60%; TP by 30%–40%), salinity (4–10 ppt), and soil temperature (8–10°C; Salmo et al., 2014). Increases in soil temperature can be attributed to the removal of canopy shading that exposes the soil to solar radiation (Bosire et al., 2003). The increases in temperature after

FIGURE 21.3 Contrasting photos of sediment losses (A) and dumping (B) in Leyte. Sediment losses (of at least ca. 0.5m surface elevation) were probably a result of scourging from storm surge, which consequently exposed the mangrove roots. In contrast, sediment dumping (mainly composed of sandy particles) altered the sediment texture.

defoliation may also cause increases in salinity due to higher evaporation and decreases in redox potential through enhanced microbial activity. The increase in nutrient availability after the typhoon may be explained by the influx of fine-grained materials during the typhoon surge (Fig. 21.3) that were deposited on the forest floor (pers. obs.) and also could be explained by the mortality of roots and release of nutrients in the biomass to the soil. The depositions of sediment materials within the forests during intense storms were observed at other sites (Smith et al., 1994; Ellison, 1998) where 50%–100% increases in TN and TP were reported (Castaneda-Moya et al., 2010). The observed reduction in soil OM in this study could be due to enhanced decomposition of OM due to increased nutrient availability (Bianchi and Allison, 2011), although this hypothesis requires further assessment. However, for sites that experienced storm surge, it is more likely to have severe reductions in OM and nutrients.

The changes to the soil environment were closely linked with the continuous decline in mangrove vegetation after the typhoon. While post-typhoon salinity (33 ppt) falls within the normal range of mangrove conditions, the 4–10 ppt sudden increase (above 20 ppt from pre-typhoon) may have caused reductions in tree growth and slowed forest recovery. Abrupt increases in soil salinity have caused massive mangrove mortality in the Caribbean coast of Colombia (Cardona and Botero, 1998). The increase in nutrient availability, particularly of P, is known to contribute to changing vegetation characteristics (Chen and Twilley, 1998; Castaneda-Moya et al., 2010), which may have long-term effects on the structure and function of forests (Smith et al., 2009). Abrupt changes in the sediments contribute to the prolonged mortality and delayed recovery.

The sediment strata revealed changes in nutrient levels with depth. A high nutrient content in the upper sediment layer may indicate an accumulation of nutrients from detritus inputs. Moreover, nutrient losses may be attributed to large-scale disturbances such as typhoons. In Vietnam, organic carbon (OC) loss was observed at 35–50 cm depth equivalent to 15 cm of sediment OM as a result of massive forest degradation (Dung et al., 2016). Similarly, sediment OC losses were observed at 0–18 cm depth in 25-year-old planted stand in Kalibo and 20–60 cm depth in Cebu.

Moreover, the responses of mangrove ecosystems to nutrient limitation are complex and are governed by ecological processes, the species composition, and the limiting nutrients (Feller et al., 2003, 2007). The N and P contents over depths revealed that the mangrove sediments had low nutritive values and that N (and therefore also N conservation) are limited in the typhoon-damaged sites (see Fig. 21.4). Under a nutrient-limited condition, it is expected that the low nutrient levels will be strongly conserved through effective nutrient recycling mechanism (Garcia, 2017).

3.3 Fauna

The impacts of typhoons on fauna are less documented (compared to vegetation and sediment). There are very few studies on the effects of typhoons on the dynamics of mangrove-dependent fauna (Smith et al., 1994; although see Diele et al., 2013). Perhaps the closest comparison that could be made is the loss of forest cover brought by deforestation in Gazi Bay, Kenya (Fondo and Martens, 1998; Bosire et al., 2003) and habitat modification in Moreton Bay, Australia (Skilleter and Warren, 2000). These studies pointed out that any level

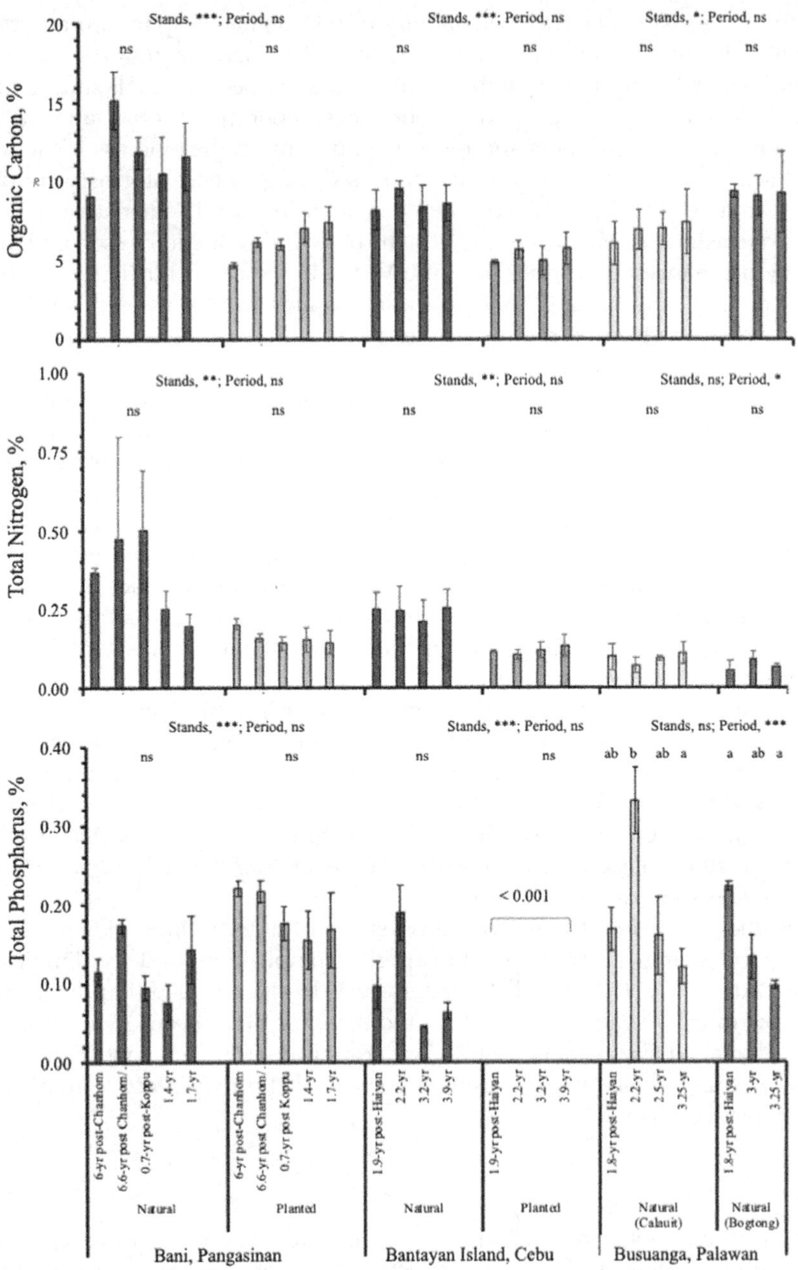

FIGURE 21.4 Contrasting patterns of changes in carbon, nitrogen, and phosphorus contents in the 20-cm sediment depths of different typhoon-damaged sites and of different post-typhoon periods. Different letters indicate significant difference between post-typhoon period (*ns*, not significant).

of disturbance may affect the diversity and abundance of fauna. The loss of forest cover and impoverished soil conditions (anoxic and warm soil) translates into loss or reduction of food and shelter for fauna. The reduction in physical structure and detritus production may also consequently reduce the macrofaunal community within its habitat. The presence, abundance, and composition of fauna are directly related to the habitat and that are provided by mangroves (Sasekumar and Chong, 1998).

Typhoons are known to cause reductions in forest cover and also OM in the soil, which may result to the reduction of habitat quality and food for both mangrove-resident and mangrove-transient species (Salmo et al., 2014). It is apparent that typhoons result to severe reductions in faunal abundance and biomass, although the magnitude may vary with the degree of dependence of the species to the mangrove vegetation and sediment. In mollusk, for example, the typhoon caused significant changes in arboreal, epifaunal, and infaunal mollusk assemblages (Salmo et al., 2019a). Significant correlations were more apparent between the mollusk assemblages and vegetation than with soil properties. Reductions in the abundance and biomass of mollusks were manifested at 2.5-month post-typhoon and do not appear to recover. Similarly, the length–biomass allometry of the dominant species in typhoon-impacted sites significantly changes after the typhoon, implying reduction of food source and desiccation (Salmo, 2011).

In Bani, Pangasinan, the mollusk species composition drastically changed after the typhoon because species that were recorded prior to the typhoon were lost. The shifts in species composition (Table 21.2) were also related to the sizes of canopy gaps where some species dominate in larger gaps while other species are more abundant under canopy (Carlos, 2014).

The combination of both vegetation and soils affected the post-typhoon changes in mollusk assemblages (Table 21.3; Salmo et al., 2019a). It is not straightforward, however, to identify a single particular variable that caused the changes since vegetation and soils are interrelated, although at some point certain variables showed stronger influence. For

TABLE 21.2 Summary results showing contrasting patterns of the species similarities (decrease in arboreal assemblage and increase in epifaunal assemblage) and the shifts in dominance in arboreal and epifaunal assemblages between pre- and post-typhoon periods.

Assemblage	Pre-typhoon	Post-typhoon
Arboreal	74%	58%
	Terebrali sulcata *Nerita polita* *Pirenella cingulata*	*Terebralia sulcata* *Littoraria scabra*
Epifauna	76%	82%
	Pirenella cingulata *Terebralia sulcata* *Latirus philberti*	*Terebralia sulcata* *Latirus philberti*

Modified from Salmo, SG, Tibbetts, I, Duke, NC. 2019. Recolonization of mollusc assemblages in mangrove plantations damaged by Typhoon Chan-hom in the Philippines. Estuarine, Coastal and Shelf Science **228***: 106365.*

TABLE 21.3 Summary of correlation results showing pre- and post-typhoon association between mollusk assemblages and vegetation and sediment variables in Lingayen Gulf. The association varied with assemblages and with periods implying different degrees of association with different parameters. The arboreal assemblage was not correlated with sand content while the epifaunal assemblage was not correlated with nitrogen content and redox.

	Arboreal		Epifauna	
Variables/Assemblages	**Pre-typhoon**	**Post-typhoon**	**Pre-typhoon**	**Post-typhoon**
Vegetation				
Leaf area index	0.48	0.47	0.47	0.49
Above-ground biomass	0.41	0.32	0.42	0.35
Sediment				
Sand	ns		0.27	0.36
Organic matter	0.60	0.36	0.59	0.34
Nitrogen	0.47	0.19	ns	
Redox	0.44	0.25	ns	
Temperature	0.20	0.32	0.60	0.45

ns, not significant.

Modified from Salmo, SG, Tibbetts, I, Duke, NC. 2019. Recolonization of mollusc assemblages in mangrove plantations damaged by Typhoon Chan-hom in the Philippines. Estuarine, Coastal and Shelf Science ***228****: 106365.*

example, in arboreal mollusk assemblage, the reductions in leaf area index (LAI) and above-ground biomass (AGB) have greater influence at 0.5-month post-typhoon. But as time progressed, it appeared that forest cover alone has more direct effect than either tree density or AGB. At 7-month and 9-month post-typhoon, both tree density and AGB continuously decline while LAI regenerate, although very slowly (Table 21.3).

Nekton communities, in general, have a less apparent pattern (as compared to mollusk). There was a change in nekton communities in typhoon-impacted sites but only at 9-month post-typhoon, which may indicate a lag period in the manifestation of typhoon effects. Surprisingly, the post-typhoon changes in nekton communities were not associated with the reductions in mangrove vegetation, but rather were more influenced by environmental variables primarily proximity to reef, tidal inundation, and fishing pressure (Salmo, 2011). The association of nekton with proximity to reef and fishing pressure may indicate that post-typhoon macrofaunal colonization will either be facilitated or derailed by the proximity and status of coral reefs and the degree of fishing pressure on the supply of nekton.

The sustained presence of fishes in typhoon-impacted sites may be explained by the fact that the dead trees are possibly still capable to offer a suitable refuge. But as these dead trees decompose, it may no longer be possible to serve as fish habitat and eventually fish abundance could be reduced. Differences in fish assemblages between vegetated and unvegetated habitats have been reported by Huxham et al. (2004) and Taylor et al. (2007) in Kenya and

Belize, respectively. The decline in crustaceans population (categorized as omnivores and mangrove-associated species) after the typhoon may have led to the slight post-typhoon increases in the relative abundance and biomass of detritivores and reef-associated species (Salmo, 2011).

The occurrence of Typhoon Chan-hom resulted to the changes in nekton communities. Fishes were not affected by the typhoon but there was greater sensitivity in the crustaceans which showed reduced abundance and biomass from 50% to 70% at 7-month post-typhoon. There was a slight increase in the abundance and biomass of crustaceans at 9-month post-typhoon (Salmo, 2011). In Bantayan Island (Cebu), crab activities in terms of burrowing, leaf tethering, and propagule consumption were at least 50% higher in natural stands than restored stands (Gallmann, 2016). The sensitivity of crustaceans to typhoon damage further strengthens proposition that crustaceans may serve as a more useful indicator of disturbance than fish.

3.4 Prospects for recovery

Reports of the recovery of mangroves after a catastrophic typhoon are highly variable (Saenger, 2002). There are some reports that mangroves are resilient against typhoons (see for example Primavera et al., 2016; Long et al., 2016). There is a general consensus, however, that recovery could be very slow or may even be uncertain (Milbrandt et al., 2006; Smith et al., 2009) as lag effect (through delayed tree mortalities) may be manifested even more than 10 years post-typhoon. As the extent and severity of damages are uneven, recoveries are also irregular and will depend on several factors. Species differ in both their susceptibility to storm damage and their recovery (Baldwin et al., 2001). Even trees that initially survived the typhoon may not recover. In Samar, for example, some mangrove sites had rapid post-typhoon recovery (1–5 year post-typhoon), stabilized at 5–6 year post-typhoon, and appeared to be steadily recovering until some trees died (Cabello and Germentil, 2020). But in general, trees with smaller diameters and shorter heights were observed to survive better. Due to higher structural complexity, diverse stands regenerate faster than monospecific plantation stands (Salmo and Gianan, 2019).

The canopy gaps (of different sizes) resulting from the typhoon will provide a space for seedling recruitment and growth. Based on regeneration pathways (*sensu* Everham and Brokaw, 1996), the canopy gaps will undergo seedling colonization, growth, and maturity until it matures and recovers the pre-typhoon canopy level. But, larger canopy gaps will take longer period to recover. Seedling colonization and growth are proposed to be the primary modes for the recovery of typhoon-damaged mangroves. Rapid seedling recruitment and growth were observed in the mangroves in Nicaragua and Dominican Republic at 17- and 18-month post-typhoon, respectively (Roth, 1992; Sherman et al., 2001). Mangrove stands that have high seedling density pre-typhoon have higher chance to recover. But, mangrove stands that does not have available seedlings and source of seedlings have lower chance to recover.

The species that will recover first is related to the type/size of seeds/propagules and its tolerance/intolerance to shade (Salmo and Juanico, 2015). Species from the *Avicenniaceae*

FIGURE 21.5 Contrasting coppicing/refoliation patterns showing less/no refoliation in *Rhizophora* mangroves in Busuanga 3 years post-typhoon (A) and rapid refoliation in *Avicennia marina* in Leyte < 6 months post-typhoon (B). Both sites were severely damaged by Super Typhoon Haiyan in November 2013.

and *Sonneratiaceae* have small and buoyant seeds that can colonize >100 m from its parent trees (<100 m). In contrast, species with larger and heavier propagules (e.g., *Rhizophora* spp.) have limited dispersal distance. Thus, monospecific plantations that are mainly composed of *Rhizophora* spp. have longer time to recover (if at all it will recover) than diverse stands. Tide inundation also assists in seed dispersal. But, it is common that coastal fringes accumulate more seeds/propagules, although its long-term survival is not assured.

Aside from seedling colonization, the other prominent prospect for recovery is refoliation, although this is also genus- and species-specific. Species from *Avicenniaceae* and *Sonneratiaceae* can coppice unlike most species from *Rhizophoraceae* (Aung et al., 2013). In fact, some Avicennia and Sonneratia trees that were completely defoliated have refoliated even within a year (Fig. 21.5). Refoliation was faster in coastal fringes and in areas with larger canopy gaps. Younger stands refoliated more rapidly because of more epicormic buds than that of the mature stands.

Canopy regeneration, seedling colonization, and refoliation may be inherently dependent on the vegetation and physiological characteristics of the mangrove species. However, these are also dependent on the sediment characteristics. Vegetation and sediment recovery are significantly correlated (Salmo et al., 2014). Anoxic sediments with low nutrients, high salinity, and high temperature will constrain seedling growth. In some cases, it is possible that there will be sudden dumping of nutrient-rich sediments in the forest floor due to the deposition and decay of organic materials. Despite the available nutrients, most of these are not used by mangroves because of physiological stress caused by salinity, temperature, and redox (Salmo et al., 2014). Sediment recovery is also dependent on vegetation recovery. Soil temperature and redox, for example, will improve when canopy shade and seedling density improves. OM will improve when litter production (and subsequently accumulation and stabilization) increases.

Faunal recovery is less certain than vegetation and sediment recovery. The presence, abundance, and assemblage of fauna are dependent on the post-typhoon improvements of the vegetation and sediment conditions. Abundance and biomass are more likely not a straightforward indicator of faunal recovery. In fact, the typhoon-damaged mangroves in Bani even have higher mollusk abundance and biomass post-typhoon than pre-typhoon. A more reliable indicator was the shift in species composition—from having more specialist species pre-typhoon to having more generalist species post-typhoon. Species that were less dominant pre-typhoon have sudden dominance (Salmo et al., 2019a).

3.5 Recovery indicators

The use of "indicators" has a long history in science. It is widely practiced by scientists, environmental managers, policy makers, and the general public. Indicators are used as proxy parameter to assess the condition of the environment, as warning signals of ecological problems, and as barometer for state of the ecosystem. The use of indicators is linked with environmental change that is tied with ecological concepts and processes occurring in an ecosystem. For biological indicator, it is assumed that the presence and absence of certain species or fluctuations in species composition and dominance reflect changes occurring at different spatial and temporal scales in the ecosystem (Weilhoefer, 2010; Noon et al., 2012).

The criteria for "viable" or "qualified" indicators have been reviewed by Niemi and McDonald (2004). Indicators could be a state indicator or response indicator and should reflect the status or change in an ecosystem. It could be a vegetation variable, sediment variable, species presence/dominance, a process, or a combination thereof. But, a critical requirement is the documentation of change (or rate of change) relative to the environmental condition before and after the disturbance. It is important that the changes, the spatial and temporal patterns of changes, and the rates of change should be consistently monitored and compared with predisturbance data.

Unless there is a total ecosystem collapse, typhoons have more patchy damages than a homogenous pattern. The differences in the distribution and sizes of patches will also influence recovery patterns, following canopy regeneration dynamics (*sensu* Duke, 2001). Recovery will happen, but the time and patterns will depend on several interdependent factors. Remnant trees may serve as initial source of recovery, but some may manifest lag effects; hence, trees that initially survived may also eventually die. The sources of seedlings will also affect the initial post-typhoon recovery. These sources may come from pre-typhoon seedling within the system or from external adjacent sources. Seedling recruitment and growth are genus/species- and location-specific.

Recovery will be uneven; hence, monitoring of "recovery indicators" will be challenging. It is not straightforward to identify recovery indicators that can be directly related to post-typhoon recovery similar to the criteria and process proposed by Niemi and McDonald (2004). A predisturbance versus postdisturbance comparison will provide a more comprehensive assessment of recovery. In most cases, pre-typhoon data are lacking. Without pre-typhoon data, the assessment and interpretation of post-typhoon recovery (relative to attainment of pre-typhoon conditions) may be precluded. The use of a space-for-time (SFT) substitution approach is a common experimental design that identifies recovery indicators (Pickett, 1989; Blois et al., 2013). The timeline of measurements should be properly documented to tract trajectory patterns. The question on until when would the monitoring be conducted will depend on the research and management needs. But, following canopy regeneration dynamics (Duke, 2001) and based on observations in typhoon-damaged mangroves in Bani, Pangasinan, it will be ideal to have measurements done at least twice a year (to account seasonal variation) and for 5–15 years (to account possible lag effects and capture full canopy recovery).

The indicators mentioned here are, *sensu lato*, based on the observations of the author. Readers are advised to interpret these parameters with caution as the assessment of spatial and temporal changes may vary with geographical location and site-specific conditions.

3.5.1 Vegetation

Refoliation and canopy diameter are one of the more obvious indicators of post-typhoon recovery. The lost of canopy (or "complete defoliation") will provide spaces (referred to as "canopy gaps") for the surviving mangroves or the recruited/new mangroves to occupy. In some cases, the initial refoliation could fully lead to canopy closure/recovery (e.g., >75% in less than a year). There are some cases, however, that the initial fast refoliation will not be sustained and therefore had delayed or stunted recovery.

Seedling recruitment, colonization, and growth are common regeneration pathways in mangroves (see for example Baldwin et al., 2001; Aung et al., 2013; Salmo et al., 2019b). In some stands, the seedlings and saplings that grow after the typhoon will have rapid growth until such time that they reached maturity size (e.g., >5 cm tree diameter; > 5 m tree height; Salmo et al., 2013). Structural complexity, tree density, and tree biomass will have rapid increase but will eventually slow as the gaps are occupied (e.g., 5–10 years). Similar to refoliation, the development of seedlings into maturity may be slowed or delayed. This could be linked to a poor nutrient status and degraded sediment condition (Fig. 21.6).

3.5.2 Sediment

The recovery indicators for vegetation are closely related to that of the sediment parameters, and vice versa. For example, the OM accumulation and stabilization are directly related to the amount of primary production and litters produced by trees. There are contrasting reports on post-typhoon changes in OM. In Florida, USA (Smoak et al., 2013), OM increased due to sudden dumping of fine sediments, which is in stark contrast to the severe depletion of OM in Pangasinan, Palawan, and Aklan due to erosion of OM-rich surface associated with the accompanying storm surge.

Aside from OM, the other possible sediment recovery indicators are temperature, salinity, and redox. An often-overlooked sediment parameters are the nutrient contents (primarily N and P). But these are also related to the state of post-typhoon forest development (Garcia,

FIGURE 21.6 Contrasting seedling densities: no/few seedling recruits even 5 years post-typhoon (A) versus high seedling density at 1.5 years post-typhoon (B) in Leyte.

2017). The nutrient contents will help support the growth of seedlings and saplings (Killingbeck, 1996). Related to the nutrient content is the nutrient resorption efficiency, which is defined as the capacity of mangroves to use conserved/retained nutrients.

Based from the range given by Killingbeck (1996), the N resorption in this study was considered complete resorption. This means that N reached the minimum level that evergreen leaves can be reduced as it senesced. The senescent leaves contained <0.7% N except for the natural mangrove stands in Bani, which had 1.04 ± 0.20% N. On the other hand, the P content of the senescent leaves were all >0.04%, indicating an incomplete P resorption, and did not fully conserve (Killingbeck, 1996). Similar result was obtained in Twin Cays, Belize, where P resorption was incomplete regardless whether the site is N-limited or P-limited (Feller et al., 2007). While phosphorus retention efficiency (PRE) was higher than nitrogen resorption efficiency (NuRE), the N in the leaves appeared to be more reduced than P. Lin and Sternberg (2007) suggested that nutrient level in the senescent leaves (a proxy indicator for complete and incomplete resorption) is a more reliable indicator of nutrient limitation.

Most studies on nutrient resorption in mangrove forests were mostly from temperate areas and thus there is lack of data for this study to compare with. Among the available reports, the obtained NuRE values in typhoon-affected sites were comparable to the reports of Lin and Sternberg (2007; Florida, USA) but were lower than those in the studies of Feller et al. (2002, 2007; Twin Cays, Belize). The PRE was lower than in the study of Feller et al. (2002). The green and senescent leaf N contents across sites were comparable with several studies (Feller et al., 2003; Lin and Sternberg, 2007). The green and senescent leaf P contents in the typhoon-damaged sites were two to seven times higher as compared with other studies (Lin and Sternberg, 2007; Feller et al., 2002, 2003). The inferior NuRE and PRE obtained in this study suggest that mangrove forests had little nutrient reserves and may be inadequate to facilitate its post-typhoon recovery.

3.5.3 *Faunal assemblage and activities*

Faunal species and assemblages are reliable recovery indicators (Macintosh et al., 2002). However, the recovery of the faunal abundance/biomass and species composition are very much dependent on the state of both vegetation and sediment recoveries. Similar to forest succession, the faunal recolonization and succession patterns will show shifts of dominance from generalist to specialist species (e.g., mollusk assemblages; Salmo et al., 2014; Salmo et al., 2019a). It is important to capture the period at which a shift is documented. The selection of indicator species will have to be based on the dependence of the faunal species to mangroves.

Faunal activities through burrowing, bioturbation, and leaf/propagule tethering are also an important recovery indicator. The burrowing/bioturbation activities by mollusks and crustaceans provide aeration to an otherwise anoxic, warm, and highly saline typhoon-damaged sediment. The occurrences and sizes of burrows are also indicative of the faunal activities (Gallmann, 2016). Leaf/propagule tethering contributes in producing and storing OM in the sediments. Typhoon-damaged mangroves that have more abundant and active fauna recovered faster than mangroves without/less fauna (for example, in Cebu; Fig. 21.7; Gallmann, 2016).

FIGURE 21.7 Crab burrowing and leaf/propagule tethering in Cebu (photos courtesy of Maria Rachelle Gallmann). Higher crab abundance and activities were observed in natural mangroves that have higher survival than planted mangroves.

4. Summary: vulnerability, research/management implications, and recommendations

The Philippines experience 20 typhoons in a year, 3–4 of which are catastrophic. The return period of catastrophic typhoons is estimated at ca. 16–32 years (Kubota and Chan, 2009), although it became more frequent and more devastating in the last 20–30 years (Holden and Marshall, 2018). Some of these catastrophic typhoons result to complete forest die-off, but some also survived. It is possible that as the mangroves recover from first impact, another catastrophic typhoon will hit the same area (note: typhoons Frank and Haiyan happened in <6 years). A delayed recovery will make mangroves more vulnerable to the damages. Those stands which survived repeated disturbances (e.g., eastern seaboard of the Philippines) have high species diversity (>25 species) and probably make them adaptive to typhoons. Higher species diversity (and therefore high structural complexity) are generally perceived to have higher resiliency against disturbance (Hughes, 2010; Beckschäfer et al., 2013).

The perturbations regularly experienced by Philippine mangroves may lead to severe depletion in the future. It is also possible that these "repeated disturbances" make the Philippine mangroves adaptive to typhoons similar to the proposition of Krauss and Osland (2020). Species-specific adaptiveness and differences in post-typhoon recovery may also be related to the high species diversity in the country (second worldwide; Duke, 2006). However, the fragmented mangrove state in the country due to anthropogenic pressures and land use modification (see for example Castillo et al., 2018) will eventually make mangroves more vulnerable to typhoons.

Typhoons just like other forms of disturbances serve as a regulatory mechanism that shapes the structure and functioning of mangroves (Krauss and Osland, 2020). Practically, the entire country is exposed to typhoon, although some regions (e.g., eastern seaboard and northern Philippines) have higher typhoon occurrences than the other regions (Cinco et al., 2016). Mangroves in some places (e.g., in selected sites in Cagayan province in north-northeastern Luzon) showed adaptiveness, but some places (e.g., Cebu, Pangasinan, etc.) show severe damages.

The differences in the adaptability (or vulnerability) of mangroves to typhoons are influenced by the state and quality of ecosystem health of mangroves. The mangroves that survived the regular typhoons in northeastern Philippines have wider extent and are more structurally complex. In contrast, mangroves that exhibited severe damages are largely fragmented with low structural complexity. The documentation and assessment of extent of typhoon damage, including inference on the rates and patterns of recovery relative to pre-typhoon conditions, are challenging. Most sites have no baseline data; therefore, it is difficult to establish how much mangrove areas including specific locations have been really affected. The effects of typhoons could be immediate, delayed, or permanent. Without a baseline data to compare with, it will be difficult to assess recovery.

The case of Philippine mangroves offers a unique case. While it is subjected to regular typhoon disturbances, some of which survived previous catastrophic typhoons. These mangroves provide an unusual opportunity to assess and track short- and long-term vulnerability or resiliency against typhoons. The Philippines has a long and rich history of several mangrove conservation and restoration programs. In the field of conservation/restoration ecology, the SFT substitution approach is a commonly used experimental design to track post-restoration (or in this case, post-disturbance) changes (Pickett, 1989). Similar approach can be used where sampling plots should be established in different conserved and restored sites ideally for the entire country.

Permanent vegetation plots should be tagged (and georeferenced) and consistently be monitored. From those monitoring plots, some (post-typhoon) recovery indicators (*sensu* Cristofoly and Mahy, 2010; Salmo et al., 2013; Zhao et al., 2016) can be identified and tracked. Structural complexity (*sensu* Gardiner and Quine, 2000) should be included as one of the key parameters to measure as it is a better indicator of vulnerability/resiliency than the usually measured tree diameter, height, and canopy. The conventional field sampling should be complemented with remote sensing or geographic information system (GIS) approach (see for example Buitre et al., 2019; Baloloy et al., 2020). Similarly, if consistently and systematically done, a long-term assessment of the adaptability, vulnerability, or resiliency of Philippine mangroves against typhoons can be documented.

The coastal communities particularly the artisanal fishers dependent on mangroves are the main sectors that are negatively affected when mangroves are damaged by typhoons. These communities are also the principal sector engaged in post-typhoon rehabilitation. An informed citizenry will lead to proactive engagement in coastal management. Local citizens have an important role in the damage and post-typhoon recovery assessments. Citizen science has been an emerging field that can complement the conventional scientific sampling approach (see for example Kobori et al., 2016). A limited, incomplete, and inaccurate information may lead to mismanagement. A case in point is that the locals think that cleaning debris

FIGURE 21.8 Clearing of typhoon debris (A) (photo courtesy of Anne Brigitte U. Lim) and planting (B and C) in still inhospitable vegetation and sediment conditions. These activities were sponsored/funded by the government in an attempt to expedite the post-typhoon recovery process.

(Fig. 21.8A) would actually help mangroves recover. However, the removal of those debris could actually dampen the sediments and lessen the structure that could otherwise provide shades and initial source of recovery. On a similar account, some government institutions planted mangrove seedlings (again using the conventional *Rhizophora* spp.; see Fig. 21.8B and C) hoping that planting could help enhance post-typhoon recovery. However, the vegetation and sediment conditions are still damaged and are most likely not suitable for planting. While the intentions are noble, the lack of appreciation of science-based intervention would actually result to missed opportunities and inappropriate management.

Lastly, policies and proper implementation of it are needed to support mangrove conservation and restoration programs. Land use change and the effects of climate change will continue to threaten Philippine mangroves. Mangrove restoration programs, if done properly, will definitely contribute in enhancing the resiliency of Philippine mangroves against typhoon. It will be ideal to maintain >500 m width of mangroves to substantially serve as buffer against typhoon, a condition that is rarely found in the country. However, if need to prioritize, then the regions that are prone to typhoons should be prioritized (e.g., eastern Philippines). Other areas that should be considered for restoration are areas where there were substantial losses due to conversion to aquaculture ponds (e.g., Manila Bay, Panay Island, Mindanao, and Lingayen Gulf).

Acknowledgments

I thank former Research Assistants, students, collaborators, community and local government partners, national agencies, NGOs, and academic institutions for the >15 yrs of lessons and experiences in mangrove research and management. The projects were funded by USAID-NSF through the Partnership for Enhanced Engagement in Research

(PEER; Project 3–191), Oscar M. Lopez Center (NP 2015–16), Asia–Pacific Network for Global Change Research (ARCP2014-14NMY(B&ES), and Commission on Higher Education (CHED)—Philippine Higher Education Research Network (PHERNet), and the Department of Science and Technology (DOST)—National Academy of Science and Technology (NAST) through Environmental Science Award grant.

References

Ancelin, P., Courbaud, B., Fourcaud, T., 2004. Development of an individual tree-based mechanical model to predict wind damage within forest stands. For. Ecol. Manag. 203, 101–121.

Aung, T.T., Mochida, Y., Than, M.M., 2013. Prediction of recovery pathways of cyclone-disturbed mangroves in the mega delta of Myanmar. For. Ecol. Manag. 293, 103–113.

Baldwin, A., Egnotovich, Ford, M., Platt, W., 2001. Regeneration in fringe mangrove forests damaged by Hurricane Andrew. Plant Ecol. 157, 149–162.

Baloloy, A.B., Blanco, A.C., Sta Ana, R.R.C., Nadaoka, K., 2020. Development and application of a new mangrove vegetation index (MVI) for rapid and accurate mangrove mapping. ISPRS J. Photogrammetry Rem. Sens. 166, 95–117.

Beckschäfer, P., Mundhenk, P., Kleinn, C., Ji, Y., Yu, D., Harrison, R.D., 2013. Enhanced structural complexity index: an improved index for describing forest structural complexity. Open J. For. 3 (1), 23–29.

Bianchi, T.S., Allison, M.A., 2011. Large-river delta-front estuaries as natural "recorders" of global environmental change. Proc. Natl. Acad. Sci. U. S. A. 106 (20), 8085–8092.

Blois, J.L., et al., 2013. Space can substitute for time in predicting climate-change effects on biodiversity. Proc. Natl. Acad. Sci. U. S. A. 110 (23), 9374–9379.

Bosire, J.O., Dahdouh-Guebas, F., Kairo, J.G., Koedam, N., 2003. Colonization of non-planted mangrove species into restored mangrove stands in Gazi Bay, Kenya. Aquat. Bot. 76, 267–279.

Breithaupt, J.L., Hurst, N., Steinmuller, H.E., Duga, E., Smoak, J.M., Kominoski, J.S., Chambers, L.G., 2019. Comparing the biogeochemistry of storm surge sediments and pre-storm soils in coastal wetlands: hurricane Irma and the Florida Everglades. Estuar. Coast 43, 1090–1103.

Buitre, M.J.C., Zhang, H., Lin, H., 2019. The mangrove forests change and impacts from tropical cyclones in the Philippines using time series satellite imagery. Rem. Sens. 11 (688).

Busby, P.E., Motzkin, G., Boose, M.R., 2008. Landscape-level variation in forest response to hurricane disturbance across a storm track. Can. J. For. Res. 38, 2942–2950.

Cabello, K.A., Germentil, M.J., 2020. Haiyan Ka Na Naman: A Post-disaster Assessment of Mangrove Forest Recovery in Lawaan-Balangiga, Eastern Samar (BSc Geodetic Engineering thesis). University of the Philippines, Diliman, Quezon City.

Cahoon, D.R., Hensel, P., Rybczyk, J., McKee, K., Proffitt, C.E., Perez, B.C., 2003. Mass tree mortality leads to mangrove peat collapse at Bay Islands, Honduras after Hurricane Mitch. J. Ecol. 91 (6), 1093–1105.

Cardona, P., Botero, L., 1998. Soil characteristics and vegetation structure in a heavily deteriorated mangrove forest in the Caribbean coast of Colombia. Biotropica 30 (1), 24–34.

Carlos, C.M., 2014. The Influence of Canopy Gaps on Post- Typhoon Regeneration of Seedlings and Mollusc Assemblage in Bangrin Mangrove Protected Area in Bani, Pangasinan (M.Sc. Environmental Science thesis). Ateneo de Manila University, Quezon City, Philippines.

Castaneda-Moya, E., Twilley, R.R., Rivera-Monroy, V.H., Zhang, K., Davis, S.E., Ross, M., 2010. Sediment and nutrient deposition associated with Hurricane Wilma in mangroves of the Florida coastal everglades. Estuar. Coast 33, 45–58.

Castillo, J.A.A., Apan, A.A., Maraseni, T.N., Salmo, S.G., 2018. Tree biomass quantity, carbon stock and canopy correlates in mangrove forest and land uses that replaced mangroves in Honda Bay, Philippines. Reg. Stud. Mar. Sci. 24, 174–183.

Chen, R., Twilley, R.R., 1998. A gap dynamic model of mangrove forest development along gradients of soil salinity and nutrient resources. J. Ecol. 86 (1), 37–51.

Cinco, T.A., de Guzman, R.G., Ortiz, A.M.D., Delfino, R.J.P., Lasco, R.D., Hilario, F.D., Juanillo, E.L., Barba, R., Ares, E.D., 2016. Observed trends and impacts of tropical cyclones in the Philippines. Int. J. Climatol. 36 (14), 4638–4650.

Cristofoli, S., Mahy, G., 2010. Ecological restoration: context, constraints, and indicators. Biotechnol. Agron. Soc. Environ. 14 (1), 203–211.

Diele, K., Tran Ngoc, D.M., Geist, S.J., Meyer, F.W., Pham, Q.H., Saint-Paul, U., Tran, T., Berger, U., 2013. Impact of typhoon disturbance on the diversity of key ecosystem engineers in a monoculture mangrove forest plantation, Can Gio Biosphere Reserve, Vietnam. Global Planet. Change 110, 236–248.

Duke, N.C., 2001. Gap creation and regenerative processes driving diversity and structure of mangrove ecosystems. Wetl. Ecol. Manag. 9, 257–269.

Duke, N.C., 2006. Australia's Mangroves. The Authoritative Guide to Australia's Mangrove Plants. The University of Queensland, Brisbane, 200 pp.

Dung, L.V., Tue, N.T.T., Nhuan, M.T., Omori, K., 2016. Carbon storage in a restored mangrove forest in Can Gio Mangrove Forest Park, Mekong Delta, Vietnam. Forest Ecol. Manag. 380, 31–40. https://doi.org/10.1016/j.foreco.2016.08.032.

Ellison, J.C., 1998. Impacts of sediment burial on mangroves. Mar. Pollut. Bull. 37 (8–12), 420–426.

Ellison, A.M., Farnsworth, E.J., 1996. Spatial and temporal variability in growth of Rhizophora mangle saplings on coral cays: links with variation in insolation, herbivory, and local sedimentation rate. J. Ecol. 84 (5), 717–731.

Everham, E.M., Brokaw, N.V.L., 1996. Forest damage and recovery from catastrophic wind. Bot. Rev. 62 (2), 113–185.

Feller, I., Ball, M., Ellis, J., Lovelock, C., Reef, R., 2017. Interactive effects of climate and nutrient enrichment on patterns of herbivory by different feeding guilds in mangrove forests. Global Ecol. BioGeograp. 26 (11), 1326–1338. https://doi.org/10.1111/geb.12644.

Feller, I., Lovelock, C., McKee, K., 2007. Nutrient addition differentially affects ecological processes of Avicennia germinans in nitrogen versus phosphorus limited mangrove ecosystems. Ecosystems 10 (3), 347–359. https://doi.org/10.1007/s10021-007-9025-z.

Feller, I., McKee, K., Whigham, D., O'Neill, J., 2002. Nitrogen vs. phosphorus limitation across an ecotonal gradient in a mangrove forest. BioGeoChem. 62 (2), 145–175.

Feller, I., Whigham, D., McKee, K., Lovelock, C., 2003. Nitrogen limitation of growth and nutrient dynamics in a disturbed mangrove forest, Indian River Lagoon, Florida. Oecologia 134 (3), 405–414. https://doi.org/10.1007/s00442-002-1117-z.

Fondo, E.N., Martens, E.E., 1998. Effects of mangrove deforestation on macrofaunal densities, Gazi Bay, Kenya. Mangroves Salt Marshes 2, 75–83.

Friess, D.A., Thompson, B.S., Brown, B., Amir, A.A., Cameron, C., Koldewey, H.J., Sasmito, S.D., Sidik, F., 2016. Policy challenges and approaches for the conservation of mangrove forests in Southeast Asia. Conserv. Biol. 30 (5), 933–949.

Gaillard, J.-C., Liamzon, C.C., Villanueva, J.D., 2007. 'Natural' disaster? A retrospect into the causes of the late-2004 typhoon disaster in Eastern Luzon, Philippines. Environ. Hazards 7, 257–270.

Gallmann, M.R., 2016. Brachyuran Herbivory and Burrowing in Typhoon Haiyan Affected Mangroves in Bantayan Island, Cebu (B.Sc. Environmental Science thesis). Ateneo de Manila University, Quezon City, Philippines.

Garcia, M.C.A., 2017. Nutrient Content and Resorption Efficiency as Recovery Indicators in Typhoon-Damaged Mangroves in the Philippines (M.Sc. Environmental Science thesis). Ateneo de Manila University, Quezon City, Philippines.

Gardiner, B.A., Quine, C.P., 2000. Management of forests to reduce the risk of abiotic damage - a review with particular reference to the effects of strong winds. For. Ecol. Manag. 135, 261–277.

Green, R.H., 1979. Sampling Design and Statistical Methods for Environmental Biologists. Wiley, New York, p. 257.

Hogarth, P., 2007. The Biology of Mangroves and Seagrasses. Oxford University Press, New York.

Holden, W.N., Marshall, S.J., 2018. Climate change and typhoons in the Philippines: extreme weather events in the Anthropocene. In: Samui, P., Kim, D., Ghosh, C. (Eds.), Integrating Disaster Science and Management: Global Case Studies in Mitigation and Recovery. Elsevier.

Hughes, A., 2010. Disturbance and diversity: an ecological chicken and egg problem. Nat. Educ. Knowl. 3 (10), 48.

Huxham, M., Kimani, E., Augley, J., 2004. Mangrove fish: a comparison of community structure between forested and cleared habitats. Estuar. Coast. Shelf Sci. 60, 637–647.

Killingbeck, K., 1996. Nutrients in senesced leaves: Keys to the search for potential resorption and resorption proficiency. Ecology 77 (6), 1716–1727. https://doi.org/10.2307/2265777.

Kobori, H., Dickinson, J.L., Washitani, I., Sakurai, R., Amano, T., Komatsu, N., Kitamura, W., Takagawa, S., Koyama, K., Ogawara, T., Miller-Rushing, A.J., 2016. Citizen science: a new approach to advance ecology, education, and conservation. Ecol. Restor. 31, 1–19.

Kozlowski, T.T., Pallardy, S.G., 1997. Physiology of woody plants, 2nd. Academic Press, San Diego, CA, USA.

Krauss, K.W., Osland, M.J., 2020. Tropical cyclones and the organization of mangrove forests: a review. Ann. Bot. 125, 213–234.

Kubota, H., Chan, C.L., 2009. Interdecadal variability of tropical cyclone landfall in the Philippines from 1902 to 2005. Geophys. Res. Lett. 36, L12802.

Lantican, R.M., 2001. The Science and Practice of Crop Production. Southeast Asian Ministers of Education Organization (SEAMEO). SEAMEO Regional Center for Graduate Study and Research in Agriculture (SEARCA), and University of the, Philippines Los Baños (UPLB), Laguna, Philippines.

Lin, Y., Sternberg, L.S.L., 2007. Nitrogen and phosphorus dynamics and nutrient resorption of Rhizophora mangle leaves in south Florida, USA. Bull. Mar. Sci. 80 (1), 159–169.

Long, J., Giri, C., Primavera, J.H., Trivedi, M., 2016. Damage and recovery assessment of the Philippines' mangroves following Super Typhoon Haiyan. Mar. Pollut. Bull. 109, 734–743.

Lovelock, C.E., Cahoon, D.R., Friess, D.A., Guntenspergen, G.R., Krauss, K.W., Reef, R., Rogers, K., Saunders, M.L., Sidik, F., Swales, A., Saintilan, N., Thuyen, L.X., Triet, T., 2015. The vulnerability of Indo-Pacific mangrove forests to sea-level rise. Nature 526, 559–563.

Macintosh, D.J., Ashton, E.C., Havanon, S., 2002. Mangrove rehabilitation and intertidal biodiversity: a study in the Ranong mangrove ecosystem, Thailand. Estuar. Coast. Shelf Sci. 55, 331–345.

McIvor, A.L., Spencer, T., Möller, I., Spalding, M., 2012. Storm surge reduction by mangroves. In: Natural Coastal Protection Series: Report 2. Cambridge Coastal Research Unit Working Paper 41. The Nature Conservancy and Wetlands International, 35 pp. ISSN 2050-7941.

Michener, W.K., 1997. Quantitatively evaluating restoration experiments: research design, statistical analysis, and data management considerations. Restor. Ecol. 5 (4), 324–337.

Milbrandt, E.C., Greenawalt-Boswell, J.M., Sokoloff, P.D., Bortone, S.A., 2006. Impact and response of Southwest Florida mangroves to the 2004 hurricane season. Estuar. Coast 29 (6a), 979–984.

NCDC IBTrACS, 2019. US NOAA National Climatic Data Center – International Best Track Archive for Climate Stewardship (IBTrACS) Project (Version 3). https://catalog.data.gov/dataset/ncdc-international-best-track-archive-for-climate-stewardship-ibtracs-project-version-3. (Accessed 1 September 2020).

Niemi, G.J., McDonald, M.E., 2004. Application of ecological indicators. Ecol. Indicat. 35, 89–111.

Noon, B.R., Bailey, L.L., Sisk, T.D., McKelvey, K.S., 2012. Efficient species-level monitoring at the landscape scale. Conserv. Biol. 26 (3), 432–441.

Pickett, S.T.A., 1989. Space-for-time substitution as an alter- native to long-term studies. In: Likens, G.E. (Ed.), Long-Term Studies in Ecology: Approaches and Alternatives. Springer, New York, pp. 110–135.

Primavera, J.H., de la Cruz, M., Montilijao, C., dela Paz, M., Rollon, R.N., Maranan, K., Samson, M.S., Blanco, A., 2016. Preliminary assessment of post-Haiyan mangrove damage and short-term recovery in Eastern Samar, central Philippines. Mar. Pollut. Bull. 109, 744–750.

Rivera-Monroy, V.H., Twilley, R.R., Medina, E., Moser, B., Botero, L., Francisco, A.M., Bullard, E., 2004. Spatial variability of soil nutrients in disturbed riverine mangrove forests at different stages of regeneration in the San Juan River estuary, Venezuela. Estuaries 27 (1), 44–57.

Romañach, S., De Angelis, D.I., Koh, H.L., Li, Y., Teh, S.Y., Barian, R.S.R., Zhai, L., 2018. Conservation and restoration of mangroves: global status, perspectives, and prognosis. Ocean Coast Manag. 154, 72–82.

Roth, L.C., 1992. Hurricanes and mangrove regeneration: effects of Hurricane Joan, October 1988, on the vegetation of Isla del Venado, Bluefields, Nicaragua. Biotropica 24 (3), 375–384.

Saenger, P., 2002. Mangrove Ecology, Silviculture and Conservation. Springer-Science.

Salmo, S.G., 2011. Early Post-typhoon Effects on the Restoration Trajectory of Planted Mangroves in the Philippines (Ph.D. dissertation). The University of Queensland, Brisbane.

Salmo, S.G., Gianan, E.L.D., 2019. Post-disturbance carbon stocks and rates of sequestration: implications on "blue carbon" estimates in Philippine mangroves. Philipp. Sci. Lett. 12 (2), 122–132.

Salmo, S.G., Juanico, D.E.O., 2015. An individual-based model of long-term forest growth and carbon sequestration in planted mangroves under salinity and inundation stresses. Int. J. Philipp. Sci. Technol. 8 (2), 31–35.

Salmo, S.G., Torio, D.D., Esteban, J.M.A., 2007. Evaluation of rehabilitation strategies and management schemes for the improvement of mangrove management programs in Lingayen Gulf. Sci. Diliman 19 (1), 24–34.

Salmo III, S., Lovelock, C., Duke, N., 2013. Vegetation and soil characteristics as indicators of restoration trajectories in restored mangroves. Hydrobiologia 720, 1–18. https://doi.org/10.1007/s10750-013-1617-3.

Salmo, S.G., Lovelock, C.E., Duke, N.C., 2014. Assessment of vegetation and soil conditions in restored mangroves interrupted by severe tropical typhoon 'Chan-hom' in the Philippines. Hydrobiologia 733, 85–102.

Salmo III, S., Tibbetts, I., Duke, N., 2019a. Recolonization of mollusc assemblages in mangrove plantations damaged by Typhoon Chan-hom in the Philippines. Estuar. Coast. Shelf Sci. 228 (106365). https://doi.org/10.1016/j.ecss.2019.106365.

Salmo, S.G., Malapit, V., Garcia, M.C.A., Pagkalinawan, H.M., 2019b. Establishing rates of carbon sequestration in mangroves from an earthquake uplift event. Biol. Lett. 15 (20180799).

Sasekumar, A., Chong, V.C., 1998. Faunal diversity in Malaysian mangroves. Global Ecol. Biogeogr. Lett. 7 (1), 57–60.

Sherman, R.E., Fahey, T.J., Martinez, P., 2001. Hurricane impacts on a mangrove forest in the Dominican Republic: damage patterns and early recovery. Biotropica 33 (3), 393–408.

Skilleter, G.A., Warren, S., 2000. Effects of habitat modification in mangroves on the structure of molluscs and crab assemblages. J. Exp. Mar. Biol. Ecol. 244, 107–129.

Smith, T.J., Robblee, M.B., Wanless, H.R., Doyle, T.W., 1994. Mangroves, hurricanes, and lightning strikes. Bioscience 44 (4), 252–262.

Smith, T.J., Anderson, G.H., Balentine, K., Tiling, G., Ward, G.A., Whelan, K.R.T., 2009. Cumulative impacts of hurricanes on Florida mangrove ecosystems: sediment deposition, storm surges and vegetation. Wetlands 29 (1), 24–34.

Soria, J.L.A., Switzer, A.D., Pilarczyk, J.E., Tang, H., Weiss, R., Siringan, F., Manglicmot, M., Gallentes, A., Lau, A.Y.A., Cheong, A.Y.L., Koh, T.W.L., 2018. Surf beat-induced overwash during Typhoon Haiyan deposited two distinct sediment assemblages on the carbonate coast of Hernani, Samar, central Philippines. Mar. Geol. 396, 215–230.

Smoak, J.M., Breithaupt, J.L., Smith, T.J., Sanders, C.J., 2013. Sediment accretion and organic carbon burial relative to sea-level rise and storm events in two mangrove forests in Everglades National Park. Catena 104, 58–66.

Takagi, H., Esteban, M., 2016. Statistics of tropical cyclone landfalls in the Philippines: unusual characteristics of 2013 Typhoon Haiyan. Nat. Hazards 80, 211–222. https://doi.org/10.1007/s11069-015-1965-6.

Taylor, D.S., Reyier, E.A., Davis, W.P., McIvor, C.C., 2007. Mangrove removal in the Belize cays: effects on mangrove-associated fish assemblages in the intertidal and subtidal. Bull. Mar. Sci. 80 (3), 879–890.

Velasco, L.R., Cabanilla, L., 2003. Assessment of Physical Resource Capability in Philippine Agriculture. Philippine APEC Study Center Network Discussion Paper No. 2003-03.

Webster, P.J., Holland, G.J., Curry, J.A., Chang, H.R., 2005. Changes in tropical cyclone number, duration, and intensity in a warming environment. Science 309, 1844–1846.

Weilhoefer, C.L., 2010. A review of indicators of estuarine tidal wetland condition. Ecol. Indicat. 11 (2), 514–525.

Zhang, C.G., Leung, K.K., Wong, Y.S., Tam, N.F.Y., 2019. Modeling risk of mangroves to tropical cyclones: a case study of Hurricane Irma. Estuar. Coast. Shelf Sci. 224, 108–116.

Zhao, Q., Bai, J., Huang, L., Lu, Q., Gao, Z., 2016. A review of methodologies and success indicators for coastal wetland restoration. Ecol. Indicat. 60, 442–452.

CHAPTER

22

Managing sediment dynamics through reintroduction of tidal flow for mangrove restoration in abandoned aquaculture ponds

Frida Sidik[1,2], *Denny Wijaya Kusuma*[1], *Bayu Priyono*[1], *Christophe Proisy*[3,4], *Catherine E. Lovelock*[2]

[1]Institute for Marine Research and Observation, Ministry of Marine Affairs and Fisheries, Bali, Indonesia; [2]School of Biological Sciences, University of Queensland, St Lucia, QLD, Australia; [3]French Institute of Pondicherry, Pondicherry, India; [4]AMAP, University of Montpellier, IRD, CIRAD, CNRS, INRAE, Montpellier, France

OUTLINE

1. Introduction 564
2. Mangrove–sediment–hydrodynamics interactions 565
3. Estuarine ecohydrology in mangrove restoration and rehabilitation 567
4. Implications of mangrove restoration: carbon abatement opportunity and climate change mitigation 568
5. Case study: hydrological recovery in abandoned ponds in Perancak, Bali 568
   5.1 Methodology and site description 568
   5.2 Tidal flow reintroduction and sediment accumulation 571
   5.3 Postplanting mangrove regeneration 574
6. Lesson learned from the case study 576

Acknowledgments 578

References 578

https://doi.org/10.1016/B978-0-12-816437-2.00004-5

1. Introduction

A growing awareness of the importance of mangroves for coastal protection, fishery productivity, and reductions in atmospheric carbon has increased efforts to restore or rehabilitate vast areas of mangrove forests that had been converted into other land uses. In these areas, the coast has lost its natural resilience, affecting the potential of mangroves as natural barriers to stabilize shorelines (Toorman et al., 2018). The most common approach used for mangrove reforestation, especially in the tropics, is the direct planting of seedlings or propagules in mangrove plantations (Iftekhar, 2008). In Southeast Asian countries, such as Indonesia and the Philippines, mangrove plantations have been the primary program for government, nongovernment organization, and communities to rehabilitate degraded coast due to aquaculture pond conversion and human settlements (Kusmana, 2014; Primavera and Esteban, 2008). Mangrove plantations are sometimes conducted with participatory approaches with expected benefits to local communities (Walton et al., 2006).

However, a number of studies have documented the failures of mangrove restoration and rehabilitation practices. These failures are mainly explained by a lack of assessment on the environmental stressors and inappropriate approaches to reintroduce mangroves (Lewis, 2005; Lewis and Gilmore, 2007; López-Portillo et al., 2017), alongside socioeconomic constraints such as unclear land tenure and perverse planting targets (Lovelock and Brown, 2019; Wodehouse and Rayment, 2019). In Sri Lanka, Kodikara et al. (2017) reported low survival rates in a 1000–1200 ha mangrove plantation made after the 2004 tsunami due to inappropriate site selection, unsuitable planted species, and lack of postplanting care and monitoring. Brown et al. (2015) also reported failures of similar projects in Aceh, Indonesia, while recovery rates in neighboring natural recruitment sites were high. In the Philippines, mangrove afforestation in seafront areas has been historically unsuccessful when the efforts took place in low intertidal areas with hydrological conditions inappropriate for young mangrove growth (Primavera and Esteban, 2008). Mangrove plantations on bare intertidal mudflats were also considered inappropriate by Erftemeijer and Lewis (1999) as the effort should be considered as "habitat conversion" rather than mangrove restoration, even if mangroves can successfully develop in some areas. In the case of abandoned pond reversion to mangroves, Proisy et al. (2018) highlighted the higher growth rates of natural mangrove over plantations, which can be explained by postplanting invasion of nonplanted native species as natural colonizers recruited through tidal inundation (Bosire et al., 2003; Walters, 2000). Most reports have showed the link between low success in restoration with oversimplified planting practices that neglect physical factors, such as hydrology, substrate, and sedimentation (Lewis, 2009; Primavera and Esteban, 2008).

Among the main causes explaining the failure of wrongly named "restoration" programs, the poor management of local hydrology ranks the first, as it is often constrained by human infrastructures built inside, upstream and/or downstream of mangrove areas (Anthony and Goichot, 2019). That is why any selection of a restoration approach must take into account the cause of mangrove degradation (Lewis, 2005). Restoring hydrology, i.e., removing any constraints on fresh and tidal water fluxes, has been argued for decades as a necessity for mangrove restoration. Very few studies, however, have explored this "nature-inspired" management through breaching aquaculture pond walls, which is low-cost and easily achievable, and described by the Community Based Ecological Mangrove Rehabilitation (CBEMR) approach developed by Mangrove Action Project, which is based on hydrological repair techniques (Brown et al., 2014).

In addition to hydrological conditions, the understanding of sedimentary processes of mangrove coastlines is also important for mangrove restoration and rehabilitation (Anthony and Goichot, 2019; Balke and Friess, 2016; Winterwerp et al., 2013). This knowledge becomes crucial when the efforts are aimed for disaster risk reduction or take place on eroding coastlines (Winterwerp et al., 2013). Several methods are suggested in the literature to improve restoration success by classifying geomorphology of the site based on geomorphic settings (Balke and Friess, 2016) and sediment flow connectivity (Anthony and Goichot, 2019). The geomorphic classification considers the functionality of two geomorphology settings that need to be addressed by different approaches such as (1) minerogenic restoration sites which require allochthonous sediment inputs to maintain accretion and remain intact in the face of erosive forces, and (2) the organogenic sites, which depend on autochthonous supply to avoid peat collapse (Balke and Friess, 2016). For practical implementation, Anthony and Goichot (2019) proposed the approaches of managing sediment flow and availability in the frame of relevant governance at four spatial scales of analysis, which in order of decreasing scale are river basins, deltas, coastal environments, and the local site scale.

This chapter aims to highlight how appropriate management of hydrology in mangrove areas fragmented by aquaculture ponds in a minerogenic site can lead to successful mangrove restoration. We begin the chapter by reviewing the importance of hydrogeomorphological regimes in mangrove establishment and early development phases that are often neglected in mangrove restoration and rehabilitation. We focus on mangrove restoration in abandoned ponds where pond walls are the main obstruction to tidal and freshwater flows. To assess evidence for the importance of geomorphic and hydrological regimes, we established a case study of restored mangroves in abandoned ponds in Bali, Indonesia, which consists of monitoring mangrove recovery in relation to hydrology and edaphic conditions. The implication of the case study for understanding restoration in minerogenic settings is discussed in the context of soil and biomass accumulations that are important for achieving blue carbon benefits and climate change mitigation.

2. Mangrove–sediment–hydrodynamics interactions

Mangroves develop in intertidal areas, from mean sea level to highest spring tide (Alongi and Brinkman, 2011). Thus, mangrove distributions are closely linked to hydrology (Alongi and Brinkman, 2011; Crase et al., 2013). The concept of interconnection between tidal hydrodynamics and mangroves has been long studied by ecologists and engineers, from the early observation of mangroves as "land builders" (Davis, 1938; Friess, 2016) to proposed models of tidal creek–mangrove systems showing tidal mixing and lateral sediment trapping in mangroves (Ridd et al., 1990; Wolanski, 1992; Wolanski and Ridd, 1986), followed by articulation of the mechanisms of trapping sediments in mangroves controlled by hydrodynamic processes (Furukawa et al., 1997; Furukawa and Wolanski, 1996). In the intertidal zone, a hydrographic pattern of tides works to export sediment from coastal areas or/and runoff to the fringing mangroves (Ridd et al., 1990; Wolanski, 1992). The suspended sediment is carried into the forest during spring flood tides and trapped in the mangrove through a mechanism that involves interactions between tidal currents and mangrove roots (Furukawa et al., 1997; Furukawa and Wolanski, 1996). Over time, this cohesive sediment, which is a dense sediment–water mixture dominated by clay, consolidates, and compacts, contributes to land accretion in the low energy zone (Alongi and Brinkman, 2011; Furukawa et al., 1997; Toorman et al., 2018).

The ability to accrete sediments depends on the sediment supply into the system and sediment trapping by mangroves (Friess et al., 2012; Furukawa et al., 1997). The elaborate aboveground roots, stems, and branches of mangrove create the friction that impedes tidal currents, creating complex flow in the form of turbulence (Furukawa and Wolanski, 1996). About 80% of suspended sediment supplied to the mangrove is trapped and settles around mangroves before slack high tide when turbulence vanishes and is then followed by sedimentation (Furukawa et al., 1997; Furukawa and Wolanski, 1996). The degree of sediment deposition derived from mangrove–hydrology interactions varies with vegetation structure and density; for example, the complex matrix roots of *Rhizophora* generates high turbulence and often results in more sedimentation trapping compared to single tree trunk of *Ceriops* or pneumatophores of *Sonneratia* (Friess et al., 2012; Furukawa and Wolanski, 1996; Krauss et al., 2003). In the absence of vegetation, such as on bare mud banks, the system can have lower friction force by a factor of 25 compared with vegetated areas (Furukawa and Wolanski, 1996), and a cycle of land accretion and erosion may occur on coasts dominated by unvegetated tidal flats (Toorman et al., 2018). The linkages among vegetation–hydrodynamics–sediment show that mangroves may not only be passive colonizers of sediment (Swales et al., 2015) but also can contribute to the development and consolidation of newly accreted sediments (Krauss et al., 2011).

When sediment deposition increases, the interaction can create positive surface elevation change that accommodates pioneer mangrove vegetation to expand, stabilizing its soil and enhancing further accretion (Willemsen et al., 2016). Not yet sufficiently considered in the literature are examples of plant–sediment interactions during natural mangrove colonization processes occurring in undisturbed areas. Proisy et al. (2009) highlighted rapid and widespread mangrove expansion over square kilometers of bare mudflats in French Guiana: when sediment supply and consolidation reached adequate levels, mangrove propagules dispersed by tidal currents may establish securely and sustainably. In contrast, erosion may occur if sediment supply is scarce, significantly reducing the sediment trapping capacity of the mangrove (Willemsen et al., 2016). As mangroves are sensitive to sediment inputs, there is a threshold of which hydrodynamics can alter the establishment and early development of mangrove, from the dispersal to recruitment of sapling stage (Friess et al., 2012; Krauss et al., 2008). For example, mangrove forests worldwide exist largely in a raised and sloped platforms above mean sea level, which are inundated for approximately 30%, or less of the time by tidal waters, and therefore natural generation occurs when suitable sediment conditions are met (Lewis, 2005). Mangrove–hydrology linkages are also visible through mangrove zonation patterns, which often reflect a range of inundation frequencies offering suitable but specific conditions for the development of distinct mangroves species (Krauss et al., 2008). Compared with forest succession in heterogeneous species-diverse rainforest, such zonation patterns may appear simple. However, mangrove ecology remains as complex due to the interwoven influence of nonstationary hydrosedimentary processes on the establishment of new seedlings or the competition between individuals of the same or distinct species (Alongi, 2009; Berger et al., 2008).

3. Estuarine ecohydrology in mangrove restoration and rehabilitation

By definition, ecosystem restoration is the recovery from degraded conditions to a state characterized by similar structure and function prior to disturbance, whereas ecosystem rehabilitation aims to bring the environment to an improved state (López-Portillo et al., 2017). Mangrove degradation over vast areas has pushed local communities and policy makers to take action toward restoring mangroves. However, many restoration programs have low success rates with high plant mortality, and most action plans continue to emphasize planting seedlings or propagules as the primary restoration method, neglecting the importance of the underlying hydrology and sediment dynamics that determine the success of this effort (Iftekhar, 2008; Lewis et al., 2016; Primavera and Esteban, 2008). These studies documented that many plantations have occurred in locations where the tidal and hydrologic conditions have been deteriorated and are no longer suitable for natural mangrove recovery, e.g., mangrove areas converted to aquaculture ponds. Over recent years, coastal managers and practitioners have become more aware of mangrove ecohydrology, which has finally become an integral part of mangrove restoration and rehabilitation programs at a number of sites (Brown et al., 2014; López-Portillo et al., 2017).

The growing role of forest plantations in fostering the recovery of natural vegetation has focused the attention of ecologists and policy makers who aspire to restore biodiversity on degraded mangrove sites (Lugo, 1997; Parrotta et al., 1997). Maintaining or increasing species richness in mangrove plantations is driven by water fluxes through propagule dispersal and the influence of varying hydroperiods on mangrove growth (Biswas et al., 2012; Crase et al., 2013; Walters, 2000). Sites planted with one or a few tree species can be rapidly occupied by many nonplanted species and thus show increasing species richness with age (Biswas et al., 2012; Proisy et al., 2018). In the case of the highly dynamic environment of mangroves, natural selection over millions of years has also affected species tolerance to changing coastal environmental conditions (Ball, 1988). Thus, the success of plantations for harboring high levels of biodiversity will be improved if suitable hydrodynamic conditions are established and maintained such that the tidal vegetation can development and mangrove forest vegetation can naturally recover (López-Portillo et al., 2017).

The implementation of hydrological knowledge into wider restoration practices gives rise to the findings that effective mangrove restoration in former mangrove area can be best achieved by eliminating environmental stressors (Krauss et al., 2008; Lewis III et al., 2016; Wolanski and Richmond, 2008). In the case of abandoned aquaculture ponds, pond walls can be regarded as physical barriers to the establishment of mangrove seedlings, which is the period between dispersal and recruitment to the sapling stage (Krauss et al., 2008). When tidal flow is restricted, stagnant water can stop the growth of planted seedlings, which progressively rot and die (Wolanski and Richmond, 2008). Indeed, pond bed level is usually near the level of the low tide, leading to planted seedling mortality during long periods of inundation (Wolanski and Richmond, 2008). In addition, hypersaline conditions in ponds due to reductions in freshwater availability can increase hypoxic or anoxic conditions, and elevated sulfide concentrations can kill mangrove seedlings and even mature trees (Lewis, 2005; Mckee and Faulkner, 2000). All these examples demonstrate that constraining water flows in any mangrove area threatens present mangroves and hampers restoration efforts (Dale et al., 2014).

4. Implications of mangrove restoration: carbon abatement opportunity and climate change mitigation

In terms of climate change mitigation, the protection of mangroves could reduce annual land use emissions caused by mangrove loss and mineralization of soil carbon as well as eliminating their roles as carbon sinks, known collectively as "blue" carbon (Nellemann et al., 2009; Pendleton et al., 2012). This mitigation potential has been acknowledged by recent studies in recognition of the valuable role of mangroves in sequestering and storing carbon, mostly in the soil, over long periods of time (Alongi, 2012; Donato et al., 2011). Soil pools store the largest carbon stocks in mangroves, accounting for up to 75% of total mangrove carbon stock in the system (Alongi, 2014). Therefore, blue carbon strategies focus on preserving and enhancing organic carbon stocks and burial capacity particularly within their soils (Donato et al., 2011; Lovelock et al., 2017). As a mangrove-rich country, Indonesia contains the highest carbon stocks in mangroves with substantial potential CO_2 emissions from soils if they are disturbed (Atwood et al., 2017; Sanderman et al., 2018).

The role of mangrove as a carbon sink is associated with biogeochemical consequences of water movements through tidal exchange (Alongi and Brinkman, 2011). There is an exchange of solid materials between and among mangroves and adjacent coastal waters that allows for the exchange of dissolved organic carbon and particulate organic carbon within the system (Alongi and Brinkman, 2011). In sites with hydrological obstacles, such as aquaculture pond walls and gates, reintroducing tidal flow to coastal ecosystems by removing barriers to tidal exchange can lead to carbon storage (abatement) through an increase in the organic carbon pools, mostly in the soil (Macreadie et al., 2017; Matsui et al., 2010). The degree of organic carbon sequestered in the soil can be estimated by measuring the surface and subsurface accumulation processes which are integrated in soil surface elevation change (Lovelock et al., 2013). Moreover, the processes controlling mangrove potential for sequestering carbon are also key to the maintenance of mangrove wetlands within the tidal frame of an appropriate position for plant growth with projected acceleration in sea level rise (Cahoon and Guntenspergen, 2010).

5. Case study: hydrological recovery in abandoned ponds in Perancak, Bali

5.1 Methodology and site description

In this chapter, we present our field study with focus on the ecophysiological and mangrove growth in the early development of mangroves in abandoned ponds in the minerogenic setting of estuary in Indonesia. Our site was within the 755 ha catchment area of the Perancak River estuary, Bali (8° 23′ 40″ S, 114° 37′ 39″ E), which has geology dominated by sedimentary limestones. A large part of the floodplain is highly fragmented into thousands of shrimp ponds (Fig. 22.1; Gusmawati et al., 2018; Proisy et al., 2018). Although aquaculture had been initiated since the mid-20th century, major mangrove conversion to

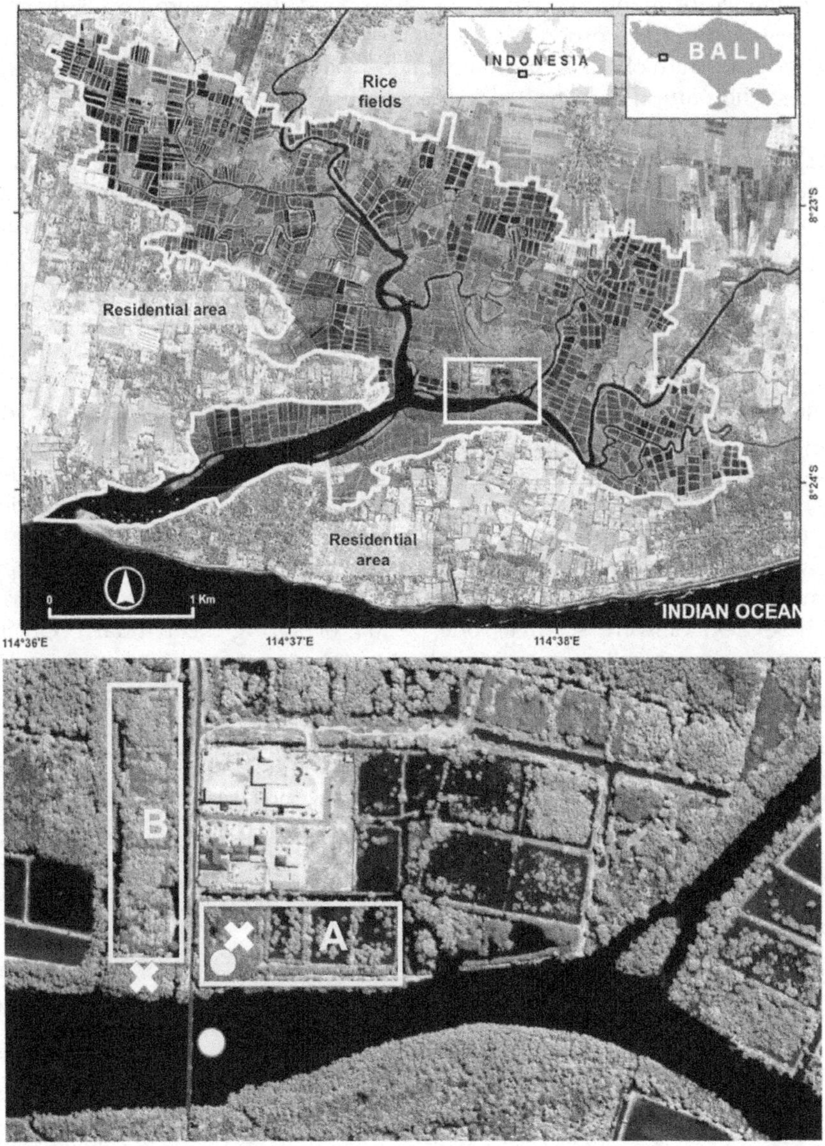

FIGURE 22.1 Location of the Perancak estuary, Southwest of Bali, Indonesia (top). A mosaic of shrimp ponds with different mangrove covers revealed in a WorldView-2 satellite image acquired in 2011 (top and bottom). (A and B) rectangular areas correspond to monitoring site of mangrove development. Water level logger (yellow [light gray in print version] circle) and rod surface elevation table–marker horizon (white cross) measurement are located.

aquaculture ponds in this area was attributed to the extensive and intensive aquaculture practices deployed in the early 1980s. The reconversion of aquaculture ponds to mangroves through restoration began in early 2000s. A number of aquaculture ponds were placed under the responsibility of IMRO (Institute of Marine Research and Observation, located at the center of the estuary) i.e., the ex-ponds for aquaculture research were transformed to "open air" monitoring stations of mangrove recovery. Some pond dykes were no longer maintained, became damaged, and have progressively allowed fresh and tidal water to flood the ponds.

Time series data were obtained and integrated from in situ and remotely sensed observations to assess the natural recovery as function of mangrove development. Tidal flow and sediment supply were monitored along with surveys of mangrove cover in two sites (Table 22.1). The monitoring of restoration site A (Site A) started from preplantation (t_0 of spatial observation) to the year after the reopening of obstructed channels (t_{final}). We set up instruments for observing hydrological and soil dynamics in Site A with reference to the river (hydrology) and intact mangroves (soil dynamics). We selected nearby ponds (Site B) as a reference site that illustrated natural mangrove regeneration in ponds with damaged walls, which exemplified nonplanted ponds across the Perancak estuary. We assessed the hydrology of a reference site in order to account for the remediation measures of changing environments in the estuary (Anthony and Goichot, 2019; Wolanski et al., 2004). Although there was no in situ measurement taken in this site, we assumed that Site B had similar hydrology and geomorphological processes as observed in Site A. The data set was classified into three groups to represent "mangrove states" during the restoration phases: (1) no mangrove, (2) beginning of plantation, and (3) beginning of nonplanted mangrove (natural) regeneration.

Hydrodynamic measurements were achieved to obtain the tidal pattern in the ponds and river by measuring water levels using pressure and temperature data using two HOBO loggers. These devices were set over one period of spring–neap tidal cycle in root of

TABLE 22.1 Time series data set used in this study including color-composite satellite images, water level, surface elevation change (SEC), and vertical accretion (VA).

	Spatial data			Hydrological data		Geomorphological data	
Year	**Time**	**Satellite**	**Kappa**	**Time**	**Location**	**Time**	**Parameters**
2001	t_0	Ikonos	−0.54				
2007	t_1	QuickBird	0.27				
2009	t_2	QuickBird	0.50				
2011						t_0	SEC, VA
2012	t_3	WorldView-2	0.61			t_1	SEC, VA
2013						t_2	SEC, VA
2014				December	River	t_3	SEC, VA
2015	t_4	WorldView-3	1.00	January	Ponds	t_4	SEC, VA

mangrove tree in the pond and under the bridge over the river (Fig. 22.1). The measurement was taken in between December 2014 and February 2015, which fell in wet season. The pressure data from the logger were converted into water level data and analyzed for tidal patterns using harmonic analysis. As each data point was taken in a different measurement period, we built harmonic recombination of observed measurements from both sites and reconstruct the tidal elevation by using the amplitude and phase of 35 tidal constituents. Tidal data from the nearby tide gauge at Pengambengan Fishing Port, 6.2 km northwest of the Perancak estuary, were also used as the oceanic tidal data for assessing the deformation of tidal wave inside estuary in the river and ponds.

Soil dynamics were analyzed as vertical accretion and soil surface elevation change, which were measured using the rod surface elevation table–marker horizon technique (RSET-MH) (Cahoon and Lynch, 1997). Measurements were undertaken every 2 months from July 2011 to March 2015, which related to the study published by Sidik et al. (2019). A set of three RSET-MH measurements was conducted in Site A and the reference site to give a mean of vertical accretion surface and elevation change for both sites. The reference site was located in intact mangroves that naturally developed along the riverbank on the south side. These measurements temporally overlapped with the water level measurements.

The expansion of planted mangroves and nonplanted mangroves was monitored using very high spatial resolution (VHSR) multispectral satellite images (Table 22.1), as presented by Proisy et al. (2018). Planted mangroves can be distinguished in VHSR satellite images from nonplanted mangroves because they correspond to a "regular organization of trees." Even recently planted seedlings can be detected, mostly *Rhizophora* spp. At the adult stage, the multispectral signature of even-aged high-density planted young *Rhizophora* stands is uniform, with a higher response in the near-infrared channel compared to mixed adult natural forests that have higher tree species diversity (Viennois et al., 2016) (Fig. 22.1). By contrast, the latter, i.e., patchy stands or individual trees of nonplanted mangroves, were discriminated by more heterogeneous spectral signatures, inside and outside (river banks, for instance) aquaculture ponds. Nonplanted mangroves were dominated by species such as *Avicennia alba, Avicennia officinalis, Avicennia marina*, and *Sonneratia alba*, with low abundance of introduced *Rhizophora* spp., which were probably dispersed from plantation areas (Proisy et al., 2018; Sidik et al., 2019). The images were analyzed for mangrove expansion by visual delineating the region of interest (Sites A and B) and calculating the area from both region of interest of planted and nonplanted mangrove. The observation started (t_0) in 2001 when the pond walls (dykes) in Sites A and B were still intact along with the water outlets that controlled the water exchange in the ponds.

5.2 Tidal flow reintroduction and sediment accumulation

Perancak River is the main distributary channel in the catchment, which receives sediment loads from upstream and channels it to the coast. To understand the hydrological pattern of the study site, we assessed the tidal regime of the Perancak River by setting up the water logger near the study site (Fig. 22.1). Measurements at this point showed that the study site is characterized as a well-mixed diurnal tidal estuary with a phase difference of 55 min between the catchment and the nearest port (Fig. 22.2A and C).

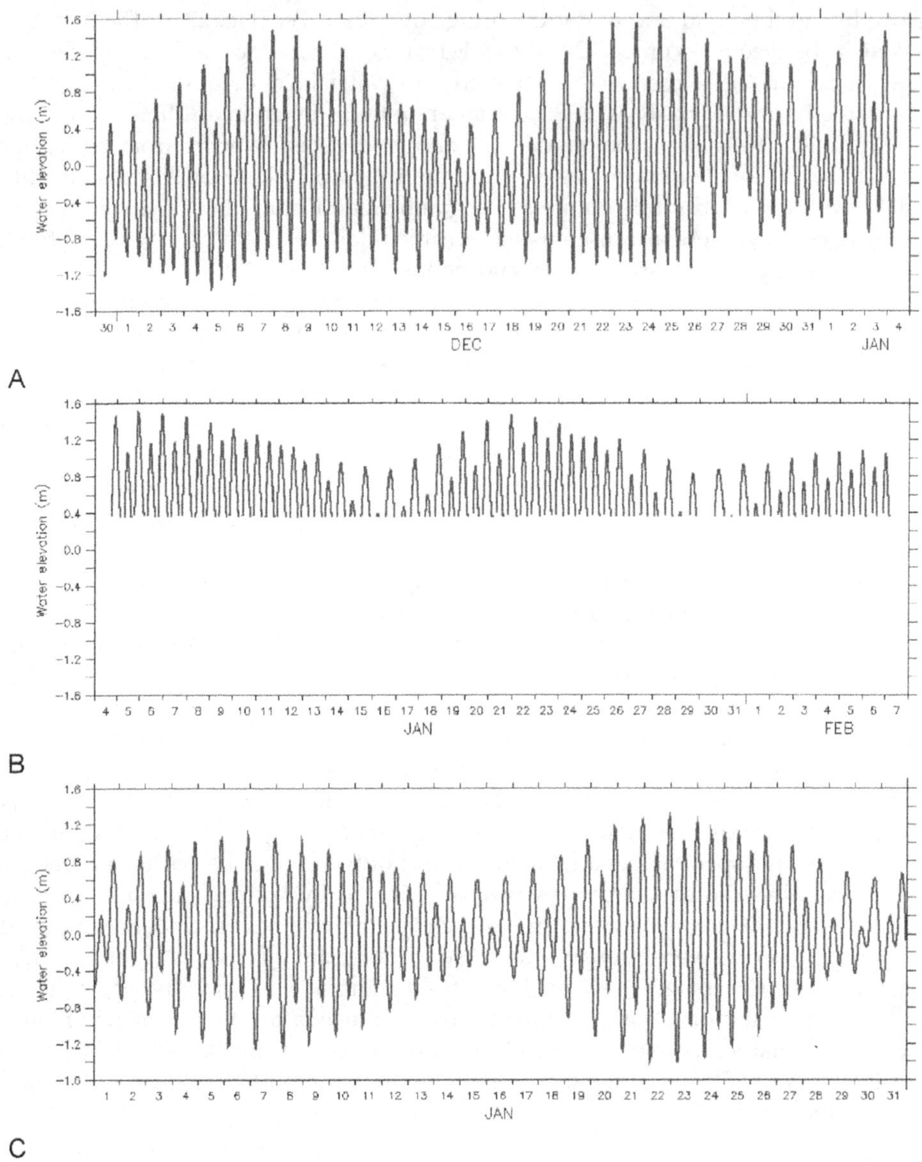

FIGURE 22.2 The comparison of tidal cycle from observation between t_3 and t_4 in Perancak River (A), inside pond (B), and Pengambengan Fishing Port (C).

Tidal patterns and periods between the river and pond showed a similar pattern with a phase difference of 2 minutes. The tidal amplitude from mean sea level at high tide reached 1.28 m at the spring tide and −1.17 m at the neap tide, allowing the water trapped in the ponds to attain a maximum height of 1.06 m during the spring tide and to be almost dry in the neap tide (Fig. 22.3). After tidal flow was reintroduced, the hydrological regime was

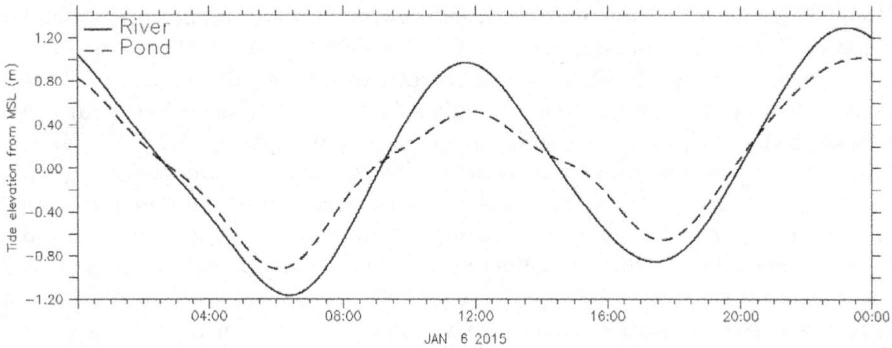

FIGURE 22.3 Tidal cycles in the river and ponds during the observation period of t_4.

restored, indicated by similar tidal cycles in both the pond and the river. This suggests that tidal flows in the pond resembled the natural pattern, with a longer hydroperiod compared to the reference site (river).

The pattern of soil dynamics in the pond showed a clear relationship with hydrodynamics. The trend of increasing elevation in the ponds coincided with reintroduction of tidal flooding in ponds. When the dykes had been removed (see t_2 of spatial observation), tidal flow started to resemble the natural pattern and followed by sediment deposition, which has been monitored since 2011 (before t_3). In 2012 (t_3 of spatial observation and t_1 of elevation change monitoring), the elevation change in pond showed similar pattern as the reference site (Fig. 22.4). The slope of elevation change in pond started to rise rapidly when entering the t_2 of the elevation change monitoring period, exceeding accumulation rates in the reference site. Over the course of the 3-year study period, the rates (slope) of surface elevation change in the pond and intact forest were as follows: pond 1.87 cm year^{-1} and forest 0.46 cm year^{-1}, respectively.

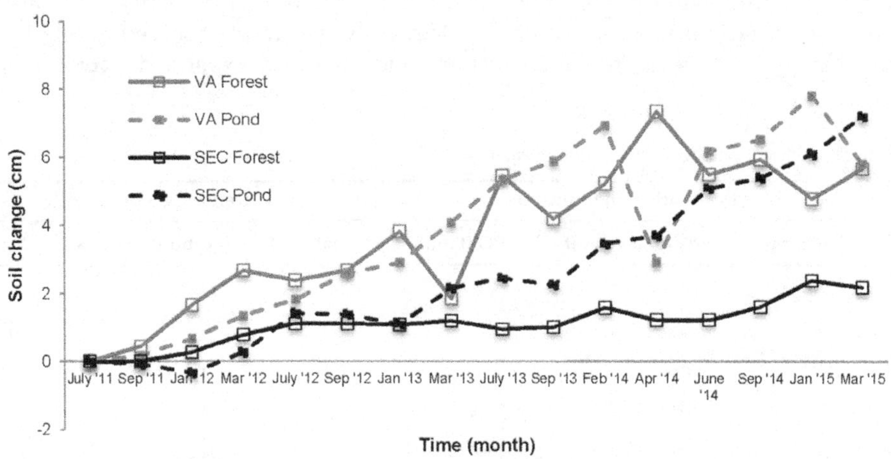

FIGURE 22.4 Soil change in pond and reference site (forest), which is represented by vertical accretion (VA) and surface elevation change (SEC). The measurement was undertaken in between July 2011 (t_0) and March 2015 (t_4).

The trend in elevation change was consistent with increasing vertical accretion. Over the observation period, which was started after the removal of tidal obstruction (see t_2 of spatial observation), sediment rapidly accumulated in both areas at a similar rate of 1.97 cm $year^{-1}$ (pond) and 1.59 cm $year^{-1}$ (forest). This is higher than the range of 0–1 cm $year^{-1}$ proposed by Furukawa et al. (1997) for early studies in Australian mangroves, but in the range of sedimentation rates in other mangrove forests in Southeast Asia and Indo-Pacific region (Alongi, 2009; Lovelock et al., 2015). As the availability of suspended sediment determined the ability of mangroves to accrete vertically, we assumed that mangroves in our study are highly impacted by suspended sediment delivered in river flows in this catchment associated with high rates of freshwater discharge (Alongi, 2009; Furukawa et al., 1997). High rates of vertical accretion over elevation change may suggest a level of subsidence or sediment compaction in both areas and possibly erosion in the forest due to variations in river discharge (Alongi, 2009; Cahoon and Guntenspergen, 2010).

5.3 Postplanting mangrove regeneration

The general trend of sediment accumulation in ponds, attributed to the reinstated hydrological regime in Site B, was likely to have triggered natural recruitment and establishment of mangrove propagules from nearby mangrove sites (Fig. 22.2). The assessment started in 2007 (t_1 of spatial observation) when Site A was established as a pilot site for a plantation and nursery of *Rhizophora* spp. In this period, no mangroves were observed in reference site (Site B) but some dykes started to reopen after t_1, with subsequent increasing water flows into this area. At the end of observation period (t_4), nonplanted species appeared to rapidly occupy the plantation site as patchy stands in plantation area and in empty ponds. This group of mangroves invaded the area with the rates of about 110 m^2 $year^{-1}$, whereas the planted mangroves developed in the plantation area and empty ponds with rates of about 325 m^2 $year^{-1}$. The plantation site was covered by 42% of planted mangrove and 20% of nonplanted mangroves (Table 22.2). This period was also marked by the reestablishment of tidal flow to natural patterns and similar sediment accumulation rates (Figs. 22.3 and 22.4).

The success of natural recovery can be explained by the trend observed in Site B, where nonplanted mangroves were recruited inside ponds (t_2) and expanded over the site (t_4)

TABLE 22.2 Mangrove forest development is shown by the change of mangrove cover in Sites A and B

	Restoration site (Site A)			Reference site (Site B)		
Year	Total (m^2)	Natural recruits	Plantation	Total (m^2)	Natural recruits	Plantation
2001 (t_0)	0	0%	0%	0	0%	0%
2007 (t_1)	795	0%	8%	0	0%	0%
2009 (t_2)	2982	0%	30%	353	4%	0%
2012 (t_3)	3678	0%	37%	8395	84%	11%
2015 (t_4)	6163	20%	42%	8395	84%	11%

with a rate of 619.22 m^2 $year^{-1}$. Plantation activities were detected during t_3, characterized by parallel rows of seedling among nonplanted mangroves (Fig. 22.5). Despite this active intervention creating a mosaic of planted and nonplanted species mixture, planted mangroves

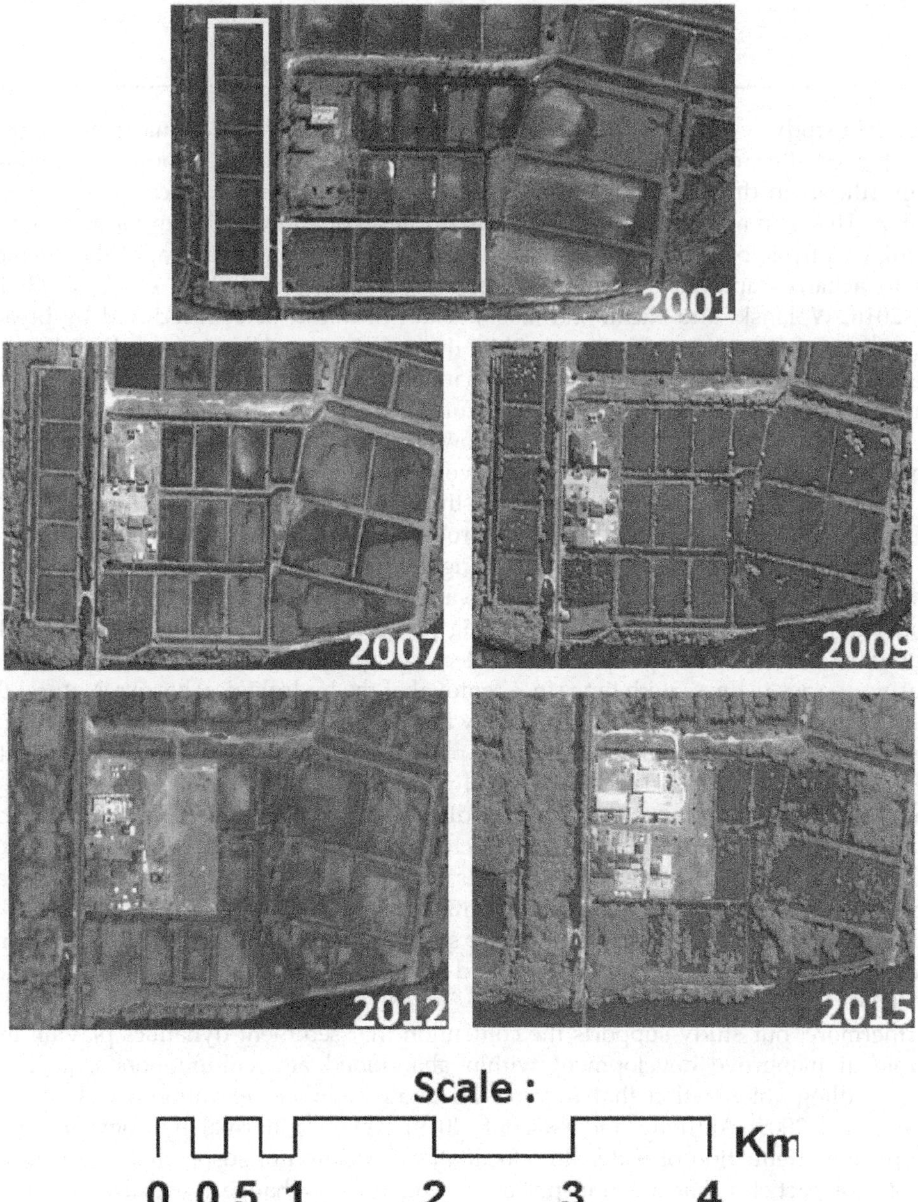

FIGURE 22.5 Mangrove development in Sites A and B (below) during the period of observation: 2001 (t_0), 2007 (t_1), 2009 (t_2), 2012 (t_3), and 2015 (t_4).

were unlikely to dominate, as their annual expansion rates were "only" about 81 m^2. The domination of species not used in plantations in nonplantation areas and their colonization within plantation areas appeared to be the process of natural dispersal and recruitment that was likely to occur associated with the dynamic environment (Ball, 1988; Walters, 2000).

6. Lesson learned from the case study

Our field study revealed that the successional processes of restored mangroves appeared to be achieved after reopening of obstructed channels, a conditional factor to restore tidal hydrology (the main driver), followed by sediment accumulation and increasing soil surface elevation. This is in agreement with previous studies that showed the importance of understanding ecophysiological characteristics of restoration sites and removal of the stressors in order to achieve rapid development of restored mangroves (Krauss et al., 2008; Lewis et al., 2016; Wolanski and Richmond, 2008). Tidal flow can be reintroduced by breaching pond walls to allow tides flooding within the pond area where plantations have been made or are planned. Opening dewatering channels between ponds was followed by colonization of nonplanted mangroves through natural recruitment during t_2 (Site B, 2009) and t_3 (Site A, 2012) of our observations. The tidal flows transported suspended sediments to ponds and also assisted in the dispersal of mangrove propagules or seeds, which is facilitated by tidal water from adjacent mangrove areas or through direct seed fall (Rabinowitz, 1978).

Mixed success can be achieved when reintroducing tidal flow into plantation areas; however, a similar trajectory of a formation of mangrove forest with multispecies stands can also be achieved without planting. After the removal of the hydrological barrier, nonplanted species of *Avicennia* spp., which is a fast-growing pioneer species (Osborne and Berjak, 1997), increased in cover more rapidly than planted mangroves and was also associated with increasing species richness with time since removal of the hydrological barrier. Natural development of mangroves occurs when hydrology is restored, showing that natural hydrological processes can lead to conditions that are a prerequisite for certain species to recruit and develop (Bosire et al., 2003). This shows that hydrological processes interplay with the plant communities through composition, species diversity, and succession to influence species composition (Zhang et al., 2012). Other factor regulating natural regeneration, such as propagule predation, was not identified in this site. The postplanting invasion of nonplanted mangroves into monospecific *Rhizophora* plantations demonstrated a tendency of "plantation" forest to give rise to forests with similar structure as natural forests which have higher species diversity, ecosystem productivity, and stability (Barnuevo et al., 2017; Bosire et al., 2006, 2003).

Furthermore, our study supports the contention that sediment dynamics play an important role in mangrove development within abandoned aquaculture ponds, particularly during seedling colonization that requires adequate sediment elevation and hydroperiod (Proisy et al., 2009; Anthony and Goichot, 2019). When a normal hydrographic pattern develops, the attenuation of water flow facilitates deposition of suspended sediment in the form of fine particles that are transported during tidal exchange (Furukawa et al., 1997; Furukawa and Wolanski, 1996). Sediment starts to consolidate and becomes more compact, allowing propagules to colonize and mangroves to grow (Proisy et al., 2009; Toorman et al.,

2018). This mangrove–sediment–hydrodynamics interaction was associated with newly formed mangrove areas inside the ponds in parallel with increasing elevation slope and tidal flows that resembled the natural pattern. Our results are consistent with the study of mangrove rehabilitation in abandoned ponds in Sulawesi, which supported evidence of the role of hydroperiod and optimal elevation in early development of mangroves (Oh et al., 2017). This earlier study found the range of optimal elevation for mangrove seedlings to establish in abandoned ponds was restricted to intertidal position of $-1.511\ m \leq x \leq 0.228\ m$ WGS 84b. Although an elevation survey was not undertaken in our study, based on our assessment of tidal amplitudes, we assumed that the elevation of ponds in our site had reached that range, as shown by the establishment of nonplanted mangrove in Sites A and B at the end of postplanting regeneration observation. Limited hydrological observations at the study site and elevation assessment in this study lead to uncertainties (Van Loon et al., 2016); thus, further investigation of inundation extent is needed to generate robust analysis of mangroves–sediment–hydrodynamics interactions.

Sediment supply is important for mangrove sustainability and water regime variation, and this driver of sediment dynamics appears to be a key factor for maintaining wetland ecosystem functioning, such as its role as a carbon sink. The valuable role of mangroves as a carbon sink advances mangrove restoration as one way to achieve carbon abatement goals (Macreadie et al., 2017). A previous study on carbon sequestration and flux in the same site found that the increasing slope of sediment surface elevation change over time was correlated with carbon accumulation (Sidik et al., 2019). After 10 years of restoration, the observations provided evidence that reintroduction of tidal flow was likely the driver of the increasing elevation of the pond floor and also rapid growth of nonplanted mangroves, suggesting the recovery of carbon sequestration function of abandoned ponds (Proisy et al., 2018; Sidik et al., 2019). Similar results were found in abandoned shrimp ponds in Sulawesi, Indonesia, and Nakorn Sri Thammarat, Thailand, showing recovery of carbon sinks after hydrologic restoration (Cameron et al., 2019; Matsui et al., 2012). These findings confirm that reintroduction of tidal flow by removing barriers to tidal inundation can lead to carbon abatement through new growth of mangrove biomass and accumulation of additional soil carbon (Kelleway et al., 2017). In addition, the removal of pond walls potentially contributes further to climate change mitigation, as the pond walls can emit CO_2 at rates three times higher than pond floors (Sidik and Lovelock, 2013).

Considering mangrove–sediment–hydrodynamic interactions and their implications for mangrove ecosystem functioning, mangrove plantations should be a management option when natural regeneration had not occurred, perhaps due to limitations in propagule abundance. Limitation in the abundance of propagules is a situation in which natural waterborne propagules are not available in a restoration area (Lewis, 2005). In other locations, planting of pioneer species may facilitate the restoration of degraded mangroves. For example, the exotic species of *Sonneratia apetala* planted in China appeared to facilitate recolonization of native species that failed to establish through natural regeneration (Ren et al., 2008). The presence of pioneer species of *Avicennia* spp. after the removal of the hydrological barrier confirms the idea that planting mangrove seedlings/propagules is the last option to restore such mangrove ecosystems (Kamali and Hashim, 2011; Lewis, 2005). In addition, a study by Bosire et al. (2006) and Salmo et al. (2014) found that plantation forests have a low threshold of disturbance tolerance, suggesting their high vulnerability to environmental change. The current study shows the importance of facilitating the natural recovery of mangroves in abandoned ponds by restoring hydrodynamic conditions. Given that aquaculture throughout the world is responsible for the substantial loss of mangroves, this recommendation should be considered by this industry for restoring unproductive ponds and enhancing management of ponds used for silvo-fisheries systems. Overall, if coastal sustainability is to be achieved, aquaculture practices, considered as the main driver of mangrove loss in Indonesia (Ilman et al., 2016), should develop management plans for disused pond conversion to mangroves as a "post" program in order to improve environmental conditions that will be beneficial to productivity in ponds still in use as well as for local communities (Stevenson, 1997).

Acknowledgments

The satellite images used in this study were acquired under the Infrastructure Development of Space Oceanography (INDESO) project. The hydrological and geomorphological measurement was part of the project Partnerships for Enhanced Engagement in Research (PEER) "Enhancement of Research for Adaptation of Wetlands in Indonesia to Projected Impacts of Sea Level Rise," performed by the first author.

References

Alongi, D.M., 2014. Carbon cycling and storage in mangrove forests. Annu. Rev. Mar. Sci. 6, 195–219. https://doi.org/10.1146/annurev-marine-010213-135020.

Alongi, D.M., 2012. Carbon sequestration in mangrove forests. Carbon Manag. 3, 313–322. https://doi.org/10.4155/cmt.12.20.

Alongi, D.M., 2009. The Energetics of Mangrove Forests. Springer Netherlands, Dordrecht. https://doi.org/10.1007/978-1-4020-4271-3.

Alongi, D.M., Brinkman, R., 2011. Hydrology and biogeochemistry of mangrove forests. In: Levia, D.F., Carlyle-Moses, D., Tanaka, T. (Eds.), Forest Hydrology and Biogeochemistry: Synthesis of Past Research and Future Direction, Ecological Studies. Springer Netherlands, Dordrecht, pp. 203–219. https://doi.org/10.1007/978-94-007-1363-5.

Anthony, E.J., Goichot, M., 2019. Sediment Flow in the Context of Mangrove and Restoration: A Rapid Assessment Guidance Manual.

Atwood, T.B., Connolly, R.M., Almahasheer, H., Carnell, P.E., Duarte, C.M., Lewis, C.J.E., Irigoien, X., Kelleway, J.J., Lavery, P.S., Macreadie, P.I., Serrano, O., Sanders, C.J., Santos, I., Steven, A.D.L., 2017. Global patterns in mangrove soil carbon stocks and losses. Nat. Clim. Change 7. https://doi.org/10.1038/NCLIMATE3326.

Balke, T., Friess, D.A., 2016. Geomorphic knowledge for mangrove restoration: a pan-tropical categorization. Earth Surf. Process. Landforms 41, 231–239. https://doi.org/10.1002/esp.3841.

Ball, M.C., 1988. Ecophysiology of mangroves. Trees 2, 129–142.

Barnuevo, A., Asaeda, T., Sanjaya, K., Kanesaka, Y., 2017. Drawbacks of mangrove rehabilitation schemes : lessons learned from the large-scale mangrove plantations. Estuar. Coast. Shelf Sci. 1–6. https://doi.org/10.1016/j.ecss.2017.02.015.

Berger, U., Rivera-Monroy, V.H., Doyle, T.W., Dahdouh-Guebas, F., Duke, N.C., Fontalvo-herazo, M.L., Hildenbrandt, H., Koedam, N., Mehlig, U., Piou, C., Twilley, R.R., 2008. Advances and limitations of individual-based models to analyze and predict dynamics of mangrove forests : a review. Aquat. Bot. 89, 260–274. https://doi.org/10.1016/j.aquabot.2007.12.015.

Biswas, S.R., Khan, M.S.I., Mallik, A.U., 2012. Invaders' control on post-disturbance succession in coastal mangroves. J. Plant Ecol. 5, 157–166. https://doi.org/10.1093/jpe/rtr050.

Bosire, J.O., Dahdouh-Guebas, F., Kairo, J.G., Koedam, N., 2003. Colonization of non-planted mangrove species into restored mangrove stands in Gazi Bay, Kenya. Aquat. Bot. 76, 267–279. https://doi.org/10.1016/S0304-3770(03)00054-8.

Bosire, J.O., Dahdouh-Guebas, F., Kairo, J.G., Wartel, S., Kazungu, J., Koedam, N., 2006. Success rates of recruited tree species and their contribution to the structural development of reforested mangrove stands. Mar. Ecol. Prog. Ser. 325, 85–91. https://doi.org/10.3354/meps325085.

Brown, B., Fadillah, R., Nurdin, Y., Soulsby, I., Ahmad, R., 2014. Case Study: community based ecological mangrove rehabilitation (CBEMR) in Indonesia. SAPIENS 7.

Brown, B., Yuniati, W., Ahmad, R., Soulsby, I., 2015. Observations of natural recruitment and human attempts at mangrove rehabilitation after seismic (Tsunami and Earthquake) events in Simeulue Island and Singkil Lagoon, Aceh, Indonesia. In: Santiago-Fandino, V., Kontar, Y.A., Kaneda, Y. (Eds.), Advances in Natural and Technological Hazards Research Post-Tsunami Hazard: Reconstruction and Restoration. Springer, Switzerland. https://doi.org/10.1007/978-3-319-10202-3_19.

Cahoon, D.R., Guntenspergen, G.R., 2010. Climate change, sea-level rise, and coastal wetlands. Natl. Wetl. Newsl. 32, 8–12.

Cahoon, D.R., Lynch, J.C., 1997. Vertical accretion and shallow subsidence in a mangrove forest of southwestern Florida, USA. Mangroves Salt Marshes 1, 173–186.

Cameron, C., Hutley, L.B., Friess, D.A., Munksgaard, N.C., 2019. Hydroperiod, soil moisture and bioturbation are critical drivers of greenhouse gas fluxes and vary as a function of landuse change in mangroves of Sulawesi , Indonesia. Sci. Total Environ. 654, 365–377. https://doi.org/10.1016/j.scitotenv.2018.11.092.

Crase, B., Liedloff, A., Vesk, P.A., Burgman, M.A., Wintle, B.A., 2013. Hydroperiod is the main driver of the spatial pattern of dominance in mangrove communities. Global Ecol. Biogeogr. 22, 806–817. https://doi.org/10.1111/geb.12063.

Dale, P.E.R., Knight, J.M., Dwyer, P.G., 2014. Mangrove rehabilitation : a review focusing on ecological and institutional issues. Wetl. Ecol. Manag. 22, 587–604. https://doi.org/10.1007/s11273-014-9383-1.

Davis, J.H., 1938. Mangrove-makers of land. Nat. Mag. 551–553.

Donato, D.C., Kauffman, J.B., Murdiyarso, D., Kurnianto, S., Stidham, M., Kanninen, M., 2011. Mangroves among the most carbon-rich forests in the tropics. Nat. Geosci. 4, 293–297. https://doi.org/10.1038/ngeo1123.

Erftemeijer, P.L.A., Lewis III, R.R., 1999. Planting mangroves on intertidal mudflats: habitat restoration or habitat conversion. In: ECOTONE-VIII Seminar "Enhancing Coastal Ecosystem Restoration for the 21st Century. Phuket.

Friess, D.A., 2016. Ecosystem services and disservices of mangrove forests: insights from historical colonial observations. Forests 7. https://doi.org/10.3390/f7090183.

Friess, D.A., Krauss, K.W., Horstman, E.M., Balke, T., Bouma, T.J., Galli, D., Webb, E.L., 2012. Are all intertidal wetlands naturally created equal? Bottlenecks, thresholds and knowledge gaps to mangrove and saltmarsh ecosystems. Biol. Rev. Camb. Phil. Soc. 87, 346–366. https://doi.org/10.1111/j.1469-185X.2011.00198.x.

Furukawa, K., Wolanski, E., 1996. Sedimentation in mangrove forests. Mangroves Salt Marshes 1, 3–10.

Furukawa, K., Wolanski, E., Mueller, H., 1997. Currents and sediment transport in mangrove forests. Estuar. Coast. Shelf Sci. 44, 301–310. https://doi.org/10.1006/ecss.1996.0120.

Gusmawati, N., Soulard, B., Selmaoui-Folcher, N., Proisy, C., Mustafa, A., Le Gendre, R., Laugier, T., Lemonnier, H., Laugier, T., Lemonnier, H., 2018. Surveying shrimp aquaculture pond activity using multitemporal VHSR satellite images – case study from the Perancak estuary, Bali Indonesia. Mar. Pollut. Bull. 131, 49–60. https://doi.org/10.1016/j.marpolbul.2017.03.059.

Iftekhar, M., 2008. Functions and development of reforested mangrove areas: a review. Int. J. Biodivers. Sci. Manag. 4, 1–14.

Ilman, M., Dargusch, P., Dart, P., Onrizal, 2016. A historical analysis of the drivers of loss and degradation of Indonesia's mangroves. Land Use Policy 54, 448–459. https://doi.org/10.1016/j.landusepol.2016.03.010.

Kamali, B., Hashim, R., 2011. Mangrove restoration without planting. Ecol. Eng. 37, 387–391. https://doi.org/10.1016/j.ecoleng.2010.11.025.

Kelleway, J.J., Serrano, O., Baldock, J., Cannard, T., Lavery, P., Lovelock, C.E., Macreadie, P., Masqué, P., Saintilan, N., Steven, A.D.L., 2017. Technical Review of Opportunities for Including Blue Carbon in the Australian Government's Emissions Reduction Fund. Australia.

Kodikara, K.A.S., Mukherjee, N., Jayatissa, L.P., Dahdouh-guebas, F., Koedam, N., 2017. Have mangrove restoration projects worked ? An in-depth study in Sri Lanka. Restor. Ecol. 25, 705–716. https://doi.org/10.1111/rec.12492.

Krauss, K.W., Allen, J.A., Cahoon, D.R., 2003. Differential rates of vertical accretion and elevation change among aerial root types in Micronesian mangrove forests. Estuar. Coast. Shelf Sci. 56, 251–259. https://doi.org/10.1016/S0272-7714(02)00184-1.

Krauss, K.W., From, A.S., Doyle, T.W., Doyle, T.J., Barry, M.J., 2011. Sea-level rise and landscape change influence mangrove encroachment onto marsh in the ten thousand Islands region of Florida, USA. J. Coast. Conserv. 15, 629–638. https://doi.org/10.1007/s11852-011-0153-4.

Krauss, K.W., Lovelock, C.E., McKee, K.L., López-Hoffman, L., Ewe, S.M., Sousa, W.P., 2008. Environmental drivers in mangrove establishment and early development: a review. Aquat. Bot. 89, 105–127. https://doi.org/10.1016/j.aquabot.2007.12.014.

Kusmana, C., 2014. Distribution and current status of mangrove forests in Indonesia. In: Faridah-Hanum, I., Latiff, A., Hakeem, K.R., Ozturk, M. (Eds.), Mangrove Ecosystems of Asia: Status, Challenges and Management. Springer, New York.

Lewis III, R.R., 2009. Knowledge overload, wisdom underload. Ecol. Eng. 35, 341–342. https://doi.org/10.1016/j.ecoleng.2008.10.006.

Lewis III, R.R., 2005. Ecological engineering for successful management and restoration of mangrove forests. Ecol. Eng. 24, 403–418. https://doi.org/10.1016/j.ecoleng.2004.10.003.

Lewis III, R.R., Gilmore, R.G., 2007. Important considerations to achieve successful mangrove forest restoration with. Optimum Fish Habitat 80, 823–837.

Lewis III, R.R., Milbrandt, E.C., Brown, B., Krauss, K.W., Rovai, A.S., Beever, J.W., Flynn, L.L., 2016. Stress in mangrove forests: early detection and preemptive rehabilitation are essential for future successful worldwide mangrove forest management. Mar. Pollut. Bull. https://doi.org/10.1016/j.marpolbul.2016.03.006.

López-Portillo, J., Lewis, R.R., Saenger, P., Rovai, A.S., Koedam, N., Farid, D.-G., Claudia, A.-H., Rivera-Monroy, H.,V., 2017. Mangrove forest restoration and rehabilitation. In: Rivera-Monroy, V.H., Lee, S.Y., Kristensen, E., Twilley, R. (Eds.), Mangrove Ecosystems : A Global Biogeographic Perspective Structure, Function, and Services. Springer, pp. 301–345.

Lovelock, C.E., Adame, M.F., Bennion, V., Hayes, M., O'Mara, J., Reef, R., Santini, N.S., 2013. Contemporary rates of carbon sequestration through vertical accretion of sediments in mangrove forests and saltmarshes of South East Queensland, Australia. Estuar. Coast 37, 763–771. https://doi.org/10.1007/s12237-013-9702-4.

Lovelock, C.E., Atwood, T., Baldock, J., Duarte, C.M., Hickey, S., Lavery, P.S., Masque, P., Macreadie, P.I., Ricart, A.M., Serrano, O., Steven, A., 2017. Assessing the risk of carbon dioxide emissions from blue carbon ecosystems. Front. Ecol. Environ. 15, 257–265. https://doi.org/10.1002/fee.1491.

Lovelock, C.E., Brown, B.M., 2019. Land tenure considerations are key to successful mangrove restoration. Nat. Ecol. Evol. 3, 1135. https://doi.org/10.1038/s41559-019-0942-y.

Lovelock, C.E., Cahoon, D.R., Friess, D.A., Guntenspergen, G.R., Krauss, K.W., Reef, R., Rogers, K., Saunders, M.L., Sidik, F., Swales, A., Saintilan, N., Thuyen, L.X., Triet, T., 2015. The vulnerability of Indo-Pacific mangrove forests to sea-level rise. Nature 526. https://doi.org/10.1038/nature15538. 559-U217.

Lugo, A.E., 1997. The apparent paradox of reestablishing species richness on degraded lands with tree monocultures. For. Ecol. Manag. 99, 9–19.

Macreadie, P.I., Nielsen, D.A., Kelleway, J.J., Atwood, T.B., Seymour, J.R., Petrou, K., Connolly, R.M., Thomson, A.C.G., Trevathan-tackett, S.M., Ralph, P.J., 2017. Can we manage coastal ecosystems to sequester more blue carbon ? Front. Ecol. Environ. 15, 206–213. https://doi.org/10.1002/fee.1484.

Matsui, N., Morimune, K., Meepol, W., Chukwamdee, J., 2012. Ten year evaluation of carbon stock in mangrove plantation reforested from an abandoned shrimp pond. Forests 3, 431–444. https://doi.org/10.3390/f3020431.

Matsui, N., Suekuni, J., Nogami, M., Havanond, S., Salikul, P., 2010. Mangrove rehabilitation dynamics and soil organic carbon changes as a result of full hydraulic restoration and re-grading of a previously intensively managed shrimp pond. Wetl. Ecol. Manag. 18, 233–242. https://doi.org/10.1007/s11273-009-9162-6.

Mckee, K.L., Faulkner, P.L., 2000. Restoration of biogeochemical function in mangrove forests. Restor. Ecol. 8, 247–259. https://doi.org/10.1046/j.1526-100x.2000.80036.x.

Nellemann, C., Corcoran, E., Duarte, C.M., Valdés, L., De Young, C., Fonseca, L., Grimsditch, G. (Eds.), 2009. Blue Carbon. A Rapid Response Assessment. GRID-Arendal: United Nations Environment Programme.

Oh, R.R.Y., Friess, D.A., Brown, B., 2017. The role of surface elevation in the rehabilitation of abandoned aquaculture ponds to mangrove forests, Sulawesi, Indonesia. Ecol. Eng. 100, 325–334. https://doi.org/10.1016/j.ecoleng.2016.12.021.

Osborne, D.J., Berjak, P., 1997. The making of mangroves: the remarkable pioneering role played by seeds of *Avicennia marina*. Endeavour 21, 143–147. https://doi.org/10.1016/S0160-9327(97)01077-6.

Parrotta, J.A., Turnbull, J.W., Jones, N., 1997. Catalyzing native forest regeneration on degrade tropical lands. For. Ecol. Manag. 11, 1–7.

Pendleton, L., Donato, D.C., Murray, B.C., Crooks, S., Jenkins, W.A., Sifleet, S., Craft, C., Fourqurean, J.W., Kauffman, J.B., Marbà, N., Megonigal, P., Pidgeon, E., Herr, D., Gordon, D., Baldera, A., 2012. Estimating global 'blue carbon' emissions from conversion and degradation of vegetated coastal ecosystems. PloS One 7. https://doi.org/10.1371/journal.pone.0043542.

Primavera, J.H., Esteban, J.M.A., 2008. A review of mangrove rehabilitation in the Philippines : successes, failures and future prospects. Wetl. Ecol. Manag. https://doi.org/10.1007/s11273-008-9101-y.

Proisy, C., Gratiot, N., Anthony, E.J., Gardel, A., Fromard, F., Heuret, P., 2009. Mud bank colonization by opportunistic mangroves: a case study from French Guiana using lidar data. Continent. Shelf Res. 29, 632–641. https://doi.org/10.1016/j.csr.2008.09.017.

Proisy, C., Viennois, G., Sidik, F., Andayani, A., Anthony, J., Guitet, S., Gusmawati, N., Lemonnier, H., Muthusankar, G., Olagoke, A., Prosperi, J., Rahmania, R., Ricout, A., Soulard, B., Suhardjono, 2018. Monitoring mangrove forests after aquaculture abandonment using time series of very high spatial resolution satellite images : a case study from the Perancak estuary, Bali, Indonesia. Mar. Pollut. Bull. 131, 61–71. https://doi.org/10.1016/j.marpolbul.2017.05.056.

Rabinowitz, D., 1978. Dispesrsal properties of mangrove propagules. Biotropica 10, 47–57.

Ren, H., Jian, S., Lu, H., Zhang, Q., Shen, W., Han, W., Yin, Z., Guo, Q., 2008. Restoration of mangrove plantations and colonisation by native species in Leizhou bay, South China. Ecol. Res. 23, 401–407. https://doi.org/10.1007/s11284-007-0393-9.

Ridd, P.V., Wolanski, E., Mazda, Y., 1990. Longitudinal diffusion in mangrove-fringed tidal creeks. Estuar. Coast. Shelf Sci. 31, 541–554. https://doi.org/10.1016/0272-7714(90)90012-G.

Salmo, S.G., Lovelock, C.E., Duke, N.C., 2014. Assessment of vegetation and soil conditions in restored mangroves interrupted by severe tropical typhoon 'Chan-hom' in the Philippines. Hydrobiologia 733, 85–102. https://doi.org/10.1007/s10750-013-1766-4.

Sanderman, J., Hengl, T., Fiske, G., Solvik, K., Adame, M.F., Benson, L., Bukoski, J.J., Carnell, P., Cifuentes-jara, M., Donato, D., Duncan, C., Eid, E.M., Ermgassen, P., Lewis, C.J.E., Macreadie, P.I., Glass, L., Gress, S., Jardine, S.L., Jones, T.G., Nsombo, E.N., Rahman, M.M., Sanders, C.J., Spalding, M., Landis, E., 2018. A global map of mangrove forest soil carbon at 30 m spatial resolution. Environ. Res. Lett. 13 https://doi.org/10.1088/1748-9326/aabe1c.

Sidik, F., Adame, M.F., Lovelock, C.E., 2019. Carbon sequestration and fluxes of restored mangroves in abandoned aquaculture ponds. J. Indian Ocean Reg. 1–16. https://doi.org/10.1080/19480881.2019.1605659.

Sidik, F., Lovelock, C.E., 2013. CO_2 efflux from shrimp ponds in Indonesia. PloS One 8. https://doi.org/10.1371/journal.pone.0066329.

Stevenson, N.J., 1997. Disused shrimp ponds: options for redevelopment of mangroves. Coast. Manag. 25, 425–435. https://doi.org/10.1080/08920759709362334.

Swales, A., Bentley Sr, S.J., Lovelock, C.E., 2015. Mangrove-forest evolution in a sediment-rich estuarine system: opportunists or agents of geomorphic change? Earth Surf. Process. Landforms 40, 1672–1687. https://doi.org/10.1002/esp.3759.

Toorman, E.A., Anthony, E.J., Augustinus, P.G.E.F., Gardel, A., Gratiot, N., Homenauth, O., Huybrechts, N., Monbaliu, J., Moseley, K., Naipal, S., 2018. Interaction of mangroves, coastal hydrodynamics and morphodynamics along the coastal fringes of the Guianas. In: Makowski, C., Finkl, C. (Eds.), Threats to Mangrove Forests. Springer, pp. 429–473. https://doi.org/10.1007/978-3-319-73016-5_20.

Van Loon, A.F., Te Brake, B., Van Huijgevoort, M.H.J., Dijksma, R., 2016. Hydrological classification, a practical tool for mangrove restoration. PloS One 11. https://doi.org/10.1371/journal.pone.0150302.

Viennois, G., Proisy, C., Féret, J.B., Prosperi, J., Sidik, F., Suhardjono, Rahmania, R., Germain, O., Gaspar, P., 2016. Multitemporal analysis of high spatial resolution satellite imagery for mangrove species mapping, Bali, Indonesia. IEEE J. Sel. Top. Appl. Earth Obs. Remote Sens. 9, 3680–3686.

Walters, B.B., 2000. Local mangrove planting in the Philippines: are fisherfolk and fishpond owners effective restorationists? Restor. Ecol. 8, 237–246. https://doi.org/10.1046/j.1526-100x.2000.80035.x.

Walton, M.E.M., Samonte-tan, G.P.B., Primavera, J.H., Edwards-jones, G., Vay, L.L.E., 2006. Are mangroves worth replanting ? The direct economic benefits of a community-based reforestation project. Environ. Conserv. https://doi.org/10.1017/S0376892906003341.

Willemsen, P.W.J.M., Horstman, E.M., Borsje, B.W., Friess, D.A., Dohmen-janssen, C.M., 2016. Geomorphology Sensitivity of the sediment trapping capacity of an estuarine mangrove forest. Geomorphology 273, 189–201. https://doi.org/10.1016/j.geomorph.2016.07.038.

Winterwerp, J.C., Erftemeijer, P.L.A., Suryadiputra, N., van Eijk, P., Zhang, L., 2013. Defining eco-morphodynamic requirements for rehabilitating eroding mangrove-mud coasts. Wetlands 33, 515–526. https://doi.org/10.1007/s13157-013-0409-x.

Wodehouse, D.C.J., Rayment, M.B., 2019. Mangrove area and propagule number planting targets produce sub-optimal rehabilitation and afforestation outcomes. Estuar. Coast. Shelf Sci. 222, 91–102. https://doi.org/10.1016/j.ecss.2019.04.003.

Wolanski, E., 1992. Hydrodynamics of mangrove swamps and their coastal waters. Hydrobiologia 247, 141–161.

Wolanski, E., Boorman, L.A., Chícharo, L., Langlois-Saliou, E., Lara, R., Plater, A.J., Uncles, R.J., Zalewski, M., 2004. Ecohydrology as a new tool for sustainable management of estuaries and coastal waters. Wetl. Ecol. Manag. 12, 235–276.

Wolanski, E., Richmond, R.H., 2008. Estuary Restoration. Elsevier.

Wolanski, E., Ridd, P., 1986. Tidal mixing swamps and trapping in mangrove. Estuar. Coast. Shelf Sci. 23, 759–771.

Zhang, L., Yin, J., Jiang, Y., Wang, H., 2012. Relationship between the hydrological conditions and the distribution of vegetation communities within the Poyang Lake National Nature Reserve, China. Ecol. Inf. 11, 65–75. https://doi.org/10.1016/j.ecoinf.2012.05.006.

CHAPTER

23

Impacts of forestry on mangrove sediment dynamics

Mériadec Sillanpää[1,2], Massimo Lupascu[1], Ruhuddien Pandu Yudha[3], Sigit D. Sasmito[1,4,5], Susan Vulpas[6], Rui Xiang Teo[2], Sartji Taberima[7]

[1]Department of Geography, National University of Singapore, Singapore; [2]Green Forest Product and Tech. Pte. Ltd., Research Department, Singapore; [3]Forestry Department, PT. Bintuni Utama Murni Wood Industries, Jakarta, Indonesia; [4]Center for International Forestry Research (CIFOR), Jalan CIFOR, Situgede, Bogor, Indonesia; [5]NUS Environmental Research Institute, National University of Singapore, Singapore; [6]Department of Environmental Science, American University, Washington D.C., United States; [7]Department of Soil Science and Land Resource, Faculty of Agriculture, Papua University, Manokwari, West Papua, Indonesia

OUTLINE

1. Introduction 584

2. Global overview of soil dynamics and recovery in mangrove harvest regimes 587

3. Postharvest impacts on mangrove soils in different harvest regimes 589

3.1 Comparison of soil conditions in different mangrove silviculture regimes 590

3.2 General trends in soil properties and forest structure of secondary mangrove forests 592

4. Forestry case study: soil dynamics in a large-scale selective logging regime in the Bintuni Bay mangrove, West Papua, Indonesia 594

4.1 General description of large-scale selective logging in Bintuni Bay 595

4.2 Soil recovery in different aged secondary and primary mangrove forests of Bintuni Bay, West Papua 595

4.3 Relationship between soil indicators and mangrove forest structure in a selective logging system 598

Dynamic Sedimentary Environments of Mangrove Coasts
https://doi.org/10.1016/B978-0-12-816437-2.00018-5

4.4 *Soil dynamics against depth in secondary and primary mangrove forests* 600
4.5 *Summary of case study* 600

5. Overview and future considerations 602

Appendix 603
I. Literature review methods 603
II. Literature review data: data from soil studies in harvested areas around the world 604
Supplementary data 604

Acknowledgments 604

References 604

Further reading 607

1. Introduction

Mangrove forests have been utilized for firewood, tannin (from bark), and construction material by local communities (Hauff et al., 2006) and for woodchip (Sillanpää et al., 2017), charcoal (FAO, 1985; Goessens et al., 2014), and possibly wood pellets by forest industries for commercial use. Through small- or large-scale harvesting, these uses have significant impacts on soil dynamics and characteristics, consequentially affecting forest structure and growth (Saenger, 2002). The sustainability of mangrove forest management regimes is scarcely studied, with the current research mostly focusing on whether the forest can regenerate back into its former structure, and whether it could be harvested multiple times over a long period without losing its volume potential (Goessens et al., 2014; Sillanpää et al., 2017; Tovilla Hernández et al., 2001). These silviculture systems have also been evaluated if they could contribute to coastal protection and preserve the carbon sequestration potential of intact mangrove ecosystems (Alongi et al., 2002; Gleason et al., 2003; Sasmito et al., 2020a).

As the rate of mangrove deforestation has been increasing globally (Hamilton and Friess, 2018), it has become essential to observe the dynamics within different silviculture systems, which, if practiced correctly, can prevent permanent loss of mangrove and associated ecosystem services such as carbon sequestration, coastline protection, and fish nurseries (Lunstrum and Chen, 2014; Aziz et al., 2015). The main harvest and forest management regimes for mangrove wood extraction could be described as either plantations or selective logging systems. These harvest regimes can be further defined as the following:

Plantations (PLs): Clearcutting of assigned mangrove areas, followed by single-species planting in a grid pattern (usually as monocultures, Fig. 23.1A.) for either (a) restoration after harvesting (illegal or legal, no subsequent logging allowed) or conversion (aquaculture, agriculture), (b) rotational harvesting by communities, or (c) commercial rotational harvesting (mostly for charcoal production) (FAO, 1985; Goessens et al., 2014). The harvesting of a forest parcel is either done once or on a strict rotation period (usually from 20 to 30 years) (Goessens et al., 2014).

FIGURE 23.1 Mangrove forests harvested and managed under the (A) plantation, (B) small-scale selective logging, and (C) large-scale selective logging silviculture regimes (locations: (A) Matang, Malaysia (Ahmad Aldrie Amir 2019 ©), (B) Baie D'Assassins, Madagascar (Daniel A. Friess 2020 ©), (C) Bintuni Bay, West Papua, Indonesia, (PT. BUMWI 2020 ©)).

Selective Logging (SL): Harvesting of selected areas or trees in a natural mangrove forest without altering the rest of the environment. The harvesting is either done during a defined rotation period or irregularly, and not necessarily in the same area. The previously harvested areas are left to regenerate naturally and may be assisted with additional planting or by leaving trees to supply propagules for forest recovery (also known as "seed trees"). This harvest method can be further categorized into the following:

Small-Scale Selective Logging (SSSL): Harvest areas are less than 1 hectare in size, and the harvesting is usually carried out by communities or individuals for self-sufficiency (firewood, poles, charcoal, construction material) (Alongi and de Carvalho, 2008; Saenger, 2002) (Fig. 23.1B).

Large-Scale Selective Logging (LSSL): Large-scale harvesting of selected mangrove forest stands of more than 1 hectare, forming either a cheese-hole pattern in a greater area or a large harvest block. This type of management regime is usually adopted for commercial purposes (for woodchip or charcoal production). The logging operations typically follow a strict rotation period (Sillanpää et al., 2017) (Fig. 23.1C).

As these different harvest practices extract wood from mangroves at various intensities, soil disturbance and the amount of wood debris left in harvest areas may vary greatly. Consequently, the physicochemical properties of the soil would also be influenced and vary according to harvest practices, which in turn affect mangrove regeneration rates.

Different soil components have different impacts not only on mangrove growth and forest structure but also on ecosystem functions such as carbon storage, water quality, and fisheries (Ewel et al., 1998a, 1998b; Alongi, 2009; Adame et al., 2012; Suárez-Abelenda et al., 2014). There is also a mutually influential relationship of soil with macrobenthos and insect communities and diversity (Cannicci et al., 2008; Feller, 1995; Feller and Chamberlain, 2007; Ozaki et al., 2000). In the maintenance of the overall mangrove ecosystem structure and function, the following soil indicators could be considered as the most critical ones:

Carbon (SOC or TC, %): Provides the essential building blocks for growth and overall productivity of mangrove forests (Alongi, 2009). It is also indicative of the carbon storage potential in the forest soil (Huang et al., 2017). Carbon can be measured either in the form of soil organic carbon (SOC, %) or total carbon (%).

Nitrogen (TN, %): An essential macronutrient that has shown to increase the photosynthetic rate or growth of mangrove seedlings (Alongi, 2009; Reef et al., 2010). It can be measured either as total nitrogen (TN) or ammonium (NH_4^+) which significantly impacts mangrove growth (Reef et al., 2010).

Phosphorus (AP, ppm, or TP, %): A macronutrient which is readily absorbed by mangrove roots and contributes to photosynthetic rates (Alongi, 2009). Phosphorus deficiency may cause limited growth and canopy development of mangroves (Lovelock et al., 2006; Chowdhury et al., 2019). It can be measured either as available phosphorus (AP) or total phosphorus (TP).

Soil Bulk Density (SBD, g cm^{-3}): A parameter that influences the establishment and anchorage of mangrove propagules (Salmo III et al., 2014) and root growth. Higher soil densities are shown to enhance root growth and thus increase water and carbon intake (Ola et al., 2018). However, a too dense soil may also impede the establishment of propagules (Yang et al., 2014).

Salinity (ppt): Usually a limiting factor in the mangrove and for any plant growth, as highly saline soils impede water intake and water balance, reducing photosynthesis potential (Alongi, 2009; Kirui et al., 2008) and, consequentially, stand growth.

Different mangrove species have varying tolerances to these soil indicators, eventually leading to diverse forests of mixed species and structures (Gleason et al., 2003; Alongi, 2009). The effects of forestry on the mangrove soils should be observed closely, especially in plantations, as soil dynamics can have a significant influence on forest growth, hydrology, erosion, and topography (Ewel et al., 1998b; Phillips et al., 2017; Ellison, 2019). In turn, soil indicators and dynamics are also significantly affected by other biotic and abiotic factors such as tidal regimes and adjacent habitats within the catchment which both act as source and sink of sediments and nutrients (Feller et al. 1999, 2002; Granek and Ruttenberg, 2008; Sasmito et al., 2020b).

A global overview of mangrove silviculture systems and associated soil dynamics is yet to be done. This chapter thus aims to evaluate the impact of various forest management practices on sediment dynamics in mangroves. The objectives are to (1) review the currently available data and literature on the impact of mangrove harvest regimes on soil properties, (2) analyze trends and any large differences in soil conditions between mangrove sites under different silviculture systems, and (3) present a case study of the effects of LSSL on mangrove sediment dynamics. General trends and gaps in research regarding the effects of harvest practices on mangrove sediment dynamics are highlighted. The findings from this chapter will be useful to guide policymakers for improving current mangrove sustainable forestry via minimizing their impacts to mangrove ecosystem services deliverability.

2. Global overview of soil dynamics and recovery in mangrove harvest regimes

Globally, there were 38 studies (out of the 147 initially found articles), carried out between 1986 and 2018, that directly or indirectly observe changes in the soil's physicochemical properties (e.g., carbon (C), nitrogen (N), phosphorus (P) contents, soil bulk density, and salinity) in harvested, damaged, or rehabilitated mangroves (Fig. 23.2, see Appendix I for literature

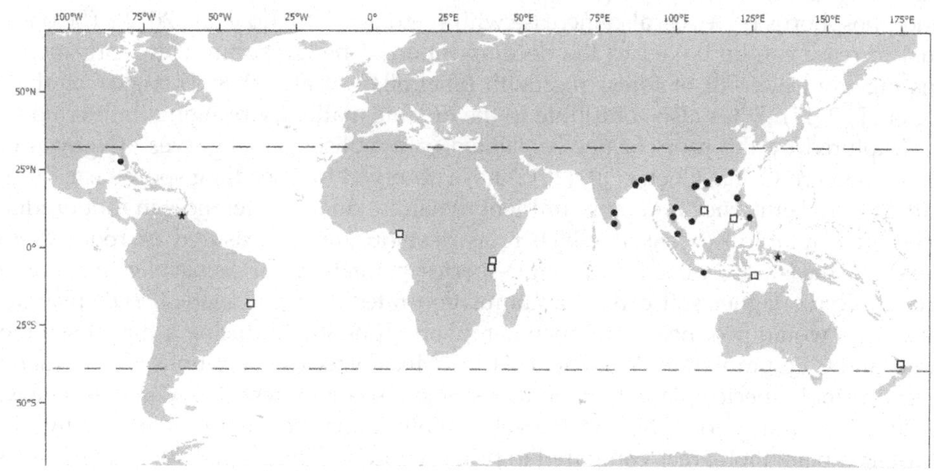

FIGURE 23.2 Global extent of selected study sites on mangroves under different harvest regimes and with available soil data (*dashed lines*: the boundary of mangrove ecosystems, world map from GADM 2018 © •: mangrove plantations (PLs), □: Small-Scale Selective Logging (SSSL), ★: Large-Scale Selective Logging (LSSL)). *Note:* Full list of studies, sites, and data in Appendix II. *According to Giri, C., Ochieng, E., Tieszen, L.L., Zhu, Z., Singh, A., Loveland, T., et al., 2011. Status and distribution of mangrove forests of the world using earth observation satellite data. Global Ecol. Biogeogr., 20(1), 154–159.*

review methods, Appendix II for datasets). Changes in soil conditions before and after harvest have been recorded and studied in plantations (Bhomia et al., 2016; Kathiresan et al., 2014; Sahu et al., 2016; Salmo III et al., 2014; Thongjoo et al., 2018) and forest plots subject to SSSL (Alongi and de Carvalho, 2008; Castillo et al., 2017; Lang'at et al., 2014; Sjöling et al., 2005). However, to date, there has been no comparison in a single mangrove area, or globally, of the impact of harvesting on soil and soil properties recovery (in C, N, P, soil bulk density, and salinity throughout time) between different harvest treatments. The great majority of the studies have only postharvest soil data, and do not present any preharvest or primary forest soil conditions for the study sites.

Currently, most of the literature on mangrove soil recovery and forest structure focus on plantation regimes (28 out of 35 articles with datasets, see Appendix II). Many sites under plantation regimes have ecosystem recovery and rehabilitation from previous land use (e.g., aquaculture) as the main objective rather than continuous wood extraction and forestry. The general observation in mangrove plantations is a gradual recovery of soil indicators postharvest to conditions similar to those seen in mature forests. Initial carbon stocks in younger stands are observed to be lower than mature stands (Chen et al., 2012), with soil bulk density increasing significantly (Grellier et al., 2017), probably due to felling activity and log transport. There is an initial C, N, and P loss following harvest that can be attributed to the loss and decomposition of root systems (Grellier et al., 2017), loss of litterfall, sediment acidification (Kathiresan et al., 2014; Shaifullah et al., 2008), and an increase in mineralization of organic matter (Kathiresan et al., 2014).

However, after this initial loss, observations show an increase in SOC with stand age (Nguyen et al., 2004; Salmo III et al., 2013) which can reach its highest levels at around 30 years postdisturbance (Pham et al., 2017). In Malaysian plantations, carbon burial efficiency in mangrove sediments (the rate at which carbon is fixed into the soil) goes "from 16% to 27% with increasing stand age" (Alongi et al., 2004; Kathiresan et al., 2014). Likewise, nitrogen (N) and phosphorus (P) levels also increase with stand age (Osland et al., 2012). The observed gradual increase can be linked to the decomposition of newly accumulated organic matter originating from litterfall of forest regrowth (Shaifullah et al., 2008; Thongjoo et al., 2018; Zhang et al., 2012). Tides also contribute to the organic matter by bringing in litter and sediment, "captured by the roots of newly planted or established mangroves" (Nguyen et al., 2004). In Southern China, Chen et al. (2012) have observed that seedling species can influence the carbon burial efficiency and other nutrient dynamics due to differences in root production efficiency (the rate and extent at which roots develop, often measured as root density or biomass). In the study, mixed and single-species plantations of *Sonneratia caseolaris* were observed to have higher soil carbon accumulation rates than *Sonneratia apetala* plantations, as the species would produce more pneumatophores, possibly inducing a higher sedimentation rate, and C accumulation (Chen et al., 2012). Mixed species plantations would, therefore, outperform single-species plantations in terms of soil recovery (especially in C levels) (Chen et al., 2012). Contrarily to C, N, and P levels, salinity initially increases significantly in soils of clear-felled plantation areas compared to primary forests. This can be attributed to the sudden opening of the canopy and sun exposure that increases evaporation (Kathiresan et al., 2014). As the canopy closes with the forest regrowth, upper soil temperatures are observed to decrease (Osland et al., 2012), which most likely results to lower salinity levels.

Studies observing impacts of SSSL on mangroves are few (7 out of 35 articles with datasets, see Appendix II), and show similar results to that of plantation systems. Initial losses in C, N, and P could be observed within the first 2 years postharvesting (Bernardino et al., 2018; Lang'at et al., 2014), followed by a gradual recovery. However, this initial loss is not always the case, as litterfall left by logging activities in Timor-Leste, for example, have induced an increase in nutrient concentrations (C, N, P compounds) in harvested mangroves areas (Alongi and de Carvalho, 2008). The size of the harvest area, which is usually small in an SSSL system, might influence this, as sources of litterfall (from the surrounding forest) are still at closer proximity (Alongi and de Carvalho, 2008). Soil salinities in mangroves under SSSL systems, such as the ones in Tanzania (Sjöling et al., 2005), do not necessarily increase postharvest, most likely because the harvested areas are smaller than plantations, keeping some canopy cover over cleared areas.

Out of the three defined harvest regimes, LSSL is the most understudied, with only one published study site in Venezuela, described by Rivera-Monroy et al. (2004). The mangrove was subjected to a completed test-harvest and not intended for a long-term forest management plan and a comparison of soil and forest conditions between forest plots aged 15–25 years and natural forests was done. There were no significant differences in measured soil indicators (C, N, P) between the regenerated and natural mangroves, suggesting a successful recovery (Rivera-Monroy et al., 2004). There was, however, a gradient of increasing nutrient concentrations from the river to inland, especially with P, indicating a significant influence of tidal inundation on the soil indicators (Rivera-Monroy et al., 2004). A potential case study for impact of LSSL on soil dynamics, demonstrated in this chapter (see Section 4), would be the PT. Bintuni Utama Murni Wood Industries (PT. BUMWI) mangrove concession (West Papua, Indonesia), where forest structure recovery and some of soil C components dynamic has been described earlier by Sillanpää et al. (2017) and Sasmito et al. (2020a), respectively.

In general, for mangrove harvest regimes, there is little long-term data on the progression of soils and soil nutrients during a mangrove recovery process of a single site, apart from studies such as Salmo III et al., 2013, Osland et al. (2012), and Ngole-Jeme et al. (2016). Most of the studies observed changes in mangrove soil conditions and nutrients only within the top 20 cm of the soil column (Alongi and de Carvalho, 2008; Bhomia et al., 2016; Ngole-Jeme et al., 2016; Sjöling et al., 2005), as the most significant changes in physical–chemical variables are expected to occur on the surface (Osland et al., 2012). Data from these soil depths are used for analysis for the subsequent sections of this chapter. However, some changes have been observed in lower soil horizons with reductions in C, N, and P concentrations (Thongjoo et al., 2018). Hydrological regimes have been rarely observed and described in the found studies, with almost no data, so a global analysis cannot be done in this chapter. Under all silvicultural systems, the major factors that influence soil indicators and nutrients are the topography, tidal regimes, the newly established mangroves (seedlings), and the surrounding forest structure of the harvest area that acts as a potential seedbank. A significant change in any of these biophysical variables can alter greatly the dynamics of the different soil indicators and recovery of the mangrove stand.

3. Postharvest impacts on mangrove soils in different harvest regimes

For nearly all soil indicators (C, N, P), we observed an immediate decrease occurring post-logging, followed by a recovery process that might take a couple of decades. The size of the harvest areas might contribute to the recovery speed as the soil in plantation systems seems to recover at a slower pace compared to other management regimes. Plant–soil interaction within the harvested mangroves can also be observed as stem density and volume correlate with SOC, AP, and salinity levels. Since all the comparisons made are driven from a global dataset with few sites and small sample populations, the ranges of values and correlations tend to be weak. Therefore, testing soil and forest structure conditions for different silviculture systems in similar sites is imperative. The global relationships of other forest structure variables such as diameters, heights, and species diversity with soil nutrients should be carried out with a larger sample population and after further studies will be published.

3.1 Comparison of soil conditions in different mangrove silviculture regimes

Overall, the impacts of harvesting for most of the soil indicators are low across three main, from recently harvested areas (0–2 years old of cleared forests), secondary forest (recovering forest in previously harvested areas), to the natural forest (unlogged stands) (Fig. 23.3). The lack of significant differences among silviculture regimes can be attributed to the scattered data distribution as well as the diversity of climatic, hydrology, and geomorphological conditions in different sites. Our global overview detailed in Fig. 23.3 suggests that there are few studies assessing soil conditions between comparable regimes. Therefore, impacts may be driven by confounding factors (i.e., hydrogeomorphic setting, Rovai et al., 2018; Sasmito et al., 2020a) rather than silviculture regimes.

SOC across the three silviculture systems shows significant differences, with selective logging regimes having a higher mean carbon content (SSSL: 7.4% ± 3.6%, LSSL: 7.1% ± 1.3%). In plantations, secondary (5.44.2% ± 6.6%) and natural forests (4.9% ± 4.4%) fall within the same range and are higher on average than postharvest areas (2.1% ± 2.3%) (Fig. 23.3A). Overall, there is a loss of SOC in all systems (between 19% and 58% decrease), which eventually recovers when the forest reaches maturity (assumed to be the monitored natural forest). The recovery success must be site-specific as some areas might take longer (≥40 years) to achieve the baseline range of soil conditions as observed in the Matang Mangrove by Adame et al. (2018).

AP levels in LSSL (Fig. 23.3B) show higher levels ($P < .05$) postharvest soils (16.78 ± 5.55 ppm), than soils of secondary (10.09 ± 3.76 ppm) and natural forests (10.05 ± 3.34 ppm), which have a lower mean AP. Higher phosphorus levels postharvest may be caused by the decomposition of leaves, branches, and stumps left by the harvesters (Alongi and de Carvalho, 2008). There is no significant difference in AP levels in primary and secondary stands in an LSSL system ($P > .05$), suggesting that phosphorus concentrations can return to regular levels due to the reaccumulation of organic material originating from mangrove regrowth (Nguyen et al., 2004). Unfortunately, AP levels between forest states could not be compared for plantations (PL) and SSSL because of small sample populations (PL: n = 0,13,2 and SSSL: n = 2,4,6 for different states).

Looking at TN levels, there is a significant difference ($P < .05$) in concentrations between forest states under the plantation system (postharvest: 0.10% ± 0.16%, secondary: 0.29% ± 0.27%, natural: 0.13% ± 0.08%) (Fig. 23.3C). Only a few studies have been reported for TN in SSSL and LSSL sites (Alongi and de Carvalho 2008; Rivera-Monroy et al., 2004). In Alongi and de Carvalho (2008), there is no significant difference in TN levels between forest conditions in a SSSL system, as there are still important N sources available, mostly from the

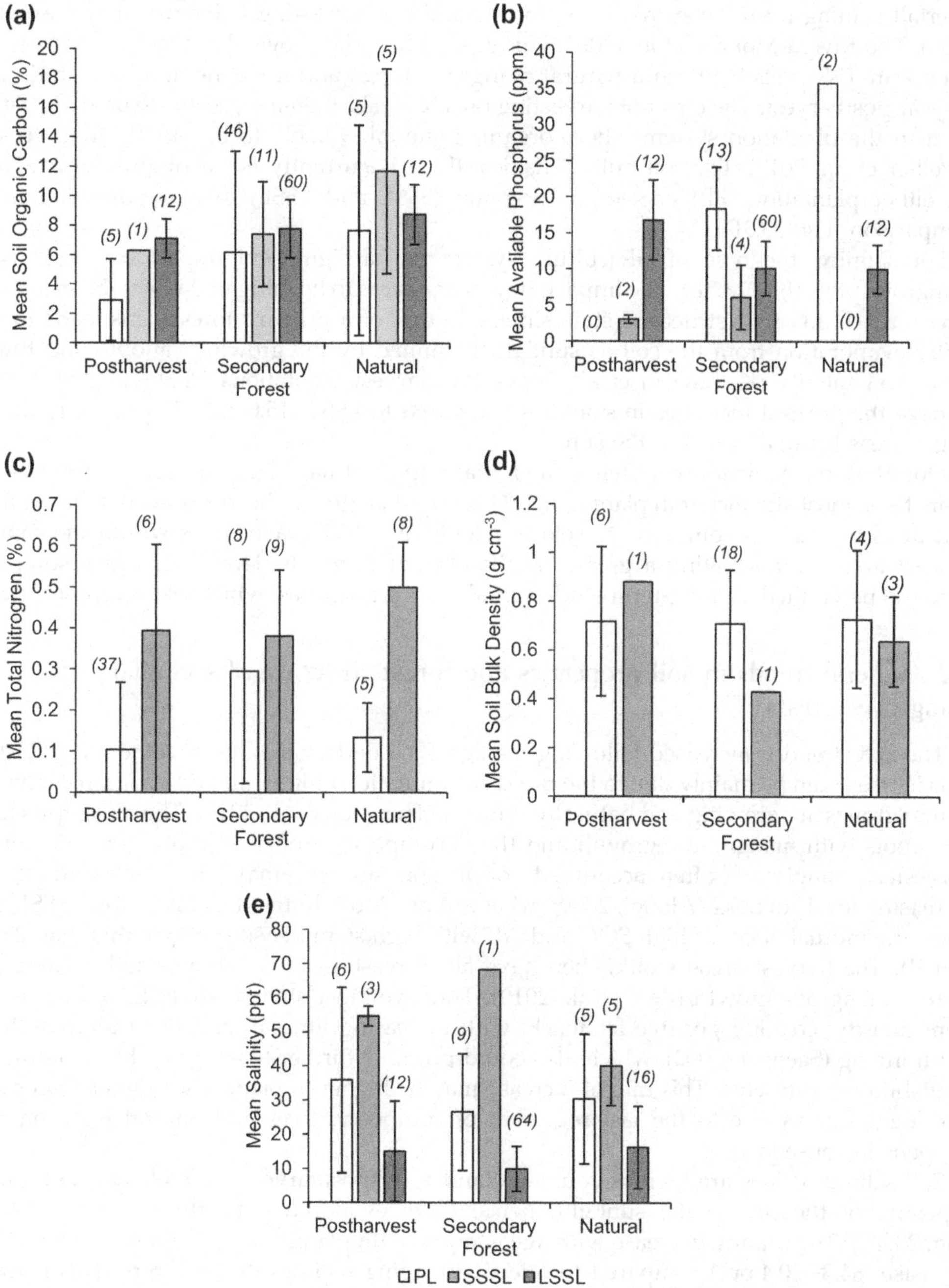

FIGURE 23.3 Global differences in soil conditions for (A) soil organic carbon, (B) available phosphorus, (C) total nitrogen, (D) soil bulk density, (E) and salinity among newly harvested forests (postharvest), mature secondary forests (≥1-year-old stands), and natural forests (never logged) under different forest management regimes (PL: Plantation, SSSL: Small-Scale Selective Logging, LSSL: Large-Scale Selective Logging). *LSSL contains secondary data from Bintuni Bay, West Papua (PT. BUMWI 2020 ©). **(E) has additional data from Venezuela (Rivera-Monroy et al., 2004, see Appendix II). Note: Number of sample areas per category is in brackets.

litterfall coming from the surrounding forests and the harvesting (Alongi and de Carvalho, 2008). The Rivera-Monroy et al. (2004) study of a LSSL site shows that there is no great difference in TN levels between a natural mangrove forest and a regenerating forest 15- and 25-year postharvest. The effect of harvesting on TN levels, therefore, seems to be more significant in the plantation system where organic material is likely to be lost through erosion (Grellier et al., 2017). For soil bulk densities, there is currently not enough data available for either plantation (PL) or selective logging (SSSL and LSSL) to make any significant comparison (Fig. 23.3D).

For salinity, the type of silviculture system has a significant impact on postharvest mangrove plots (Fig. 23.3E). The impact, however, seems to be mitigated when the secondary forests reach a certain structural state similar to that of a primary forest. This leads to less water evaporation from the soil as sunlight is limited by the growing canopy, and thus a lower soil salinity (Kathiresan et al., 2014). Postharvest plantations (35.9 ± 27.2 ppt) seem to have the greatest increases in salinities compared to LSSL (15.0 ± 14.7 ppt), likely due to larger areas being exposed to the sun.

Out of all the silviculture systems, the initial impact of harvesting on soil parameters appears to be most significant in plantations. This might be due to the lower availability of litter and dead biomass as compared to selective logging, which leaves trees within the smaller harvest areas, thus creating a greater availability of nutrients. However, this assumption needs to be verified by long-term studies in all harvest regimes, which are severely lacking.

3.2 General trends in soil properties and forest structure of secondary mangrove forests

The SOC trend is increased following the age of secondary mangrove stands (Fig. 23.4A). This increase can be mainly due to the rapid accumulation of litter and dead organic material as mangroves are growing and naturally thinned (Thongjoo et al., 2018). The development of new roots with mangrove regrowth and the decomposition of dead roots from previously harvested mangroves, often accounted for in soil surveys, may also contribute to the increasing levels of SOC (Alongi, 2009; Adame et al., 2018). In the data taken from SSSL systems, the mutual effect of high SOC and AP with high stem density reflects this (Fig. 23.5A and B). The harvest areas would then have an increase in N and P availability, inducing further mangrove growth (Reef et al., 2010). Thus, younger stands would have high initial stem density, creating positive feedback, with increasing litterfall and dead biomass from self-thinning (Saenger, 2002), which also could promote further forest growth due to higher availability of nutrients. This rate of increase may be higher in plantations rather than selective logged areas due to the fast regrowth of competing trees and natural thinning that happens in consequence.

Soil salinity values are high at younger stand ages (postharvest). It is likely due to more exposure of the soil to the sunlight, hence more evaporation (Kathiresan et al., 2014) (Fig. 23.4C). The salinity decrease with age is rapid, with plantations exhibit more significant decrease (64.3–20.4 ppt) compared to selective logging regimes (54.7–33.8 ppt). This trend suggests that forest development stages may control the degree of soil surface sunlight exposure and, thus, soil salinity.

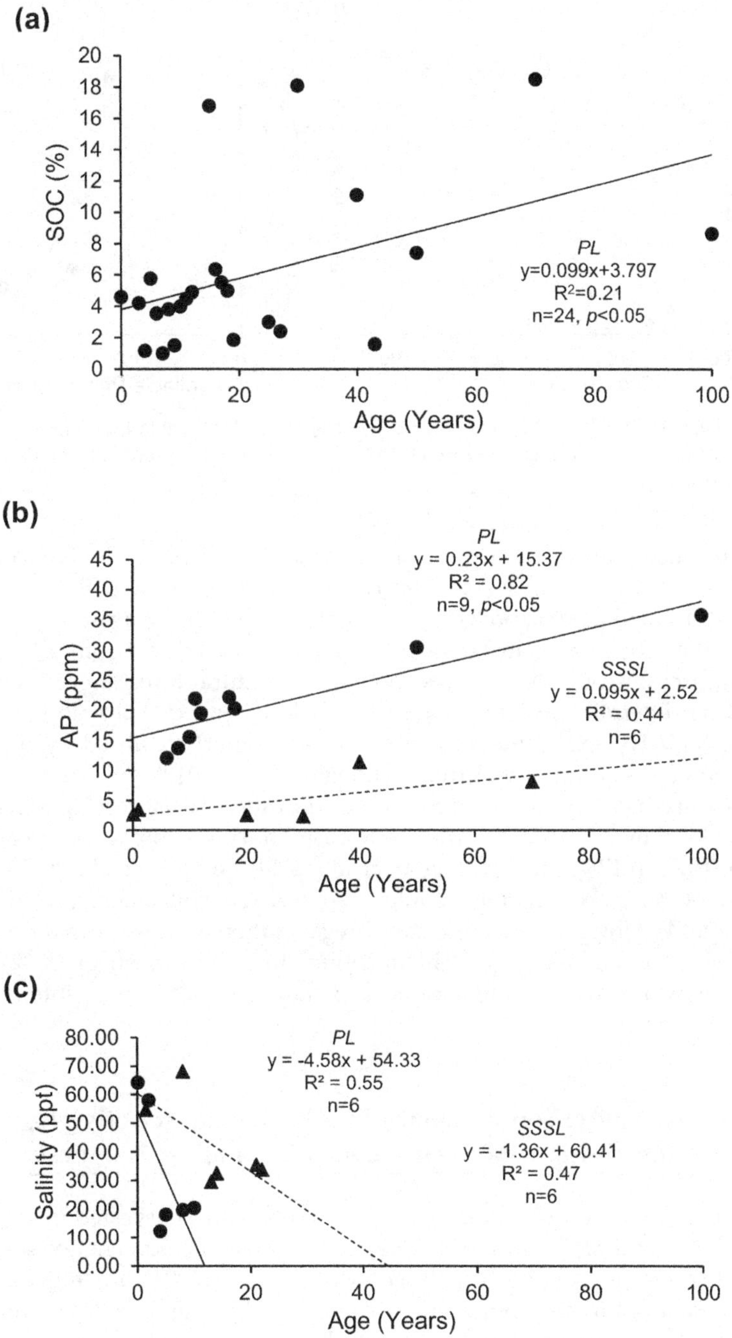

FIGURE 23.4 Global changes in soil characteristics in different aged forest stands subjected to plantation (PL, •, –) or small-scale logging (SSSL, ▲, —); (A) Mean Soil Organic Carbon (%), (B) Mean Available Phosphorus (ppm), and (C) Mean Salinity (ppt).

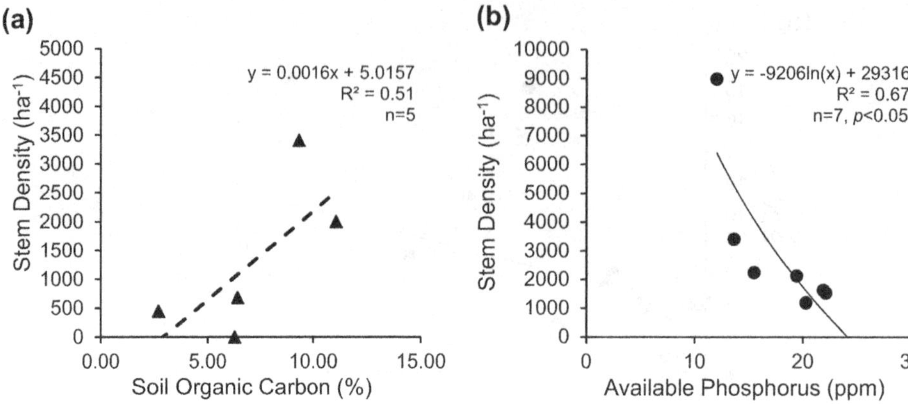

FIGURE 23.5 Global Influence of Soil Organic Carbon (%) (A) and Available Phosphorus (ppm) (B) on stem density in postharvest mangrove forests under different forest management regimes; (PL (•): Plantation, SSSL (▲): Small-Scale Selective Logging.).

For nitrogen concentrations, no significant conclusions can be made due to the scarcity of the data in different silviculture systems. In two studies on SSSL and PL systems, however, Alongi and de Carvalho (2008) and Thongjoo et al. (2018) observe a rise in N concentrations postharvest. In the SSSL system in Timor-Leste, the increase in TN (from 0.47% to 0.53%), as seen in the comparisons of harvested, secondary, and natural forest areas (Fig. 23.3C), most likely comes from litterfall caused by harvesters (Alongi and de Carvalho, 2008). In a Thai plantation, the high TN soil could be attributed to the litterfall caused by the self-thinning and competition between individual trees (Thongjoo et al., 2018).

Soil bulk density has been scarcely monitored in different harvest systems, but the few studies available showed an immediate increase in soil compaction postfelling (from 0.80 ± 0.08 to 0.97 ± 0.17 g cm^{-3}) (Lang'at et al., 2014; Grellier et al., 2017). The gradual decomposition of roots and stumps could also weaken soil stability, inducing erosion (Lang'at et al., 2014). However, given to rapid regeneration and mangrove propagule establishment, the system would likely be resilient enough to recover (Saenger, 2002). Bulk density would likely follow a recovery similar to SOC as they are both heavily interlinked (Phillips et al., 2017).

4. Forestry case study: soil dynamics in a large-scale selective logging regime in the Bintuni Bay mangrove, West Papua, Indonesia

Most of the current literature on mangrove soil dynamics in harvested mangroves concentrates on plantation and SSSL sites (Fig. 23.1). By contrast, assessments of soil dynamics in large selective logging systems, such as Rivera-Monroy et al. (2004), are limited. Furthermore, observations of changes in soil conditions over time and within a soil column in harvested mangroves of such systems have been understudied. In this section, we therefore present a case study of a LSSL system in West Papua, Indonesia, not only looking at the soil indicators over various stand ages but also up to 1m depth, which beyond what most of the literature concentrates on (20–30 cm soil depth, regardless of management system), except for Adame et al., 2018.

4.1 General description of large-scale selective logging in Bintuni Bay

In the southern part of Bintuni Bay, West Papua, Indonesia, which has one of the largest contiguous mangrove forests in the world (Giri et al., 2011; Pribadi, 1998; Sillanpää et al., 2017; Simard et al., 2019), selective logging has been carried out for the past 30 years for woodchip production by a local forestry company (PT. Bintuni Utama Murni Wood Industries). The rotational harvesting is conducted within an 82,120-hectare concession area, following a keyhole harvest method (Pribadi, 1998; Sillanpää et al., 2017). Through this method, forests are harvested in keyhole-shaped areas of 1 to 5-hectare size, which is mostly due to the regulations of greenbelts (from waterbodies and other harvest areas), and operational capacity of manual harvesting. The shape of harvest areas might differ as of 2019, as the company has developed mechanical logging operations where skidders are used.

Timber harvesting is targeted to areas with *Rhizophora apiculata, Bruguiera gymnorrhiza,* and *Bruguiera parviflora* which are the dominant mangrove species (Sillanpää et al., 2017). Secondary forest structure and diversity have been predicted to recover to similar levels to that of natural, untouched forests within a 30- to 50-year period (Sillanpää et al., 2017). The soil in the natural mangrove environment can be classified as silty clay (texture), acidic, and regularly water-logged. The concession is thus composed of a mix of different forest cohorts based on densities, diameters, and species, as they are at varying stages of regeneration post-harvest (as explained by Sillanpää et al., 2017). The following sections show observations made by the PT. BUMWI forestry department on soils in forest plots of different secondary forests impacted by manual selective logging and baseline forests (Fig. 23.6).

4.2 Soil recovery in different aged secondary and primary mangrove forests of Bintuni Bay, West Papua

In the secondary mangrove forests, SOC means (%) was altered during the first years following harvesting (falling to 7.4% ± 1.3% at 1-year postharvest), following by a recovery at around 25 years postharvest, with levels (8.8% ± 1.6%) resembling those of primary forests (8.9% ± 1.9%) (Fig. 23.7A). The decrease of SOC means from top-70 cm soil can be linked to two different reasons, (1) harvested and primary forest sites are located in different hydrogeomorphic settings and thus leading to an SOC variation (Sasmito et al., 2020a), (2) soil surface at harvested forest site may have experienced some physical mixing followed by fluvially driven erosion (Arias-Ortiz et al., 2020), specifically when the sites are located between two big tidal creeks (Fig. 23.6). The immediate loss of SOC in the topsoil could be attributed to a combination of erosion (Granek and Ruttenberg, 2008) and the establishment of seedlings that develop a new root system that further break down the remaining organic matter (Nguyen et al., 2004). The SOC dynamics might then follow the same trend as observed by Nguyen et al. (2004) in mangrove plantations, where SOC quickly recovers with new accumulation of organic matter coming from the vegetation regrowth. Indeed, plots in 10-year-

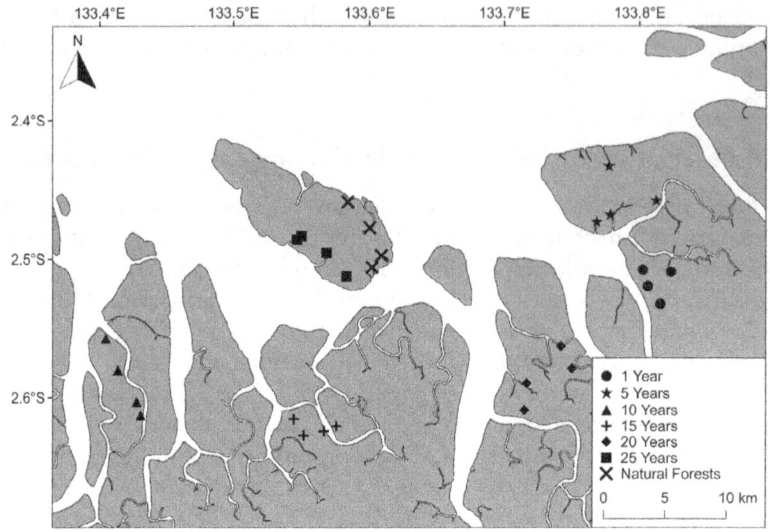

FIGURE 23.6 Mangrove Forest Survey plots in the PT. BUMWI Concession, Bintuni Bay, West Papua, Indonesia (Sillanpää et al., 2017).

old secondary forests have already the same range of SOC (with a mean of 8.4% ± 1.9%) to that of primary forests (8.9% ± 1.9%). This could be explained by the high litterfall and dead biomass originating from competing young trees (self-thinning observed by Sillanpää et al. (2017)) and decomposing stumps and dead seed trees left from the previous harvest in the first 15 years. Primary forests have a large range of structures and densities, even if they are within the same tidal range (intertidal). The reasons for this large range even within the same tidal regime might be the regeneration potential (availability of propagules, soil nutrient availability), variance interaction between topography and hydrology (hydrogeomorphic setting).

AP appears to increase significantly ($P < .05$) postharvest (15.3 ± 3.6 ppm in 1-year-old stands) and then decrease to levels that are not significantly different from primary forests (12.4 ± 2.4 ppm in 25-year-old forests, and 10.9 ± 2.8 ppm in primary forests) (Fig. 23.7B). The variation can be attributed to the new accumulation leaves, branches, bark, stumps, and roots of harvested mangroves, which provide a "flush" in phosphorus and other nutrients (Alongi and de Carvalho, 2008).

There is a large range of NH_4^+ concentrations (ranging 15.79–43.29 ppm) regardless of stand age in secondary forests, which suggest that the logging impact is limited to the first 5 years postharvest (Fig. 23.7C). It is worth to note that the widest range of value is within the 1-year postharvest, suggesting an influence of increasing organic matter from logging waste (increase in nitrogen) (Alongi and de Carvalho, 2008) and the possible influence of erosion (resulting in loss of nitrogen) (Salmo III et al., 2013).

The immediate impact of harvest can be seen in salinity as well, with a large range of high salinities in 1-year-old secondary stands (15.0 ± 14.7 ppt), which eventually stabilize to a range (9.9 ± 1.7 ppt in 25-year-old forest) akin to primary forests (9.9 ± 4.3 ppt) (Fig. 23.7D). The high

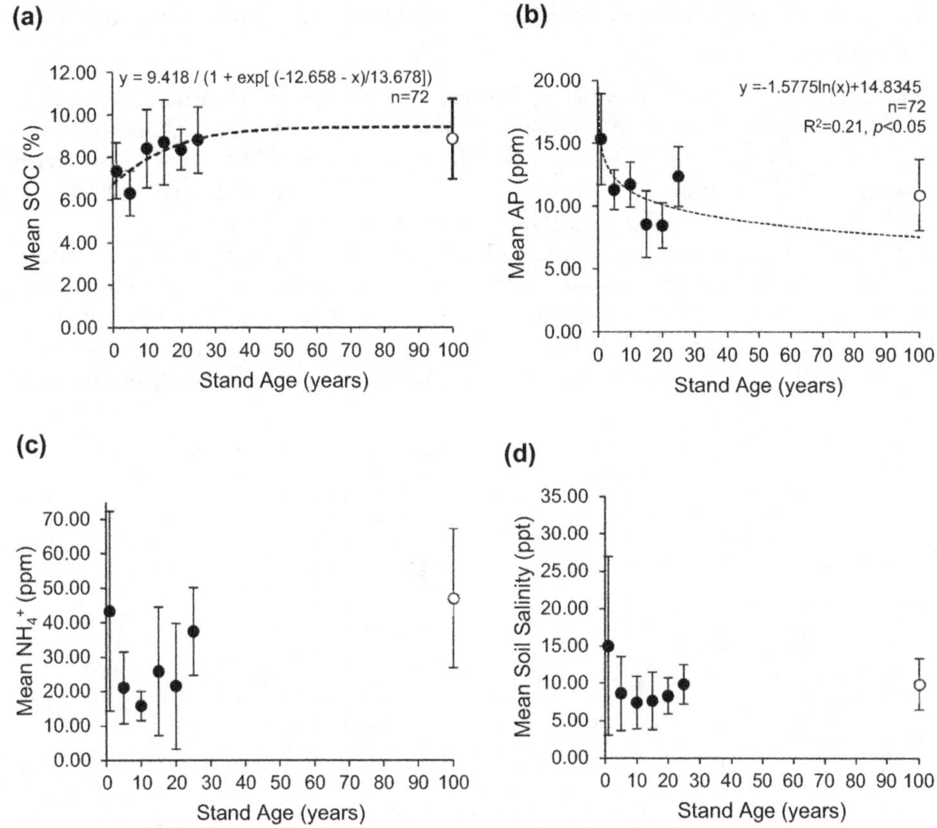

FIGURE 23.7 Changes in soil characteristics in the topsoil (30 cm) of different aged secondary (•) and primary (o) natural forest stands in Bintuni Bay, West Papua; (A) Mean Soil Organic Carbon (%), (B) Available Phosphorus (ppm), (C) Ammonium (ppm), (D) Mean Soil Salinity (ppt) (PT. BUMWI 2020 ©). Note: Natural forest stands are primary "untouched" stands that are assumed to be 100+ years old.

salinities in younger stands could be caused by the open canopy and sunlight, which dries the soil and seawater (Alongi and de Carvalho, 2008; Kathiresan et al., 2014; Salmo III et al., 2013). The high variance in salinities in the 1-year-old stands might be due to the spatial variance in disturbed hydrology, elevations, and remaining vegetation present within a harvested area (Rivera-Monroy et al., 2004).

Overall, the observed impact of logging in Bintuni is significant on most soil indicators (SOC, AP, and salinity). Despite sudden changes in physicochemical conditions postharvest, all soil indicators in maturing secondary forests return to a range of concentrations that are equal to what is observed in natural forests within a 25-year period. The mangrove trees and propagules seem well adapted to the changes in the soil as the seedling establishment and vegetation regeneration is considered successful (Sillanpää et al., 2017).

4.3 Relationship between soil indicators and mangrove forest structure in a selective logging system

When observing possible relationships between mangrove forest structure and soils, significant trends appeared between total stem densities (all trees ≥5 cm DBH) and total tree volumes. There is a large range of high tree densities at lower SOC concentrations (at 4% SOC, mean ± SE here after, 7247 ± 6857 trees ha^{-1}), however, at higher SOC levels (8%, 10%, and 12%) the tree densities and their range are lower (5347 ± 1227, 3532 ± 302, and 4073 ± 1582 trees ha^{-1}, respectively) (Fig. 23.8A). The same trend is seen with ammonium as forest soils with large stem densities seem to have low NH_4^+ concentrations in their soils (7130 ± 1457 and 7738 ± 1306 ha^{-1}, at 10 and 20 ppm, respectively). Stem densities seem to achieve a stable range (between 1042.8 and 5346.6 trees ha^{-1}) at higher NH_4^+ concentrations (30 ppm onward) (Fig. 23.8B). For phosphorus, the pattern is the same, although there is a

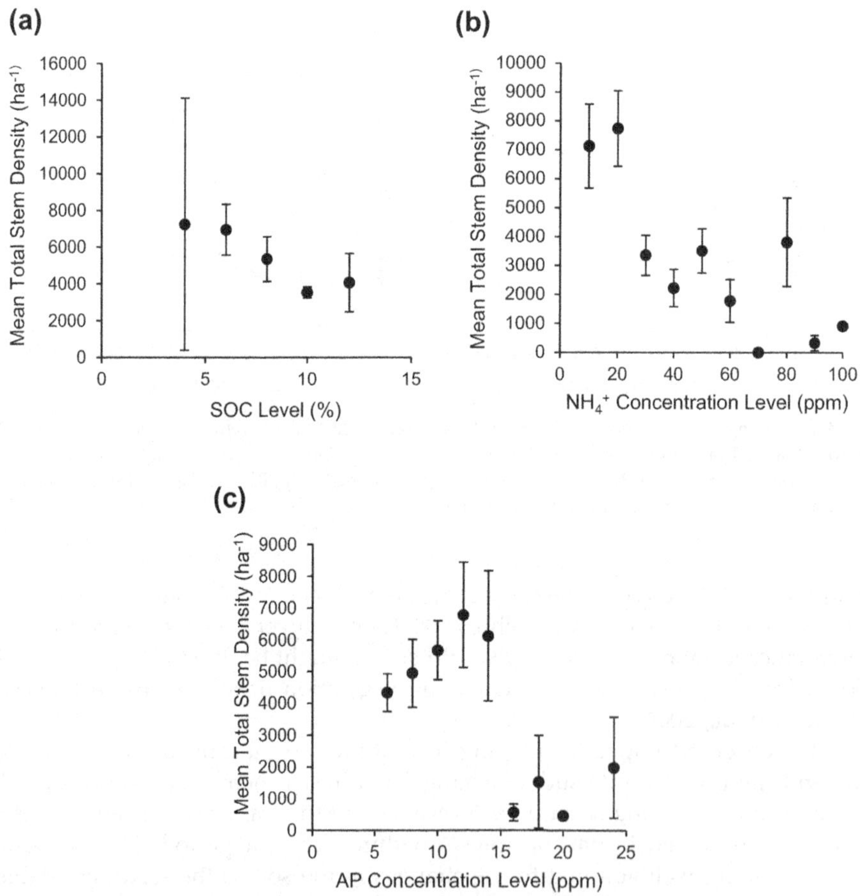

FIGURE 23.8 Correlations between mean Stem Density of mangrove stands and (A) Mean Soil Organic Carbon (%), (B) Ammonium (ppm), and (C) Available Phosphorus (ppm) (with Standard Errors, S.E.) in topsoil (up to 30 cm depth) of Bintuni Bay, West Papua, Indonesia. (PT. BUMWI 2020 ©).

distinct difference between low stem density and high stem density forests, with ~15 ppm AP being the breaking point (Fig. 23.8C). From this point, there are high stem densities at lower AP levels (between 3744 and 8435 trees ha^{-1} at the 5–15 ppm range) and lower stem densities (between 65 and 3575 trees ha^{-1}) at higher AP levels (15–25 ppm).

The trends in C, N, and P could be explained by lower density forests being the older secondary forests that have accumulated organic matter through litterfall and soil accretion, hence SOC and NH_4^+, on the topsoil layer over time (Alongi et al., 2002; Nguyen et al., 2004; Salmo III et al., 2013; Kathiresan et al., 2014; Thongjoo et al., 2018). High stem density, younger secondary forests, would still be influenced by the lack of organic matter and sediment acidification following harvest, which in turn produces phosphorus deficiency (Kathiresan et al., 2014). No distinct patterns were observed between salinity concentrations in soil and forest structure.

Like stem density, the same relationship between soil indicators and forest volumes is observed. Total volume seems to increase with SOC (153.0 ± 37.3 and 250.7 ± 34.3 m^3 ha^{-1} at 6% and 10% SOC, respectively) (Fig. 23.9A). A higher above-ground biomass, and therefore carbon, could increase the below-ground biomass, and vice versa. Total volume of a forest may

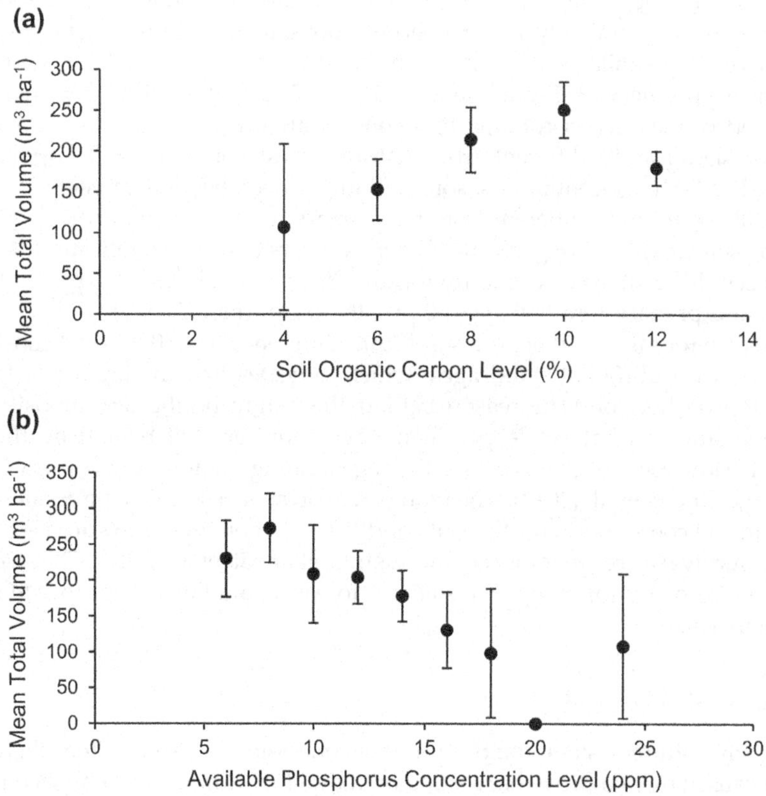

FIGURE 23.9 Change in Total Volume in different levels of (A) Mean Soil Organic Carbon (%) and (B) Available Phosphorus (ppm) (with Standard Errors, S.E.) concentration levels in topsoil (up to 30 cm depth) of Mangrove stands of Bintuni Bay, West Papua, Indonesia (PT. BUMWI 2020 ©).

decrease with a higher AP and vice versa (272.7 ± 48.2 and 130.7 ± 53.1 m^3 ha^{-1} at 8 and 16 ppm AP, respectively) (Fig. 23.9B). Lower volume plots are again from the younger stands that may produce a lot of biomass through self-thinning. The accumulation of SOC over time is then reflected in the older stands with higher volumes (and higher topsoil values).

The ranges of stem densities volumes observed in different soil nutrient levels demonstrate the adaptability of mangrove forest to various soil conditions. Stem density and volume are the only forest structure variables that currently demonstrate any pattern with the designated soil indicators. The major conclusion we can currently draw from the plant–soil observation is a lower variation in forest structure in higher SOC concentrations, which reflects the condition of a maturing secondary stand or untouched natural forest.

4.4 Soil dynamics against depth in secondary and primary mangrove forests

When observing SOC, AP, NH_4^+, and salinity concentrations with increasing depth, there is not much significant variance within each forest stand age, except for the topsoil (first measurement between 0 and 15 cm depth) ($P < .05$). There is, however, a significant difference ($P < .05$) in soil conditions between different stand ages throughout all soil segments for almost all soil indicators. This is the case for SOC where 5-year-old stands (e.g., 0–15 cm depth, 6.25% ± 0.34%) have an initially lower concentrations than natural forests (8.74% ± 0.59%) but then recover to a similar state in 25 years (8.54% ± 0.49%) (Fig. 23.10A and C). Yet, AP seems to increase postharvest (5-year stands: 10.5 ± 0.7 ppm at 0–15 cm depth) and stabilizes around the end of the harvest rotation to a concentration (25 year stands: 12.3 ± 1.3 ppm at 0–15 cm) not significantly different from natural forest soils (10.1 ± 0.9 ppm at 0–15 cm) (Fig. 23.10B). For NH_4^+ concentrations, soils still might need beyond 25 years to recover fully as they are still significantly different from the primary forest (except for the 0–15 cm soil horizon that has similar values) (Fig. 23.10C). There appears to be no significant difference in soil salinity between different aged secondary forests (25-year stand: 9.9 ± 0.5 ppt, at the 0–15 cm soil horizon) and primary forests (9.9 ± 1.2 ppt, 0–15 cm) (Fig. 23.10D).

Overall, except for salinity, there is a significant difference ($P < .05$) in soil conditions along all horizons between different stand ages, which indicates that the impact of forestry goes beyond the topsoil horizon. The reason behind this might be the decomposition of roots, bark, and tree stumps (that go deeper than 30 cm) that are left behind by the harvesters and act like a "slow release fertilizer," for the regenerating mangrove, as observed by Alongi et al. (2000) and Zhang et al. (2012). The mangrove forest also seems to be resilient to the current change in soil conditions, and the soil conditions do not appear to surpass the threshold of mangrove survival as regeneration is successful (Sillanpää et al., 2017). The strict condition for a successful recovery for mangrove soil and forest would be an appropriate rotation age and harvest area size.

4.5 Summary of case study

In a LSSL silviculture system, the Bintuni Bay concession has shown that there is a significant impact caused by mangrove harvesting on the soil. This impact can be seen immediately for all soil indicators within the first 5 years postharvest. However, measurements presented previously show that, given a correct rotation, the mangrove soil recovers eventually to a

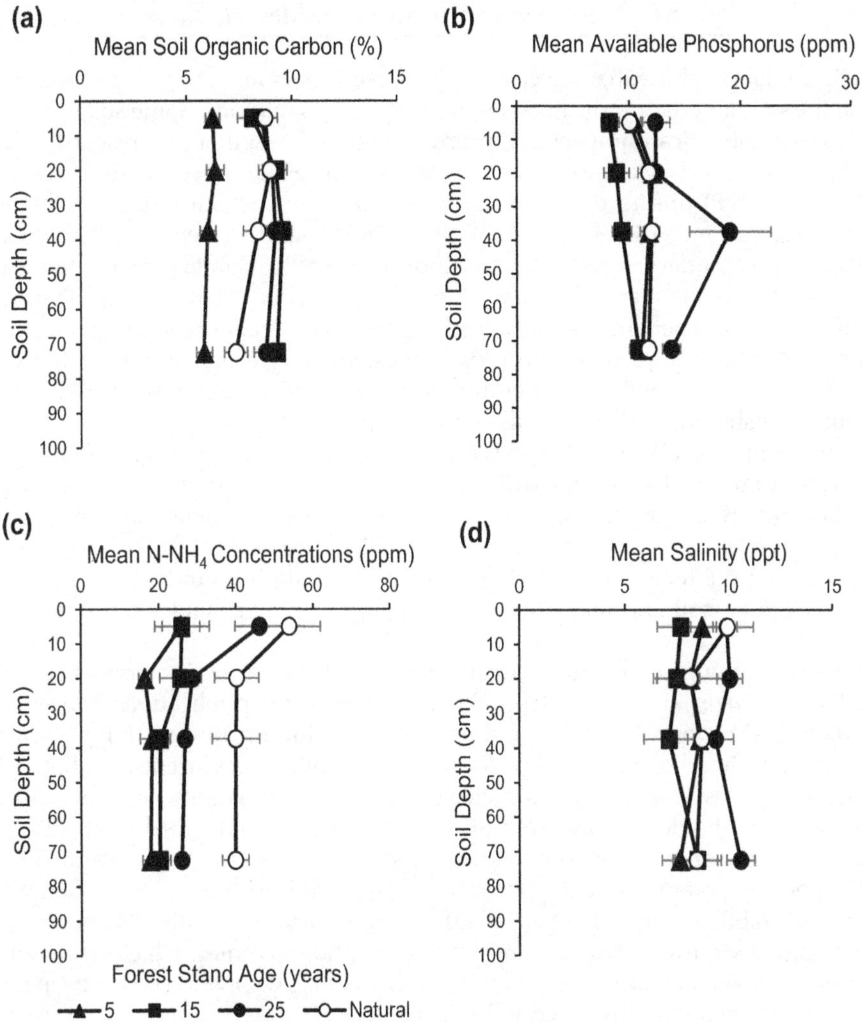

FIGURE 23.10 Change in Mean Soil characteristics with increasing soil depth among different aged stands in Bintuni Bay, West Papua, Indonesia (A): Soil Organic Carbon (%), (B) Available Phosphorus (ppm), (C) Ammonium (ppm), and (D) Salinity (ppt) (with Standard Errors, S.E.) (PT. BUMWI 2020 ©).

similar range of conditions as observed in primary forests. Direct correlations between forest structure and soil conditions are difficult due to the vast range of forest structures present in the concession (Erftemeijer et al., 1989; Sillanpää et al., 2017). The mutual influence of forest structure and soil conditions in harvested areas has not been studied and is seldomly studied in general. Further measurements would be required to isolate the causes of such changes to specific forest structures and by creating a "forest structure classification" to see any significant differences in soil conditions. In addition, the impact of harvesting on mangrove soil could be seen up to 1m depth, way beyond the usual 20–30 cm analyzed in previous studies.

5. Overview and future considerations

Overall, we can conclude that selective logging regimes in mangrove may cause a reduced impact on the soils and their indicators, as compared to plantations. Some soil indicators such as salinity have a significant impact in limiting mangrove growth and forest structure within all three harvest regimes. Mangrove forest soils experience an immediate decrease in soil nutrients (such as CNP) due to decomposition caused by oxygenation of soil (Adame et al., 2018), and decrease in sediment accumulation (Granek and Ruttenberg, 2008). Generally, in more mature secondary forests the soil indicator would stabilize or fluctuate within a certain range that is equivalent of primary forests in each site. We also realize that mangrove harvesting impact goes beyond the 20 cm soil depth that is usually monitored. Due to decomposition of roots and stumps that are left by harvesters, soil nutrient concentrations can fluctuate for at least 25 years with a regenerating forest. It is recommended that all soil sampling for soil nutrients should be done at least down to 1 m depth.

One of the biggest challenges in making comparisons between mangrove forest stands is the scarcity of data availability for different harvest regimes, be it plantation or selective logging data (for SSSL and LSSL). There have been many articles and locations under different regimes with unpublished data (most likely because the data are not deemed important for the published article). A key recommendation would be for the authors to publish all the data (soil and forest structure) in supplementary tables that would complement their article.

Many research groups choose to measure only one of the various components of each soil nutrient for their own study objective. For example, many publications collect data and analyze data on N concentration, TN, ammonium (including Bintuni Bay) or nitrates, to quantify nitrogen fluxes or storage. For future studies, other measurements should be done to estimate soil quality and conditions such as SOC, NH_4^+, AP, salinity, and soil bulk density. All these variables should be measured at least to 1 m depth (Howard et al., 2014; Adame et al., 2018). To further monitor soil conditions, soil grain size should be studied further as it is an important parameter that affects mangrove growth due to its influences on soil anchorage and stability (Salmo III et al., 2013). Soil redox has shown also to impact mangrove growth (Alongi, 2009; Ewel et al., 1998b). In addition, tidal regimes, which heavily influence sediment deposition and nutrient availability (Feller et al., 2002), have rarely been studied in the context of soil dynamics in harvested mangroves.

The lack of observed forest structure–soil correlations in this chapter further shows that global trends are hard to determine as forests dominated by different species and sites have different tolerances to soil conditions (e.g., *Avicennia marina* can grow in highly saline conditions compared to *B. gymnorrhiza*). Hence, regional studies, once there are enough data, would perhaps be more indicative than a global review. A great area for such a study would be Southeast Asia, as the region's mangroves are subjected to various human activities, especially logging. We would further recommend continuing mangroves research on different harvest and management regimes within sites, especially in locations such as Micronesia, Thailand, and Vietnam, where there are both selective logging and plantation systems.

Appendix

I. Literature review methods

The literature review on soil recovery in harvested mangrove areas was conducted following Ilstedt et al.'s (2007) method, using search strings on online databases and then sieving through the produced articles (Figs. 23.2 and 23.3). Article searches were done on the NUS Library (National University of Singapore), Web of Science, SCOPUS, PubMed, and Google Scholar databases by first using the "mangrove harvesting soils" string, followed by the topic strings (coded as "TS" or "TITLE-ABS-KEY" in SCOPUS) below:

String 1

TS = ((logging OR deforest * OR harvest * OR plantation *) AND mangrove * near/2 soil * AND(soil carbon OR carbon sequestration OR sediment * OR nutrient))

String 2

TS = ((logging OR deforest * OR harvest * OR plantation *)AND ((mangrove * near/2 soil *) OR (mangrove * near/2 sediment *)) AND (soil carbon OR carbon sequestration OR sediment * OR nutrient *)

String 3

TS = ((logging OR deforest * OR harvest * OR plantation *)AND ((mangrove * near/2 soil *) OR (mangrove * near/2 sediment *))

String 4

TS = ((logging OR deforest * OR harvest * OR plantation *) AND mangrove AND (soil AND carbon OR carbon AND sequestration OR sediment * OR nutrient))

There are multiple strings due to the limitations each database search engine imposes on the string equation (for example, SCOPUS search does not allow to search with the "near/2" function). Articles were then categorized by database, and the literature search was considered finished once the same article results repeatedly appeared on string searches from other databases. Each articles' abstract, title, and key words are then reviewed according to how appropriate it was to the current study. Next, the appropriate soil data (any value regarding soil carbon, nitrogen, phosphorus, soil bulk density, salinity, and soil composition) with the corresponding forest structure data (mean DBH, height, tree density, above-ground biomass) are compiled into a single table. The data were categorized per article, country, dominant mangrove tree species, and forest management regime (PL, SSSL, and LSSL). To add additional data entries to the soil organic carbon (%) variable, organic matter was converted using the following equation from Howard et al. (2014):

$$\text{Carbon Concentration } (\%) = 0.415\ (\text{Soil Organic Matter } (\%)) + 2.8857 \quad (R^2 = 0.5916)$$

Data that were converted from another measurement are highlighted in red in the tables of Supplementary Materials I (Appendix II).

II. Literature review data: data from soil studies in harvested areas around the world

Supplementary data

Supplementary data related to this article can be found online at https://doi.org/10.1016/B978-0-12-816437-2.00018-5

Acknowledgments

Part of the figures and data appearing in this chapter and this chapter's case study (LSSL) are secondary data (2016) provided by the PT. Bintuni Utama Murni Wood Industries Forestry Department (Head of Department: Ruhuddien Pandu Yudha).

References

Adame, M.F., Reef, R., Herrera-Silveira, J.A., Lovelock, C.E., 2012. Sensitivity of dissolved organic carbon exchange and sediment bacteria to water quality in mangrove forests. Hydrobiologia 691 (1), 239–253.

Adame, M.F., Zakaria, R.M., Fry, B., Chong, V.C., Then, Y.H.A., Brown, C.J., Lee, S.Y., 2018. Loss and recovery of carbon and nitrogen after mangrove clearing. Ocean Coast Manag. 161, 117–126.

Alongi, D.M., Tirendi, F., Trott, L.A., Xuan, T.T., 2000. Benthic decomposition rates and pathways in plantations of the mangrove *Rhizophora apiculata* in the Mekong delta, Vietnam. Mar. Ecol. Prog. Ser. 194, 87–101.

Alongi, D.M., Trott, L.A., Wattayakorn, G., Clough, B.F., 2002. Below-ground nitrogen cycling in relation to net canopy production in mangrove forests of southern Thailand. Mar. Biol. 140 (4), 855–864.

Alongi, D.M., Wattayakorn, G., Tirendi, F., Dixon, P., 2004. Nutrient capital in different aged forests of the mangrove *Rhizophora apiculata*. Bot. Mar. 47, 116–124.

Alongi, D.M., de Carvalho, N.A., 2008. The effect of small-scale logging on stand characteristics and soil biogeochemistry in mangrove forests of Timor Leste. For. Ecol. Manag. 255 (3–4), 1359–1366.

Alongi, D.M., 2009. The Energetics of Mangrove Forests. Springer Science & Business Media.

Arias-Ortiz, A., Masqué, P., Glass, L., et al., 2020. Losses of soil organic carbon with deforestation in mangroves of Madagascar. Ecosystems. https://doi.org/10.1007/s10021-020-00500-z.

Aziz, A.A., Dargusch, P., Phinn, S., Ward, A., 2015. Using redd+ to balance timber production with conservation objectives in a mangrove forest in Malaysia. Ecol. Econ. 120, 108–116.

Bhomia, R.K., MacKenzie, R.A., Murdiyarso, D., Sasmito, S.D., Purbopuspito, J., 2016. Impacts of land use on Indian mangrove forest carbon stocks: implications for conservation and management. Ecol. Appl. 26 (5), 1396–1408.

Bernardino, A.F., Gomes, L.E.O., Hadlich, H.L., Andrades, R., Correa, L.B., 2018. Mangrove clearing impacts on macrofaunal assemblages and benthic food webs in a tropical estuary. Mar. Pollut. Bull. 126, 228–235.

Cannicci, S., Burrows, D., Fratini, S., Smith III, T.J., Offenberg, J., Dahdouh-Guebas, F., 2008. Faunal impact on vegetation structure and ecosystem function in mangrove forests: a review. Aquat. Bot. 89 (2), 186–200.

Castillo, J.A.A., Apan, A.A., Maraseni, T.N., Salmo III, S.G., 2017. Soil C quantities of mangrove forests, their competing land uses, and their spatial distribution in the coast of Honda Bay, Philippines. Geoderma 293, 82–90.

Chen, L., Zeng, X., Tam, N.F.Y., Lu, W., Luo, Z., Du, X., Wang, J., 2012. Comparing carbon sequestration and stand structure of monoculture and mixed mangrove plantations of *Sonneratia caseolaris* and *S. apetala* in Southern China. For. Ecol. Manag. 284, 222–229.

Chowdhury, R., Sutradhar, T., Begam, M.M., Mukherjee, C., Chatterjee, K., Basak, S.K., Ray, K., 2019. Effects of nutrient limitation, salinity increase, and associated stressors on mangrove forest cover, structure, and zonation across Indian Sundarbans. Hydrobiologia 842 (1), 191–217.

Ellison, J.C., 2019. Biogeomorphology of mangroves. In: Coastal Wetlands. Elsevier, pp. 687–715.

Ewel, K.C., Zheng, S., Pinzon, Z.S., Bourgeois, J.A., 1998a. Environmental effects of canopy gap formation in high-rainfall mangrove forests. Biotropica 30 (4), 510–518.

Erftemeijer, P., Allen, G., Zuwendra, 1989. Preliminary Resource Inventory of Bintuni Bay and Recommendations for Conservation and Management. Direktorat Perlindungan Hutan Dan Pelestanan Alam Departement Kehutanan Republik Indonesia. Asian Wetland Bureau Bogor.

Ewel, K.C., Bourgeois, J.A., Cole, T.G., Zheng, S., 1998b. Variation in environmental characteristics and vegetation in high-rainfall mangrove forests, Kosrae, Micronesia. Global Ecol. Biogeogr. Lett. 7, 49–56.

Feller, I.C., 1995. Effects of nutrient enrichment on growth and herbivory of dwarf red mangrove (*Rhizophora mangle*). Ecol. Monogr. 65 (4), 477–505.

Feller, I.C., Whigham, D.F., O'Neill, J.P., McKee, K.L., 1999. Effects of nutrient enrichment on within-stand cycling in a mangrove forest. Ecology 80 (7), 2193–2205.

Feller, I.C., McKee, K.L., Whigham, D.F., O'Neill, J.P., 2002. Nitrogen vs. phosphorus limitation across an ecotonal gradient in a mangrove forest. Biogeochemistry 62, 145–175.

Feller, I.C., Chamberlain, A., 2007. Herbivore responses to nutrient enrichment and landscape heterogeneity in a mangrove ecosystem. Oecologia 153 (3), 607–616.

FAO, 1985. Mangrove Management in Thailand, Malaysia, and Indonesia. Food and Agriculture Organization of the United Nations, Rome.

Giri, C., Ochieng, E., Tieszen, L.L., Zhu, Z., Singh, A., Loveland, T., et al., 2011. Status and distribution of mangrove forests of the world using earth observation satellite data. Global Ecol. Biogeogr. 20 (1), 154–159.

Gleason, S.M., Ewel, K.C., Hue, N., 2003. Soil redox conditions and plant–soil relationships in a Micronesian mangrove forest. Estuar. Coast Shelf Sci. 56 (5–6), 1065–1074.

Goessens, A., Satyanarayana, B., Van der Stocken, T., Zuniga, M.Q., Mohd-Lokman, H., Sulong, I., Dahdouh-Guebas, F., 2014. Is Matang Mangrove Forest in Malaysia sustainably rejuvenating after more than a century of conservation and harvesting management? PLoS One 9 (8), e105069.

Granek, E., Ruttenberg, B.I., 2008. Changes in biotic and abiotic processes following mangrove clearing. Estuar. Coast Shelf Sci. 80 (4), 555–562.

Grellier, S., Janeau, J.L., Nhon, D.H., Cuc, N.T.K., Quynh, L.T.P., Thao, P.T.T., et al., 2017. Changes in soil characteristics and C dynamics after mangrove clearing (Vietnam). Sci. Total Environ. 593–594, 654–663.

Hamilton, S.E., Friess, D.A., 2018. Global carbon stocks and potential emissions due to mangrove deforestation from 2000 to 2012. Nat. Clim. Change 8 (3), 240.

Howard, J., Hoyt, S., Isensee, K., Telszewski, M., Pidgeon, E. (Eds.), 2014. Coastal Blue Carbon: Methods for Assessing Carbon Stocks and Emissions Factors in Mangroves, Tidal Salt Marshes, and Seagrasses. Conservation International, Intergovernmental Oceanographic Commission of UNESCO, International Union for Conservation of Nature, Arlington, Virginia, US.

Hauff, R.D., Ewel, K.C., Jack, J., 2006. Tracking human disturbance in mangroves: estimating harvest rates on a Micronesian Island. Wetl. Ecol. Manag. 14 (2), 95–105.

Huang, X., Wang, X., Li, X., Xin, K., Yan, Z., Sun, Y., Bellerby, R., 2017. Distribution pattern and influencing factors for soil organic carbon (SOC) in mangrove communities at Dongzhaigang, China. J. Coast Res. 34 (2), 434–442.

Ilstedt, U., Malmer, A., Verbeeten, E., Murdiyarso, D., 2007. The effect of afforestation on water infiltration in the tropics: a systematic review and meta-analysis. For. Ecol. Manag. 251 (1–2), 45–51.

Kathiresan, K., Gomathi, V., Anburaj, R., Saravanakumar, K., 2014. Impact of mangrove vegetation on seasonal carbon burial and other sediment characteristics in the Vellar-Coleroon estuary, India. J. For. Res. 25 (4), 787–794.

Kirui, B.Y.K., Huxham, M., Kairo, J., Skov, M., 2008. Influence of species richness and environmental context on early survival of replanted mangroves at Gazi bay, Kenya. Hydrobiologia 603, 171–181.

Lang'at, J.K.S., Kairo, J.G., Mencuccini, M., Bouillon, S., Skov, M.W., Waldron, S., Huxham, M., 2014. Rapid Losses of surface elevation following tree girdling and cutting in tropical mangroves. PLoS One 9 (9), e107868.

Lovelock, C.E., Ball, M.C., Choat, B., Engelbrecht, B.M., Holbrook, N.M., Feller, I.C., 2006. Linking physiological processes with mangrove forest structure: phosphorus deficiency limits canopy development, hydraulic conductivity and photosynthetic carbon gain in dwarf Rhizophora mangle. Plant Cell Environ. 29 (5), 793–802.

Lunstrum, A., Chen, L., 2014. Soil carbon stocks and accumulation in young mangrove forests. Soil Biol. Biochem. 75, 223–232.

Ngole-Jeme, V.M., Fonge, B.A., Tabot, P.T., Mumbang, C., 2016. Impact of logging activities in a tropical mangrove on ecosystem diversity and sediment heavy metal concentrations. J. Coast Conserv. 20 (3), 245–255.

Nguyen, H.T., Yoneda, R., Ninomiya, I., Harada, K., Dao, T.V., Sy, T.M., Phan, H.N., 2004. The effects of stand-age and inundation on carbon accumulation in mangrove plantation soil in Namdinh, Northern Vietnam. Tropics 14 (1), 21–37.

Ola, A., Schmidt, S., Lovelock, C.E., 2018. The effect of heterogeneous soil bulk density on root growth of field-grown mangrove species. Plant Soil 432 (1–2), 91–105.

Osland, M.J., Spivak, A.C., Nestlerode, J.A., Lessmann, J.M., Almario, A.E., Heitmuller, P.T., et al., 2012. Ecosystem development after mangrove Wetland creation: plant–soil change across a 20-year chronosequence. Ecosystems 15 (5), 848–866.

Ozaki, K., Takashima, S., Suko, O., 2000. Ant predation suppresses populations of the scale insect *Aulacaspis marina* in natural mangrove forests. Biotropica 32 (4), 764–769.

Pham, V.H., Luu, V.D., Nguyen, T.T., Koji, O., 2017. Will restored mangrove forests enhance sediment organic carbon and ecosystem carbon storage? Reg. Stud. Mar. Sci. 14, 43–52.

Phillips, D.H., Kumara, M.P., Jayatissa, L.P., Krauss, K.W., Huxham, M., 2017. Impacts of mangrove density on surface sediment accretion, belowground biomass and biogeochemistry in Puttalam Lagoon, Sri Lanka. Wetlands 37 (3), 471–483.

Pribadi, R., 1998. The Ecology of Mangrove Vegetation in Bintuni Bay, Irian Jaya, Indonesia (Ph.D. thesis). University of Stirling, Scotland, pp. 1–225.

Reef, R., Feller, I.C., Lovelock, C.E., 2010. Nutrition of mangroves. Tree Physiol. 30 (9), 1148–1160.

Rivera-Monroy, V.H., Twilley, R.R., Medina, E., Moser, E.B., Botero, L., Francisco, A.M., Bullard, E., 2004. Spatial variability of soil nutrients in disturbed riverine mangrove forests at different stages of regeneration in the San Juan River Estuary, Venezuela. Estuaries 27 (1), 44–57.

Rovai, A.S., Twilley, R.R., Castañeda-Moya, E., Riul, P., Cifuentes-Jara, M., Manrow-Villalobos, M., Horta, P.A., Simonassi, J.C., Fonseca, A.L., Pagliosa, P.R., 2018. Global controls on carbon storage in mangrove soils. Nat. Clim. Change 8 (6), 534–538.

Saenger, P., 2002. Mangrove Ecology, Silviculture and Conservation. Springer Science & Business Media.

Sahu, S.C., Kumar, M., Ravindranath, N.H., 2016. Carbon stocks in natural and planted mangrove forests of Mahanadi mangrove Wetland, East coast of India. Curr. Sci. 110 (12), 2253–2261.

Salmo III, S.G., Lovelock, C., Duke, N.C., 2013. Vegetation and soil characteristics as indicators of restoration trajectories in restored mangroves. Hydrobiologia 720 (1), 1–18.

Salmo III, S.G., Lovelock, C.E., Duke, N.C., 2014. Assessment of vegetation and soil conditions in restored mangroves interrupted by severe tropical typhoon 'Chan-hom' in the Philippines. Hydrobiologia 733 (1), 85–102.

Sasmito, S.D., Sillanpää, M., Hayes, M.A., Bachri, S., Saragi-Sasmito, M.F., Sidik, F., et al., 2020a. Mangrove blue carbon stocks and dynamics are controlled by hydrogeomorphic settings and land-use change. Global Change Biol. 26 (5), 3028–3039.

Sasmito, S.D., Kuzyakov, Y., Lubis, A.A., Murdiyarso, D., Hutley, L.B., Bachri, S., Friess, D.A., Martius, C., Borchard, N., 2020b. Organic carbon burial and sources in soils of coastal mudflat and mangrove ecosystems. Catena 187, 104414.

Shaifullah, K.M., Mezbahuddin, M., Sujauddin, M., Haque, S.M.S., 2008. Effects of coastal afforestation on some soil properties in Lakshmipur coast of Bangladesh. J. For. Res. 19 (1), 32–36.

Sillanpää, M., Vantellingen, J., Friess, D.A., 2017. Vegetation regeneration in a sustainably harvested mangrove forest in West Papua, Indonesia. For. Ecol. Manag. 390, 137–146.

Simard, M., Fatoyinbo, L., Smetanka, C., Rivera-Monroy, V.H., Castañeda-Moya, E., Thomas, N., Van der Stocken, T., 2019. Mangrove canopy height globally related to precipitation, temperature and cyclone frequency. Nat. Geosci. 12 (1), 40–45.

Sjöling, S.M., Mohammed, S.M., Lyimo, T.J., Kyaruzi, J.J., 2005. Benthic bacterial diversity and nutrient processes in mangroves: impact of deforestation. Estuar. Coast Shelf Sci. 63 (3), 397–406.

Suárez-Abelenda, M., Ferreira, T.O., Camps-Arbestain, M., Rivera-Monroy, V.H., Macías, F., Nóbrega, G.N., Otero, X.L., 2014. The effect of nutrient-rich effluents from shrimp farming on mangrove soil carbon storage and geochemistry under semi-arid climate conditions in northern Brazil. Geoderma 213, 551–559.

Thongjoo, C., Choosak, S., Chaichana, R., 2018. Soil fertility improvement from commercial monospecific mangrove forests (*Rhizophora apiculata*) at Yeesarn Village, Samut Songkram Province, Thailand. Trop. Ecol. 59 (1), 91–97.

Tovilla Hernández, C., de la Lanza Espino, G., Orihuela Belmonte, D.E., 2001. Impact of logging on a mangrove swamp in South Mexico: cost/benefit analysis. Rev. Biol. Trop. 49 (2), 571–580.

Yang, S.C., Riddin, T., Adams, J.B., Shih, S.S., 2014. Predicting the spatial distribution of mangroves in a South African estuary in response to sea level rise, substrate elevation change and a sea storm event. J. Coast Conserv. 18 (4), 459–469.

Zhang, J.-P., Shen, C.-D., Ren, H., Wang, J., Han, W.-D., 2012. Estimating change in sedimentary organic carbon content during mangrove restoration in southern China using carbon isotopic measurements. Pedosphere 22 (1), 58–66.

Further reading

Bulmer, R.H., Lundquist, C.J., Schwendenmann, L., 2015. Sediment properties and CO_2 efflux from intact and cleared temperate mangrove forests. Biogeosciences 12 (20), 6169–6180.

Chen, G.-C., Ye, Y., Lu, C.-Y., 2007. Changes of macro-benthic faunal community with stand age of rehabilitated *Kandelia candel* mangrove in Jiulongjiang Estuary, China. Ecol. Eng. 31 (3), 215–224.

Datta, D., Deb, S., 2017. Forest structure and soil properties of mangrove ecosystems under different management scenarios: experiences from the intensely humanized landscape of Indian Sunderbans. Ocean Coast Manag. 140, 22–33.

Ellegaard, M., Nguyen, N.T.G., Andersen, T.J., Michelsen, A., Nguyen, N.L., Doan, N.H., et al., 2014. Temporal changes in physical, chemical and biological sediment parameters in a tropical estuary after mangrove deforestation. Estuar. Coast Shelf Sci. 142, 32–40.

Gillis, L.G., Belshe, E.F., Narayan, G.R., 2017. Deforested mangroves affect the potential for carbon linkages between connected ecosystems. Estuar. Coast 40 (4), 1207–1213.

Ha, T.H., Marchand, C., Aimé, J., Dang, H.N., Phan, N.H., Nguyen, X.T., Nguyen, T.K.C., 2018. Belowground carbon sequestration in a mature planted mangroves (Northern Viet Nam). For. Ecol. Manag. 407, 191–199.

Kauffman, J.B., Donato, D.C., 2012. Protocols for the Measurement, Monitoring and Reporting of Structure, Biomass, and Carbon Stocks in Mangrove Forests. CIFOR, Bogor, Indonesia, p. 50.

Leung, J.Y.S., Tam, N.F.Y., 2013. Influence of plantation of an exotic mangrove species, *Sonneratia caseolaris* (L.) Engl., on macrobenthic infaunal community in Futian Mangrove National Nature Reserve, China. J. Exp. Mar. Biol. Ecol. 448, 1–9.

Leung, J.Y.S., 2015. Habitat heterogeneity determining the macrobenthic infaunal community in a mangrove swamp in South China: implication for plantation and plant invasion. J. Coast Res. 31 (3), 624–633.

López-Hoffman, L., Monroe, I.E., Narváez, E., Martínez-Ramos, M., Ackerly, D.D., 2006. Sustainability of mangrove harvesting: how do harvesters' perceptions differ from ecological analysis? Ecol. Soc. 11 (2), 14.

Putz, F.E., Chan, H.T., 1986. Tree growth, dynamics, and productivity in a mature mangrove forest in Malaysia. For. Ecol. Manag. 17 (2–3), 211–230.

Ren, H., Chen, H., Li, Z.A., Han, W., 2009. Biomass accumulation and carbon storage of four different aged *Sonneratia apetala* plantations in Southern China. Plant Soil 327 (1–2), 279–291.

Salmo III, S.G., Duke, N.C., 2010. Establishing mollusk colonization and assemblage patterns in planted mangrove stands of different ages in Lingayen Gulf, Philippines. Wetl. Ecol. Manag. 18 (6), 745–754.

Sukardjo, S., Yamada, I., 1992. Biomass and productivity of a *Rhizophora mucronata* Lamarck plantation in Tritih, Central Java, Indonesia. For. Ecol. Manag. 49 (3–4), 195–209.

Weyl, P.K., 1964. On the change in electrical conductance of seawater with temperature. Limnol. Oceanogr. 9 (1), 75–78.

Conclusions

Daniel A. Friess[1], *Frida Sidik*[2]

[1]Department of Geography, National University of Singapore, Singapore; [2]Institute for Marine Research and Observation, Ministry of Marine Affairs and Fisheries, Bali, Indonesia

Introduction

Sedimentary environments underpin >90% of the world's tropical, subtropical, and warm temperate mangrove forests (Worthington et al., 2020), and as such, an effective understanding of the processes underpinning their management is of global concern. Mangrove sedimentary environments are particularly at risk of human disturbance; sedimentary settings such as deltas and estuaries have experienced rapid population growth, particularly in the tropics, with knock on impacts on mangrove forests along these coastlines. Indeed, the largest areas of mangrove loss between 1996 and 2016 occurred on deltas such as the Rakhine River delta (Myanmar), the Mahakam and Kayan deltas (Indonesia), and the deltaic coastline of northern Brazil (Worthington et al., 2020). At smaller scales, other stressors such as pollution contribute to mangrove degradation, and individual management actions such as harvesting also impact the sedimentary status of mangrove forests (Sillanpää et al., 2020). The myriad threats facing mangroves have important implications for carbon storage and dynamics in mangroves, from large-scale carbon emissions (Atwood et al., 2017) to smaller scale disturbance, peat collapse, and the release of dissolved organic and inorganic matter (e.g., Sippo et al., 2020). This ultimately impacts the ability of mangrove forests to contribute to broader climate change mitigation objectives.

Alongside direct anthropogenic threats such as land cover change, mangrove sedimentary environments are also highly vulnerable to the longer term process of sea-level rise due to natural subsidence and compaction processes that reduce soil surface elevations relative to the tidal frame. In many coastal landscapes, these natural sedimentary processes operate alongside human-exacerbated reductions in sediment supply from many river basins due to damming and other hydrological modifications (Besset et al., 2019). Importantly, not all mangrove systems are equally vulnerable to sea-level rise, and vulnerability is spatially variable. Geomorphic and sedimentary settings are key determinants of the overall vulnerability of mangrove forests to sea-level rise (McKee et al., 2020), and vulnerability assessments must take into account the diversity of sedimentary environments within which mangrove forests are found. Ultimately, a range of biotic, abiotic, and human factors come together to determine the overall vulnerability and resilience of sedimentary mangrove forests to global change (Ong and Ellison, 2020).

At the same time, mangrove sedimentary coasts provide a range of future opportunities and benefits to human populations. Economic and conservation opportunities can be found particularly in the long-term accumulation and storage of blue carbon (*sensu* Lovelock and Duarte, 2019). Blue carbon has created strong national and international policy interest to use mangrove forests as a nature-based climate solution that can contribute to country-level targets for international climate change agreements (Herr and Landis, 2016). While their small extent means that their impact on global atmospheric carbon concentrations is small (Taillardat et al., 2018), their contribution at national scales can be substantial (Sidik et al., 2018). Mangroves are viewed as a particularly cost-effective nature-based solution to climate change at smaller scales compared with other ecosystems; this is because they show high rates of carbon sequestration relative to the cost of their restoration (Taillardat et al., 2020).

Our conceptual and applied knowledge of sedimentary mangrove environments is now strong

This book shows that we now have a strong fundamental knowledge of many of the key processes shaping and affecting dynamic mangrove sedimentary environments. Studies and reviews have clearly characterized key biogeomorphological processes (Le Minor et al., 2020; Swales et al., 2020) and how they relate to ongoing and future challenges facing mangroves, such as sea-level rise (Ward and Lacerda, 2020). The research field's knowledge of mangrove blue carbon is now particularly advanced, as shown in the number of chapters in this book dedicated to this topic. Recent international policy interest and the work presented here build upon decades of research that sought to understand carbon cycling in mangroves (e.g., Twilley et al., 1992). This early work spurred huge research efforts in estimating the carbon stocks held by mangrove forests and their soils (Donato et al., 2011; Banda et al., 2020). When discussing the future of blue carbon science, Macreadie et al. (2019) posed 10 key research questions that remain to be answered to advance this research field, and it is clear that mangrove researchers are already contributing to filling these knowledge gaps. We now have a stronger understanding on organic matter dynamics (Zimmer and Helfer, 2020) and how natural disturbance impacts blue carbon and mangrove status (Salmo, 2020). Researchers are also now asking more nuanced questions around greenhouse gas fluxes (Ulumuddin et al., 2020) and how mangrove blue carbon stocks and processes change within ecotones and at latitudinal ecosystem boundaries (Banda et al., 2020; Chen et al., 2020).

Academic research on mangrove sedimentary environments has a long and rich history (Friess and McKee, 2020), but it is not a purely scientific endeavor; it is crucial that we apply this research to real-world contexts if we are to better manage, conserve, and restore our mangrove-fringed shorelines. Sustainable forest management relies on research and synthesis to help managers understand how to minimize the impact of harvesting practices on soil carbon stocks (Sillanpää et al., 2020). Successful mangrove restoration must be underpinned by a sound understanding of hydrodynamic and geomorphological processes (Lewis, 2005), and Sidik et al. (2020) show how working with, instead of against, these processes can facilitate higher rates of restoration success in projects in Indonesia. Successful restoration can ultimately lead to an increase in ecosystem function and service provision, including increased carbon sequestration and a return to predisturbance levels of carbon dynamics (Sidik et al., 2019). The conceptual and empirical evidence base highlighted in this book is having real-world application and is linking researchers to managers and other coastal stakeholders who are managing dynamic and complex shorelines.

We need to embrace the complexity of sedimentary mangrove environments

Mangrove forests are inherently complex and dynamic environments, and they provide a model ecosystem through which to study issues of biocomplexity at different scales (Feller et al., 2010). This book has shown that mangrove sedimentary environments and their processes must be viewed and characterized across a range of scales. Spatially, we have to be aware of processes occurring from the scale of an individual seedling (Vovides et al., 2020) to the global scale, which may drive, for example, regional- or global-scale variation in mangrove blue carbon stocks (Mackenzie et al., 2020; Ray et al., 2020; Rovai and Twilley, 2020). Spatial scale is also important to help researchers and managers incorporate processes occurring in landscapes external to, but connected to and influencing mangroves (Jennerjahn, 2020). Temporally, processes affecting mangrove sedimentary environments may occur from short-term (Le Minor et al., 2020) to geological timescales (Saintilan et al., 2020; Sefton et al., 2020). Understanding the diversity of spatial and temporal scales over which processes affect mangrove sedimentary environments is the first step in characterizing those sedimentary processes and understanding how to manage them effectively.

The chapters contained in this book have made clear that understanding the complexity of mangrove sedimentary environments, solving the challenges that mangroves face, and realizing their blue carbon opportunities require an inherently multidisciplinary approach. No single research discipline is able to adequately study the range and complexity of processes and scales found in the mangrove ecosystem. Research presented in this book covers disciplines as varied as geology (Saintilan et al., 2020), geomorphology (Swales et al., 2020), hydrology (Macklin et al., 2020), biogeochemistry (Ulumuddin et al., 2020), remote sensing (Proisy et al., 2020), forestry (Sillanpää et al., 2020), microbiology (Helfer and Hassenrück, 2020), resilience theory (Ong and Ellison, 2020), and the history of science (Friess and McKee, 2020). All of these disciplines bring different techniques and philosophies that together can help us understand variable sedimentary and biogeomorphological processes and their influence on different components of the mangrove environment. It is this rich diversity of disciplinary outlooks that makes mangrove research such an exciting field of study. It is this disciplinary diversity that equips researchers with the wide range of techniques required to understand mangrove sedimentary environments at multiple scales.

The work of all the authors who contributed to this book represents the current state of conceptual and applied knowledge of mangrove sedimentary environments across a range of settings and countries. Through this work we now have a clearer understanding of the multidisciplinary and multiscalar nature of mangrove shorelines. Mangrove sedimentary environments provide an exciting and dynamic research field which is built on an increasingly strong conceptual and empirical foundation. This knowledge is helping us to increasingly turn the tide on mangrove forest loss and degradation, so that we can secure these forests and their important ecosystem services long into the future.

References

Atwood, T.B., Connolly, R.M., Almahasheer, H., Carnell, P.E., Duarte, C.M., et al., 2017. Global patterns in mangrove soil carbon stocks and losses. Nat. Clim. Change 7, 523–528.

Banda, S.B., Adams, J.B., Rajkaran, A., Johnson, J.-L., Raw, J.L., 2020. Blue carbon storage comparing mangroves with salt marsh and seagrass habitats at a warm temperate continental limit. In: Sidik, F., Friess, D.A. (Eds.), Dynamic Sedimentary Environments of Mangrove Coasts, first ed. Pgs tbc.

Besset, Manon, Anthony, Edward J., Bouchette, Frederic, 2019. Multi-decadal variations in delta shorelines and their relationship to river sediment supply: an assessment and review. Earth-sci. Rev. 193, 199–219.

Chen, L., Chen, Y., Zhang, Y., Feng, H., 2020. Mangrove carbon sequestration and sediment deposition changes under cordgrass invasion. In: Sidik, F., Friess, D.A. (Eds.), Dynamic Sedimentary Environments of Mangrove Coasts, first ed. Pgs tbc.

Donato, D.C., Kauffman, J.B., Murdiyarso, D., Kurnianto, S., Stidham, M., Kanninen, M., 2011. Mangroves among the most carbon-rich forests in the tropics. Nat. Geosc. 4, 293–297.

Feller, I.C., Lovelcok, C.E., Berger, U., McKee, K.L., Joye, S.B., Ball, M.C., 2010. Biocomplexity in mangrove ecosystems. Annu. Rev. Mar. Sci. 2, 395–417.

Friess, D.A., McKee, K.L., 2020. The history of surface-elevation paradigms in mangrove biogeomorphology. In: Sidik, F., Friess, D.A. (Eds.), Dynamic Sedimentary Environments of Mangrove Coasts, first ed. Pgs tbc.

Helfer, V., Hassenrück, C., 2020. Microbial communities in mangrove sediments. In: Sidik, F., Friess, D.A. (Eds.), Dynamic Sedimentary Environments of Mangrove Coasts, first ed. Pgs tbc.

Herr, D., Landis, E., 2016. Coastal blue carbon ecosystems: opportunities for nationally determined contributions. Int. Union Conserv. Nat., Nat. Conserv. 28p.

Jennerjahn, T.C., 2020. Relevance of allochthonous input from an agriculture-dominated hinterland for 'blue carbon' storage in mangrove sediments in Java, Indonesia. In: Sidik, F., Friess, D.A. (Eds.), Dynamic Sedimentary Environments of Mangrove Coasts, first ed. Pgs tbc.

Le Minor, M., Zimmer, M., Helfer, V., Gillis, L.G., Huhn, K., 2020. Flow and sediment dynamics around structures in mangrove ecosystems – a modelling perspective. In: Sidik, F., Friess, D.A. (Eds.), Dynamic Sedimentary Environments of Mangrove Coasts, first ed. Pgs tbc.

Lewis, R.R., 2005. Ecological engineering for successful management and restoration of mangrove forests. Ecol. Eng. 24, 403–418.

Lovelock, C.E., Duarte, C.M., 2019. Dimensions of blue carbon and emerging perspectives. Biol. Lett. 15, 20180781.

Mackenzie, R., Sharma, S., Rovai, A., 2020. Environmental drivers of blue carbon burial and soil carbon stocks in mangrove forests. In: Sidik, F., Friess, D.A. (Eds.), Dynamic Sedimentary Environments of Mangrove Coasts, first ed. Pgs tbc.

Macklin, P., Rosentreter, J., Arifanti, V.B., Suryaputra, I.G., 2020. Groundwater research in mangrove coastal ecosystems – new prospects. In: Sidik, F., Friess, D.A. (Eds.), Dynamic Sedimentary Environments of Mangrove Coasts, first ed. Pgs tbc.

Macreadie, P.I., Anton, A., Raven, J.A., Beaumont, N., Connolly, R.M., et al., 2019. The future of blue carbon science. Nat. Commun. 10, 3998.

McKee, K.L., Krauss, K.W., Cahoon, D.R., 2020. Does geomorphology determine vulnerability of mangrove coasts to sea-level rise? In: Sidik, F., Friess, D.A. (Eds.), Dynamic Sedimentary Environments of Mangrove Coasts, first ed. Pgs tbc.

Ong, W.J., Ellison, J., 2020. A framework for the quantitative assessment of mangrove resilience. In: Sidik, F., Friess, D.A. (Eds.), Dynamic Sedimentary Environments of Mangrove Coasts, first ed. Pgs tbc.

Proisy, C., Walcker, R., Blanchard, E., Gardel, A., Anthony, E.J., 2020. Mangroves: a natural early warning system of erosion on open muddy coasts in French Guiana. In: Sidik, F., Friess, D.A. (Eds.), Dynamic Sedimentary Environments of Mangrove Coasts, first ed. Pgs tbc.

Ray, R., Sharma, S., Jana, T.K., 2020. State of biogeochemical blue carbon in South Asian mangroves. In: Sidik, F., Friess, D.A. (Eds.), Dynamic Sedimentary Environments of Mangrove Coasts, first ed. Pgs tbc.

Rovai, A.S., Twilley, R.R., 2020. Gaps, challenges and opportunities in mangrove blue carbon research: a biogeographic perspective. In: Sidik, F., Friess, D.A. (Eds.), Dynamic Sedimentary Environments of Mangrove Coasts, first ed. Pgs tbc.

Saintilan, N., Kelleway, J., Rogers, K., 2020. Australian mangroves through the Holocene: interactions between sea-level, mangrove extent and carbon sequestration. In: Sidik, F., Friess, D.A. (Eds.), Dynamic Sedimentary Environments of Mangrove Coasts, first ed. Pgs tbc.

Salmo, S.G., 2020. Assessment of typhoon impacts and post-typhoon recovery in Philippines mangroves: lessons and challenges for adaptive management. In: Sidik, F., Friess, D.A. (Eds.), Dynamic Sedimentary Environments of Mangrove Coasts, first ed. Pgs tbc.

Sefton, J., Woodroffe, S., Ascough, P., 2020. Radiocarbon dating of mangrove sediments. In: Sidik, F., Friess, D.A. (Eds.), Dynamic Sedimentary Environments of Mangrove Coasts, first ed. Pgs tbc.

Sidik, F., Supriyanto, B., Krisnawati, H., Muttaqin, M.Z., 2018. Mangrove conservation for climate change mitigation in Indonesia. WIREs Clim. Change 9, e529.

Sidik, F., Adame, M.F., Lovelock, C.E., 2019. Carbon sequestration and fluxes of restored mangroves. J. Indian Ocean Reg. 15, 177–192.

Sidik, F., Kusuma, D.W., Priyono, B., Priosy, C., Lovelock, C.E., 2020. Managing sediment dynamics through reintroduction of tidal flow for mangrove restoration in abandoned aquaculture ponds. In: Sidik, F., Friess, D.A. (Eds.), Dynamic Sedimentary Environments of Mangrove Coasts, first ed. Pgs tbc.

Sillanpää, M., Lupascu, M., Sasmito, S., Vulpas, S., Teo, R.X., Taberima, S., 2020. Impacts of forest on mangrove sediment dynamics. In: Sidik, F., Friess, D.A. (Eds.), Dynamic Sedimentary Environments of Mangrove Coasts, first ed. Pgs tbc.

Sippo, J.Z., Sanders, C.J., Santos, I.R., Jeffrey, L.C., Call, M., et al., 2020. Coastal carbon cycle changes following mangrove loss. Limnol. Oceanogr. In press, doi: https://doi.org/10.1002/lno.11476.

Swales, A., Pritchard, M., McBride, G.B., 2020. Biogeomorphic evolution and expansion of mangrove forests in New Zealand's sediment-rich estuarine systems. In: Sidik, F., Friess, D.A. (Eds.), Dynamic Sedimentary Environments of Mangrove Coasts, first ed. Pgs tbc.

Taillardat, P., Friess, D.A., Lupascu, M., 2018. Mangrove blue carbon strategies for climate change mitigation are most effective at the national scale. Biol. Lett. 14, 20180251.

Taillardat, P., Thompson, B.S., Garneau, M., Trottier, K., Friess, D.A., 2020. Climate change mitigation potential of wetlands and the cost-effectiveness of their restoration. Interface Focus 10, 20190129.

Twilley, R.R., Chen, R.H., Hargis, T., 1992. Carbon sinks in mangroves and their implications to carbon budget of tropical coastal ecosystems. Water Air Soil Pollut. 64, 265–288.

Ulumuddin, Y.I., Beavis, S., Roderick, M., Eggins, S., Sugoro, I., Sukardjo, S., 2020. Potential carbon loss in sediment through methane production during early development stage of mangrove regeneration in restored mangroves. In: Sidik, F., Friess, D.A. (Eds.), Dynamic Sedimentary Environments of Mangrove Coasts, first ed. Pgs tbc.

Vovides, A.G., Berger, U., Balke, T., 2020. Morphological plasticity and survival thresholds of mangrove plants growing in active sedimentary environments. In: Sidik, F., Friess, D.A. (Eds.), Dynamic Sedimentary Environments of Mangrove Coasts, first ed. Pgs tbc.

Ward, R.D., Lacerda, L.D., 2020. Responses of mangrove ecosystems to sea level change. In: Sidik, F., Friess, D.A. (Eds.), Dynamic Sedimentary Environments of Mangrove Coasts, first ed. Pgs tbc.

Worthington, T.A., zu Ermgassen, P.S.E., Friess, D.A., Krauss, K.W., Lovelock, C.E., et al., 2020. A global biophysical typology of mangroves and its relevance for ecosystem structure and deforestation. Sci. Rep. 10, 14652.

Zimmer, M., Helfer, V., 2020. Quantity and quality of organic matter in mangrove sediments. In: Sidik, F., Friess, D.A. (Eds.), Dynamic Sedimentary Environments of Mangrove Coasts, first ed. Pgs tbc.

Index

'*Note*: Page numbers followed by "f" indicate figures, "t" indicate tables and "b" indicate boxes'

A

Above-ground biomass (AGB), 300t–306t, 307, 449, 458–460, 549–550
Aboveground plant macrofossils, 208
Absorptive capacity, 516
Acanthus, 183
Accelerator mass spectrometry (AMS), 201
Accommodation space, 248
Acidification process, 452–453
Active methanogenesis, 146–147
Active sedimentary environments, 122
  colonization and biogeomorphic succession, 124–127
  plastic responses to sediment dynamics in mature forests, 134–135
  root responses for mechanical stability, 129–134
  wind resistance and mechanical stability in unstable sediments, 127–129
Adaptive capacity, 515
Aegiceras corniculatum, 218, 223, 282–284, 343
Aerial thermal infrared imaging (ATI), 75
Aerobic microbial communities, 146
Aerobic respiration, 146–147
Algae, 286
Aliphatic compounds, 376b
Allochromatium, 149–150
Allochthonous
  allochthonous C sources, 277, 480
  contributions, 405–406
  OM sources, 408–410
Alongshore mud-bank migration rates, 60–61
Amazon mud, 49
Ammonia (NH_3), 147–148
Ammonia-oxidizing archaea (AOA), 147–148
Ammonia-oxidizing bacteria (AOB), 147–148
Ammonification, 148
Anaerobic degradation processes, 146
Anaerobic sulfate reduction, 146–147
Anoxygenic photosynthesis. *See* Photoautotrophic oxidation of sulfide
Anthropogenic CO_2, 319–321
Apicuns, 247–248
Aquaculture
  conversion of mangrove forest to, 71–72
  pond walls, 564–565
    estuarine ecohydrology in mangrove restoration and rehabilitation, 567–568
    hydrological recovery in abandoned ponds, 568–576
    mangrove–sediment–hydrodynamics interactions, 565–567
Aquatic molds, 143
Arabian Gulf coast, 239, 241–242
Arafura Sill, 221
Archaea, 143
Aromatic compounds, 376b
Atlantic East Pacific region (AEP region), 295–297, 307
Australian mangroves, 218
  21st century sea-level rise acceleration and challenges of coastal planning, 227–229
  contemporary patterns of mangrove–saltmarsh interaction, 226–227
  early phases of postglacial marine transgression, 219–221
  estuary infill in mid-late holocene, 224–226
  late phases of postglacial marine transgression, 221–223
Autochthonous
  autochthonous C sources, 277, 480
  contributions, 405–406
  OM sources, 408–410
Automated ribosomal intergenic spacer analysis (ARISA), 159
Autotrophic microorganisms, 143
Available phosphorus (AP), 543
Avicennia, 85, 94–96, 103, 130, 183, 223
  A. alba, 103, 124–127, 181
  A. germinans mangroves, 52, 131, 246
  A. officinalis, 106
Avicennia marina, 6–7, 9, 93, 126–127, 218, 223, 228, 237, 282–284, 343, 384–385

B

Bacteria, 143
Balance trees, 145
Barrier-enclosed estuaries, 13
Basal area (BA), 419–420
Below-ground biomass (BGB), 307–310, 309t–310t, 458–460
Bintuni Bay
  large-scale selective logging in, 595
  soil recovery in different aged secondary and primary mangrove forests, 595–597
Biodiversity Conservation Action Plan of China, 475
Biofouling, 105–107
Biogenic mangroves, 268
Biogeochemical C exchange flux, 353–355
  forest–atmosphere coupled system, 353–354
  forest–sediment coupled system, 354
  sediment–atmosphere coupled system, 354
  sediment–water coupled system, 354–355
Biogeochemical processes, 154–155
Biogeography of blue carbon research in mangroves, 307–315
Biogeomorphic succession, 124–127
Biogeomorphological descriptions, 183
Biogeomorphology, 4–5
Biomarkers, 209, 378–381, 381b
Biomass, 297–298, 336
  allocation and productivity, 485–486
  analysis, 453–454
  carbon, 416
  collection, 453
Biotic interactions, 144
Bioturbation, 150, 207, 555
Bioturbators, 207
Blue carbon (Blue C), 218, 336–337, 394, 448
  analytical methods, 398–399
  biogeography of blue carbon research in mangroves, 307–315
    AGB, 307
    BGB, 307–310
    SOC, 310–315
  budget and fate of, 355–356
  coastal ecosystems, 70
  ecosystems, 276, 298
  global trends in blue C in mangroves, 315–325
    biogeographic and regional controls, 315–325
  grain-size composition of mangrove sediments, 400
  habitats, 449
  material and methods, 395–399
  organic carbon and total nitrogen content, 399–400
  sampling, 398
  sensitivity of blue C to climate change, 356–357
  sink, 395
  stable isotope composition
    mangrove sediments, 400
    of plants and soils, 399–400
    of surface sediments, 400
  study area, 395–396
Blue forests, 448
Boosted-regression trees method (BRT method), 23–24, 26
Bruguiera spp., 183
  B. gymnorrhiza, 94–96, 131, 262–263, 283–284, 384–385
  B. parviflora, 131
Bulk sediment approaches, 205
Burrowing, 555
Buttresses, 130–132

C

Calcium carbonate ($CaCO_3$), 353
Candidatus genera, 149–150
Canopy gaps, 554
Cararma Inlet, 224–226
Carbohydrate-active enzyme (CAZyme), 156–157
Carbon (C), 145, 200, 276, 586
  abatement opportunity, 568
  burial, 280
    in mangrove sediment, 337–338
  C-cycle, 146–147
  cycle in mangrove ecosystems, 68–71
  cycling processes, 72
  density, 453
  dynamics in mangroves and saltmarsh ecotones, 498–500
  emission from sediment to atmosphere, 339–341
  export from sediment to water, 338–339
  import and export, 207
  sequestration, 299, 387b
  source changes, 487
  stocks, 299, 387b
^{13}C isotope technique, 487
Carbon dioxide (CO_2), 69–70, 336, 416
Carbon hydrogen and nitrogen (CHN), 453
CARD-FISH, 163
Carnivory, 374b
Casuarina, 223
  C. glauca, 224–226
Catastrophic typhoons, 540
Ceiba pentandra, 131
Ceriops spp., 183
Ceriops tagal, 131, 237, 343
Chemolithoheterotrophs, 149–150
Chemoorganoheterotrophs, 149–150
Chlorobium, 149–150
Chronology, 204–205
Classical fluid dynamic theory, 489

Clean Development Mechanism, 359
Climate change, 229, 268, 394, 514
  mitigation, 568
Climate variability, 226
Climate-based model, 277
Climatic drivers, 279–280
"Climax" community, 187
Clone libraries, 158–159
CMEMS. *See* Copernicus Marine Environment Monitoring Service (CMEMS)
Coarse wood debris (CWD), 345–346
Coastal
  21st century sea-level rise acceleration and challenges of coastal planning, 227–229
  accretion, 48
  environmental settings, 297–298
  habitats, 336, 448
  instability, 50–52, 51f
  modeling, 49
  morphology, 281
  wetland biogeomorphology, 5
Coastal vulnerability indices (CVIs), 58
Cohesive sediment, 565–566
Colonization, 124–127
Colorless sulfur bacteria, 149–150
Community Based Ecological Mangrove Rehabilitation (CBEMR), 564–565
Community fingerprinting, 159
Compound-specific approaches, 209
Contact metagenomics, 160–161
Continuous Global Mangrove Forest Cover for the 21st Century (CGMFC-21), 327
Conventional radiocarbon dating, 201
Cooks River in Sydney, 228
Copernicus Marine Environment Monitoring Service (CMEMS), 48–49
Cordgrass (*Spartina alterniflora*), 246, 474
  and biological characteristics, 477–478
  biological factors affecting sediment carbon flux, 480–488
  conceptual model and evaluation of regional carbon storage, 496–500
  invasive status along Chinese mangroves, 478–480
  vegetation, 488–496
Cosmic rays, 200
Crabs, 207, 286–287
Cross-shore mangrove shoreline fluctuations, 56–58
Cultivation-dependent approaches, 158
Currambene Creek, 224–226
Cylinder arrays, 99–101

D
Decay, 370b–371b, 373–374
Decomposition, 370b–371b
Deep subsidence, 28
Deltaic mangroves, 281
Denaturing gradient gel electrophoresis (DGGE), 159
Denitrification, 148
Desulfovibrio desulfodismutans, 150
Detached Eddy Simulation, 96–97
Detritivory, 374b
Diazotrophs, 147
Digital Shoreline Analysis System, 55–56
Dinitrogen gas (N_2), 147–148
Direct Numerical Simulation (DNS), 96–97, 104–105
Dissimilatory nitrate reduction to ammonia (DNRA), 147
Dissimilatory sulfate reduction (DSR), 148–149
Dissolved inorganic carbon (DIC), 339
Dissolved organic carbon (DOC), 205
Dissolved organic matter (DOM), 146, 371, 373
DNA, 161–162
  metagenomics, 381–382
Drought, 238
Dry bulk density (DBD), 452
Dynamic elevation models, 219

E
Ecogeomorphology. *See* Biogeomorphology
Ecohydraulics, 98
Ecological resilience, 514
Ecophysiological responses, 123
Ecosystem
  ecosystem-level C stocks, 298–299
  resilience, 514–515
  restoration, 567
  robustness components, 526–527
El Niño-Southern Oscillation, 11
Element cycling
  drivers of microbial element cycling, 150–151
  microbial communities, 143–151
Elevation, 187
  capital, 30
Engineering resilience, 514
Enteromorpha intestinalis, 286
Environmental DNA (eDNA), 346, 382b
  metabarcoding, 381–382
Environmental drivers of blue carbon burial and soil carbon stocks, 276
  abiotic factors, 279–282
    climatic drivers, 279–280
    geomorphology and nutrients, 281–282

Environmental drivers of blue carbon burial and soil carbon stocks (*Continued*)
salinity, 282
storms, 280
biotic factors, 283–287
algae, 286
crabs, 286–287
forest age, 284–286
tree diversity, 283–284
Environmental interactions, 144
Environmental modeling, 242
Epibionts, 106
Epifaunal assemblages, 543–544
Erosional processes, 495
Estuarine ecohydrology in mangrove restoration and rehabilitation, 567–568
Estuary, 30–35
evolution, 11–13
infill in mid-late holocene, 224–226
infilling, 9
Eubacteria, 143
European Center for Medium-Range Weather Forecasts (ECMWF), 48–49
Eutrophication, 475
Exchangeable DOC (EDOC), 339
Excoecaria agallocha, 181, 283–284
Exploratory/abstracted modeling approaches, 5

F
Facultative chemolithoautotrophic sulfide oxidizers, 149–150
Fauna, 547–551
Faunal assemblages, 543–544
and activities, 555
Ficus, 131
Findable, accessible, interoperable, and reusable data (FAIR data), 162
Fine root necromass, 430
Firth of Thames mesotidal estuarine embayment, 15
Flocs, 88–89
Flooding, 181–182
Flora, 145
Flow dynamics, 92
Flume experiments, 99, 101, 106
Flux method, 499
Fluxes, 388
Flying buttresses, 130, 132–134
Forest
age, 284–286
forest–atmosphere coupled system, 353–354
forest–sediment coupled system, 354
Fractionation factor, 202
French Guiana, 48
mud dynamics and vulnerability of wave-exposed mangrove coast of, 49–52
Freshwater inputs, 68
Friction coefficients, 97
Fringe forest, 8
Fulvic acids (FAs), 352–353, 375–376, 376b
Fungi, 143

G
Gain–loss method, 499
Gas chromatography (GC), 420–421
Generalized additive models (GAM), 23–24
Geomorphology, 281–282
Glacial low sea-level deposits, 219
Global warming potential (GWP), 488
Grain-size composition of mangrove sediments, 400
Gray mangrove (*Avicennia marina*), 6–7, 9, 93, 126–127, 218, 223, 228, 237, 282–284, 343, 384–385
Greenhouse gas emissions (GHG emissions), 416, 487–488
Groundwater, 69
importance of, 68
significance, 68–69

H
Hawkesbury River estuary, 224–226
Heavy metals, 70–71
Herbivory, 374b
Heriteria litoralis, 131
Holocene
climatic optimum, 223
mangrove peat deposits, 221–222
Humic acids (HAs), 352–353, 375–376, 376b
Humic substances, 375, 376b
Hurricanes, 280
Hybrid short-and long-read sequencing, 160–161
Hydro-and sediment dynamics in mangrove ecosystems, 86–89
state-of-the-art numerical modeling on, 89–107
modeled sediment dynamics at forest scale, 89–94
sediment dynamics around idealized cylinders, 98–102
sediment dynamics at microtopography scale, 103–107
sediment dynamics at root or seedling scale, 102–103
sediment dynamics at tree scale, 94–98
Hydrochloric acid (HCL), 452–453
Hydrodynamic
forcing, 122
modifications by vegetation, 489–493
Hydrogen peroxide (H_2O_2), 149–150

Hydrogeological
cycle in mangrove ecosystems, 68–71
mapping, 72–73
Hydrogeology, 70
Hydrographic methods, 72–73
Hydrological
pathways, 72
recovery in abandoned ponds, 568–576
postplanting mangrove regeneration, 574–576
tidal flow reintroduction and sediment accumulation, 571–574
regime, 565
Hydrometric analysis, 72–73
Hydroperiod (Hp), 26

I

Impacts of typhoons, 540, 543
In situ surface water temperature measurements, 75
Indian Sundarbans
environmental settings, 349t
mangrove blue C in, 348–357
model mangrove site in South Asia, 348–350
organic and inorganic C storage, 350–353
Indo West Pacific region (IWP region), 295–297, 307
Initial platform height (IPH), 33
Inorganic carbon (IC), 338, 454. *See also* Organic carbon (OC)
analysis, 452–453
as carbonate in sediment, 353
storage, 350–353
Institute of Marine Research and Observation (IMRO), 568–570
Intact mangroves, 540
Integrated Coastal Zone Management (ICZM), 527–528
Inter-bank phase, 55
Inter-decadal Pacific Oscillation, 11
Intergovernmental Panel on Climate Change (IPCC), 394
Inundation, 182, 187
optimum, 181
Iron (Fe), 145

J

Juncus kraussii, 224–226

K

Kaipara Harbour, 9
Kandelia spp., 131

L

Lagoon, 396
Lagrangian approach, 105
Laguncularia racemosa, 131, 155–156, 283–285
Land builders, mangroves as, 190–191, 193–194
Land consolidators, mangroves as, 191–194
Landward migration in arid and semiarid areas, 238–242
Large-Eddy Simulation (LES), 96–97
Large-scale selective logging (LSSL), 586, 589
Leaf area index (LAI), 549–550
Leaf/propagule tethering, 555
Legacy sediments, 35
Lenticels, 104
Lignin, 346
lignin-derived phenols, 346
Lipids, 379–381
Liquid or gas chromatography and high-resolution mass spectrometry (LG/MS or GC/MS), 161–162
Litterfall, 347–348
Local Early Action Planning and Management Planning, 521
Loss on ignition (LOI), 374–375, 452
Low-lying coastal wetlands, 256
Lumnitzera racemosa, 183, 186

M

Macroecology, 298
Macrofossils, 208
Mahakam Delta, 72
Management capacity, 517–518, 521
Mānawa, 9
Manganese (Mn), 145
Mangrove Action Project, 564–565
Mangrove Forests of the World (MFW), 327
Mangrove harvest regimes
forestry case study, 594–602
future considerations, 602–603
global overview of soil dynamics and recovery, 587–589
literature review methods, 603–604
postharvest impacts on mangrove soils, 589–594
comparison of soil conditions in mangrove silviculture regimes, 590–592
general trends in soil properties and forest structure of secondary mangrove forests, 592–594
Mangrove(s), 48, 84–85, 122, 276, 295–297, 336–337, 448, 474, 540
approaches used in mangrove research, 164–165
as blue carbon sinks, 298–300
challenges and opportunities to improving mangrove blue C inventories and mapping, 325–328
of China, 475–477, 476t–477t

Mangrove(s) (*Continued*)
condition, 520
contemporary patterns of mangrove–saltmarsh interaction, 226–227
crowns, 128–129
deforestation, 72, 584–586
ecohydrology, 567
ecology, 566–567
ecosystems, 371–373
erosion, 48
forests, 4, 68, 236, 394, 584
degradation and loss, 68, 71–72
hydrology, 68
implications for mangrove management, 387–388
mangrove-based early warning of coastal vulnerability to mud erosion, 55–61
mangrove-dominated wetlands, 256
mangrove-habitat expansion, 9–10
mangrove–sediment–hydrodynamics interactions, 565–567
microbial communities in mangroves sediments, 142
migration, 242
new biogeochemical setting, 242–243
plantations, 564
propagules, 124–125
quantitative descriptions of mangrove-elevation linkages, 184–186
regeneration in restored mangroves, 416
effects of seasons and mangrove regrowth on dissolved CH_4, 427–435
indication of CH_4 releases through mangrove trees, 434–435
physicochemical parameters, 423–424
potential mangrove contribution to CH_4 emissions, 431–434
relationships between porewater CH_4 and stand basal area, 424–427
sample collection and analytical methods, 420–421
stand basal area, porewater dissolved CH_4, and CH_4 fluxes, 422–423
study area and sampling sites, 418–420
rehabilitation projects, 182
restoration, 564
implications, 568
roots, 205–206
sediments, 199–200, 337, 369–370
organic matter in, 369–374
radiocarbon dating of, 204–209
seedlings, 103, 567–568
shoreline fluctuations, 48–49
species, 84
status and dynamics of, 52–55
tree loss, 72
vegetation, 418–419
worldwide, 382–388
Māori land use activities, 9
Marine sediments, 145
Mass failure, 496
Matang Mangrove Forest Reserve, 524, 528
Mean air temperature, 26
Mean sea level (MSL), 10, 419
Mechanical stability
root responses for, 129–134
in unstable sediments, 127–129
Melaleuca forest, 222–223, 226
Meltwater pulse 1A (MWP1A), 220
Metabarcoding, 159–160
Metabolomics, 161–162
Metagenomics, 160–161
Metaproteomics, 161–162
Metatranscriptomics, 161
Methane (CH_4), 69–70, 146–147, 416–417
fluxes, 422–423, 487–488
from tree stem and soil surface in literature, 438
Microbes, 142
Microbial cells, 146
Microbial communities
drivers, 144–145
drivers, and role in element cycling, 143–151
in mangroves, 151–157
knowledge on, 152–157
sediments, 142
members, 143–144
peculiarities of mangrove ecosystems and constraints, 151
techniques to depict microbial communities and processes, 157–165
Microbiome, 142
Microenvironmental heterogeneity, 358–359
Microfauna, 207
Micronesian Conservation Trust, 528
Microorganisms, 142–143
Microphytobenthos (MPB), 340–341
Migration inland, 238
Mineralization, 370b–371b
Minerogenic
mangrove settings, 268
process, 180, 192
restoration sites, 565
Mini-metagenomics, 160–161
Model(s), 4
calibration, 93
Molecular biomarkers, 346
Mollusks, 207
Morphological plasticity, 121–122

Mud bank, 49
phase, 54–55
Mud dynamics and vulnerability of wave-exposed mangrove coast, 49–52
Mud-dominated mangroves, 192
Myxomycetes, 143

N

Nahoon Estuary, 449
Nationally determined contributions (NDC), 359–360
Natural coastal wetlands, 488
Natural tracers, 73, 75
Nature Reserve of Kaw, 56
Nekton communities, 550
Net primary productivity (NPP), 279–280, 356, 430
New South Wales (NSW), 218
New Zealand estuarine system, 5
New Zealand's mangrove forests, 4, 6–11, 30–35
biogeomorphology, 4–5
biophysical drivers, 10–11
elevation capital and future fate, 30
estuaries, 30–35
estuary evolution, 11–13
geographic distribution, environmental settings, and forest types, 6–8
historical patterns, 9–10
legacy sediments, 35
sand flat to mangrove forest, 14–30
Next-generation sequencing (NGS), 157–158
Nitrate (NO_3^-), 147
respiration, 148
Nitrification, 147–148
Nitrite (NO_2^-), 147
Nitrogen (N), 145, 148, 281–282, 586
fixation, 147
N-cycle, 147–148
transformation in mangrove ecosystems, 148
Nitrogen resorption efficiency (NuRE), 555
Nitrous oxide (N_2O), 148
Numerical models/modeling, 72–73, 93, 99–101
Nutrient(s), 70–71, 242, 281–282, 370b–371b
resorption efficiency, 554–555
Nxaxo Estuary, 450
Nypa, 183

O

Obligate chemolithoautotrophs, 149–150
Oceanic data, 60–61
Oceanic inputs, 68
Ocypodid crabs, 286–287
"Old-wood" effect, 208
Omics
challenges of omics data analysis, 162
environmental characterization, 163
metabarcoding, 159–160
metagenomics, 160–161
metaproteomics and metabolomics, 161–162
metatranscriptomics, 161
and related approaches, 159–162
supplementing omics with quantitative information, 163–164
visualization of microbes at microscale, 163
Optical method, 97
Organic carbon (OC), 336, 454, 547. *See also* Inorganic carbon (IC)
analysis, 452–453
stocks, 382–383
storage, 350–353
Organic compounds, 376b
Organic concentrates, 208–209
Organic matter (OM), 369–370, 394, 452, 540
dynamics of organic matter turnover, 373–374
in mangrove sediments, 369–374
sources of, 400–404
sources or sinks, 371–373
turnover, 142–143
Organogenic sites, 565
Oxygen exposure time, 405–406
Oxygenated sediments, 207

P

Paris Agreement, 394
Partial dependency (PD), 23–24
Particulate organic matter (POM), 371, 373
Pedogenesis, 369–374
Pelliciera rhizophorae, 131
Perancak River, 571
Phenotypic plasticity. *See* Morphological plasticity
Philippine mangroves, 540
fauna, 547–551
field assessment, 543–544
materials and methods, 541–544
prospects for recovery, 551–553
recovery indicators, 553–555
results, 544–555
sediment damage, 546–547
site description, 541
typhoon history, 542–543
vegetation damage, 544–546
vulnerability, research/management implications, and recommendations, 556–558
Phosphorous (P), 145, 281–282, 586
Phosphorus retention efficiency (PRE), 555
Photoautotrophic oxidation of sulfide, 149–150
Photogrammetric measurements, 106–107
Physical processes, 188–189

Plant
productivity, 243–244
traits, 122–123
Plantations (PLs), 584, 590
Plasmodial slime molds, 143
Plastic responses, 122, 124
to sediment dynamics in mature forests, 134–135
Plasticity, 121–122
Plate root systems, 130–131
Pneumatophore(s), 134
density, 122–123
height, 122–123
Pohnpei Watershed Forest Reserve and Mangrove Protection Act of (1987), 528
Point model, 30–31
Pollen concentrates, 208–209
Pollutants, 242–243
Polysaccharides, 161–162
Populus deltoides, 434–435
Porewater, 69
dissolved CH_4, 420–423
importance of, 68
physicochemistry, 421
salinity, 68–69
significance, 68–69
variables, 543
Post-depositional effects, 203
Postglacial marine transgression
early phases of, 219–221
late phases of, 221–223
Postplanting mangrove regeneration, 574–576
Potential evapotranspiration (PET), 319
Precipitation, 279–280
Process-based numerical simulations, 94–96
Progradation, 28
Proteins, 161–162
Protists, 143
Protozoa, 143
PT. Bintuni Utama Murni Wood Industries (PT. BUMWI), 589

Q

Quantitative dating approaches, 209–210
Quantitative PCR (qPCR), 163–164

R

Radiocarbon age calibration and corrections, 201–202
Radiocarbon dating, 200
of mangrove sediments, 204–209
principles, 200–204
quantitative dating approaches, 209–210
Radioisotope dating methods, 245
Radiometric radiocarbon dating, 201
Radon (^{222}Rn), 73
and ATI methodologies, 76–77
as naturally occurring SGD tracer, 74–75
Rainfall, 279
Randhir's model, 530
Ranking, 521
Recovery
indicators, 553–555
of mangroves, 551–553
Red mangrove (*Rhizophora mucronata*), 94–96, 184–185, 283–284, 343
Reducing Emissions from Deforestation and Forest Degradation (REDD+), 359–360
Reef-dominated mangroves, 192
Regional carbon storage
carbon dynamics in mangroves and saltmarsh ecotones, 498–500
changes after cordgrass invasion in China, 496–498
conceptual model and evaluation of, 496–500
Relative sea-level rise (RSLR), 15, 256, 515. *See also* Sea level rise (SLR)
calculation of relative sea-level rise, vertical land development, and deficits, 262
Remote sensing, 75
image analyses, 497–498
remote sensing–based methods, 53–54
Representative Concentration Pathways (RCPs), 31–33
Reservoir effect, 203
Resilience, 514–515
absorptive capacity, 516
adaptive capacity, 515
assessment, 516–521
in coastal areas, 513–514
dimensions, 517–518
improvement, 529–530
management components, 527–528
methods, 517–521
objective, 517
results, 521–530
study areas, 517
transformative capacity, 516
Reynolds numbers, 97
Reynolds-Averaged Navier–Stokes model (RANS model), 96–97
Rhizophora, 183, 220, 222–223
prop root system, 88–89, 99
R. apiculata, 237, 262–263, 284–285
R. mangle, 132–133, 183–184, 186, 262–263, 283–285, 379
R. stylosa, 85, 94–96, 133–134, 218
trees, 87, 94–96, 95f

Rhizophora mucronata. *See* Red mangrove (*Rhizophora mucronata*)
Rhizophoraceae, 218
River discharge, 318
Riverine
forest, 8
mangroves, 318
RNA, 161–162
Rod surface elevation table–marker horizon technique (RSET-MH), 571
Rod Surface Elevation Tables (RSET), 19–20, 23, 245–246
Root(s), 85–86, 277
production, 279–280
responses for mechanical stability, 129–134
buttresses, 131–132
flying buttresses and stilt roots, 132–134
plate root systems, 130–131
Roughness density (RD), 106

S

Sabkha environments, 241–242
Saccharides, 379–381
Salicornia spp., 460–461
S. quinqueflora, 464–465
Salinity, 282, 586
Saltmarsh(es), 218, 276, 448, 474
grasses, 488–489
plants, 453–454
Sand flat to mangrove forest, 14–30
biogeomorphic evolution, 18–23
environmental setting, 15–16
mangrove-seedling recruitment, 16–18
surface-elevation dynamics, 23–30
Sand-dominated mangroves, 192
Satellite images, 54–55
Science for the Protection of Indonesian Coastal Ecosystems program (SPICE program), 396
Scour, 496
Sea Level Affecting Marshes Model (SLAMM), 218–219, 246
Sea level rise (SLR), 4, 180, 218, 220, 236
biogenic *vs.* minerogenic settings, 266–268
differential responses of mangrove types to, 245–246
factors contributing to mangrove resilience to, 237–238
global dataset, 263–266
management considerations, 268
measuring mangrove resilience to, 244–245
methods, 260–263
regions with mangroves under threat from, 246–248
Africa, 248
Indo-Pacific, 247
North America, 246
South America, 247–248
surface elevation change and component variables, 261–262
surmountable challenges in determining vulnerability to, 269
Sea-level stabilization, 224
Seagrass(es), 448
meadows, 276
Seaward–landward transect in Southeast Asia, 84, 84f
Sediment, 188, 370b–371b, 554–555
accretion, 239–241, 244
accumulations, 480–485
availability in arid and semiarid areas, 238–242
CO_2 fluxes, 341
damage, 546–547
deposition, 495
dynamics, 567
plastic responses to sediment dynamics in mature forests, 134–135
properties, 101–102
sediment–atmosphere coupled system, 354
sediment–water coupled system, 354–355
settling and trapping by vegetation, 493–495
stability associated with carbon release from sediments, 495–496
supply, 519
transport, 86, 93
Sediment accumulation rates (SARs), 9, 244–245, 247
Sediment organic carbon (SOC), 370
Sediment organic matter (SOM), 370
origin, 376–382
quantity *vs.* quality of, 374–376
Sedimentation, 256, 284
Seed trees, 585–586
Seedlings, 122
Segara Anakan Lagoon (SAL), 395, 397f, 398t
Selective logging (SL), 585–586
Sensitivity of blue C to climate change, 356–357
Sequence stratigraphy, 22
Sequestration, 298–300
Sesarmid crabs, 286–287
Shallow
coastal habitats, 336
subsidence rates, 26, 28
Shoreline stabilization, 85
Shotgun sequencing, 160–161
Shrimp, 207
Silviculture, 584
Single cylinder, 98–99
Single-cell genomics, 160–161
Sinks of organic matter, 371–373

Small-scale selective logging (SSSL), 586
Smoothed Particle Hydrodynamics method, 105
Soil, 370b–371b
  analysis, 452–453
  carbon, 455–458
    stocks in mangroves, 277, 281
  CO_2
    efflux, 340–341
    flux, 487–488
  collection, 451
  dynamics against depth in secondary and primary mangrove forests, 600
  indicators and mangrove forest structure in selective logging system, 598–600
  recovery in different aged secondary and primary mangrove forests, 595–597
  soil–root plate, 130
  surface CH_4 fluxes, 421, 436–438
  surface elevation, 256
Soil bulk density (SBD), 586
Soil organic carbon (SOC), 310–315, 311t–314t, 427–428, 487
Sonneratia, 85, 94–96, 183, 222
  S. alba, 94–96, 126–127, 181, 237, 262–263, 283–284
  S. caseolaris, 126–127
Sources of organic matter, 371–373
South Africa, 449
South Asian mangroves, 341
  blue C, 336–337
  distribution of mangrove blue C, 341–343
  mangrove blue C in Indian Sundarbans, 348–357
  past and current research in, 341–348
  research gaps and future directions, 357–360
    gaps and challenges, 359–360
    laboratory-based data, 357
    microenvironmental heterogeneity, 358–359
    research on fine root production, 358
  state of C stock and sources in South Asian countries, 343–348
    litterfall, 347–348
    sources of OC in sediment, 346–347
    total blue C stock in mangroves, 343–346
Southeast Asia (SE Asia), 218
Space-for-time (SFT), 553
Spartina alterniflora. *See* Cordgrass (*Spartina alterniflora*)
Spatial resolution, 327
Spatial scales, 85–86, 85f
  disparate treatment, 107–108
Spatial subsidies, 373b
Species-specific allometric equations, 449
Sporobolus virginicus, 224–226
Stable isotope signatures, 376–378
Sterols, 379–381
Stilt roots, 130, 132–134
Stocks, 298–300
  stock–difference method, 499
Storms, 244, 280
Stressors, 516
  components, 522–524
Subaerial erosion, 496
Submarine groundwater discharge (SGD), 68–71
  land conversions, 71f
  potential interactions, 70f
  techniques in SGD research, 72–77
Subsurface processes, 188–189
Subtropical mangrove forests, 180
Succession, 188
Successional theory, 187
Sulfate (SO_4^{2-}), 148–149
  reduction, 146–147, 149, 242–243
Sulfide oxidation, 149–150
Sulfur (S), 145
  disproportionation, 150
  S-cycle, 148–150
Sundarbans mangroves, 336–337
Supratidal zone, 369–370, 370b–371b
Surface
  processes, 188–189
  texture, 104–105
  water hydrology, 70
Surface elevation table-marker horizon system (SET-MH system), 188–189, 256–260, 480–481
Surface elevation tables (SETs), 218, 261–262, 520–521
Surface elevations, 180
  dynamics, 23–30
  gain rates, 28
  mangrove species distribution and growth, 181–188
    historical origins, 182–184
    importance of, 181–182
    inundation and elevation, 187
    quantitative descriptions of mangrove-elevation linkages, 184–186
  mangroves modifying, 188–194
    elevation adjustment importance by mangroves, 188–190
    land building or consolidating depends on biogeomorphic setting, 192–193
    mangroves as land builders, 190–191, 193–194
    mangroves as land consolidators, 191–194
  trends, 25–26
Surface-elevation change (SEC), 237–238, 480–481, 483t
  and component variables, 261–262

Suspended particulate matter (SPM), 61
Suspended sediment concentrations (SSCs), 31
Swartkops Estuary, 449

T
Tanakeke Island, 418
Tanjung Panjang Nature Reserve, 524
Temperature, 279–280
Temporal variations of organic matter composition, 406–408
Tephrochronological approach, 209–210
Terminal restriction fragment length polymorphisms (TRFLP), 159
Thermal imagery, 76
Thraustochytrids, 156–157
Tidal
amplitude, 318
cycle, 181
flow reintroduction and sediment accumulation, 571–574
inundation, 180, 318–319
prism, 13
pumping, 69–71, 84–85
saltmarshes, 474
Tides, 281–282
TOC
in mangrove plant, 350–352
in mangrove sediment, 352–353
Total nitrogen (TN), 543
Total phosphorus (TP), 543
Total soil carbon, 453
Transformative capacity, 516
Tree
crowns, 127–128
diversity, 283–284
Tropical cyclones, 280
Tropical mangrove forests, 180
Turbid fringe, 27–28
Turbulence models, 96–97
Typhoons, 280, 540, 549
history, 542–543

U
Ulva pertusa, 286
Unicellular algae, 143
United Arab Emirates (UAE), 427
United Nations Framework Convention on Climate Change (UNFCCC), 359–360
Untargeted environmental metabolomics, 378–381

V
Vegetated coastal habitats, 394
Vegetation, 488–496, 554
damage, 544–546
hydrodynamic modifications by, 489–493
sediment settling and trapping by, 493–495
Vertical accretion (VA), 480–481, 483t
Vertical land motion (VLM), 23
Very high spatial resolution (VHSR), 571
VORTEX model, 98

W
Water-soluble aliphatic compounds, 375
Water-soluble aromatic compounds, 375
Watson's inundation classes, 185–186
Western Indian Ocean region (WIO region), 448
Wetland, 180
Wind
drag, 129
resistance and mechanical stability in unstable sediments, 127–129
Window of Opportunity (WoO), 17–18, 124–125
Written scientific records, 183

X
Xylocarpus
X. granatum, 131
X. mekongensis, 131

Z
Zero-dimensional model (0-D model), 30–31
Zhangjiang Estuary, 479, 499–500
Zonation, 187
Zostera capensis, 463, 465

Made in the USA
Monee, IL
27 September 2025

30710751R00359